Contents

Contributors

Numbers in parentheses indicate the pages on which the author's contributions begin.

Wenbing Guo (495), School of Energy Science and Engineering, Henan Polytechnic University, Jiaozuo, China

Xiangxin Liu (399), School of Mining Engineering, North China University of Science and Technology, Tangshan, China

Hossein Masoumi (1), Department of Civil Engineering, Faculty of Engineering, Monash University, Melbourne, VIC, Australia

Joung Oh (259), School of Minerals and Energy Resources Engineering, The University of New South Wales, Sydney, NSW, Australia

Shuren Wang (399,495), School of Civil Engineering, Henan Polytechnic University, Jiaozuo, China

Sheng Zhang (145), School of Energy Science and Engineering, Henan Polytechnic University, Jiaozuo, China

About the authors

Shuren Wang is the distinguished professor of Henan Province and an adjunct professor at the University of New South Wales, Sydney, Australia. He mainly focuses on the challenging areas of mining engineering, geotechnical engineering, rock mechanics, and numerical simulation analysis. His research projects have been supported by the National Natural Science Foundation of China (51774112; U1810203; 51474188; 51074140; 51310105020) and the International Cooperation Project of Henan province (162102410027; 182102410060), etc. He is the recipient of nine state- and province-level awards and the 2015 Endeavour Research Fellowship provided by the Australian government. He has published more than 97 articles in peer-reviewed journals as well as nine monographs and textbooks. He has received 22 patents in China.

Hossein Masoumi obtained his Ph.D. at the University of New South Wales in 2012. He mainly focuses on rock mechanics with particular interest in the size/scale and shape effects of intact rock; constitutive modeling of intact rocks exhibiting size/scale behaviors under different test conditions; shaly-sandstone microscale characterization; and nonpersistent jointed rock characterization. He has more than 10 research projects that have been supported by the Australian Coal Association Research Program (ACARP) and the Mining Education Australia (MEA) Collaborative Research Project. He has published more than 30 academic articles.

Joung Oh earned his Ph.D. in Civil Engineering from the University of Illinois in the United States in 2012. Now, he combines his deep knowledge of geotechnical engineering and mine geomechanics to improve safety, productivity, and sustainability in the Australian resources sector. He is currently involved in several ACARP (Australian Coal Association Research Program) projects and is the lead investigator on one project that is focused on understanding the fundamentals of rock failure mechanisms under different geotechnical environments. He has published more than 29 journal articles with important academic influence, 13 conference papers, and 1 report.

Sheng Zhang, during September 2012 through September 2013, spent time at the University of Kentucky in the United States. He mainly engaged in rock fracture mechanics and engineering, safety mining, and roadway support under complex and difficult conditions. He has presented lectures, such as on rock mechanics and engineering, mine pressure and strata control, and other professional courses. He has presided over and participated in 8 National Natural Science Foundation of China projects as well as more than 40 enterprise projects. He has published more than 60 papers in academic journals and 1 book. He is the recipient of 5 province-level awards and has received 30 patents in China.

Preface

The scale effect objectively exists. The scale effect of rock refers to the dependence of the change of the rock's mechanical properties on the size of the sampling grid. The scale is the spatial dimension and time dimension of the object or process. The spatial scale refers to the area size of the study unit or the spatial resolution level of the smallest information unit, and the time scale is the time interval of its dynamic change. There are great mechanical differences in the strength and deformation characteristics of rocks of different sizes. The strength and deformation characteristics of rocks of a certain size cannot be directly applied to geotechnical engineering design and the establishment of constitutive relations. Therefore, rock scale analysis and scale effect are important to the engineering.

Rock mass differs from the general continuous medium in that there are various structural planes in the rock mass. Also, the rock mass structure, composed of the structural plane and the rock created by the structural plane, control the mechanics and mechanical properties of the rock mass. The influence of the rock mass structure on the mechanical properties of the rock mass is called the structural effect of rock mass mechanical properties. Due to the loading and unloading processes of engineering loads, structural loads, temperature loads, and underground fluid infiltration, the stability of rock engineering is a very prominent area of research. It has become a research hotspot of geotechnical engineering to study the structural effects of rock mass.

This book summarizes and enriches the latest research results on the scale-size and structural effects of rock materials, including test methods, innovative technologies and their applications in indoor tests, rock mechanics, and rock engineering. The book is divided into five chapters: Chapter 1: Size Effect of Rock Samples (Hossein Masoumi); Chapter 2: Rock Fracture Toughness (Sheng Zhang); Chapter 3: Scale Effect of Rock Joint (Joung Oh); Chapter 4: Microseismic Monitoring and Application (Shuren Wang 1–3, Xiangxin Liu 4–6), and Chapter 5: Structural Effect of Rock Blocks (Shuren Wang 1–6, Wenbing Guo 7). This book is innovative, practical, and rich in content. It will be of great use and interest to researchers undertaking various rock tests, geotechnical engineering, and rock mechanics as well as for teachers and students in related universities and onsite technical people.

The material presented in this book contributes to the expansion of knowledge related to rock mechanics and engineering. Through their extensive fundamental and applied research over the past decade, the authors cover a diverse range of topics, including the scale-size and structural effects of rock materials through the interaction of large-scale rock masses and engineering practices; the mechanics of rock cutting; techniques to improve the strength and integrity of rock structures in surface and underground excavations; and improvements in approaches to modeling techniques used in engineering design.

Shuren Wang[a]
Hossein Masoumi[b]
Joung Oh[c]
Sheng Zhang[d]
[a]*Ph.D. Professor in School of Civil Engineering, Henan Polytechnic University, Jiaozuo, China*
[b]*Ph.D. Senior Lecturer in Department of Civil Engineering, Faculty of Engineering, Monash University, Melbourne, VIC, Australia*
[c]*Ph.D. Senior Lecturer in School of Minerals and Energy Resources Engineering, The University of New South Wales, Sydney, NSW, Australia*
[d]*Ph.D. Professor in School of Energy Science and Engineering, Henan Polytechnic University, Jiaozuo, China*

Acknowledgments

The authors are pleased to acknowledge the support received from various organizations, including the National Natural Science Foundation of China (51774112; U1810203; 51674190; 51474188; 51074140; 51310105020), the Natural Science Foundation of Hebei Province of China (E2014203012), the China Scholarship Council (CSC) and the Hebei Provincial Office of Education (2010813124), the 2015 Endeavor Research Fellowship and Program for Taihang Scholars, the International Cooperation Project of the Henan Science and Technology Department (182102410060; 162102410027), the Doctoral Fund of Henan Polytechnic University (B2015-67), the International Joint Research Laboratory of Henan Province for Underground Space Development and Disaster Prevention, the Henan Key Laboratory for Underground Engineering and Disaster Prevention, the Collaborative Innovation Center of Coal Work Safety, the Provincial Key Disciplines of Civil Engineering of Henan Polytechnic University, and the School of Minerals and Energy Resources Engineering at the University of New South Wales, Sydney, Australia.

While it is not possible to name everyone, the authors are particularly thankful to Prof. Manchao He, Prof. Meifeng Cai, Prof. Ji'an Wang, Prof. Youfeng Zou, Prof. Xiaolin Yang, Prof. Xiliang Liu, Prof. Zhaowei Liu, Prof. Zhengsheng Zou, Prof. Yanbo Zhang, Prof. Yahong Ding, Prof. Paul Hagan, Prof. Bruce Hebblewhite, Prof. Ismet Canbulat, Prof. Fidelis Suorineni, Prof. Serkan Saydam, Prof. Kodama Jun-ichi, and related persons. A number of postgraduate students, namely, Yanhai Zhao, Danqi Li, Chunliu Li, Yingchun Li, Jianhang Chen, Chengguo Zhang, Chen Cao, Xiaogang Wu, Jiyun Zhang, Chunyang Li, and others assisted in the design, construction, and commissioning of the test facility as well as the experimentation; their contributions to the book are acknowledged and appreciated. The untiring efforts of Mr. Kanchana Gamage, Dr. Mojtaba Bahaaddini, Dr. Wenxue Chen, and Dr. Faqiang Su during the equipment design phase and during laboratory testing programs are gratefully appreciated. Thanks and apologies to others whose contributions we have overlooked.

Chapter 1

Size effect of rock samples

Hossein Masoumi
Department of Civil Engineering, Faculty of Engineering, Monash University, Melbourne, VIC, Australia

Chapter outline

1.1 Size effect law for intact rock

1.1.1 Introduction

In general, the term "size effect" refers to the influence of sample size on mechanical characteristics such as strength. Size effects in rock engineering have been of particular interest over the last four decades and many studies have been undertaken to understand the phenomenon. Size effects are not limited to rock. Almost all quasibrittle and brittle materials such as concrete, ceramic, and ice have shown some form of size effect.

To date, among different quasibrittle and brittle materials, the most comprehensive size effect studies have been undertaken on concrete samples. Bazant (Bazant and Planas, 1998) and Van Mier (1996) have been the leading experts in this respect. Bazant improved the knowledge of the size effect from a theoretical perspective, whereas Van Mier concentrated on experimental studies. Unfortunately, earlier size effect investigations on rock are not as comprehensive as for concrete. Nevertheless, Hoek and Brown (1980a, b, 1997) attempted to develop an understanding of size effects in rock from an experimental viewpoint. Prior to Hoek and Brown, there were a number of published papers (Mogi, 1962; Bieniawski, 1968; Koifman, 1969; Pratt et al., 1972) providing size effect data, which were collected and used by Hoek and Brown in order to propose their well-known empirical size effect model.

Apart from the advantages of the Hoek and Brown (1980a, b, 1997) studies, there were two important shortcomings that were unavoidable. First, their focus was only on the size effect of the uniaxial compressive test and nothing was reported on other experiments such as the point load test. Indeed, there is very limited research in the literature that has investigated simultaneously size effects under different testing conditions (e.g. Bieniawski, 1975; Wijk et al., 1978; Panek and Fannon, 1992; Kramadibrata and Jones, 1993). Second, according to some past observational studies (Hiramatsu and Oka, 1966; Hoskins and Horino, 1969; Abou-Sayed and Brechtel, 1976) as well as recent ones (Hawkins, 1998; Darlington and Ranjith, 2011), in the uniaxial compressive test, the size effect behavior of small samples does not follow a commonly assumed size effect model in which the strength reduces as the sample size increases. This important observation was not discussed by Hoek and Brown (1980a, b, 1997) and there has been no comprehensive investigation that has assessed this behavior from an analytical viewpoint.

In view of these knowledge gaps, this paper will propose a unified size effect law for intact rock, building on work by Bazant (1997). The model is applied to experimental data obtained from Gosford sandstone for uniaxial compressive and point load results as well as data reported in the literature. The impact of surface flaws on the size effect behavior is also discussed. It is shown that the fractal fracture theory and surface flaws play key roles in sample failure for both uniaxial compressive and point load tests, and that point load results are best interpreted by considering the condition of the load contact area to properly explore size effects.

1.1.2 Background

The existing size effect models can be divided into two major categories: descending and ascending types. The descending models can be classified into four subcategories: those based on statistics, fracture energy, multifractals as well as that includes empirical and semi-empirical models.

1.1.2.1 Descending models

Statistical models

A statistical explanation for the size effect in materials was initially proposed by Weibull (1939), which later become known as the weakest link model. Weibull (1939) postulated that every solid consists of preexisting flaws (microcracks) that play a significant role in determining the strength of the material. Based on this explanation, for two samples with different sizes but identical shapes (e.g., cylinders), the probability of failure in the larger sample that has more flaws is higher than that of the smaller sample. Therefore, the larger sample fails at a lower strength in comparison to the smaller sample. In other words, an increase in size causes a decrease in strength according to:

$$P_f(\sigma) = 1 - \exp\left[-\frac{V}{V_r} P_1(\sigma)\right] \tag{1.1}$$

where V is the volume of the sample, V_r represents the volume of one element in the sample, $P_f(\sigma)$ is the material strength, and $P_1(\sigma)$ is the strength of the representative sample. Eq. (1.1) is the initial statistical model proposed by Weibull (1939) and later Weibull (1951) introduced a more general form of Eq. (1.1) through:

$$m \log\left(\frac{P_f(\sigma)}{P_1(\sigma)}\right) = \log\left(\frac{V}{V_r}\right) \tag{1.2}$$

where m is a material constant introduced for better simulation of the size effect behavior ($m=1$ was assumed in Eq. (1.1)). In Eq. (1.2), V and V_r can be substituted by any characteristic measure of volume such as length or sample diameter. For example, in the case of cylindrical samples with identical shapes and constant length-to-diameter ratios instead of volume, the diameter can be substituted.

Two modified formulations of Eq. (1.2) were proposed by Brook (1980, 1985) and Hoek and Brown (1980a, b) to predict the size effect in point load and uniaxial compressive tests, respectively, as follows:

$$\frac{I_s}{I_{s50}} = \left(\frac{50}{d}\right)^{k_1} \tag{1.3}$$

$$\frac{\sigma^c}{\sigma^c_{50}} = \left(\frac{50}{d}\right)^{k_2} \tag{1.4}$$

where, in Eq. (1.3), I_s is the point load strength index, I_{s50} is the characteristic point load strength index measured on a sample with a characteristic size of 50 mm, d is the sample characteristic size, and k_1 is a positive constant controlling the statistical decay of the strength with an increase in size, which is also related to m in Eq. (1.2). Similarly, in Eq. (1.4), the measured uniaxial compressive strength (UCS) σ^c is a function of sample characteristic size d (in mm), $\sigma c50$ is the characteristic UCS measured on a sample with a characteristic size of 50 mm, and k_2 is a positive constant. In uniaxial compressive and point load tests, the characteristic size d is usually taken to be the sample core diameter. The schematic trends of Eqs. (1.3) or (1.4) for different values of k are presented in Fig. 1.1.

Fracture energy model

Size effect fracture theory originated from Griffith (1924), who indicated that in brittle materials, a crack grows and propagates only when the total potential energy of the system of applied forces and material reduces or remains constant with an increase in crack length. Later, other researchers (Hunt, 1973; Bazant, 1984; Bazant and Kazemi, 1990; Kim and Eo, 1990; Bazant and Xi, 1991; Bazant et al., 1991; Smith, 1995; Huang and Detournay, 2008; Van Mier and Man, 2009; Villeneuve et al., 2012) extended this theory by proposing some modifications for better applicability of the original concept to quasibrittle and brittle behavior.

It is known that during a compressive or tensile test, at the same stress level, the stored elastic energy in a larger sample is more than that of the smaller one (Hudson et al., 1972). Therefore, a higher energy release can be expected from a

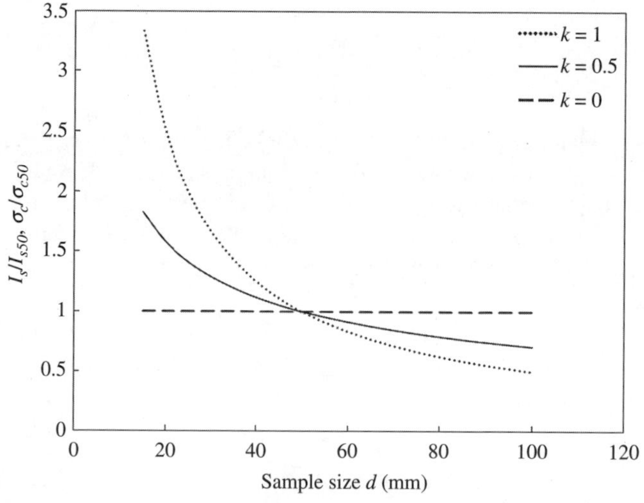

FIG. 1.1 Schematic representation of the statistical size-effect model at three different k values, where k can be k_1 or k_2.

larger sample at the commencement of the crack propagation. As a result, this higher energy release rate leads to lower crack initiation stress in a larger sample in comparison with a smaller one (Abou-Sayed and Brechtel, 1976). In other words, as is the case with the statistical model, an increase in size causes the failure stress to reduce.

Bazant (1984) was the first to define a size effect model using fracture energy theory, known as the size effect law (SEL). Bazant (1984) was suitable for quasibrittle and brittle materials such as rock and concrete. Bazant (1984) took into account the role of energy for quantification of the crack growth and propagation. The final expression can be written in a manner that does not explicitly show the fracture energy term:

$$\sigma_N = \frac{Bf_t}{\sqrt{1 + (d/\lambda d_0)}} \tag{1.5}$$

where σ_N is a nominal strength, B and λ are dimensionless material constants, f_t is a strength for a sample with negligible size that may be expressed in terms of an intrinsic strength, d is the characteristic sample size, and d_0 is the maximum aggregate size. In order to understand the influence of the defined constants of Eq. (1.5), a number of schematic size effect trends are plotted in Figs. 1.2 and 1.3.

Figs. 1.2 and 1.3 illustrate that the strength parameter Bf_t mostly controls the upward or downward movement of the size effect trend, whereas λd_0 mainly influences the rate of strength change with size.

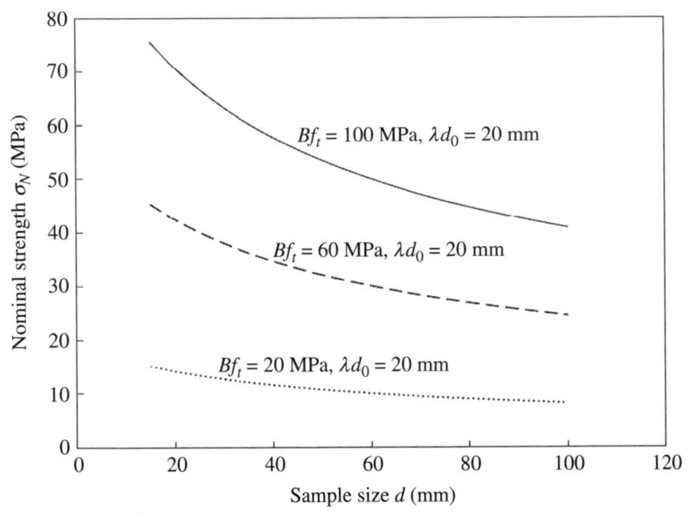

FIG. 1.2 Schematic representation of SEL at different Bf_t values and identical λd_0 values.

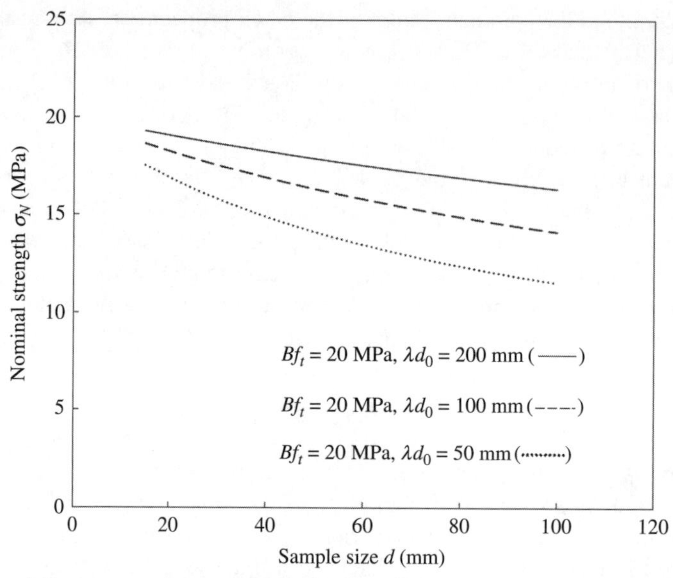

FIG. 1.3 Schematic representation of SEL at different λd_0 values and identical Bf_t values.

Fractal and multifractal models

Fractals have been utilized to explore different properties in rocks. According to Carpinteri (1994), Carpinteri and Ferro (1994), and Borodich (1999), the self-similar properties of multiple fractures can appear in a wide range of material sizes, thus making them fractal. Yi et al. (2011) used the concept of fractal theory to explore the damage evolution in mortar under triaxial conditions and Cnudde et al. (2011) utilized this theory for pore size distribution and the packing efficiency of Ferruginous sandstone. Other researchers such as Thompson (1991), Muller and Mccauley (1992), and Radlinski et al. (2004) applied this theory to explore other characteristics of sedimentary rocks.

Multifractality was initially proposed by Mandelbort (1982), who indicated that in many physical realities, a material under peak load can be considered as multifractal. Carpinteri et al. (1995) adopted the topological concept of geometrical multifractality, which is an extension of self-similarity, to explain the size effect in quasibrittle and brittle materials. Carpinteri et al. (1995) pointed out that such a fractal set can be observed via two different ranges of fractal dimensions, namely local and global. The local dimension applies in the limit of scale tending to zero and has a noninteger value, whereas the global dimension corresponds to the large scale and can only attain an integer value. As a result, according to the concept of multifractality, Carpinteri et al. (1995) proposed a size effect model known as the multifractal scaling law (MFSL) with the following analytical expression:

$$\sigma_N = f_c\sqrt{1+\frac{l}{d}} \tag{1.6}$$

where σ_N is the nominal strength, l is a material constant with unit of length, f_c is the strength of a sample with an infinite size that may be expressed in terms of an intrinsic strength, and d is the characteristic sample size. The fractal dimension in Eq. (1.6) was assumed as 1. In principle, the MFSL acts similar to the SEL and Weibull model in which the strength reduces as size increases. The significant advantage of MFSL versus the Weibull model and SEL is the ability to estimate the realistic strength of a very large sample with infinite size. The impacts of two MFSL constants on the final size effect trend are demonstrated in Figs. 1.4 and 1.5, which show that the strength parameter f_c mostly controls the upward or downward movement of the size effect trend while l influences the rate of strength change with size.

In principle, Eqs. (1.3)–(1.6) are similar in that the strength reduces with an increase in size. However, the statistical size effect models rely on only one material constant, whereas the two others require two different constants. A careful observation of the SEL and MFSL models shows that the pair of constants in one criterion has similar characteristics as that in the other. Both models have one strength parameter and one index size coefficient. In the MFSL, the term under the square root reveals that when the sample size approaches infinity, l/d tends to zero and eventually the nominal strength of the sample is equal to f_c. In the SEL, when the size approaches infinity, the term under the square root tends to infinity and eventually the nominal strength approaches zero. The statistical size effect model is similar to SEL in that zero

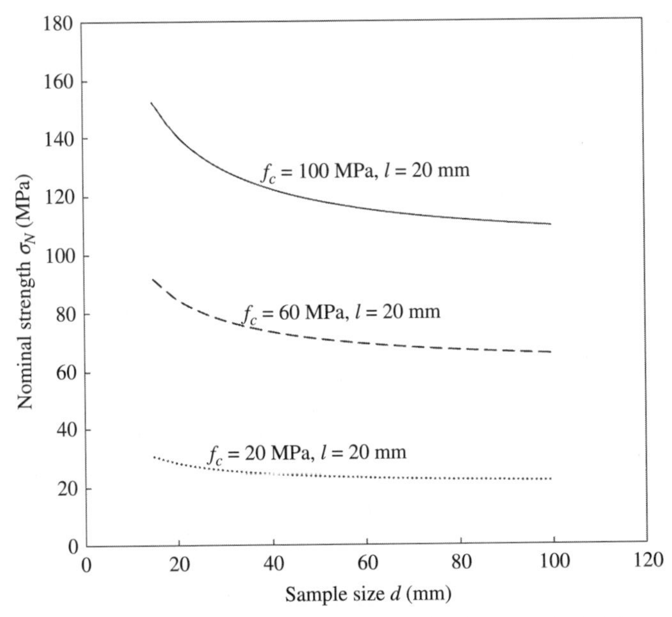

FIG. 1.4 Schematic representation of MFSL at different f_c values and identical l values.

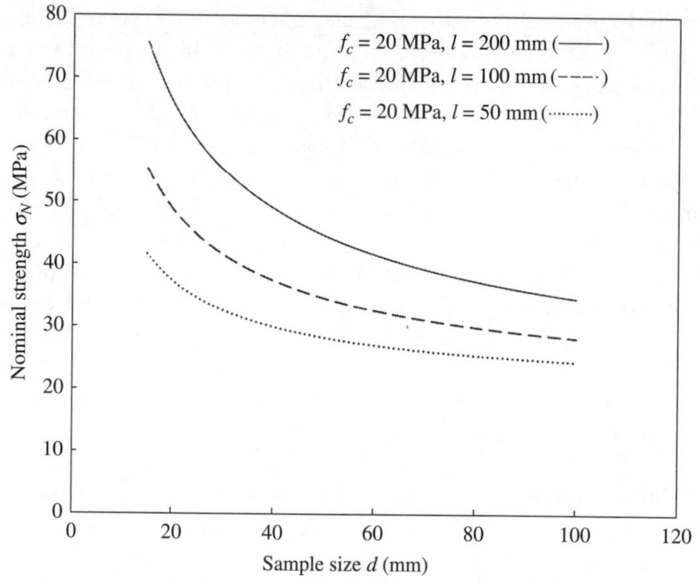

FIG. 1.5 Schematic representation of MFSL at different l values and identical f_c values.

strength is also predicted for infinitely large samples. Perhaps the significant advantage of MFSL over the statistical and SEL models is the ability to estimate nonzero strength for very large sample sizes.

For linking the SEL and MFSL to the experimental data obtained from the intact rock, it is feasible to relate the parameters of these models to a number of characteristics of rock samples. Since Bazant (1984) introduced the SEL for size effect in concrete samples, d_0 was defined as the maximum aggregate size. For intact rock, this definition may be linked to the maximum grain size of the rock sample. Similarly for MFSL, the l value can be linked to the maximum grain size if it is represented in the form of a coefficient multiplied by this maximum grain size, such as $\beta l_0 = l$ where l_0 is the maximum grain size.

Bazant (1984) defined f_t as a strength for a sample with negligible size, which may be expressed in terms of an intrinsic strength and so this negligible size can be referred to the strength of the grain with the maximum size as defined earlier. Unfortunately, to date, there is no such equipment with the ability to gain only one grain out of the sandstone properly in order to conduct the uniaxial compressive test on it and therefore it is not feasible to verify this definition for f_t. It is important to state that Bazant (1984) defined f_t as the tensile strength of the concrete sample for dimensional purposes, but did not specify any particular size. For MFSL, f_c is the strength of the sample with infinite size and so it can be linked to the rock mass, which represents the very large sample with infinite diameter.

Empirical and semiempirical models

The majority of the empirical and semiempirical size effect models originated from or have a similar form to the statistical size effect model. These models (Mogi, 1962; Dey and Halleck, 1981; Silva et al., 1993; Adey and Pusch, 1999; Castelli et al., 2003; Yoshinaka et al., 2008; Darlington and Ranjith, 2011; Zhang et al., 2011) resulted from curve fitting with similar logarithmic equations. All these models follow a commonly assumed size effect concept in which the strength reduces as the size increases.

1.1.2.2 Ascending model

Bazant (1997) incorporated the concept of fractals into fracture energy and proposed the fractal fracture size effect law (FFSEL). It was argued that within a certain range of sizes, the fracture surfaces in a number of materials such as rock, concrete, and ceramics exhibit fractal characteristics in some way. Fractal characteristics were captured through the fractal dimension, d_f, which takes a positive value as follows: (1) $d_f=1$ for nonfractal characteristics; (2) $d_f \neq 1$ for fractal characteristics.

Bazant (1997) derived the FFSEL model for nominal strength according to:

$$\sigma_N = \frac{\sigma_0 d^{(d_f-1)/2}}{\sqrt{1+(d/\lambda d_0)}} \tag{1.7}$$

where σ_0 is the strength for a sample with negligible size that may be expressed in terms of an intrinsic strength, d_f is a fractal dimension, and other constants are the same as those defined for SEL (Eq. 1.5). In general, the structures of the SEL and FFSEL models are very similar. In order to obtain $\sigma_0$and d_f in FFSEL, it is required to initially attain λd_0 from SEL and then obtain σ_0 and d_f for FFSEL. For those sizes that exhibit nonfractal characteristics, $d_f=1$, FFSEL becomes the same as SEL, in which $Bf_t=\sigma_0$. To demonstrate the influence of d_f on FFSEL, a number of trends are presented in Figs. 1.6 and 1.7 with various d_f values.

1.1.3 Experimental study

1.1.3.1 Rock sample selection

Gosford sandstone from Gosford Quarry, Somersby, New South Wales, Australia, was used to conduct a number of point load and uniaxial compressive tests according to International Society for Rock Mechanics (ISRM) suggested methods (ISRM, 2007). Homogenous samples were carefully selected, having no macro defects (see Fig. 1.8). Sufian and Russell (2013) conducted an X-ray CT scan on the same batch of Gosford sandstone at a resolution of 5 μm, and estimated the porosity to be 18.5%. X-ray diffraction results showed that the sandstone comprises 86% quartz, 7% illite, 6% kaolinite, and 1% anatase.

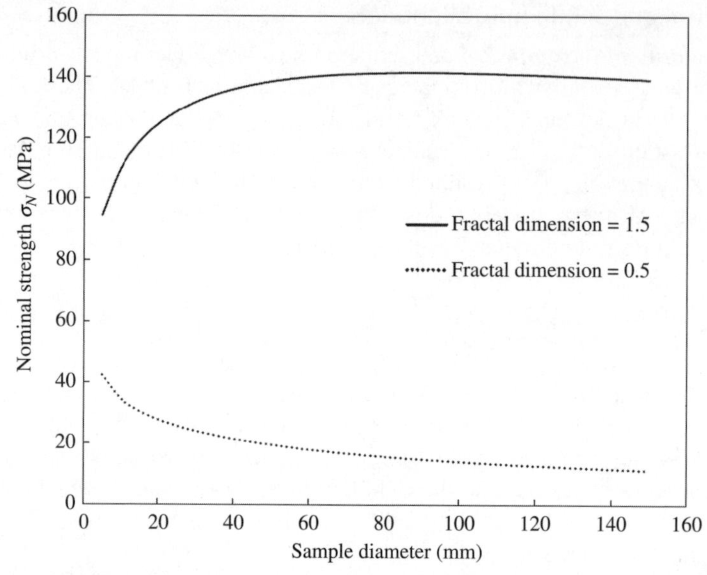

FIG. 1.6 Fractal fracture size effect law trends with various d_f values (0.5 and 1.5) and constants σ_0=65 MPa and λd_0=90 mm.

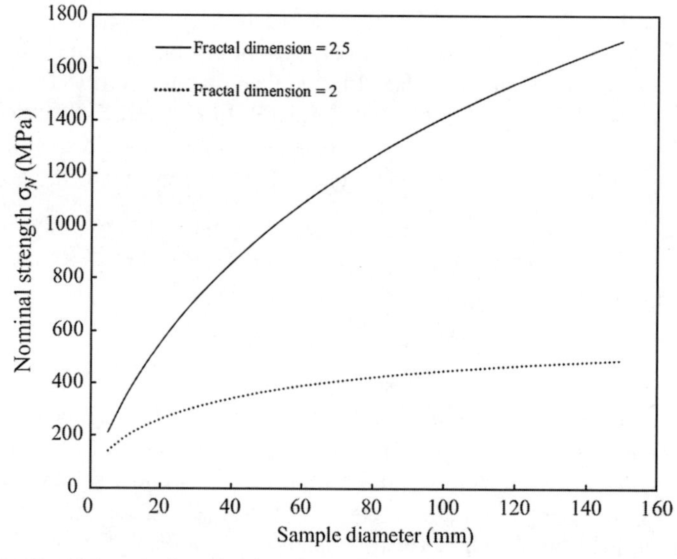

FIG. 1.7 Fractal fracture size effect law trends with various d_f values (2 and 2.5) and constants σ_0=65 MPa and λd_0=90 mm.

FIG. 1.8 Cross-sectional area of Gosford sandstone sample with 50-mm diameter.

1.1.3.2 UCS results

The uniaxial compressive tests were undertaken on core samples with 19, 25, 31, 50, 65, 96, 118, and 146 mm diameters. These diameters were selected according to Hawkins (1998) to cover a wide range of sample sizes for model verification process. The constant length-to-diameter ratio of two was selected as specified by ISRM (2007). Multiple tests as suggested by ISRM (2007) were performed at each diameter using an INSTRON loading frame with a 300-t maximum loading capacity (see Figs. 1.9 and 1.10) to enable an average strength to be determined and account for scatter in the data. More experiments were conducted on small samples in comparison with larger ones. The typical fracture patterns that were observed from the uniaxial compressive tests on Gosford sandstone are presented in Fig. 1.11.

Table 1.1 lists the mean UCS values obtained from Gosford sandstone at different sizes and Fig. 1.12 presents all UCS results.

The experimental data from Fig. 1.12 clearly shows that the resulting size effect behavior is not in agreement with a commonly assumed size effect concept (Weibull, 1939). This observation also is in conflict with the Hoek and Brown (1980a, b) size effect model (see Fig. 1.13).

These results however, are in good agreement with the study from Hawkins (1998), who conducted a significant number of uniaxial compressive tests on

FIG. 1.9 Setup of a 31-mm diameter sample for a uniaxial compressive test using a servo-controlled testing frame.

FIG. 1.10 Setup of a 118-mm diameter sample for a uniaxial compressive test using a servo-controlled testing frame.

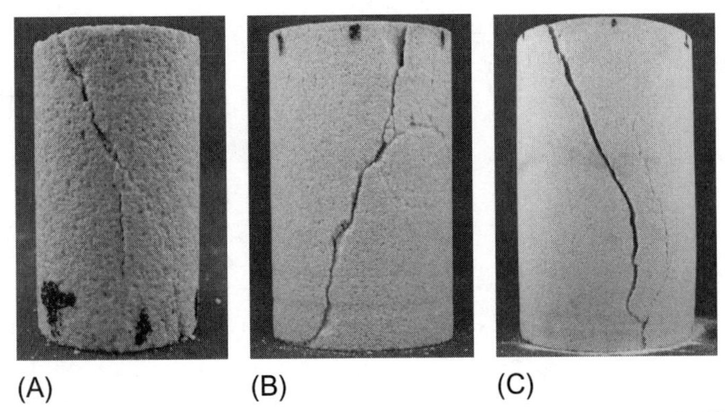

FIG. 1.11 Typical fracture patterns for samples of Gosford sandstone: (A) 25; (B) 50; (C) 96 mm.

TABLE 1.1 Mean UCS results and number of repetitions at different diameters for Gosford sandstone

Sample diameter (mm)	Number of tests	Average of UCS (MPa)	SD	
			MPa	%
19	6	34.6	5.2	14.9
25	5	36.5	3.2	8.8
31	6	42.3	4.0	9.4
50	6	52.3	2.7	5.2
65	7	58.8	2.8	4.8
96	5	56.1	2.7	4.8
118	4	54.7	1.5	2.8
145	2	54.2	0.4	0.7

different sedimentary rocks at variable sizes (see Fig. 1.14). These demonstrated that, with an increase in sample size up to a characteristic diameter, the UCS increases and then above this characteristic diameter, the UCS reduces as the sample size increases.

1.1.3.3 Point load results

A number of point load tests were conducted in both diametral and axial loadings, where the coring was performed perpendicular to the bedding.

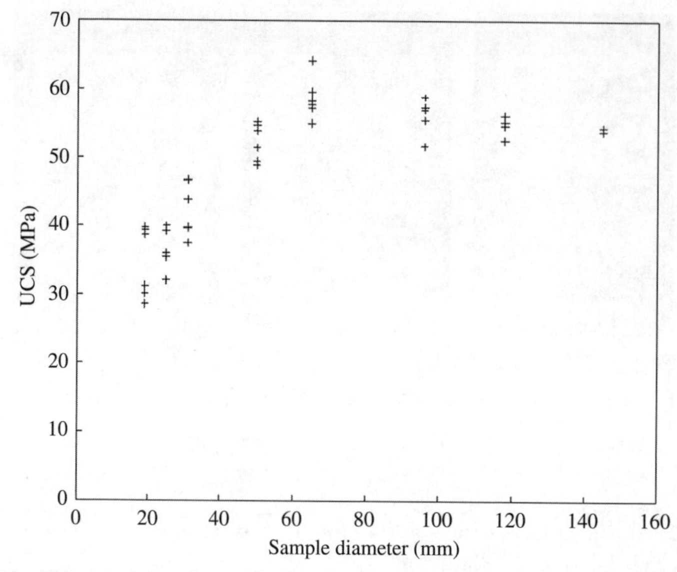

FIG. 1.12 Uniaxial compressive strength results at different diameters for Gosford sandstone.

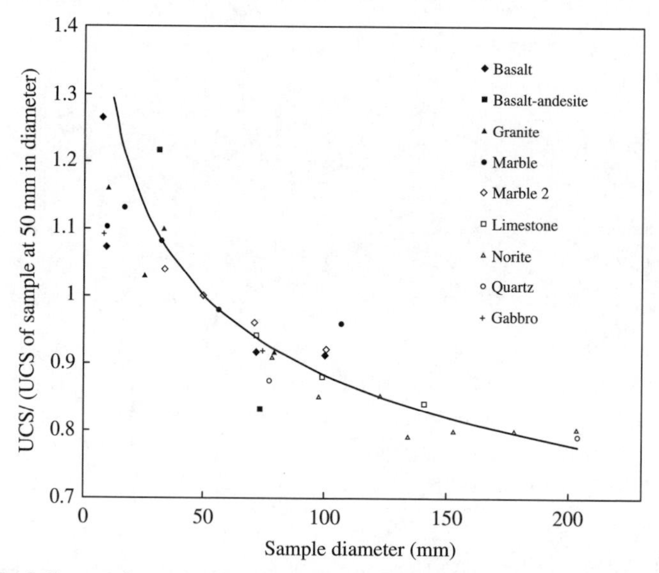

FIG. 1.13 Influence of sample size on the strength of intact rock. *(Reprinted from Hoek, E., Brown, E.T., 1997. Practical estimates of rock mass strength. Int. J. Rock Mech. Min. Sci., 34 (8), 1165-1186, Copyright (1997), with permission from Elsevier.)*

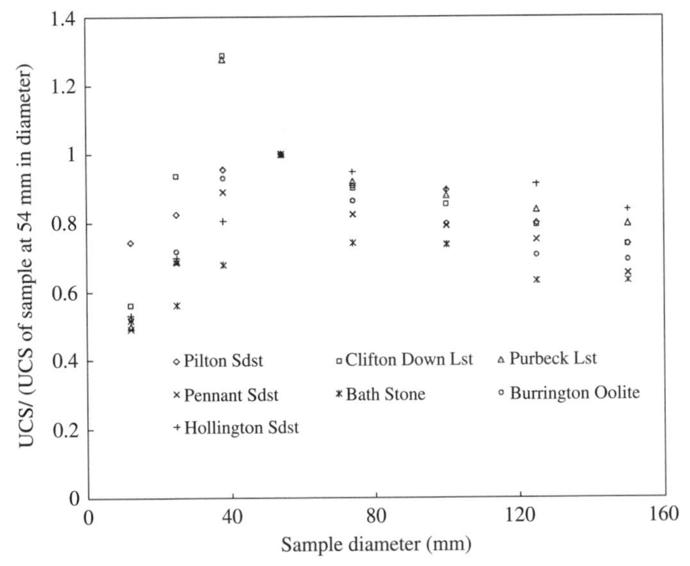

FIG. 1.14 Uniaxial compressive strength results obtained from seven different sedimentary rocks reported by Hawkins (1998).

Diametral loading

For the diametral test, the sample length should be greater than or equal to the diameter and the load should be applied (location of the conical platens) at least $0.5D$ from the ends of the sample as specified by the ISRM (Franklin, 1985) (see Fig. 1.15). The diameter of samples ranged between 19 and 65 mm.

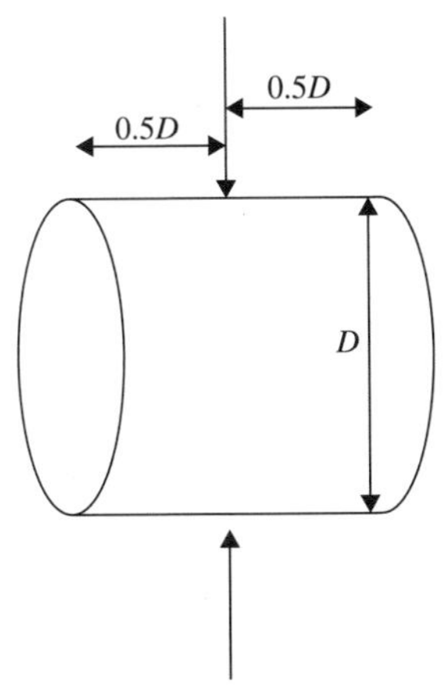

FIG. 1.15 Diametrically loaded cylinder.

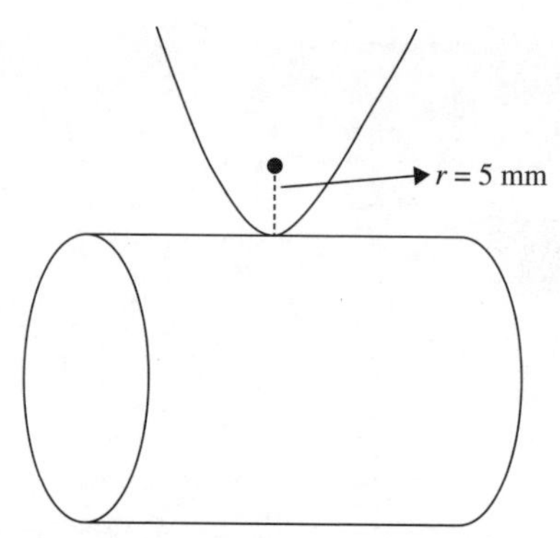

FIG. 1.16 Conical platen is loading diametrically.

According to the Broch and Franklin (1972) study, the standard radius of the conical platens of the point load test machine is 5 mm (see Fig. 1.16). This radius causes some practical limitations when testing large samples (being greater than 65 mm diameter) diametrically. It was observed that when the sample approached the state of failure, it started twisting between the top and lower conical platens. Eventually, only a fragment broke off, leading to an unacceptable outcome according to the ISRM (Franklin, 1985; Franklin, 2007). This behavior was observed at the 96 mm diameter several times and all results had to be abandoned.

It is convention (ISRM, 2007) to define a diametral point load strength index as:

$$I_s = \frac{P}{D^2} \tag{1.8}$$

where P is the peak load measured during the test and D is the distance between two pointers (conical platens), which is also the diameter of the sample (Fig. 1.16). The typical diametral fracture patterns from Gosford sandstone are presented in Fig. 1.17.

The mean diametral point load strength indices as well as the number of repetitions for each sample are given in Table 1.2.

Fig. 1.18 presents the variation of the diametral point load indices versus the sample diameters. Despite the number of tests that were reported in Table 1.2, fewer symbols can be observed in Fig. 1.18 due to the closeness of some point load results (particularly at small diameters) that overlapped. It is evident that with an increase in size, the strength reduces. This agrees with the observations of earlier researchers (Brook, 1980; Greminger, 1982; Brook, 1985; Thuro et al., 2001a, b).

Axial loading

For the axial test, ISRM (Franklin, 1985; Franklin, 2007) suggests that the ratio of length over diameter can range between 0.3 and 1. This ratio varied between

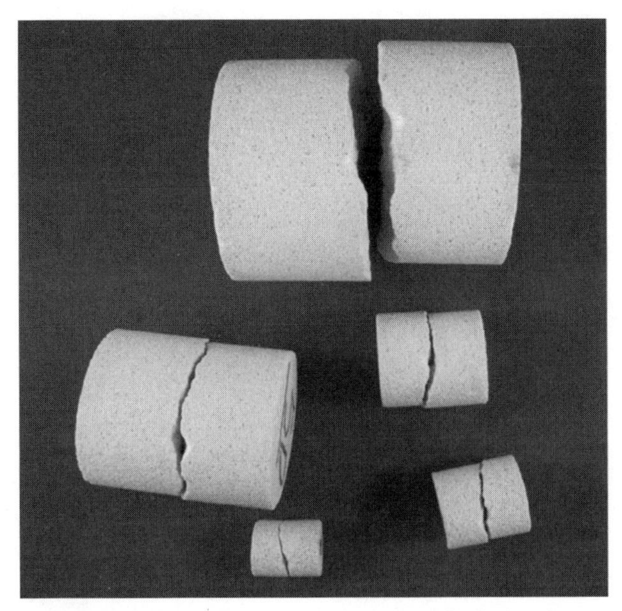

FIG. 1.17 Fracture patterns from a diametral point load test on Gosford sandstone at diameters of 19, 25, 31, 50, and 65 mm.

TABLE 1.2 Mean diametral point-load strength indices and number of repetitions at different sizes for Gosford sandstone

Sample diameter (mm)	Number of tests	Average of point-load strength index (MPa), I_S	SD	
			MPa	%
19	10	4.5	0.3	6.7
25	10	4.4	0.3	6.8
31	10	3.9	0.1	2.6
50	5	3.3	0.1	3.0
65	3	3.0	0.2	6.7

0.3 and 1 in the tests conducted here. The loading is applied (location of the conical platens) at the center of the end surfaces (see Fig. 1.19), axially. The selected sample sizes for axial tests on Gosford sandstone varied between 19 and 96 mm diameters.

For this test, the point load strength index (ISRM, 2007) is obtained using:

$$I_s = \frac{P}{4LD/\pi} \tag{1.9}$$

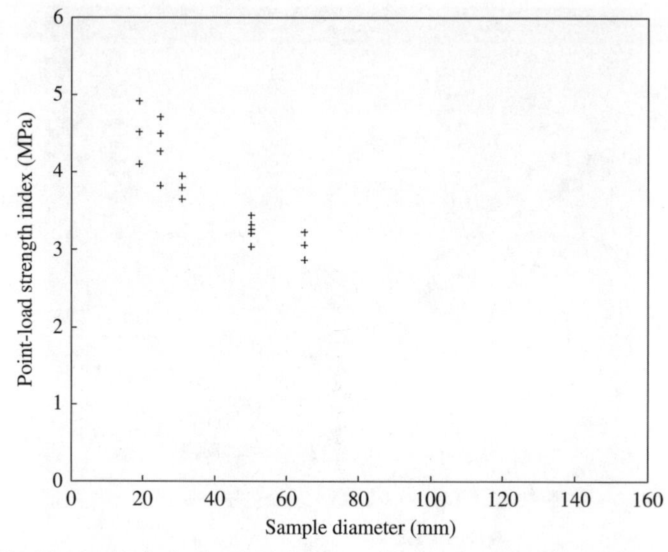

FIG. 1.18 Diametral point load strength indices at different sizes from Gosford sandstone.

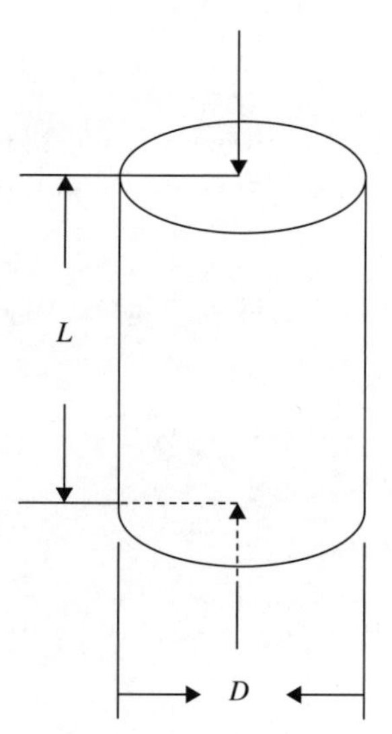

FIG. 1.19 Axially loaded cylinder.

where LD is the minimum cross-sectional area of a plane through the pointers (see Fig. 1.19) and P is the applied load. Fig. 1.20 depicts the typical fracture patterns from the axial point load test on Gosford sandstone.

Table 1.3 lists the characteristic parameters obtained from the axial point load test and Fig. 1.21 illustrates the size effect results from the axial point load

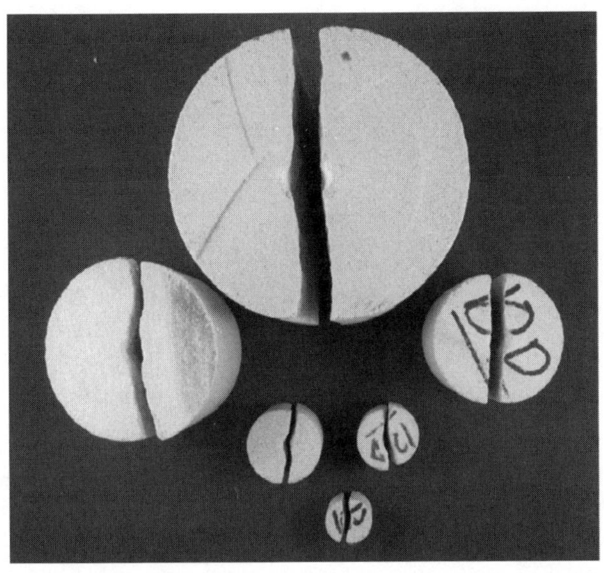

FIG. 1.20 Typical fracture patterns from the axial point load test on Gosford sandstone at diameters of 19, 25, 31, 50, 65, and 96 mm.

TABLE 1.3 Mean axial point-load strength indices and number of repetitions at different sizes for Gosford sandstone

Sample diameter (mm)	Number of tests	Average of point-load strength index (MPa), I_S	SD	
			MPa	%
19	10	4.1	0.3	7.3
25	10	3.6	0.2	5.6
31	10	3.5	0.1	2.9
50	6	3.1	0.4	12.9
65	3	2.6	0.1	3.8
96	4	1.9	0.1	5.3

tests, which are in agreement with a commonly assumed size effect concept (Weibull, 1939).

1.1.4 Unified size effect law

It is identifiable from background studies presented earlier that neither of the existing size effect models on their own can predict the size effect behavior

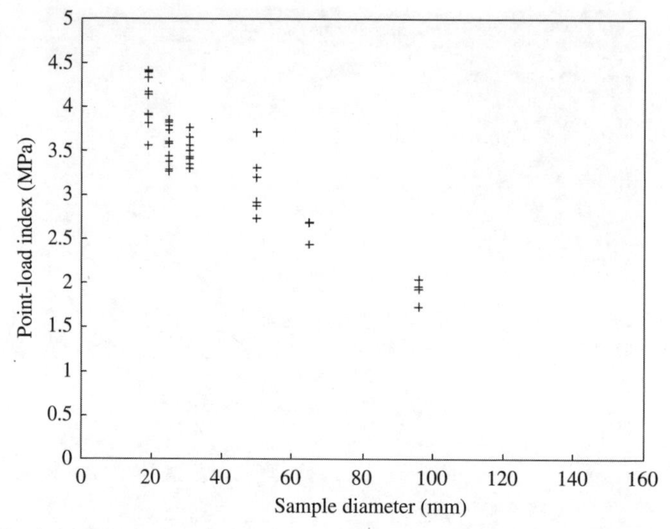

FIG. 1.21 Axial point-load strength indices at different sizes from Gosford sandstone.

of the UCS data across a wide range of diameters (see Figs. 1.12 and 1.14), thus a unified law is required. The law should be able to reproduce the size effect behaviors of different testing conditions and capture increasing and decreasing strengths with size.

Based on the Bazant (1997) argument, it is always the minimum strength predicted by the SEL and FFSEL models that represents the nominal strength of a material at any size. Therefore, the combination of these two models represents a unified size effect law (USEL), capturing ascending and descending strength zones as depicted in Fig. 1.22.

The intersection between SEL and FFSEL models occurs when:

$$d_i = \left(\frac{Bf_t}{\sigma_0}\right)^{2/(d_f-1)} \tag{1.10}$$

The USEL verification involved using the UCS data from Gosford sandstone as well as UCS data for five other sedimentary rock types reported by Hawkins (1998). The range of sample sizes should be sufficiently wide below and above the intersection diameter where the maximum strength is observed.

Initially, SEL was fitted to the UCS data above the intersection diameter to obtain Bf_t and λd_0. Then, the FFSEL was fitted to the UCS data below the intersection diameter using the same λd_0 as that resulting from SEL to attain the σ_0 and fractal dimension (d_f). The complete USEL captures both ascending and descending strength zones.

Estimating the exact location of the intersection diameter was the concern and it should be addressed that whether the sample size exhibiting the maximum

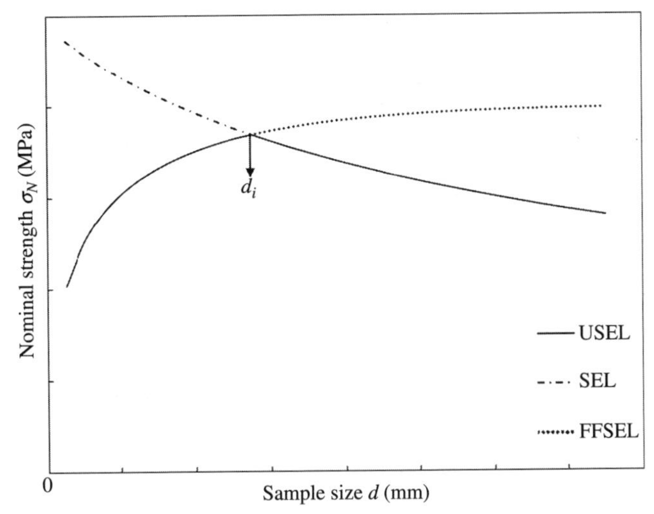

FIG. 1.22 Depiction of USEL, SEL, and FFSEL.

UCS among the data be included as part of the SEL or FFSEL. To address this issue, three cases were considered. First, the maximum UCS was only included in SEL. Second, it was only included in FFSEL. Finally it was simultaneously included in both SEL and FFSEL.

The obtained parameters for the three cases and six rock types are listed in Tables 1.4–1.6 and the selected case for each rock type is shaded. Only the

TABLE 1.4 Obtained USEL parameters for different rocks where sample with maximum UCS included in SEL

Rock sample	Diameter of sample with maximum UCS (mm)	USEL parameters				Diameter at intersection (mm)
		Bf_t *(MPa)*	λd_0 *(mm)*	σ_0 *(MPa)*	d_f	
Gosford sandstone	65	60.79	532.7	8.19	1.97	62.37
Pilton sandstone (Hawkins, 1998)	**54**	**241.73**	**75.82**	**60.26**	**1.70**	**52.93**
Pennant sandstone (Hawkins, 1998)	54	109	75.09	9.83	2.26	45.55

Continued

TABLE 1.4 Obtained USEL parameters for different rocks where sample with maximum UCS included in SEL—cont'd

Rock sample	Diameter of sample with maximum UCS (mm)	USEL parameters				Diameter at intersection (mm)
		Bf_t *(MPa)*	λd_0 *(mm)*	σ_0 *(MPa)*	d_f	
Bath stone (Hawkins, 1998)	54	19.18	93.64	1.71	2.28	43.69
Burrington oolite limestone (Hawkins, 1998)	54	205.34	49.75	17.80	2.27	47.05
Hallington sandstone (Hawkins, 1998)	**54**	**37.66**	**249.76**	**5.39**	**1.97**	**55.06**

TABLE 1.5 Obtained USEL parameters for different rocks where the sample with maximum UCS included in FFSEL

Rock sample	Diameter of sample with maximum UCS (mm)	USEL parameters				Diameter at intersection (mm)
		Bf_t *(MPa)*	λd_0 *(mm)*	σ_0*(MPa)*	d_f	
Gosford sandstone	65	64.26	322.65	8.04	1.99	66.62
Pilton sandstone (Hawkins, 1998)	54	244.03	73.50	61.76	1.68	56.90
Pennant sandstone (Hawkins, 1998)	54	151.44	29.45	7.67	2.52	50.64

TABLE 1.5 Obtained USEL parameters for different rocks where the sample with maximum UCS included in FFSEL—cont'd

Rock sample	Diameter of sample with maximum UCS (mm)	USEL parameters				Diameter at intersection (mm)
		Bf_t *(MPa)*	λd_0 *(mm)*	σ_0*(MPa)*	d_f	
Bath stone (Hawkins, 1998)	54	71.38	3.77	3.61	2.35	83.20
Burrington oolite limestone (Hawkins, 1998)	**54**	**253.74**	**28.24**	**13.61**	**2.51**	**48.18**
Hallington sandstone (Hawkins, 1998)	54	38.88	207.65	6.66	1.84	66.74

TABLE 1.6 Obtained USEL parameters for different rocks where the sample with maximum UCS included in SEL and FFSEL

Rock sample	Diameter of sample with maximum UCS (mm)	USEL parameters				Diameter at intersection (mm)
		Bf_t *(MPa)*	λd_0 *(mm)*	σ_0 *(MPa)*	d_f	
Gosford sandstone	**65**	**64.26**	**322.65**	**7.80**	**2.01**	**65.10**
Pilton sandstone (Hawkins, 1998)	54	244.03	73.50	59.93	1.70	55.24
Pennant sandstone (Hawkins, 1998)	**54**	**151.44**	**29.45**	**8.29**	**2.47**	**52.07**

Continued

TABLE 1.6 Obtained USEL parameters for different rocks where the sample with maximum UCS included in SEL and FFSEL—cont'd

Rock sample	Diameter of sample with maximum UCS (mm)	USEL parameters				Diameter at intersection (mm)
		Bf_t (*MPa*)	λd_0 (*mm*)	σ_0 (*MPa*)	d_f	
Bath stone (Hawkins, 1998)	**54**	**71.38**	**3.77**	**1.51**	**2.91**	**56.70**
Burrington oolite limestone (Hawkins, 1998)	54	253.74	28.24	N	N	N
Hallington sandstone (Hawkins, 1998)	54	38.88	207.65	5.29	1.99	56.24

resulting parameters from this case were used for model simulation presented in Figs. 1.23–1.28. For three rock types, the successful cases were obtained when the sample with maximum UCS was included in both SEL and FFSEL.

1.1.5 Reverse size effects in UCS results

The ascending and descending trend of UCS data at different sizes was initially reported by Hoskins and Horino (1969). Later, Vutukuri et al. (1974) argued that there are two mechanisms that influence size effects in unaxial compressive test simultaneously. The first mechanism is a commonly assumed size effect concept similar to that that causes the descending strength zone in which the strength reduces when size increases, consistent with the descending models presented earlier. The second mechanism is associated with surface flaws or surface imperfections created during sample preparation, which leads to the ascending strength zone in which, with an increase in size, the strength rises. Most likely, the flaws exist on the end surfaces where axial loads are applied and were produced during the sample cutting and/or grinding procedure when it is attempted to make the sample ends flat and square. There may also be some surface flaws on the sides of the sample. However, during a uniaxial compressive test, the sample is loaded through the end surfaces using flat platens, and the role of the end surface flaws becomes more important.

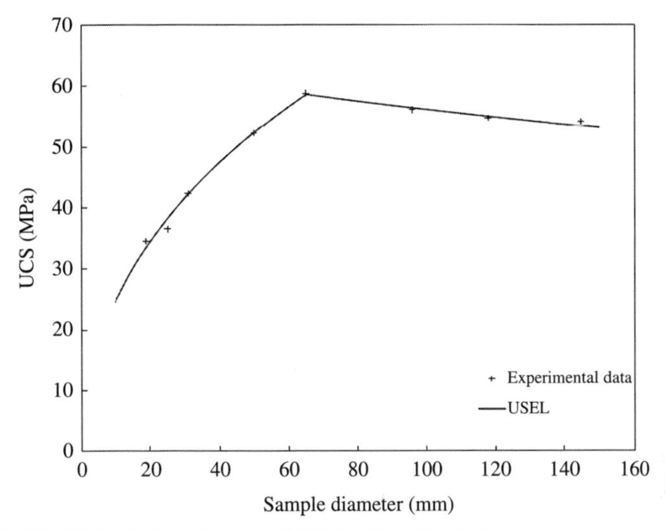

FIG. 1.23 Model simulation using mean UCS data from Gosford sandstone.

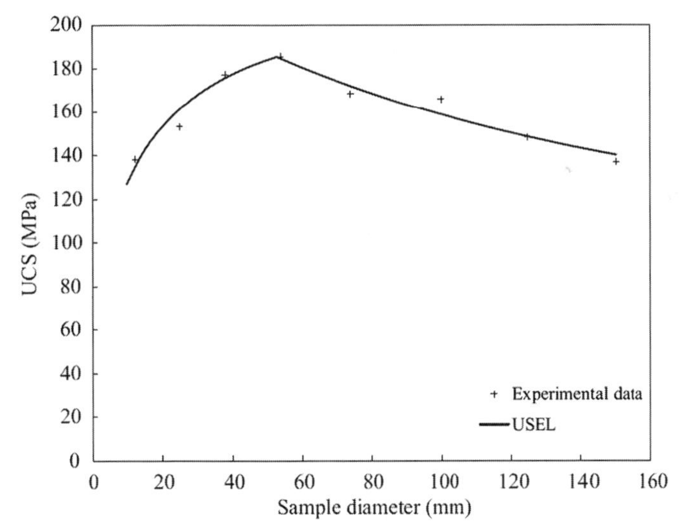

FIG. 1.24 Model simulation using UCS data from Pilton sandstone. *(Data from Hawkins, A.B., 1998. Aspects of rock strength. Bull. Eng. Geol. Environ. 57, 17–30.)*

The surface flaw idea of Vutukuri et al. (1974) was proposed before the fractal fracture energy-based ascending strength model of Bazant (1997). It may be that both surface flaws and fractal fracture contribute to the strength increase with size for small samples. If the surface flaw idea has validity, then the surface flaw effects may be eliminated (or reduced) by polishing the sample end

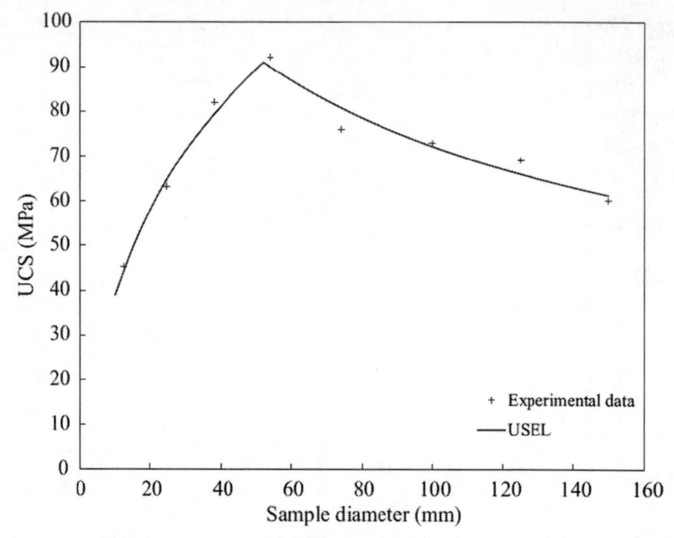

FIG. 1.25 Model simulation using UCS data from Pennant sandstone. *(Data from Hawkins, A.B., 1998. Aspects of rock strength. Bull. Eng. Geol. Environ. 57, 17–30.)*

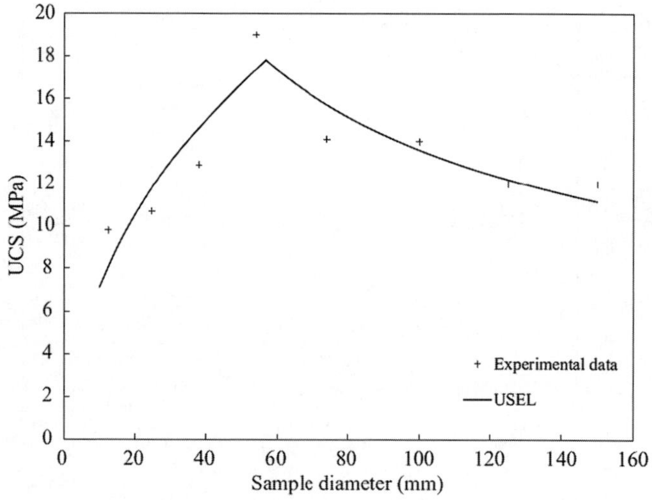

FIG. 1.26 Model simulation using UCS data from Bath stone. *(Data from Hawkins, A.B., 1998. Aspects of rock strength. Bull. Eng. Geol. Environ. 57, 17–30.)*

surfaces. A strength increase should result from polishing, and this is what is seen here. Two samples of Gosford sandstone, with 25 mm diameters, were carefully polished and tested under uniaxial compression. The comparison between the UCS results from unpolished and polished samples is presented in Fig. 1.29.

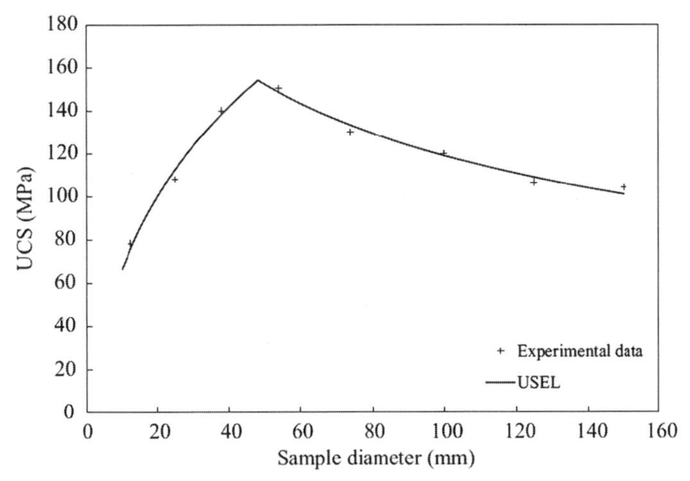

FIG. 1.27 Model simulation using UCS data from Burrington oolite limestone. *(Data from Hawkins, A.B., 1998. Aspects of rock strength. Bull. Eng. Geol. Environ. 57, 17–30.)*

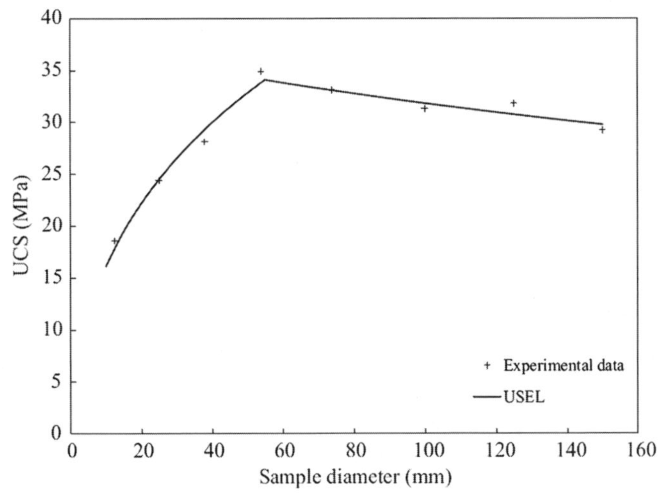

FIG. 1.28 Model simulation using UCS data from Hollington sandstone. *(Data from Hawkins, A. B., 1998. Aspects of rock strength. Bull. Eng. Geol. Environ. 57, 17–30.)*

The increase in UCS of the samples with 25 mm diameter due to polishing is quite evident from Fig. 1.29. However, it should be noted that it was difficult to perfectly polish a sedimentary rock due to its cemented structure with lots of pores, thus it was not possible to totally eliminate the effects of surface flaws on failure.

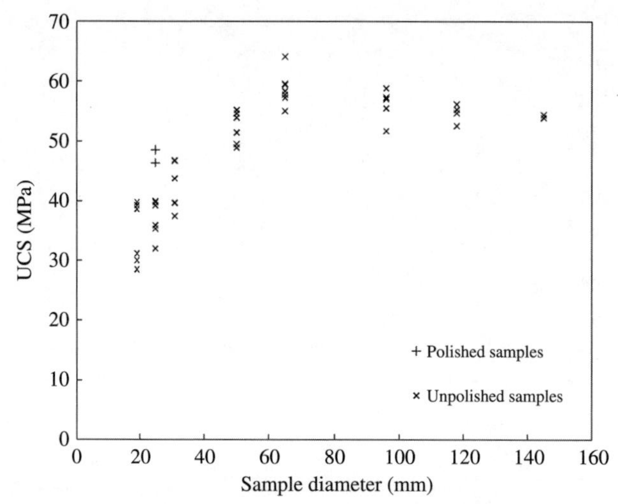

FIG. 1.29 Comparison between the UCS of polished and unpolished Gosford sandstone samples with diameters of 25 mm.

Even after polishing, it seems that other mechanisms are at work in influencing the strength of the rock as the polished samples were still weaker than the 65 mm diameter samples (see Fig. 1.29). Therefore, it remains necessary to apply an ascending strength model like that of Bazant (1997). The fractal characteristics of Gosford sandstone seem to be the primary mechanism that causes the strength ascent and surface flaws can be considered the secondary mechanism.

1.1.6 Contact area in size effects of point load results

In this section, an approach for obtaining a new point load strength index is proposed. The approach is novel in the way it incorporates the load contact area. The results were compared with the conventional method. The size effect trends of point load results obtained through the new method are presented and reproduced using USEL.

1.1.6.1 Conventional approach to highlight size effects

It is clear that neither Eq. (1.8) nor Eq. (1.9) includes a parameter that represents the contact surface area between the pointer and the sample. This is of concern because Russell and Wood (2009) demonstrated that load contact area controls the stress intensity immediately below the contact points, and it is near the load contact points that failure initiates. As a result, plotting I_s versus sample size may not enable the true size effects to be observed. It is also noted that, as shown in Fig. 1.30, I_s always reduces with increasing size, even for very small samples,

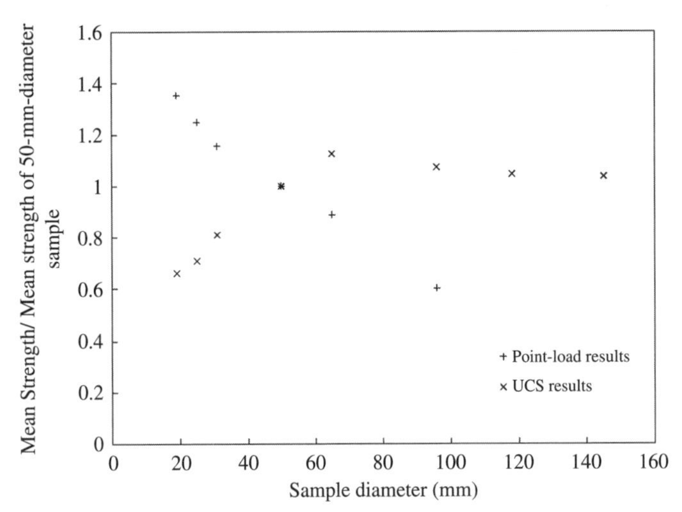

FIG. 1.30 Comparison between the size-effect trends of point load and UCS results for Gosford sandstone.

contrary to UCS data. This contrasting behavior has never been explored or properly understood.

1.1.6.2 A new approach incorporating contact area

Expressions for obtaining the contact area between two elastic spheres with different properties and radii are given by Timoshenko and Goodier (1951). The contact area is circular with a radius of:

$$r = \sqrt[3]{\frac{3\pi P(k_1 + k_2)R_1R_2}{4 \quad R_1 + R_2}} \tag{1.11}$$

where P is the contact load, R_1 and R_2 are the radii of two spheres, and k_1 and k_2 are obtained through:

$$k_1 = \frac{1 - \nu_1^2}{\pi E_1} \quad \text{and} \quad k_2 = \frac{1 - \nu_2^2}{\pi E_2} \tag{1.12}$$

in which ν_1 and ν_2 are the Poisson's ratios and E_1 and E_2 are the Young's moduli of the materials making up the two spheres. For simplicity, the surface roughness between pointer and sample (as discussed by Russell and Wood, 2009) is ignored. Typically, the pointers are made of tungsten carbide or hardened steel with a smooth and spherically curved tip of $R_1 = 5$ mm, as indicated by ISRM (Franklin, 1985; Franklin, 2007). Also, the elastic modulus and Poisson's ratio of tungsten carbide are about 700 GPa and 0.25, respectively (Russell and Wood, 2009).

It is noted that in a diametral point load test, as a pointer with a spherical tip pushes on a cylindrical surface, the contact area has an elliptical shape. However, for typical elastic properties, the ratio of major diameter over minor diameter in the ellipse is very close to unity and the contact area can be treated as circular for simplicity.

In an axial point load test, due to the pointer pushing on a flat surface, the contact area is circular and Eq. (1.11) is applicable with $R_2 \gg R_1$. Eq. (1.11) then simplifies to:

$$r = \sqrt[3]{\frac{3\pi P(k_1 + k_2)R_1}{4}} \tag{1.13}$$

The schematic trends of Eq. (1.13) for different values of E_2 and ν_2 are presented in Figs. 1.31 and 1.32, which show that both parameters control the upward or downward movement of the trend. Also, they demonstrate that the greater values of E_2 and ν_2 lead to lower trends of change in the radius of contact area during the axial loading.

A new point load strength index is defined as:

$$I_{st} = \frac{P}{A} \tag{1.14}$$

where A is the contact area. The resulting elastic properties of Gosford sandstone ($E = 15.5$ GPa and $\nu = 0.2$) were used for contact area calculations, leading to a new plot of I_{st} against a sample size for axial and diametral tests (see Fig. 1.33).

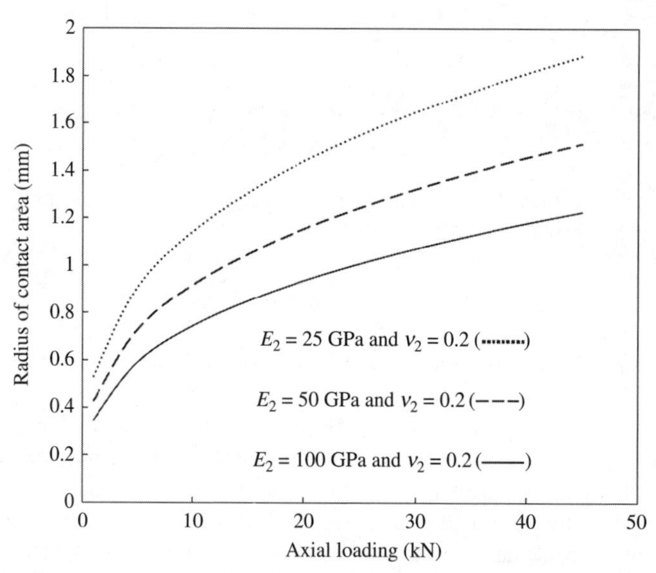

FIG. 1.31 Schematic representation of Eq. (1.13) at different E_2 values and identical ν_2 values.

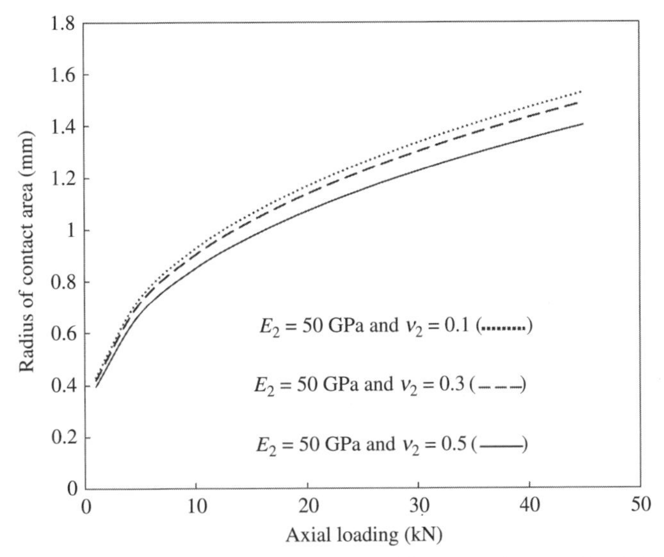

FIG. 1.32 Schematic representation of Eq. (1.13) at different ν_2 values and identical E_2 values.

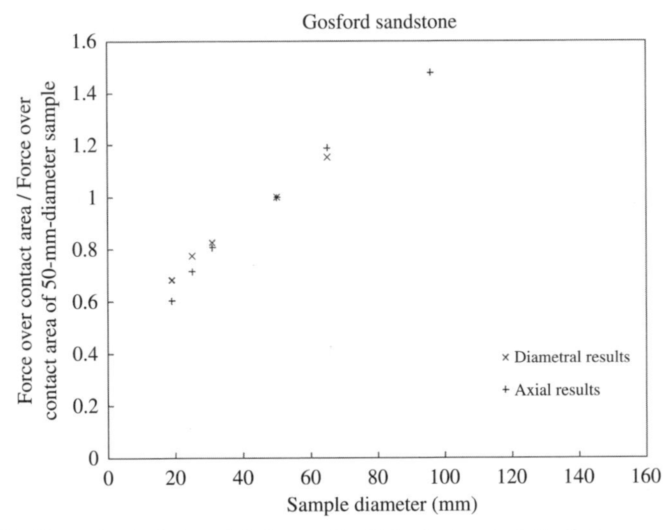

FIG. 1.33 New point load strength index results obtained using elasticity theory.

Fig. 1.33 shows the opposite trend to that observed in Fig. 1.30. Including contact area in the point load strength causes an ascending strength with size for both diametral and axial load configurations. The point load strength variation with size is simulated well using USEL (see Figs. 1.34 and 1.35).

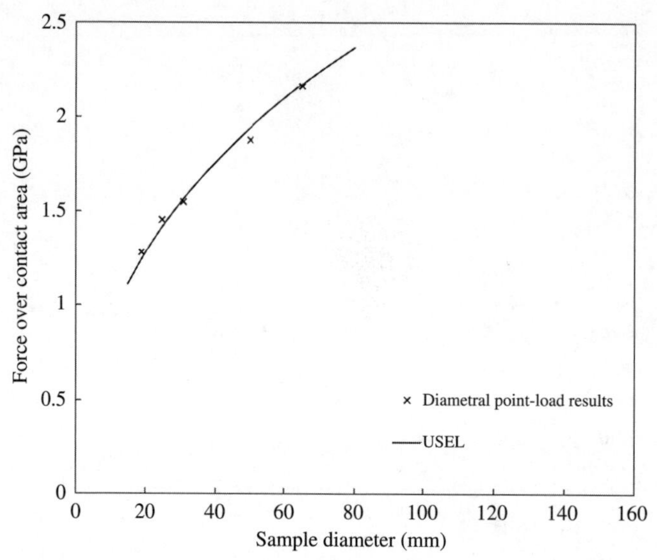

FIG. 1.34 Comparison between the diametral point load results obtained through the new approach and USEL.

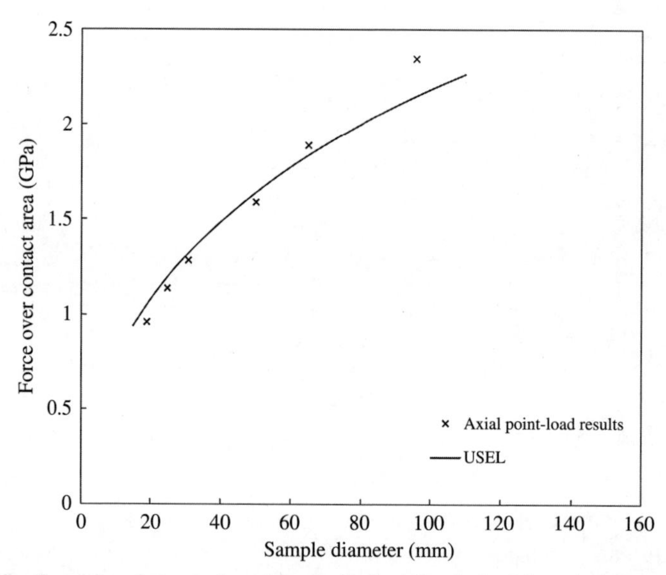

FIG. 1.35 Comparison between the axial point load results obtained through the new approach and USEL.

TABLE 1.7 The resulting constants of USEL for UCS, diametral, and axial results obtained from Gosford sandstone samples

Testing conditions	σ_0 (MPa)	d_f	λd_0 (mm)
UCS	7.8	2.01	322.65
Diametral PLT	289.25	2.01	322.65
Axial PLT	244.43	2.01	322.65

Note that due to a lack of descending strength data, the obtained $\lambda d_0 = 322.65$ mm from UCS data for Gosford sandstone was used for model simulations of point load results. Also, the same fractal dimension $d_f = 2.01$ as that attained from UCS results was utilized for the fitting process. Table 1.7 compares the resulting characteristic strengths σ_0 for UCS as well as diametral and axial point load data. The small difference between the obtained σ_0 for axial and diametral point load data is associated with the anisotropy of the samples and testing conditions, similar to that observed from the conventional method.

Similar to the UCS results, the author believes that perhaps two mechanisms cause the ascending strengths. The primary one is the fractal fracture theory, which underpins the model used in the simulations presented in Figs. 1.34 and 1.35. The secondary mechanism is the surface flaws effects.

Franklin (1985) conducted an extensive investigation using a number of rock types to formulate the difference between the resulting characteristic strengths from uniaxial compressive and point load tests, leading to the following relationship:

$$UCS = KI_s \tag{1.15}$$

where K is a correlation factor, UCS is uniaxial compressive strength, and I_s is the conventional point load strength index obtained from Eqs. (1.8), (1.9) for diametral and axial data, respectively. Later, Eq. (1.15) has been adopted by ISRM (2007) suggested methods for indirect estimation of UCS of an intact rock from the conventional point strength index. Based on the new approach, Eq. (1.14) was used to attain an alternate point load strength index in which the contact area is incorporated. Therefore, Eq. (1.15) can be written:

$$\text{For diametral}: UCS = K\frac{A}{D^2}I_{st} \quad \text{For axial}: UCS = K\frac{A}{4LD/\pi}I_{st} \tag{1.16}$$

where A is contact area and D and L are the same as those defined in Eqs. (1.8), (1.9). Now, the resulting characteristic strengths from UCS and point load data presented in Table 1.7 may be substituted into Eq. (1.16), and after some rearrangement:

$$\text{For diametral}: \frac{\sigma 0_{UCS}}{\sigma 0_{I_{st}}} = K\frac{A}{D^2} \quad \text{For axial}: \frac{\sigma 0_{UCS}}{\sigma 0_{I_{st}}} = K\frac{A}{4B/\pi} \tag{1.17}$$

where$\sigma 0_{UCS}$ is the characteristic uniaxial compressive strength and $\sigma 0_{I_{st}}$ is the characteristic point load strength index obtained from the new approach. Using Eq. (1.17), for different sample diameters, K values were estimated and are presented in Tables 1.8 and 1.9 for diametral and axial point load results, respectively. These tables also compare the resulting correlation factors from the new approach to those obtained from the conventional method.

Note that in Eq. (1.17), A represents the average of the load contact area for a given sample size. Tables 1.6 and 1.7 show that the resulting K values from both methods are similar and the difference for the diametral condition is less than that for the axial condition.

1.1.7 Conclusions

A unified size effect law (USEL) was introduced for intact rock in order to model both the ascending and descending strength zones. The USEL was verified against the UCS results from Gosford sandstone as well as five other rock types reported by Hawkins (1998). It was shown that there was good agreement between the model outputs and the experimental data. The influence of surface flaws on the sample failure was identified as a significant mechanism for strength ascending behavior in UCS results. This impact was ranked as the secondary mechanism while the fractal characteristics seemed to be the primary one.

TABLE 1.8 Comparison between estimated correlation factors *K* at different diameters obtained from new and conventional approaches for diametral point-load results

Sample diameter (mm)	*K* obtained from new approach	*K* obtained from conventional approach
19	8.1	7.7
25	9.0	8.7
31	10.8	10.8
50	15.9	15.8
65	18.5	19.6

TABLE 1.9 Comparison between the estimated correlation factors K at different diameters obtained from new and conventional approaches for axial point-load results

Sample diameter (mm)	K obtained from new approach	K obtained from conventional approach
19	6.5	8.4
25	8.6	10.7
31	9.9	12.1
50	14.2	16.9
65	20.7	22.6
96	33.0	29.5

The inclusion of contact area in size effects in point load results was investigated. It was shown that the resulting size effect trend from the new approach was quite different to that of the conventional method. Using the new approach, the point load strength index (I_{st}) increased as size increases for small sample sizes. Using a conventional approach to interpret point load data makes the results seem to always exhibit a strength reduction with increasing size, contrary to UCS data. Finally, it was demonstrated that the difference between the UCS results and the point load strength indices obtained from the new approach was approximately the same as the resulting difference between the UCS data and the point load indices obtained from the conventional method.

1.2 Length-to-diameter ratio on point load strength index

1.2.1 Introduction

The point load test is widely used within geotechnical and rock engineering in the classification of rock. Due to the low cost and portability of the test unit, it is commonly used within the mining industry to classify intact rock strength for use in rock mass classification systems such as the Rock Mass Rating (RMR) or Q systems (Brady and Brown, 2006). Knowledge of intact rock strength and the rock mass classification is a fundamental input for geotechnical design and also is a parameter for selecting the mining method. Accurate classification of intact rock strength is critical for the accurate classification of rock mass, which is necessary for the development of safe geotechnical and mine designs.

Since the introduction of the point load test, a size effect has been observed in the test results where the measured rock strength varies increasingly with sample size. This effect can have a significant impact on the classification of intact rock strength. Considerable research has been conducted to investigate the size effect in the point load test, with previous research finding that point load results vary with sample diameter and sample length-to-diameter ratio (Broch and Franklin, 1972; Brook, 1980; Greminger, 1982; Forster, 1983; Thuro et al., 2001a, b). This research has led to the introduction of sample size requirements for the point load test.

The size effect of sample size on strength and other mechanical properties has been observed in most strength tests of rock and other quasibrittle materials. The cause of the size effect has been studied extensively with numerous theories, including those based on statistics, fracture energy, and multifractals having been proposed to explain the cause of the size dependency of material strength (Masoumi, 2013).

While there has been considerable research to investigate the size effect in the point load test and the cause of the size effect, there has been limited research to apply the theories of the causes of size effect to the size effect observed in the point load test results. Also, few investigations have applied size effect theories to point load test results with varying length-to-diameter ratios.

1.2.2 Background

1.2.2.1 Point load test size effect

Broch and Franklin (1972) provided a comprehensive investigation of the impact of sample size on the point load strength index to define an accepted unified size dimension for conducting point load tests. They found that as the diameter increased, the point load strength index decreased for both axial and diametral testing conditions. This relationship between diameter and the point load strength has been observed by multiple studies (Brook, 1980; Greminger, 1982; Forster, 1983; Thuro et al., 2001a, b; Masoumi, 2013; Brook, 1985).

A number of investigations have also investigated the impact of the length-to-diameter ratio of samples on the point load strength index (Broch and Franklin, 1972; Greminger, 1982; Forster, 1983). For the axial point load test, the point load strength index is observed to decrease as the sample length-to-diameter ratio increases. For the diametral point load test, a very different relationship between the point load strength index and the sample length-to-diameter ratio was observed by Broch and Franklin (1972). They found that as sample length-to-diameter ratio increased, the point load strength index increased up until the length-to-diameter ratio was approximately equal to one. Beyond this value, the point load strength index was independent of the sample length-to-diameter ratio.

1.2.2.2 Size effect models

A number of models have been proposed to explain and predict the size effect. Darlington and Ranjith (2011) categorized size effect models into three categories: statistical, fracture energy, and multifractal models.

The most common size correction method used in the point load test is the one proposed by Brook (1985). The size corrected strength index to a standard core diameter of 50 mm ($I_{s(50)}$) can be obtained by incorporating a size correction factor (F), as shown in Eqs. (1.18), (1.19) where P is the peak load and D_e is the equivalent diameter. The exponent, a, has been experimentally determined to be approximately 0.5.

$$I_{s(50)} = F\frac{P}{D_e^{\ 2}} \tag{1.18}$$

$$F = \left(\frac{D_e}{50}\right)^a \tag{1.19}$$

Bazant's (1984) size effect law (SEL) was the first size effect model using fracture energy theory. Bazant (1984) derived the SEL by assuming a certain level of energy (also known as fracture energy) that is required to extend a crack band along a unit length. This level of energy is assumed to be a material constant. By quantifying the energy of crack propagation and solving for the nominal stress, Bazant (1984) derived the SEL size effect model according to Eq. (1.20), where σ_N is a nominal strength, B and λ are material constants, f_t is the strength of a sample with negligible size, d is the sample size, and d_0 is the maximum aggregate size.

$$\sigma_N = \frac{Bf_t}{\sqrt{1+d/\lambda d_0}} \tag{1.20}$$

Cracked zones under peak load have been observed to behave as multifractals. This behavior means that the cracks are self-similar at different sizes where their distribution and shape are the same at different microscales. Carpinteri et al. (1995) proposed a new size effect law called the multifractal scaling law (MFSL) based on the fractal behavior of quasibrittle materials. The MFSL is shown in Eq. (1.21), where σ_N is the nominal strength, f_c is the strength of a sample with infinite size, l is a material constant, and d is the characteristic sample size.

$$\sigma_N = f_c\sqrt{1+\frac{l}{d}} \tag{1.21}$$

After the introduction of MFSL, Bazant and Planas (1998) combined fracture energy and multifractals, leading to the proposal of a new size effect model known as the fractal fracture size effect law (FFSEL). This model assumed that quasibrittle materials display fractal characteristics only within a certain range of sizes. By expressing and equating the energy dissipated using a fractal crack to the energy required to extend a fractal crack, Bazant and Planas (1998) derived the equation for FFSEL. The FFSEL equation is shown in Eq. (1.22), where σ_N is the nominal strength, σ_0 and λd_0 are material constants, and d_f is the fractal dimension. Bazant and Planas (1998) provided no experimental results to verify the FFSEL.

$$\sigma_N = \frac{\sigma_0 d^{(d_f-1)/2}}{\sqrt{1+d/\lambda d_0}} \tag{1.22}$$

1.2.3 Methodology

Axial and diametral point load tests on more than 375 samples were conducted to investigate the size effect of the point load strength index on Gosford sandstone samples. The diameters varied between 17 and 95 mm with length-to-diameter ratios ranging between 0.3 and 2, so that the impact of the diameter and the length-to-diameter ratio could be observed. A GCTS point load testing system with a maximum load of 100kN was used for all point load tests. This system used a digital force gauge to eliminate human error in recording peak force. With the exception of sample dimensions, the ISRM (2007)-recommended guidelines for axial and diametral point load tests were followed during the sample preparation and point load testing.

For axial and diametral testing, samples were prepared with diameters of 17, 26, 39, 51, 67, and 95 mm diameters. The samples were prepared with length-to-diameter ratios of 0.3, 0.5, 1, 1.5, and 2. These length-to-diameter ratios were chosen, as they are within the suggested range by ISRM (2007) and also beyond the recommended range for both axial and diametral tests.

With six diameters and five length-to-diameter ratios used for testing under axial and diametral conditions, a total of 60 different testing scenarios could be created. However, no diametral test results with a diameter of 95 mm were obtained due to sample twisting during the experiment leading to invalid test results (Masoumi, 2013). Axial testing of 95 mm diameter samples with a length-to-diameter ratio of 2 was also excluded due to equipment limitations.

Masoumi (2013) conducted a number of point load tests using the same batch of Gosford sandstone and the results from his investigation were used in this study. Sample preparation was completed in three stages: coring, cutting, and drying. The samples were oven dried for at least 24 h at 105°C before testing.

The ISRM (2007) guidelines for point load tests were followed during testing. Samples were placed into the point load testing frame as shown in Fig. 1.36. A total of 205 axial and 169 diametral point load tests were conducted.

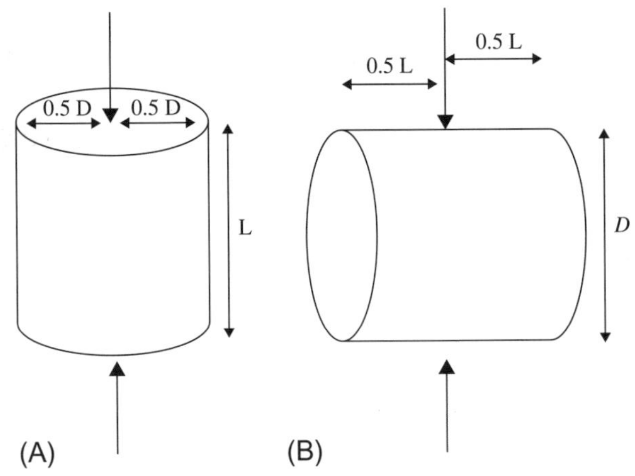

FIG. 1.36 Demonstration of placement of samples into the point load testing frame (A) axial and (B) diametral tests.

1.2.4 Valid and invalid failure modes

As outlined by the ISRM (2007), if the fracture surface after failure only passes through one loading pointer, the test is invalid. Valid failure in axial and diametral testing requires that the failure surface passes through both loading pointers. Examples of invalid modes are shown in Fig. 1.37. During the experiments, it was observed that the percent of invalid samples was affected by the sample length-to-diameter ratio.

1.2.4.1 Failure mode in axial testing

In axial testing, it was observed that for samples with a high length-to-diameter ratio, few samples failed validly. The percentage of failure at each sample length-to-diameter ratio for axial testing is shown in Table 1.10. It demonstrates that for a sample with a length-to-diameter ratio less than or equal to 1, all samples failed validly. For samples with a length-to-diameter ratio of 1.5, less than

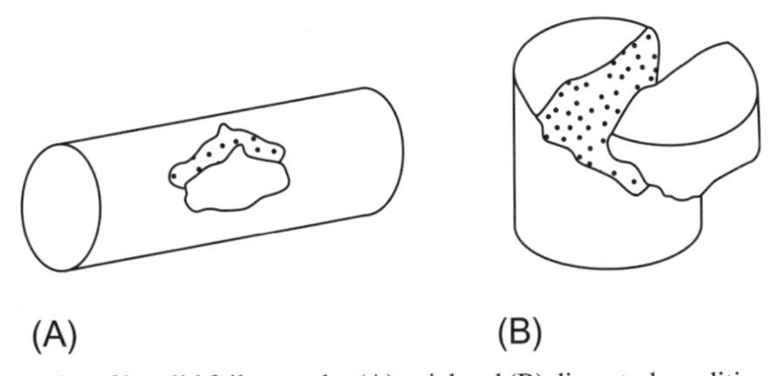

FIG. 1.37 Examples of invalid failure under (A) axial and (B) diametral conditions (ISRM, 2007).

TABLE 1.10 Number of tests performed at all length-to-diameter ratios for axial testing and the number of valid failures

L/D	Number tested	Valid failures	Valid failure percentage (%)
0.3	45	45	100
0.5	45	45	100
1.0	32	32	100
1.5	44	19	43
2.0	39	2	5

TABLE 1.11 Number of tests performed at all length-to-diameter ratios for diametral testing and the number of valid failures

L/D	Number tested	Valid failures	Valid failure percentage (%)
0.3	40	23	58
0.5	40	31	78
1.0	11	11	100
1.5	39	39	100
2.0	39	39	100

half the samples failed in a valid manner. However, for the samples with a length-to-diameter ratio of 2, approximately 5% failed validly.

1.2.4.2 Failure mode in diametral testing

In diametral testing, the opposite trend compared to axial testing was observed. The percentage of failure at each sample length-to-diameter ratio for diametral testing is shown in Table 1.11. This table shows that for samples with a length-to-diameter ratio of 1 or greater, all samples failed validly. For samples with a length-to-diameter ratio of 0.5 and 0.3, the number of samples that failed validly was 78% and 58%, respectively.

For samples with 17 and 26 mm diameters, 100% of samples with a length-to-diameter ratio of 0.3 or 0.5 failed validly. Samples with a length-to-diameter ratio of 0.3 having 51 and 67 mm diameters had no valid sample failures. In general, it was observed that for diametral testing of a sample with a length-to-diameter ratio of 0.5 or less, samples with a small diameter still failed validly.

1.2.4.3 Impact of stress distribution on failure mode

The failure mode results show a very different relationship between the valid failure and the length-to-diameter ratio for axial and diametral testing conditions. It is speculated that the reason behind length-to-diameter ratio impacting the failure mode for both axial and diametral testing is the stress distribution within the sample during the loading. The stress distribution of a point load can be approximated by Boussinesq's formula for point loading shown in Eq. (1.23). Boussinesq established that the vertical stress at a point P (σ_z) is dependent on the size of the applied load (Q) through the pointer and the vertical (z) and horizontal (r) distance of the pointers from the point load. This relationship is simplified with the introduction of the Boussinesq stress coefficient (I_B).

$$\sigma_z = \frac{3Q}{2\pi z^2}\frac{1}{\left[1+\left(\frac{r}{z}\right)^2\right]^{\frac{5}{2}}} = I_B\frac{Q}{z^2} \tag{1.23}$$

The application of this formula to solids requires a number of assumptions. These assumptions include that the solid is elastic, isotropic, homogeneous, and semiinfinite and that the solid is weightless such that there is no stress within the solid due to supporting its own weight (Murthy, 2002).

For axial testing of samples with large length-to-diameter ratios, Boussinesq's formula indicates that before failure at the sample ends, there would be high stresses right up to the sample edges due to the low horizontal distance of the sample edge from the loading pointer. However, in the center of the sample, the stresses would be low due to the large vertical distance from the pointers. At failure for a sample with a large length-to-diameter ratio, failure will occur in the regions with high stress, resulting in the failure surface starting at the pointer where the stress is highest. Failure will then propagate toward the sample edge where the stress is high rather than propagating down the middle of the sample where stress is low. This results in the formation of a chip rather than splitting the sample through the middle, leading to an invalid failure.

A similar view of stress distribution and failure propagation can be used to explain the formation of chips during the failure of samples with a low length-to-diameter ratio during diametral testing. The difference in the impact that length-to-diameter ratio has on axial and diametral testing can simply be explained due to diameter being the horizontal distance in axial testing but the vertical distance in diametral testing while length is the vertical distance in axial testing but the horizontal distance in diametral testing.

Stress distribution from Boussinesq's formula suggests that as long as the sample shape stays constant, the stress distribution would be the same regardless of scale. However, in reality there is a significant difference for the point load test as scale changes. The loading by the conical pointers is applied over a contact area and is therefore not a true point load. For large sample sizes, this can be approximated as a point. For small samples, the contact area is significant when compared to the size of the sample and therefore the applied load cannot be approximated as a point load, resulting in Boussinesq's formula becoming nonapplicable.

1.2.5 Conventional point load strength index size effect

1.2.5.1 Axial and diametral point load strength index results

The axial and diametral point load strength index was calculated using the ISRM's (2007) suggested formula. Shown in Eq. (1.24) is the formula used for calculating the axial point load strength index and shown in Eq. (1.25) is the formula used for calculating the diametral point load strength index.

$$I_s = \frac{P}{4A/\pi} \tag{1.24}$$

$$I_s = \frac{P}{D^2} \tag{1.25}$$

Summaries of the axial and diametral point load strength index results are presented in Figs. 1.38 and 1.39. These graphs display the average point load index for all tested samples having different length-to-diameter ratios and valid failure modes. Within each graph, the results have been grouped by sample length-to-diameter ratio, showing how the point load strength index changes with the diameter.

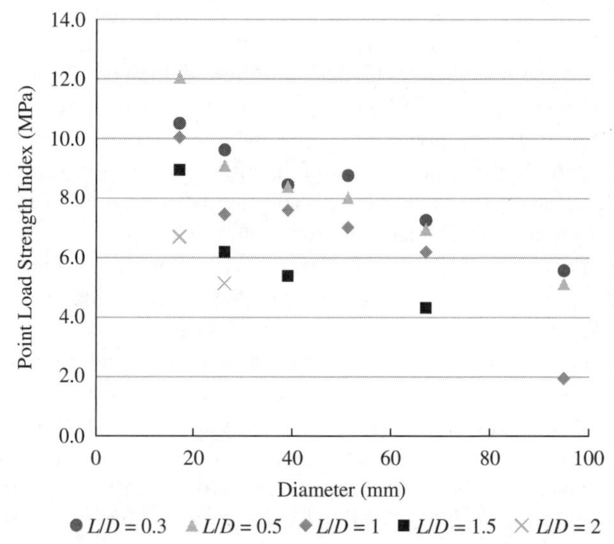

FIG. 1.38 Summary of average axial point load test results as diameter changes for all diameters and length-to-diameter ratios.

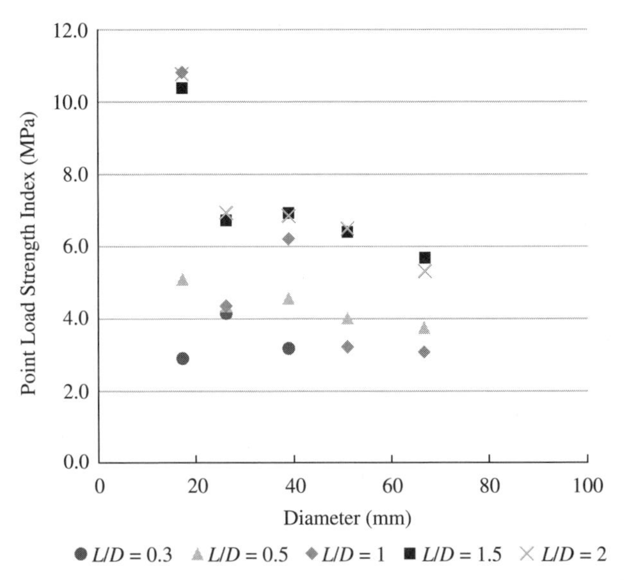

FIG. 1.39 Summary of average diametral point load test results as diameter changes for all diameters and length-to-diameter ratios.

Figs. 1.38 and 1.39 further illustrate that as the sample diameter increases, the point load strength index decreases. This trend corresponds to the relationship between point load strength index and sample diameter observed by other researchers (Greminger, 1982; Masoumi, 2013; Brook, 1985). It is evident from Figs. 1.38 and 1.39 that the size effect can be observed across a range of length-to-diameter ratios, both inside and outside the range recommended by the ISRM (2007) for both axial and diametral testing. The size effect is not limited to only a number of length-to-diameter ratios but can be observed in all ratios that were tested and had sufficient valid results. This size effect is in agreement with the basic size effect theory (Bazant, 1984; Weibull, 1939). The size effect cannot be adequately determined for axial testing with a length-to-diameter ratio of 2 and diametral testing with a length-to-diameter ratio of 0.3 due to the limited available results caused by invalid failure.

Also, for a number of sample diameters such as 26 mm and 39 mm with length-to-diameter ratios of 1 in both axial and diametral testing, the point load strength index was observed to increase as the diameter decreases. The cause of this reversal trend between some data is speculated to be due to data scatter.

1.2.5.2 Applicability of existing size effect models

The applicability of the three main size effect models to point load test results was examined using an analytical approach. The data fitting process was carried out on the mean point load index of each sample diameter with different length-to-diameter ratios.

The size effect models used in the fitting process consisted of the SEL, the MFSL, and the Brook Model (Brook, 1985). The results from the data fitting process are presented in Table 1.12. The Brook Model could not be applied to all length-to-diameter ratios as a requirement of the Brook Model is the value of $I_{s(50)}$, which was not measured in some instances due to invalid failure resulting in no results. For all other length-to-diameter ratios, the point load strength index of the 51 mm diameter samples was used as an approximation of $I_{s(50)}$. No models could be fitted to axial results with a length-to-diameter ratio of 2 or to diametral results with a length-to-diameter ratio of 0.3 as few valid results were collected.

From Table 1.12 it can be seen that for axial results, the coefficient of correlation (R^2) of SEL and MFSL is very similar for most length-to-diameter ratios. Using the coefficient of correlation as a measure of fit, SEL fits better to the length-to-diameter ratio of 0.3 in axial results while SEL and MFSL fit similar to other axial results. The Brook Model performed poorly compared to SEL and MFSL. This is likely due to the existence of two constants in SEL and MFSL, which allowed for better fitting of the models to the experimental results.

The coefficients of correlation (R^2) for diametral results shown in Table 1.12 show that for most length-to-diameter ratios, all three models perform very similarly. This is also illustrated in Fig. 1.40. One exception is the samples with length-to-diameter ratio of 1, where SEL and MFSL performed poorly and the Brook Model performed significantly better. The differences in the model predictions for this scenario can be clearly observed in Fig. 1.40, which also illustrates that the reason for the poor performance of the SEL and MFSL models on diametral tests with a length-to-diameter ratio of 1 is that the observed size effect in the experimental data does not follow the conventional size effect concept. As discussed previously, this reversal trend is possibly due to the data scatter.

1.2.6 Size effect of point load strength index

Masoumi (2013) proposed a new point load strength index (I_{st}) that included the contact area between the conical load pointers and the sample. This strength index shown in Eq. (1.26) is equal to the peak load at failure (P) over the contact area (A). This definition of the point load strength index of force over area is similar to the definition of stress, which is also force over area. Area is assumed to be circular with radius (r) calculated using Eq. (1.27) where R_1 is the radius of the point load conical platen. The values of k_1 and k_2 are obtained from Eq. (1.28), where v_1 and v_2 are the Poisson's ratios, and E_1 and E_2 are the Young's moduli of the conical platens and the tested samples, respectively. For simplicity, it is assumed that the contact area between the conical platen and the sample surface is close to circular in shape. Surface roughness between the conical platen and sample has also been ignored.

TABLE 1.12 Fitting constants and coefficients of correlation for SEL, MFSL, and Brook Model obtained from axial and diametral testing conditions

		SEL			MFSL			Brook model	
	L/D	λd_0 *(mm)*	Bf_t*(MPa)*	R^2	f_c*(MPa)*	l *(mm)*	R^2	K	R^2
Axial	0.3	13.8	24.3	0.92	5.13	61.45	0.82	0.22	0.60
	0.5	23.2	5.81	0.94	3.09	233	0.94	0.36	0.89
	1.0	40.4	1.11	0.74	0.31	18,503	0.74	0.37	0.58
	1.5	91.3	0.14	0.94	0.31	12,249	0.94	No 50 mm result	
	2.0	Insufficient data			Insufficient data			No 50 mm result	
Diametral	0.3	Insufficient data			Insufficient data			No 50 mm result	
	0.5	5.76	51.03	0.84	3.42	20.79	0.82	0.21	0.83
	1.0	187	0.032	0.61	0.11	88,618	0.61	1.06	0.77
	1.5	22.67	3.87	0.79	3.21	136	0.82	0.38	0.79
	2.0	100	0.18	0.85	2.07	386	0.86	0.40	0.81

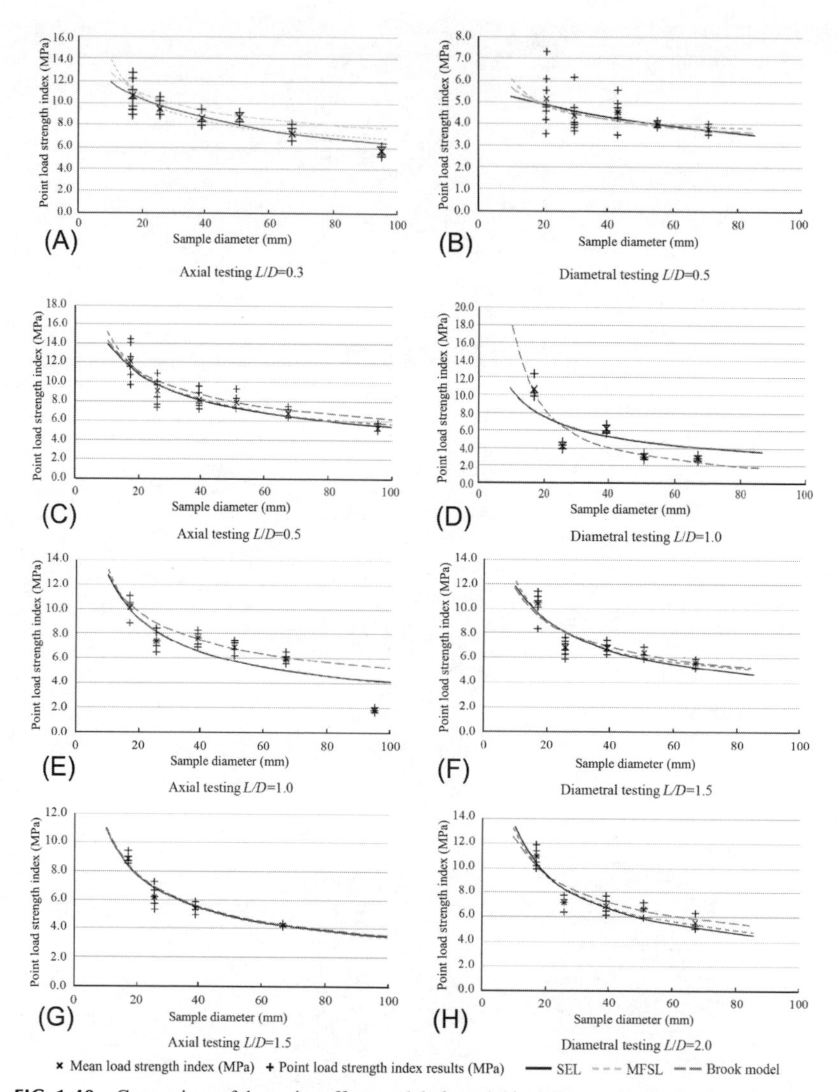

FIG. 1.40 Comparison of three size effect models for axial and diametral point load results over a range of diameters and length-to-diameter ratios.

$$I_{st}=\frac{P}{A} \tag{1.26}$$

$$r=\sqrt[3]{\frac{3\pi P(k_1+k_2)R_1}{4}} \tag{1.27}$$

$$k_1=\frac{1-v_1^2}{\pi E_1} \text{ and } k_2=\frac{1-v_2^2}{\pi E_2} \tag{1.28}$$

The radius of the tungsten carbide conical pointer (R_1) was assumed to be 5 mm as indicated by the ISRM (2007), and the Young's Modulus (E_1) and Poisson's Ratio (v_1) of tungsten carbide were assumed to be 700 GPa and 0.25, respectively (Russell and Wood, 2009). The Young's Modulus (E_2) and Poisson's Ratio (v_2) for Gosford sandstone were assumed to be 12.1 GPa and 0.14, according to Masoumi (2013) study. The size dependency of the Young's Modulus and Poisson's Ratio were ignored for simplicity.

1.2.6.1 Axial and diametral point load strength index

Summaries of the axial and diametral point load strength index results incorporating contact area are presented in Figs. 1.41 and 1.42. These graphs display the average I_{st} for all tested samples having varying diameters and length-to-diameter ratios. Within each graph, the results have been grouped by length-to-diameter ratios.

Figs. 1.41 and 1.42 illustrate that as sample diameter increases, I_{st} increases too. An important observation from these graphs is that there is a trend of size effect across a range of length-to-diameter ratios for both axial and diametral testing. The size effect is not limited to only some length-to-diameter ratios, but can be observed in all ratios included in this study. This size effect is in agreement with the size effect reported by Masoumi (2013). The size effect cannot be adequately determined for axial testing with a length-to-diameter ratio of 2 nor for diametral testing with a length-to-diameter ratio of 0.3 due to the limited results available caused by invalid failure.

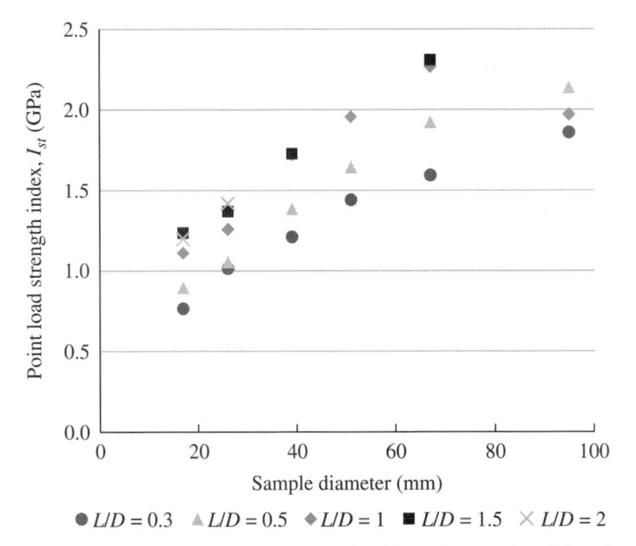

FIG. 1.41 Summary of average axial I_{st} results obtained from the samples with various diameters and length-to-diameter ratios.

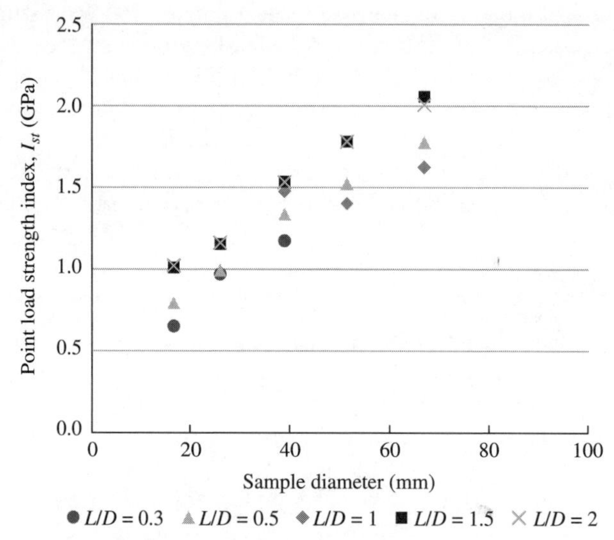

FIG. 1.42 Summary of average diametral I_{st} results obtained from the samples with various diameters and length-to-diameter ratios.

1.2.6.2 Applicability of existing size effect models

The results from the data fitting process of FFSEL to the axial and diametral I_{st} results at a range of length-to-diameter ratios are presented in Table 1.13. This shows the values determined for σ_0 and d_f through the size effect fitting process. The λd_0 values determined during SEL data fitting of the conventional axial point load strength index are shown in Table 1.12. The FFSEL model could not be applied to axial results with a length-to-diameter ratio of 2 nor to diametral results with a length-to-diameter ratio of 0.3 due to the availability of the limited data.

TABLE 1.13 Fitting constants and coefficient of correlation for FFSEL from axial and diametral point load tests

		FFSEL			
	L/D	λd_0 *(mm)*	σ_0 *(MPa)*	d_f	R^2
	0.3	13.8	0.10	2.75	0.99
	0.5	23.2	0.10	2.71	0.99
Axial	1.0	40.4	0.24	2.24	0.82
	1.5	91.3	0.20	2.28	0.97
	2.0	Insufficient data			
	0.3	Insufficient data			
	0.5	5.76	0.08	3.06	0.99

TABLE 1.13 Fitting constants and coefficient of correlation for FFSEL from axial and diametral point load tests—cont'd

		FFSEL			
	L/D	λd_0 *(mm)*	σ_0 *(MPa)*	d_f	R^2
Diametral	1.0	187	0.26	1.94	0.83
	1.5	22.67	0.11	2.73	0.99
	2.0	100	1.58	2.33	0.99

Table 1.13 indicates good correlation with the R^2 values range from 0.82 to 0.99. The length-to-diameter ratio of 1 has a lower R^2 value. The fitted models to the results are shown in Fig. 1.43. In general, it is true to state that FFSEL can accurately predict the size effect of axial I_{st} for samples with different length-to-diameter ratios.

The values of the coefficient of correlation (R^2) for FFSEL for the diametral results are also shown in Table 1.13. The R^2 values range from 0.83 to 0.99, indicating that the FFSEL size effect model can accurately model and predict the observed size effect of diametral I_{st} for all length-to-diameter ratios. Similar to the axial results, the length-to-diameter ratio of 1 has a lower R^2 than others. Overall, the FFSEL can accurately predict the size effect behavior of diametral I_{st} for samples with various length-to-diameter ratios.

1.2.7 Conclusions

A sample length-to-diameter ratio was found to have a significant impact on the failure mode and the validity of the failure modes for both axial and diametral point load testing. Axial testing on the samples with a length-to-diameter ratio equal to or greater than 1.5 and diametral testing using samples with a length-to-diameter ratio equal or less than 0.5 resulted in the failure of few samples under valid conditions. It is thought that the reason for the invalid failures of axial samples with a large length-to-diameter ratio and diametral samples with a small length-to-diameter ratio is due to the uneven internal stress distribution in the sample at the time of failure.

Results from the conventional point load strength index demonstrated that as the sample diameter increased, the point load strength index decreased. Results from this investigation further conclude that this trend is observable at all tested length-to-diameter ratios. Data fitting of three size effect models to the collected results found that the SEL and MFSL better predict the size effect trend compared to the Brook Model for various length-to-diameter ratios.

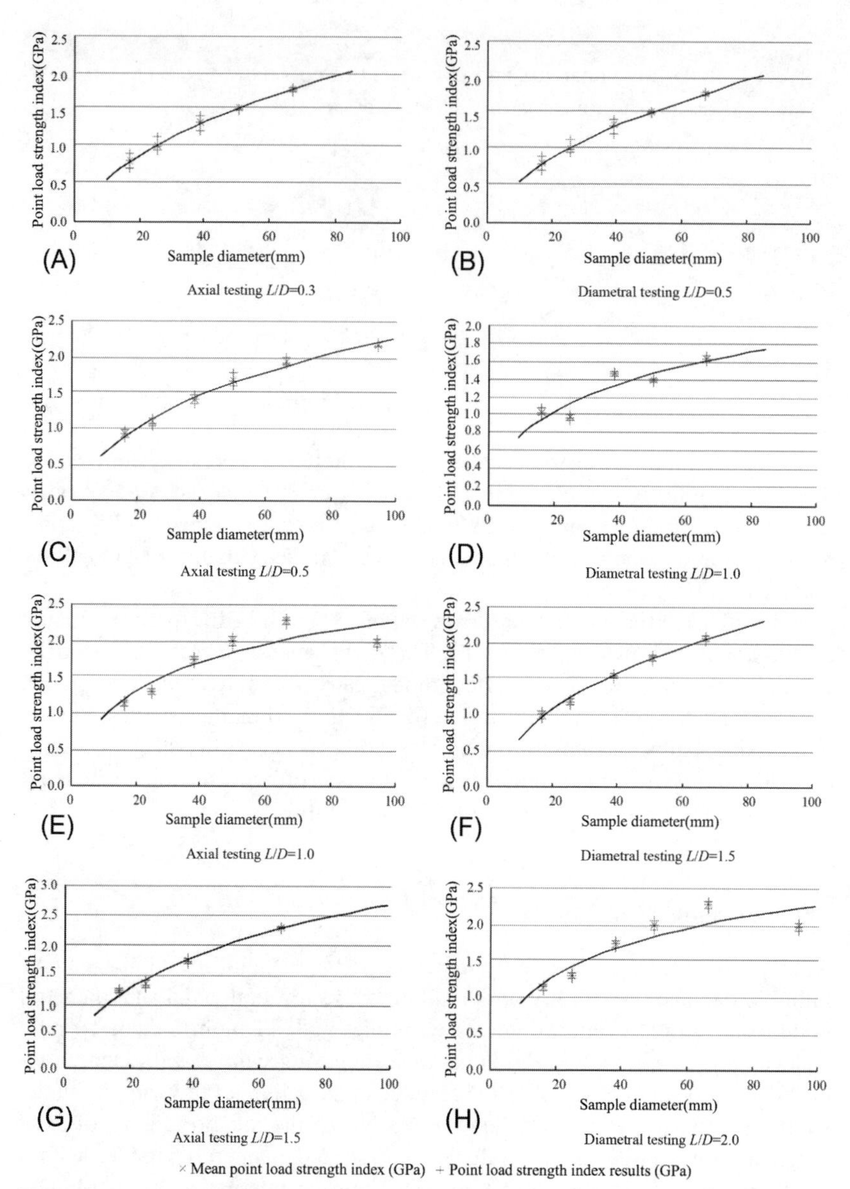

FIG. 1.43 Comparison between axial I_{st} results and FFSEL over a range of length-to-diameter ratios for axial and diametral testing conditions.

Results from the point load strength index incorporating contact area demonstrated that as the sample diameter increased, the new point load strength index also increased. With the results from this study, it is possible to also conclude that the increasing size effect trend is present over the full range of length-to-diameter ratios investigated here. Data fitting was also carried out to fit the FFSEL model to the axial and diametral size effects for different length-to-diameter ratios that had sufficient valid results. The data fitting process found that the FFSEL size effect model can accurately model and predict the size effect of the axial and diametral point load strength index incorporating contact area for all length-to-diameter ratios.

1.3 Plasticity model for size-dependent behavior

1.3.1 Introduction

The strength, stiffness, and stress-strain behavior of rock are affected by the size of the rock subjected to loading. The size dependence is observed in laboratory tests, for example in unconfined compressive strength tests (Mogi, 1962; Lundborg, 1967; Hoskins and Horino, 1969; Nishimatsu et al., 1969; Dhir and Sangha, 1973; Baecher and Einstein, 1981; Kramadibrata and Jones, 1993; Pells, 2004; Simon and Deng, 2009; Darlington and Ranjith, 2011), point load tests (Broch and Franklin, 1972; Bieniawski, 1975; Brook, 1980; Greminger, 1982; Brook, 1985; Hawkins, 1998; Thuro et al., 2001b), and indirect tension tests or Brazilian disc tests (Mellor and Hawkes, 1971; Wijk et al., 1978; Thuro et al., 2001a). A similar size effect is also observed in concrete (e.g., Van Mier, 1996; Bazant and Planas, 1998).

The size dependence is important to civil and mining engineering practice. One example relates to the design of structures on or within a rock mass for which an estimate of the strength of the intact rock blocks within the mass is required. These blocks can be many orders of magnitude greater in size than the laboratory samples tested. Another example relates to two mining methods, the room-and-pillar and bord-and-pillar methods, which rely on the strengths of pillars to support underground openings. However, pillar sizes and thus their strength and stiffness can vary significantly. There is a need to be able to correct laboratory strength measurements made on small samples so they can be suitably applied to the design of larger rock structures. There is also a need to be able to describe the stress-strain behavior of rock in a way that recognizes its size dependency.

Size effects in rock engineering have been of interest for several decades with many studies focusing on the phenomenon. A number of models have been produced that define strength at different sizes. Most describe a descending strength with increasing sample size. They are based on either statistics, fracture energy, or fractals or are empirical in origin. Many empirical models have appeared in the literature (e.g., Mogi, 1962; Dey and Halleck, 1981; Yoshinaka et al., 2008; Darlington and Ranjith, 2011). Hoek and Brown (1980a, b, 1997) considered experimental results (Mogi, 1962; Pratt et al., 1972) and proposed a model in which strength is related to sample size in a power law. The power law can be derived using the statistical analysis presented by Weibull (1939) in which it was postulated that a solid consists of preexisting flaws (microcracks). The larger the sample of the solid, the greater the number of flaws it contains and the greater the likelihood of failure under a given load. Bazant (1984) derived a theoretically based descending strength model by considering the strain energy or fracture energy dissipated at failure and during crack propagation. Carpinteri et al. (1995) considered the geometries of fractures that have self-similar fractal properties (Carpinteri, 1994; Carpinteri and Ferro, 1994; Borodich, 1999) to propose a model for descending strength. However, many researchers point out that a descending strength with increasing sample size may not be applicable for small sizes (Hiramatsu and Oka, 1966; Hoskins and Horino, 1969; Abou-Sayed and Brechtel, 1976; Hawkins, 1998; Darlington and Ranjith, 2011). For small sample sizes, an ascending strength is more realistic and one model that describes this was developed by Bazant and Planas (1998) by integrating the concepts of fractals and fracture energy.

While size-dependent rock strength has been studied extensively, there are very few constitutive models that capture the size-dependent stress-strain behavior (Aubertin et al., 1999; Aubertin et al., 2000). Most of the constitutive models in the literature were adapted from models originally developed for soil (Cividini, 1993). Most have been formulated using the conventional elastic-plastic framework, in which a purely elastic response occurs until yield followed by elastic-plastic deformation (e.g. Lade, 1977; Desai, 1980; Desai and Faruque, 1984; Lade and Nelson, 1987; Kim and Lade, 1988; Khan et al., 1991; Khan et al., 1992; Weng et al., 2005). A typical feature of models developed using the conventional elastic-plastic framework is the sharp transition at yield from elastic behavior to elastic-plastic behavior, which is rarely an accurate reflection of true rock behavior. Also, models of this framework are generally poor in simulating the transition from brittle to ductile postpeak behavior as confining pressure increases. Models developed using the bounding surface plasticity theory framework have a versatility that may overcome these limitations. Bounding surface plasticity models developed for soils produce highly accurate simulations of stress-strain behavior. Most notably, they produce a smooth transition from an initially near-linear behavior to a highly nonlinear behavior as peak strength is mobilized and postpeak deformation occurs. Examples for soils include Crouch et al. (1994), Gajo and Wood (1999), Morvan et al. (2010), and Wong et al. (2010).

The aim of this study is to present a new constitutive model for rock that recognizes its size-dependent behavior. An input parameter to the model is the unconfined compressive strength, which is made to depend on sample size. Through this single property, the stress-strain responses for the triaxial load path also become size-dependent. The model is formulated using bounding surface plasticity to avoid the sharp transition at the change from elastic to elastic-plastic behavior. Some ingredients of the model are similar to those of Gajo and Wood's (1999) Severn-Trent model for sand, in which simple linear Mohr-Coulomb type loading and bounding surfaces are defined.

The model is validated against a series of monotonically loaded unconfined compression test results and triaxial test results conducted on Gosford sandstone. Test samples for unconfined compression range in size from 19 mm diameter to 145 mm diameter and for triaxial compression from 25 mm diameter to 96 mm diameter.

Using a single set of equations, the strength and complete stress-strain behavior from initial loading to large shear strains are modeled well. The stress level-dependent transition from a brittle behavior toward a ductile behavior is captured. Most notably, for all sample sizes, the initial near-linear response and then gradual transition to a highly nonlinear response as peak strength is mobilized and postpeak softening occurs are simulated well.

1.3.2 Notation and unified size effect law

Conventional triaxial notation is used in which p' is the mean effective stress and q is the deviator stress. A prime indicates the stress is effective. The work conjugate strain variables are the volumetric strain ε_p and shear (deviatoric) strain ε_q. These are related to axial and radial stresses (σ_1' and $\sigma_3' = \sigma_2'$) and strains (ε_1 and $\varepsilon_2 = \varepsilon_3$) according to:

$$p' = \frac{\sigma_1' + 2\sigma_3'}{3} \quad q = \sigma_1' - \sigma_3' \quad \varepsilon_p = \varepsilon_1 + 2\varepsilon_3 \quad \varepsilon_q = \frac{2}{3}(\varepsilon_1 - \varepsilon_3) \tag{1.29}$$

where the subscripts 1 and 3 symbolize axial and radial components, respectively. Total strain increments are the sum of elastic and plastic strain increments:

$$\begin{bmatrix} \delta\varepsilon_p \\ \delta\varepsilon_q \end{bmatrix} = \begin{bmatrix} \delta\varepsilon_p^e \\ \delta\varepsilon_q^e \end{bmatrix} + \begin{bmatrix} \delta\varepsilon_p^p \\ \delta\varepsilon_q^p \end{bmatrix} \tag{1.30}$$

where the superscripts e and p denote the elastic and plastic components, respectively. The pairs of stresses and strains may be written in vector form:

$$\boldsymbol{\sigma} = [p', q]^{\mathrm{T}} \text{ and } \varepsilon = [\varepsilon_p, \varepsilon_q]^{\mathrm{T}} \frac{-b \pm \sqrt{b^2 - 4ac}}{2a} \tag{1.31}$$

A unified size effect law is adopted here that describes an ascending then descending unconfined compressive strength (σ_{UCS}) with increasing sample size.

The part of the law that describes a descending strength with sample size was derived by Bazant (1984). It considers the role of energy on crack growth and propagation. The descending strength is defined as:

$$\sigma_{UCS} = \frac{Bf_t}{\sqrt{1 + (d/\lambda d_0)}} \tag{1.32}$$

where d is the sample diameter, d_0 is a characteristic size sometimes taken to be the maximum aggregate size, λ and B are dimensionless material parameters, and f_t is a characteristic intrinsic strength.

The part of the law that describes an ascending strength with sample size was derived by Bazant (1997). It extends the work of Bazant (1984) by accounting for the fractal shape of a fracture and then linking strength to fracture energy. The ascending strength is defined as:

$$\sigma_{UCS} = \frac{\sigma_0 d^{(D_f-1)/2}}{\sqrt{1+(d/\lambda d_0)}} \tag{1.33}$$

where σ_0 is a characteristic strength and D_f is the fractal dimension of the fracture surface ($D_f \neq 1$ for fractal surfaces and $D_f = 1$ for nonfractal surfaces).

In the unified size affect law, σ_{UCS} is taken to be the lower of the strengths predicted by the two parts (Fig. 1.44):

$$\sigma_{UCS} = \text{Min}\left(\frac{\sigma_0 d^{(D_f-1)/2}}{\sqrt{1+(d/\lambda d_0)}}, \frac{Bf_t}{\sqrt{1+(d/\lambda d_0)}}\right) \tag{1.34}$$

The intersection between the two parts occurs at a sample diameter d_i defined as:

$$d_i = \left(\frac{Bf_t}{\sigma_0}\right)^{2/(D_f-1)} \tag{1.35}$$

The σ_{UCS} at the intersection point (σ_{UCSi}) represents the maximum that can be observed and is defined as:

$$\sigma_{UCSi} = \frac{Bf_t}{\sqrt{1+\left((Bf_t/\sigma_0)^{2/(D_f-1)}/\lambda d_0\right)}} \tag{1.36}$$

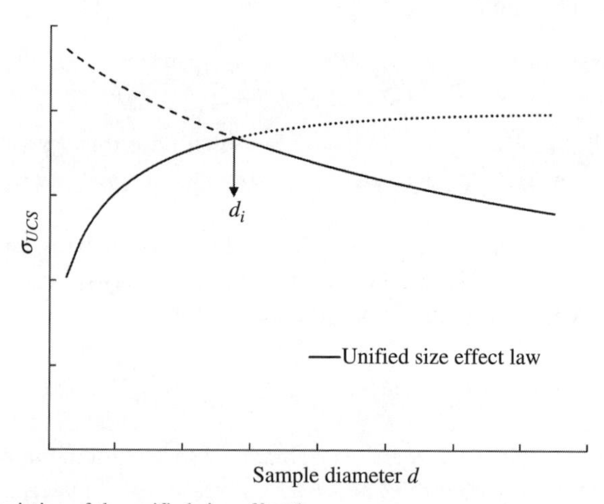

FIG. 1.44 Depiction of the unified size effect law.

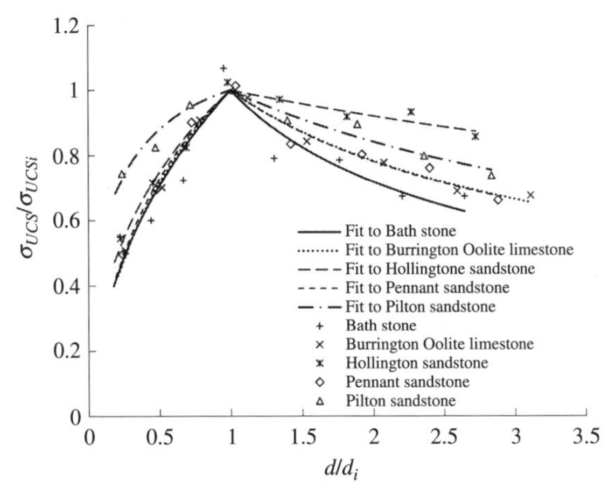

FIG. 1.45 The unified size effect law fitted to unconfined compressive strength data for five sedimentary rocks (Hawkins, 1998).

The law is fitted to the unconfined compressive strengths measured for five sedimentary rock types (Hawkins, 1998) in Fig. 1.45. The fitting parameters are listed in Table 1.14.

1.3.3 Bounding surface plasticity

One appealing feature of bounding surface plasticity models is the ability to abandon the notion of a purely elastic region so that the entire stress-strain response is smooth and defined by a single set of elastic-plastic constitutive laws.

TABLE 1.14 Parameters of the unified size effect law for different sedimentary rocks

Rock name	Bf_t (MPa)	λd_0 (mm)	σ_0 (MPa)	D_f	d_i (mm)	R^2
Bath stone	71.4	3.77	1.5	2.91	56.7	0.82
Burrington Oolite limestone	254	28.2	13.6	2.51	48.2	0.98
Hollington sandstone	37.7	250	5.4	1.97	55.1	0.94
Pennant sandstone	151	29.5	8.3	2.47	52.1	0.92
Pilton sandstone	242	75.8	60.3	1.7	52.9	0.9

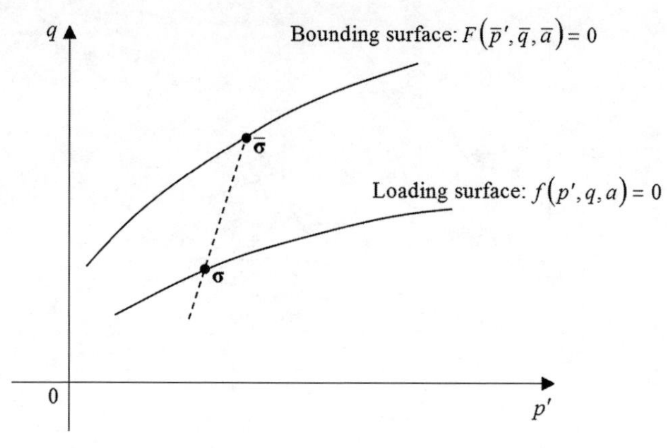

FIG. 1.46 Representation of the loading and bounding surfaces in the q-p' plane.

Bounding surface plasticity models, in general, require the definition of two surfaces (Fig. 1.46). One surface is a loading surface on which the current stress state $\boldsymbol{\sigma}$ always lies. The function f defines the loading surface:

$$f(p', q, a) = 0 \tag{1.37}$$

where a represents a hardening parameter controlling movement of the loading surface. The generic term "hardening" is used in this paper to mean hardening or softening. The other surface is the bounding surface, which also evolves as hardening occurs. An image point, $\bar{\boldsymbol{\sigma}} = [\bar{p}', \bar{q}]^{\mathrm{T}}$, is always located on the bounding surface. $\bar{\boldsymbol{\sigma}}$ is the image of $\boldsymbol{\sigma}$ defined according to a mapping rule. The function F defines the bounding surface:

$$F(\bar{p}', \bar{q}, \bar{a}) = 0 \tag{1.38}$$

where $\bar{a}$ is a hardening parameter that has dimensions of stress.

It is usual to define the ingredients of a bounding surface plasticity model so that the loading surface never intersects the bounding surface, but approaches it as infinitely large shear strains develop. In models for soils, for example, it is only at the critical state that the loading and bounding surfaces become one and $\boldsymbol{\sigma} = \bar{\boldsymbol{\sigma}}$.

When formulating a bounding surface plasticity model, the usual elastic relations may be adopted:

$$\delta\boldsymbol{\sigma} = \mathbf{D}^e \delta\boldsymbol{\varepsilon}^e \tag{1.39}$$

where $\mathbf{D}^e$ is the elastic matrix. For triaxial conditions, $\mathbf{D}^e$ is related to the tangential bulk (K) and shear (G) moduli according to:

$$\mathbf{D}^e = \begin{bmatrix} K & 0 \\ 0 & 3G \end{bmatrix} \tag{1.40}$$

The relationship between the incremental stresses and plastic strains is of the general form:

$$\delta \boldsymbol{\varepsilon}^p = \frac{1}{H}\left(\mathbf{n}^{\mathrm{T}} \delta \boldsymbol{\sigma}\right)\mathbf{m} \tag{1.41}$$

which has the expanded form:

$$\begin{bmatrix} \delta\varepsilon_p^p \\ \delta\varepsilon_q^p \end{bmatrix} = \frac{1}{H}\begin{bmatrix} n_p m_p & n_q m_p \\ n_p m_q & n_q m_q \end{bmatrix}\begin{bmatrix} \delta p' \\ \delta q \end{bmatrix} \tag{1.42}$$

in which $\mathbf{n} = [n_p, n_q]^{\mathrm{T}}$ is the unit normal vector that controls the direction of loading, $\mathbf{m} = [m_p, m_q]^{\mathrm{T}}$ is the unit normal vector controlling plastic flow, and H is the hardening modulus. It is usual to split H into two components:

$$H = H_b + H_f \tag{1.43}$$

where H_b is the part of the modulus related to the movement of $\overline{\boldsymbol{\sigma}}$ on F and H_f is the part controlling the movement of $\boldsymbol{\sigma}$ on f toward $\overline{\boldsymbol{\sigma}}$ on F.

Combining Eqs. (1.39), (1.41) leads to the elastic-plastic stress-strain relationship:

$$\delta \boldsymbol{\sigma} = \left(\mathbf{D}^e - \frac{\mathbf{D}^e \mathbf{m}\mathbf{n}^{\mathrm{T}} \mathbf{D}^e}{H + \mathbf{n}^{\mathrm{T}} \mathbf{D}^e \mathbf{m}}\right)\delta \boldsymbol{\varepsilon} \tag{1.44}$$

1.3.4 Model ingredients

The model presented here describes, in a very simple way using a single set of equations, the complete stress-strain behavior from initial loading to large shear strains. The stress level-dependent transition from brittle to ductile behavior is captured. The initial near-linear response and the gradual transition to a highly nonlinear response as peak strength is mobilized and postpeak deformation occurs are also captured. This represents an improvement over other models developed in the conventional elastic-plastic modeling framework (Cividini, 1993), which typically exhibit a sharp transition from elastic deformation to elastic-plastic deformation.

1.3.4.1 Elasticity

G and K are linked to E (tangential Young's modulus) and ν (Poisson's ratio) in the standard way through:

$$K = \frac{E}{3(1-2\nu)} \tag{1.45}$$

$$G = \frac{E}{2(1+\nu)} \tag{1.46}$$

Triaxial results in the literature (Chiarellia et al., 2003; Sulem and Ouffroukh, 2006a; Sulem and Ouffroukh, 2006b; Corkum, 2007) show that sedimentary rocks exhibit an elastic behavior that is stress level-dependent. Therefore, the following formulation is adopted for G:

$$\frac{G}{G_{UCS}} = \left(b\sigma_3' + 1\right)^z \tag{1.47}$$

where G_{UCS} is the shear modulus obtained for $\sigma_3' = 0$ (as in unconfined compression tests) and b and z are positive material constants (b has units of stress^{-1}). The resulting G is then used to compute E from Eq. (1.46) and subsequently compute K through Eq. (1.45) using an assumed constant for Poisson's ratio.

The use of Eq. (1.47) and a constant Poisson's ratio means that energy will not be conserved during a closed cycle of purely elastic loading when σ_3' varies. Even so, it is reasonable to use laws of this type in which stiffness depends on stress state or strain state without concern for the thermodynamic consequences, provided that the stress paths or strain paths to which the rock is being subjected are not very repeatedly cyclic (Wood, 2004).

1.3.4.2 Bounding surface and image point

It is assumed that deformation includes both elastic and plastic components from first loading. The bounding surface is defined using the simple Mohr-Coulomb criterion as:

$$F = \overline{q} - M\overline{p}' - \overline{a} = 0 \tag{1.48}$$

in which M is a dimensionless material constant representing the slope of the surface. $\overline{a}$ represents a kinematic hardening parameter that controls the movement of the surface in the stress space while it maintains constant slope and represents the intersection of the bounding surface with the $\overline{q}$ axis (see Figs. 1.47 and 1.48).

$\overline{\boldsymbol{\sigma}}$ is defined using a simple vertical mapping rule such that a straight vertical line in the $q - p'$ plane through $\boldsymbol{\sigma}$ on the loading surface intersects the bounding surface at $\overline{\boldsymbol{\sigma}}$ (see Fig. 1.47). According to this mapping rule:

$$\frac{p'}{\overline{p}'} = 1 \tag{1.49}$$

The ratio between q and $\overline{q}$ is denoted s:

$$\frac{q}{\overline{q}} = s \tag{1.50}$$

Another mapping rule may have been adopted for which $p' \neq \overline{p}'$ but vertical mapping was preferred here due to its simplicity.

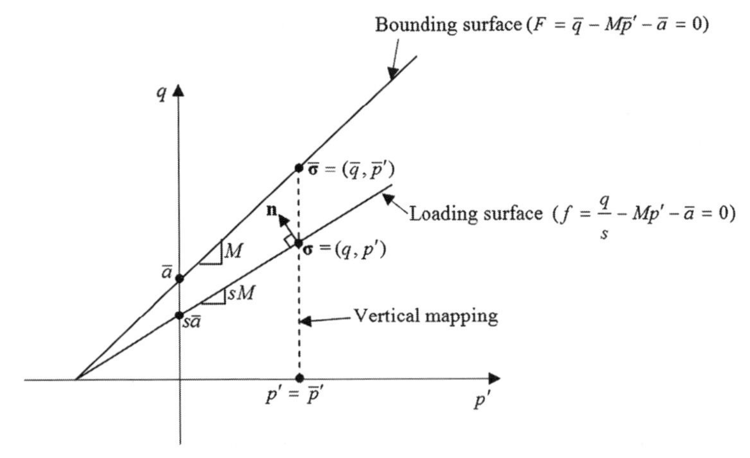

FIG. 1.47 Loading surface, bounding surface, vertical mapping line, and image point in the q-p' plane.

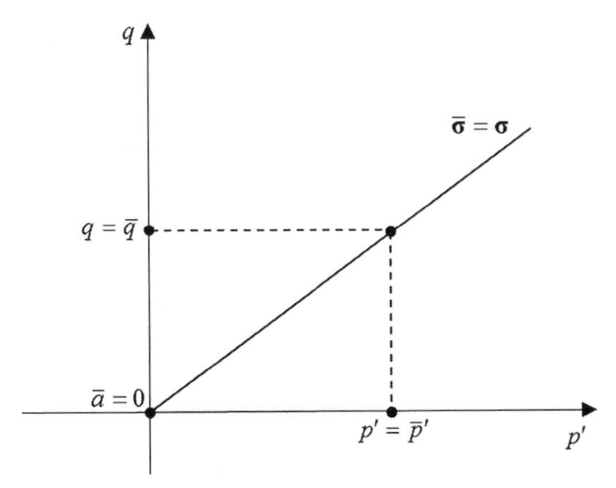

FIG. 1.48 Assumed overlapping of loading and bounding surfaces at infinite large shear strains in the q-p' plane.

It follows that the loading surface is defined as:

$$f = \frac{q}{s} - Mp' - \bar{a} = 0 \tag{1.51}$$

The loading surface is a line that intersects the q axis at $s\bar{a}$ and has a slope of sM (Fig. 1.47). s is thus a hardening variable controlling the changing slope of the loading surface.

1.3.4.3 Hardening law

The hardening law may be a function of ε_p^p and/or ε_q^p (Roscoe and Schofield, 1963; Khan et al., 1992; Shah, 1997; Weng et al., 2005). A unique strength at large shear strains is usually observed in triaxial results of rock samples having a single size (e.g., Wawersik and Fairhurst, 1970; Besuelle et al., 2000). In particular, at very large shear strains, samples are usually damaged to an extent that distinct failure planes (strain localizations) form, along which shearing occurs. The shear strength on the failure planes is purely frictional as there is an absence of cohesion. In other words, at very large shear strains, when $\overline{\boldsymbol{\sigma}}=\boldsymbol{\sigma}$, strength is controlled entirely by M and $\overline{a}=0$ (see Fig. 1.5).

A simple hardening law that ensures $\overline{a}\to 0$ as $\varepsilon_q^p\to\infty$ is:

$$\overline{a}=\frac{\overline{a}_0}{\exp\left[\left(\frac{\varepsilon_q^p}{\varepsilon_{ref}}\right)^c\right]} \tag{1.52}$$

in which $\overline{a}_0$ is a material constant and is the amount of strength $\overline{a}$ available (although not mobilized) at the commencement of loading, ε_{ref} is another constant that controls the plastic shear strain level at which $\overline{a}$ has reduced to a value of $\overline{a}_0/\mathrm{e}$ (where e is the natural number), and c is another material constant that controls the rate of decay of $\overline{a}$ at ε_{ref}. The sudden decrease of strength during softening (e.g., in an unconfined compressive strength test or triaxial test with low confining pressure) can be simulated using this definition for $\overline{a}$. Note that it has been assumed here for simplicity that hardening occurs only with ε_q^p.

The evolution of $\boldsymbol{\sigma}$ toward $\overline{\boldsymbol{\sigma}}$ is assumed to be controlled by a simple hyperbolic law similar to that used by Gajo and Wood (1999) and Wood (2004). Specifically, s is defined as:

$$s=\frac{q}{\overline{q}}=\frac{\varepsilon_q^p}{r+\varepsilon_q^p} \tag{1.53}$$

where r is a positive material constant. Here it is evident how the vertical mapping simplifies the hardening, as only the movement of q toward $\overline{q}$ needs definition because $p'=\overline{p}'$ always.

It is noted that strain localizations begin to develop around the peak strength and are very pronounced at large shear strains. Even so, it is possible to capture the general constitutive behavior of the rock by treating it as a homogenous continuum as above and as also done by others (e.g., Shah, 1997; Weng et al., 2005).

1.3.4.4 *Plastic potential and elastic-plastic matrix*

A nonassociated flow rule is assumed in which dilatancy d is defined as in Cam-Clay type models (Roscoe and Schofield, 1963):

$$d=\frac{\delta\varepsilon_p^p}{\delta\varepsilon_q^p}=\frac{\partial g/\partial p'}{\partial g/\partial q}=M-\eta \tag{1.54}$$

where $\eta=q/p'$ is the mobilized stress ratio.

The components of the elastic-plastic matrix (Eq. 1.44) including **n**, **m,** and H are now defined in expanded form as follows:

$$\mathbf{n}=\left[\frac{-M}{\sqrt{M^2+\left(\frac{1}{s}\right)^2}},\frac{\frac{1}{s}}{\sqrt{M^2+\left(\frac{1}{s}\right)^2}}\right]^T \tag{1.55}$$

$$\mathbf{m}=\left[\frac{d}{\sqrt{d^2+1}},\frac{1}{\sqrt{d^2+1}}\right]^T \tag{1.56}$$

The differential form of the function defining the bounding surface (Eq. 1.48) is:

$$\frac{\partial F}{\partial \overline{p}'}\delta \overline{p}'+\frac{\partial F}{\partial \overline{q}}\delta \overline{q}+\frac{\partial F}{\partial \overline{a}}\delta \overline{a}=0 \tag{1.57}$$

in which the differential form of $\overline{a}$is:

$$\delta \overline{a}=\frac{\partial \overline{a}}{\partial \varepsilon_q^p}\delta \varepsilon_q^p \tag{1.58}$$

The differential form of s is:

$$\delta s=\frac{\partial s}{\partial \varepsilon_q^p}\delta \varepsilon_q^p \tag{1.59}$$

The hardening modulus H is then:

$$H=\frac{-\left(\frac{\partial F}{\partial \overline{a}}\frac{\partial \overline{a}}{\partial \varepsilon_q^p}+\frac{\partial F}{\partial \overline{q}}\frac{\partial \overline{q}}{\partial s}\frac{\partial s}{\partial \varepsilon_q^p}\right)m_q}{\sqrt{\left(\frac{\partial F}{\partial \overline{p}'}\frac{\partial \overline{p}'}{\partial p'}\right)^2+\left(\frac{\partial F}{\partial \overline{q}}\frac{\partial \overline{q}}{\partial q}\right)^2}} \tag{1.60}$$

which is the sum of the two components H_b and H_f:

$$H_b=\frac{-\left(\frac{\partial F}{\partial \overline{a}}\frac{\partial \overline{a}}{\partial \varepsilon_q^p}\right)m_q}{\sqrt{\left(\frac{\partial F}{\partial \overline{p}'}\frac{\partial \overline{p}'}{\partial p'}\right)^2+\left(\frac{\partial F}{\partial \overline{q}}\frac{\partial \overline{q}}{\partial q}\right)^2}}=\frac{-1}{\sqrt{M^2+\left(\frac{1}{s}\right)^2}}\left(\frac{a_0c\left(\frac{\varepsilon_q^p}{\varepsilon_{ref}}\right)^c}{\varepsilon_q^p e^{\left(\frac{\varepsilon_q^p}{\varepsilon_{ref}}\right)^c}}\right)\frac{1}{\sqrt{d^2+1}} \tag{1.61}$$

$$H_f=\frac{-\left(\frac{\partial F}{\partial \overline{q}}\frac{\partial \overline{q}}{\partial s}\frac{\partial s}{\partial \varepsilon_q^p}\right)m_q}{\sqrt{\left(\frac{\partial F}{\partial \overline{p}'}\frac{\partial \overline{p}'}{\partial p'}\right)^2+\left(\frac{\partial F}{\partial \overline{q}}\frac{\partial \overline{q}}{\partial q}\right)^2}}=\frac{-1}{\sqrt{M^2+\left(\frac{1}{s}\right)^2}}\left(-\frac{q}{s^2}\left(\frac{r}{(r+\varepsilon_q^p)^2}\right)\right)\frac{1}{\sqrt{d^2+1}} \tag{1.62}$$

As is common when using bounding surface plasticity (Crouch et al., 1994; Russell and Khalili, 2004; Morvan et al., 2010; Wong et al., 2010), part of the hardening modulus, H_b, is derived directly from the bounding surface.

A unique characteristic of this simple model when applied to a conventional triaxial load path is that q may be written as a function of ε_q^p, σ_3', and constants M, $\overline{a}_0$, ε_{ref}, c, and r. To do so, initially, Eq. (1.25) is substituted into Eq. (1.20), then the resulting formulation is substituted into Eq. (1.53) in order to expand q as a function of M, $\overline{a}_0$, ε_{ref}, c, r, $\overline{p}'$, and ε_q^p. If p' is expressed as a function of q and σ_3' based on Eq. (1.1), then the final formulation of q as a function of M, $\overline{a}_0$, ε_{ref}, c, r, ε_q^p, and σ_3' is:

$$q = -\frac{3\left(M\sigma_3' \exp^{\left(\frac{\varepsilon_q^p}{\varepsilon_{ref}}\right)^c} + \overline{a}_0\right)\varepsilon_q^p}{\exp^{\left(\frac{\varepsilon_q^p}{\varepsilon_{ref}}\right)^c}\left(-3r - 3\varepsilon_q^p + M\varepsilon_q^p\right)} \tag{1.63}$$

1.3.4.5 Model outputs and parameter sensitivity

In order to show the influence of $\overline{a}_0$, c, ε_{ref}, and r on the stress-strain behavior, a number of model outputs are presented in Figs. 1.49–1.52 for a range of different values. These constants only control the plastic stress-strain behavior and therefore plots of $q/p' - \varepsilon_q^p$ are used in the illustrations.

The peak strength is mostly controlled by $\overline{a}_0$, as shown in Fig. 1.49. c and r have a slight influence on the peak strength, according to Figs. 1.59 and 1.52.

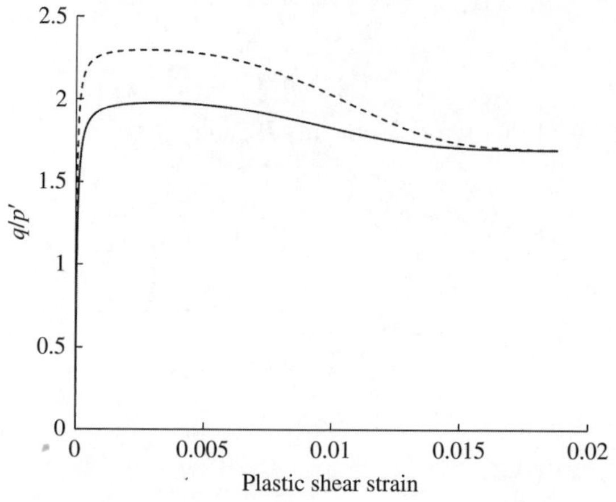

FIG. 1.49 Model simulation of q/p' versus plastic shear strain ε_q^p for $\overline{a}_0 = 28\,\text{MPa}$ (—) and $\overline{a}_0 = 84\,\text{MPa}$ (- - -) and the constants $M = 1.7$, $c = 3$, $\varepsilon_{ref} = 0.01$, $r = 5 \times 10^{-5}$, and $\sigma_3' = 30\,\text{MPa}$.

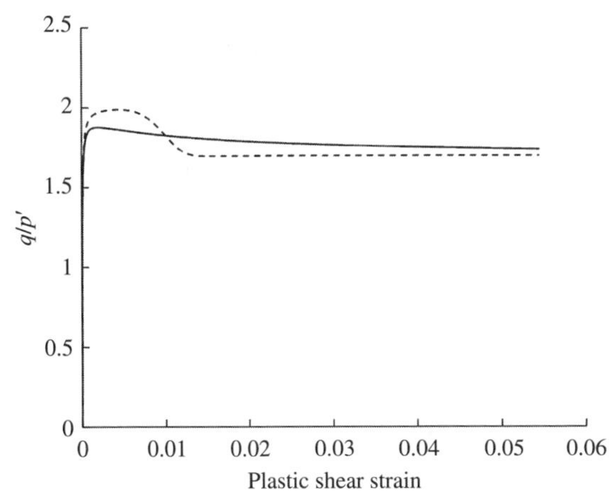

FIG. 1.50 Model simulation of q/p' versus plastic shear strain ε_q^p for $c=0.5$ (——) and $c=5$ (- - -) and the constants $M=1.7$, $\overline{a}_0=28\,\text{MPa}$, $\varepsilon_{ref}=0.01$, $r=5\times10^{-5}$, and $\sigma_3'=30\,\text{MPa}$.

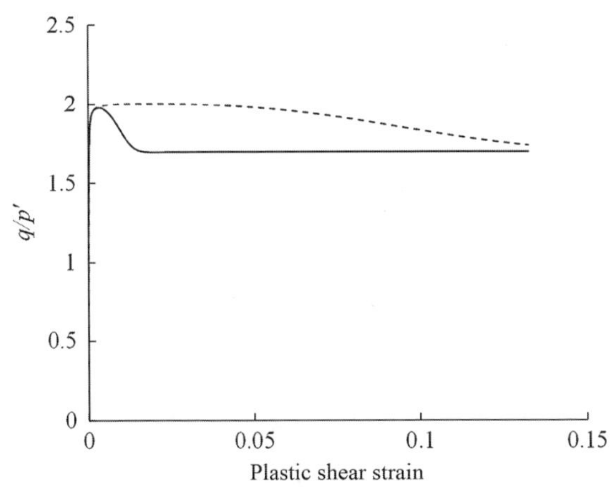

FIG. 1.51 Model simulation of q/p' versus plastic shear strain ε_q^p for $\varepsilon_{ref}=0.01$ (——) and $\varepsilon_{ref}=0.1$ (- - -) and the constants $M=1.7$, $\overline{a}_0=28\,\text{MPa}$, $c=3$, $r=5\times10^{-5}$, and $\sigma_3'=30\,\text{MPa}$.

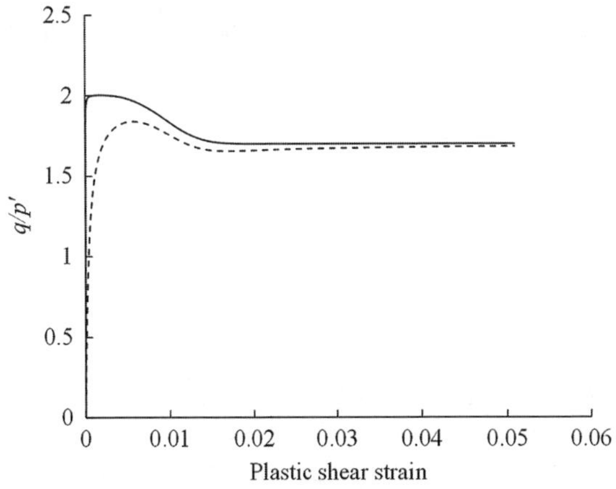

FIG. 1.52 Model simulation of q/p' versus plastic shear strain ε_q^p for $r=5\times10^{-6}$ (——) and $r=5\times10^{-4}$ (- - -) and the constants $M=1.7$, $\overline{a}_0=28\,\text{MPa}$, $\varepsilon_{ref}=0.01$, $c=3$, and $\sigma_3'=30\,\text{MPa}$.

The strain level at peak strength does not vary when $\overline{a}_0$ is altered whereas a change in c and r has significant impact on the strain level at peak strength.

From Figs. 1.50 and 1.52, it can be seen that the rate of decay of strength postpeak is controlled by c and ε_{ref}. The role of ε_{ref} is more dominant than c in this decay.

1.3.4.6 Initial stiffness

Fig. 1.52 shows that r controls the initial plastic stiffness. More specifically, for $\sigma_3'= 0$, the initial plastic stiffness is given by:

$$\lim_{\varepsilon_q^p \to 0}\left(\frac{\partial q}{\partial \varepsilon_q^p}\right) = \frac{\overline{a}_0}{r} \tag{1.64}$$

When $\sigma_3' \neq 0$, the initial plastic stiffness depends on σ_3' in addition to $\overline{a}_0$ and r and cannot be expressed in closed form. Eq. (1.36) shows that ε_{ref} and c have no impact on the initial plastic stiffness.

The influence of $\overline{a}_0$ and r relative to G on the initial total stiffness is very small. For example, for an unconfined compression test, the initial plastic and elastic shear stiffnesses are:

$$\lim_{\varepsilon_q^p \to 0}\left(\frac{\partial q}{\partial \varepsilon_q^q}\right) = \frac{\overline{a}_0}{r} \text{ and } \lim_{\varepsilon_q^e \to 0}\left(\frac{\partial q}{\partial \varepsilon_q^e}\right) = 3G \tag{1.65}$$

resulting in an initial total stiffness of:

$$\lim_{\varepsilon_q \to 0}\left(\frac{\partial q}{\partial \varepsilon_q}\right) = \frac{3G}{3Gr/\overline{a}_0 + 1} \tag{1.66}$$

If $\overline{a}_0/r$ is much larger than $3G$, then $3Gr/\overline{a}_0$ is much smaller than unity and can be neglected in Eq. (1.38), meaning the initial total stiffness is dominated by the elastic part and may be assumed equal to $3G$ when interpreting laboratory data.

For example, for an unconfined compression test conducted on 96 mm diameter Gosford sandstone, with values of G_{UCS}=7.1 GPa, $\overline{a}_0$ = 28MPa, and $r = 5 \times 10^{-5}$, the plastic, elastic, and total initial stiffness are:

$$\lim_{\varepsilon_q^p \to 0}\left(\frac{\partial q}{\partial \varepsilon_q^p}\right) = 560\text{GPa}, \quad \lim_{\varepsilon_q^e \to 0}\left(\frac{\partial q}{\partial \varepsilon_q^e}\right) = 21.3\text{GPa}, \quad \lim_{\varepsilon_q \to 0}\left(\frac{\partial q}{\partial \varepsilon_q}\right) = 20.52\text{GPa} \tag{1.67}$$

In which case, the impact of the plastic component is less than 4% on the overall initial shear stiffness.

1.3.4.7 Incorporating size effects

For unconfined compression tests, q is equal to σ_{UCS} and is directly proportional to $\overline{a}_0$. The unified size effect law fitted to unconfined compressive strength data can then be used to scale $\overline{a}_0$. $\overline{a}_0$ is defined as:

$$\overline{a}_0 = \frac{\overline{a}_{0_{100}} \sigma_{UCS}}{\sigma_{UCS_{100}}} \tag{1.68}$$

in which $\overline{a}_{0_{100}}$ and $\sigma_{UCS_{100}}$ are values of $\overline{a}_0$ and the unconfined compressive strength for a sample with 100 mm diameter, respectively.

Eq. (1.52) can be adapted to include size effects:

$$\overline{a} = \frac{\dfrac{\overline{a}_{0_{100}} \sigma_{UCS}}{\sigma_{UCS_{100}}}}{\exp\left[\left(\dfrac{\varepsilon_q^p}{\varepsilon_{ref}}\right)^c\right]} \tag{1.69}$$

All other parameters controlling plastic deformation are unchanged and are unaffected by size.

1.3.5 Model calibration

When modeling samples of a single size, five parameters—M, $\overline{a}_0$, ε_{ref}, c, and r—plus four elastic constants—G_{UCS}, b, ν and z—are needed to describe stress-strain behavior from first loading to large shear strains. When incorporating size effects, it is necessary to make the unconfined compressive strength and thus the $\overline{a}_0$ sample size-dependent. In this case, $\overline{a}_0$ may be replaced with parameters λd_0, Bf_t, σ_0, and D_f.

The rock type used here is Gosford sandstone, obtained from Gosford Quarry, Somersby, New South Wales, Australia. Homogenous samples were carefully selected to have no macrodefects. Sufian and Russell (2013) conducted an X-ray CT scan on the same batch of Gosford sandstone at a resolution of 5 μm. They characterized the microstructure and estimated the porosity to be 18.5%. X-ray diffraction results showed that the sandstone comprises 86% quartz, 7% illite, 6% kaolinite, and 1% anatase.

1.3.5.1 Fitting the unified size effect law

Samples of different sizes were tested to determine their unconfined compressive strengths. The mean of the unconfined compressive strengths for each sample size was used to obtain the size effect law parameters via a standard nonlinear multiple determination method to produce maximum coefficients of determination R^2. Table 1.15 lists the mean strengths obtained from Gosford sandstone at different sizes.

The strengths are shown in Fig. 1.53 along with the fitted unified size effect law defined by parameters $\lambda d_0 = 323$ mm, $Bf_t = 64.3$ MPa, $\sigma_0 = 7.8$ MPa, and $D_f = 2.01$. The intersection size and maximum strength are $d_i = 65.1$ mm and $\sigma_{UCSi} = 58.6$ MPa, respectively.

The shear modulus observed in unconfined compression (G_{UCS}) and ν were found to vary with sample size. These constants were estimated based on the

TABLE 1.15 Mean unconfined compressive strengths at different diameters for Gosford sandstone

Name		Values							
Sample diameter (mm)		19	25	31	50	65	96	118	145
Number of tests		6	5	6	6	7	5	4	2
Mean unconfined compressive strength (MPa)		34.6	36.48	42.3	52.3	58.8	56.1	54.7	54.2
Standard deviation	MPa	5.2	3.2	4.0	2.7	2.8	2.7	1.5	0.4
	%	14.9	8.8	9.4	5.2	4.8	4.8	2.8	0.7

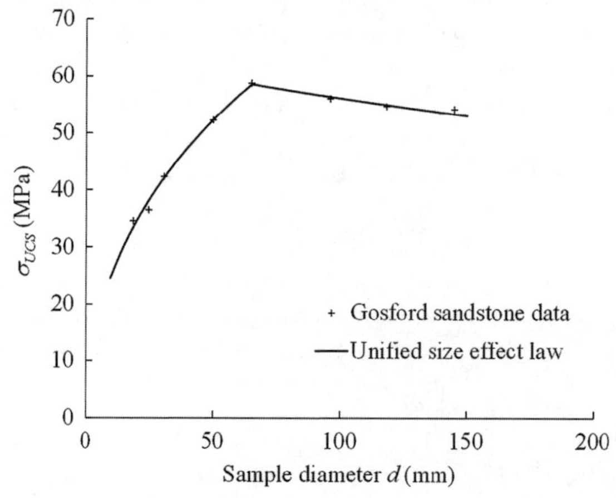

FIG. 1.53 Unified size effect law fitted to the unconfined compressive strength data for Gosford sandstone.

suggested method by ISRM (1979). The other two constants that influence the elastic response, b and z, were obtained using a standard nonlinear multiple determination method. There was no discernible sample size dependence. The best fit b and z values for all sample sizes were 0.73 MPa^{-1} and 0.12, respectively. Table 1.16 lists G_{UCS} and ν values for 96, 50, and 25 mm diameter samples. Suitable values for b and z were found to be 0.73 MPa^{-1} and 0.12, respectively, and independent of sample size.

TABLE 1.16 Lists of applied G_{UCS} and ν for different sample sizes

Sample diameter (mm)	G_{UCS} (GPa)	ν
96	7.1	0.14
50	7.7	0.13
25	5.8	0.055

1.3.5.2 Simulation for 96-mm diameter samples

In the triaxial tests, for each confining pressure and sample size, a minimum of three tests were carried out. The test that produced the stress-strain data most typical of all those conducted at that confining pressure and sample size was used for model calibration.

Considered here are the samples tested that had diameters of 96 mm and lengths of 192 mm. The large strain q and p' values for each confining pressure are plotted in Fig. 1.54. The slope of the line of best fit through the origin gives $M=1.7$. A standard linear multiple determination method was employed to obtain the best linear fit to the data.

Plots of a nondimensional mobilized strength, denoted V, versus ε_q^p, obtained from the experimental data, are illustrated in Fig. 1.55. V is defined as:

$$V = \frac{(\eta - \eta_r)}{(\eta_p - \eta_r)} \tag{1.70}$$

In computing V for any one set of stress-strain data, η_p and η_r are the stress ratios (q/p') at peak and large shear strains, respectively. It so happens that this

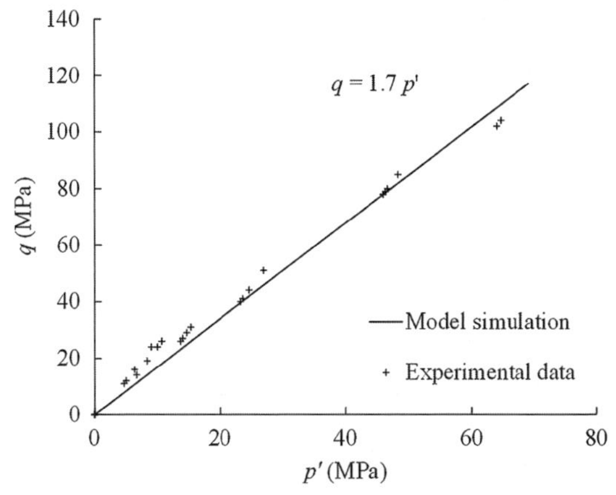

FIG. 1.54 Comparison between the experimental data for 96 mm diameter samples and model simulation for the large shear strain strengths in the q-p' plane.

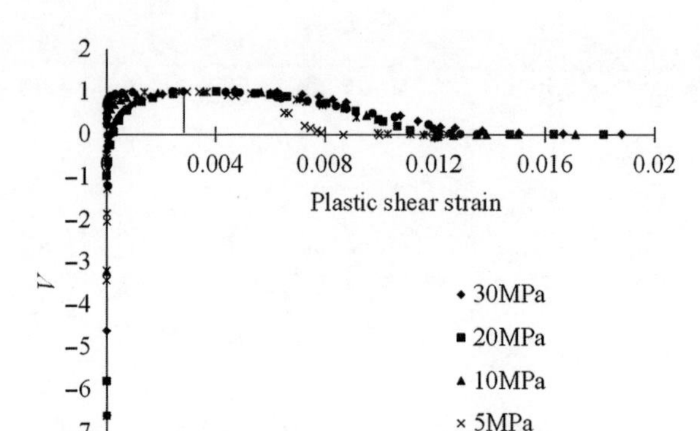

FIG. 1.55 Normalized triaxial data for Gosford sandstone.

type of plot results in the overlapping of the triaxial data within a narrow range. As shown in Fig. 1.55, the maximum possible value of V is 1, and occurs when $\eta=\eta_p$. The minimum possible value of V occurs when $\varepsilon_q^p=0$ and has a value of:

$$V_{\min}=\frac{-(\eta_r)}{(\eta_p-\eta_r)} \tag{1.71}$$

From the experimental data in Fig. 1.55, it can be seen that when V is at its maximum, ε_q^p is about 0.0025 for all confining pressures. This can assist the selection of a suitable set of model parameters. $V=1$ at the peak strength occurs when:

$$\frac{\partial q}{\partial \varepsilon_q^p}=0 \tag{1.72}$$

which has the expanded form:

$$\frac{\partial q}{\partial \varepsilon_q^p}=\frac{3}{\left(-3r-3\varepsilon_q^p+M\varepsilon_q^p\right)}\left(\begin{array}{l}\dfrac{\left(M\sigma_3'\exp\left(\dfrac{\varepsilon_q^p}{\varepsilon_{ref}}\right)^c+\bar{a}_0\right)}{\exp\left(\dfrac{\varepsilon_q^p}{\varepsilon_{ref}}\right)^c}\left(\dfrac{\varepsilon_q^p(-3+M)}{\left(-3r-3\varepsilon_q^p+M\varepsilon_q^p\right)}+\left(\dfrac{\varepsilon_q^p}{\varepsilon_{ref}}\right)^c c-1\right)\\ -M\sigma_3'\left(\dfrac{\varepsilon_q^p}{\varepsilon_{ref}}\right)^c c\end{array}\right)=0 \tag{1.73}$$

For unconfined compression, this simplifies to:

$$\frac{\partial q}{\partial \varepsilon_q^p} = \frac{3\bar{a}_0}{\exp\left(\frac{\varepsilon_q^p}{\varepsilon_{ref}}\right)^c (-3r - 3\varepsilon_q^p + M\varepsilon_q^p)} \left(\left(\frac{\varepsilon_q^p}{\varepsilon_{ref}}\right)^c c + \varepsilon_q^p(-3+M) - 1\right) = 0 \tag{1.74}$$

Noting that $M = 1.7$ has been defined already, and ε_q^p was visually estimated to be 0.0025 from Fig. 1.55, therefore $V = 1$ when:

$$\left(-1 + \left(\frac{0.0025}{\varepsilon_{ref}}\right)^c c + 0.0025(-3+1.7)\right) = 0 \tag{1.75}$$

As a result, by trial and error, c and ε_{ref}^p can then be computed by satisfying Eq. (1.75). Suitable values were found to be 3 and 0.01, respectively. Subsequently, the model outputs were fitted to the triaxial data. Again, by trial and error, the best fit values for $\bar{a}_0$ and r were estimated, being 28 MPa and 5×10^{-5}, respectively. The model predicted stress conditions at peak strength are studied using Eq. (1.73) and are compared with the experimental data in Table 1.17.

The peak strengths are also presented graphically in Fig. 1.56, where it can be seen that the model produces a peak strength criterion of $q = 1.7p' + 27$ MPa and agrees well with experimental data.

The complete simulations of stress-strain behavior are presented in Fig. 1.57 in the $q - \varepsilon_q$ and $\varepsilon_p - \varepsilon_q$ planes. The model simulations are represented by continuous lines while experimental results are represented by symbols. There is a

TABLE 1.17 Comparison between peak strengths from experiments and model simulations

Confining pressure	Experimental data (average values)		Model simulations	
σ_3' (MPa)	q (MPa)	p′ (MPa)	q (MPa)	p′ (MPa)
0	56.1	18.7	60.8	20.3
1	65.6	22.9	64.5	22.5
2	71.2	25.7	68.3	24.8
5	86.2	33.7	79.6	31.5
10	106	45.2	98.4	42.8
20	136	65.4	136	65.4
30	169	86.4	174	88.0

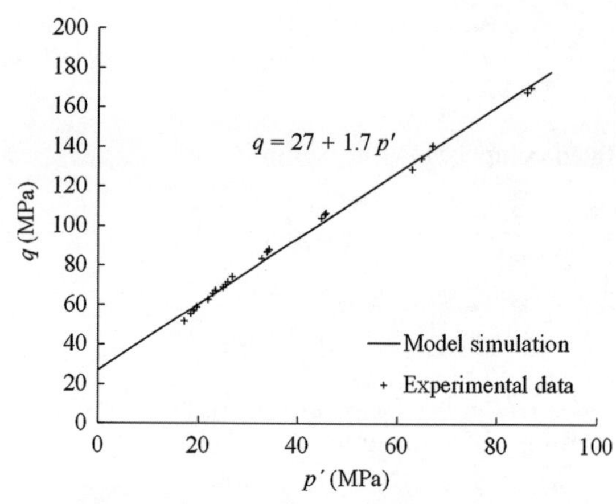

FIG. 1.56 Comparison between the experimental data for 96 mm diameter samples and model simulation for peak strengths in the q-p' plane.

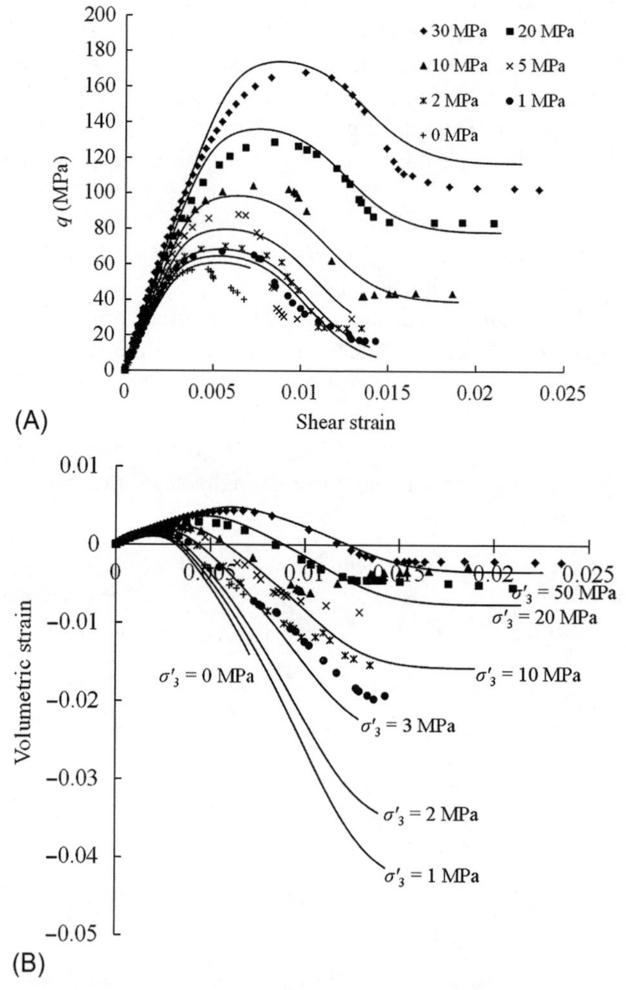

FIG. 1.57 Model simulation of the triaxial results for 96 mm diameter Gosford sandstone samples with constants $\nu=0.14$, $G_{UCS}=7.1$ GPa, $b=0.73$, $z=0.12$, $M=1.7$, $\bar{a}_0=28\,\text{MPa}$, $\varepsilon_{ref}=0.01$, $c=3$, and $r=5\times10^{-5}$.

good agreement between the model simulations and the experimental data, in particular at high confining pressures.

1.3.5.3 Simulation for 50-mm diameter samples

The large shear strain and peak strength conditions for 50-mm diameter and 100-mm long samples are presented graphically in Figs. 1.58 and 1.59. The large shear strain strengths did not show a significant size effect and the M value is assumed constant for different sizes and confining pressures. It can be seen

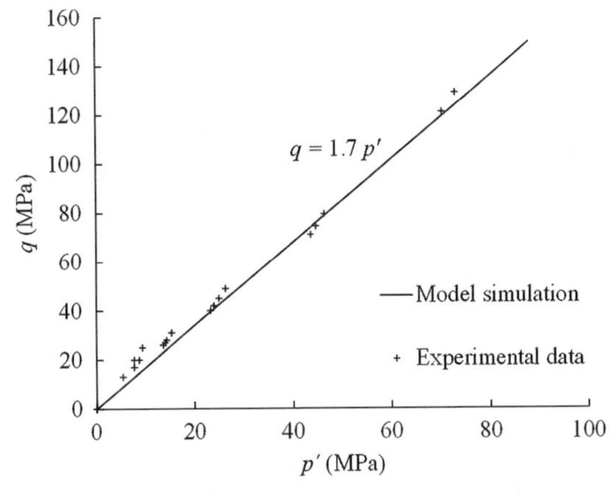

FIG. 1.58 Comparison between the experimental data for 50 mm diameter samples and model simulation for the large shear strain strengths in the q-p' plane.

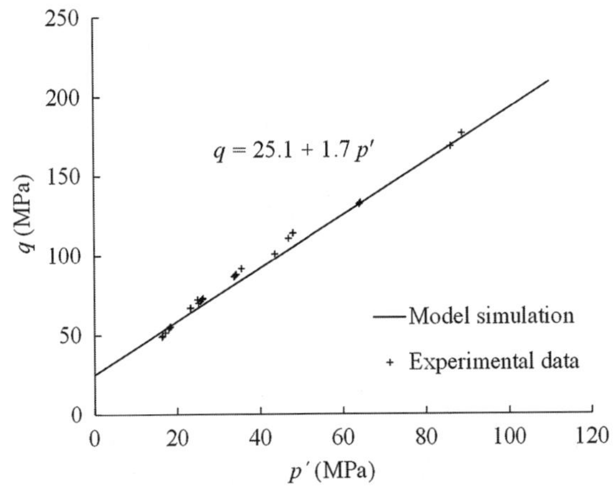

FIG. 1.59 Comparison between the experimental data for 50 mm diameter samples and model simulation for peak strengths in the q-p' plane.

that the model produces a peak strength criterion of $q = 1.7p' + 25.1$ MPa (Fig. 1.59) and agrees well with experimental data.

The unconfined compressive strengths of Gosford sandstone for 96 mm and 50 mm diameter samples are 56.2 and 52.3 MPa, respectively (using Eq. 1.34). $\bar{a}_0$ for 50 mm diameter samples was estimated to be 26.1 MPa (using Eq. 1.68 and assuming that $\bar{a}_0$ is equal to 28 MPa and applies to both 96 and 100 mm diameter samples).

The model simulations for 50 mm diameter samples are presented in Fig. 1.60 along with experimental data. In general, there is a good agreement between the model simulations and experimental data. Only for the test at 30 MPa confining pressure is there a poor agreement between experiment and simulation in the volumetric strain versus shear strain plane.

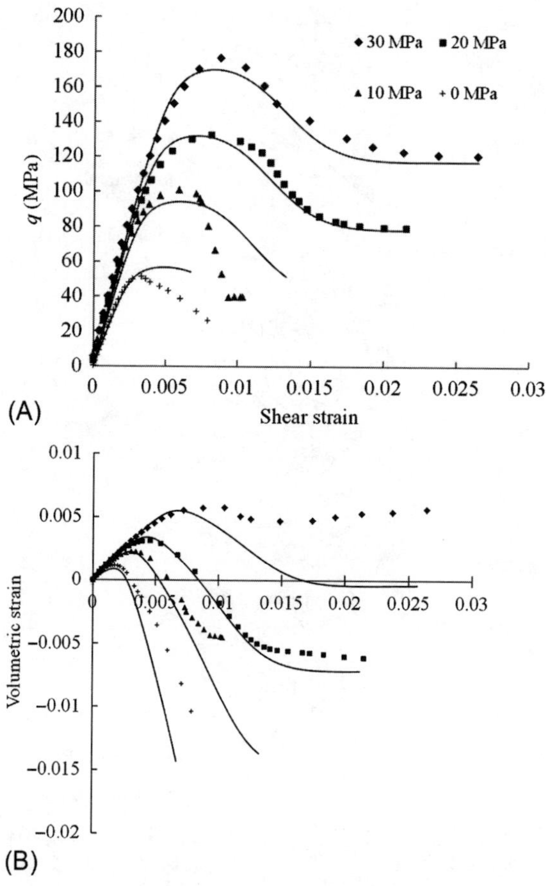

FIG. 1.60 Model simulation of the triaxal results for 50 mm diameter Gosford sandstone samples with constants $\nu = 0.13$, $G_{UCS} = 7.7$ GPa, $b = 0.73$, $z = 0.12$, $M = 1.7$, $\bar{a}_0 = 26.1$MPa, $\varepsilon_{ref} = 0.01$, $c = 3$, and $r = 5 \times 10^{-5}$.

1.3.5.4 Simulation for 25 mm diameter samples

Figs. 61 and 1.62 present the large shear strain and peak strengths. Again M appears unaffected by sample size. The model produces a peak strength criterion of $q = 1.7p' + 18.1$ MPa (Fig. 1.62) and agrees well with experimental data.

The unconfined compressive strength for 25 mm diameter and 50 mm long samples was taken to be 38.2 MPa, and thus $\overline{a}_0$ was determined to be 19.0 MPa.

The model simulations as well as the experimental data are presented in Fig. 1.63. This includes a plot of deviatoric stress versus axial strain because

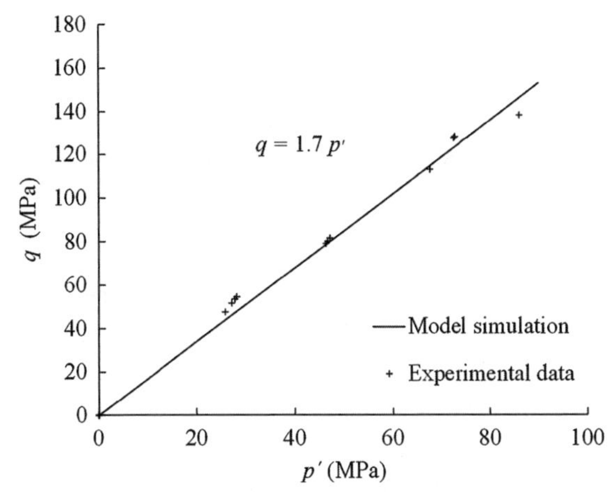

FIG. 1.61 Comparison between the experimental data for 50 mm diameter samples and model simulation for the large shear strain strengths in the q-p' plane.

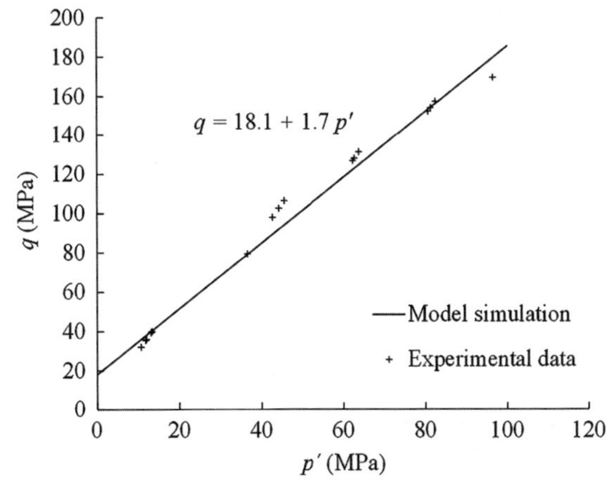

FIG. 1.62 Comparison between the experimental data for 25 mm diameter samples and model simulation for peak strengths in the q-p' plane.

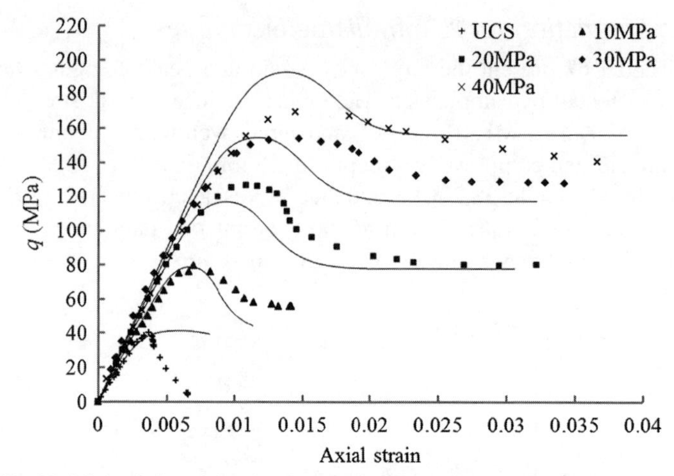

FIG. 1.63 Model simulation of the triaxal results for 25 mm diameter Gosford sandstone samples with constants $\nu=0.055$, $G_{UCS}=5.8$ GPa, $b=0.73$, $z=0.12$, $M=1.7$, $\overline{a}_0=19.04$MPa, $\varepsilon_{ref}=0.01$, $c=3$, and $r=5\times10^{-5}$.

only axial strain was recorded during the experiments. There is a reasonable agreement between the model simulations and experimental data.

1.3.5.5 Comparing models for different diameter samples

The stress-strain model simulations for unconfined compression as well as triaxial compression with 30 MPa confining pressures are plotted and compared in Figs. 1.64 and 1.65 for 96, 50, and 25 mm diameter samples. These include the graphs of deviatoric stress versus shear strain and volumetric strain versus shear strain.

The size dependencies of the peak strength and initial stiffness are recognizable from the graphs. The graphs of volumetric strain versus shear strain show that with a decrease in size, the volumetric strain tended to be more compressive. Also, the resulting simulations of 50 mm and 96 mm diameter samples for both unconfined and triaxial conditions were very similar.

1.3.6 Conclusions

A new bounding surface plasticity model for intact rock has been presented. It adopts linear Mohr-Coulomb loading and bounding surfaces, a vertical mapping rule, hardening with plastic shear strains, and a Cam clay flow rule.

Unconfined compression and triaxial compression test results for Gosford sandstone performed on samples with three different diameters were used to calibrate the model and demonstrate its simulative capabilities. The sample size-dependent unconfined compressive strength of the sandstone was defined as the minimum predicted by two strength criteria.

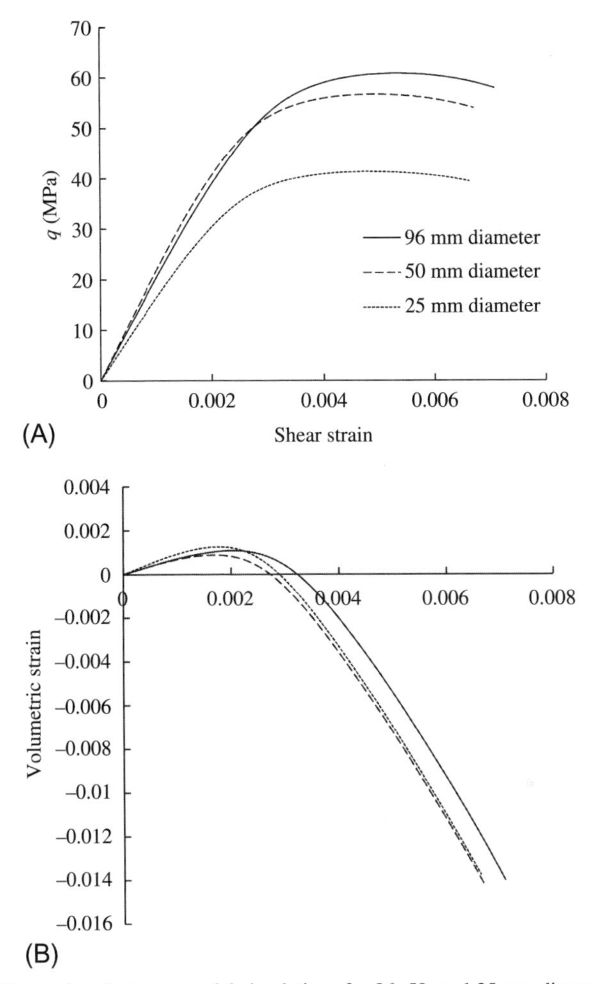

FIG. 1.64 Comparison between model simulations for 96, 50, and 25 mm diameter samples in unconfined compression.

Using a single set of equations, the complete stress-strain behavior from initial loading to large shear strains was simulated well. The stress level-dependent transition from a brittle behavior toward a ductile behavior was captured. Most notably, the initial near linear response and then gradual transition to a highly nonlinear response as peak strength was mobilized and postpeak deformation occurred were simulated very well. Many earlier studies have not attempted to model the post peak region (e.g., Khan et al., 1991; Khan et al., 1992; Shah, 1997; Weng et al., 2005). While the model presented here captures many of the postpeak characteristics reasonably well, there is a space for

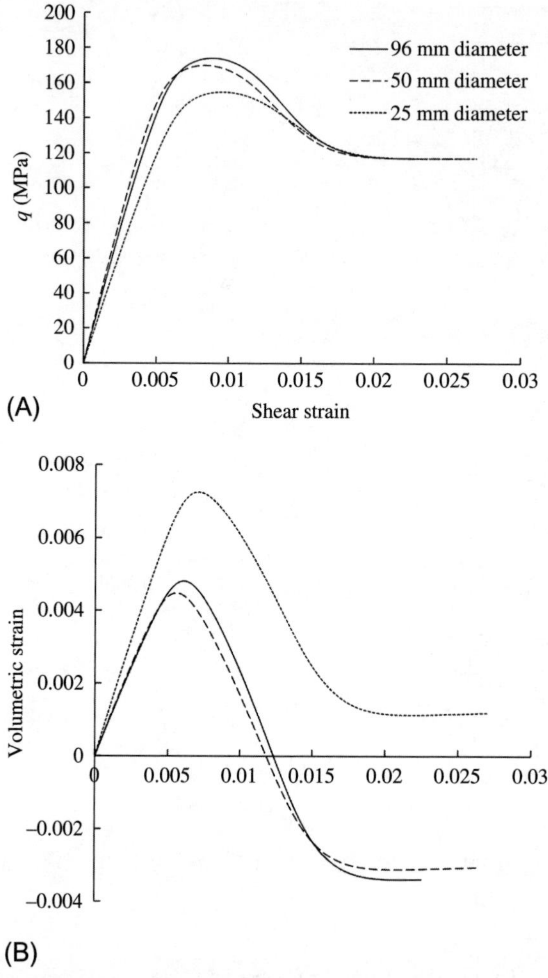

FIG. 1.65 Comparison between model simulations for 96, 50, and 25 mm diameter samples in triaxial compression with $\sigma_3' = 30\,\text{MPa}$.

improvement, in particular around the transition between a declining strength and the attainment of a constant strength at large shear strains.

The sample size-dependent unconfined compressive strength was a model input parameter. The elastic shear modulus and Poisson's ratio were also size-dependent. All other model input parameters were size-independent. The calibration of this model was simple, owing to: (i) the negligible influence plastic deformation had on initial stiffness and (ii) the plastic shear strain magnitude at which peak strength was mobilized was constant, irrespective of stress level.

The model has been presented using q-p' notation. Further work is needed to extend the model to a more general 3D stress state once data from more varied (nontriaxial) load paths becomes available. The model in its current form is intended for monotonic loading, although bounding surface models are routinely extended to cyclic loading (Cremer et al., 2001, Muir Wood, 2004), so that the average stiffness in any cycle reflects the strain level at which the direction of loading is reversed. An extension of this model to cyclic loading will be the subject of future work.

1.4 Scale-size dependency of intact rock

1.4.1 Introduction

Scale-size dependency of the mechanical properties of intact rock has been the central focus of various studies over the last four decades. Both terms (scale and size) have been used to refer to the influence of three-dimensional volumetric changes on the mechanical characteristics of intact rock.

A large portion of size effect studies on intact rock have been conducted under uniaxial and triaxial compression (Mogi, 1962; Dhir and Sangha, 1973; Hunt, 1973; Baecher and Einstein, 1981; Hawkins, 1998; Aubertin et al., 2000; Thuro et al., 2001a; Yoshinaka et al., 2008; Darlington and Ranjith, 2011; Masoumi et al., 2012; Masoumi et al., 2014; Masoumi et al., 2016a; Roshan et al., 2016a; Masoumi et al., 2017a; Quiñones et al., 2017; Roshan et al., 2017a). Some have considered point loading (Broch and Franklin, 1972; Bieniawski, 1975; Brook, 1977; Wijk, 1978; Brook, 1980; Greminger, 1982; Brook, 1985; Hawkins, 1998; Thuro et al., 2001b; Masoumi et al., 2012; Forbes et al., 2015) and only a few have considered tensile testing (Wijk, 1978; Andreev, 1991a; Andreev, 1991b; Butenuth, 1997; Thuro et al., 2001a; Çanakcia and Pala, 2007). Due to the complexity of direct tensile testing, indirect tensile testing known as the Brazilian test is a common alternative. It is noteworthy that some numerical modeling studies have been conducted to assess the size effects in rocks, particularly under uniaxial compression (Zhang et al., 2011; Ding et al., 2014; Gao et al., 2014; Wong and Zhang, 2014; Zhang et al., 2015; Bahrani and Kaiser, 2016). However, a limited number of numerical studies have included size-dependent behavior of tensile strength data (Yu et al., 2006; Wang et al., 2014) and the size effect in point load strength data has remained unexplored from the numerical point.

Most of the studies on size dependency of intact rock have attempted to address this problem from an experimental viewpoint while the need for analytical models is essential for practical applications. In general, for brittle and quasibrittle materials such as rock and concrete, three major size effect models have been proposed based on statistics (Weibull, 1939), fracture energy (Bazant, 1984), and multifractals (Carpinteri et al., 1995). All these models indicate that with an increase in size, the strength decreases. Darlington and

Ranjith (2011) assessed the applicability of these models to a set of uniaxial compressive strength (UCS) data from different geological origins to identify the one that fits each set of rock types the best. However, there is no similar study to that of Darlington and Ranjith (2011) for point load and tensile strength data.

The determination of UCS is an oft-used parameter in most rock mechanics projects while rock failure in tension is also important in practice (Hoek and Brown, 1980a, b; Goodman, 1989; Brady and Brown, 2006). In both tensile and point load testing, the intact rock fails in tension as demonstrated by Russell and Wood (2009) and Masoumi et al. (2016b). Hence, consideration was given to both these testing regimes in assessing the problem of size dependency of intact rock under tension.

A comprehensive set of laboratory experiments consisting of point loading and Brazilian tests was conducted on different rock types from various geological origins, including sedimentary, igneous, and metamorphic. The experiments were conducted according to the relevant International Society for Rock Mechanics (ISRM, 2007) suggested methods. The applicability of different size effect models to point load and tensile strength data was assessed similar to that carried out by Darlington and Ranjith (2011) for UCS data. The need for such an analysis is essential for a proper design process where the point load and tensile strength data are used. Upscaling of point load strength data from laboratory to field scale is often conducted based on the ISRM (2007) suggested method using only a statistical model (Brook, 1980; Brook, 1985). It is important to assess the applicability of other available size effect theories based on the fracture energy (Bazant, 1984) and the multifractals (Carpinteri et al., 1995) for upscaling of point load strength laboratory data to field condition that can lead to identification of the most efficient size correction methodology. Such an analysis for tensile strength data has even been overlooked in rock engineering (e.g., costly hydraulic fracturing operations) and thus a systematic analysis is required to determine the best approach for upscaling the lab tensile strength data to field condition.

1.4.2 Rock types

Six different rock types were selected for experimental investigation. Gosford sandstone was studied using point load and indirect tensile (Brazilian) tests. Other rock types—Bentheim sandstone, Gambier limestone, granite, and marble—were also examined, but through point loading only.

Bentheim sandstone was sourced from the Gildehausen quarry near the village of Bentheim in Germany. The rock type is similar to the reservoir rock from the Schoonebeek oil field with approximately 23% porosity. It is relatively homogenous (Fig. 1.66A) with approximately 95% quartz, 3% kaolinite, and 2% orthoclase. The quartz grain size varies between 0.05 and 0.5 mm (Klein et al., 2001; Baud et al., 2004). The mean UCS of the Bentheim sandstone used

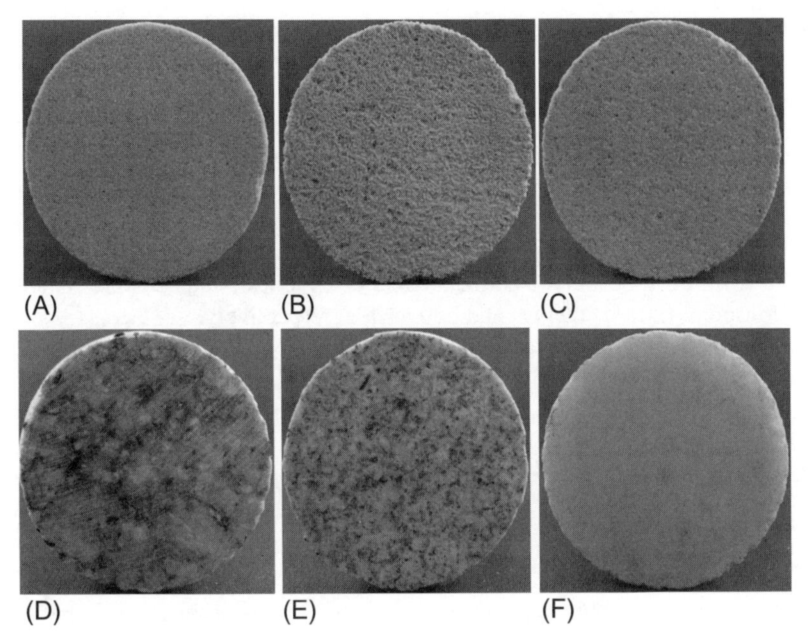

FIG. 1.66 Cross-sectional areas of (A) Bentheim sandstone, (B) Gambier limestone, (C) Gosford sandstone, (D) Granite A, (E) Granite B, and (F) Marble with 50 mm diameters.

in this study with a 50-mm diameter and a length-to-diameter ratio of 2 was measured at 41.4 MPa. The bulk density of Bentheim sandstone was estimated to be 1983 kg/m^3.

Gambier limestone (Fig. 1.66B) was extracted from the Mount Gambier coastal region in South Australia. The rock is the product of shoreline sedimentation during successive interglacial maxima and various interstadials (Murray-Wallace et al., 1999). Two studies have investigated the geological and micro structure of this limestone (Allison and Hughes, 1978; Melean et al., 2009). Melean et al. (2009) reported the porosity of this rock to be approximately 50%. The mean UCS of Gambier limestone having a 50-mm diameter and a length-to-diameter ratio of 2 was 4.7 MPa. Also, its bulk density was estimated to be 1213 kg/m^3.

Gosford sandstone was obtained from the Gosford Quarry, Somersby, New South Wales, Australia (Fig. 1.66C). Samples were carefully selected to be as homogeneous as possible. The maximum grain size of this sandstone was reported by Masoumi et al. (2016c) to be 0.6 mm. The detailed microstructural information and petrography of Gosford sandstone have been reported by Roshan et al. (2016b) and Masoumi et al. (2016c), respectively. The UCS of Gosford sandstone was reported by Masoumi et al. (2016c) at 52.3 MPa. It is noteworthy that in this study, only Brazilian tests were conducted on Gosford sandstone and the point load strength indices reported by Masoumi et al. (2016c) were included for analytical size effect investigation.

The crystal size was used to differentiate between two granite types (Figs. 1.66D and E). The granite with larger crystals (ranged between 3 and 5 mm) was labeled A while the small crystal sample ranging between 1 and 2 mm was labeled Granite B. Both granites consist of quartz, plagioclase feldspar, biotite mica, and finely disseminated iron oxides (Savidis, 1982). The average bulk densities of Granites A and B were estimated to be 2766 and 2588 kg/m^3, respectively. The mean UCS of Granites A and B were reported by Masoumi (2013) at 217.9 and 250.8 MPa, respectively.

The marble samples used in this study came from Wombeyan, New South Wales, Australia (see Fig. 1.1F). Based on the Savidis (1982) study, this marble mostly consists of calcite and its average bulk density was estimated to be 2758 kg/m^3. Masoumi (2013) reported the average UCS of this marble at 72.2 MPa.

1.4.3 Experimental procedure

The samples were prepared in accordance with the ISRM (2007)-suggested methods and tested in a completely dry condition (oven dried for 24 h at 105°C temperature).

1.4.3.1 Point load testing

All the point load tests (PLT) were conducted under axial loading conditions according to the ISRM (Franklin, 1985)-suggested method where the ratio of length over diameter can vary from 0.3 to 1 ($0.3 \leq L/D \leq 1$). In the axial point loading, the force was applied at the center of the end surfaces and the point load strength index (I_s) was estimated using the following equation:

$$I_s = \frac{P}{4A/\pi} \tag{1.76}$$

where A is the minimum cross-sectional area of a plane through the platen contact points and P represents the axial force.

For Bentheim sandstone and Gambier limestone, the sample sizes ranged from 19 to 65 mm diameters and more experiments were conducted on small diameters (Tables 1.18 and 1.19). Considering the standard deviations (SD) and coefficient of variations (CV) at different sizes, it is evident that the scatter of the Bentheim sandstone data was significantly less than that of the Gambier limestone. However, it should be noted that the measured I_s values from the Gambier limestone were approximately half that of the Bentheim sandstone. Higher homogeneity in the Bentheim sandstone potentially caused less scatter in the data. Figs. 1.67 and 1.68 present the size effect trends of Bentheim sandstone and Gambier limestone, respectively, which show the decrease in strength

TABLE 1.18 Mean axial point-load strength indices and different sizes for Bentheim sandstone

Sample diameter (mm)	Number of tests	Average of point-load index [I_s (MPa)]	SD (MPa)	CV (%)
19	9	2.4	0.2	8.3
25	10	1.9	0.2	10.5
38	7	1.8	0.1	5.6
50	5	1.6	0.1	6.3
65	4	1.4	0.1	7.1

TABLE 1.19 Mean axial point-load strength indices and different sizes for Gambier limestone

Sample diameter (mm)	Number of tests	Average of point-load index [I_s (MPa)]	SD (MPa)	CV (%)
19	6	1.2	0.1	8.3
25	5	1.0	0.2	20.0
38	5	0.9	0.1	11.1
50	5	0.8	0.1	12.5
65	5	0.7	0.1	14.3

with an increase in size. Figs. 1.69 and 1.70 illustrate the typical fracture patterns from the axial point load tests on Bentheim sandstone and Gambier limestone, respectively. A single fracture that led to two broken pieces was the most dominant failure pattern in the point load tests conducted on Bentheim sandstone and Gambier limestone at different sizes.

The sample diameters of Granite A varied between 19 and 50 mm diameters. For Granite B, diameters ranged from 18 to 96 mm. Tables 1.20 and 1.21 list the average point load strength indices, the number of tests at different sizes, and the SD and CV obtained from the point load testing on these two granite samples. The resulting CVs from both granites were approximately less than 22% for almost all sizes. Figs. 1.71 and 1.72 show the graph of point load indices versus sample diameter, which follows the generalized size effect concept where strength decreases with an increase in size. Also, the typical fracture patterns from the axial PLT on Granites A and B are presented in Figs. 1.73 and 1.74.

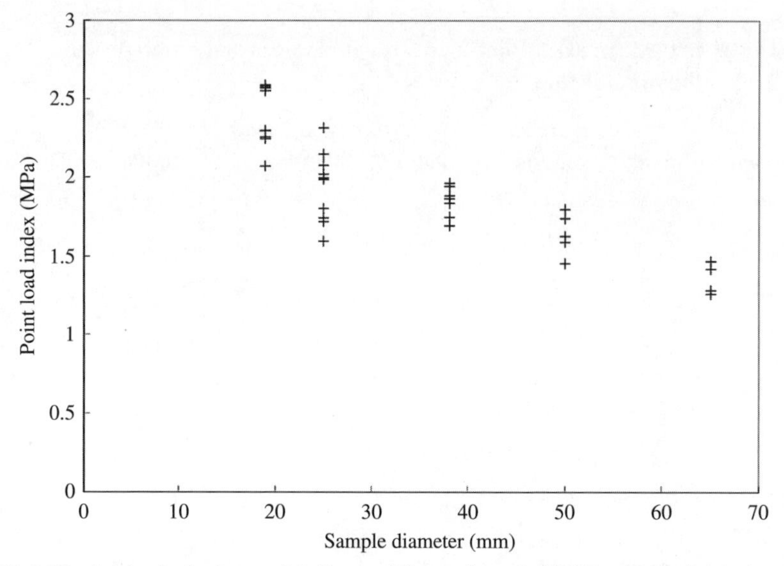

FIG. 1.67 Axial point load strength indices at different sizes obtained from Bentheim sandstone.

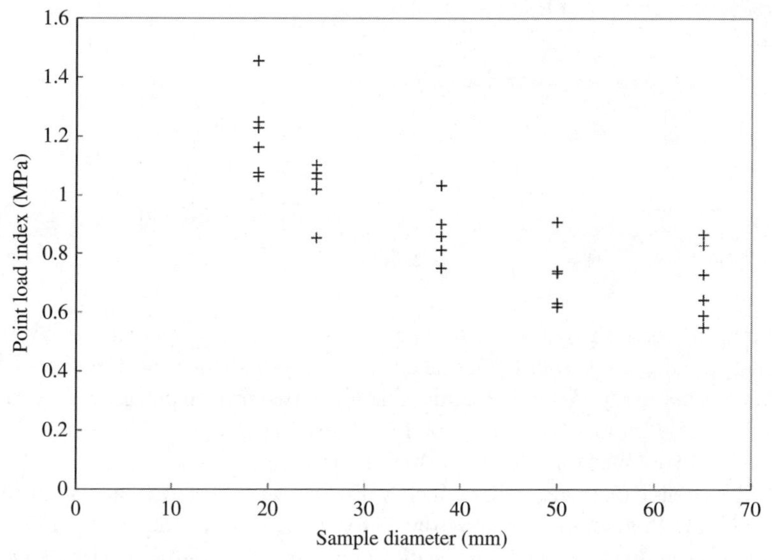

FIG. 1.68 Axial point load strength indices at different sizes obtained from Gambier limestone.

Marble was the only metamorphic rock that was tested here under axial point loading. The sample sizes for this rock varied from 19 to 96 mm diameter; Table 1.5 lists the resulting experimental data from the point load tests. From Table 1.22, it is evident that the CVs are greater than 20% for 19, 50, and 65 mm diameters. It is believed that this is associated with the existence of a weathered

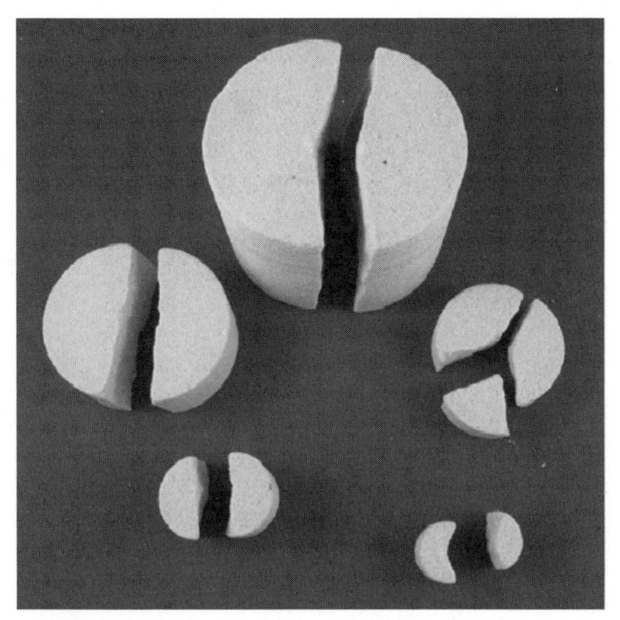

FIG. 1.69 Typical fracture patterns from axial PLT on Bentheim sandstone having 19, 25, 38, 50, and 65 mm diameters.

FIG. 1.70 Typical fracture patterns from axial PLT on Gambier limestone having 19, 25, 38, 50, and 65 mm diameters.

TABLE 1.20 Mean axial point-load strength indices and different sizes for Granite A

Sample diameter (mm)	Number of tests	Average of point-load index [I_s (MPa)]	SD (MPa)	CV (%)
19	6	18.2	3.4	18.7
25	4	17.3	2.2	12.7
38	6	14.4	0.9	6.3
50	5	10.7	1.0	9.3

TABLE 1.21 Mean axial point-load strength indices and different sizes for Granite B

Sample diameter (mm)	Number of tests	Average of point-load index [I_s (MPa)]	SD (MPa)	CV (%)
19	11	16.3	1.9	11.7
25	5	15.8	1.0	6.3
38	8	12.6	1.2	9.5
50	7	10.5	0.6	5.7
65	6	8.8	1.9	21.6
96	2	7.1	0.0	0.0

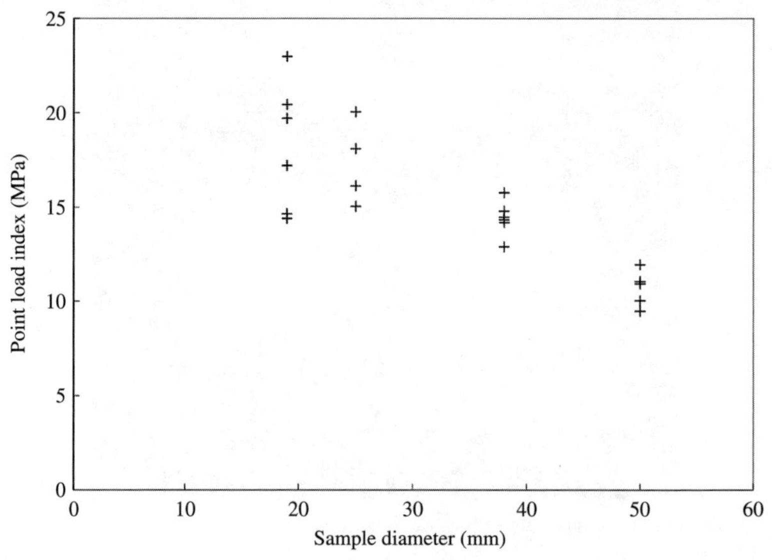

FIG. 1.71 Axial point load strength indices at different sizes obtained from Granite A.

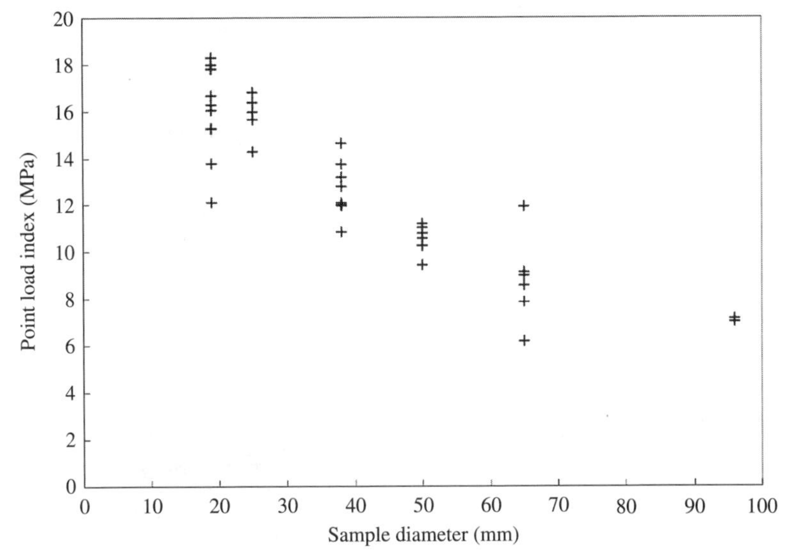

FIG. 1.72 Axial point load strength indices at different sizes obtained from Granite B.

FIG. 1.73 Typical fracture patterns from axial PLT on Granite A having 25, 38, and 50 mm diameters.

FIG. 1.74 Typical fracture patterns from axial PLT on Granite B having 38, 50, 65, and 96 mm diameters.

TABLE 1.22 Mean axial point-load strength indices and different sizes for marble

Sample diameter (mm)	Number of tests	Average of point-load index [I_s (MPa)]	SD (MPa)	CV (%)
19	5	2.9	0.6	20.7
25	9	2.7	0.4	14.8
38	8	1.7	0.3	17.6
50	9	1.7	0.4	23.5
65	8	1.6	0.4	25.0
96	3	1.5	0.1	5.3

plane (defect) randomly spreading across the marble block before the coring process. Identification of such a weak plane was very difficult through only observation, but attempts were made to avoid any coring from the affected zones. It could still, however, affect the results at the microscale. The point load strength indices obtained from marble (Fig. 1.75) samples are in agreement with

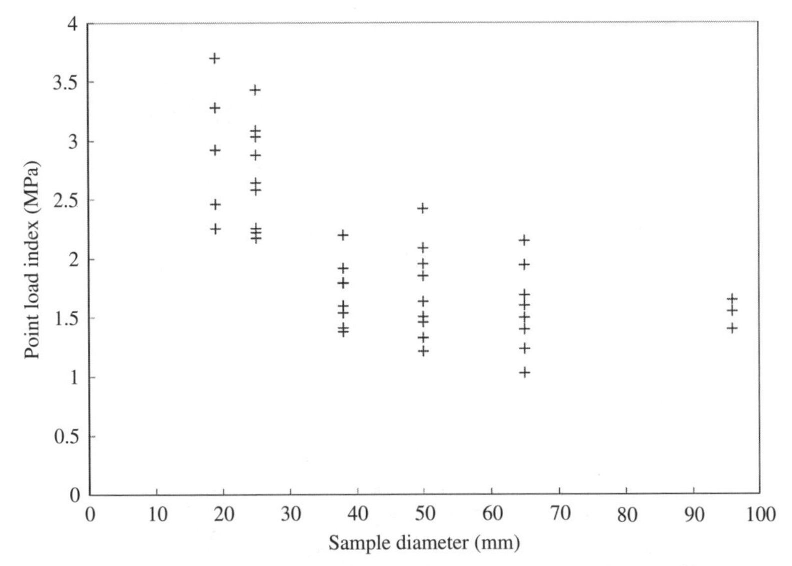

FIG. 1.75 Axial point load strength indices at different sizes obtained from marble.

the basic size effect theory (Weibull, 1951). There were, however, two diameters (38 and 96 mm) where the mean strength indices placed slightly below and slightly above the expected trend. Fig. 1.76 illustrates the typical fracture patterns of marble samples tested under axial point loading. Single fracture was the common failure pattern in this rock.

It has been reported by different researchers (Lama and Vutukuri, 1978; Carmichael, 1982; Kwasniewski, 1989; Jizba, 1991; Wong et al., 1997; Chang et al., 2006; Zoback, 2007) that an increase in porosity leads to a decrease in strength of the rocks. Such a reverse correlation is clearly recognizable in this study in which the Gambier limestone with maximum porosity exhibited the minimum point load strength at all tested diameters. It is noteworthy that Roshan et al. (2017b) demonstrated that in the sedimentary rocks (e.g. shaly-sandstones) apart from porosity, the clay content also has a substantial impact on the strength of rocks.

1.4.3.2 Indirect tensile (Brazilian) testing

For the Brazilian test, ISRM (Bieniawski and Hawkes, 1978; ISRM, 2007) proposed length-to-diameter ratios between 0.3 and 0.6, thus 0.5 has been selected as an appropriate ratio in this study. A total of 40 tests were conducted on Gosford sandstone according to the ISRM (2007)-suggested method. The sample sizes ranged from 19 to 145 mm diameters. The tensile strength was calculated using the following formula (Bieniawski and Hawkes, 1978, ISRM, 2007):

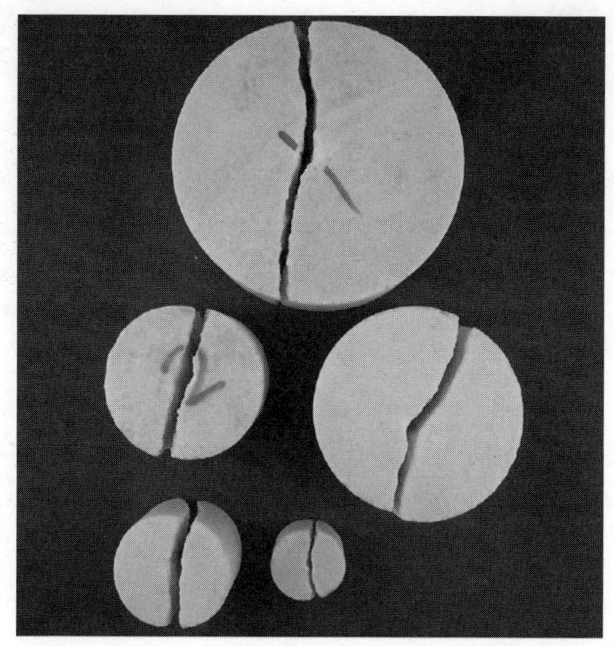

FIG. 1.76 Typical fracture patterns from axial PLT on the marble at 25, 38, 50, 65, and 96 mm diameters.

$$\sigma_t = \frac{2P}{\pi Dt} \tag{1.77}$$

where P is the peak load at failure, D is the diameter, and t is the thickness of the sample measured at the center.

The mean tensile strengths for Gosford sandstone samples as well as the number of repetitions for each size are given in Table 1.23. The overall scatter for Brazilian results was less than 15% in all sizes, except for the 25 mm diameter that resulted in about a 24% coefficient of variation. Fig. 1.77 shows the variation of the tensile strength data versus the sample diameter for Gosford sandstone. In general, a descending trend is observable from the tensile strength data ranging from 19 to 118 mm diameters where the increase in size leads to a decrease in strength. By contrast, the resulting tensile strength data for 145 mm diameter samples does not follow this trend and lies significantly above the mean strengths of almost all other sizes. This is an interesting observation that is reported here for the first time. It is believed to be associated with the fracture pattern of 145 mm diameter samples under the Brazilian test.

According to the ISRM (2007)-suggested method, only a single axial fracture at the center of the sample is a valid failure pattern for the Brazilian test. Fig. 1.78 highlights the difference between the failure patterns resulting from 145 mm diameter samples and those obtained from other sizes. Fig. 1.78 shows

TABLE 1.23 Mean tensile strengths for different sizes of Gosford sandstone samples

Sample diameter (mm)	Number of tests	Average of point-load index [I_s (MPa)]	SD (MPa)	CV (%)
19	5	3.4	0.4	12.4
25	5	3.3	0.8	24.2
38	5	3.2	0.3	9.4
50	5	3.2	0.3	9.4
65	5	2.8	0.4	14.3
96	5	2.5	0.2	8.0
118	5	2.4	0.3	12.5
145	5	3.4	0.1	2.9

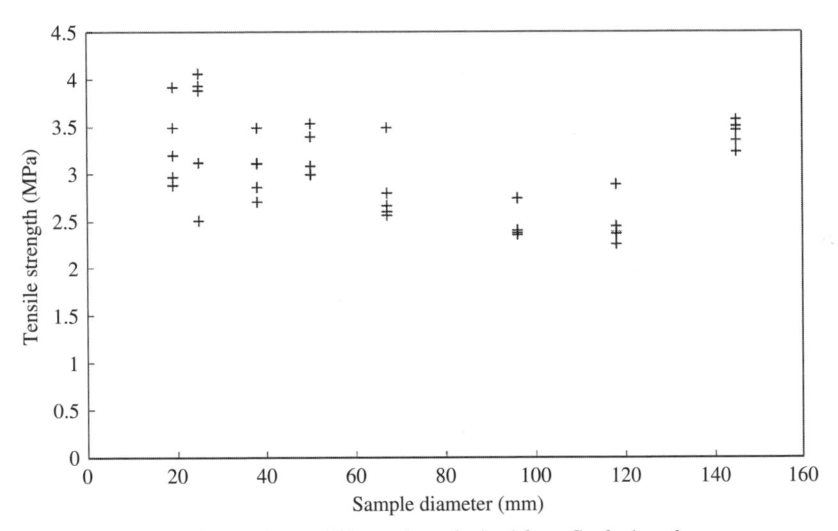

FIG. 1.77 Tensile strength data at different sizes obtained from Gosford sandstone.

that for the largest samples, an axial fracture at the center along with the extra shearing zones developed at the contact points between the samples and the compressive loading frame. This behavior was studied by Serati et al. (2015) through an extensive analytical investigation. They argued that the concentration of shear stresses developed in the vicinity of contacts can interfere with the tensile breakage of disc samples through the formation of inverse shear conical

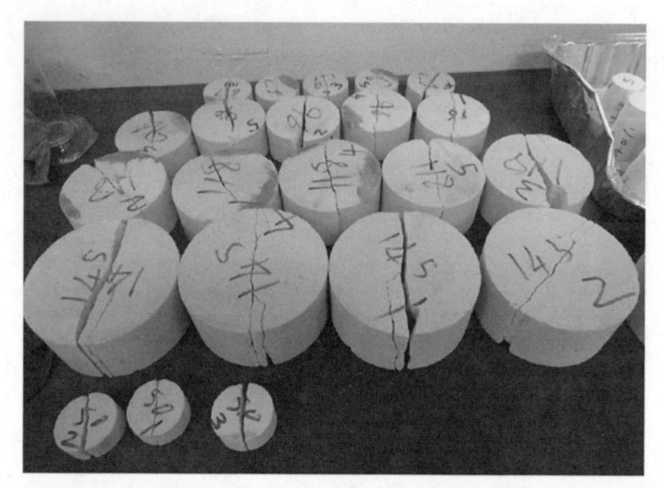

FIG. 1.78 Comparing the resulting fracture patterns of Gosford sandstone samples with various diameters after Brazilian tests.

plugs and multiple cracking. It has been reported by different researchers (Erarslan and Williams, 2012; Serati, 2014; Komurlu and Kesimal, 2015; Serati et al., 2015) that in hard and brittle materials, it is impractical to follow the standard Brazilian test methodology in its entirety (Fig. 1.79). Deviations from the standard test method arise mainly from the violation of the accepted

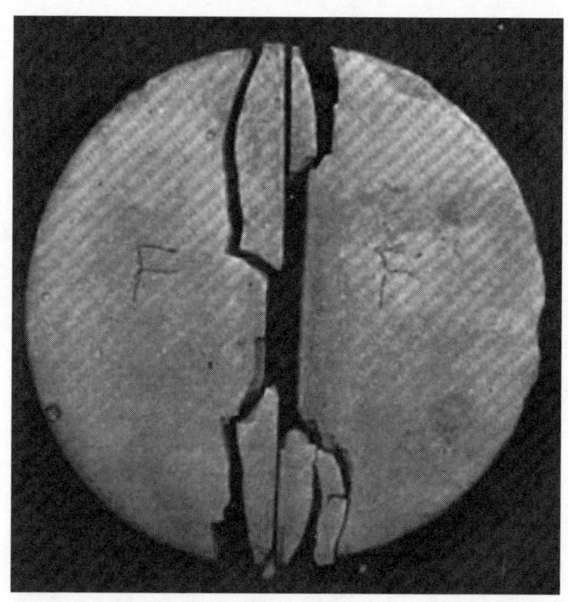

FIG. 1.79 The resulting fracture pattern from the Brazilian test on a graphite sample with 50 mm diameter reported by Serati et al. (2015). The UCS of this sample was measured at about 100 MPa.

boundary conditions in a conventional Brazilian test. The flexural tensile strength measured at the center of a solid Brazilian disc, therefore, could easily lead to erroneous estimations of the actual tensile strength of the material if the test method employed deviates significantly from the standard test method. Thus, it is likely that with an increase in sample size, the stored energy inside the Gosford sandstone sample becomes higher and therefore the behavior of rock becomes similar to hard rock materials, leading to an inaccurate measurement of tensile strength. As a result, the tensile strength data from 145 mm diameter samples were excluded for the size effect modeling process. It is noteworthy that such a phenomenon has been further investigated by Masoumi et al. (2017b) in which a number of Brazilian tests were conducted on hard rock samples with high brittleness at different diameters. Interestingly, it was found that with a decrease in sample size, the level of stored energy inside the hard rock samples reduced, leading to a valid single tensile crack at the center of samples.

1.4.4 Comparative study

In this section, the experimental results obtained from different sizes are compared with well-known size effect models based on statistics, fracture energy, and multifractals to investigate their applicability to point load and tensile strength data and identify the one that has the best fit to the experimental data. The sample diameter is selected as the size-dependent representative parameter rather than the total volume of the sample because it is intuitive to conceive that the diameter of the sample is related to its volumetric change and the fact that such a practice has been already adopted by ISRM (2007) leading to a so well-known Brook (1980, 1985) size effect model for upscaling of point load strength data from laboratory to field condition based on statistical theory e.g. where the diameter is a representative factor for size correction. Furthermore, Carpinteri (1994), Carpinteri and Ferro (1994), and Carpinteri et al. (1995) conducted size-effect studies on the tensile strength data of concrete samples obtained from Brazilian tests based on multifractal theory where the sample diameter was the representative size effect parameter. As a result, in this study the size effect analysis is based on sample diameter and to account for proper volumetric change in size-effect analysis, the length-to-diameter ratio of the samples was kept constant so that the change in sample sizes can be comparable to that of volume at different diameters. Subsequently, the length-to-diameter ratio of the critical zones where the fracture is initiated under point load and indirect tensile testing conditions was also constant in all samples with different sizes.

Masoumi et al. (2016c) has extensively reviewed the existing size effect models and elaborated their features through comprehensive discussion as well as some graphical representations. Below is a brief summary from the Masoumi et al. (2016c) study on these models.

1.4.4.1 Size effect models

Statistical model

The statistical theory was initially proposed by Weibull (1939), followed by an improvement to the original concept by Weibull (1951). The theory is commonly known as the weakest link theory and states that the probability of failure under any load can be attributed to structural flaws inherent in solids. If the probability of an element to survive under load is assumed as $1-P$ where P is the probability of the element to fail, then the survivability of the entire chain of n elements is the accumulative probability, which equals to $(1-P)^n$ or $1-P_f$ where P_f is the probability of failure of the chain and therefore:

$$1-P_f=(1-P)^n \tag{1.78}$$

then:

$$\ln(1-P_f)=n\ln(1-P) \tag{1.79}$$

In practice, P is very small and thus $\ln(1-P)\approx -P$. Hence:

$$\ln(1-P_f)=-nP \tag{1.80}$$

The number of elements, n, in a chain situation can be replaced by the representative volume of the material, V/V_r, as shown below:

$$\ln(1-P_f)=-(V/V_r)P \quad \text{or} \quad P_f=1-\exp\left[-\frac{V}{V_r}P\right] \tag{1.81}$$

A general form of Eq. (1.81) was proposed by Weibull (1939) and then an additional variable, m, was added for better model simulation (Weibull, 1951) according to:

$$m\log\left(\frac{P_f(\sigma)}{P(\sigma)}\right)=\log\left(\frac{V}{V_r}\right) \tag{1.82}$$

In Eq. (1.82), V and V_r can be substituted by any characteristic measure of size such as length or sample diameter. A modified formulation of the statistical model (Eq. 1.82) was proposed by Brook (1980, 1985) to predict the size effect in point load tests as follows:

$$\frac{I_s}{I_{s50}}=\left(\frac{50}{d}\right)^{k_1} \tag{1.83}$$

where, I_s is the point load strength index, I_{s50} is the characteristic point load strength index obtained from a sample with the characteristic size of 50 mm, d is sample size, and k_1 is a positive constant controlling the statistical decay of the strength with a raise in size, which is also related to m in Eq. (1.82).

Fracture energy model

A size effect model based on the fracture energy theory was proposed by Bazant (1984), who postulated that a zone of microcracking that precedes the fracture energy would occur in most brittle and quasibrittle materials such as rock and concrete. Bazant (1984) then quantified the fracture energy and incorporated this energy into the formulation of the size effect law (SEL) according to:

$$\sigma_N = \frac{Bf_t}{\sqrt{1+(d/\lambda d_0)}} \tag{1.84}$$

where σ_N is a nominal strength (e.g., uniaxial compressive strength, point load strength, and tensile strength) B and λ are dimensionless material constants, f_t is a strength for a sample with a very small size, d is the sample size, and d_0 is the maximum aggregate size. Generally, Bf_t and λd_0 are determined through a curve fitting process as instructed by Bazant (1984). Masoumi et al. (2016c) demonstrated the process of determination of Bf_t and λd_0 for UCS data of six different rock types with various sizes.

Multifractal model

Fractals have been implemented to study different properties in rocks. Carpinteri (1994) and Carpinteri and Ferro (1994) argued that the self-similar properties of multiple cracks can appear in a wide range of material sizes, therefore making them fractal. Carpinteri et al. (1995) proposed the multifractal scaling law (MFSL) based on the topological concept of geometrical multifractality with the following analytical expression:

$$\sigma_N = f_c\sqrt{1+\frac{l}{d}} \tag{1.85}$$

where σ_N is the nominal strength (e.g., uniaxial compressive strength, point load strength, and tensile strength), l is a material constant with unit of length, f_c is the strength of a sample with an infinite size that may be expressed in terms of an intrinsic strength, and d is the sample size. Similar to SEL, the model parameters of MFSL are determined through the curve fitting process as demonstrated by Carpinteri et al. (1995) and Masoumi et al. (2016c).

1.4.4.2 Existing size effect models to point load

Here, the applicability of the three main size effect models (statistical, SEL, and MFSL) to the point load results obtained from sedimentary, igneous, and metamorphic rocks is examined using an analytical approach.

The point load results for Gosford sandstone reported by Masoumi et al. (2016c) under axial loading were also included here for extensive analytical study. It is aimed to identify which size effect model provides the best fit to the experimental data based on the coefficient of multiple determination (R^2) values.

For sedimentary rocks, the resulting size effect comparative analysis is presented in Tables 1.24 and 1.25. Also, Figs. 1.80–1.82 illustrate the comparison between the model fits and the experimental data. As can be seen from Figs. 1.80–1.82 and Tables 1.24 and 1.25, the SEL and MFSL generally show a more accurate fit to the data compared to the statistical model. The presence of two fitting parameters in SEL and MFSL gives more flexibility to the mathematical form to fit better to the experimental data.

The applicability of existing size effect models to the point load results from granites and marble samples was investigated. As these rocks have a crystalline structure, they were grouped together. The resulting fitting constants are listed in Tables 1.26 and 1.27, followed by graphical representations in Figs. 1.83–1.85. In Fig. 1.85, almost all three model fits are alike whereas in Figs. 1.83 and 1.85, the SEL and MFSL better fit the data compared to the statistical model.

The normalized mean axial point load strength indices for all rock samples grouped together for the best fits of the statistical model, SEL, and MFSL as shown in Fig. 1.86. All the axial point load strength indices were nondimensionalized using the average 50 mm diameter point load strength index. As a result, some modifications were required in SEL and MFSL, respectively, according to:

$$\frac{\sigma_N}{\sigma_{N50}} = \frac{\sqrt{1 + (50/\lambda d_0)}}{\sqrt{1 + (d/\lambda d_0)}} \tag{1.86}$$

TABLE 1.24 List of fitting constants for statistical model, SEL, and MFSL obtained from mean point-load strength indices of sedimentary rocks

	Statistical model (Eq. 1.83)		SEL			MFSL		
Sample	k_1	R^2	Bf_t (MPa)	λd_0 (mm)	R^2	f_c (MPa)	l (mm)	R^2
Bentheim sandstone	0.38	0.92	4.30	7.61	0.92	0.82	133.23	0.93
Gambier limestone	0.39	0.97	2.31	6.57	0.97	0.38	161.15	0.98
Gosford sandstone	0.31	0.82	6.85	10.51	0.95	1.25	195.37	0.92

TABLE 1.25 Range of fitting constants estimated for sedimentary rocks at 95% and 99% confidence intervals for statistical model, SEL, and MFSL

	95% confidence intervals					99% confidence intervals				
	Eq. (1.83)	*SEL*		*MFSL*		*Eq. (1.83)*	*SEL*		*MFSL*	
Sample	k_1	Bf_t *(MPa)*	λd_0 *(mm)*	f_c *(MPa)*	l *(mm)*	k_1	Bf_t *(MPa)*	λd_0 *(mm)*	f_c *(MPa)*	l *(mm)*
Bentheim sandstone	±0.11	±4.85	±21.64	±0.81	±327.36	±0.18	±8.91	±39.72	±1.49	±600.83
Gambier limestone	±0.06	±1.75	±12.19	±0.26	±261.74	±0.11	±3.21	±22.37	±0.48	±480.39
Gosford sandstone	±0.15	±4.04	±16.55	±1.25	±459.28	±0.24	±6.69	±27.45	±2.08	±761.62

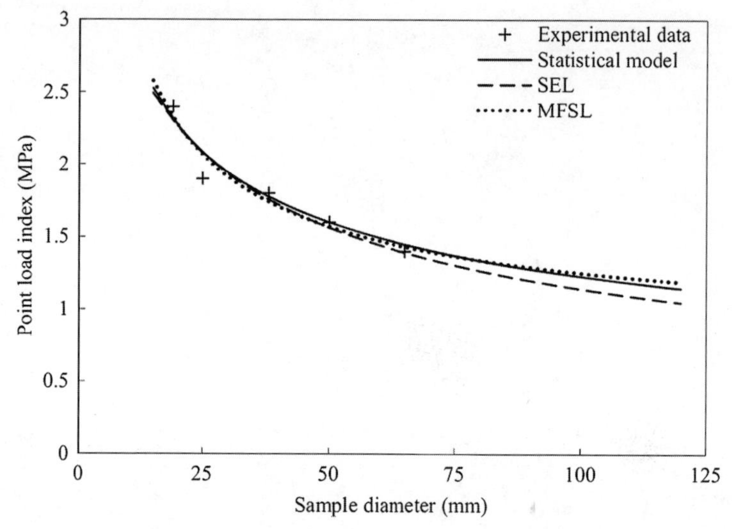

FIG. 1.80 Comparing three size effect models using the mean axial point load results from Bentheim sandstone.

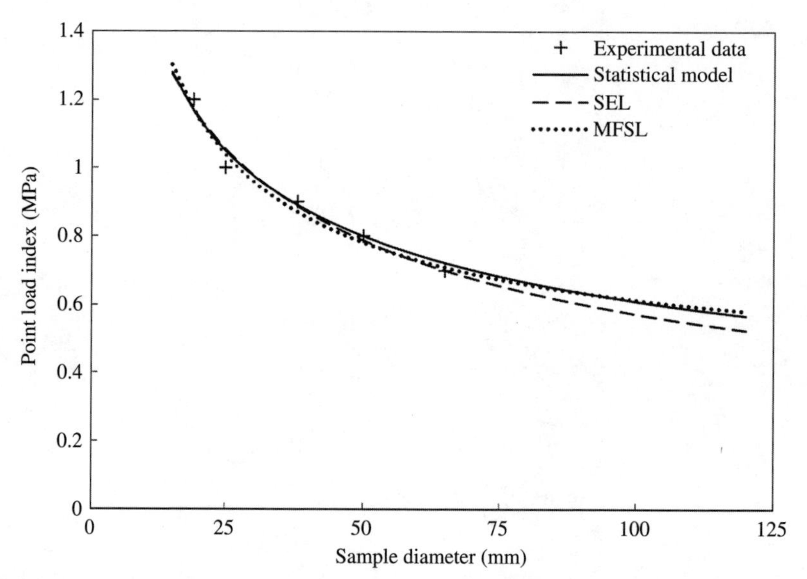

FIG. 1.81 Comparing three size effect models using the mean axial point load results from Gambier limestone.

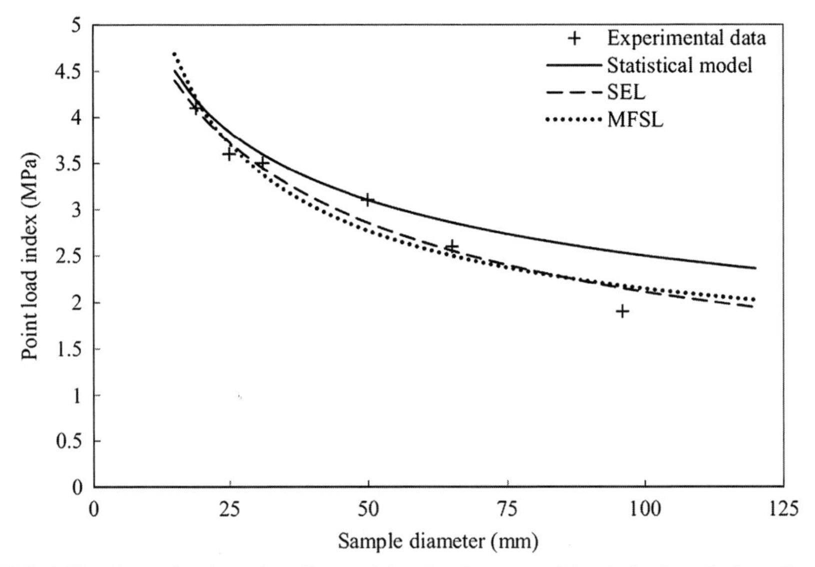

FIG. 1.82 Comparing three size effect models using the mean axial point load results from Gosford sandstone reported by Masoumi et al. (2016c).

TABLE 1.26 List of fitting constants for statistical model, SEL, and MFSL obtained from mean point-load strength indices of crystalline rocks

	Statistical model (Eq. 1.83)		SEL			MFSL		
Sample	k_1	R^2	Bf_t *(MPa)*	λd_0 *(mm)*	R^2	f_c *(MPa)*	l *(mm)*	R^2
Granite A	0.60	0.85	64.32	1.77	0.91	1.23	4475.42	0.91
Granite B	0.51	0.97	114.31	0.43	0.97	0.37	39,724.53	0.97
Marble	0.54	0.88	36.58	0.12	0.89	0.49	632.15	0.90

$$\frac{\sigma_N}{\sigma_{N50}} = \frac{\sqrt{1 + \dfrac{l}{d}}}{\sqrt{1 + \dfrac{l}{50}}} \tag{1.87}$$

Tables 1.28 and 1.29 list the fitting constants for the statistical model, the modified SEL (Eq. 1.86), and the modified MFSL (Eq. 1.87). It is evident from

TABLE 1.27 Range of fitting constants estimated for crystalline rocks at 95% and 99% confidence intervals for statistical model, SEL, and MFSL

	95% confidence intervals					99% confidence intervals				
	Eq. (1.83)	*SEL*		*MFSL*		*Eq. (1.83)*	*SEL*		*MFSL*	
Sample	k_1	Bf_t *(MPa)*	λd_0 *(mm)*	f_c *(MPa)*	l *(mm)*	k_1	Bf_t *(MPa)*	λd_0 *(mm)*	f_c *(MPa)*	l *(mm)*
Granite A	±0.19	±536.09	±31.44	±88.27	±645,705	±0.35	±1237	±72.51	±203.6	±1,489,397
Granite B	±0.08	±1227.6	±9.35	±47.06	±10,041,235	±0.13	±2036	±15.51	±78.05	±16,651,365
Marble	±0.16	±2571.0	±16.77	±1.89	±5190.6	±0.25	±4263	±27.82	±3.13	±8607

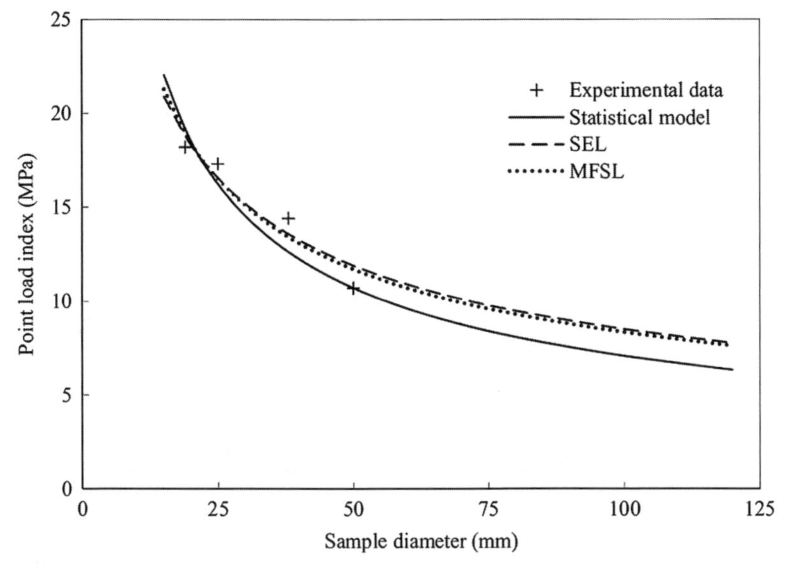

FIG. 1.83 Comparing three size effect models using the mean axial point load results from Granite A.

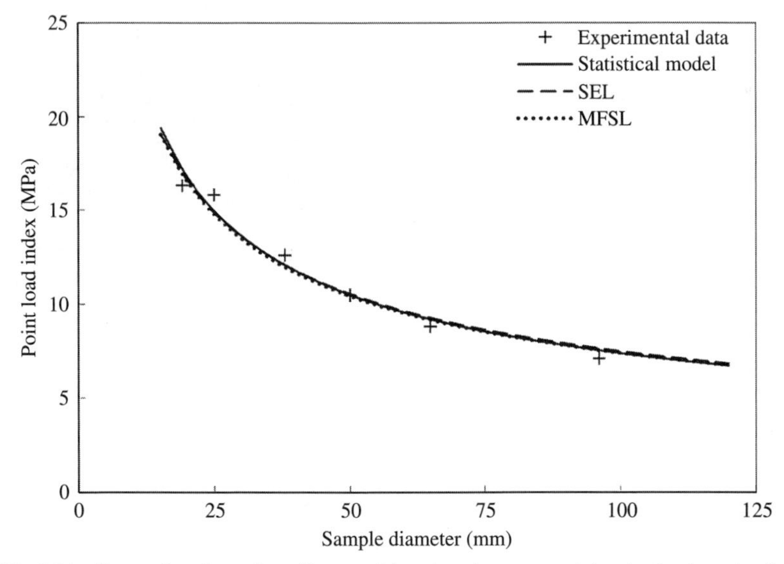

FIG. 1.84 Comparing three size effect models using the mean axial point load results from Granite B.

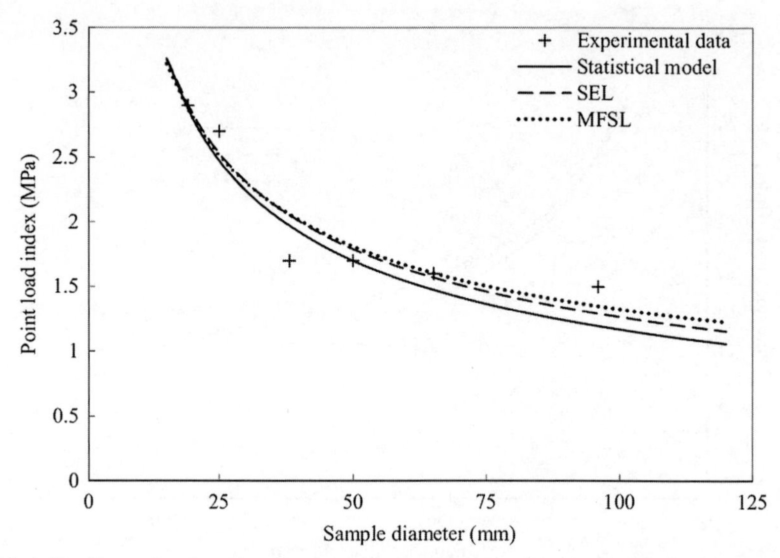

FIG. 1.85 Comparing three size effect models using the mean axial point load results from marble.

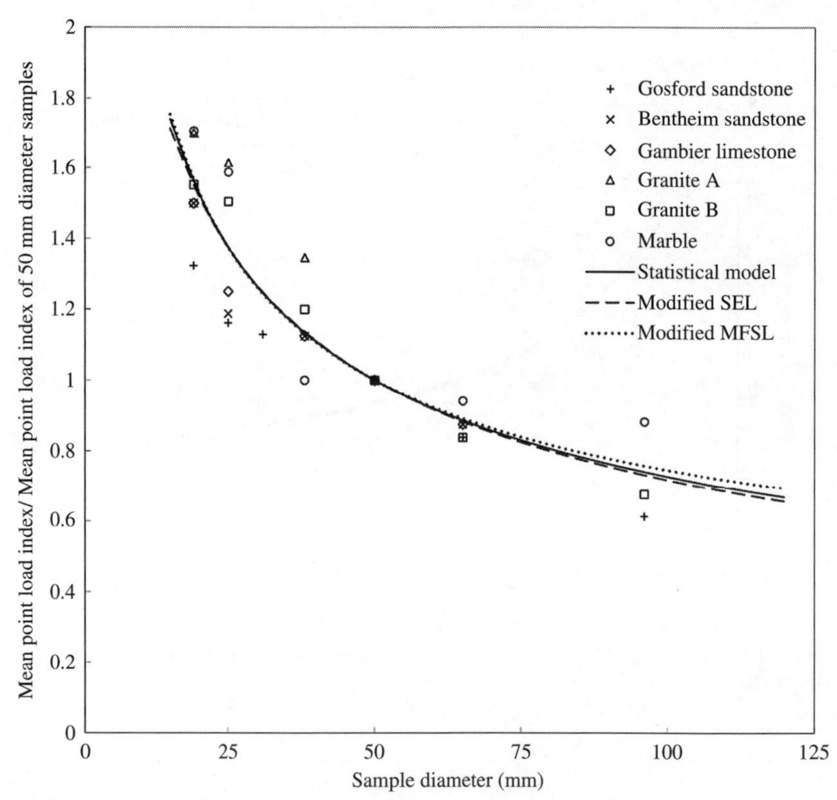

FIG. 1.86 Comparing three size effect models using the mean axial point load results from all rock samples.

TABLE 1.28 List of fitting constants for statistical model, modified SEL, and modified MFSL obtained from mean point-load strength indices of all rock samples

	Statistical model (Eq. 1.83)		Modified SEL		Modified MFSL	
Sample	k_1	R^2	λd_0 *(mm)*	R^2	l *(mm)*	R^2
All rock samples	0.46	0.84	3.084	0.84	400.1	0.84

Tables 1.28 and 1.29 that the resulting R^2 from the statistical model and the modified SEL and MFSL are the same when all rock samples were grouped together. This could be associated with the modification process applied to SEL and MFSL (Eqs. 1.86, 1.87) where one of the constants during the normalization process was eliminated. Also, it is evident that the resulting k_1 for the statistical model is approximately the same as that suggested by ISRM (Brook, 1985; ISRM, 2007) "$k_1 = 0.45$" when all rock samples were included. However, the resulting k_1 values for individual rock types were different to that suggested by ISRM (Brook, 1985, ISRM, 2007).

1.4.4.3 Existing size effect models to tensile strength

The tensile strength results from Gosford sandstone were used to assess which size effect model provides the best fit to the experimental data. For the statistical model, a modified version of Eq. (1.82) similar to that suggested by Brook (1985) for point load strength index is proposed:

$$\frac{\sigma_t}{\sigma_{t50}} = \left(\frac{50}{d}\right)^{k_2} \tag{1.88}$$

where σ_t is the tensile strength, $\sigma_{t50}$is the characteristic tensile strength obtained from a sample with the characteristic size of 50 mm, d is sample size, and k_2 has the same role as that for k_1 in Eq. (1.83), which controls the statistical decay of the strength with an increase in size. A summary of the results of model calibration is presented in Tables 1.30 and 1.31. Also, Fig. 1.87 compares the resulting model fits by different size effect models against the experimental data. From Fig. 1.87 as well as Tables 1.30 and 1.31, it is clear that SEL provided the best fit to the tensile strength data while the statistical model resulted in the poorest fit and MFSL lies in between. This is an ideal example to highlight the benefit of using a model with double constants over the one with a single constant. SEL and MFSL have two constants for the fitting process while the statistical model relies only on one constant, which is the strength of

TABLE 1.29 Range of fitting constants estimated for all rock samples at 95% and 99% confidence Intervals for statistical model, modified SEL, and modified MFSL

	95% confidence intervals			99% confidence intervals			
	Eq. (1.83)	*Modified SEL*	*Modified MFSL*	*Eq. (1.83)*	*Modified SEL*	*Modified MFSL*	
Sample	k_1	λd_0 *(mm)*	l *(mm)*	k_1	λd_0 *(mm)*		l *(mm)*
All rock samples	±0.05	±4.18	±581.86	±0.07	±5.62		±782.86

TABLE 1.30 Statistical model, SEL, and MFSL obtained from mean tensile strengths of Gosford sandstone

	Statistical model (Eq. 1.88)		SEL			MFSL		
Sample	k_1	R^2	Bf_t *(MPa)*	λd_0 *(mm)*	R^2	f_c *(MPa)*	l *(mm)*	R^2
Gosford sandstone	0.16	0.41	3.85	77.50	0.94	2.36	24.59	0.77

characteristic size (in here 50 mm). Thus, the whole statistical model shifted upward, leading to the least best fit compared to SEL and MFSL.

1.4.5 Conclusion

A set of point-load and indirect tensile (Brazilian) tests was conducted on six different rock types having various geological origins over a range of sizes. It was demonstrated that all rock types follow the generalized size effect trend where an increase in size leads to a decrease in strength. Also, it was noted that in the Brazilian test, an increase in size causes the failure of intact rock transfers from pure tensile failure to a combination of shear and tensile failure.

The applicability of three well-known size effect models based on statistics, fracture energy, and multifractals to point load and tensile strength data was investigated. It was confirmed that, in addition to the statistical size effect model, the fracture energy and multifractal size effect models can suitably predict the size effect behavior of point load results. Also, it was demonstrated that the fracture energy size effect model provided the best fit to the tensile strength data while the model fit based on the statistical concept provided the least best fit.

1.5 Scale effect into multiaxial failure criterion

1.5.1 Introduction

The scale effect is a significant characteristic in brittle and quasibrittle media such as rock. Many studies have explored the scale effect with regard to the uniaxial compressive test in different rock types (Mogi, 1962; Pratt et al., 1972; Hoek and Brown, 1980a, b; Baecher and Einstein, 1981; Dey and Halleck, 1981; Tsur-Lavie and Denekamp, 1982; Silva et al., 1993; Kramadibrata and Jones, 1993; Hawkins, 1998; Arioglu, 1999; Thuro et al., 2001a, b; Pells, 2004; Yoshinaka et al., 2008; Darlington and Ranjith, 2011). Some research

TABLE 1.31 Constants estimated for Gosford sandstone for statistical mode, SEL, and MFSL

	95% confidence intervals					99% confidence intervals				
	Eq. (1.88)	*SEL*		*MFSL*		*Eq. (1.83)*	*SEL*		*MFSL*	
Sample	k_1	Bf_t *(MPa)*	λd_0 *(mm)*	f_c *(MPa)*	l *(mm)*	k_1	Bf_t *(MPa)*	λd_0 *(mm)*	f_c *(MPa)*	l *(mm)*
Gosford sandstone	±0.14	±0.37	±38.37	±0.47	±24.31	±0.21	±0.58	±60.19	±0.74	±38.14

has also investigated the scale effect for different stress paths such as point load and tensile tests (Broch and Franklin, 1972; Bieniawski, 1975; Wijk et al., 1978; Brook, 1980; Greminger, 1982; Forster, 1983; Brook, 1985; Panek and Fannon, 1992; Prakoso and Kulhawy, 2011; Thuro et al., 2001a, b). On the other hand, investigations into the mechanical behavior of intact rock have resulted in various failure criteria (Lade and Duncan, 1975; Desai, 1980; Desai and Faruque, 1984; Kim and Lade, 1984; Lade and Nelson, 1987; Kim and Lade, 1988; Khan et al., 1991). Some criteria (Lade and Duncan, 1975; Iwan, 1967; Desai, 1980; Liu et al., 2005) are extensions of those developed for soils, for which the scale effect has not been incorporated. Perhaps the most widely used criterion is that of Hoek and Brown (1997), and scale effect has been incorporated into it by assigning a scale dependence to the uniaxial compressive strength term that appears in its definition. However, there was no clear analytical justification for this approach to incorporate scale effect, and its suitability to capture scale effect for rocks brought to failure in paths other than uniaxial compression such as point loading, uniaxial tension, and pure shearing remains unverified.

An alternate multiaxial failure criterion that incorporates scale effect is presented here. It is an extension of the simple two-parameter multiaxial failure criterion for brittle materials proposed by Christensen (2000), which is modified to include scale effect. With this modification, the scale effect under different stress paths such as uniaxial compression, point loading, uniaxial tension, and

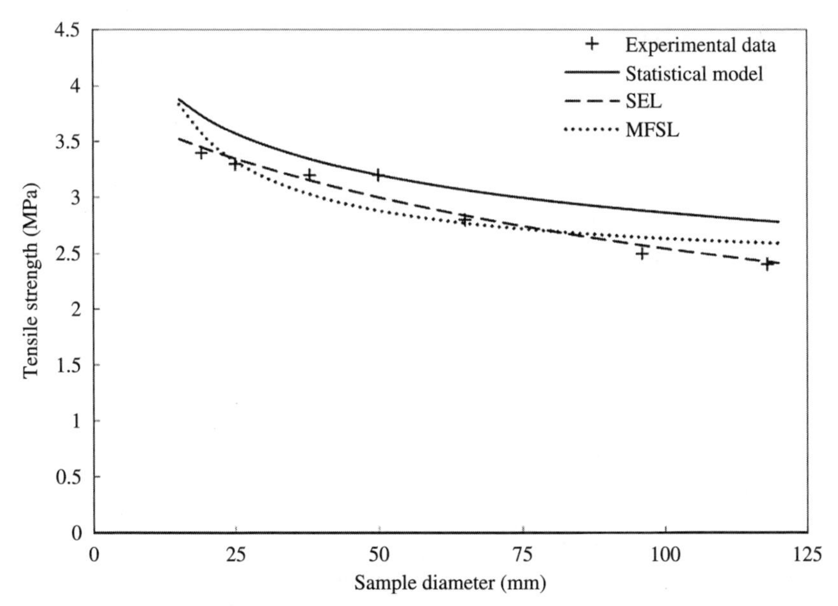

FIG. 1.87 Comparing three size effect models using the mean tensile strength results from Gosford sandstone.

pure shearing can be suitably captured. Finally, a parametric study is carried out to investigate the scale dependency of the proposed parameters for the modified multiaxial failure criterion.

1.5.2 Background

The original criterion of Christensen (2000), with no account of scale, will be used as a basis in this study. It states that a material is not at failure when:

$$\frac{\chi\kappa}{\sqrt{3}}I_1+(1+\chi)^2\left(\frac{I_1^{\ 2}}{3}-I_2\right)<\frac{\kappa^2}{1+\chi}\text{ or }\frac{\chi\kappa}{\sqrt{3}}I_1+(1+\chi)^2J_2<\frac{\kappa^2}{1+\chi} \tag{1.89}$$

where I_1 and I_2 are the first and second invariants of the stress tensor and $J_2=I_1^{\ 2}/3-I_2$ is the second invariant of the deviatoric stress tensor, with tensile stresses being taken as positive. Writing the stress tensor as:

$$\boldsymbol{\sigma}=\begin{bmatrix}\sigma_x & \tau_{xy} & \tau_{xz}\\ \tau_{yx} & \sigma_y & \tau_{yz}\\ \tau_{zx} & \tau_{zy} & \sigma_z\end{bmatrix} \tag{1.90}$$

leads to the definitions:

$$I_1=\sigma_x+\sigma_y+\sigma_z\text{and}I_2=\sigma_x\sigma_y+\sigma_x\sigma_z+\sigma_y\sigma_z-\tau_{xy}^{\ 2}-\tau_{xz}^{\ 2}-\tau_{yz}^{\ 2} \tag{1.91}$$

For uniaxial compression, where σ^c is the applied uniaxial compressive strength, $I_1=\sigma^c$, $I_2=0$, and $J_2=(\sigma^c)^2/3$. It is clear from the criterion that for a particular material, two parameters χ and κ are required to characterize its strength. χ is a dimensionless shape parameter and represents the ratio between the characteristic uniaxial compressive and tensile strengths, $\sigma^c$and σ^t, respectively, through:

$$\chi=\frac{|\sigma^c|}{\sigma^t}-1 \tag{1.92}$$

κ is a strength parameter giving the criterion and the dimensions of stress, and is defined as:

$$\kappa=\frac{1+\chi}{\sqrt{3}}|\sigma^c| \tag{1.93}$$

Clearly, uniaxial compressive and tensile tests would be sufficient to evaluate χ and κ, although as will be demonstrated later, other combinations of tests involving different stress paths would suffice.

If $\chi=0$, then the material behavior at failure is of the von Mises type as the criterion in Eq. (1.89) simplifies to $J_2<\kappa^2$ and κ individually controls failure. When $\chi=0$, the criterion defines a failure surface in the three-dimensional stress state as a right circular cylinder, symmetrical about the hydrostatic axis with radius$\sqrt{2}\kappa$. If $\chi>0$, then Eq. (1.89) defines a failure surface in the three-dimensional stress space that is a revolved paraboloid symmetric about the hydrostatic axis. When $\chi \rightarrow \infty$ the material is unable to sustain load of any kind and disintegration occurs. Christensen (2000) suggested that κ is a measure of the strength of the material having no microstructural damage and must be related to atomic scale properties. Furthermore, Christensen (2000) suggested that χ represents the effects of microstructural deviations from the ideal sample with no microdamage. However, the attribution of microstructural origins to κ and χ has not actually been supported by targeted experimental investigations, although it is aimed in this research to provide some support to Christensen's ideas using the data gathered below.

Goodman (1989) showed compressive and tensile strengths for a range of rock types indicating that, if Eqs. (1.92), (1.93) apply, $1+\chi$ varies from about 10 to 170. It is therefore sensible to assume for all rock types that $\chi > 9$.

The challenge is to define κ and χ as functions of scale, as the criterion in Eq. (1.89) would then become scale-dependent. The reason that only these two constants should be scale-dependent is that they describe the material characteristics while the other parameters such as I_1 and J_2 are stress-dependent. For example, in a rock sample with the same material characteristics, I_1 and J_2 can attain different values under various stress conditions.

1.5.3 Scale and Weibull statistics into strength measurements

The Weibull (1951) probability distribution function was proposed to describe the survival probability of a block of volume V contained within a larger volume of material V_r. Bazant et al. (1991) described the mathematical principles of the statistical model in a simple and elegant way. According to Bazant et al. (1991), in a chain, if the failure probability of an element (link) is assumed P_1, then the chance of survival would be $(1-P_1)$ and therefore, in the case of many connecting elements, the survival probability would be as follows:

$$(1-P_1)(1-P_1)(1-P_1)(1-P_1)\ldots(1-P_1)\,\text{or}\,(1-P_1)^N = 1-P_f \tag{1.94}$$

where P_f is the failure probability of the chain. So,

$$N\ln(1-P_1) = \ln\left(1-P_f\right) \tag{1.95}$$

In practice, P_1 has a very small value. This leads to $\ln(1-P_1) \approx -P_1$. Therefore:

$$\ln\left(1-P_f\right) = -NP_1 \tag{1.96}$$

Now, by setting $N=V/V_r$, Eq. (1.6) would be:

$$\ln\left(1-P_f\right) = -(V/V_r)P_1\,\text{or}\,P_f(\sigma) = 1-\exp\left[-\frac{V}{V_r}P_1(\sigma)\right] \tag{1.97}$$

where V is the volume of the sample, V_r represents the volume of one element in the sample, $P_f(\sigma)$ is the material strength, and $P_1(\sigma)$ is the strength of the representative sample. Eq. (1.97) is the initial statistical model proposed by Weibull (1951); later, he introduced a more general form of Eq. (1.9) through:

$$m\log\left(\frac{P_f(\sigma)}{P_1(\sigma)}\right) = \log\left(\frac{V}{V_r}\right) \tag{1.98}$$

where m is a material constant introduced for better simulation of the size effect behavior ($m=1$ was assumed in Eq. (1.97)). In Eq. (1.98), V and V_r can be substituted by any characteristic measure of volume such as length^3 or sample diameter^3. For example, in the case of cylindrical samples with identical shapes and constant length-to-diameter ratios, instead of volume the diameter can be substituted.

It will now be demonstrated that Eq. (1.98) can be applied to strengths observed in rock when subjected to different stress paths.

1.5.3.1 Scale effect in uniaxial compressive strength

The uniaxial compressive strength (UCS) of rock samples measured in a laboratory is well known to be scale-dependent and obey Eq. (1.98) written in a slightly different form:

$$\frac{\sigma^c}{\sigma^c_{50}} = \left(\frac{d}{50}\right)^{-k_1} = \beta_1 \tag{1.99}$$

where the measured UCS (σ^c) is a function of sample diameter d (in millimeters) and σ^c_{50} is the characteristic UCS measured of a sample with a diameter of 50 mm. Hoek and Brown (1980a, b) collected UCS results from different rock types and suggested that the value of k_1 is 0.18 (Fig. 1.88).

For the uniaxial compressive test:

$$\sigma^c = -\frac{\sqrt{3}\kappa}{\chi+1} \quad \text{and} \quad \sigma^c_{50} = -\frac{\sqrt{3}\kappa_{50}}{\chi_{50}+1} \tag{1.100}$$

where κ_{50} and χ_{50} are material properties for 50 mm diameter samples, and κ and χ are material properties for samples of diameter d. An expression linking $\kappa_{50}, \chi_{50}, \kappa$, and χ to β_1 (or $d/50$) is then obtained by substituting Eq. (1.100) into Eq. (1.99):

$$\frac{\kappa(\chi_{50}+1)}{\kappa_{50}(\chi+1)} = \left(\frac{d}{50}\right)^{-k_1} = \beta_1 \tag{1.101}$$

1.5.3.2 Scale effect in point load strength index

Scale effect is also observed in strength measured using the point load test (PLT). Franklin (1985) showed that the point load strength index $I_s = f/d^2$ (where f is the force required to fail a sample of characteristic diameter d) is a function of d according to:

$$\frac{I_s}{I_{s50}} = \left(\frac{d}{50}\right)^{-k_2} = \beta_2 \tag{1.102}$$

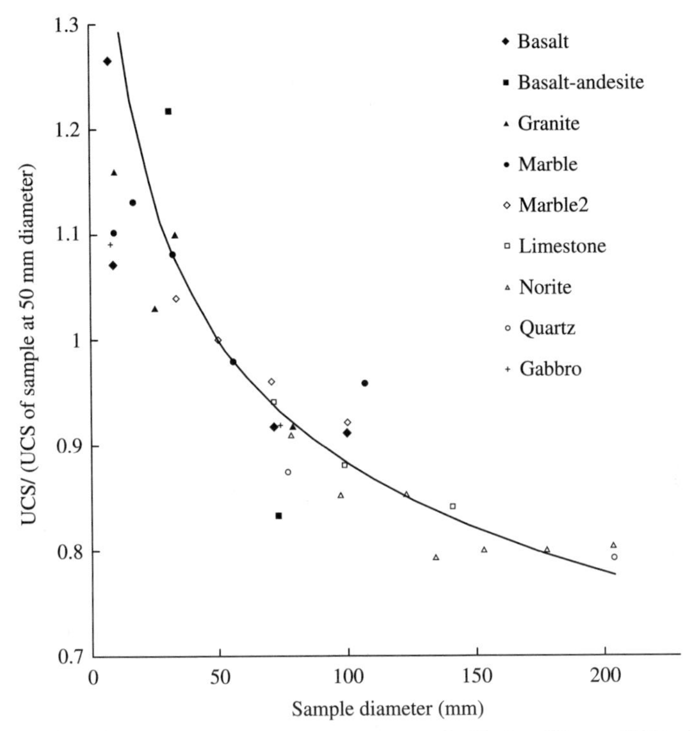

FIG. 1.88 Scale effect trend fitted to different rock types by Hoek and Brown (1980a, b).

where I_{s50} corresponds to a sample with $d=50$ mm. Franklin (1985), based on the earlier studies on scale effect in point load testing (Greminger, 1982; Forster, 1983), suggested that the value of k_2 is 0.45.

Russell and Wood (2009) showed that in PLT when the sample subjected to the force f that acts normal to sphere surface on an area defined by the angle θ_0 (see Fig. 1.89), failure initiates near the point load contact where the stresses satisfy:

$$I_1=\left(2-\sqrt{2}\right)(1+\upsilon)\rho \text{ and } J_2=\left(\frac{3}{32}+\frac{\sqrt{2}}{24}+\left(\frac{\sqrt{2}}{12}-\frac{1}{4}\right)\upsilon+\left(\frac{1}{2}-\frac{\sqrt{2}}{3}\right)\upsilon^2\right)\rho^2 \tag{1.103}$$

in which υ is the Poisson's ratio, $\rho=\frac{4f}{\pi d^2 \sin^2\theta_0}$, and θ_0 controls the radius of the contact area between the pointer and sample.

The expressions in Eq. (1.103) can be introduced into the criterion and combined with Eq. (1.102) to demonstrate that:

$$\frac{\kappa\left(\chi C_1\sqrt{3}-C_4\right)(1+\chi_{50})^2}{\kappa_{50}\left(\chi_{50}C_1\sqrt{3}-C_3\right)(1+\chi)^2}=\left(\frac{d}{50}\right)^{-k_2}=\beta_2 \tag{1.104}$$

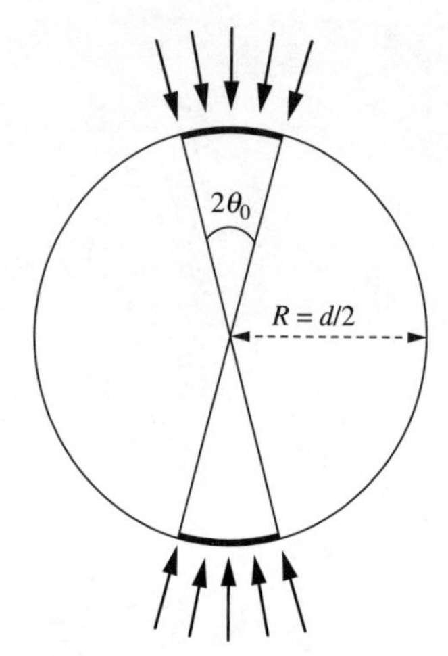

FIG. 1.89 Illustration of force that acts normal to the sphere surface (Russell and Wood, 2009).

in which $C_1 = (2 - \sqrt{2})(1 + \mu)$, $C_2 = \left(\frac{3}{32} + \frac{\sqrt{2}}{24} + \left(\frac{\sqrt{2}}{12} - \frac{1}{4}\right)\mu + \left(\frac{1}{2} - \frac{\sqrt{2}}{3}\right)\mu^2\right)$, $C_3 = \left(3{\chi_{50}}^2{C_1}^2 + 36\chi_{50}C_2 + 36C_2\right)^{\frac{1}{2}}$, and $C_4 = \left(3\chi^2{C_1}^2 + 36\chi C_2 + 36C_2\right)^{\frac{1}{2}}$.

1.5.3.3 Scale effect in tensile strength

In a tensile test, the stress condition for a 50-mm diameter sample at failure is:

$$\sigma_{50}^t = \frac{\sqrt{3}\kappa_{50}}{(1 + \chi_{50})^2} \tag{1.105}$$

This only applies to direct tensile testing. For indirect tensile testing (known as the Brazilian test) due to different stress conditions at failure compared to that for direct tensile testing that has not been sufficiently explored, Eq. (1.105) is not applicable.

The scale-dependent tensile strength is assumed to obey Eq. (1.98) so that:

$$\frac{\sigma^t}{\sigma_{50}^t} = \left(\frac{d}{50}\right)^{-k_3} = \beta_3 \tag{1.106}$$

After some rearrangements, it follows that:

$$\frac{\kappa(\chi_{50} + 1)^2}{\kappa_{50}(\chi + 1)^2} = \left(\frac{d}{50}\right)^{-k_3} = \beta_3 \tag{1.107}$$

1.5.3.4 Scale effect in pure shear strength

A pure shear test brings a sample to failure when $I_1 = 0$. The pure shear strength for a 50-mm diameter sample is defined as (Christensen, 2000):

$$\sigma_{50}^{y} = \frac{\kappa_{50}}{(1+\chi_{50})^{3/2}} \tag{1.108}$$

Again, the scale-dependent pure shear strength is supposed to obey Eq. (1.98) so that:

$$\frac{\sigma^{y}}{\sigma_{50}^{y}} = \left(\frac{d}{50}\right)^{-k_4} = \beta_4 \tag{1.109}$$

After some rearrangements, Eq. (1.110) can be derived as follows:

$$\frac{\kappa(\chi_{50}+1)^{3/2}}{\kappa_{50}(\chi+1)^{3/2}} = \left(\frac{d}{50}\right)^{-k_4} = \beta_4 \tag{1.110}$$

1.5.4 The modified failure criteria

Any two of Eqs. (1.101), (1.104), (1.107), and (1.110) may be combined to obtain scale-dependent definitions for χ and κ in terms of χ_{50}, κ_{50}, and two β values (or $d/50$ and two k values). Perhaps the most useful combination is the equations for uniaxial compressive and point load tests, as these tests have been the subject of many experimental studies of scale effect. Combining Eqs. (1.101), (1.104) leads to:

$$\frac{\beta_1(1+\chi_{50})}{\beta_2(1+\chi)} = \frac{(\chi_{50}C_1\sqrt{3}-C_3)}{(\chi C_1\sqrt{3}-C_4)} \tag{1.111}$$

$$\kappa = \frac{\beta_1\kappa_{50}(1+\chi)}{(1+\chi_{50})} \tag{1.112}$$

Using these expressions, minor approximations need to be introduced to make χ and κ the subjects:

$$\chi \cong \chi_{50}\frac{\beta_1}{\beta_2} \tag{1.113}$$

and

$$\kappa \cong \kappa_{50}\frac{\left(1+\dfrac{\beta_1}{\beta_2}\chi_{50}\right)}{1+\chi_{50}}\beta_1 \tag{1.114}$$

Eq. (1.114) can be simplified even further for large values of χ, which can vary from about 9 to 169 based on the study by Goodman (1989):

$$\kappa \cong \kappa_{50} \frac{{\beta_1}^2}{\beta_2} \tag{1.115}$$

The accuracy of the approximations introduced in χ and κ can be verified numerically, as illustrated in Table 1.32. Solving Eq. (1.111) for χ leads to the exact value while Eq. (1.113) provides an approximate value. Similarly, Eq. (1.112) results in the exact value for κ whereas Eq. (1.115) leads to the approximate value.

A comparison between the calculated exact and approximate values for χ and κ with various input parameters confirms the validity of the proposed approximations. As a result, the scale-dependent failure criterion (which includes a small approximation) becomes:

$$\frac{\chi_{50}\kappa_{50}}{\sqrt{3}} \frac{{\beta_1}^3}{{\beta_2}^2} I_1 + \left(1 + \chi_{50}\frac{\beta_1}{\beta_2}\right)^2 J_2 < \frac{{\kappa_{50}}^2}{\left(1 + \chi_{50}\frac{\beta_1}{\beta_2}\right)} \frac{{\beta_1}^4}{{\beta_2}^2} \tag{1.116}$$

which may be rewritten in terms of $d/50$ and k_1 and k_2 as:

$$\frac{\chi_{50}\kappa_{50}}{\sqrt{3}} \left(\frac{d}{50}\right)^{2k_2-3k_1} I_1 + \left(1 + \chi_{50}\left(\frac{d}{50}\right)^{k_2-k_1}\right)^2$$
$$J_2 < \frac{{\kappa_{50}}^2}{\left(1 + \chi_{50}\left(\frac{d}{50}\right)^{k_2-k_1}\right)} \left(\frac{d}{50}\right)^{2k_2-4k_1} \tag{1.117}$$

Sample diameter clearly has an influence on the failure criterion, as do the magnitudes of k_1 and k_2. A few schematic surfaces are drawn in the p and q plane in Figs. 1.90–1.92 (where $p = I_1/3$ is the mean stress and $q = \sqrt{3J_2}$ is the deviatoric stress) for different values of d and different combinations of k_1 and k_2 (e.g., k_1, $k_2 \geq 0$). The value of $\chi_{50} = 9$ was assumed, which is much larger than unity, and it was required to satisfy the approximations used in the derivation of Eq. (1.117) having negligible influence. The assumed value of χ_{50} resulted in $\kappa_{50} = 173.21\,\text{MPa}$ if $\sigma_{50}^c = 30\,\text{MPa}$ is also assumed.

When $k_1 = 0$ and $k_2 = 0.4$ (see Fig. 1.91), the surfaces intersect at the point where the uniaxial compressive test would reach the failure surface (that is, when $q = 3p$), as due to the $k_1 = 0$ condition, scale effect has no influence on UCS. If a stress path causes the failure surface to be reached at a p value lower than 10 MPa, then the strength would be reduced as d increases. Conversely, if a stress path causes the failure surface to be reached at a p value larger than 10 MPa, then the strength would be increased as d increases.

TABLE 1.32 Exact and approximate values for χ and κ based on the various input parameters

Input parameters						χ		κ (GPa)	
d (mm)	υ	k_1	k_2	χ_{50}	κ_{50} (GPa)	*Exact from Eq. (1.101)*	*Approximate from Eq. (1.113)*	*Exact from Eq. (1.112)*	*Approximate from Eq. (1.115)*
25	0.15	0.09	0.22	25	5	22.77	22.84	4.87	4.86
100	0.15	0.09	0.22	25	5	27.44	27.36	5.14	5.14
200	0.15	0.09	0.22	25	5	30.11	29.94	5.28	5.29
200	0.15	0.09	0.22	25	5	30.07	29.94	5.27	5.29
200	0.15	0.09	0.22	25	5	30.04	29.94	5.27	5.29
200	0.35	0.18	0.45	25	5	36.60	36.35	5.63	5.66
200	0.35	0.36	0.90	25	5	53.45	52.85	6.36	6.42
200	0.35	0.36	0.90	50	10	106.30	105.70	12.77	12.83
200	0.35	0.36	0.90	100	20	211.00	211.40	25.61	25.67

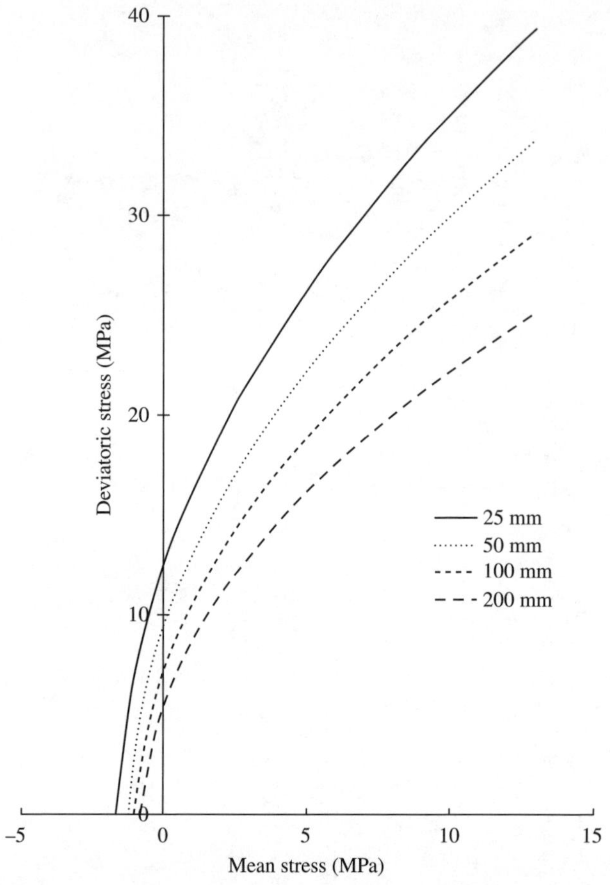

FIG. 1.90 Schematic representation of modified failure criterion for different sample diameters when $k_1 = k_2$, $\chi_{50} = 9$, and $\sigma_{50}^c = 30\,\text{MPa}$.

When $k_1 = 0.4$ and $k_2 = 0$ (Fig. 1.92), the surfaces intersect at the point where PLT would reach the failure surface, as due to the $k_2 = 0$ condition, scale effect has no influence on the point load strength index. The stress condition at this point on a failure surface is the same as Eq. (1.103) that:

$$p = I_1/3 = \frac{(2-\sqrt{2})}{3}(1+\upsilon)I_s \text{ and } q = \sqrt{3J_2}$$
$$= \sqrt{3}\left(\frac{3}{32} + \frac{\sqrt{2}}{24} + \left(\frac{\sqrt{2}}{12} - \frac{1}{4}\right)\upsilon + \left(\frac{1}{2} - \frac{\sqrt{2}}{3}\right)\upsilon^2\right)^{\frac{1}{2}} I_s \tag{1.118}$$

where I_s is the point load strength index, which is constant at different diameters, and υ is assumed 0.25. If a stress path causes the failure surface to be reached at a p value larger than -1.031 MPa (the intersection between the modified failure criteria at different scales according to Fig. 1.92), then the strength

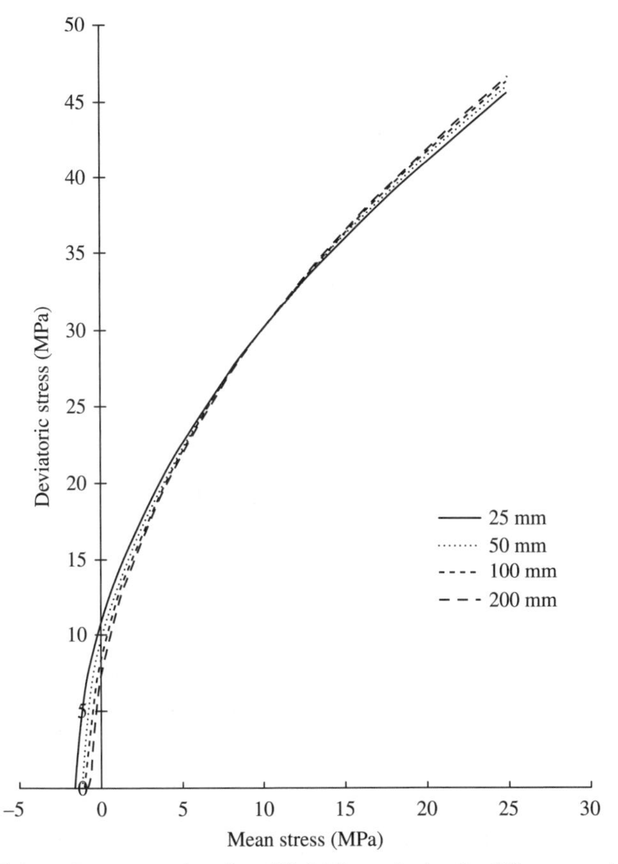

FIG. 1.91 Schematic representation of modified failure criterion for different sample diameters when $k_1=0$, $k_2=0.4$, $\chi_{50}=9$, and $\sigma^c{}_{50}=30\text{MPa}$.

would be reduced as d increases. Conversely, if a stress path causes the failure surface to be reached at a p value lower than -1.031 MPa, then the strength would be increased as d increases.

From Eq. (1.115), if $k_2=2k_1$, that is, when $\beta_1{}^2=\beta_2$, then κ would be scale-independent, and in support of the definition of Christensen (2000) that κ is a measure of an intrinsic strength dependent on atomic scale properties. From Eq. (1.113), if $k_1=k_2$, that is, when $\beta_1=\beta_2$, then χ would be scale-independent.

To ensure that the failure envelopes at different diameters do not intersect, Eq. (1.88) is rearranged in the plain of p and q as follows:

$$p=-\frac{\sqrt{3}}{9}\underbrace{\frac{(1+\chi_{50})^2}{\chi_{50}\kappa_{50}}}_{a_1}q^2+\frac{\sqrt{3}}{3}\underbrace{\frac{\kappa_{50}}{\chi_{50}(1+\chi_{50})}}_{c_1} \tag{1.119}$$

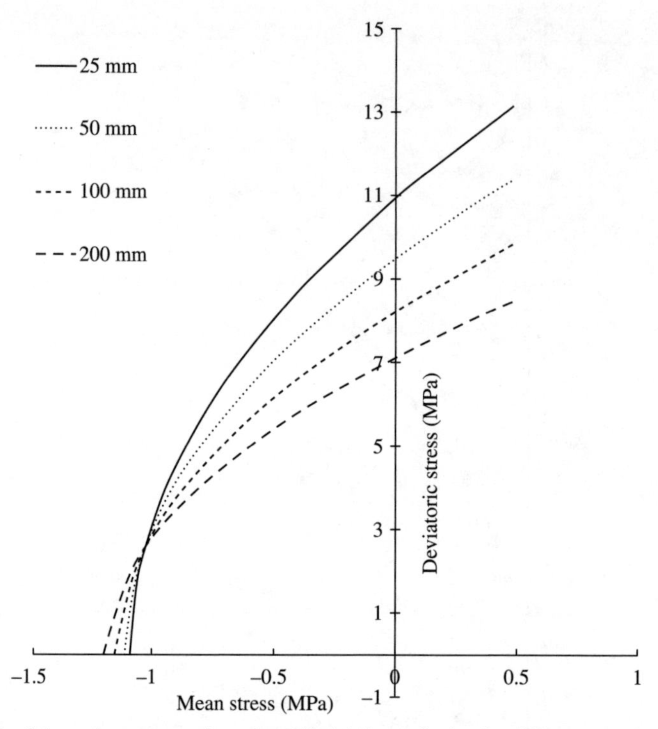

FIG. 1.92 Schematic representation of modified failure criterion for different sample diameters when $k_1=0.4$, $k_2=0$, $\chi_{50}=9$, and $\sigma_{50}^c=30\,\text{MPa}$.

Similarly, the rearrangement of the modified failure criterion (Eq. 1.117) in the same plain would be as follows:

$$p=-\frac{\sqrt{3}}{9}\underbrace{\frac{\left(1+\chi_{50}\left(\dfrac{d}{50}\right)^{k_2-k_1}\right)^2}{\chi_{50}\kappa_{50}\left(\dfrac{d}{50}\right)^{2k_2-3k_1}}}_{a_2}q^2+\frac{\sqrt{3}}{3}\underbrace{\frac{\kappa_{50}}{\chi_{50}\left(\dfrac{d}{50}\right)^{k_1}\left(1+\chi_{50}\left(\dfrac{d}{50}\right)^{k_2-k_1}\right)}}_{c_2} \tag{1.120}$$

In Eqs. (1.119), (1.120), if $a_1 \leq a_2$ and $c_1 > c_2$, then the failure envelopes that result from different diameters do not intersect. To obtain this, the following relationships as functions of k_1, k_2, d, and χ_{50} should be true:

$$(1+\chi_{50})^2\left(\frac{d}{50}\right)^{2k_2-3k_1} \leq \left(1+\chi_{50}\left(\frac{d}{50}\right)^{k_2-k_1}\right)^2 \text{and}\,(1+\chi_{50}) < \left(\frac{d}{50}\right)^{k_1}\left(1+\chi_{50}\left(\frac{d}{50}\right)^{k_2-k_1}\right) \tag{1.121}$$

If the scale effect observed in uniaxial compressive and uniaxial tensile tests as well as point load and uniaxial tensile tests were combined, then the following expressions for χ and κ are obtained:

For uniaxial compressive and uniaxial tensile tests:

$$\chi = \frac{\beta_1(1+\chi_{50})}{\beta_3} - 1 \tag{1.122}$$

and

$$\kappa = \kappa_{50}\frac{{\beta_1}^2}{\beta_3} \tag{1.123}$$

leading to the alternate criterion:

$$\begin{aligned}&\frac{\kappa_{50}}{\sqrt{3}}\left(\frac{d}{50}\right)^{k_3-2k_1}\left((1+\chi_{50})\left(\frac{d}{50}\right)^{k_3-k_1}-1\right)I_1\\&+\left((1+\chi_{50})\left(\frac{d}{50}\right)^{k_3-k_1}\right)^2 J_2 < \frac{{\kappa_{50}}^2}{(1+\chi_{50})}\left(\frac{d}{50}\right)^{k_3-3k_1}\end{aligned} \tag{1.124}$$

For point load and uniaxial tensile tests, similar to Eqs. (1.109), (1.110), the following terms are resulted to acquire χ and κ:

$$\frac{\beta_3}{\beta_2} = \frac{(\chi_{50}C_1\sqrt{3} - C_3)}{(\chi C_1\sqrt{3} - C_4)} \tag{1.125}$$

$$\kappa = \frac{\beta_3\kappa_{50}(1+\chi)}{(1+\chi_{50})} \tag{1.126}$$

Due to the complexity of Eq. (1.126), it is not possible to propose a suitable approximation for χ and κ, thus the exact values should be applied.

In the same vein, combining the scale effect observed in uniaxial compressive and pure shear tests would lead to:

$$\begin{aligned}&\frac{\kappa_{50}}{\sqrt{3}}\left(\frac{d}{50}\right)^{2k_4-3k_1}\left((1+\chi_{50})\left(\frac{d}{50}\right)^{2k_4-2k_1}-1\right)I_1\\&+(1+\chi_{50})^2\left(\frac{d}{50}\right)^{4k_4-4k_1} J_2 < \frac{{\kappa_{50}}^2}{(1+\chi_{50})}\left(\frac{d}{50}\right)^{2k_4-4k_1}\end{aligned} \tag{1.127}$$

1.5.5 Comparison with experimental data

The modified failure criterion was correlated against the experimental data obtained from uniaxial compressive and point load tests for a range of rock types. The authors conducted an extensive literature review (Mogi, 1962; Lundborg, 1967; Bieniawski, 1968; Pratt et al., 1972; Bieniawski, 1975; Abou-Sayed and Brechtel, 1976; Panek and Fannon, 1992; Silva et al., 1993; Kramadibrata and Jones, 1993; Simon and Deng, 2009; Darlington and Ranjith, 2011) in order to include additional UCS data to that provided by

Hoek and Brown (1980a, b). These results were normalized against the UCS of the sample having a 50-mm diameter and then were combined to plot Eq. (1.99) to this new dataset, leading to the proposed value of 0.2 for k_1 with an acceptable coefficient of determination ($R^2=0.76$).

Similarly, for the point load strength index, a comprehensive literature review (Broch and Franklin, 1972; Bieniawski, 1975; Wijk et al., 1978; Brook, 1980; Greminger, 1982; Brook, 1985; Hawkins, 1998; Thuro et al., 2001a, b) was carried out on scale effect in point loading. The point load strength indices were normalized against the sample having a 50-mm diameter and then grouped together to fit Eq. (1.102) to the resulting dataset, leading to the proposed value of 0.4 for k_2 with $R^2=0.84$.

Wijk et al. (1978) were the only researchers who conducted an experimental scale effect study on Bohus granite under direct tensile testing where the results were inconclusive and so no scale effect trend was recognizable. No reported investigation could be found that included the experimental scale effect study under direct pure shear testing. This is mainly due to the complexity of the setup and test procedure.

As a result, according to the UCS and point load strength index data presented earlier, it is true to state that $2k_1 = k_2$ leading to ${\beta_1}^2 \cong \beta_2$, suggesting that κ is scale-independent as hypothesized by Christensen (2000). This would be even further supported if ${\beta_1}^2=\beta_3$ and ${\beta_1}^3={\beta_4}^2$ while due to the lack of data, it remains unverified. Relying on the concluding values for k_1 and k_2 to suggest the scale-independency of κ, it can be assumed that $k_3 = 0.4$ and $k_4 = 0.3$ to satisfies ${\beta_1}^2=\beta_3$ and ${\beta_1}^3={\beta_4}^2$. Thus, the resulting modified multiaxial failure criterion parameters for different pairs of testing conditions and various sample diameters can be summarized in Table 1.32 if $\chi_{50} = 9$ and $\sigma_{50}^c = 30\,\text{MPa}$ are assumed.

The failure envelopes for four different sample diameters associated with the parameters presented in Table 1.33 are demonstrated in Fig. 1.93.

It is evident from Fig. 1.93 that an increase in scale leads to a decrease in material strength. There are four positions (A, B, C, and D) in Fig. 1.6 that mark the intersection of the uniaxial compressive stress path with the modified multiaxial failure envelopes for the samples having 25, 50, 100, and 200 mm diameters, respectively. The slope of the UCS stress path is 3 based on the relation between deviatoric (q) and mean (p) stresses in which $q=3p$. The coordinates of the points A, B, C, and D are listed in Table 1.34, which demonstrates for both the deviatoric and mean stresses, an increase in sample diameter leads to a decrease in strength.

It is important to note that the above parametric study was conducted to assess the sensitivity of the proposed modified failure criterion for different possible input parameters. This is mainly due to the fact that the available scale effect data under different stress conditions are very limited. The analytical solution presented here has been based on a logical mathematical process that confirms its validity and thus the experimental data was used to verify the potential validity of the Christensen hypotheses on the definitions of χ and κ.

TABLE 1.33 Resulting modified multiaxial failure criterion parameters for different pairs of testing conditions and various sample diameters where $\chi_{50} = 9$ and $\sigma_{50}^{c} = 30\,\text{MPa}$

	UCS-Point load				UCS-Tensile				UCS-Pure shear			
Diameter (50 mm)	$\frac{\kappa}{\sigma_{50}^{c}}$	$\frac{\chi}{\chi_{50}}$	β_1	β_2	$\frac{\kappa}{\sigma_{50}^{c}}$	$\frac{\chi}{\chi_{50}}$	β_1	β_2	$\frac{\kappa}{\sigma_{50}^{c}}$	$\frac{\chi}{\chi_{50}}$	β_1	β_2
0.5	5.77	0.87	1.15	1.32	5.77	0.86	1.15	1.32	5.77	0.86	1.15	1.23
1.0	5.77	1.00	1.00	1.00	5.77	1.00	1.00	1.00	5.77	1.00	1.00	1.00
2.0	5.77	1.15	0.87	0.76	5.77	1.17	0.87	0.76	5.77	1.17	0.87	0.81
4.0	5.77	1.32	0.76	0.57	5.77	1.36	0.76	0.57	5.77	1.36	0.76	0.66

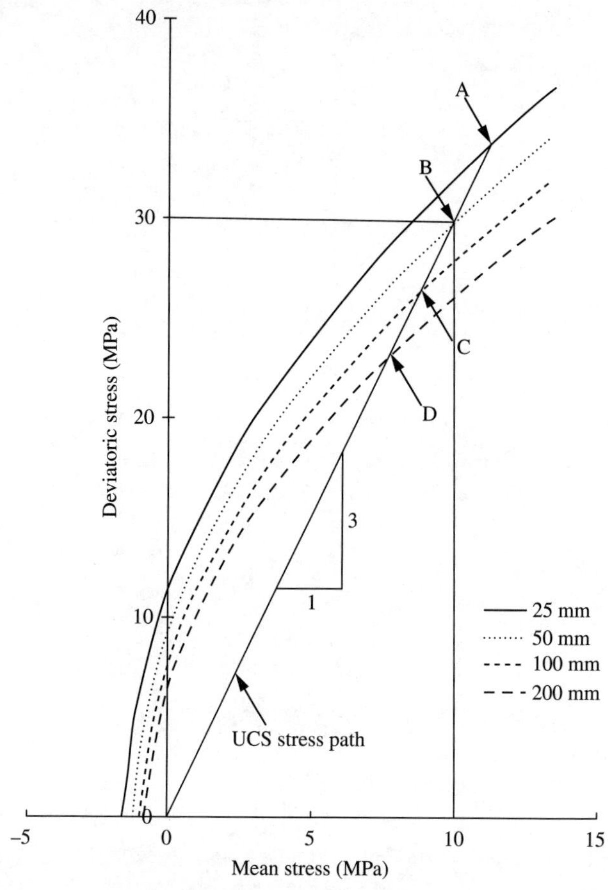

FIG. 1.93 Demonstration of the modified multiaxial failure envelopes for different sample diameters associated with the parameters presented in Table 1.33.

TABLE 1.34 The UCS values at different scales

Sample diameter (mm)	Deviatoric stress (MPa)	Mean stress (MPa)
25 (A)	33.9	11.3
50 (B)	30.0	10.0
100 (C)	26.4	8.8
200 (D)	23.1	7.7

1.5.6 Conclusions

A modified multiaxial failure criterion including scale effect was developed. The criterion was initially proposed by Christensen (2000) as the general failure criterion applicable to different brittle materials. The scale effect under different stress paths such as uniaxial compression, point loading, uniaxial tension, and pure shearing were incorporated, resulting in the modified multiaxial failure criterion.

Because the uniaxial compressive and point load tests have been the subject of many studies on scale effect, the modified multiaxial failure criterion was calibrated against the experimental data obtained from these two tests for a range of rock types. The scale dependency of the proposed parameters for the modified multiaxial failure criterion was assessed using the results from these two tests. It was concluded that one of the modified failure criterion parameters (κ) is scale-independent as hypothesized by Christensen (2000) while the other parameter (χ) is scale-dependent.

1.6 Size-dependent Hoek-Brown failure criterion

1.6.1 Introduction

Size effect is an important characteristic in brittle and quasibrittle media such as rock. A significant amount of research has been conducted to better understand the influence of size on the mechanical behavior of intact rock under different stress conditions, including uniaxial compressive (Mogi, 1962; Pratt et al., 1972; Baecher and Einstein, 1981; Panek and Fannon, 1992; Thuro et al., 2001a; Darlington and Ranjith, 2011; Masoumi et al., 2014), point load (Broch and Franklin, 1972; Brook, 1980; Greminger, 1982; Hawkins, 1998; Thuro et al., 2001b; Forbes et al., 2015), and indirect tensile (Andreev, 1991a; Andreev, 1991b; Carpinteri et al., 1995; Butenuth, 1997; Elices and Rocco, 1999; Thuro et al., 2001a; Çanakcia and Pala, 2007) testing. However, there have been limited investigations into the size effect under triaxial conditions (Singh and Huck, 1972; Hunt, 1973; Medhurst and Brown, 1998; Aubertin et al., 2000). On the other hand, investigation into the mechanical behavior of intact rocks has resulted in various failure criteria (Lade and Duncan, 1975; Desai, 1980; Desai and Faruque, 1984; Kim and Lade, 1984; Lade and Nelson, 1987; Kim and Lade, 1988; Khan et al., 1991; Khan et al., 1992). Perhaps the most widely used criterion in rock engineering is that of Hoek and Brown (1997). The size effect has been later incorporated into the Hoek-Brown criterion (Medhurst and Brown, 1998) by assigning a generalized size effect model based on statistical theory (Weibull, 1939) in which the strength reduces with an increase in sample size.

Masoumi et al. (2015) conducted an extensive size effect study on intact rocks and consequently introduced the unified size effect law (USEL). USEL significantly enhances the prediction of the size-dependent behavior of intact rocks. They investigated size effects in a number of intact rocks (particularly sedimentary rock types) and concluded that their size effect trends follow an ascending-descending behavior, as opposed to the generalized size effect concept (Weibull, 1939) that assumes only a descending trend. Despite the verification of the applicability of USEL to uniaxial compressive strength (UCS) and point load strength data (Masoumi et al., 2015), further investigation is required to assess its applicability to triaxial test data.

The Hoek-Brown failure criterion (Hoek and Brown, 1997) is modified to include USEL, leading to a size-dependent Hoek-Brown failure criterion. A suite of laboratory triaxial experiments was performed on Gosford sandstone samples having diameters of 25, 50, and 96 mm under a range of confining pressures (σ_3) between 1 and 40 MPa. It is shown that the size-dependent Hoek-Brown failure criterion provides a good estimation of the size effect behavior of Gosford sandstone under a triaxial condition where the samples stay within the brittle regime.

1.6.2 Background

1.6.2.1 Analytical study

There are only a limited number of studies in the literature that have investigated the size effect in relation to triaxial tests.

Hoek and Brown (1980a, b) introduced their well-known size effect model based on the statistical concept (Weibull, 1939) as follows:

$$\frac{\sigma^c}{\sigma_{50}^c} = \left(\frac{d}{50}\right)^{-0.18} \tag{1.128}$$

where the measured uniaxial compressive strength σ^c is a function of sample size d (in millimeters) and σ_{50}^c is the characteristic uniaxial compressive strength measured on a sample with a characteristic size of 50 mm. Medhurst and Brown (1998) used the Hoek and Brown (1980a, b) size effect model as the starting point to estimate the triaxial compressive strength of coal samples. They showed that the size effect behavior in uniaxial and triaxial conditions is similar.

Aubertin et al. (2000) developed the only notable size effect model that considered triaxial confinement, which follows the generalized size effect concept in a very complex form according to:

$$\sigma_N = \sigma_s - x_1(\sigma_s - \sigma_L)\left\langle\frac{d_N - d_s}{d_L - d_s}\right\rangle^{m_1} \tag{1.129}$$

where σ_N is the nominal strength of a sample with size (d_N), σ_s is the maximum strength of a representative volume element of isotropic material with minimum size d_s, σ_L is the minimum strength of a material at a very large size (d_L), m_1 is a positive constant controlling the statistical decay of the strength with increase in size, and x_1 was defined as a function of confining pressure according to:

$$x_1 = \exp\left(x_0\sigma_3/T_0\right) \tag{1.130}$$

in which T_0 is the uniaxial tensile strength (negative value $= -\sigma_t$) and x_0 is a material constant. The above equations were then incorporated into a failure criterion known as the Mises-Schleicher and Drucker-Prager unified ($MSDP_u$) (Aubertin et al., 1999).

Aubertin et al. (2000) argued that with an increase in confining pressure, the brittle behavior of intact rock changes to ductile behavior and therefore less size effect is expected. They also stated that the closure of microcracks at high confining pressures reduces size dependency. Based on this hypothesis, $MSDP_u$ predicts approximately the same peak stress for two samples having different sizes at very high confinement. Unfortunately, the use of the $MSDP_u$ model is limited due to the complexity in obtaining its required coefficients.

Masoumi et al. (2015) divided the existing size effect models into two major categories: descending and ascending types. The descending models can be classified into four subcategories: statistics, fracture energy, multifractals, and empirical and semiempirical models. Neither of the existing size effect models on their own can predict the size effect behavior of the UCS data across a wide range of diameters and thus USEL was introduced (Masoumi et al., 2015). In the USEL, UCS is the lower strength predicted by the following two functions (Fig. 1.94):

$$\text{Strength} = \text{Min}\left(\frac{\sigma_0 d^{(D_f-1)/2}}{\sqrt{1+(d/\lambda d_0)}}, \frac{Bf_t}{\sqrt{1+(d/\lambda d_0)}}\right) \tag{1.131}$$

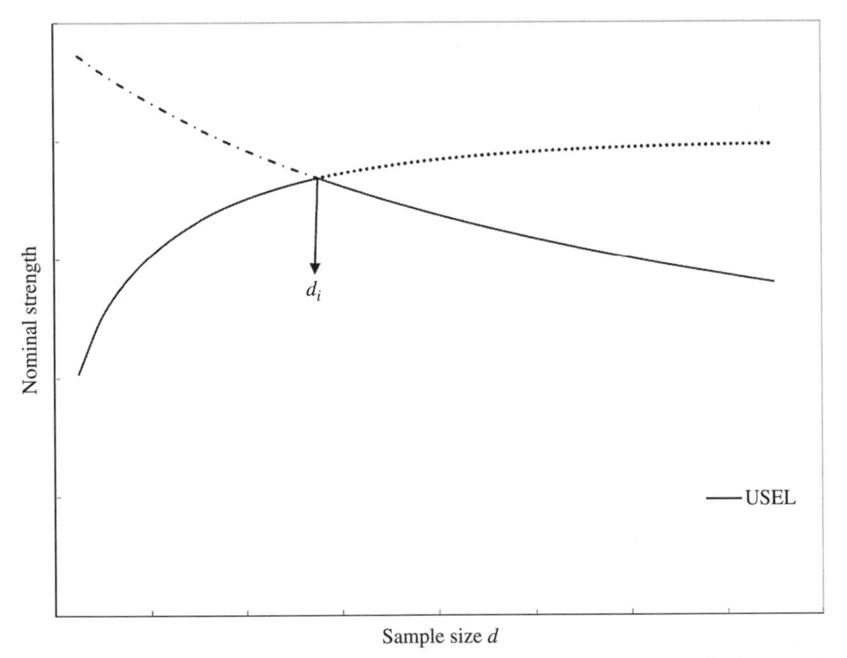

FIG. 1.94 Ascending-descending strength zones. *(Reprinted from Masoumi, H., Saydam, S., Hagan, P.C., 2015. Unified size-effect law for intact rock. Int. J. Geomech. doi: 10.1061/(ASCE) GM.1943-5622.0000543.)*

where d is the sample diameter, d_0 is a characteristic size sometimes taken to be the maximum aggregate size, λ and B are dimensionless material parameters, f_t is a characteristic intrinsic strength for the descending zone, σ_0 is the characteristic strength for the ascending zone, and D_f is the fractal dimension of the fracture surface ($D_f \neq 1$ for fractal surfaces and $D_f = 1$ for nonfractal surfaces).

The intersection between the two functions occurs at a sample diameter (d_i) defined as:

$$d_i = \left(\frac{Bf_t}{\sigma_0}\right)^{2/(D_f-1)} \tag{1.132}$$

The maximum strength at the intersection point is defined as:

$$\text{Strength} = \frac{Bf_t}{\sqrt{1 + \left((Bf_t/\sigma_0)^{2/(D_f-1)}/\lambda d_0\right)}} \tag{1.133}$$

1.6.2.2 Experimental study

Large-scale triaxial tests on cylindrical rock samples were initially conducted by Singh and Huck (1972) using Charcoal Black granite and Indiana limestone. The sizes of the granite samples were 50, 100, and (approximately) 800 mm in diameter. The first two small samples were tested up to about 40 MPa confining pressure while the largest sample was tested at about 10 MPa confinement. The final results showed that the recorded peak stresses for the 50 and 100 mm diameter samples at different confining pressures were of similar values. For the largest sample, the reported strength was significantly less than the smaller samples at identical confinements. It should be noted that the largest sample was not completely intact due to some macrodefects.

The Indiana limestone samples were tested at 50, 100, and (approximately) 300 mm diameters (Singh and Huck, 1972). The largest sample (which also contained macrodefects) gave lower strengths in comparison with the smaller samples at every confinement. For the first two smaller samples, the outcome was similar to that reported for granite samples.

In another investigation, Hunt (1973) carried out a number of uniaxial and triaxial tests on gypsum samples with approximately 25, 38, and 50 mm diameters at different confining pressures up to 10 MPa. The obtained UCS and triaxial results from the Hunt (1973) study showed some form of ascending-descending size effect trend (see Fig. 1.95).

Medhurst and Brown (1998) performed a number of triaxial tests on coal samples. The samples were prepared at 61, 100, 146, and 300 mm diameters. The largest sample was tested up to only 1 MPa confining pressure. The other samples were tested up to about 5 MPa confinement except the 61 mm diameter sample, which was tested up to 10 MPa confining pressure. According to the final results, the size effect behavior from the triaxial tests was similar to the

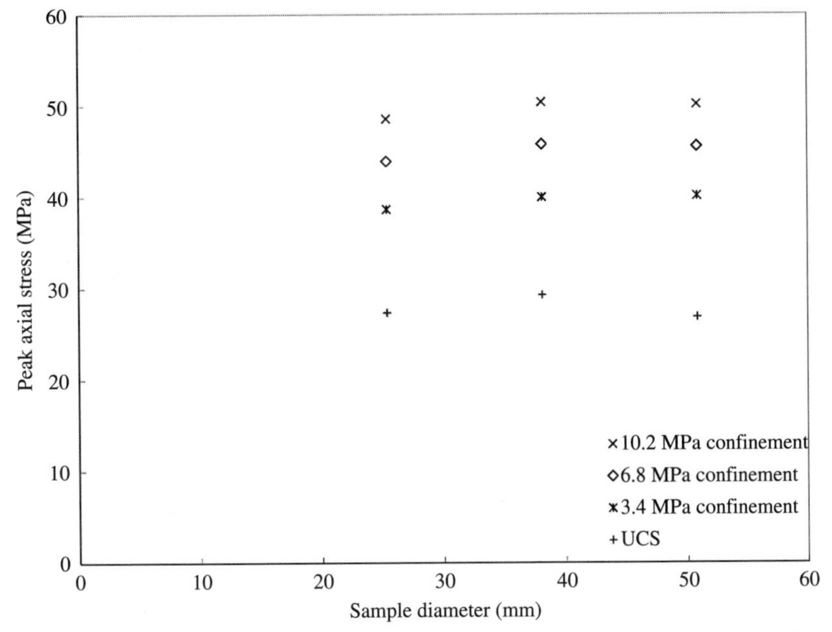

FIG. 1.95 Peak axial stress versus sample diameter at different confining pressures. *(Data from Hunt, D.D., 1973. The Influence of Confining Pressure on Size Effect. Master of Science, Massachusetts Institute of Technology, USA.)*

uniaxial compressive tests. It is important to note that in general, coal behaves similar to a rock mass and not an intact rock.

1.6.2.3 Testing procedure

Gosford sandstone from Gosford Quarry, Somersby, New South Wales, Australia, was used to conduct a number of triaxial compressive tests according to the International Society for Rock Mechanics (ISRM)-suggested methods (ISRM, 2007). Homogenous samples were carefully selected, having no visible sign of macrodefects. The maximum grain size of this sandstone was estimated to be 0.6 mm (Masoumi, 2013). Detailed information regarding the petrography and microstructure of Gosford sandstone can be found in Sufian and Russell (2013) and Masoumi et al. (2015).

The laboratory triaxial experiments were conducted on samples with diameters of 25, 50, and 96 mm. The constant length-to-diameter ratio of two was selected. A servo-controlled loading frame system with a 300-t maximum loading capacity was used to perform the experiments, along with a triaxial cell having a 200-t maximum axial loading capacity and 70 MPa confining pressure (see Fig. 1.96).

The triaxial cell came with three sets of platens at 100, 50, and 25 mm diameters. A manual hydraulic pump with maximum pressure capacity of 100 MPa

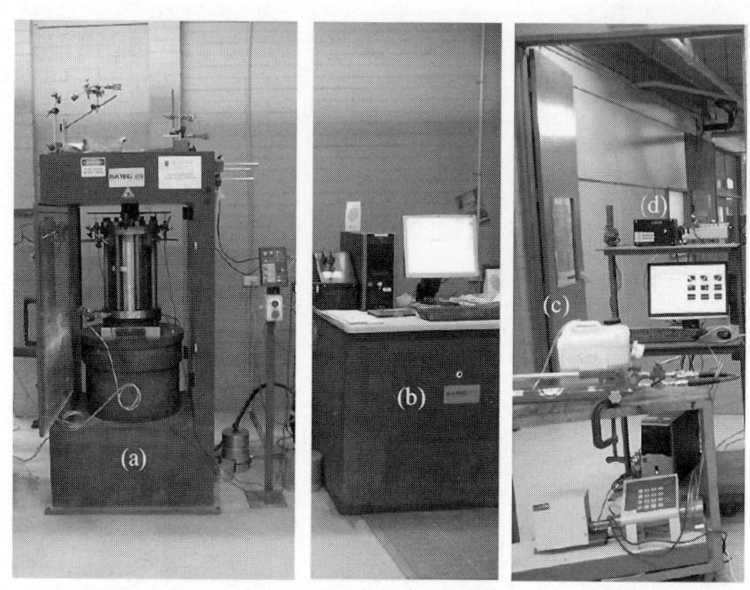

FIG. 1.96 Linkage of all elements during a triaxial experiment: (A) loading frame with triaxial cell; (B) loading frame control system; (C) hydraulic pump to provide and control confining pressure; and (D) computer for data acquisition.

was utilized to provide the confining pressure. An additional digital gauge with an accuracy of ±0.01 MPa was used to control the confining pressure during the experiment. Several experiments were conducted at different sizes and confining pressures to account for possible scatter.

1.6.2.4 Experimental results

The peak axial stresses (σ_1) are extracted from the experiments for this analysis. Table 1.35 lists the mean peak axial stresses obtained from the triaxial tests at different sizes and confining pressures while Fig. 1.97 presents all triaxial data for the samples with 96, 50, and 25 mm diameters. As expected, the peak axial stress increases with an increase in confining pressure with a slight variation at each confinement for all sample diameters.

1.6.3 Size-dependent Hoek-Brown failure criterion

1.6.3.1 Model development

If the mean peak axial stresses obtained from Gosford sandstone at different confining pressures and sizes are combined with the UCS data of this rock (Fig. 1.98), the same trend as that reported for UCS (Masoumi et al., 2015) is observable in the triaxial condition. Fig. 1.98 demonstrates that the UCS of the 50 and 96 mm diameter samples are very close to each other (UCS of

TABLE 1.35 Mean peak axial stresses obtained from Gosford sandstone at different sizes and confining pressures

Confining pressure (MPa)	Mean peakaxial stress (MPa)	Number of tests	Standard deviation	
			MPa	%
96-mm diameter samples				
1	66.6	4	2.1	3.2
2	73.2	4	2.4	3.3
5	91.2	4	2.0	2.3
10	115.6	4	1.1	1.1
20	156.1	4	5.8	4.3
30	199.1	2	1.8	1.0
50-mm diameter samples				
1	69.6	3	3.0	4.4
2	73.6	4	1.4	1.9
5	93	5	2.1	2.4
10	119.9	4	6.2	5.7
20	152.6	3	0.6	0.4
30	202.4	2	5.8	3.4
25-mm diameter samples				
10	106.7	4	11.92	12.32
20	148.8	3	2.3	1.8
30	184.5	3	2.7	1.7
40	209.6	1	–	–

96 mm is slightly higher) but considerably higher than that of the 25 mm diameter sample. These results are in good agreement with the study by Hunt (1973), who first showed that under the triaxial condition, with an increase in sample size up to a characteristic diameter, the strength increases and then, above this characteristic diameter, the strength reduces as the sample size increases. Based on these similarities, it is suggested that the triaxial size effect trends can also follow the ascending and then descending behavior. Masoumi et al. (2015) showed that the influences of surface flaws and fractal characteristics on the sample failure are the important mechanisms for strength ascending behavior, followed by the descending trend based on fracture energy theory.

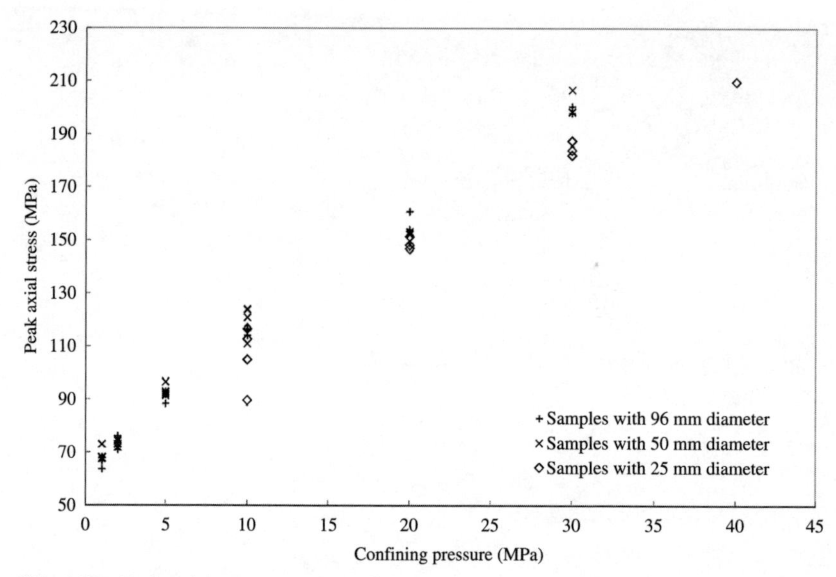

FIG. 1.97 Peak axial stress versus confining pressure for samples with 96, 50, and 25 mm diameters.

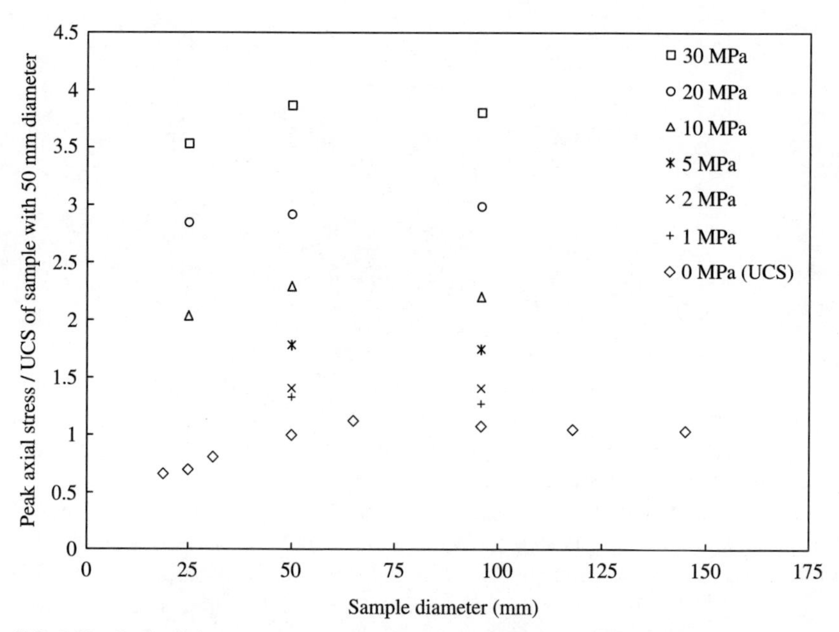

FIG. 1.98 Peak axial stress versus sample diameter at different confining pressures.

The data from Fig. 1.98 indicate that the resulting size effect trends from uniaxial and triaxial tests are quite similar. This is in agreement with the Medhurst and Brown (1998) study, which concluded that the size effect behaviors under uniaxial and triaxial conditions are similar and then included a descending size effect model into the Hoek and Brown (1997) failure criterion. The form of the original Hoek-Brown failure criterion for intact rock is:

$$\sigma_1 = \sigma_3 + \sigma_{ci}\left(m\frac{\sigma_3}{\sigma_{ci}} + 1\right)^{0.5} \tag{1.134}$$

where σ_1 and σ_3 are the major (peak axial stress) and minor (confining pressure) principal stresses at failure, m is the value of the Hoek-Brown constant, and σ_{ci} is the uniaxial compressive strength of the intact rock. In the Hoek-Brown failure criterion, m mainly influences the curvature of the criterion or rate of strength change with confining pressure (see Fig. 1.99), whereas σ_{ci} controls the upward or downward movement of the criterion (see Fig. 1.100).

Medhurst and Brown (1998) rearranged Eq. (1.128) and incorporated it into Eq. (1.134) to include the size effect in the Hoek-Brown failure criterion according to:

$$\sigma_1 = \sigma_3 + \sigma_{c50}\left(\frac{50}{d}\right)^{0.18}\left(m\frac{\sigma_3}{\sigma_{c50}\left(\frac{50}{d}\right)^{0.18}} + 1\right)^{0.5} \tag{1.135}$$

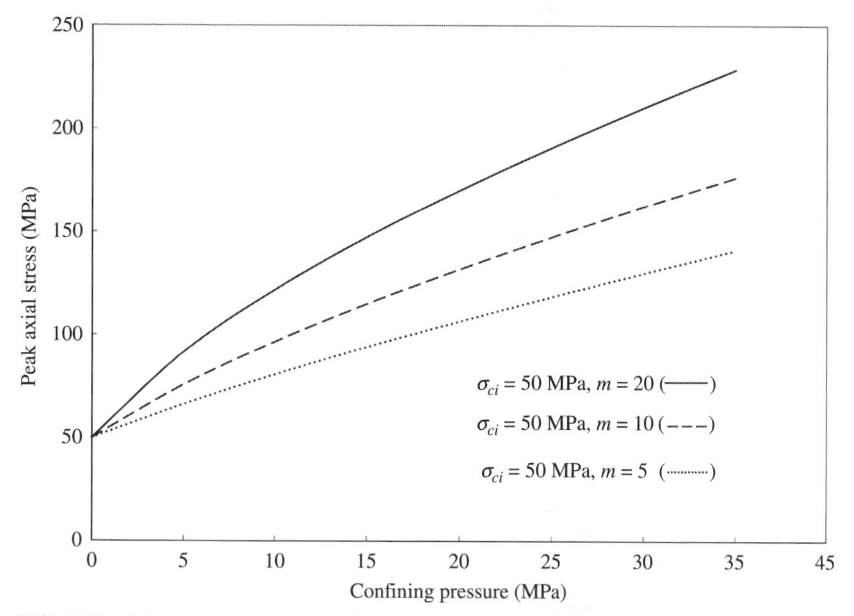

FIG. 1.99 Schematic representation of Hoek-Brown failure criterion at different m values and identical σ_{ci} values.

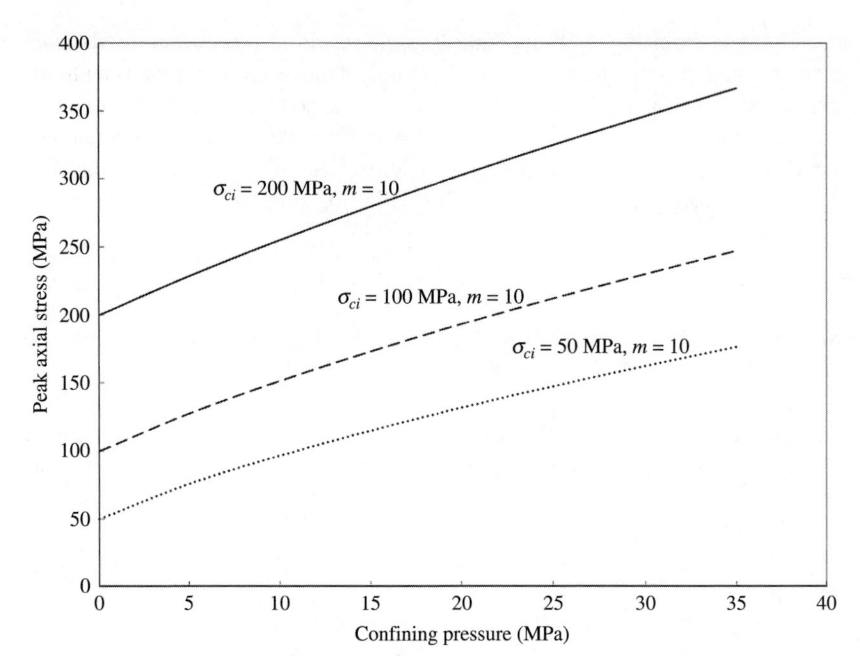

FIG. 1.100 Schematic representation of Hoek-Brown failure criterion at different σ_{ci} values and identical m values.

where the size-dependent behavior of the UCS and triaxial data have been assumed similar. On the other hand, as discussed earlier, Eq. (1.128) follows the generalized size effect concept in which the strength reduces with an increase in sample size. The size effect trends of the UCS and triaxial data from Gosford sandstone, however, exhibit ascending-descending behavior. Thus, USEL is incorporated into the original Hoek-Brown failure criterion, leading to a size-dependent Hoek-Brown failure criterion that can comprehensively capture the ascending and descending strength behavior of rock under uniaxial and triaxial conditions. As a result, σ_{ci} in the Hoek-Brown failure criterion is substituted by USEL (Eq. 1.131) according to:

$$\sigma_1 = \sigma_3 + \mathrm{Min}\left(\frac{\sigma_0 d^{(D_f-1)/2}}{\sqrt{1+(d/\lambda d_0)}}, \frac{Bf_t}{\sqrt{1+(d/\lambda d_0)}}\right)\left(m\frac{\sigma_3}{\mathrm{Min}\left(\frac{\sigma_0 d^{(D_f-1)/2}}{\sqrt{1+(d/\lambda d_0)}}, \frac{Bf_t}{\sqrt{1+(d/\lambda d_0)}}\right)}+1\right)^{0.5} \tag{1.136}$$

This is a simple but versatile size-dependent form of the Hoek-Brown failure criterion that will be validated by the triaxial data obtained from Gosford sandstone at different sizes. The generality and versatility of the USEL have been confirmed by Masoumi et al. (2015) using six different rock types such as Gosford sandstone as well as that reported by Hawkins (1998).

1.6.3.2 Model calibration

From USEL (Eq. 1.121), the predicted strength parameters (σ_{ci}) for the samples with 96, 50, and 25 mm diameter are 56.4, 52.1, and 38.2 MPa, respectively. To obtain the values of USEL parameters, a simple calibration process based on UCS data is sufficient. Masoumi et al. (2015) calibrated USEL based on UCS results obtained from Gosford sandstone and the resulting parameters are listed in Table 1.36.

The resulting $m = 16.7$ for Gosford sandstone obtained from the triaxial experiments at different sizes and confining pressures is used for model predictions. The applicability of the proposed model in this study (Eq. 1.136) to triaxial data is compared with that introduced by Medhurst and Brown (1998) for Gosford sandstone samples with 96, 50, and 25 mm diameters as demonstrated in Figs. 1.101–1.103. Note that in Eq. (1.135), $\sigma_{c50} = 52.3\,\text{MPa}$ as reported by Masoumi et al. (2015).

Comparison between the model predictions resulted from size-dependent Hoek-Brown and Medhurst and Brown (1998) criteria demonstrates that for the samples with 50 mm diameter, both predictions deliver similar results with good agreement with the experimental data. For 96 mm diameter samples, the proposed model predicts the experimental data more precisely than the Medhurst and Brown (1998) model, particularly at low confining pressures. Despite the good prediction by both models at relatively higher confinement, it seems that the Medhurst and Brown (1998) criterion and the size-dependent Hoek-Brown criterion slightly under- and overestimate the experimental data,

TABLE 1.36 List of obtained USEL parameters for Gosford sandstone

Parameter	Value
Diameter of sample with maximum UCS (mm)	65
USEL parameters	
Bf_t (MPa)	64.26
λd_0 (mm)	322.65
σ_0 (MPa)	7.8
d_f	2.01
Predicted intersection (mm)	65.1

Datafrom Masoumi, H., Saydam, S., Hagan, P.C., 2015. Unified size-effect law for intact rock. Int. J. Geomech., https://doi.org/10.1061/(ASCE)GM.1943-5622.0000543.

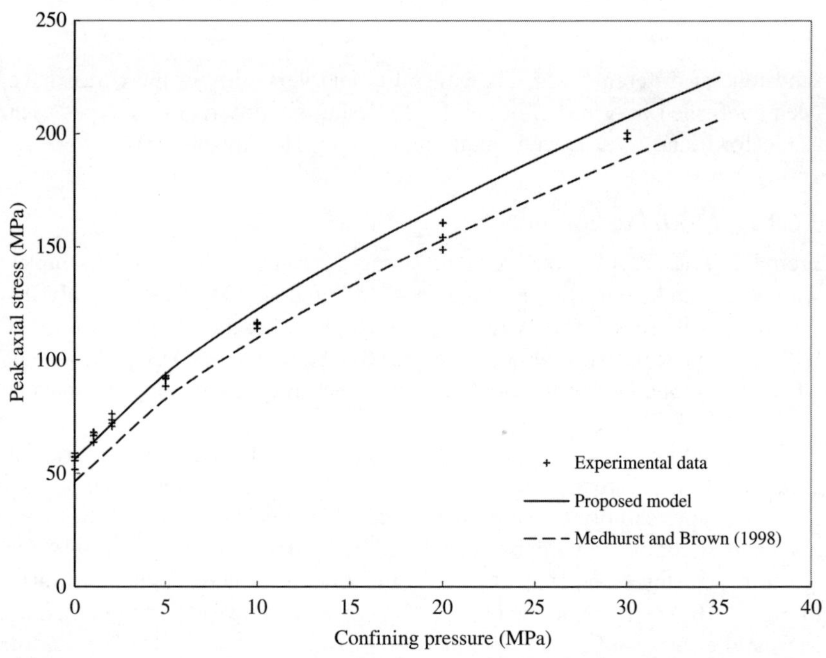

FIG. 1.101 Comparison between applicability of the proposed model and the Medhurst and Brown (1998) model to the peak axial stresses obtained from 96 mm diameter samples at different confining pressures.

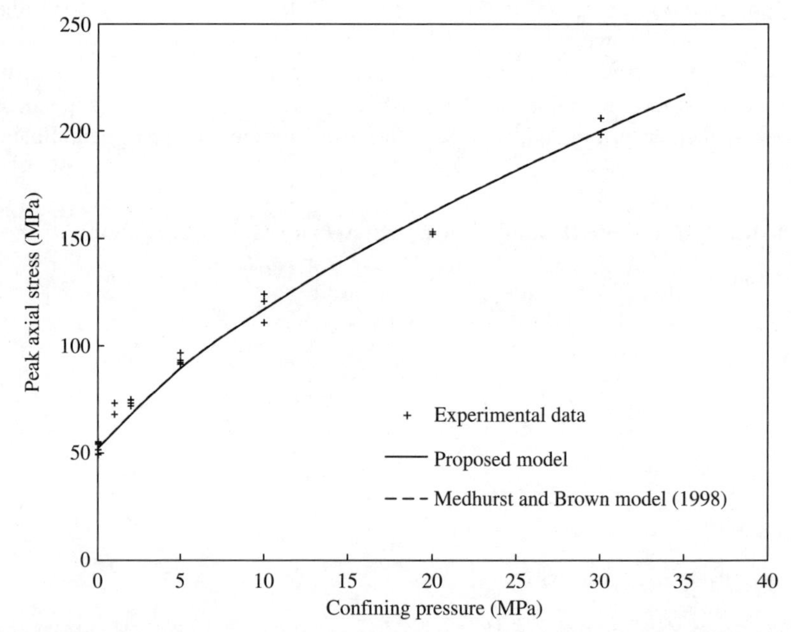

FIG. 1.102 Comparison between applicability of the proposed model and the Medhurst and Brown (1998) model to the peak axial stresses obtained from 50 mm diameter samples at different confining pressures.

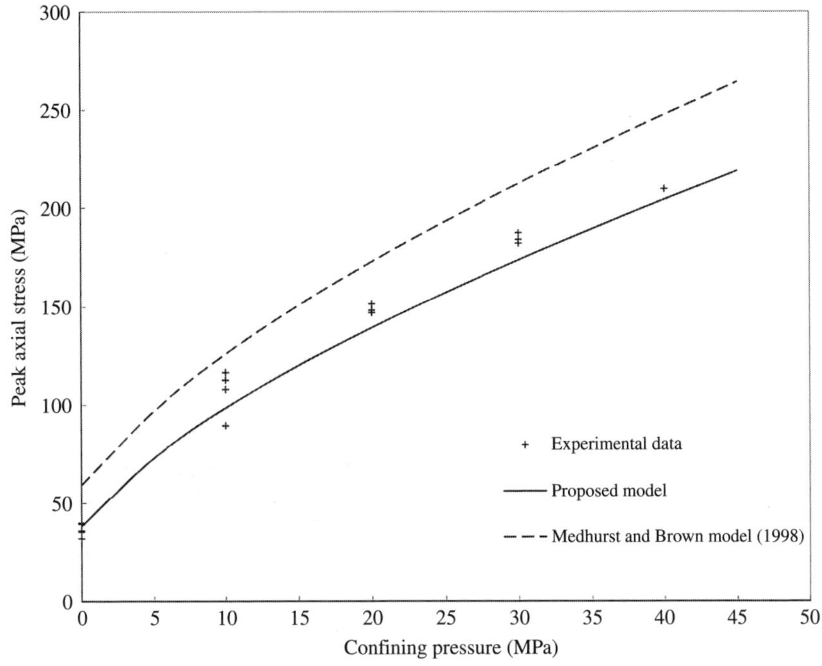

FIG. 1.103 Comparison between applicability of the proposed model and Medhurst and Brown (1998) model to the peak axial stresses obtained from 25 mm diameter samples at different confining pressures.

respectively. Moreover, for samples with 25 mm diameter, the size-dependent Hoek-Brown criterion can suitably fit the experimental results as opposed to the Medhurst and Brown (1998) model, which considerably overpredicts the peak axial stresses at almost all confinements.

In order to highlight the differences between the predictability of the proposed model and the Medhurst and Brown (1998) model, the predicted peak axial stresses versus confining pressure and sample diameter are simultaneously plotted in the three-dimensional figures using Gosford sandstone parameters. As seen from Fig. 1.104, both models predict similar results for samples above approximately 50 mm diameter. However, the predicted trends of both models below this characteristic size start to deviate.

The deviation is significant at small diameters. For instance, in the predictions by both models for samples with 10 and 200 mm diameters (Figs. 1.105 and 1.106), the difference is highly noticeable at the 10 mm diameter sample where the estimation by the Medhurst and Brown (1998) model is much higher than that of the size-dependent Hoek-Brown model. In fact, based on the model verification process presented earlier, at small diameters the Medhurst and Brown (1998) model exhibits gross overprediction while for the larger sizes,

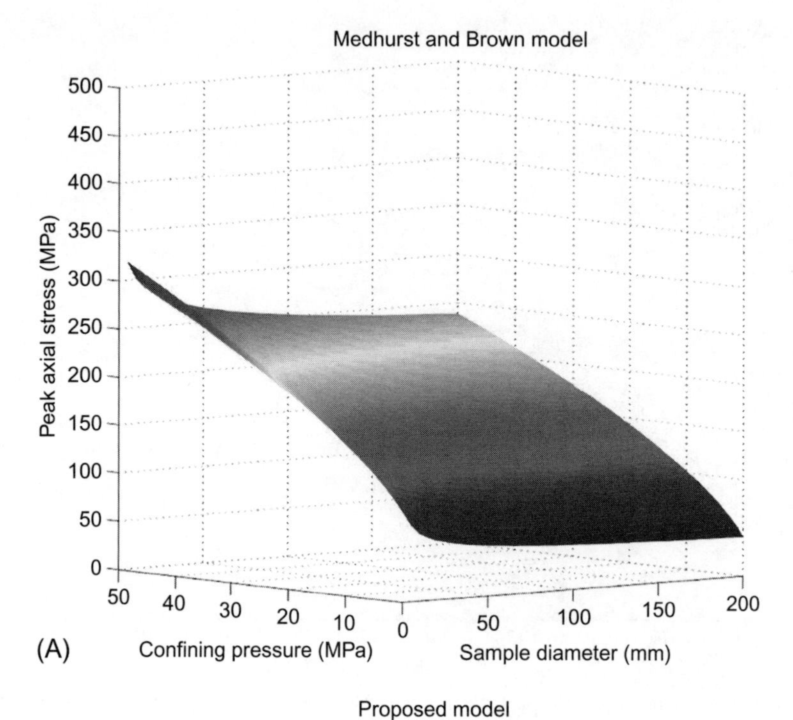

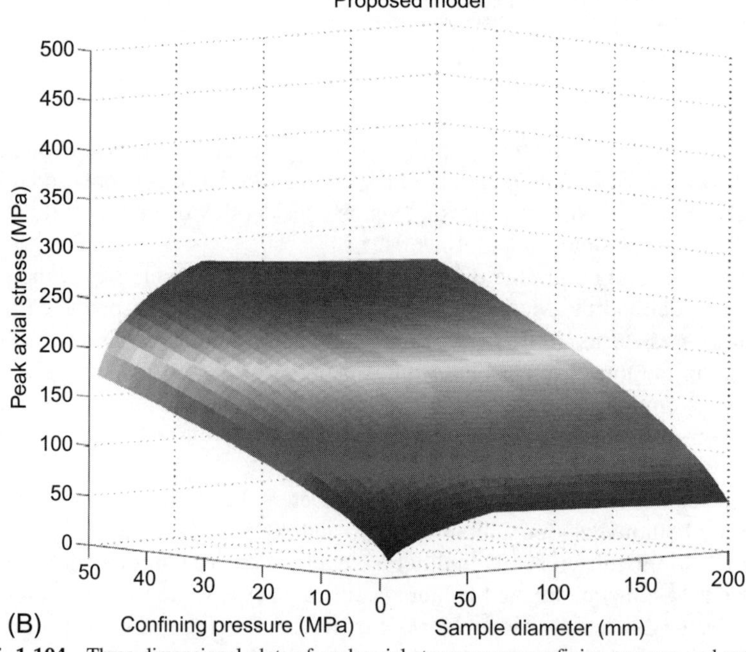

FIG. 1.104 Three-dimensional plots of peak axial stress versus confining pressure and sample diameter obtained from (A) the Medhurst and Brown (1998) model and (B) the proposed model using Gosford sandstone parameters.

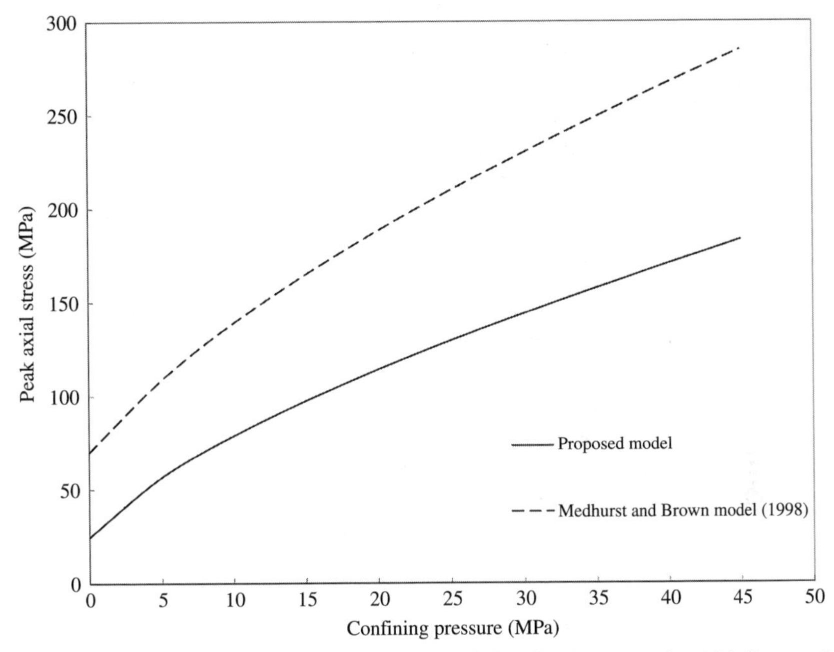

FIG. 1.105 Schematic comparison between the predictions by the proposed and Medhurst and Brown (1998) failure criteria for the samples with 10 mm diameter.

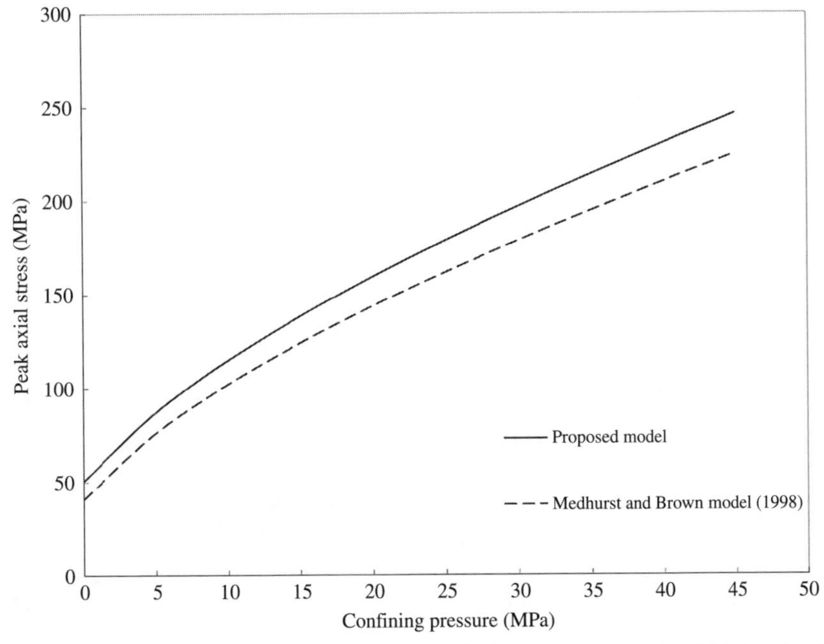

FIG. 1.106 Schematic comparison between the predictions by the proposed and Medhurst and Brown (1998) failure criteria for the samples with 200 mm diameter.

to some extent it underpredicts. Recently, it has become common to conduct an X-ray computed tomography (CT) scan for microstructural analysis on small core samples with less than 10 mm diameter under triaxial conditions, particularly in petroleum engineering. The result from such a study is then extrapolated to field conditions and thus the application of the proposed size effect model becomes very important for accurate size effect modeling.

1.6.4 Example of application

A simple practical example of using the size-dependent Hoek-Brown failure criterion for predicting the strengths of an intact rock under triaxial condition with various sizes is presented.

In the majority of the rock testing programs within the academic and industrial disciplines, the uniaxial compressive test is the most common used. It is, in fact, often conducted as a basic experiment to deploy important mechanical properties of intact rock such as UCS, Elastic Modulus, and Poisson's ratio. This test in some instances is also complemented by a set of triaxial experiments using a Hoek cell (ISRM, 2007). The cell is available in the range of sizes from 30 mm (AX) to about 54 mm diameter (NX) where ISRM (2007) suggested the use of a 54-mm diameter Hoek cell for the triaxial experiment. Thus, in the case of a Hoek cell of a single size (e.g., NX) and a suitable loading frame, the following steps can be taken to predict the strength of an intact rock under triaxial condition with various sizes:

(a) A number of uniaxial compressive tests are conducted on different sample diameters where the range of sizes is sufficiently wide below and above the intersection diameter (e.g., 65 mm for Gosford sandstone).
(b) The USEL (Eq. 1.121) parameters are then calibrated using this set of UCS data.
(c) A suite of triaxial testing is performed on single size samples at different confining pressures using a 54-mm diameter Hoek cell (suggested by ISRM, 2007).
(d) The strength parameter of the size-dependent Hoek-Brown failure criterion is then predicted using USEL (Eq. 1.121) for a 54-mm diameter sample.
(e) The size-dependent Hoek-Brown failure criterion is then calibrated versus the resulting triaxial data to achieve the m value.
(f) This will then lead to determination of all parameters of the size-dependent Hoek-Brown failure criterion to be used for model predictions.

If it is possible to perform a number of triaxial experiments on different sizes, then it is recommended that the average of the m values from the calibration processes of the different sizes is used similar to that implemented for Gosford sandstone. Note that the proposed model was verified only for the case where the samples exhibit brittle behavior and thus further investigation is required for the ductile response of rock.

1.6.5 Conclusions

An advanced set of triaxial experiments was conducted on Gosford sandstone samples having 96, 50, and 25 mm diameters. The unified size effect law was incorporated into the Hoek-Brown failure criterion (Hoek and Brown, 1997), leading to the development of the size-dependent Hoek-Brown failure criterion.

The proposed model was verified against the triaxial data obtained from Gosford sandstone with various sizes. It was confirmed that the size-dependent Hoek-Brown failure criterion can capture the size effect behavior of intact rock with high accuracy under triaxial conditions where they exhibit strength ascending-descending behavior. In addition, it was demonstrated that the proposed model is suitable for the stress conditions where the rock samples exhibit brittle behavior and further investigation is required for ductile response.

References

Abou-Sayed, A.S., Brechtel, C.E., 1976. Experimental investigation of the effects of size on the uniaxial compressive strength of cedar city quartz diorite. In: The 17th U.S. Symposium on Rock Mechanics, US.

Adey, R.A., Pusch, R., 1999. Scale dependency in rock strength. Eng. Geol. 53, 251–258.

Allison, G.B., Hughes, M.W., 1978. The use of environmental chloride and tritium to estimate total recharge to an unconfined aquifer. Aust. J. Soil Res. 16 (2), 181–195.

Andreev, G.E., 1991a. A review of the Brazilian test for rock tensile strength determination. Part I: calculation formula. Min. Sci. Technol. 13 (3), 445–456.

Andreev, G.E., 1991b. A review of the Brazilian test for rock tensile strength determination. Part II: contact conditions. Min. Sci. Technol. 13 (3), 457–465.

Arioglu, E., 1999. Discussion on, Aspect of rock strength, by A. B. Hawkins, Bull EngGeolEnv (1998) 57:13-30. Bull. Eng. Geol. Environ. 57, 319–320.

Aubertin, M., Li, L., Simon, R., Khalfi, S., 1999. Formulation and application of a short-term strength criterion for isotropic rocks. Can. Geotech. J. 36, 947–960.

Aubertin, M., Li, L., Simon, R., 2000. A multiaxial stress criterion for short and long term strength of isotropic rock media. Int. J. Rock Mech. Min. Sci. 37, 1169–1193.

Baecher, G.B., Einstein, H.H., 1981. Size effect in rock testing. Geophys. Res. Lett. 8 (7), 671–674.

Bahrani, N., Kaiser, P.K., 2016. Numerical investigation of the influence of specimen size on the unconfined strength of defected rocks. Comput. Geotech. 77, 56–67.

Baud, P., Klein, E., Wong, T.F., 2004. Compaction localization in porous sandstone: spatial evolution of damage and acoustic emission activity. J. Struct. Geol. 26, 603–624.

Bazant, Z.P., 1984. Size effect in blunt fracture: concrete, rocks and metal. Eng. Mech. 110 (4), 518–535.

Bazant, Z.P., 1997. Scaling of quasibrittle fracture: hypotheses of invasive and lacunar fractality, their critique and Weibull connection. Int. J. Fract. 83, 41–65.

Bazant, Z.P., Kazemi, M.T., 1990. Determination of fracture energy, process zone length and brittleness number from size effect, with application to rock and concrete. Int. J. Fract. 44 (1), 111–131.

Bazant, Z.P., Planas, J., 1998. Fracture and Size Effect in Concrete and Other Quasibrittle Materials. CRC Press, London, UK.

Bazant, Z.P., Xi, Y., 1991. Statistical size effect in quasi-brittle structures: II. Nonlocal theory. J. Eng. Mech. 117 (11). Paper No. 26347.

Bazant, Z.P., Xi, Y., Deid, S.G., 1991. Statistical size effect in quasi-brittle structures: I. Is weibull theory applicable? J. Eng. Mech. 117 (11). Paper No. 26346.

Besuelle, P., Desrues, J., Raynaud, S., 2000. Experimental characterisation of the localisation phenomenon inside a Vosges sandstone in a triaxial cell. Int. J. Rock Mech. Min. Sci. 37, 1223–1237.

Bieniawski, Z.T., 1968. The effect of specimen size on the compressive strength of coal. Int. J. Rock Mech. Min. Sci. 6, 325–335.

Bieniawski, Z.T., 1975. The point load test in geotechnical practice. Eng. Geol. 9, 1–11.

Bieniawski, Z.T., Hawkes, I., 1978. Suggested methods for determining tensile strength of rock materials. Int. J. Rock Mech. Min. Sci. 15 (3), 99–103.

Borodich, F.M., 1999. Fractals and fractal scaling in fracture mechanics. Int. J. Fract. 95, 239–259.

Brady, B.H.G., Brown, E.T., 2006. Rock Mechanics for Underground Mining. Springer.

Broch, E., Franklin, J.A., 1972. The point-load strength test. Int. J. Rock Mech. Min. Sci. 9, 669–697.

Brook, N., 1977. A Method of Overcoming Both Shape and Size Effects in Point Load Testing. Rock Engineering, Geotech Society, University of Newcastle, London, UK, pp. 53–70.

Brook, N., 1980. Size correction for point load testing. Int. J. Rock Mech. Min. Sci. 17, 231–235.

Brook, N., 1985. The equivalent core diameter method of size and shape correction in point load testing. Int. J. Rock Mech. Min. Sci. 22 (2), 61–70.

Butenuth, C., 1997. Comparison of tensile strength values of rocks determined by point load and direct tension tests. Rock Mech. Rock. Eng. 30 (1), 65–72.

Çanakcia, H., Pala, M., 2007. Tensile strength of basalt from a neural network. Eng. Geol. 94 (1-2), 10–18.

Carmichael, R.S., 1982. Handbook of Physical Properties of Rocks. CRC Press, Boca Raton, FL.

Carpinteri, A., 1994. Fractal nature of material microstructure and size effects on apparent mechanical properties. Mech. Mater. 18, 89–101.

Carpinteri, A., Ferro, G., 1994. Size effects on tensile fracture properties: a unified explanation based on disorder and fractality of concrete microstructure. Mater. Struct. 27, 563–571.

Carpinteri, A., Chiaia, B., Ferro, G., 1995. Size effects on nominal tensile strength of concrete structures: multifractality of material ligaments and dimensional transition from order to disorder. Mater. Struct. 28, 311–317.

Castelli, M., Saetta, V., Scavia, C., 2003. Numerical study of scale effects on the stiffness modulus of rock masses. Int. J. Geomech. 3 (2), 160–169.

Chang, C., Zoback, M.D., Khaksar, A., 2006. Empirical relations between rock strength and physical properties in sedimentary rocks. J. Pet. Sci. Eng. 51 (3-4), 223–237.

Chiarellia, A.S., Shaoa, J.F., Hoteitb, N., 2003. Modeling of elastoplastic damage behaviour of a Claystone. Int. J. Plast. 19, 23–45.

Christensen, R.M., 2000. Yield functions, damage states and intrinsic strength. Math. Mech. Solids. 5, 285–300.

Cividini, A., 1993. Constitutive behavior and numerical modeling. In: Hudson, J.A. (Ed.), Comprehensive Rock Engineering, Principles, Practice and Projects. vol. 1. BPCC Wheatons Ltd., London.

Cnudde, V., Boone, M., Dewanckele, J., Dierick, M., Van Hoorebeke, L., Jacobs, P., 2011. 3D Characterization of sandstone by means of X-Ray computed tomography. Geosphere. 7 (1), 54–61.

Corkum, A.G., 2007. The mechanical behaviour of weak mudstone (opalinus Clay) at low stresses. Int. J. Rock Mech. Min. Sci. 44, 196–209.

Cremer, C., Pecker, A., Davenne, L., 2001. Cyclic macro-element for soil-structure interaction: material and geometrical nonlinearities. Int. J. Numer. Anal. Methods Geomech. 25 (13), 1257–1284.

Crouch, R.S., Wolf, P.J., Dafalias, Y.F., 1994. Unified critical-state bounding surface plasticity model for soil. J. Eng. Mech. 120 (11), 2251–2270.

Darlington, W.J., Ranjith, P.G., 2011. The effect of specimen size on strength and other properties in laboratory testing of rock and rock-like cementitious brittle materials. Rock Mech. Rock. Eng. 44, 513–529.

Desai, C.S., 1980. A general basic for yield, failure and potential functions in plasticity. Int. J. Numer. Anal. Methods Geomech. 4, 361–375.

Desai, C.S., Faruque, M.O., 1984. Constitutive model for (geological) materials. J. Eng. Mech. 110 (9), 1391–1408.

Dey, T., Halleck, P., 1981. Some aspects of size-effect in rock failure. Geophys. Res. Lett. 8 (7), 691–694.

Dhir, R.K., Sangha, C.M., 1973. Relationships between size, deformation and strength for cylindrical specimens loaded in uniaxial compression. Int. J. Rock Mech. Min. Sci. 10, 699–712.

Ding, X., Zhang, L., Zhu, H., Zhang, Q., 2014. Effect of model scale and particle size distribution on PFC3D simulation results. Rock Mech. Rock. Eng. 47, 2139–2156.

Elices, M., Rocco, C., 1999. Size effect and boundary conditions in Brazilian test: theoretical analysis. Journal of Material Structures. 32, 437–444.

Erarslan, N., Williams, D.J., 2012. Experimental, numerical and analytical studies on tensile strength of rocks. Int. J. Rock Mech. Min. Sci. 49, 21–30.

Forbes, M., Masoumi, H., Saydam, S., Hagan, P., 2015. Investigation into the effect of length to diameter ratio on the point load strength index of Gosford sandstone. In: 49th US Rock Mechanics/Geomechanics Symposium, American Rock Mechanics Association, San Francisco, US, 28 June–1 July, pp. 478–488.

Forster, I.R., 1983. The influence of core sample geometry on the axial point-load test. Int. J. Rock Mech. Min. Sci. 20 (6), 291–295.

Franklin, J.A., 1985. Suggested method for determining point load strength. Int. J. Rock Mech. Min. Sci. 22, 51–60.

Gajo, A., Wood, M., 1999. Severn-Trent sand: a kinematic-hardening constitutive model: the q-p formulation. Geotechnique. 49 (5), 595–614.

Gao, F., Stead, D., Kang, H., 2014. Numerical investigation of the scale effect and anisotropy in the strength and deformability of coal. Int. J. Coal Geol. 136, 25–37.

Goodman, R.E., 1989. Introduction to Rock Mechanics. Wiley.

Greminger, M., 1982. Experimental studies of the influence of rock anisotropy and size and shape effects in point-load testing. Int. J. Rock Mech. Min. Sci. 19, 241–246.

Griffith, A.A., 1924. The theory of rupture. In: 1st International Congress of Applied Mechanics, Delf, Netherlands.

Hawkins, A.B., 1998. Aspects of rock strength. Bull. Eng. Geol. Environ. 57, 17–30.

Hiramatsu, Y., Oka, Y., 1966. Determination of the tensile strength of rock by a compression test of an irregular test piece. Int. J. Rock Mech. Min. Sci. 3, 89–99.

Hoek, E., Brown, E., 1980a. Underground Excavations in Rock. Institution of Mining and Metallurgy, Spon Press, Hertford, London.

Hoek, E., Brown, E.T., 1980b. Underground Excavations in Rock. Institute of Mining and Metallurgy.

Hoek, E., Brown, E.T., 1997. Practical estimates of rock mass strength. Int. J. Rock Mech. Min. Sci. 34 (8), 1165–1186.

Hoskins, J.R., Horino, F.G., 1969. Influence of Spherical Head Size and Specimen Diameters on the Uniaxial Compressive Strength of Rocks. US department of the Interior, Bureau of Mines, Washington, US.

Huang, H., Detournay, E., 2008. Intrinsic length scales in tool-rock interaction. Int. J. Geomech. 8, 39–44. SPECIAL ISSUE: Excavation Geomechanics for Energy, Environmental, and Transportation Infrastructrue.

Hudson, J.A., Crouch, S.L., Fairhurst, C., 1972. Soft, stiff and servo-controlled testing machines: a review with reference to rock failure. Eng. Geol. 6 (3), 155–189.

Hunt, D.D., 1973. The Influence of Confining Pressure on Size Effect. Master of Science, Massachusetts Institute of Technology, USA.

ISRM, 1979. Suggested methods for determining the uniaxial compressive strength and deformability of rock materials. Int. J. Rock Mech. Min. Sci. 16 (2), 135–140.

ISRM, 2007. The complete suggested methods for rock characterization, testing and monitoring: 1974–2006 ISRM. In: Ulusay, R., Hudson, J.A. (Eds.), Suggested Methods Prepared by the Commission on Testing Methods. Compilation arranged by the ISRM Turkish National Group, Kozanofset, Ankara.

Iwan, W.D., 1967. On class of models for yielding behaviour of continuous and composite systems. J. Appl. Mech. 34, 612–617.

Jizba, D. 1991. Mechanical and Acoustical Properties of Sandstones and Shales. PhD dissertation, Stanford University.

Khan, A.S., Xiang, Y., Huang, S., 1991. Behaviour of Berea sandstone under confining pressure part I: yield and failure surfaces and nonlinear elastic response. Int. J. Plast. 7, 607–624.

Khan, A.S., Xiang, Y., Huang, S., 1992. Behaviour of Berea sandstone under confining pressure part II: elastic-plastic response. Int. J. Plast. 8, 209–230.

Kim, J.K., Eo, S.H., 1990. Size effect in concrete specimens with dissimilar initial cracks. Mag. Concr. Res. 42 (153), 233–238.

Kim, M.K., Lade, P.V., 1984. Modelling rock strength in three dimensions. Int. J. Rock Mech. Min. Sci. 21 (1), 21–33.

Kim, M.K., Lade, P.V., 1988. Single hardening constitutive model for frictional materials: I. Plastic potential function. Comput. Geotech. 5, 307–324.

Klein, E., Baud, P., Reuschle, T., Wong, T.F., 2001. Mechanical behaviour and failure mode of Bentheim sandstone under triaxial compression. Phys. Chem. Earth. 26 (1-2), 21–25.

Koifman, M.I., 1969. Investigation of the effect of specimen dimensions and anisotropy on the strength of some coals in Donets and Kuznetsk Basin. In: Mechanical Properties of Rock, pp. 118–129.

Komurlu, E., Kesimal, A., 2015. Evaluation of indirect tensile strength of rocks using different types of jaws. Rock Mech. Rock. Eng. 48, 1723–1730.

Kramadibrata, S., Jones, I.O., 1993. Size effect on strength and deformability of brittle intact rock. In: Scale Effect in Rock Masses. Balkema, Lisbon, Portugal, pp. 277–284.

Kwasniewski, M., 1989. Laws of brittle failure and of B-D transition in sandstones. In: Rock at Great Depth, ISRM-SPE International Symposium, Balkema, Elf Aquitaine, Pau, France.

Lade, P.V., 1977. Elasto-plastic stress-strain theory for cohesionless soil with curved yield surfaces. Int. J. Solids Struct. 13, 1019–1035.

Lade, P.V., Duncan, J.M., 1975. Elasto-plastic stress-strain theory for cohesionless soil. J. Geotech. Eng. 101 (GT10), 1037–1053.

Lade, P.V., Nelson, R.B., 1987. Modeling the elastic behaviour of granular materials. Int. J. Numer. Anal. Methods Geomech. 11, 521–542.

Lama, R., Vutukuri, V., 1978. Handbook on Mechanical Properties of Rock. Trans Tech Publications, Clausthal, Germany.

Liu, X., Cheng, X.H., Scarpas, A., Blaauwendraad, J., 2005. Numerical modelling of nonlinear response of soil. Part 1: Constitutive model. Int. J. Solids Struct. 42, 1849–1881.

Lundborg, N., 1967. The strength-size relation of granite. Int. J. Rock Mech. Min. Sci. 4, 269–272.

Mandelbort, B.B., 1982. The Fractal Geometry of Nature. W. H. Freeman, New York, USA.

Masoumi, H. 2013. Investigation Into the Mechanical Behaviour of Intact Rock at Different Sizes. PhD Thesis, UNSW Australia, Sydney, Australia.

Masoumi, H., Douglas, K., Saydam, S., Hagan, P., 2012. Experimental study of scale effects of rock on UCS and point load tests. In: Proceedings of the 46th US Rock Mechanics Geomechanics Symposium (ARMA), Chicago, US, 24–27 June, pp. 2561–2568.

Masoumi, H., Bahaaddini, M., Kim, G., Hagan, P., 2014. Experimental investigation into the mechanical behavior of Gosford sandstone at different sizes. In: 48th US Rock Mechanics/Geomechanics Symposium, ARMA, Minneapolis, USA, 1–4 June, pp. 1210–1215.

Masoumi, H., Saydam, S., Hagan, P.C., 2015. Unified size-effect law for intact rock. Int. J. Geomech. https://doi.org/10.1061/(ASCE)GM.1943-5622.0000543.

Masoumi, H., Douglas, K., Russell, A.R., 2016a. A bounding surface plasticity model for intact rock exhibiting size-dependent behaviour. Rock Mech. Rock. Eng. 49 (1), 47–62.

Masoumi, H., Saydam, S., Hagan, P.C., 2016b. Incorporating scale effect into a multiaxial failure criterion for intact rock. Int. J. Rock Mech. Min. Sci. 83, 49–56.

Masoumi, H., Saydam, S., Hagan, P.C., 2016c. Unified size-effect law for intact rock. Int. J. Geomech. 16(2).

Masoumi, H., Roshan, H., Hagan, P.C., 2017a. Size-dependent Hoek-Brown failure criterion. Int. J. Geomech. 17(2).

Masoumi, H., Serati, M., Williams, D.J., Alehossein, H., 2017b. Size dependency of intact rocks with high brittleness: a potential solution to eliminate secondary fractures in Brazilian test. In: 51st US Rock Mechanics/Geomechanics Symposium (ARMA), San Francisco, USA, 25–28 June.

Medhurst, T.P., Brown, E.T., 1998. A study of the mechanical behaviour of coal for pillar design. Int. J. Rock Mech. Min. Sci. 35 (8), 1087–1105.

Melean, Y., Washburn, K.E., Callaghan, P.T., Arns, C.H., 2009. A numerical analysis of NMR pore-pore exchange measurements using micro X-ray computed tomography. Diff. Fundament. 10 (30), 1–3.

Mellor, M., Hawkes, I., 1971. Measurement of tensile strength by diametral compression of discs and annuli. Eng. Geol. 5 (3), 173–225.

Mogi, K., 1962. The influence of dimensions of specimens on the fracture strength of rocks. Bull. Earth Res. Inst. 40, 175–185.

Morvan, M., Wong, H., Branque, D., 2010. An unsaturated soil model with minimal number of parameters based on bounding surface plasticity. Int. J. Numer. Anal. Methods Geomech. 34, 1512–1537.

Muller, J., Mccauley, J., 1992. Implication of fractal geometry for fluid flow properties of sedimentary rocks. Transp. Porous Media. 8 (2), 133–147.

Murray-Wallace, C.V., Belperio, A.P., Bourman, R.P., Cann, J.H., Price, D.M., 1999. Facies architecture of a last interglacial barrier: a model for Quaternary barrier development from the Coorong to Mount Gambier Coastal Plain, southeastern Australia. Mar. Geol. 158, 177–195.

Murthy, V.N.S., 2002. Geotechnical Engineering Principles and Practices of Soil Mechanics and Foundation Engineering. Marcel Dekker, New York.

Nishimatsu, Y., Yamagushi, U., Motosugi, K., Morita, M., 1969. The size effect and experimental error of the strength of rocks. J. Soc. Mater. Sci. Jpn. 18, 1019–1025.

Panek, L.A., Fannon, T.A., 1992. Size and shape effects in point load tests of irregular rock fragments. Rock Mech. Rock. Eng. 25, 109–140.

Pells, P., 2004. On the absence of size effects for substance strength of Hawkesbury sandstone. Aust. Geomech. 39 (1), 79–83.

Prakoso, W.A., Kulhawy, F.H., 2011. Effects of testing conditions on intact rock strength and variability. Geotech. Geol. Eng. 29, 101–111.

Pratt, H.R., Black, A.D., Brown, W.S., Brace, W.F., 1972. The effect of specimen size on the mechanical properties of unjointed diorite. Int. J. Rock Mech. Min. Sci. 9, 513–529.

Quiñones, J., Arzúa, J., Alejano, L.R., García-Bastante, F., Mas Ivars, D., Walton, G., 2017. Analysis of size effects on the geomechanical parameters of intact granite samples under unconfined conditions. Acta Geotech. https://doi.org/10.1007/s11440-017-0531-7.

Radlinski, A., Ioannidis, M., Hinde, A., Hainbuchner, M., Baron, M., Rauch, H., 2004. Angestrom-to-millimeter characterisation of sedimentary rock microstructure. J. Colloid Interface Sci. 274, 607–612.

Roscoe, K.H., Schofield, A.N., 1963. Mechanical behaviour of an idealised 'wet' clay. In: 2nd European Conference SMFE, Wiesbaden, pp. 47–54.

Roshan, H., Masoumi, H., Hagan, P., 2016a. On size-dependent uniaxial compressive strength of sedimentary rocks in reservoir geomechanics. In: 50th US Rock Mechanics/Geomechanics Symposium, ARMA, Houston, US, 26–29 June, pp. 2322–2327.

Roshan, H., Sari, M., Arandiyan, H., Hu, Y., Mostaghimi, P., Sarmadivaleh, M., Masoumi, H., Veveakis, M., Iglauer, S., Regenauer-Lieb, K., 2016b. Total porosity of tight rocks: a welcome to heat transfer technique. Energy Fuel. 30 (12), 10072–10079.

Roshan, H., Masoumi, H., Regenauer-Lieb, K., 2017a. Frictional behaviour of sandstone: a sample-size dependent triaxial investigation. J. Struct. Geol. 94, 154–165.

Roshan, H., Masoumi, H., Zhang, Y., Al-Yaseri, A.Z., Iglauer, S., Lebedev, M., Sarmadivaleh, M., 2017b. Micro-structural effects on mechanical properties of Shaly-sandstone. J. Geotech. Geoenviron. 144.

Russell, A.R., Khalili, N., 2004. A bounding surface plasticity model for sands exhibiting particle crushing. Can. Geotech. J. 41, 1179–1192.

Russell, A.R., Wood, D.M., 2009. Point load tests and strength measurements for brittle spheres. Int. J. Rock Mech. Min. Sci. 46, 272–280.

Savidis, G.M. 1982. Fundamental Studies on the Excavation of Rock by Button Cutter. PhD thesis, The University of New South Wales, Sydney, Australia.

Serati, M. 2014. Stiffness and Strength of Rock Cutting and Drilling Tools—Drag Bit and Roller Disc Cutters. PhD thesis, University of Queensland, Brisbane, Australia.

Serati, M., Alehossein, H., Williams, D.J., 2015. Estimating the tensile strength of super hard brittle materials using truncated spheroidal specimens. J. Mech. Phys. Solids. 78, 123–140.

Shah, K.R., 1997. An elasto-plastic constitutive model for brittle-ductile transition in porous rocks. Int. J. Rock Mech. Min. Sci. 34 (3-4). 283.e1–e13.

Silva, A., Azuaga, L., Hennies, W.T., 1993. A methodology for rock mass compressive strength characterization from laboratory tests. In: Scale Effects in Rock Mass. Balkema, Lisbon, Portugal, pp. 217–224.

Simon, R., Deng, D., 2009. Estimation of scale effects of intact rock using dilatometer tests results. In: 62nd Canadian Geotechnical Conference, Halifax, pp. 481–488.

Singh, M.M., Huck, P.J., 1972. Large scale triaxial tests on rock. In: The 14th US Symposium on Rock Mechanics Penn State University, Pennsylvania, USA, pp. 35–60.

Smith, E., 1995. Critical dimensions for scaling effects in the fracture of quasi-brittle materials. In: International Union of Theoretical and Applied Mechanics (IUTAM) Symposium on Size-Scale Effects in the Failure Mechanisms of Materials and Structures, Politecnico di Torino, Turin, Italy.

Sufian, A., Russell, A.R., 2013. Microstructural pore changes and energy dissipation in Gosford sandstone during pre-failure loading using X-ray CT. Int. J. Rock Mech. Min. Sci. 57, 119–131.

Sulem, J., Ouffroukh, H., 2006a. Hydromechanical behaviour of Fontainebleau sandstone. Rock Mech. Rock. Eng. 39 (3), 185–213.

Sulem, J., Ouffroukh, H., 2006b. Shear banding in drained and undrained triaxial tests on saturated sandstone: porosity and permeability evolution. Int. J. Rock Mech. Min. Sci. 43, 292–310.

Thompson, A., 1991. Fractals in rock physics. Annu. Rev. Earth Planet. Sci. 19, 237–262.

Thuro, K., Plinninger, R.J., Zah, S., Schutz, S., 2001a. Scale effect in rock strength properties. Part 1: Unconfined compressive test and Brazilian test. In: Rock Mechanics—A Challenge for Society. Zeitlinger, Switzerland, pp. 169–174.

Thuro, K., Plinninger, R.J., Zah, S., Schutz, S., 2001b. Scale effect in rock strength properties. Part 2: Point load test and point load strength index. In: Rock Mechanics—A Challenge for Society. Zeitlinger, Switzerland, pp. 175–180.

Timoshenko, S.P., Goodier, J.N., 1951. Theory of Elasticity. McGraw-Hill, New York, US.

Tsur-Lavie, Y., Denekamp, S.A., 1982. Comparison of size effect for different types of strength tests. Rock Mech. Rock. Eng. 15, 243–254.

Van Mier, J.G.M., 1996. Fracture Processes of Concrete. CRC Press, Boca Raton, FL.

Van Mier, J.G.M., Man, H.K., 2009. Some notes on microcracking, softening, localization, and size effects. International Journal of Damage Mechanics. 18, 283–309.

Villeneuve, M., Diederichs, M., Kaiser, P., 2012. Effects of grain scale heterogeneity on rock strength and the chipping process. Int. J. Geomech. 12, 632–647. SPECIAL ISSUE: Advances in Modeling Rock Engineering Problems.

Vutukuri, V.S., Lama, R.D., Saluja, S.S., 1974. Handbook on Mechanical Properties of Rocks. Trans Tech Publications, Bay Village.

Wang, S.Y., Sloan, S.W., Tang, C.A., 2014. Three-dimensional numerical investigations of the failure mechanism of a rock disc with a central or eccentric hole. Rock Mech. Rock. Eng. 47, 2117–2137.

Wawersik, W.R., Fairhurst, C., 1970. A study of brittle rock fracture in laboratory compression experiments. Int. J. Rock Mech. Min. Sci. 7, 561–575.

Weibull, W., 1939. A Statistical Theory of the Strength of Materials. Royal Swedish Academy of Engineering Science, Stockholm, Sweden, pp. 1–45.

Weibull, W., 1951. A statistical distribution of function of wide applicability. J. Appl. Mech. 18, 293–297.

Weng, M.C., Jeng, F.S., Huang, T.H., Lin, M.L., 2005. Characterizing the deformation behavior of Tertiary sandstone. Int. J. Rock Mech. Min. Sci. 42, 388–401.

Wijk, G., 1978. Some new theoretical aspects of indirect measurements of the tensile strength of rocks. Int. J. Rock Mech. Min. Sci. 15, 149–160.

Wijk, G., Rehbinder, G., Logdstrom, G., 1978. The relation between the uniaxial tensile strength and the sample size for Bohus granite. Rock Mech. Rock. Eng. 10, 201–219.

Wong, L.N.Y., Zhang, X.P., 2014. Size effects on cracking behavior of flaw-containing specimens under compressive loading. Rock Mech. Rock. Eng. 47 (5), 1921–1930.

Wong, T.F., David, C., Zhu, W., 1997. The transition from brittle faulting to cataclastic flow in porous sandstones: mechanical deformation. J. Geophys. Res. Solid Earth. 102 (B2), 3009–3025.

Wong, H., Morvan, M., Branque, D., 2010. A 13-parameter model for unsaturated soil based on bounding surface plasticity. J. Rock Mech. Eng. 2 (2), 135–142.
Wood, D.M., 2004. Geotechnical Modelling. Spon Press, New York.
Yi, C., Zhu, H., Liu, L., 2011. Damage analysis for quasi-brittle materials of CT images under uniaxial compression. Adv. Mater. Res. 170, 373–379.
Yoshinaka, R., Osada, M., Park, H., Sasaki, T., Sasaki, K., 2008. Practical determination of mechanical design parameters of intact rock considering scale effect. Eng. Geol. 96 (3-4), 173–186.
Yu, Y., Yin, J., Zhong, Z., 2006. Shape effects in the Brazilian tensile strength test and a 3D FEM correction. Int. J. Rock Mech. Min. Sci. 43 (4), 623–627.
Zhang, Q., Zhu, H., Zhang, L., Ding, X., 2011. Study of scale effect on intact rock strength using particle flow modeling. Int. J. Rock Mech. Min. Sci. 48, 1320–1328.
Zhang, X.P., Wong, L.N.Y., Wang, S., 2015. Effects of the ratio of flaw size to specimen size on cracking behavior. Bull. Eng. Geol. Environ. 74, 181–193.
Zoback, M.D., 2007. Reservoir Geomechanics. University Press, Cambridge, United Kingdom.

Further reading

Desai, C.S., Somasundaram, S., Frantziskois, G.N., 1986. A hierarchical approach for constitutive modelling of geologic materials. Int. J. Numer. Anal. Methods Geomech. 10, 225–257.
Frantziskonis, G., Desai, C.S., Somasundaram, S., 1986. Constitutive model for nonassociative behaviour. J. Eng. Mech. 112, 932–945.
Nieble, C.M., de Cornides, A.T., 1973. Analysis of point load test as a method for preliminary geotechnical classification of rocks. Bull. Eng. Geol. Environ. 7, 37–52.

Chapter 2

Rock fracture toughness

Sheng Zhang
School of Energy Science and Engineering, Henan Polytechnic University, Jiaozuo, China

Chapter outline

2.1 Fracture toughness of splitting disc specimens

2.1.1 Introduction

Rock fracture toughness is used to characterize the ability of rock materials to resist crack propagation (Yu, 1991). There are two difficulties in the testing process: it is difficult to prefabricate cracks in rock specimens and critical crack lengths are difficult to measure. Therefore, simply applying the shape and test specification of the metal specimen (ASTM, 1997) does not work, and new specimens and new methods must be developed. At present, there is not a unified standard for the mode I (opening mode) rock fracture toughness test in the world. Only two recommended methods (ISRM, 1988; ISRM, 1995) have been issued by the International Society of Rock Mechanics (ISRM). Three types of specimens were recommended: the short rod (SR) specimen and the chevron-notched bend (CB) specimen were recommended in 1988, and the cracked chevron notched Brazilian disc (CCNBD) specimen was proposed in 1995. The CCNBD specimen splitting test method is simpler and has been gradually used.

In addition, Guo et al. (1993) promoted the use of tensile strength obtained through ordinary Brazilian disc test brittle materials, and determined the rock fracture toughness using the two peak phenomena on the load-displacement curve obtained by the experiment (1993). Later, Wang and Xing (1999) focused on the key problem of the disc splitting test, that is, the center of the disc should be destroyed first. The configuration and loading method of the specimen were improved, and the fracture toughness of the rock was determined by the platform loading method using the flattened Brazilian disc (FBD). In order to better eliminate the influence of stress concentration at the loading end, the literature (Yang et al., 1997) used the holed-flattened Brazilian disc (HFBD). At the same time, the Brazilian disc developed into a specimen with a straight-though crack

(Tang et al., 1996; Fischer et al., 1996) or with a chevron notched crack (ISRM, 1995). According to the configuration of the crack, it can be divided into a cracked straight-through Brazilian disc (CSTBD), a cracked chevron notched Brazilian disc (CCNBD), and a slotted type holed-cracked Brazilian disc (HCBD). Different shapes of disc specimens have been used to determine the fracture toughness of rock, but the methods and determination results for different discs are not correspondingly compared.

It is very difficult to process specimens with a crack width of less than 1 mm (Chen and Zhang, 2004; Li et al., 2004). In the literature (Chen and Zhang, 2004), the width of the grooving has been studied. It is pointed out that the crack tip radius of the straight cut groove of the specimen shall be at least 0.4 mm, as the static fracture toughness of the rock is determined by using a disc with a diameter of not more than 100 mm. If it exceeds this range, the fracture mechanics analysis is not applicable. The grooving width of the CCNBD disc specimen recommended by ISRM requires no more than 1.5 mm, as for the unique advantages of herringbone grooving.

In this test, a diamond cutter with a diameter of 60 mm and a thickness of 1.2 mm was used, and the groove width of the machined CCNBD specimen was about 1.3 mm. Two kinds of marble disc specimens of CSTBD and HCFBD were studied, (HCFBD is the abbreviation for holed-cracked Brazilian flattened disc, and the platform loading surface is fabricated on HCBD specimen and the fabrication problem of slot width less than 1 mm was successfully solved. The formula of fracture toughness was also studied by finite element analysis combined with experiments. The formula for determining the fracture toughness of each disc was obtained. Finally, the advantages and disadvantages of the rock fracture toughness of five discs were compared.

2.1.2 Preparation of disc specimens

The specimen is white marble, collected from Ya'an, Sichuan, with a Poisson's ratio of 0.3, an elastic modulus of 16.3 GPa, and a density of 2.73 g/cm^3. The disc specimens were 80 mm in diameter and 32 mm in thickness. The FBD, HFBD, CCNBD, CSTBD, and CFBD disc specimens are shown in Fig. 2.1.

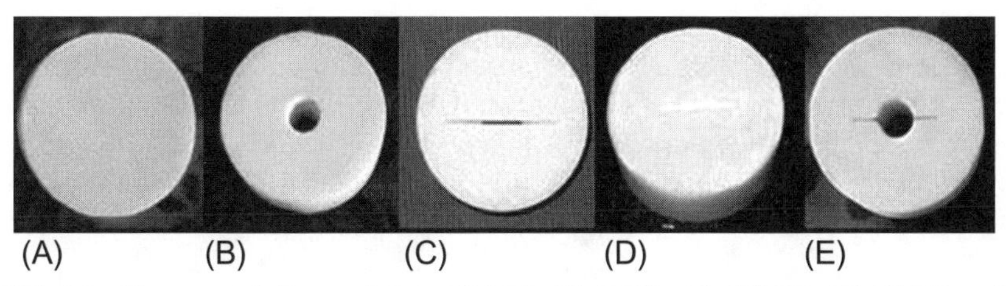

FIG. 2.1 Five types of disc specimens. (A) FBD, (B) HFBD, (C) CCNBD, (D) CSTBD, and (E) HCFBD.

According to the referenced method (Li et al., 2004), CSTBD processing is carried out by first drilling a small hole in the center, and then making a straight-through groove. Specifically, first drill a small hole in the center of the disc with a drill bit of 2 mm in diameter, and a wire with a diameter of 0.6 mm and a saw-tooth shape is cut through the small hole to make a groove. The groove is machined with a thinned steel saw blade (around 0.4 mm in thickness). Because the cutting amount of the steel wire to the rock is small, and the swing in the lateral direction makes the grooving wider, the tip is finely machined with a fine steel wire with 0.2 mm in thickness so that the width of the grooving tip is less than 0.3 mm.

In addition, the HCFBD specimen is obtained by drilling a large center hole in the center of the disc and making a groove and a platform. The advantage of HCFBD is that it does not require steel wire. It can be directly grooved with a thin steel saw blade, and specimens with different groove lengths and a minimum groove of 0.6 mm in width can be obtained. In this paper, the HCFBD specimen has a center hole diameter of 16 mm and a groove of 40 mm in length. The FBD, HFBD, and HCFBD specimens have loading platforms with a loading angle of $2\beta = 20$ degrees. All fabricated platforms and center holes are done on the drilling and milling machine.

2.1.3 Fracture toughness of five types of specimens

In the fracture toughness formula, the dimensionless stress intensity factor related to the geometry of the specimen must be determined first (China Aviation Research Institute, 1993). This paper uses the finite element method to calibrate it. The ANSYS finite element software was used to build a model for each type of disc. According to the symmetry of the disc, a 1/4 disc model was used. Also, a plane 8 node 4 edge isoparametric unit is used, and the "1/4 node" technology is used about the grooving tip.

2.1.3.1 Fracture toughness formula of FBD and HFBD

Figs. 2.2 and 2.3 show the geometry of the FBD and HFBD specimens. R is the disc radius, 2β is the platform loading angle, t is the disc thickness, AB is the loading diameter in Fig. 2.3, and d is the diameter of the center hole. The FBD and HFBD specimens are cracked at the center or center hole after being loaded, and the idealized cracks generated thereafter expand symmetrically along the direction of the vertical loading diameter AB to form a central crack.

Let the combined force of the uniform pressure of the specimen be P, the radius of the disk is R, and α is the ratio of the center crack length $2a$ to the disc diameter $2R$ generated during crack propagation, $\alpha = a/R$, ratio of the center radius of the hole to the radius of the disk, $\rho = r/R$, then the stress intensity factor of the two specimens can be expressed as

$$K_{\mathrm{I}} = \frac{P}{\sqrt{R}t} Y(\beta, \rho, \alpha) \tag{2.1}$$

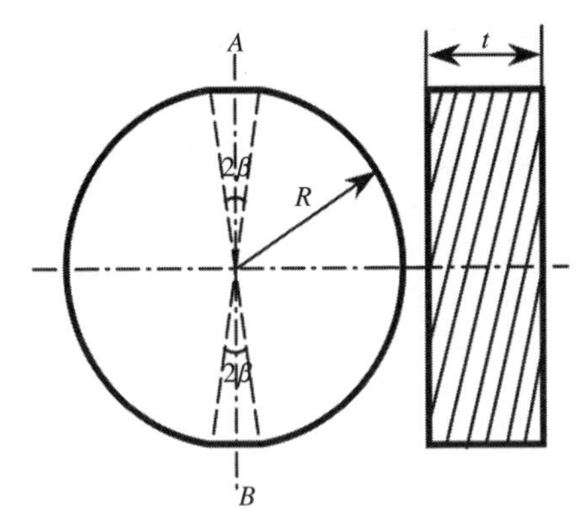

FIG. 2.2 FBD specimen.

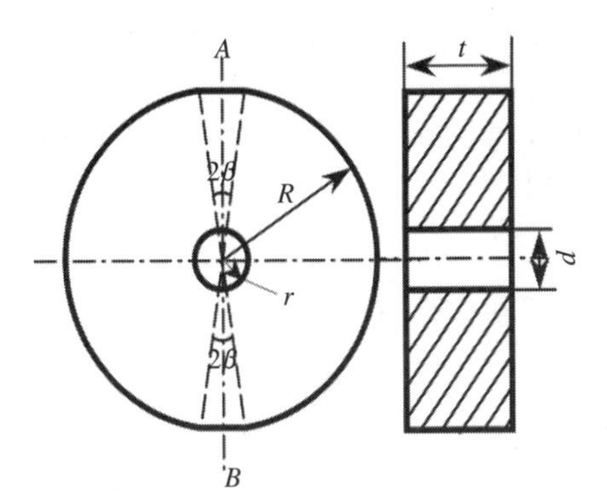

FIG. 2.3 HFBD specimen.

Because the fracture toughness K_{IC} is a material constant, according to the fracture principle $K_I = K_{IC}$, any load P during crack propagation and its corresponding dimensionless crack length α can be calculated by Formula (2.1), but this is cumbersome and even unrealistic. However, the variation of the two disk loads with α corresponds to the variation of the dimensionless stress intensity factor Y (Wang and Xing, 1999). The maximum value of the dimensionless stress intensity factor Y_{max} corresponds to the load on the load-displacement curve falling from the maximum value to a local minimum load P_{min} (at this point, $\alpha_c = a_c/R$, the critical crack length corresponds to Y_{max}). Therefore, this minimum load P_{min} and Y_{max} can be substituted into Formula (2.1) to calculate K_{IC}.

$$K_{IC} = \frac{P_{min}}{\sqrt{R}t} Y_{max} \tag{2.2}$$

In the formula, the maximum value Y_{max} is calculated by the finite element, thus the fracture toughness formula for the FBD specimen with a loading angle of $2\beta = 20$ degrees is obtained:

$$K_{IC} = \frac{P_{min}}{\sqrt{Rt}} \times 0.80 \tag{2.3}$$

From the finite element calculation, the fracture toughness formula of the HFBD specimen with a loading angle of $2\beta = 20$ degrees and a central aperture ratio of $\rho = 0.2$ is:

$$K_{IC} = \frac{P_{min}}{\sqrt{Rt}} \times 1.04 \tag{2.4}$$

2.1.3.2 Fracture toughness formula of CCNBD

Fig. 2.4 shows the geometry of the CCNBD specimen, where t is the thickness of the specimen, D is the disc diameter, R is the disc radius, α_0 is the dimensionless initial crack length ($\alpha_0 = a_0/R$), α_1 is the dimensionless maximum groove length ($\alpha_1 = a_1/R$), α_t is the thickness of the dimensionless specimen ($\alpha_t = t/R$), and P is the load.

The calculation formula for determining the fracture toughness K_{IC} of rock using CCNBD specimen is:

$$K_{IC} = \frac{P_{max}}{\sqrt{2Rt}} Y^*_{min} \tag{2.5}$$

In the formula, P_{max} is the maximum load and Y^*_{min} is the minimum dimensionless stress intensity factor. Y^*_{min} is an important and critical parameter in Formula (2.5), determined by the dimensionless parameters α_0, α_1, and α_t.

In Formula (2.5), Y^*_{min} can be obtained by numerical calculation before the experiment, and its value directly affects the accuracy of the calculation of the fracture toughness value. The expression of Y^*_{min} is given in (Wu et al., 2004a, b):

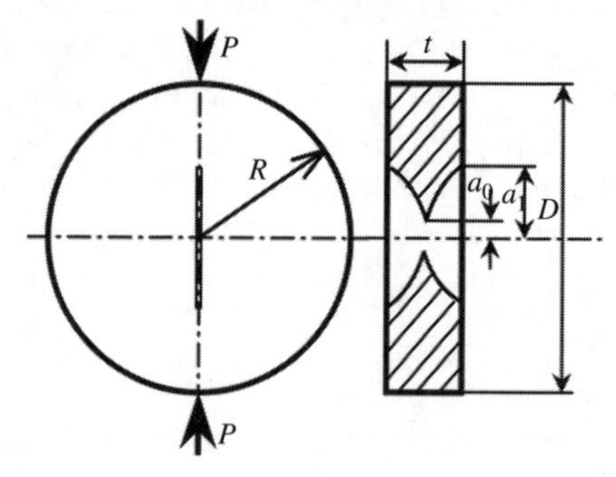

FIG. 2.4 CCNBD specimen.

$$Y_{\min}^{*}=ue^{va_1} \tag{2.6}$$

In the formula, u and v are the constants determined by α_0 and α_t, respectively; the values of u and v are given by Eqs. (2.7), (2.8).

$$u(\alpha_0,\alpha_t)=0.2553+0.0925\alpha_0+0.0327\alpha_t+0.1929\alpha_0^2+0.3473\alpha_0^3-0.9695\alpha_0^4 \tag{2.7}$$

$$v(\alpha_0,\alpha_t)=2.4404+0.8582\alpha_0+1.2698\alpha_t+0.469\alpha_0\alpha_t+0.7345\alpha_0^2-0.4819\alpha_t^2 \tag{2.8}$$

2.1.3.3 Fracture toughness formula of CSTBD and HCFBD

Figs. 2.5 and 2.6 show the geometry of the CSTBD and HCFBD specimens, respectively, and $2a_0$ is the groove length.

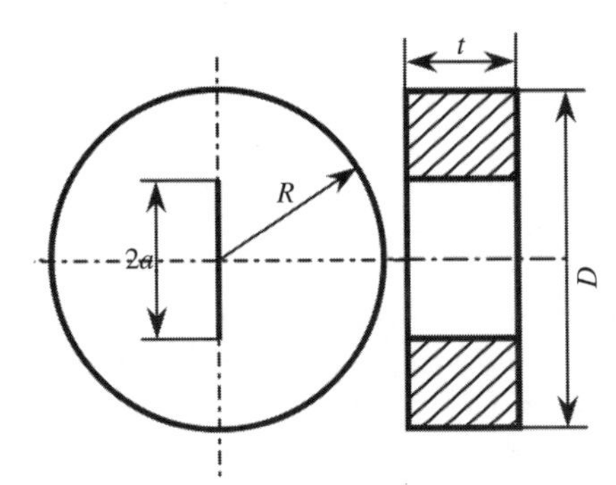

FIG. 2.5 CSTBD specimen.

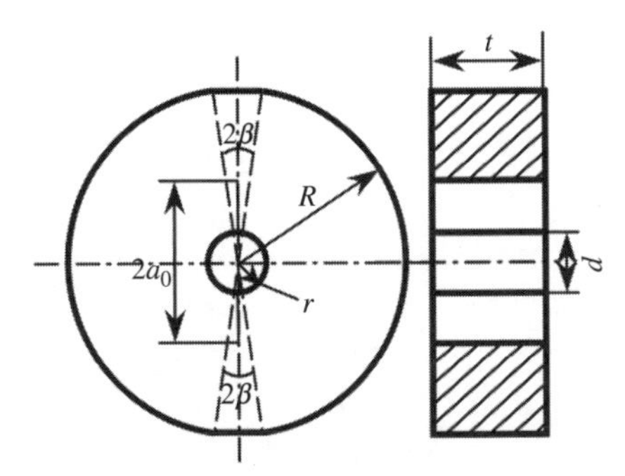

FIG. 2.6 HCFBD specimen.

The theoretical formula can be used to calculate the fracture toughness of a material within a reasonable crack tip radius. According to the research (Chen and Zhang, 2004), for the disc specimen with a diameter of 100 mm, the fracture mechanics are still applicable in the range of crack width not more than 0.8 mm, and the change of the crack tip radius has little effect on the fracture toughness. The CSTBD and HCFBD disc specimens in this paper have initial grooves, but the grooving width is controlled below 0.8 mm and the tip of the groove is further processed with a fine steel wire 0.2 mm in diameter. After processing, the produced crack width is already very small. This paper does not consider the effect of crack width on the test value. In the finite element analysis, the crack width is assumed to be 0.

The fracture toughness formula of the CSTBD specimen is:

$$K_{IC} = \frac{P_{\mathrm{cr}}}{\sqrt{R}t} Y(2\beta, \rho, \alpha_0) \tag{2.9}$$

In the formula, Y is the dimensionless stress intensity factor, which can be determined by finite element calculation; $\rho = r/R$ is the ratio of the diameter of the central hole to the radius of the disk; $\alpha_0 = a_0/R$ is the ratio of the initial slot length to the disk radius; and P_{cr} is the cracking load measured by the test, and it is the first peak load value on the load-displacement curve with a straight-cut specimen. The reason is that if the initial grooving length of the specimen is less than the critical crack length, the loading-displacement curves of the CSTBD and HCFBD specimens also have a secondary peak phenomenon as in the FBD and HFBD specimens, that is, the load will undergo a process of decreasing first and then rising. The load drops from the maximum and reaches a local minimum load, at which point the crack length reaches the critical crack length a_c and the load continues to rise. In this case, the cracking load is not necessarily the maximum load, but the first peak load. If the length of the specimen is so long that it exceeds the critical crack length, then the cracking load is the maximum load. Therefore, in determining the formula for the rock fracture toughness of the CSTBD and HCFBD specimens, the load takes the first peak load and names the load as the cracking load P_{cr}.

The fracture toughness formula of the CSTBD specimen with the groove length ratio $\alpha_0 = 0.45$ calculated by the finite element method is:

$$K_{IC} = \frac{P_{\mathrm{cr}}}{\sqrt{R}t} \times 0.50 \tag{2.10}$$

Calculated by the finite element method, the fracture toughness formula of the HCFBD specimen with the central aperture ratio $r/R = 0.2$ and the slot length ratio $\alpha = 0.5$ is

$$K_{IC} = \frac{P_{\mathrm{cr}}}{\sqrt{R}t} \times 0.75 \tag{2.11}$$

2.1.4 Load-displacement curve of disc splitting test

The experiment was carried out on the RMTS-150 material testing machine of Sichuan University Hydropower College. The load cell model is a BLR-1/5000 tension and pressure sensor with a measuring range of 0–50 kN. The experiment is controlled by a displacement loading rate of 0.02 mm/min, and the disc is directly subjected to radial splitting loading. Fig. 2.7 shows the loading of the HCFBD specimen. The typical loading-displacement curves of five types of disc specimens are shown in Fig. 2.8.

The loading-displacement curve of a disc specimen is closely related to its fracture pattern. For the loading-displacement curves of FBD and HFBD specimens, the literature (Wang and Xing, 1999; Yang et al., 1997) has been studied (Fig. 2.8). During the entire loading process, the load will undergo a process of decreasing first and then rising. At the same time, it is observed that the crack originated from the middle of the specimen or the center circular hole, and propagated in the diameter direction of the load until the final complete fracture. For the convenience of description, the crack that starts from the tip of the groove and then propagates along the loading diameter is called the primary crack.

FIG. 2.7 Loading system of the HCFBD specimen.

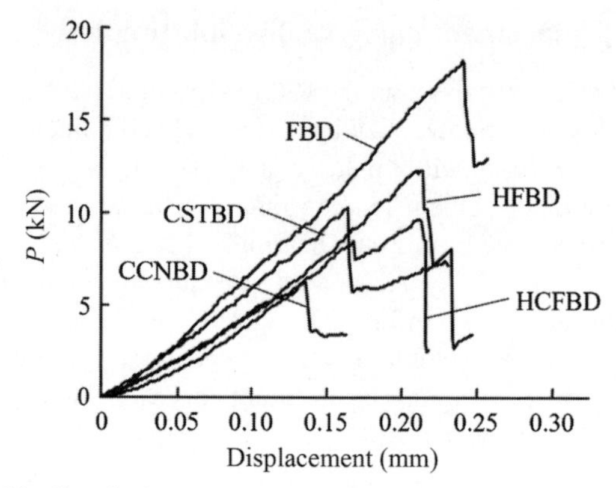

FIG. 2.8 Typical loading-displacement curves of disc specimens.

Cracks that crack and expand at other locations are called incidental cracks. The loading diameter refers to the connection of the concentrated loading of the upper and lower ends when the ordinary Brazilian disc is concentrated. During the rupture process, it is possible to observe that an incidental crack is initiated from the intersection of the arc near the platform and the outer surface. When and how this incidental crack occurs directly affects the shape of the loading-displacement curve and fundamentally affects the validity of the experiment. An effective loading-displacement curve is an incidental crack after the primary crack has fully propagated.

According to the fracture principle, the fracture toughness of the material can be calculated by any load in the crack propagation process and its corresponding crack length, but the crack length corresponding to a certain load is difficult to measure. Both the FBD and HFBD specimens have two inversions. According to the variation law of their special load-displacement curve and the dimensionless stress intensity factor in the corresponding crack propagation process, the fracture toughness of the material can be determined, and do not need to know the extended length of the crack. For a precracked CCNBD specimen, the fracture toughness is determined only by the maximum load. For the CSTBD and HCFBD specimens, because the initial groove length is known, the key is to determine the cracking load. The ideal state for the fracture of the three discs is to first crack from the end of the grooving and then expand along the direction of the load, eventually becoming two. In the load-displacement curve, the load increases almost linearly until the breaking load is reached, at which point the fracture at the tip of the grooving is followed by a drop in load. Later, as new cracks are likely to occur, the load curve may also appear as multiple rises and falls.

It is worth noting that the cracking and expansion of the specimen are determined by at least the following three influencing factors: (1) changes in local strength due to material nonuniformity; (2) changes in stress field caused by the distribution of microvoids and inclusions; and (3) the influence on processing accuracy of specimens.

These factors can cause some specimens not to crack at the center of the disc, near the center hole, or at the tip of the crack, or they will start to crack and expand in other parts after the cracking shortly. The author believes that it is ineffective for tests that do not perform initial crack initiation and expansion on the loading diameter. If the initial damage of the specimen is mainly caused by the main crack, the test is considered to be effective. During the test, it was found that the nongrooved disc was greatly affected by the loading end, and there were more ineffective tests. The cracking of the grooved disc basically starts from the tip of the grooving, which is easy to obtain an effective test curve.

2.1.5 Comparison of disc splitting test results

The fracture process of the disc is subject to crack initiation, stable crack propagation, and unsteady expansion of the crack until the final failure of the specimen. According to the phenomenon in the specimens, the cracking and crack propagation of various disc splitting specimens are analyzed and compared. The effective test discs are first cracked from the center of the disc, which cannot be observed with the naked eye for FBD specimens without pregrooving. But as the crack continues to expand, the width of the central crack will gradually increase. Fig. 2.9A shows a photo of the crack growth process of the FBD specimen captured by a digital camera before failure. From Fig. 2.9, the crack on the loading diameter of the FBD specimen is expanding from the center to the two loading ends. Fig. 2.9B is a photograph of the crack propagation process before the failure of the HFBD specimen. It can be clearly seen that the first cracking of the HFBD specimen splitting test occurs from the intersection of the loading diameter and the central circular orifice. The crack also spreads toward the two loading ends after cracking at the edge of the hole. For pregrooving disc specimens, the initiation crack basically starts from the two tips of the grooving, and then the crack spreads along the loading diameter toward the two loading ends until the crack spread near the two loading platforms of the disc specimen, the specimen completely failed. Fig. 2.9C is a photograph of the crack

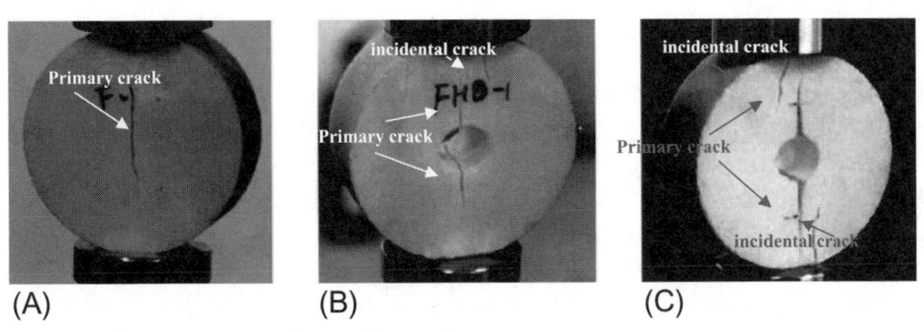

FIG. 2.9 Crack growth of different disc specimens.

propagation process before the failure of the HCFBD specimen, and the pregrooving width becomes larger due to the continuous expansion of the crack. From Fig. 2.9, the stress concentration at the loading end of the disc when the crack propagates to the vicinity of the loading end is manifested, that is, the crack is again induced along the intersection of the arc from the approaching platform and the outer surface, thereby causing an expansion of the incidental crack. However, the expansion of the accidental crack is generated after the expansion of the main crack is completed, and the critical load at this time has been determined, so the effectiveness of the test is not affected. Conversely, the test that the incidental crack occurred before the main crack is invalid.

The measured specimen geometry and the critical load P obtained in the effective test are substituted into Formulas (2.3)–(2.5) and Formulas (2.10) and (2.11) to obtain the mode I fracture toughness measured by the test, as shown in Table 2.1. The mean fracture toughness determined by the five types of discs is shown in Fig. 2.10.

From Table 2.1 and Fig. 2.10, the average fracture toughness determined by all discs is about 0.91 MPa $m^{1/2}$. The test values of the FBD and HFBD specimens without grooves are higher than those of the CCNBD, CSTBD, and HCFBD specimens with grooving. The test value of the HFBD specimens is the highest, 1.04 MPa $m^{1/2}$, and the test value of the CSTBD specimens is the lowest, 0.78 MPa $m^{1/2}$. The fracture toughness determined by the HFBD specimens is 1.33 times that of the CSTBD specimens. The test values for the HCFBD specimens are closest to the average fracture toughness determined for all discs. The cracks in the grooving specimens are mostly extended by the grain boundary of the artificial grooving tip. The reason why the fracture toughness values measured by these specimens are low may be that the initial slit length is used in calculating the fracture toughness, and the microcrack zone generated before the crack extension is ignored. In addition, in this test, try to make a specimen with a small grooving width, in the finite element calculation, the artificial grooving width is regarded as 0, and the influence of the grooving width on the fracture toughness test value is not considered. Actually, this effect exists.

It is also inaccurate if it is considered that the specimen without a groove determines that the fracture toughness value of the rock is more accurate and more conducive to test. Because the dispersion of FBD and HFBD specimens was observed to be large during the test, there were more ineffective tests, mainly due to the heterogeneity of the specimens. The first cracking of the grooving is always difficult to occur all at the center ideally. The critical load used to determine the fracture toughness is also highly susceptible to other factors, and the dispersion is large. FBD and HFBD specimens without grooves are easy to process; the grooving of CCNBD, CSTBD, and HCFBD discs is difficult. The machining of CCNBD is relatively easy, and it can be produced by directly cutting the diamond cutter on the machine tool. However, its complicated three-dimensional configuration cannot be analyzed according to the

TABLE 2.1 Experimental data of disc specimens

Specimen	R (mm)	t (mm)	2β (degree)	r/R	α_0	Y	P (kN)	Test value K_{IC} (MPa m$^{1/2}$)	Average value $K_{IC}^{(AV)}$ (MPa m$^{1/2}$)
FBD-1	40	32.00	20	0.2	0	0.8	7.15	0.89	1.01
FBD-2	40	32.30	20	0.2	0	0.8	10.60	1.31	
FBD-3	40	32.20	20	0.2	0	0.8	8.54	1.06	
FBD-4	40	32.30	20	0.2	0	0.8	6.18	0.77	
HFBD-1	40	32.30	20	0.2	0	1.04	7.12	1.15	1.04
HFBD-2	40	32.20	20	0.2	0	1.04	6.79	1.10	
HFBD-3	40	32.40	20	0.2	0	1.04	5.48	0.88	
CCNBD-1	40	32.00	0	0.2	0.49	1.21	6.24	0.83	0.83
CCNBD-2	40	32.40	0	0.2	0.33	1.06	6.98	0.81	
CCNBD-3	40	32.20	0	0.2	0.43	1.16	6.61	0.84	
CSTBD-1	40	32.00	0	0.2	0.45	0.5	10.27	0.80	0.78
CSTBD-2	40	32.30	0	0.2	0.45	0.5	9.83	0.76	
HCFBD-1	40	32.40	20	0.2	0.5	0.75	7.58	0.88	0.91
HCFBD-2	40	32.30	20	0.2	0.5	0.75	8.01	0.93	
HCFBD-3	40	32.20	20	0.2	0.5	0.75	8.23	0.96	
HCFBD-4	40	32.30	20	0.2	0.5	0.75	7.74	0.90	
HCFBD-5	40	32.50	20	0.2	0.5	0.75	7.63	0.88	

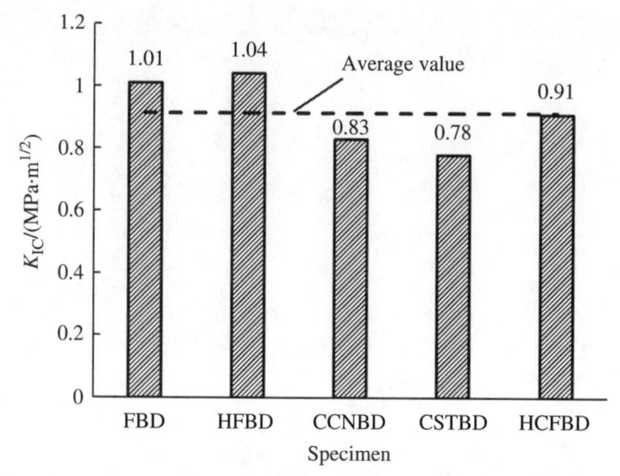

FIG. 2.10 Experimental data and their average of marble fracture toughness from disc specimens.

plane problem. There are major difficulties in the analysis process, especially in the case of composite or dynamic fracture analysis. In addition, the grooving of the CCNBD specimen is also limited by the width of the cutter. The geometry of the CSTBD specimen is the simplest, but the most difficult to process. Relatively speaking, the HCFBD specimen has a greater advantage. The disk specimen is relatively easy to process, and the groove width of the processing is small. The disk specimen with different groove lengths can be produced according to requirements, and can be analyzed according to the plane problem. However, the influence of some shape parameters (the size of the central aperture and the width of the groove, etc.) of the disc on the experimental values needs to be studied in depth.

2.1.6 Conclusions

This paper introduces the method of making the center grooving disc specimen, that is, first drilling the hole in the center of the disc, and then making the grooving. The minimum grooving width of the CSTBD and HCFBD specimens with grooving is 0.6 mm.

The machining difficulty of the five types of disc specimens is compared. The ungrooved FBD and HFBD specimens are easier to process, and the grooving of the CCNBD, CSTBD, and HCFBD discs is difficult. For disc specimens with grooves, CCNBD is relatively easy to machine, but the configuration is complex and difficult to analyze. The CSTBD specimen is the simplest to construct and the most difficult to machine. Relatively speaking, the HCFBD specimen is easy to make a smaller groove in width, and can produce a disc specimen with different grooves in length as required, which has greater advantages.

The rock fracture toughness values determined by the five types of disc specimens are between 0.78 and 1.04 MPa $m^{1/2}$, and the average value of all the disc specimens is about 0.91 MPa $m^{1/2}$. The test values of the FBD and HFBD specimens without grooves are higher than those of the CCNBD, CSTBD, and HCFBD specimens with grooves while the test values of the HFBD specimens are the highest and the test values of the CSTBD specimens are the lowest. The fracture toughness determined for the HFBD specimen is 1.33 times the test value of the CSTBD specimen.

For FBD and HFBD specimens without grooves, because the critical loads are highly susceptible to other factors, the fracture toughness determined by them is more discrete. For the specimens with grooving, the crack always occurs first from the tip of the grooving, and the test value of the fracture toughness is less discrete.

2.2 Fracture toughness of HCFBD

2.2.1 Introduction

Rock fracture toughness can be used to characterize the ability of rock to resist crack propagation and analyze rock fracture problems encountered in mine drilling, blasting, etc. (Tang, 1993; Li et al., 2003) At present, the study of rock fracture toughness is dominated by type I (open type) fractures (Zhang et al., 2009a, b; Cui et al., 2009). Three specimens for testing rock fracture toughness were recommended by the International Society for Rock Mechanics and Rock Engineering (ISRM) in 1988 (ISRM, 1988) and 1995 (ISRM, 1995). Among them, the cracked chevron notched Brazilian disc (CCNBD) recommended in 1995 has the advantages of small size, convenient loading, and low requirements on test equipment; it has been widely used. However, the CCNBD specimens are complex in configuration and cannot be analyzed according to the plane problem. The accurate calibration of the Y_{min} value in the fracture toughness formula is very difficult. Some researchers (Jia and Wang, 2003; Dai and Wang, 2004; Fan et al., 2011) have carried out a large number of calibrations, and such work is still in progress. Dai and Wang (2004) first considered the influence of the groove width on the calibration value of the minimum dimensionless stress intensity factor Y_{min} of CCNBD specimens by means of three-dimensional (3D) boundary element analysis. They made a conclusion that the larger the slot width, the larger the calibration value Y_{min}. For the fact that the CCNBD complex 3D configuration Y_{min} value is not easy to calibrate, some straight-disc disk specimens with relatively simple configurations were used to test the rock fracture toughness (Zhang and Wang, 2009; Fischer et al., 1996), such as cracked straight-through Brazilian disc (CSTBD) (Zhang and Wang, 2009) and holed-cracked flattened Brazilian disc (HCFBD) (Zhang and Wang, 2009; Fischer et al., 1996; Zhang and Wang, 2006). However, it is difficult to prefabricate straight cracks with a crack width of less than 1 mm in the rock specimen, and the influence of the prefabricated crack width on the test value is currently insufficiently studied. Chen Mian et al. (2004) The influence of crack width on the cracked specimens with water pressure in the center was studied by Zhang et al. (2002) and Chen and Zhang (2004), and they pointed out that there will be great errors when the artificially produced crack width was more than 0.8 mm. It is recommended to minimize the width of the prefabricated cracks. Dong and Xia (2004) studied the stress intensity factor of different crack widths and central aperture CSTBD specimens under combined loading conditions by the finite element method, theoretically improving the analytical value of the stress intensity factor of the specimen relative to the ideal central crack disk specimen. Obviously, an in-depth study of the effect of artificial prefabricated crack width is of great significance for the accurate testing of type I rock fracture toughness using straight cracked disc specimens.

In view of this, the HCFBD disc specimens were fabricated using marble, and the radial loading test and finite element calculation were carried out. The influences of prefabricated crack width on the fracture toughness of the HCFBD specimens were analyzed.

2.2.2 Test method and principle

The loading and crack parameters of the HCFBD specimen before and after crack fracturing are shown in Fig. 2.11. The specimen is subjected to a uniform pressure of P, the loading angle of the disc platform is 2β, the radius of the disc is R, the thickness of the disc is t, the radius of the center hole of the disc is r, the width of the artificial prefabricated crack is $2b$, and the length of the artificial prefabricated crack is $2a_0$. Fig. 2.11A shows that there is no propagation crack in the disk specimen before the HCFBD specimen is subjected to the pressure P, and only precracks exist. Fig. 2.1B shows the high stress concentration caused

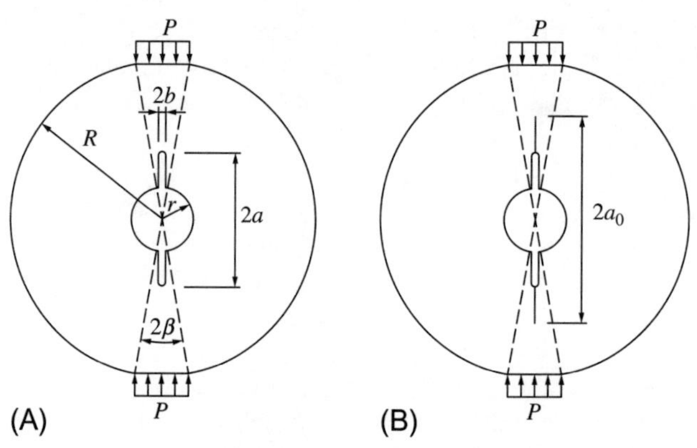

FIG. 2.11 Loading mode and parameters of the HCFBD specimen before and after the fracture process. (A) Before crack fracturing and (B) crack propagation process.

by the artificial prefabrication of the crack tip along the loading diameter line AB when the pressure P acts radially. As the load increases, the crack first intensifies at the crack tip stress concentration with a certain width, and then expands in a form of symmetric expansion along the loading diameter line AB direction.

When the influence of the prefabricated crack width is not considered, the stress intensity factor K_I of the crack front of the HCFBD specimen after being subjected to crack initiation can be expressed as Formula (2.12).

$$K_I = \frac{P}{\sqrt{R}t} Y\left(\beta, \alpha_0, \frac{r}{R}\right) \tag{2.12}$$

Where Y is the dimensionless stress intensity factor related to the geometry of the disk specimen. $\alpha_0 = a_0/R$ is the dimensionless crack length. Based on the principle of fracture mechanics, when the stress intensity factor of the crack front reaches a certain critical value, the crack will propagate. The corresponding critical value of the stress intensity factor is called the fracture toughness K_{IC}. If it is known that any load P in the crack propagation process of the loaded material and its corresponding crack length are calibrated, the dimensionless stress intensity factor can be calculated from Formula (2.12) to determine the rock fracture toughness.

When the artificial prefabricated crack width is small enough, we usually calculate the rock fracture toughness by the nondimensional stress intensity factor corresponding to the peak load and the artificial prefabricated crack length. However, when the artificial prefabricated crack width is large, if the width of the crack is still assumed to be an ideal zero-width crack, it will bring a large error to the test value. Therefore, how to determine the nondimensional stress intensity factor of a certain load and its corresponding extended crack during crack propagation is a key problem to consider the influence of crack width. Because the HCFBD specimen with the expansion of the artificial prefabricated crack tip crack, the dimensionless stress intensity factor will experience a process of rising to the maximum and falling again. The maximum dimensionless stress intensity factor Y_{max} corresponds exactly to the local minimum load P_{min} on the loading curve (Critical crack length corresponding to it) (Fan et al., 2011). Therefore, for HCFBD specimens where the prefabricated crack width cannot be ignored, the rock fracture toughness K_{IC} can be determined by the local minimum load P_{min} and the maximum dimensionless stress intensity factor Y_{max}, seen from Formula (2.13).

$$K_{IC} = \frac{P_{min}}{\sqrt{R}t} Y_{max}\left(\beta, \alpha_0, \alpha_B, \frac{r}{R}\right) \tag{2.13}$$

Where $\alpha_B = b/R$ is expressed as dimensionless crack width. The local minimum load P_{min} can be obtained by experiment, usually the minimum load that occurs after the first peak load. The maximum dimensionless stress intensity factor Y_{max} is related to the geometry of the disc specimen, and the influence of the prefabricated crack width of the specimen was considered.

2.2.3 HCFBD specimens with prefabricated cracks

The specimen material is white marble. It was collected from Nanyang, Henan province, China. The rock is a crystallite crystal structure. The grain size is between 0.1 and 0.3 mm. The main mineral composition is calcite, dolomite, and magnetite. The Poisson's ratio is 0.3, the Young's modulus of elasticity is 7.5 Gpa, and the density is 2.71 g/cm^3.

When the specimen is produced, the homogeneous stone is first made into a cylinder 84.4 mm in diameter and cut into a Brazilian disc rock sample with a thickness of 30 mm. Then, through the center drilling end face smoothing, the artificial sewing platform is made in four steps, as shown in Fig. 2.12. First, the centers of the disc specimens are drilled using a C260 lathe with a 15 mm diameter drill bit. Second, the SHM-200B double-face grinding machine is used to smooth the two end faces of the disc specimen to ensure that the parallelism deviation of the two end faces of the specimen is not more than 0.02 mm. Then, different saw blades are used to process cracks on the disc specimen. At the same time, in order to eliminate the influence of the artificial prefabricated crack tip, the crack tip is finely machined into an arc shape using a thin steel

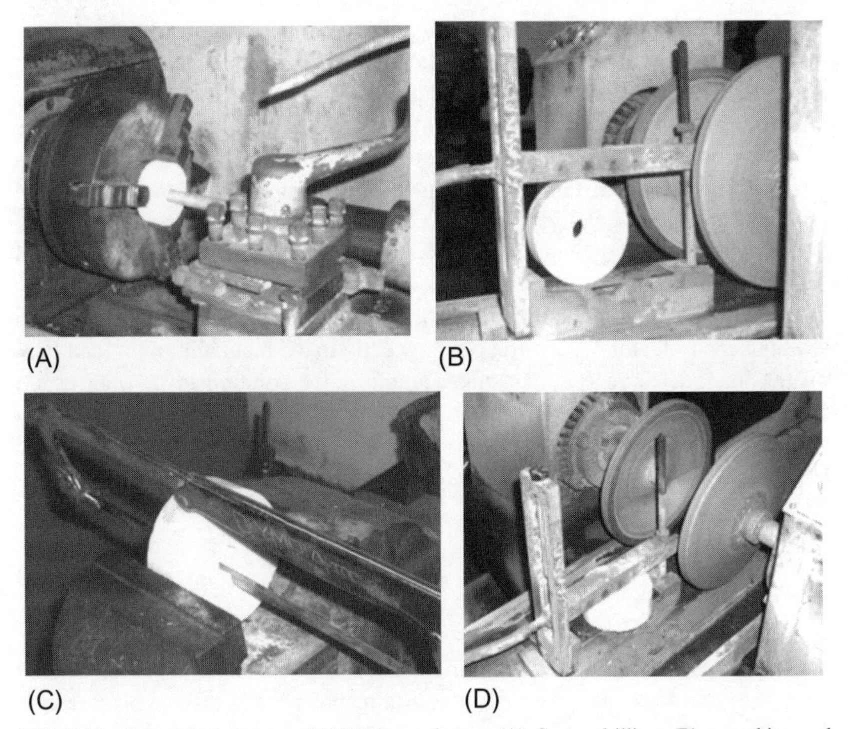

FIG. 2.12 Preparation process of HCFBD specimens. (A) Center drilling, (B) smoothing end faces, (C) artificial seaming, and (D) making platform.

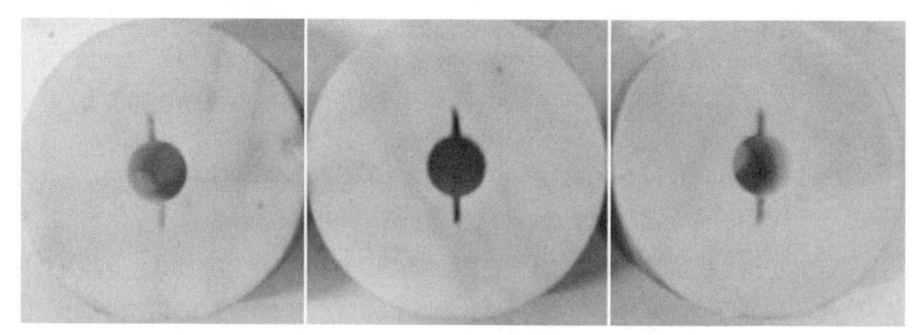

FIG. 2.13 HCFBD specimens having different prefabricated crack widths.

wire. Finally, the platform production of the disc specimen is completed by the SHM-200B double-face grinding machine.

The processed specimen is shown in Fig. 2.13. The center hole radius r of the HCFBD specimen is designed as 7.9 mm, and the platform loading angle is 20 degrees. The dimensionless crack length α_0 is 0.4, the prefabricated crack widths $2b$ are 0.55 mm, 0.64 mm, 1.08 mm, 1.3 mm, 1.42 mm, 1.94 mm, and 2.02 mm, respectively.

2.2.4 Calibration of maximum dimensionless SIF Y_{max}

The dimensionless stress intensity factors of the specimens with different artificial crack lengths under certain artificial prefabricated crack width and length conditions were calculated using ANSYS finite element software. The diameter of the disc specimen R is 42.2 mm, the center hole radius r is 7.9 mm, the platform loading angle is 20 degrees, the dimensionless crack length is 0.4, and it contains different prefabricated crack widths. The results of finite element calibration and curve fitting are shown in Fig. 2.14. Formula (2.14) gives the

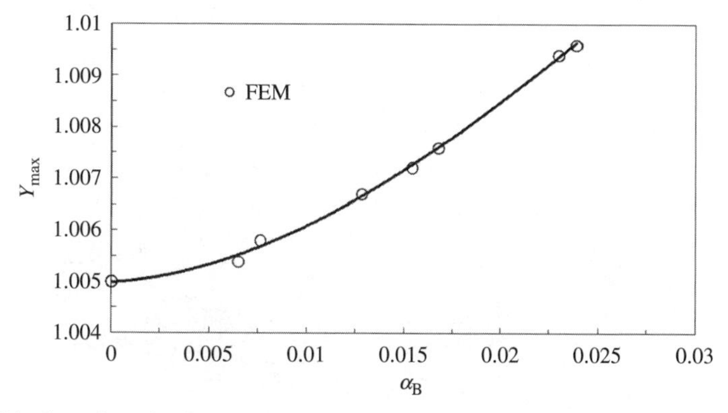

FIG. 2.14 Stress intensity factors of HCFBD specimens with different prefabricated crack widths.

relationship between the maximum dimensionless stress intensity factors corresponding to HCFBD specimens with different crack widths.

$$Y_{\max} = -100.82\alpha_B^3 + 9.5\alpha_B^2 + 0.0246\alpha_B + 1.005 \quad (2.14)$$

It can be seen from Formula (2.14) and Fig. 2.14 that the maximum dimensionless stress intensity factor $Y_{\max}$ increases with the increase of the artificial prefabrication crack width of the HCFBD specimen. Therefore, if the load of a specimen has been determined, and the influence of the width of the prefabricated crack is neglected, the $Y_{\max}$ value will inevitably be too small so that the test value of the rock fracture toughness determined by Formula (2.13) is too small.

2.2.5 Results and analysis

The loading test was carried out on the RMT-150B rock mechanics test machine in the Energy College of Henan Polytechnic University. The disc specimens were loaded by the radial splitting loading test using a displacement loading rate of 0.0005 mm/s. The test machine is shown in Fig. 2.15.

The load-displacement curve of the disk specimen is closely related to its crack propagation and failure mode. During the test of rock fracture toughness, three typical load-displacement curves are shown in Fig. 2.16.

In Fig. 2.16, case 2 corresponds to the case where the crack width is small. At this time, the crack is close to the ideal crack, and the tip expands rapidly after the crack occurs, causing the crack to completely break through the

FIG. 2.15 Loading system of the HCFBD specimen.

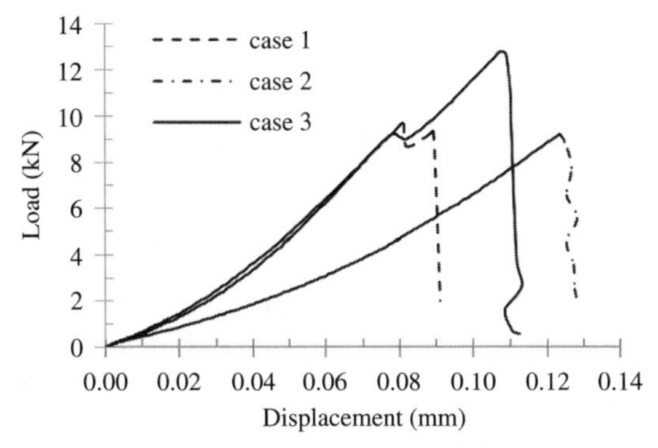

FIG. 2.16 The typical load-displacement curve of the HCFBD specimen.

specimen. In this case, there is only one peak load P_{max} and no local minimum load P_{min}. Formula (2.12) is used to determine rock fracture toughness. The load curves appearing in case 3 in Fig. 2.16 are mostly related to the poor quality of the specimen and the eccentricity of the specimen caused by natural defects during the loading process. This type of curve is not acceptable for the invalid test curve.

The validity of the experiment needs to be judged in combination with the load loading curve and the failure form of the HCFBD sample. The damaged form of the specimen is shown in Fig. 2.17. The crack of the specimen damaged by the effective test expands along the direction of the precrack, as shown in Fig. 2.17A. The failure crack of the invalid test deviates from the direction of the artificial precrack, as shown in Figs. 2.17B and C.

When the rock fracture toughness is determined, it is necessary to make the artificial prefabricated crack width less than 1 mm and refine the crack tip. At this time, the prefabricated crack is treated as an ideal crack, and the length of the crack is the length of the artificial crack. Also, the corresponding load is the first peak load loaded by the specimen. The rock fracture toughness is

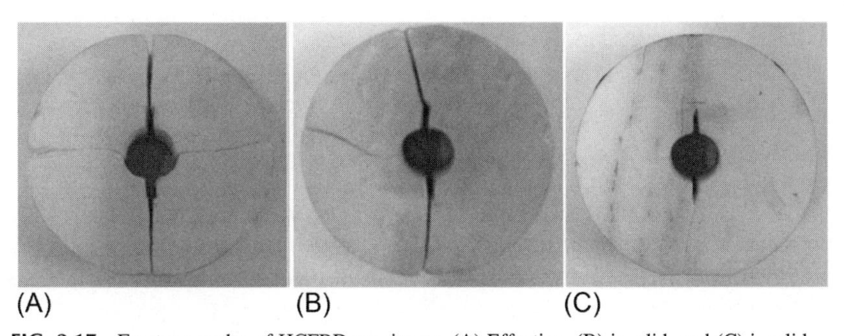

FIG. 2.17 Fracture modes of HCFBD specimens. (A) Effective, (B) invalid, and (C) invalid.

determined by the first peak load combined with the ideal crack numerical calculation results. Formula (2.12) can be used. However, it is not difficult to find out from the previous analysis of the load-displacement curves of HCFBD specimens. Formula (2.12) is only applicable in the case that the crack width is small (prefabricated crack width is less than 0.64 mm). When the prefabricated crack width is greater than 1 mm, the load-displacement curve shows a second inversion phenomenon. According to the previous analysis, the influence of the crack width must be considered. The effective P_{min} value of the HCFBD specimen is obtained by the splitting experiment. The value of Y_{max} calibrated with the finite element soft ANSYS is used to calculate the test value of the rock fracture toughness using Formula (2.13).

Table 2.2 shows the dimensions of the disc specimens used in the experiment, the critical loads measured by two methods, the dimensionless stress intensity factors, and the measured fracture toughness values. The specimen M indicates that the rock material is marble. A and G indicate specimens of different crack width types, and the number 1 indicates the number 1 of this type of specimen.

The rock fracture toughness values determined by Formulas (2.12) and (2.13) and their average values are shown in Fig. 2.18. The abscissa indicates the dimensionless crack width, and the ordinate indicates the fracture toughness value and the average fracture toughness value of the test rock calculated by Formulas (2.12) and (2.13). In Fig. 2.18, ${K_{IC}}^{(1)}$ represents the fracture toughness value calculated using Formula (2.12). ${K_{IC}}^{(2)}$ indicates the fracture toughness value calculated by Formula (2.13). ${K_{IC}}^{avg(1)}$ indicates the fracture toughness average value calculated by Formula (2.12). ${K_{IC}}^{avg(2)}$ indicates the fracture toughness average value measured by Formula (2.13).

It can be seen from Fig. 2.18 that the average fracture toughness value obtained by Formula (2.12) is 1.116 MPa $m^{1/2}$. The maximum fracture toughness value is 1.339 MPa $m^{1/2}$. The minimum fracture toughness value is 0.87 MPa $m^{1/2}$. Using Formula (2.13) to consider the calculation method of the influence of different crack widths, the average fracture toughness value is 1.447 MPa $m^{1/2}$, the maximum fracture toughness value is 1.60 MPa $m^{1/2}$, and the minimum fracture toughness value is 1.299 MPa $m^{1/2}$. It is easy to conclude from the test results that when the crack width of the specimen is constant, ${K_{IC}}^{(1)} > {K_{IC}}^{(2)} > {K_{IC}}^{avg(1)} > {K_{IC}}^{avg(2)}$. Therefore, if the artificial crack width is treated as an ideal crack, the test value of rock fracture toughness will be low. Especially when the artificial prefabricated crack width is greater than 0.64 mm, the value calculated by Formula (2.12) is 35% smaller than the value calculated by Formula (2.13) and the data is more discrete. When making HCFBD specimens, it is recommended that the width of the prefabricated cracks should not exceed 2 mm because the increase in crack width will affect the concentration of cracks and the discrete type of test data will be increased.

TABLE 2.2 Geometry size and experimental values of HCFBD specimens

Specimen no.	Specimen size					Formula (2.12)			Formula (2.13)		
	R (mm)	r (mm)	t (mm)	a_0	a_B	P_{max} (kN)	Y	$K_{IC}^{(1)}$ (MPa $m^{1/2}$)	P_{min} (kN)	Y_{max}	$K_{IC}^{(2)}$ (MPa $m^{1/2}$)
M-a1	42.2	7.9	30.58	0.4	0.0065	13.33	0.631	1.339	–	1.0054	–
M-a2	42.2	7.9	33.92	0.4	0.0065	14.27	0.631	1.292	–	1.0054	–
M-b1	42.2	7.9	33.78	0.4	0.0076	13.56	0.631	1.233	–	1.0058	–
M-b2	42.2	7.9	35.34	0.4	0.0128	11.93	0.631	1.037	10.06	1.0067	1.396
M-c1	42.2	7.9	28.94	0.4	0.0154	9.14	0.631	0.970	8.75	1.0072	1.482
M-d1	42.2	7.9	33.14	0.4	0.0168	11.56	0.631	1.072	10.81	1.0076	1.600
M-f1	42.2	7.9	31.66	0.4	0.0230	11.53	0.631	1.118	9.41	1.0094	1.460
M-g1	42.2	7.9	36.22	0.4	0.0239	10.26	0.631	0.870	9.57	1.0096	1.299

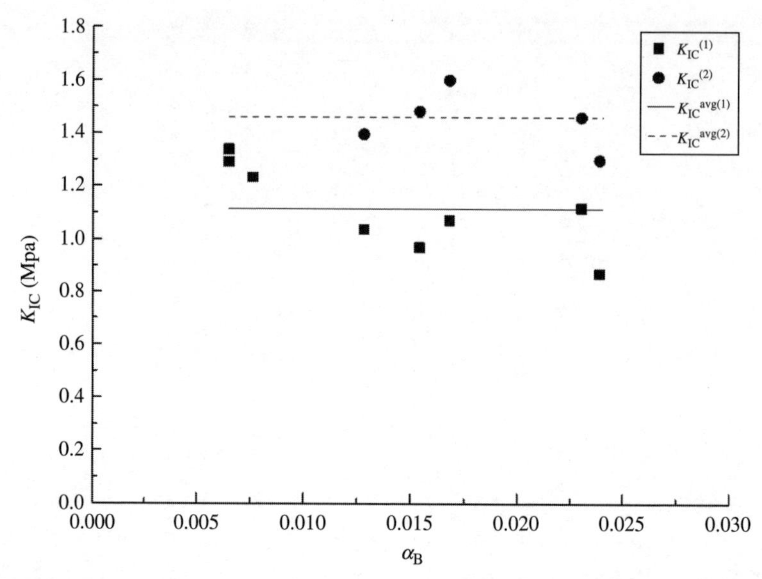

FIG. 2.18 Rock fracture toughness determined with disc specimens with different crack widths.

2.2.6 Conclusions

For HCFBD specimens having artificial prefabricated crack widths from 0.55 mm to 2.02 mm, based on the finite element analysis and the radial load curve and fracture form of the specimen, two different formulas are used to calculate the test value of rock fracture toughness. Research indicates:

For HCFBD disc specimens used in this study, when the prefabricated crack width is less than 0.64 mm, it is reasonable to use Formula (2.12) to determine the rock fracture toughness.

The finite element numerical analysis shows that the maximum dimensionless stress intensity factor Y_{max} of the HCFBD specimen increases with the increase of the artificial prefabrication crack width, as shown in Fig. 2.11 and Formula (2.14). When the radial loading test of the HCFBD specimen is carried out, if the load-displacement curve shows a secondary inversion phenomenon, the prefabricated crack cannot be treated as an ideal crack at this time, and the rock fracture toughness should be determined by Formula (2.13).

As the artificial crack with a certain width is used as the ideal crack to determine the rock fracture toughness, its test value will be lower than the actual value. A new method is suggested to determine the rock fracture toughness using Formula (2.13), which makes up for the deficiency of Formula (2.12) that cannot consider the influence of prefabricated crack width.

2.3 Crack length on dynamic fracture toughness

2.3.1 Introduction

In geotechnical engineering, crack initiation and propagation occur dynamically in many cases, such as the seismic resistance of buildings and structures, rock bursts in high stress areas of large hydropower stations and underground mines, etc. The dynamic fracture toughness of rock is an extremely important parameter when analyzing problems related to crack initiation and propagation in rock mass (Xie and Chen, 2004; Yu, 1992). Because of the convenience of loading, disc specimens can avoid the disadvantage of "disengagement" between specimens and supports under dynamic loading, such as three-point bending. It has been applied in the testing of rock fracture toughness in recent years (Fowell et al., 2006). However, the CCNBD (Fowell, 1995) specimens recommended by the International Society of Rock Mechanics (ISRM) in 1995 have complex 3D configurations and are not easy for carrying out dynamic analysis. Nakano and Kishida (1994) conducted a composite fracture test of ceramics and glass with a single-rod impact device using a disc specimen with a central crack. Lambert and Ross (2000) carried out impact tests on concrete and rock on the SHPB device with slot combined disc specimens. Li et al. (2006) measured the dynamic fracture toughness of rock by a side notch disc specimen. Zhang and Wang (2006) determined the dynamic fracture toughness of marble by using HCFBD, a circular hole platform with a central crack. HCFBD specimens are easy to fabricate, easy to test, and can avoid the inconvenience of making auxiliary devices for wedge loading for edge-notched disc specimens. It has been used to study the size effect of dynamic fracture toughness of rocks (Zhang et al., 2008).

At present, there is a lack of relevant research on the influence of the length of prefabricated cracks on the dynamic fracture toughness test values of rocks. However, there are still many problems in the shape and testing method of the disc specimen for testing the dynamic fracture toughness of rock reasonably, such as the reasonable matching relationship between the disc specimen and the compression bar loading system (Dong et al., 2006). This paper focuses on the effect of HCFBD specimens with different length cracks on the measured values of rock dynamic fracture toughness. Through a static test, the SHPB dynamic impact test, and a specimen fracture mode, the influence of fracture length on rock fracture toughness is analyzed.

2.3.2 Dynamic impact splitting test

2.3.2.1 Configuration and dimensions of specimens

The white marble is from Ya'an, Sichuan province, China. The rock sample is homogeneous and the structure is compact. According to petrographic analysis,

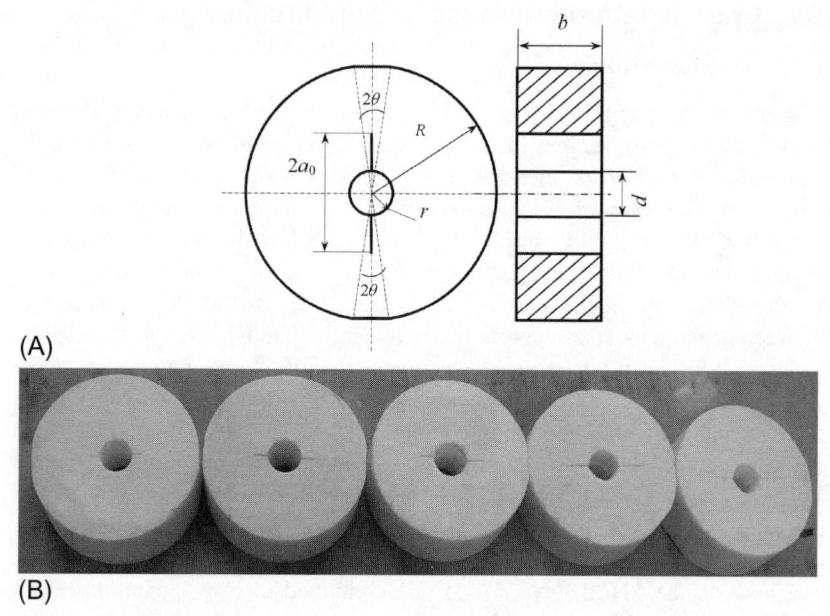

FIG. 2.19 Basic parameters and material objects of HCFBD specimens. (A) Basic parameters and (B) physical map.

the main compositions include calcite at about 95%, forsterite at about 4%, and magnetite at about 1%. Its maximum particle size is 1.2, 0.4, and 0.3 mm, respectively. The elastic modulus is 16.3 GPa, the Poisson's ratio is 0.3, and the density is 2730 kg/m^3.

The basic parameters and physical objects of the HCFBD specimen are shown in Fig. 2.19. In addition to the crack length $2a$, other nominal dimensions are taken. The disc radius R is 40 mm, the thickness t is 32 mm, the diameter d is 16 mm of the central hole, and the loading platform angle 2β is 20 degrees. Some of the specimens used in the test are shown in Fig. 2.19B, and the crack width of the specimen is less than 1 mm.

2.3.2.2 Dynamic test process

The dynamic impact splitting test of the HCFBD specimen was carried out on an SHPB system with a diameter of 100 mm. The schematic diagram of the SHPB system and the sample loading model are shown in Fig. 2.20. The elastic modulus of the compression bar is 210 GPa, the Poisson's ratio is 0.25, and the density is 7850 kg/m^3. The length of the incident bar is 450 cm, the length of the transmission bar is 250 cm, the length of the cylindrical projectile is 188 mm, the diameter is 100 mm, and the distance between the projectile and muzzle is 1.4 m.

During the test, a wave shaper was attached to the end of the projectile impact rod. Strain gauges were attached to an incident rod 1000 mm from

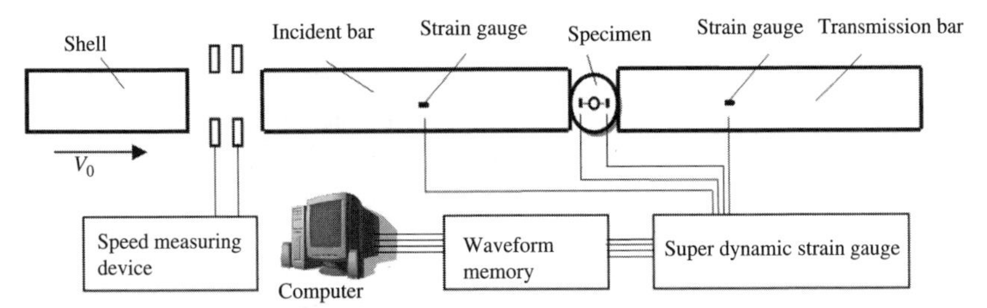

FIG. 2.20 Sketch of SHPB system and loading model.

the loading end of the specimen and a transmission rod 800 mm from the other loading end of the specimen. The strain gauge is used to record the incident strain signal ε_i, the reflected strain signal ε_r, and the transmitted strain signal ε_t produced by the stress wave action on the incident rod. The resistance strain gauge of BF120-3AA is pasted at the crack tip of both sides of the specimen to determine the initiation time t_f of the specimen. Reference is made to the study (Zhang and Wang, 2006) on the storage, archiving, related test parameters, and determination of rock dynamic fracture toughness K_{Id} for all strain signals.

2.3.3 Results and discussion

2.3.3.1 Comparison of dynamic and static fracture toughness

Typical strain gauge signals are shown in Fig. 2.21A. The incident wave strain signal ε_i, the reflected wave strain signal ε_r, and the transmitted wave strain signal ε_t are measured on the compression bar under radial impact. The fast Fourier transform method is used to filter the original strain wave curve. The filtered waveforms are shown in Fig. 2.21B.

The peak value of the transmission wave of the specimens with 155 mm diameter was nearly one order of magnitude larger than that of the specimens with 42 mm diameter (Zhang et al., 2008). However, the waveforms of the incident wave, the reflected wave, and the transmitted wave measured by the same size (diameter $D = 80$ mm) specimens with different crack lengths in this paper were basically the same, with little difference. This relatively small difference has great influence on the determination of dynamic fracture toughness. The load of the two ends of the specimen was calculated by the filtered waveform. Finally, the dynamic fracture toughness of rock was determined by combining the finite element and the strain gauge test. During dynamic loading, the projectile emission pressure was 0.15 MPa. The defined dynamic loading rate is equal to K_{Id}/t_f. The test values of initiation time and dynamic fracture toughness of the HCFBD specimens at different loading rates are shown in Table 2.3. From Table 2.3, it can be seen that when the projectile launching pressure is 0.15 MPa and other conditions are certain, when the ratio of the length to diameter of the central crack is $2a/D \in [0.31, 0.51]$, the rock dynamic fracture toughness test values are relatively discrete.

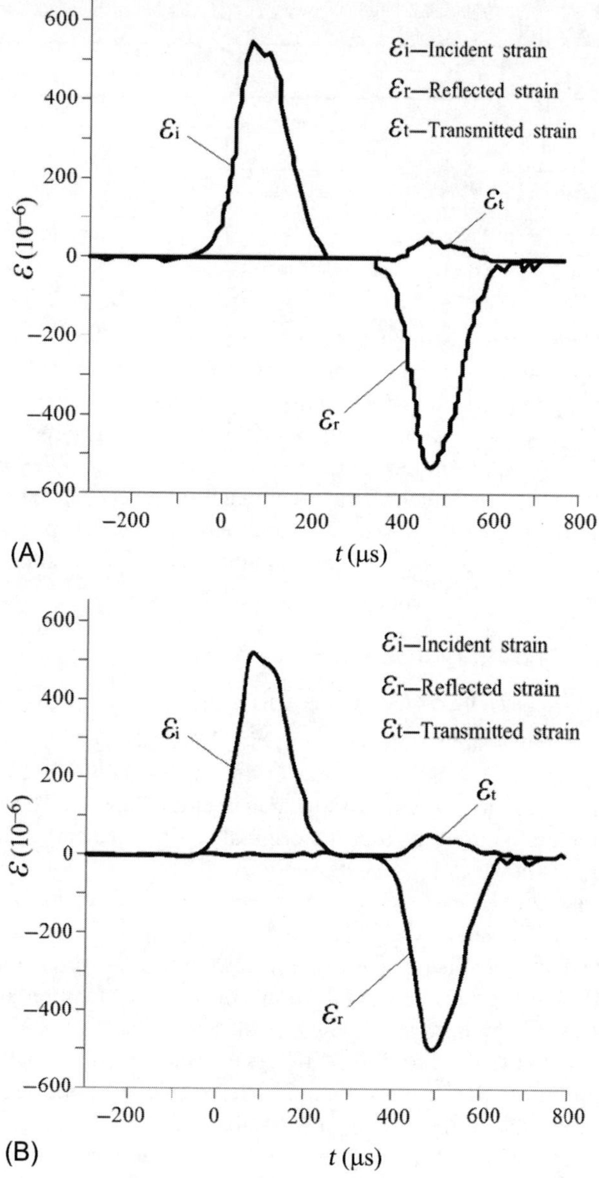

FIG. 2.21 Strain gauge signals before and after filtration. (A) Before filtering and (B) after filtering.

With the increase of the crack length, the crack initiation time of the specimen decreases slightly. When the average dynamic loading rate is 2.96×10^4 MPa m$^{1/2}$, the average dynamic fracture toughness (DFT) of marble tested by all specimens is 2.41 MPa m$^{1/2}$.

When Zhang and Wang (2007) used single-size specimens to study the size effect of rock static fracture toughness, the specimens with the same disc diameter

TABLE 2.3 Variation range of diameter and crack width parameter of disc specimen

Specimen number	Disc diameter *D* (mm)	2*a*/*D*	Cracking time t_f (μs)	Loading rate(K_{Id}/t_f) (10^4MPa $m^{1/2}$ s^{-1})	Dynamic fracture toughness K_{Id} (MPa $m^{1/2}$)
MII-01	80	0.50	76.0	3.88	2.95
MII-02	80	0.50	72.0	3.33	2.40
MII-03	80	0.50	72.0	3.07	2.21
MII-04	80	0.49	80.0	3.28	2.62
MII-05	80	0.39	85.0	3.44	2.92
MII-06	80	0.35	94.5	2.81	2.66
MII-07	80	0.36	93.0	2.23	2.07
MII-08	80	0.31	91.0	2.46	2.24
MII-09	80	0.34	81.0	2.31	1.87
MII-10	80	0.37	93.0	2.76	2.57
MII-11	80	0.39	76.0	3.11	2.36
MII-12	80	0.45	72.0	3.04	2.19
MII-13	80	0.51	85.0	2.73	2.32
Average value	**80**	**0.42**	**82.4**	**2.96**	**2.41**

and different crack lengths of marble were studied. The size of the specimen is shown in Fig. 2.19A, and the relative length of cracks is $2a/D \in [0.28, 0.51]$. The displacement loading method of 0.02 mm/min is used to load the platform directly. Fig. 2.22 compares the measured values of dynamic and static fracture toughness of marble with different crack lengths. From Fig. 2.22, it can be seen that the dynamic fracture toughness of rock is 2.6 times the mean static fracture toughness, and the dynamic fracture toughness is obviously greater than the static fracture toughness. In addition, the degree of dispersion of static fracture toughness is much less than that of the dynamic state. In the range of different crack lengths, the dispersion degree of dynamic fracture toughness values is very large. With the increase of fracture length, the test value of dynamic fracture toughness does not change significantly while the static fracture toughness obviously decreases with the increase of fracture length.

It should be pointed out that the dynamic fracture toughness in this paper is actually a constant representing the dynamic initiation and propagation of cracks in rock materials, that is, the initiation toughness, without considering

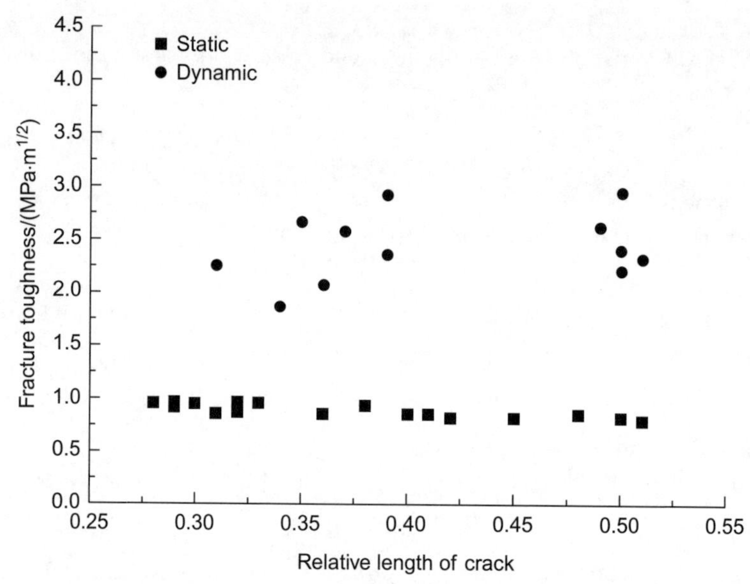

FIG. 2.22 Dynamic and static fracture toughness of marble samples of different crack lengths.

the effect of crack growth rate on the dynamic fracture toughness. In addition, the properties of rock materials, the fabrication accuracy of specimens, the dynamic loading rate, the temperature, the testing methods, and even the coupling factors between them may have a greater impact on the test results of rock dynamic fracture toughness, which should be paid attention to.

2.3.3.2 Fracture mode of specimens

The typical fracture modes of dynamic fracture and static splitting of specimens with different crack lengths are shown in Fig. 2.23. From Fig. 2.23, it can be seen that the failure of HCFBD specimens under static loading occurs along the loading diameter (diameter parallel to the direction of disk loading), and the fracture is basically symmetrical. The failure of HCFBD specimens is caused by tensile failure. Under dynamic loading, all specimens undergo macrofracture on the loading diameter plane, and the cracks propagate through strain gauges pasted at the crack tip. However, unlike static loading, when the disc specimen first breaks from the loading diameter into almost two equal parts, more fragments may be formed in the two parts, and the damage degree is relatively serious. Generally, under dynamic loading, the specimen is broken into 4–6 parts. In fact, the failure direction of other fracture surfaces except the loading diameter has no obvious law, that is, the direction of dynamic impact is not necessarily the same as that of other sections except the loading diameter.

The static fracture surface of the loading diameter is different from that of the dynamic fracture surface. The static fracture surface is smoother, and there are no bifurcations on the fracture surface. The macroscopic crack bifurcations

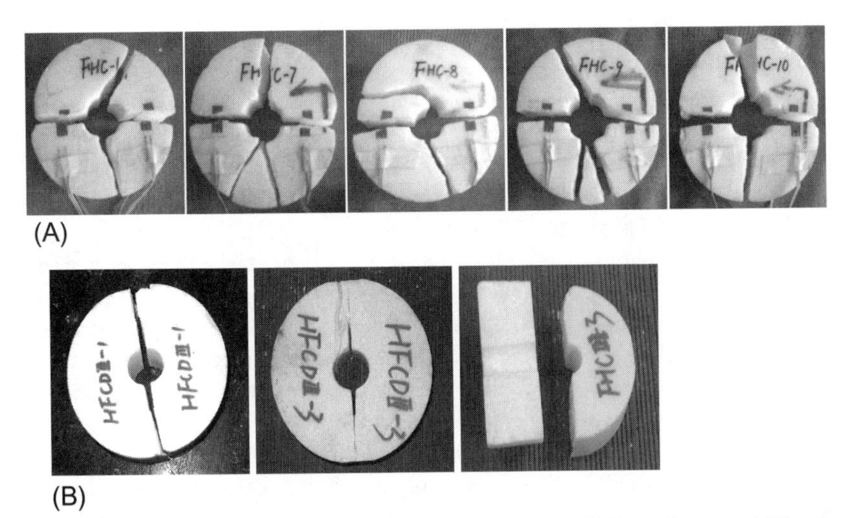

FIG. 2.23 Typical dynamic and static fracture modes of samples. (A) Dynamic case and (B) static state case.

are common near the dynamic fracture surface. The damage degree of the vertical section of the dynamic fracture surface is obviously greater than that of the static fracture. As for the reason why the dynamic fracture toughness of rock is greater than the static fracture toughness, Zhang and Yu (1995) gave a good explanation. When static loading is applied, the loading speed is small, and the main energy of the static fracture of rock is used for the expansion of the main crack. With dynamic loading, the loading rate is high, and some cracks will merge into the development of component fork cracks or two cracks. Because the generation and propagation of these bifurcated cracks and microcracks consume more energy than the static state, the energy of the main crack propagation is reduced, which hinders the growth of the main crack. This is also the reason why the dynamic fracture toughness of rock is greater than the static fracture toughness.

2.3.4 DFT irrespective of configuration and size

From the differences of mechanical parameters measured by different specimens, laboratory small specimens, and the performance of large-scale components in the field, it can be seen that the problems of specimen configuration and size effect are still pending (Bazant and Chen, 1997). Bazant and Chen (1997) believed that there exists a large fracture growth zone at the crack tip of quasibrittle materials such as rocks, which had a stable growth. Especially, the gradual release of storage energy due to the redistribution of stress, the existence of macrocracks, and the existence of large microcracks was the main reason for the size effect of quasibrittle material specimens.

For disc specimens with the same diameter and different crack lengths, there is actually no commonly considered size effect due to the different size of specimens, but the difference between specimens' configurations. In fact, under static loading, specimens with different local configurations can also be analyzed by Bazant's size effect theory. For example, Zhang and Wang (2007) used single-size specimens with variable cracks, and Wu et al. (2004a, b) used different configurations of CNCBD specimens to analyze the size effect. The real values of static fracture toughness of marble independent of size were obtained. Under dynamic loading, the fracture toughness test is more complex under the influence of specimen configuration and size effect. At present, there is very little research on the dynamic size effect. The dynamic size effect of concrete strength was tested by A and B. The causes of the dynamic size effect and the methods of eliminating the size effect were not thoroughly analyzed. Proposed a method to determine the dynamic fracture toughness of rocks in the space-time domain. It was pointed out that under dynamic loading, not only the influence of the fracture process area but also the influence of the incubation time related to the fracture process area should be considered. Unfortunately, did not propose an effective method to determine the fracture process zone and incubation time of rocks. Whether the established criteria can eliminate the size effect of the dynamic fracture toughness of rocks remains to be verified. The effect on the structure and size effect of rock dynamic fracture toughness needs extensive and in-depth research.

2.3.5 Conclusions

For specimens with a diameter of 80 mm, when the dynamic average loading rate is 2.96×10^4 MPa $m^{1/2}$ s^{-1} and the ratio of crack length to diameter is $2a/D \in [0.31, 0.51]$, the dynamic fracture toughness test value of rock does not change significantly with the increase of the central crack length. The static fracture toughness decreases with the increase of crack length.

The mean dynamic fracture toughness of marble tested by all specimens is 2.41 MPa·$m^{1/2}$, which is about 2.6 times the mean static fracture toughness. The dynamic fracture toughness is obviously greater than its static fracture toughness value, and the experimental value under dynamic loading is more discrete than that under static loading.

In static loading, the failure of specimens occurs along the loading diameter (diameter parallel to the direction of disk loading), and the basic failure is symmetrical. In the case of dynamic loading, besides macrofractures on the loading diameter surface, more fracture surfaces are generated, and there are crack bifurcations on the macrofracture surface, which is relatively rough. The main reason why the dynamic fracture toughness of rock is greater than the static fracture toughness is that the relative propagation energy of the macroscopic main crack decreases under dynamic loading.

Under dynamic loading, the effect of specimen configuration and size on the fracture toughness test is more complex. The influence of incubation time related to the fracture process zone should be considered as well as the range of the fracture process zone.

2.4 Crack width on fracture toughness

2.4.1 Introduction

With the ongoing development of human society, the demand for geothermal energy and other energy sources (oil and gas) is increasing. In these energy projects, hydraulic fracturing and blasting techniques are often used. Therefore, the evaluation of rock fracture toughness is becoming increasingly important (She and Lin, 2014; Deng et al., 2012). Experimental values of rock fracture toughness are known to be susceptible to the influences of many factors, such as the specimen geometry (Wu et al., 2004a, b), test loading conditions (Hua et al., 2016), test loading states (Hao et al., 2016), temperature (Li, 2013), test loading rate (Yin et al., 2015), and specimen configuration (Zhang et al., 2009a, b). Among the geometric factors affecting a specimen, the width of the prefabricated crack is a key factor that determines whether the fracture toughness can be accurately measured (Cui et al., 2009). To this end, researchers have carried out numerous studies. Zhang et al. (2002) studied the influence of the crack width on the stress intensity factor using experimental and finite element analyses and presented a numerical method for determining the relationship between the dimensionless stress intensity factor and the crack tip radius that addresses the problem of the zero crack width assumption in the analytical method. Zhang and Liang (2013) proposed a new method for determining rock fracture toughness using the minimum load and maximum dimensionless stress intensity factor. The method can eliminate the influence of the crack width on the fracture toughness of a disc specimen. Dai and Wang (2004) analyzed the influence of the crack width on the dimensionless stress intensity factor of cracked chevron notched Brazilian disc (CCNBD) specimens and concluded that the larger the crack width, the larger the dimensionless stress intensity factor. Zhou et al. (2005) compared the fracture behavior and toughness of straight and sharp center notched disc specimens. The tested values of the fracture toughness of straight crack-tip specimens were considered to be relatively high while the crack usually initiated at the corner of the notch end and the test results were scattered. Cui et al. (2015) used a Hopkinson pressure bar to perform dynamic impact tests of center notched disc specimens containing nonideal cracks. The results showed that when the width of the prefabricated crack was less than 1 mm, it was feasible to replace an ideal disc specimen with a non-ideal crack-disc specimen with a center notch. Some scholars (Ouinas et al., 2009; Bouiadjra et al., 2007) studied the stress intensity factor and crack propagation behavior, respectively, of composite plates with notched cracks.

Kolhe et al. (1998) studied the influence of the prefabricated crack width on chevron notched three-point bend test specimens (CBs). They found that when the crack width is sufficiently large, the dimensionless stress intensity factor does not exhibit a local minimum as the crack length changes.

In 2014, the International Society for Rock Mechanics (ISRM) recommended a new method for static fracture toughness testing. The method includes a new fracture toughness test specimen with a notched semicircular bend (NSCB) (Kuruppu et al., 2014). Because the specimen has a simple configuration and is easy to manufacture, it has been widely used in fracture toughness testing (Chong and Kuruppu, 1984; Dai et al., 2013; Gong et al., 2016).

In summary, research on the influence of the width of a prefabricated crack on the tested value of rock fracture toughness has mostly concentrated on disc specimens with a single fixed size, whereas research on the impact of the crack width on the fracture toughness of new NSCB specimens is lacking.

In view of this lack of research, a series of NSCB specimens with different radii containing prefabricated cracks with different widths was fabricated using limestone sheets with a uniform texture and fine grains to study the effect of the crack width on the experimental value of the fracture toughness of limestone for different specimen radii, with the goal of further improving the fracture toughness test method for NSCB samples.

2.4.2 NSCB three-point flexural test

2.4.2.1 Specimen preparation

Limestone from Jiaozuo, Henan Province, China, was used as the specimen material. The main content of the rock is calcite, the hardness is relatively low, the rock contains few natural microcracks, the structure is dense, the texture is uniform, and the mineral grains are small. The rock material density is $\rho = 2.785$ g/cm^3, the Young's modulus is $E = 63.938$ GPa, the Poisson's ratio $\nu = 0.27$, the longitudinal wave velocity is $C_d = 5.674 \times 103$ m/s, the compressive strength is 169.35 MPa, and the tensile strength is 5.47 MPa.

The specimen preparation process is divided into the following three steps, as shown in Fig. 2.24.

(1) *Core drilling:* as shown in Fig. 2.24A, cores with diameters of 50 mm, 75 mm, 100 mm, and 150 mm were drilled with sleeves of different diameters, and the cores were cut into Brazilian disc specimens with a thickness-to-diameter ratio of 0.3.

(2) *Specimen grinding:* as shown in Fig. 2.24B, the front and rear end faces of the cut Brazilian disc specimens were ground to a roughness of less than ± 0.02 mm, and the disc specimens were then cut into two half discs. The end surfaces of the half-disc specimens were ground again in the thickness direction to achieve a roughness of approximately 0.01 mm.

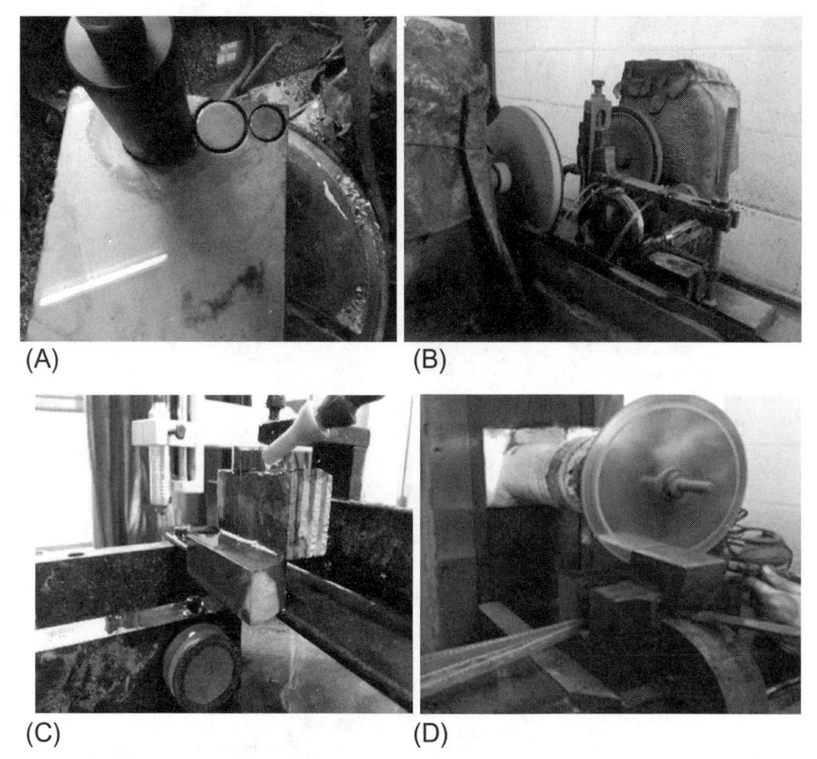

FIG. 2.24 Specimen preparation process. (A) Core drilling, (B) specimen grinding, (C) crack wire, and (D) crack disc.

(3) *Artificial cracking:* as shown in Fig. 2.24C and D, this test required the preparation of NSCB specimens with prefabricated cracks with widths varying from 0.25 mm to 2.15 mm. Because the widths of the prefabricated cracks varied considerably, the artificial cracking was performed using a method combining wire cutting and disc cutting. The wire cutting method was used for cracks with a width of 0.25 mm, and a cutting disc of a corresponding thickness was used for cracks with a width of more than 0.25 mm.

The prepared NSCB specimens are shown in Fig. 2.25. The specimen radii are 25 mm, 37.5 mm, 50 mm, and 75 mm, and the widths of the prefabricated cracks are 0.25 mm, 0.65 mm, 1.30 mm, 1.65 mm, and 2.15 mm. There are three specimens for each specification, for a total of 60 specimens.

2.4.2.2 Test equipment and test plan

The three-point flexural tests were carried out using an RMT-150B testing machine. The test loading device is shown in Fig. 2.26. The maximum vertical

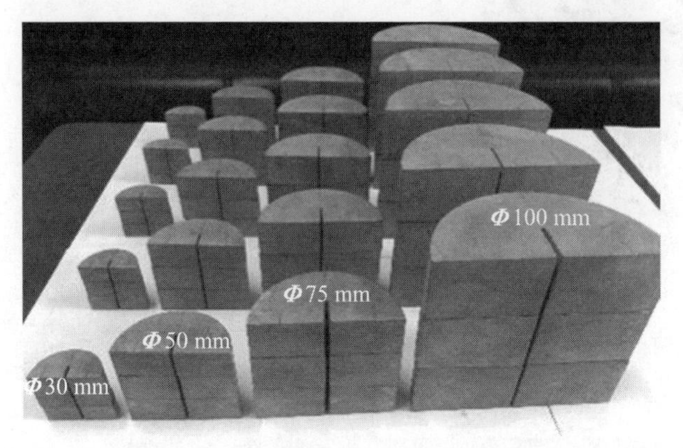

FIG. 2.25 NSCB specimens.

FIG. 2.26 NSCB specimen loading.

load of the tester is 1000 kN, and the maximum compression deformation is 20 mm; these values satisfy all of the measurement range and precision requirements of this test.

Displacement loading control is used to obtain a relatively stable load-displacement curve; to make the crack propagate sufficiently and stably along the prefabricated crack surface and obtain a more effective fracture toughness value, a relatively low loading rate of 0.0002 mm/s is employed.

2.4.3 Width influence on prefabricated crack

The load-displacement curves of sample NSCB specimens are shown in Fig. 2.27, where $2b$ is the width of the prefabricated crack and R denotes the

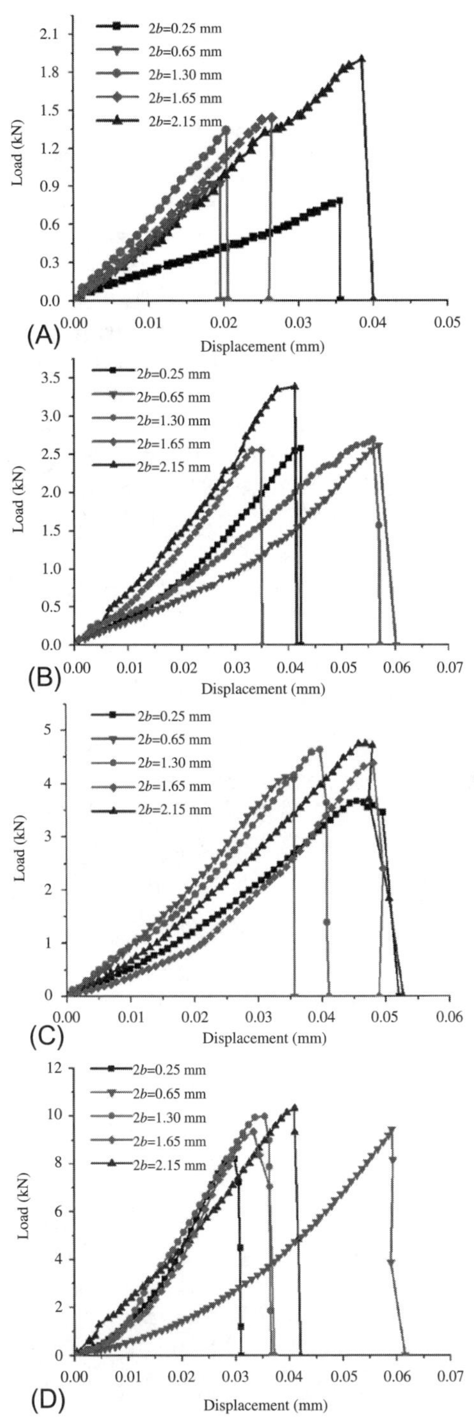

FIG. 2.27 Load-displacement curves of NSCB specimens. (A) $R = 25$ mm, (B) $R = 37.5$ mm, (C) $R = 50$ mm, and (D) $R = 75$ mm.

radius of the NSCB specimen. Sub-figures (A) to (D) show the influence of the crack width on the load-displacement curves of NSCB specimens with four different radii.

The results in Fig. 2.27A–D show the following:

(1) As the displacement increases gradually, the load-displacement curves of most specimens undergo a process of compaction, linear elasticity, and peak drop. After the load reaches the peak value, it decreases abruptly, and the specimen loses its load-bearing capacity in a short time, which demonstrates the brittle fracture characteristics of the specimen.
(2) For each specimen with a given size (radius), the width of the artificial prefabricated crack has a significant effect on the peak load of the NSCB specimen when it fractures. When the crack width increases from $2b = 0.25$ mm to $2b = 2.15$ mm, the peak loads of the specimens with $R = 25$ mm, 37.5 mm, 50 mm, and 75 mm increase from the original values of 0.788, 2.584, 3.652, and 8.212 kN to 1.9, 3.384, 4.748, and 10.33 kN, respectively, which represent increases of 141.12%, 30.96%, 30.01%, and 25.79%, respectively. Therefore, the specimen with $R = 25$ mm is affected the most by the crack width, and the specimen with $R = 75$ mm is affected the least. To investigate the influence of the crack width on the trend of the load-displacement curve, the linear elastic phases of the load-displacement curves are linearly fitted. The slopes of the curves for the specimens with the four radii are 20.69–90.46, 32.97–86.93, 93.03–215.27, and 173.48–333.58, respectively, and the corresponding maximum and minimum deviations are 77.13%, 62.07%, 56.78%, and 47.99%, respectively. Fig. 2.5 shows the tested and average values of the peak load P of the NSCB specimens with radii of 25 mm, 37.5 mm, 50 mm, and 75 mm, respectively.

As shown in Fig. 2.28, the width of the prefabricated crack has a significant effect on the peak load P of the NSCB specimen:

(1) Except for the specimens with a radius $R = 37.5$ mm, the peak load P of the NSCB specimens increases with increasing crack width. When the crack width increases from $2b = 0.25$ mm to $2b = 2.15$ mm, the average peak loads P of the specimens with $R = 25$ mm, 37.5 mm, 50 mm, and 75 mm increase from the original values of 0.986, 2.606, 4.229, and 8.35 kN to 1.544, 2.442, 4.791, and 9.231 kN, respectively, and the corresponding average peak loads P increase by 56.59%, –6.30%, 13.29%, and 10.55%, respectively.
(2) The larger the radius of a specimen, the less the average peak load P is affected by the width of the prefabricated crack. As the radius of the specimen increases from 25 mm to 7 mm, the percentage influence of the crack width decreases from 56.59% to 10.55%.
(3) For specimens with a radius $R = 37.5$ mm, the peak load P at the time of fracturing fluctuates slightly but does not increase steadily.

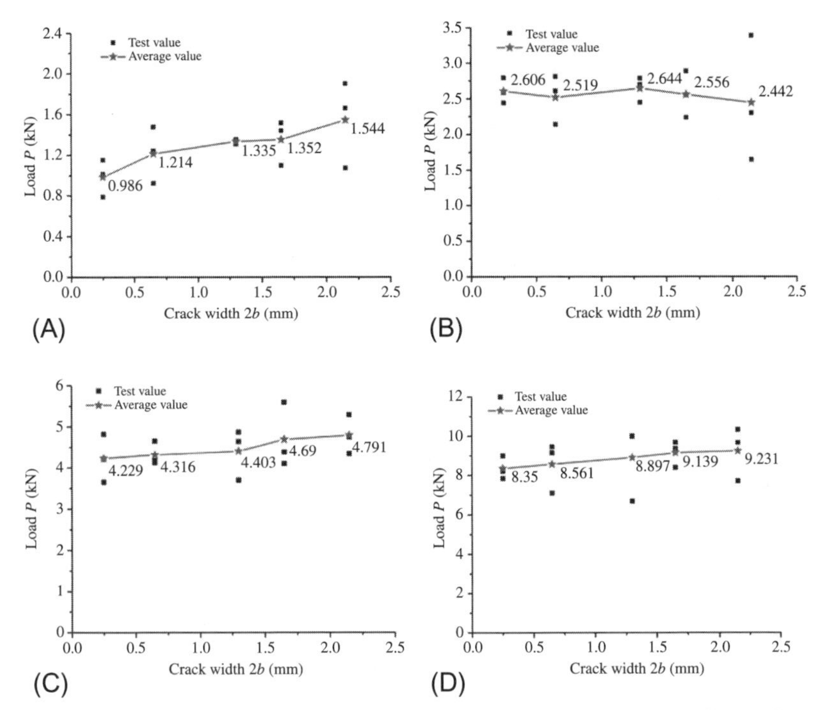

FIG. 2.28 Peak loads of the NSCB specimens. (A) $R = 25$ mm, (B) $R = 37.5$ mm, (C) $R = 50$ mm, and (D) $R = 75$ mm.

2.4.4 Width influence of cracks on tested fracture toughness

The average peak load of an NSCB specimen can be obtained by performing a three-point flexural test on it. The obtained average peak load is then combined with the dimensionless strength factor obtained using (Formula 2.16) (Table 2.4), which considers the influence of the crack width, to obtain the fracture toughness.

Fig. 2.29 shows the variations of the tested values of the fracture toughness of NSCB specimens with four radii as functions of the width of the prefabricated crack. The following observations can be made from the results.

(1) As with the average peak load, the fracture toughness of the specimens with the three radii other than $R = 37.5$ mm increase with increasing crack width. Specifically, when $R = 25$ mm, as the crack width $2b$ increases from 0.25 mm to 2.15 mm, the fracture toughness increases from 0.892 MPa m$^{1/2}$ to 1.464 MPa m$^{1/2}$, which is an increase of 64.13%. Similarly, the tested values of the fracture toughness of the specimens with $R = 50$ mm and $R = 75$ mm increase from 1.339 MPa m$^{1/2}$ and 1.435 MPa m$^{1/2}$ to 1.547 MPa m$^{1/2}$ and 1.628 MPa m$^{1/2}$, respectively, which are increases of 15.53% and 13.45%, respectively. The fracture surface analysis shows that

TABLE 2.4 Value of parameters in Formula (2.16)

Support spacing	Parameter value					
	k_0	k_1	k_2	k_3	k_4	k_5
$\alpha_S=0.5$	5.9	−2.776	−18.806	−7.413	28.427	11.109
$\alpha_S=0.6$	7.355	−3.683	−22.66	−6.03	34.191	14.464
$\alpha_S=0.7$	8.814	−5.088	−26.525	−1.74	39.977	18.582
$\alpha_S=0.8$	10.104	−5.278	−29.693	−1.507	45.064	20.545

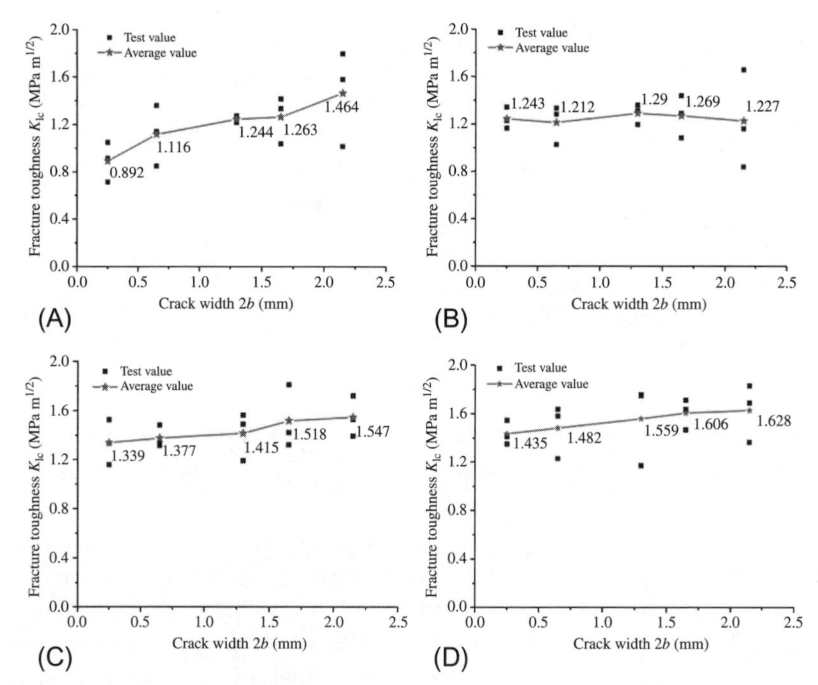

FIG. 2.29 Relationships between the tested fracture toughness and the width of the prefabricated crack. (A) $R = 25$ mm, (B) $R = 37.5$ mm, (C) $R = 50$ mm, and (D) $R = 75$ mm.

the NSCB specimen with a crack width of 0.25 mm formed a smooth shiny surface, and the fracture surface was nearly parallel to the surface of the prefabricated crack. In comparison, the fracture surfaces that formed in the NSCB specimens with crack widths of 0.75 mm or larger were rough, the crack propagation paths were relatively tortuous, and the fracture surfaces and the prefabricated crack planes formed angles of approximately 10 degrees. This may be because the grains in the limestone used in this test have a diameter of 0.3 mm. In the NSCB specimens with crack widths less than 0.25 mm, the crack widths are smaller than the grain diameters, and most of the fractures are transgranular. In contrast, the fractures that formed in the specimens with crack widths of 0.65 mm or larger were intergranular, which caused rough fracture surfaces.

(2) When the specimen radius R is 37.5 mm, the fracture toughness does not increase steadily with the increasing width of the prefabricated crack; rather, it fluctuates with an error between the maximum and minimum values of 6.44%. The author believes that this phenomenon can be attributed to the inaccuracy in the processing of NSCB specimens with $R = 37.5$ mm, which is compounded by the dispersion of the rock itself "masking" the effect of the prefabricated crack width on the fracture toughness. The measured fracture toughness values of each specimen show that prefabricated cracks with greater widths have greater dispersion of the tested fracture toughness values.

2.4.5 Method for eliminating influence of crack width

The ISRM has calibrated the dimensionless stress intensity factors of NSCB specimens with different crack lengths and dimensionless support spacing. The formula for the dimensionless stress intensity factor of an NSCB specimen is obtained by the curve fitting method, as shown in Formula (2.15). Clearly, Formula (2.15) only considers the influences of the length of the prefabricated crack and the support spacing on the dimensionless stress intensity factor. The author presented a new formula that considers not only the crack length and the support spacing, but also the influence of the crack width, as shown in Formula (2.16).

$$Y = -1297 + 9.516\alpha_S - (0.47 + 16.457\alpha_S)\alpha_0 + (1.071 + 34.401\alpha_S)\alpha_0^2 \tag{2.15}$$

$$Y = k_0 + k_1\alpha_{2b} + k_2\alpha_0 + k_3\alpha_{2b}^2 + k_4\alpha_0^2 + k_5\alpha_{2b}\alpha_0 \tag{2.16}$$

To determine whether the influence of the crack width on the tested value of the fracture toughness of the NSCB specimen should be considered, the dimensionless stress intensity factors of each NSCB specimen considering and not considering the influence of the crack width are obtained using Formulas (2.15) and (2.16), respectively. Combined with the peak load and the geometric parameters of the specimen, the fracture toughness in both cases is obtained, as shown in Fig. 2.30.

In Fig. 2.30, the curves of groups a, b, c, and d are the variations of the tested values of the fracture toughness of the NSCB specimens with radii of 25 mm, 37.5 mm, 50 mm, and 75 mm, respectively, with and without the influence of the crack width. The data in Fig. 2.30 show that:

(1) Regardless of whether Formula (2.15), which is recommended by the ISRM, or Formula (2.16), which is the corrected formula from reference 1 that considers the influence of the crack width, is used to calculate the dimensionless stress intensity factor, the variations of the tested values

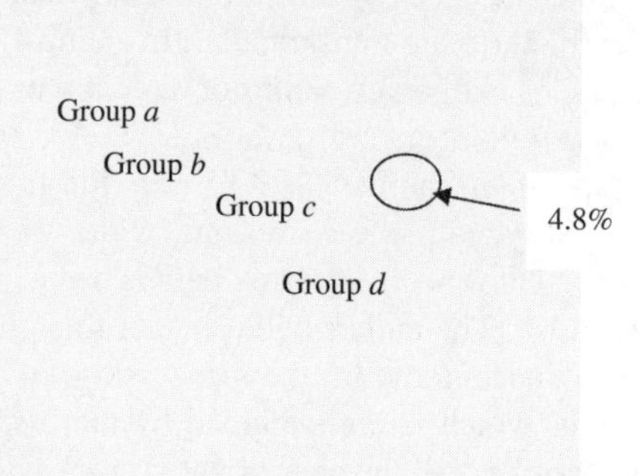

FIG. 2.30 Tested values of the fracture toughness calculated using the two formulas.

of the fracture toughness are consistent. The tested values of the fracture toughness of the three sizes of NSCB specimens other than those with a radius of 37.5 mm increase with increasing crack width.

(2) The fracture toughness value obtained using the dimensionless stress intensity factor from Formula (2.15) is smaller than the value obtained using Formula (2.16); that is, the test value of the fracture toughness tends to be smaller when the influence of the crack width is not considered. The maximum difference between the two can be up to 4.8%.

(3) The tested values of the fracture toughness show an obvious size effect. When the width of the prefabricated crack is less than 1.3 mm, the values of the fracture toughness increase continuously with increasing specimen radius. As the crack width increases, the size effect tends to weaken; that is, the difference between the tested values of the fracture toughness of specimens with different sizes decreases gradually.

(4) When the width of the prefabricated crack is greater than 1.3 mm, the tested values of the fracture toughness of the NSCB specimens with a radius of 25 mm (shown by the group *d* curves in Fig. 2.29) increase continuously and exceed the tested values of the fracture toughness of the specimens with a radius of 37.5 mm (group *c* curves). The tested values of the fracture toughness of the specimens no longer increase steadily as the specimen size increases. The author believes that the reason for this phenomenon is that the increases in the specimen size and the crack width increase the tested value of the fracture toughness. For specimens with small dimensions and large crack widths, the crack width relative to the specimen size becomes the dominating factor that affects the tested value of the fracture toughness. Therefore, when the crack width is greater than 1.65 mm, the tested values of the fracture toughness of the NSCB specimens with a radius of 25 mm are larger than those of the specimens with a radius of 37.5 mm.

To quantitatively compare the differences between the results of the fracture toughness obtained using Formula (2.15) and Formula (2.16), the values of the fracture toughness calculated using Formula (2.15) are used as a baseline.

TABLE 2.5 Relative errors between the results of Formulas (2.15) and (2.16)

Prefabricated crack width $2b$ (mm)	Specimen diameter d (mm)			
	25	37.5	50	75
0.25	0.90%	0.65%	0.60%	0.49%
0.65	1.73%	1.17%	0.95%	0.82%
1.3	3.07%	2.14%	1.65%	1.30%
1.65	3.69%	2.59%	2.02%	1.45%
2.15	4.80%	3.28%	2.52%	1.81%

The relative error between the values of the fracture toughness calculated by Formulas (2.15) and (2.16) can be obtained by subtracting the fracture toughness obtained using Formula (2.15) from that obtained using Formula (2.16) and then dividing by the value obtained using Formula (2.15), as shown in Table 2.5.

The results in Table 2.5 show that when the influence of the crack width is not considered, the tested values of the fracture toughness obtained using Formula (2.15) exhibit different degrees of error. If 1% is used as the upper limit of the relative error, it is recommended that NSCB specimens with a radius of not less than 50 mm and a crack width of not more than 0.65 mm be used to test the rock fracture toughness.

2.4.6 Conclusions

Three-point flexural tests were performed on NSCB specimens with different prefabricated crack widths. The tested values of the fracture toughness of the NSCB specimens increase with increasing crack width, and the maximum increase is 64.13%.The NSCB specimens with a radius $R = 25$ mm are affected the most by the change of crack width, and the NSCB specimens with a radius $R = 75$ mm are affected the least.

Increases in the prefabricated crack width and the radius of the specimen have significant but different influences on the tested values of the fracture toughness of NSCB specimens. An increase in the width of the prefabricated crack mainly affects the dimensionless stress intensity factor in the formula for calculating the tested value, whereas the increase in specimen size is reflected more by the size effect of the fracture toughness due to the difference in the fracture zones near the crack tip. This size effect exists in certain ranges of the specimen size and crack width. In this study, the specimen radius of 37.5 mm and the prefabricated crack width of 1.3 mm are found to form the boundary between the two mutually dominating factors.

2.5 Loading rate effect of fracture toughness

2.5.1 Introduction

Rock fracture toughness is a parameter used to characterize the ability of materials to resist crack initiation and propagation, including type I, type II, and type III fracture toughness, of which the type I fracture is the most common fracture mode. The fracture characteristics of rocks with different sizes are different, and there is an obvious size effect (Bazant, 2000; Zhang and Wang, 2007; Zhang et al., 2009a, b; Zhang et al., 2008; Feng et al., 2009). In addition, the loading rate is also one of the important factors affecting rock fracture toughness (Li, 1995; Wang et al., 2011; Zhang and Gao, 2012; Su et al., 2014; Liu et al., 2007; Zhou et al., 2013). A lot of research results have been achieved in the field of rock size effect and loading rate, both at home and abroad. Carried out a dynamic impact test on the Hopkinson pressure bar system using an HCFBD marble specimen with variable crack length and similar geometry. The influence of crack length and geometric similarity on the test value of dynamic fracture toughness was also analyzed. Feng et al. (2009) tested the size effect of rock dynamic fracture toughness by using a CSTFBD marble specimen, and the test results showed that the rock dynamic fracture toughness was affected by both the loading rate effect and the size effect. Some scholars (Li, 1995; Wang et al., 2011; Zhang and Gao, 2012) obtained results showing that the peak compressive strength increased logarithmically with the increase of loading rate through uniaxial compression tests of red sandstone at different loading rates. Su et al. (2014) has carried out uniaxial compression tests on sandstone after high temperatures at different loading rates. The results showed that the peak strength and strain of samples had an obvious loading rate effect and obeyed a positive linear relationship. Liu et al. (2007) carried out direct tension and Brazilian splitting tests at different loading rates using a self-developed test system, and the results showed that the tensile strength of rock increases with the increase of strain rate, and the critical tensile strain was positively correlated with strain rate. Zhou et al. (2013) obtained the relationship between the load-displacement curve, the tensile strength, the fracture morphology, and the loading rate by a hard and brittle marble Brazilian test and a scanning electron microscope test at different loading rates (0.001, 0.01, 0.1, 1.0, 10 kN/s). Meng et al. (2016) analyzed the variation of rock strength, deformation, and acoustic emission characteristics with specimen size and strain rate, and discussed the internal relationship between energy and specimen size. In summary, a lot of research achievements have been made on the size effect and loading rate of rock, but the influence of loading rate and specimen size on the fracture toughness of rock type I has still not been studied in depth. The research object is mainly concrete, and the rock samples used are mostly round disc specimens or beam specimens. In 2014, the International Society of Rock Mechanics (ISRM) recommended the use of the notched semicircular bend (NSCB) to test the fracture toughness of rocks. This specimen was first proposed by Kuruppu and Chong (Kuruppu et al., 2014; Chong and Kuruppu, 1984) in 1984. Because of its simple structure, easy processing, and convenient test operation, it has been widely used in the fracture toughness test of brittle materials (Dai et al., 2011; Kuruppu and Chong, 2012; Zhou et al., 2012; Li et al., 1993), and has been popularized to the recommended method under dynamic loading (Zhou et al., 2012). In this paper, a new fracture toughness test method proposed by the International Association of Rock Mechanics is used to test the fracture toughness of limestone under different sizes of NSCB specimens and loading rate ranges. The effect of the size and loading rate on the test value of the fracture toughness of limestone is revealed.

2.5.2 Specimen preparation

The test material is limestone, which is dark grey white and belongs to the typical sedimentary rock in the Jiaozuo area. In order to avoid the dispersion of the test results, the specimens were drilled from the same rock block along the vertical bedding direction and processed according to the recommended method of NSCB specimens. The flatness of the disc specimen was within 10 wires. The disc samples were cut into two semidiscs using the improved rock slitting machine, which had a cutting piece that was 200 mm in diameter and 0.3 mm in thickness. Then, a cutting piece with a diameter of 110 mm and a thickness of 0.15 was used to cut a central straight crack with a radius 0.2 times longer on the semidisc, whose width of the central straight crack was about 0.3 mm. The processes of sample processing are shown in Fig. 2.31A–D, including drilling, cutting, grinding, and slit four processes. The processing of different specimen sizes is shown in Fig. 2.31E.

2.5.3 Test process and data processing

2.5.3.1 Test method

The experiment was completed on the RMT-150B rock mechanics test system developed by the Wuhan Geotechnical Institute, Chinese Academy of Sciences. The system can realize uniaxial compression, triaxial compression, compression-shear failure, and a direct or indirect tension test. It can also carry out displacement (stroke) and load control mode. The load control loading rate is in the range of 0.001–100 kN/s and there are 12 levels of optional. Five geometrically similar NSCB specimens with diameters of 30, 50, 75, 100, and 150 mm were used in the experiment. Twenty specimens were made in each diameter, four of which were in a group. Each diameter sample was divided into five groups. The loading rates set by the five groups were 0.002, 0.02, 0.2, 2, and 10 kN/s, respectively.

Fig. 2.32 is a sample loading diagram in which the y sensor is used to monitor the vertical displacement of the sample. There are five pairs of dental grooves on the base for placing steel wire pads. Each pair of dental grooves is symmetrical with respect to the prefabricated cracks of NSCB specimens. The distance between the grooves (support spacing s) can be set to 18, 30, 45, 60, and 90 mm, respectively.

FIG. 2.31 Diagram of NSCB specimen processing. (A) Core drill, (B) rock cutting machine, (C) rock grinder, (D) improved rock cutting machine, and (E) NSCB specimen.

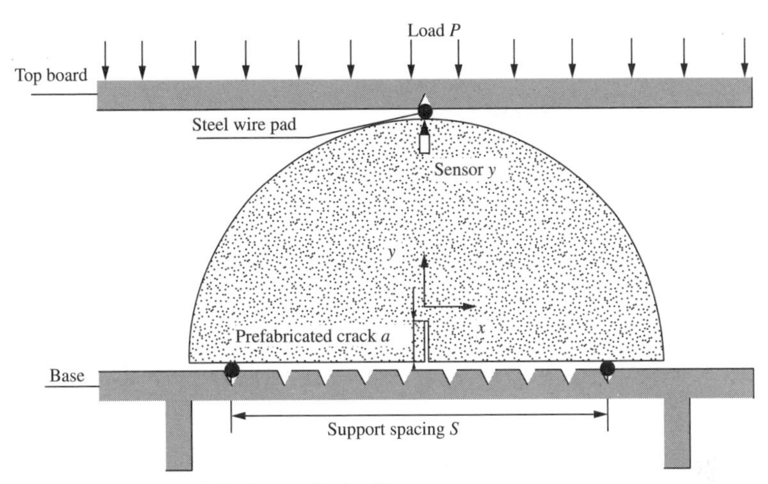

FIG. 2.32 Diagram of NSCB three-point bending test.

2.5.3.2 Fracture toughness calculation formula

The test data were calculated and processed according to a recommendation method of ISRM (Kuruppu et al., 2014). The fracture toughness K_{IC} of the NSCB specimen is calculated by Formula (2.17):

$$K_{IC} = \frac{P_{max}\sqrt{\pi a}}{2RB} Y' \tag{2.17}$$

In Formula (2.17):

$$\begin{aligned} Y' = & -1.297 + 9.516[S/(2R)] - \{0.47 + 16.457[S/(2R)]\}\beta \\ & + \{1.071 + 34.401[S/(2R)]\}\beta^2 \end{aligned} \tag{2.18}$$

$$\beta = a/R \tag{2.19}$$

In Formula (2.19), P_{max} is the peak load corresponding to specimen failure, Y' is a dimensional stress intensity factor, a is the length of the prefabricated crack for the NSCB specimen, R is the radius of the NSCB specimen, B is the thickness of the NSCB specimen, and S is the distance between the three-point bending tests and the two supporting points. In this paper, $S/(2R)$ is equal to 0.6, and β is equal to 0.2, which items are adopted according to the standard of reference [16].

2.5.4 Results and analysis

The results of the three-point bending fracture test for NSCB specimens at different loading rates and different sizes are shown in Table. 2.6. P_{max} is the peak load corresponding to the specimen failure. No. is the NSCB sample number. Y' is a dimensional stress intensity factor. K_{IC} is the fracture toughness of specimens. D is the sample diameter, and P_v is the actual loading rate of the test. It should

TABLE 2.6 Test results

No.	Y'	P_{max} (kN)	K_{IC} (MPa $m^{1/2}$)	No.	Y'	P_{max} (kN)	K_{IC} (MPa $m^{1/2}$)	No.	Y'	P_{max} (kN)	K_{IC} (MPa $m^{1/2}$)	No.	Y'	P_{max} (kN)	K_{IC} (MPa $m^{1/2}$)
A1	3.212	15.016	1.186	B6	3.210	8.202	1.185	C11	3.212	6.418	1.414	D16	3.208	3.868	1.584
A2	3.213	16.318	1.271	B7	3.212	9.332	1.255	C12	3.211	6.150	1.344	D17	3.206	3.588	1.488
A3	3.210	16.340	1.310	B8	3.211	9.512	1.399	C13	3.212	7.312	1.609	D18	3.212	3.586	1.460
A4	3.212	17.014	1.340	B9	3.212	9.992	1.441	C14	3.212	6.976	1.548	D19	3.212	3.240	1.366
A5	3.213	17.912	1.400	B10	3.213	8.990	1.289	C15	3.212	6.244	1.378	D20	3.212	3.240	1.352
A6	3.211	18.258	1.434	B11	3.209	9.218	1.397	C16	3.213	7.542	1.662	E1	3.260	0.572	0.501
A7	3.212	16.220	1.296	B12	3.211	9.396	1.410	C17	3.215	7.764	1.691	E2	3.363	0.684	0.564
A8	3.212	16.826	1.324	B13	3.212	10.888	1.585	C18	3.211	7.528	1.663	E3	3.252	0.826	0.720
A9	3.213	18.520	1.449	B14	3.212	9.196	1.354	C19	3.212	7.476	1.647	E4	3.252	0.670	0.578
A10	3.213	19.288	1.515	B15	3.213	11.602	1.668	C20	3.211	7.736	1.718	E5	3.279	0.696	0.577
A11	3.212	19.866	1.565	B16	3.213	10.924	1.605	D1	3.211	2.600	1.070	E6	3.295	0.894	0.744
A12	3.213	21.046	1.648	B17	3.214	12.546	1.816	D2	3.212	3.362	1.362	E7	3.331	0.990	0.832
A13	3.211	22.188	1.767	B18	3.211	11.502	1.681	D3	3.214	2.572	1.050	E8	3.268	0.828	0.726
A14	3.214	20.930	1.662	B19	3.213	10.824	1.610	D4	3.212	2.442	1.007	E9	3.256	1.008	0.861
A15	3.213	18.506	1.456	B20	3.211	11.032	1.626	D5	3.207	2.968	1.225	E10	3.338	0.942	0.771
A16	3.211	21.164	1.688	C1	3.213	5.354	1.183	D6	3.210	3.034	1.265	E11	3.257	0.788	0.696

A17	3.213	21.422	1.684	C2	3.212	5.172	1.135	D7	3.212	2.782	1.127	E12	3.357	1.092	0.878
A18	3.211	22.392	1.757	C3	3.212	5.142	1.134	D8	3.213	2.810	1.161	E13	3.314	1.088	0.910
A19	3.212	21.504	1.703	C4	3.212	5.550	1.221	D9	3.211	3.382	1.378	E14	3.332	1.112	0.900
A20	3.213	25.032	1.970	C5	3.213	5.708	1.250	D10	3.226	3.338	1.342	E15	3.349	1.092	0.911
B1	3.212	9.104	1.309	C6	3.212	6.478	1.433	D11	3.211	3.060	1.252	E16	3.255	1.196	1.039
B2	3.212	8.170	1.180	C7	3.213	5.464	1.198	D12	3.212	3.222	1.318	E17	3.307	1.074	0.922
B3	3.212	8.798	1.272	C8	3.213	5.548	1.217	D13	3.218	3.268	1.333	E18	3.324	1.588	1.274
B4	3.212	7.534	1.121	C9	3.213	6.262	1.371	D14	3.206	2.900	1.204	E19	3.395	1.344	1.045
B5	3.215	9.002	1.297	C10	3.213	6.302	1.388	D15	3.219	3.144	1.27	E20	3.284	1.462	1.245

be noted that the loading rates were set to be 0.002, 0.02, 0.2, 2, and 10 kN/s, but the average loading rates were 0.002, 0.02, 0.2, 1.96, and 9.52 kN/s. The average loading rates for the specimens with a diameter of 30 mm were 1.83 and 6.86 kN/s under the setting loading rates of 2 and 10 kN/s.

2.5.4.1 Load-displacement curve

Fig. 2.33A shows the load-displacement curves of the first group of four NSCB specimens with a diameter of 150 mm at a loading rate of 0.002 kN/s. It can be seen that the slope and peak load of the load-displacement curves of four specimens with the same diameter and loading rate are different to some extent, which is due to the heterogeneity of the rock itself and shows a certain degree of discreteness. The peak loads of A1, A2, A3, and A4 specimens are 15.016, 16.318, 16.34, and 17.014 kN, respectively, and the dispersion coefficient is 4.47%.

Fig. 2.33B is a load-displacement curve corresponding to 5 diameter NSCB specimens at a loading rate of 0.002 kN/s. It can be seen that the phase changes of the load-displacement curves of specimens with different sizes are different under certain loading rates. With the increase of specimen size, the load-displacement curve of the specimen changes from two stages of linear elasticity failure to three stages of compaction-linear elasticity failure. The brittleness of small specimens is strong. The large specimens exhibit certain plasticity at the initial stage, and there is a certain stage of microcrack compaction.

The load-displacement curve corresponding to different loading rates under different sizes of NSCB specimens is shown in Fig. 2.33C. It can be seen that the load-displacement curve of the NSCB specimen with a diameter of 30 mm is only two stages of cable elastic failure. It can be seen from Fig. 2.33D that most of the NSCB specimens with 50 mm only pass through two stages of linear elasticity failure, except for the load-displacement curves of individual specimens. However, with the size of specimens from small to large, the compaction stage begins to appear, and the proportion of the compaction stage in the curve before the peak load point tends to be higher. From Fig. 2.33E–G, it can be seen that the load-displacement curves of the NSCB specimens with 75 mm, 100 mm, and 150 mm have undergone three stages of compaction-linear elasticity failure under different loading rates. The NSCB specimen breaks suddenly after reaching the peak load, showing typical brittle failure. The postpeak failure time of the 30 mm NSCB specimen is shorter, which is quite different from the deformation characteristics of standard rock specimens in the uniaxial compression process. In the three-point bending test of NSCB, the crack propagates rapidly and breaks down after initiation of the prefabricated crack tip, showing stronger brittleness.

It can also be seen from Fig. 2.33 that the slope of the load-displacement curve in the linear elastic stage does not change regularly with the change of sample size. The slope of the load-displacement curve in the linear elastic stage varies with the loading rate in a certain size, but the slope of the curve increases with the increase of the loading rate in terms of the minimum and maximum loading rates, and there is a certain correlation between them.

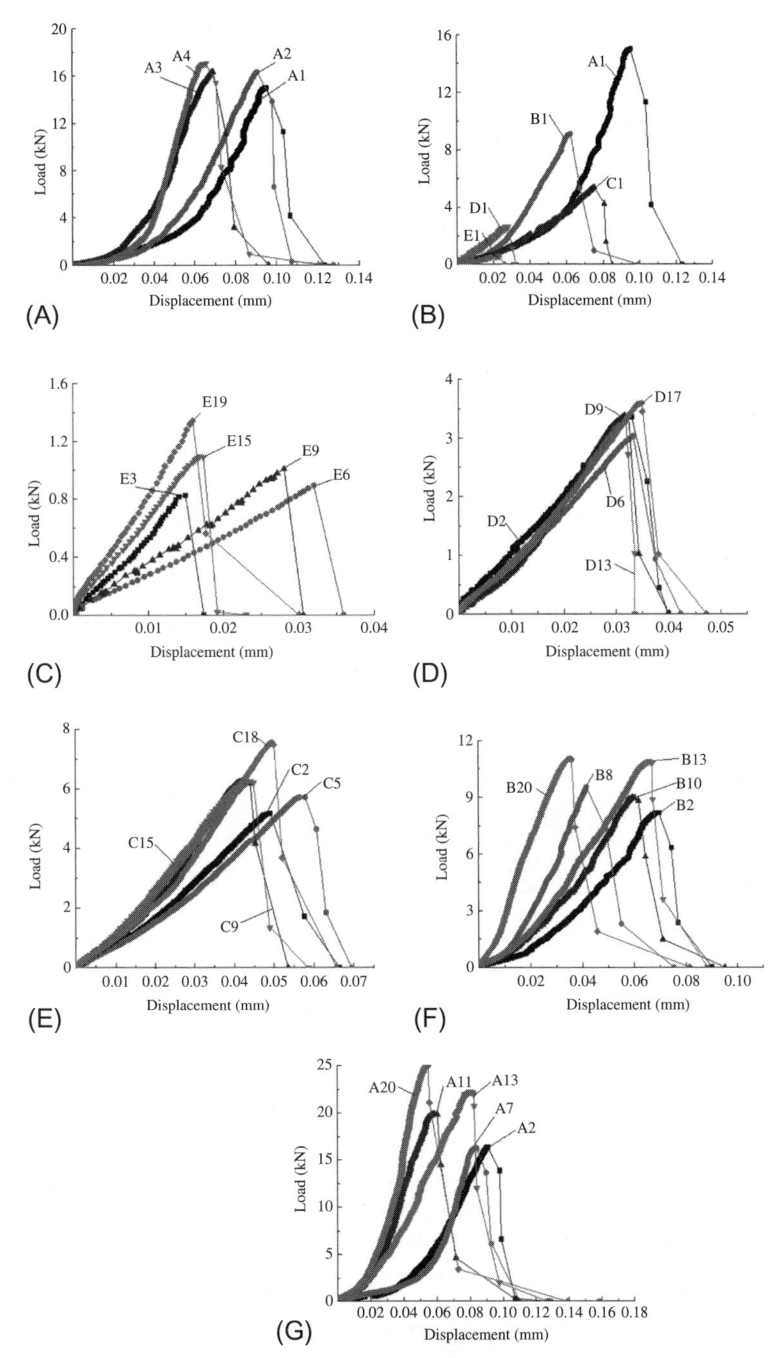

FIG. 2.33 Load-displacement curves. (A) Φ150mm $P_v = 0.002$ kN/s, (B) $P_v = 0.002$ kN/s, (C) Φ30 mm, (D) Φ50 mm, (E) Φ75 mm, (F) Φ100 mm, (G) Φ150 mm.

It can also be seen from Fig. 2.33 that the maximum vertical displacement of the NSCB specimen tends to increase as the size increases. The maximum vertical displacement of Φ30 mm, Φ50 mm, Φ75 mm, Φ100 mm, and Φ15.0 mm specimens is in the range of 0.015–0.037 mm, 0.03–0.05 mm, 0.05–0.07 mm, 0.07–0.09 mm, and 0.09–0.14 mm, respectively. It can be understood that the maximum vertical displacement of large specimens is large because of their strong bearing capacity and large deformation before failure. However, the maximum vertical displacement of NSCB specimens is not related to the loading rate when the specimen size is fixed.

2.5.4.2 Fracture toughness test value

Loading rate effect on fracture toughness

Fig. 2.34 shows the relationship between the fracture toughness test value and the loading rate of the NSCB specimen under different sizes. It can be seen from Fig. 2.34 and Table 2.6 that the fracture toughness test values of NSCB specimens are closely related to the loading rate, and the characteristics of the fracture toughness test values under the loading rate can be characterized by the average values of the test results of four specimens. From the overall rule of the test results, the fracture toughness test value of NSCB specimens is positively correlated with the loading rate, and increases logarithmically with the increase of loading rate. However, the fitting logarithm formula is different under different sizes, which is affected by the size of the NSCB specimens.

Formula (2.34) is used to calculate the average value of the fracture toughness test of NSCB specimens at different loading rates under certain sample sizes. Compared with the average value of the fracture toughness test at a loading rate of 0.002 kN/s, the increase e is shown in Table 2.7.

$$e = \frac{\overline{K}_{IC}^{i} - \overline{K}_{IC}^{0.002}}{\overline{K}_{IC}^{0.002}} \tag{2.20}$$

In Formula (2.20), $\overline{K}_{IC}^{0.002}$ represents the average fracture toughness at a loading rate of 0.002 kN/s, and $\overline{K}_{IC}^{i}$ represents the average fracture toughness test values at different loading rates.

The relationship between the average increment e of the fracture toughness test value and the loading rate for specimens of different sizes is shown in Fig. 2.35. From Table 2.7 and Fig. 2.35, it can be seen that different loading rates have different effects on the fracture toughness test values of NSCB specimens under certain sample sizes. When the loading rate (0.02, 0.2 kN/s) is low, the increase e is small, and the increase e increases logarithmically with the increase of the loading rate. It can be seen from Fig. 2.35 that the relationship between the increase of e and the loading rate is affected by the size of the sample. The correlation coefficient R^2 of the logarithmic function is larger when the sample size is larger (Φ150 mm, 100 mm, 75 mm), and the three formulas are similar. It shows that the relationship between the increase e and the loading rate

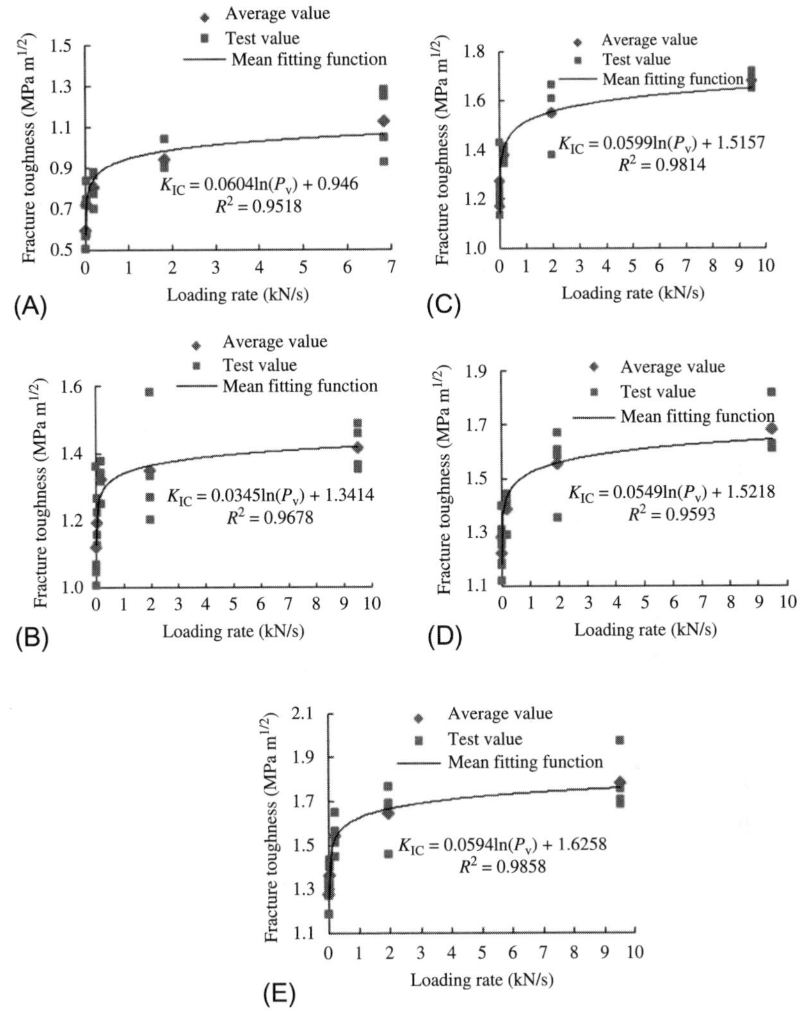

FIG. 2.34 Relationship between fracture toughness and loading rate. (A) *Φ*30 mm, (B) *Φ*50 mm, (C) *Φ*75 mm, (D) *Φ*100 mm, and (E) *Φ*150 mm.

is less affected by the sample size when the sample size is larger. When the sample size is small (*Φ*50 mm, *Φ*30 mm), the correlation coefficient R^2 of the logarithmic function is small, and the relationship between the increase e and the loading rate is greatly affected by the size.

Size effect on fracture toughness

The relationship between the fracture toughness test values of NSCB specimens and their sizes under different loading rates is shown in Fig. 2.36. It can be seen

TABLE 2.7 The amplification coefficient e under different loading rates

Diameter (mm)	150	100	75	50	150	100	75	50	Diameter (mm)	30	30
Loading rate (kN/s)	Mean fracture toughness value (MPa m$^{1/2}$)				Increase e (%)				Loading rate (kN/s)	Mean fracture toughness value (MPa m$^{1/2}$)	Increase e (%)
002	1.28	1.22	1.17	1.12	–	–	–	–	0.002	0.59	–
0.02	1.36	1.28	1.27	1.19	6.78	5.21	9.08	6.44	0.02	0.72	21.82
0.20	1.54	1.38	1.38	1.32	20.96	13.43	18.07	17.88	0.20	0.80	35.69
1.96	1.64	1.55	1.55	1.35	28.71	27.24	32.61	20.07	1.83	0.94	59.12
9.52	1.78	1.68	1.68	1.42	39.29	37.89	43.79	26.20	6.86	1.12	89.87

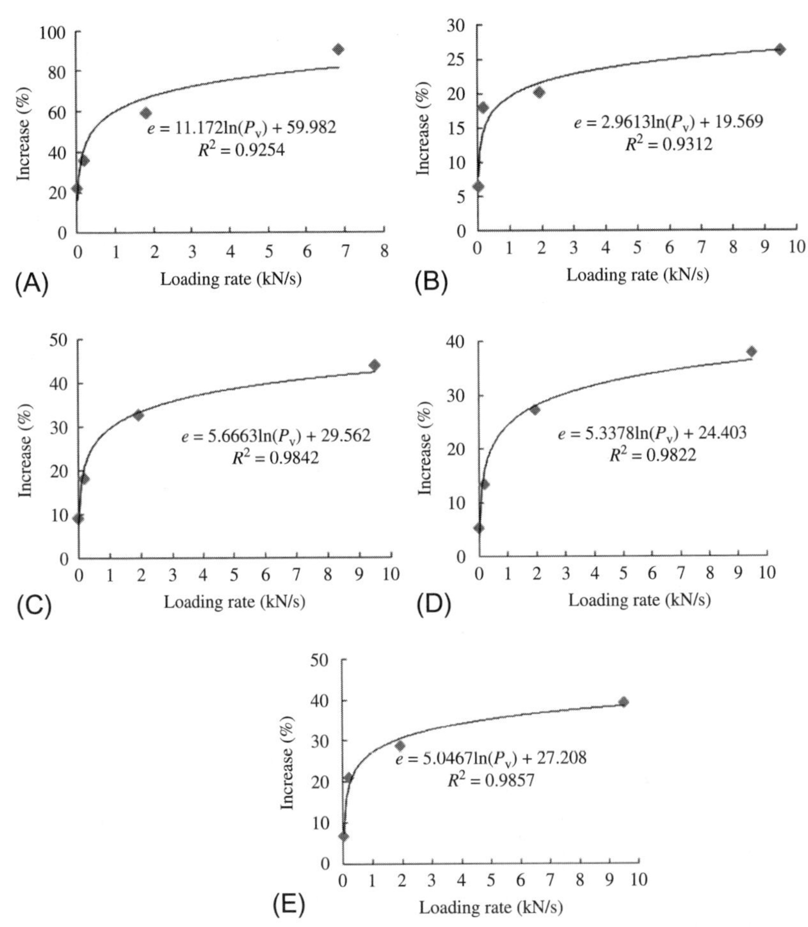

FIG. 2.35 Relationship between amplification coefficient of fracture toughness e and loading rate. (A) Φ30 mm, (B) Φ50 mm, (C) Φ75 mm, (D) Φ100 mm, and (E) Φ150 mm.

from Fig. 2.36 that the test values of the fracture toughness of NSCB specimens are closely related to the size. Under a certain loading rate, the fracture toughness test value of NSCB specimens increases with the increase of specimen size, but the increasing trend is different under different loading rates, which is affected by the loading rate.

In order to investigate the influence of different specimen sizes on the fracture toughness test values of NSCB specimens under different loading rates, the average value of the fracture toughness test value calculated by Formula (2.21) is compared with the average value test of fracture toughness with a specimen diameter of 30 mm or 50 mm, and the increased rate of fracture toughness f is obtained. The results are shown in Table 2.8.

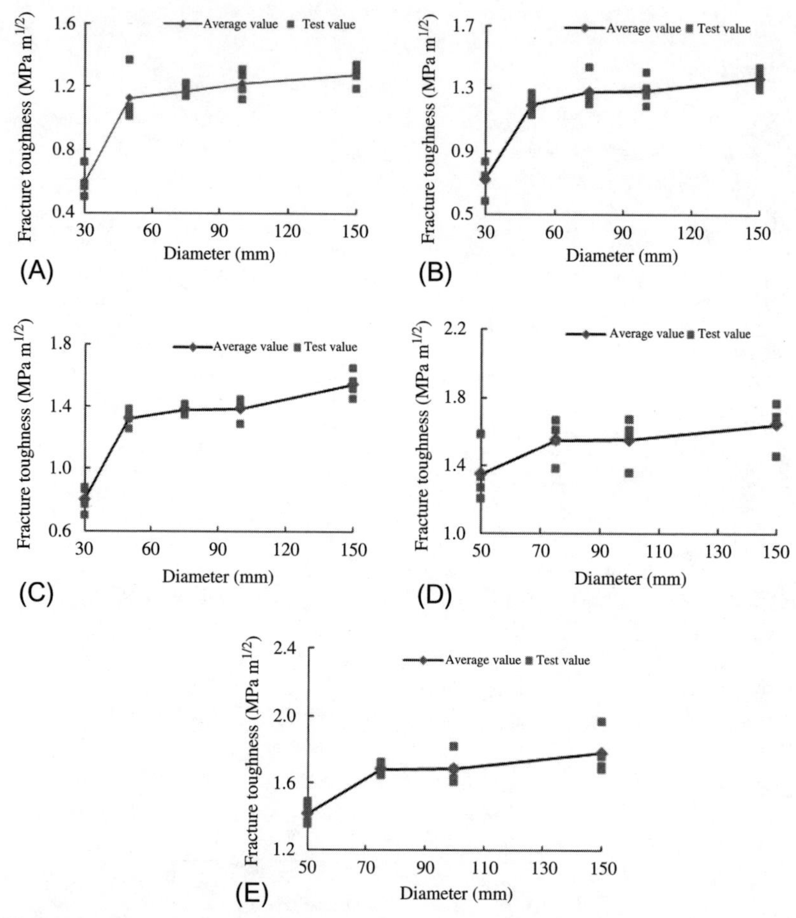

FIG. 2.36 Relationship between fracture toughness and specimen size. (A) $P_v = 0.002$ kN/s, (B) $P_v = 0.02$ kN/s, (C) $P_v = 0.2$ kN/s, (D) $P_v = 1.96$ kN/s, and (E) $P_v = 9.52$ kN/s.

$$f = \frac{\overline{K}_{IC}^{j} - \overline{K}_{IC}^{D}}{\overline{K}_{IC}^{D}} \tag{2.21}$$

In Formula (2.21), $\overline{K}_{IC}^{D}$ represents the average value of fracture toughness test values at loading rates of 0.002, 0.02, and 0.2 kN/s for a specimen diameter of 30 mm. When the loading rate is 1.96 and 9.52 kN/s, the mean value of the fracture toughness test value for the specimen diameter of 50 mm is indicated. $\overline{K}_{IC}^{j}$ indicates the average fracture toughness under different sizes. It is worth noting that the loading rates are 0.002, 0.02, 0.2, 2, and 10 kN/s, but the average loading rates are 0.002, 0.02, 0.2, 1.96, and 9.52 kN/s. The actual average loading rates of specimens with a diameter of 30 mm are 1.83 kN/s and 6.86 kN/s,

TABLE 2.8 The amplification coefficient f under different sizes

Loading rate (kN/s)	0.002	0.02	0.20	1.96	9.52	0.002	0.02	0.20	1.96	9.52
Diameter (mm)	Mean fracture toughness value (MPa $m^{1/2}$)					Increase f (%)				
30	0.59	0.72	0.80	–	–	–	–	–	–	–
50	1.12	1.19	1.32	1.35	1.42	89.96	65.98	65.03	–	–
75	1.17	1.27	1.38	1.55	1.68	97.76	77.09	72.07	14.98	18.61
100	1.22	1.28	1.38	1.55	1.68	106.61	78.44	72.72	15.26	18.83
150	1.28	1.36	1.54	1.64	1.78	116.13	89.44	92.66	21.96	25.56

respectively, under the setting loading rates of 2 and 10 kN/s. It can be seen that when the loading rate is high, the specimen with a diameter of 30 mm has two loading rates, which are different from the other four sizes.

The relationship between the average increment F of the fracture toughness test value and the specimen size at different loading rates is shown in Fig. 2.37. From Table 2.8 and Fig. 2.37, it can be seen that the effect of specimen size on the fracture toughness test value of NSCB specimens varies with the loading rate, and the amplification f increases linearly with the increase of specimen size. It can also be seen from Fig. 2.7 that the relationship between the

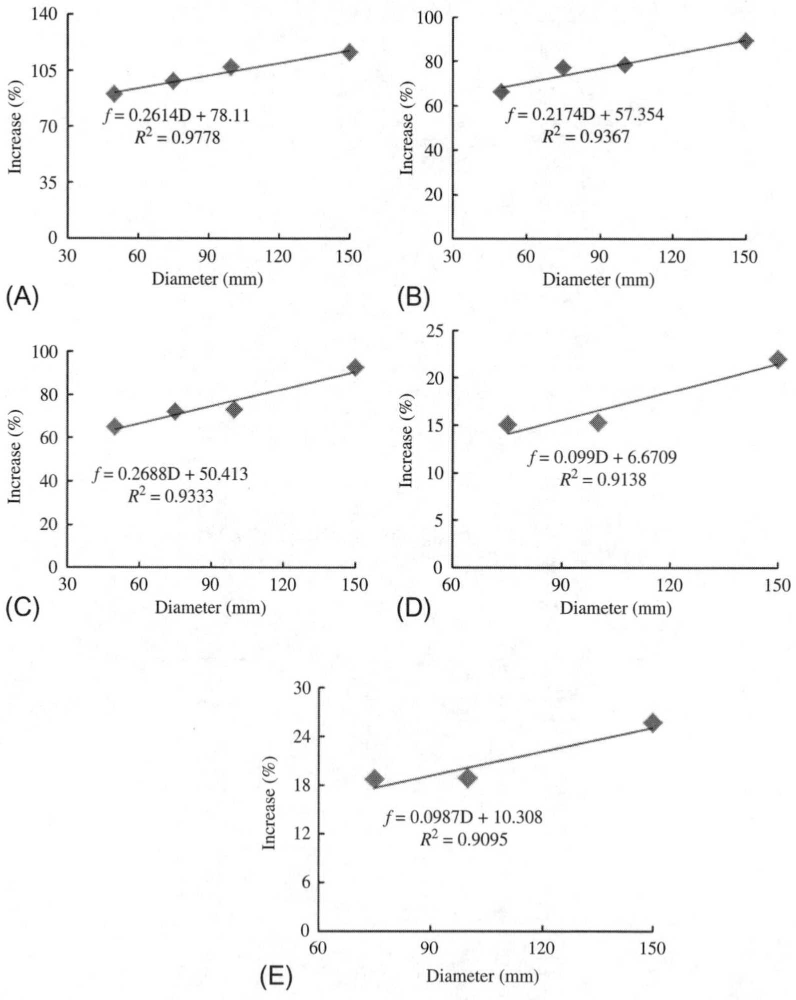

FIG. 2.37 Relationship between the amplification coefficient f of fracture toughness and the specimen size. (A) $P_v = 0.002$ kN/s, (B) $P_v = 0.02$ kN/s, (C) $P_v = 0.2$ kN/s, (D) $P_v = 1.96$ kN/s, and (E) $P_v = 9.52$ kN/s.

amplification f and the specimen size is affected by the loading rate, and the larger the loading rate, the greater the influence of the specimen size on the amplification f.

Discussion on loading rate and size effects

Bazant and Chen (1997) believed that quasibrittle materials such as rocks had a large fracture growth zone at the crack tip and undergo steady growth. Especially the gradual release of storage energy due to the redistribution of stress, the existence of macrocracks and large microcracks is the main reason for the size effect of quasibrittle materials. When the loading rate is constant for specimens with the same configuration and similar geometry, the stored energy before fracture is larger for the large specimens because of the enhanced load-carrying capacity of the specimens themselves. The specimen has more energy released at the moment, and both the fracture energy and the fracture toughness are larger.

The author considers that with the increase of the loading rate, the incubation time of microcracks in the specimen decreases gradually and the internal microcracks cannot propagate when the size of the specimen is fixed. The energy used for microcrack rupture decreases gradually and accumulates until the instantaneous release at the moment of failure. Therefore, at a higher loading rate, the energy released by the specimen at the time of failure is more, and the specimen needs to consume more fracture energy. Zhang et al. (1996) believed that under static or quasistatic loading conditions, the fracture toughness of general rocks increases slightly with the increase of loading rate, but when the loading rate exceeds a certain value (such as general impact loading), they will increase significantly with the increase of loading rate. The loading rate under static loading had less gradient, so the trend of increasing fracture toughness of rock under static loading was not obvious (Zhang et al., 1996). In fact, under the conditions of a static low loading rate, setting multiple loading rate gradients, and making the minimum value of the gradient small enough, the fracture toughness of rock increases with the increase of the loading rate. The increase is logarithmic, which has a strong regularity. The analysis of the coupling effect between sample size and loading rate is more easily affected by the homogeneity and discreteness of the test data. These problems should be noticed.

In addition, large NSCB specimens show a certain compaction stage in the initial stage of the load-displacement curves while the curves of small NSCB specimens have no obvious plastic characteristics in the initial stage. The reason should be related to the number of microcracks in the specimens and the deformation range provided by the size of the specimens. Many researcher believe that the size of small specimens is limited while the number of internal microcracks and the compacted space are very limited. So, the compact linear elasticity stage is immediately entered without more obvious deformation caused by the microcrack compression. However, large specimens are just the opposite, and they have more microcracks and defects. In the initial stage of loading, the microcracks of specimens along the direction of loading are compacted, which can produce a wider range of compaction deformation, thus showing the plastic characteristics of microcracks compacted in the initial stage on the load-displacement curve. Due to the dense material, the overall deformation is small, and no compaction stage can be observed at the macroscopic level.

2.5.5 Conclusions

The load-displacement curves of different specimen sizes are different at different stages. With the increase of specimen size, the load-displacement curve of the specimen changes from two stages of linear elasticity failure to three stages of compaction-linear elasticity failure, and the plastic properties of the large specimen are more obvious in the initial stage. After reaching the peak load, the specimens with different sizes all exhibit typical brittle failure characteristics. The slope of the linear elastic stage before the load-displacement curve has nothing to do with the size of the specimen.

Under a certain specimen size, the fracture toughness of the NSCB specimen is positively correlated with the loading rate. It increases logarithmically with the increase of the loading rate, and the amplification increases logarithmically with the increase of the loading rate.

Under a certain loading rate, the fracture toughness test value of the NSCB specimen increases with the increase of specimen size. There is a critical dimension. When the specimen size is smaller than this critical dimension, the increase of the fracture toughness test value is larger. On the contrary, when the critical size is larger than this, the increase of the fracture toughness test value is small and linear.

2.6 Hole influence on dynamic fracture toughness

2.6.1 Introduction

The phenomenon of fracture is related to the presence of cracks, notches, and voids in the material or structure, which can result in discontinuity of the macroscopic material and cause fracture failure induced by stress concentration near the defect under the action of an external force. Because the fracture of rock and concrete brittle materials only undergoes significant plastic deformation within a small range of the crack tip, there is almost no sign before the fracture, but it may often have catastrophic consequences. Studying the fracture phenomena of materials, especially the dynamic fracture mechanism and crack propagation law of materials under a high strain rate, has an important practical engineering significance for humans to scientifically evaluate the structural damage resistance of earthquakes and dynamic external forces such as explosion stress wave destructive ability.

The dynamic fracture of rock materials is a fairly complex process. It is generally considered to be influenced by factors such as the physical properties of the material, the dynamic applied load history, the geometry of the structure, the random distribution of defects, and the effects of inertia and strain rate (Li et al., 2006). During the experiment, it was found that the dynamic mechanical response curve obtained by dynamic experiments is diversified, so it is difficult to accurately determine the dynamic fracture of rock and understand the dynamic failure mechanism of rock. A herringbone slotted Brazilian disc was proposed to test rock fracture toughness in 1995 by ISRM (International Society for Rock Mechanics and Rock Engineering). This interested researchers in testing rock mechanics parameters using disc specimens.

In order to solve the problem that prefabricated cracks are difficult to make, Tang et al. (1996) proposed a central circular hole crack Brazilian disc specimen combining round holes and cracks. Wang (2004) proposed the use of a method for processing a disc specimen, which not only ensures center cracking but also reduces the stress concentration of the contact surface. Zhou et al. (2005) proposed to change the shape of the straight groove crack tip into a sharp groove specimen in order to reduce the influence of the preformed crack width of the specimen on the fracture test value of the disc. Used five kinds of disc specimens to test the fracture toughness of rock, and used the same size of the holed-cracked flattened Brazilian disc (HCFBD) to study the effect of crack length on the rock dynamic toughness test value. The central circular hole of the HCFBD specimen makes it easier to preform the crack on the rock. However, the influence of the diameter of the central circular orifice on the dynamic fracture toughness value of the tested rock has not been studied. There is no reasonable suggestion range for the selection of the diameter of the central circular orifice. Therefore, this paper uses the test method (Zhang et al., 2009) to focus on the effect of the central hole diameter change of the HCFBD specimen on the rock dynamic fracture toughness test value and its fracture mode. This provides a reference for the promotion of HCFBD specimens to test the dynamic fracture toughness of rock.

2.6.2 Dynamic cleaving specimens and equipment

The rock specimens are white marble with good homogeneity and a grain size of 0.3–1.2 mm. The mechanic parameters of the marble are a Poisson's ratio of 0.3, a density of 2730 kg/m^3, and an elasticity modulus of 16.3 GPa.

The basic parameters and physical map of the HCFBD specimen are shown in Fig. 2.38. The other nominal dimensions of the specimen are fixed except for the central bore aperture r_0. The loading platform 2α is 20 degrees, the specimen thickness b is 32 mm, the disc radius R is 40 mm, the prefabricated crack length $2a_0$ is 40 mm, and the preformed crack width of the specimen is less than

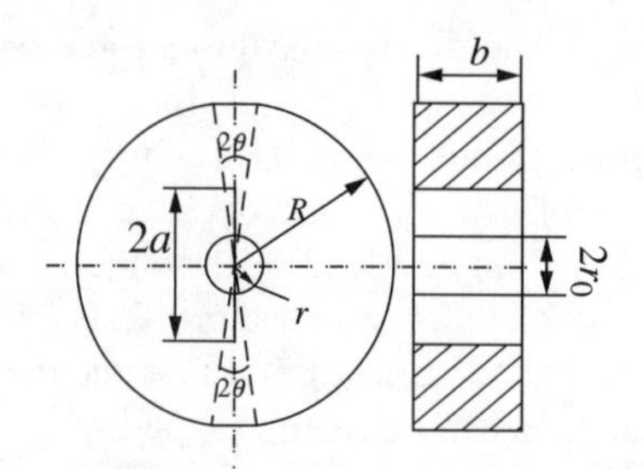

(A)

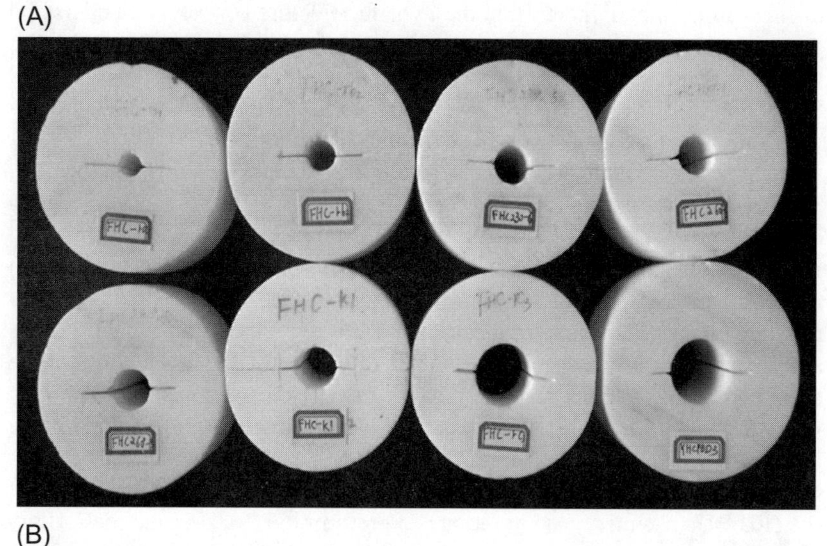

(B)

FIG. 2.38 Geometric parameters and photos of HCFBD specimens. (A) Basic parameters of HCFBD specimen and (B) photos of HCFBD marble specimens.

0.65 mm. To more easily analyze the effect of the central aperture r_0 on the dynamic test value, define the dimension of the central aperture radius parameter γ as equal to r_0/R. Considering the length of the central crack and the crack-propagation ligament, γ is between 0.10 and 0.30.

The dynamic impact splitting experiment of the HCFBD specimen was performed on a 75 mm diameter Hopkinson pressure bar system from the Central Impact Laboratory of Central South University. The SHPB pressure bar loading device is shown in Fig. 2.39. A deformed projectile with a length of about 540 mm is fired by a compressed air cannon, then collides with the incident bar coaxially to generate a compressive stress wave on the incident bar. When the compressive stress wave propagates to the contact surface of the incident bar and the disc specimen, one part of it returns to the incident bar, the other part is transmitted to the disc specimen, and an impact load is applied to the disc specimen to cause the specimen to break. The resistance strain gauge of

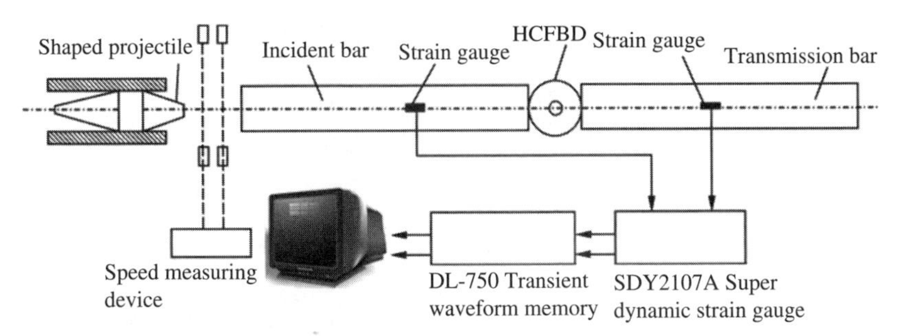

FIG. 2.39 Diagram of the SHPB load system.

BX120-10AA was adhered to the upper and lower surfaces of the specimen at a distance of 3 mm from the crack tip to determine the fracture initiation time t_f of the HCFBD specimen. The BX120-2AA resistance strain gauges are attached to the incident and transmission bars to record the voltage signals on the incident and transmission bars. The distance between the strain gauge on the elastic pressure bar and the contact end of the pressure bar and the specimen is 1004 mm. The selection of the strain gauge model mainly considers the size of the material, the adhesion area, the radius of curvature, and the installation conditions. During the experiment, the speed of the projectile was calculated by recording the time when the projectile passed the parallel light sources of the two speed detectors. The ultra dynamic strain gauge model is the SDY2107A dynamic strain gauge with automatic balancing. The DL-750 digital waveform oscilloscope has a sampling frequency of 1 MHz, a sampling length of 10 K, and a sampling delay of −2 K. Before the test, the reliability and precision of the SHPB system device were tested by adding no specimen between the incident bar and the transmission bar.

2.6.3 SHPB test and data record

2.6.3.1 Pulse signal on elastic pressure bar

Using a 540 mm long-range projectile, the strut was coaxially struck at a pressure of 0.32 MPa, and the strain gauge attached to the incident bar obtained an incident wave and a reflected wave with a rise time of about 150 μs, as shown in Fig. 2.40.

The special-shaped projectile can generate a half-sinusoidal stress wave that can better achieve the constant strain rate loading of the specimen, thereby obtaining a pulse signal with better repeatability that is unaffected by the shape and size of the waveform shaper. It can be seen from the strain signal on the pressure bar in Fig. 2.40 that the strain pulse signal obtained on the pressure bar has small oscillation, the typical transmission strain pulse waveform is steep, and its amplitude is only about 1/7 of the incident pulse waveform. This

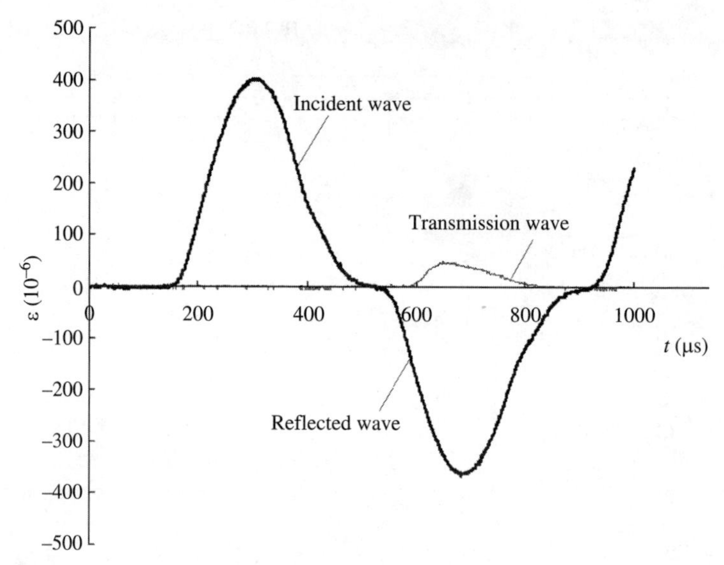

FIG. 2.40 Strain waves recorded on the pressure bar.

is due to the mismatch between the wave impedance and the contact surface of the rock specimen and the steel bar during the stress wave propagation.

2.6.3.2 Determination of cracking time

In this experiment, the strain gauge method was used to determine the fracture initiation time of the crack tip of the marble specimen. During the high-speed expansion of the crack of the HCFBD specimen, the strain and displacement fields at the crack tip are affected by the inertia effect, and the change trend has a certain hysteresis. However, when the crack starts to crack, the strain rate changes sharply (decreased or increased). Therefore, the maximum crack rate of the crack tip strain signal can be used to determine the fracture initiation time. Fig. 2.41 shows the strain signal of specimen N_1, which is directly derived to obtain the strain derivative waveform, as shown in Fig. 2.42. The time at which the incident pulse first reaches the specimen is the time starting point, and the time corresponding to the maximum value of the crack tip strain derivative on the HCFBD specimen is t_{gmax}.

The strain gauge is not stuck to the crack tip (Feng et al., 2009). And in the dynamic loading process, the crack initiation is not simultaneously cracking at all points of the crack front in the thickness direction of the disc specimen; it is generally considered that the crack initiation is started at the center point of the crack tip. Therefore, determining the fracture initiation time must also take into account the influence of the adhesion position of the strain gauge and the thickness of the specimen. After obtaining t_{gmax} by the method shown in Fig. 2.42,

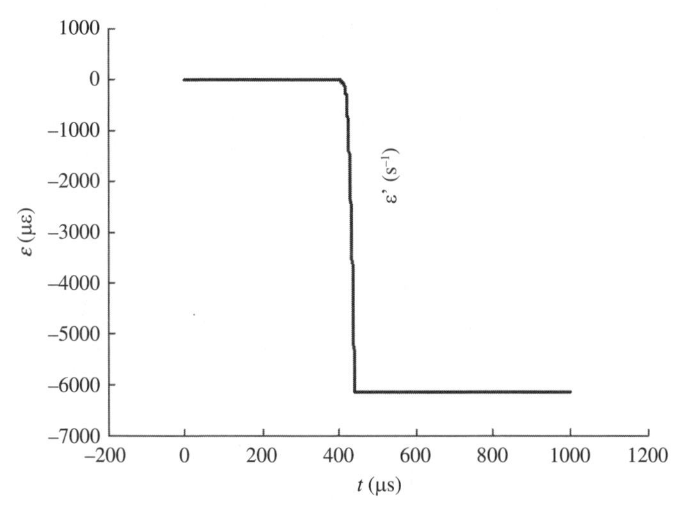

FIG. 2.41 Strain signal on specimen N_1.

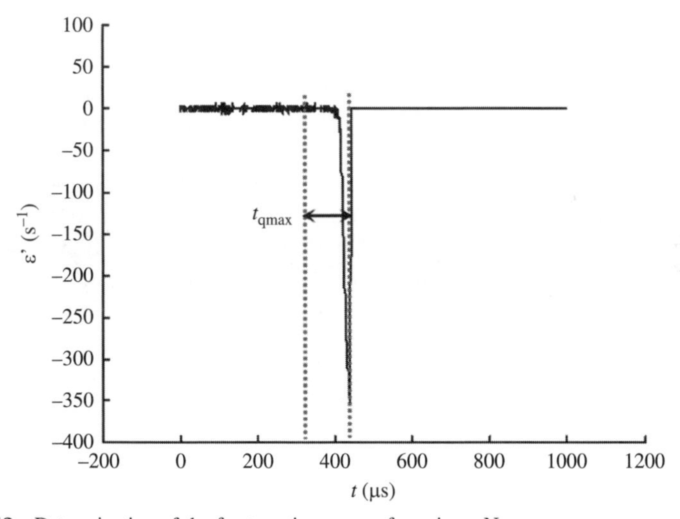

FIG. 2.42 Determination of the fracture time t_{gmax} of specimen N_1.

the fracture initiation time t_{f} of the crack tip of the specimen is determined by the following Formula (2.22).

$$t_f = t_{\text{gmax}} - \sqrt{\left(\frac{b}{2}\right)^2 + e^2}/c \tag{2.22}$$

In Formula (2.22), b is the thickness of the HCFBD specimen, e is the linear distance from the crack tip on the upper and lower surfaces of the disc specimen to the center of the strain gauge, and c is the wave velocity of the stress wave propagating in the specimen.

2.6.4 Dynamic finite element analysis

2.6.4.1 Load determination of model

The HCFBD specimen was simplified to a two-dimensional plane problem for analysis. Considering the attenuation and energy consumption of the stress pulse in the specimen, the dynamic load applied to the specimen is the average of the incident bar end load $F_L(t)$ and the transmission bar end load $F_R(t)$. The end face load of the loading platform determined by the transmission bar needs to shift the time τ of the stress wave passing through the specimen to the left, and the load after translation ε_t^* is equal to $\varepsilon_t(t-\tau)$. This eliminates the spatial and temporal nonuniformity of the two platform end faces. According to the strain signal obtained by the elastic pressure bar, the dynamic load $P(t)$ acting on both ends of the HCFBD specimen can be calculated by the following Formula (2.23).

$$P(t)=\frac{F_{\mathrm{L}}(T)+F_{\mathrm{R}}(T)}{2}=\frac{EA}{2}\left[\varepsilon_{\mathrm{i}}(t)+\varepsilon_{\mathrm{r}}(t)+\varepsilon_{\mathrm{t}}^{*}(t)\right] \tag{2.23}$$

In Formula (2.23), $\varepsilon_i(t)$ and $\varepsilon_r(t)$ are the incident and reflected waveforms of the measured pulse of the incident bar through the alignment and translation of the wave head, respectively; E and A are the elastic modulus and cross-sectional area of the elastic pressure bar, respectively.

Considering that the incident bar is a dynamic splitting load on the disc specimen through the loading platform of the HCFBD specimen, the dynamic load $P(t)$ determined by Formula (2.23) needs to be converted into the load $\sigma(t)$ applied to the loading surface of the HCFBD specimen, as shown in the following Formula (2.24).

$$\sigma(t)=\frac{P(t)}{2Rb\sin\theta} \tag{2.24}$$

2.6.4.2 Dynamic loading of model

The finite element model for the dynamic loading of HCFBD specimens is shown in Fig. 2.43. Due to the symmetry of the HCFBD specimen, a half-model of the specimen was taken for modeling analysis. Applying a dynamic load $\sigma(t)$ to the left end of the numerical model to simulate the loading process of the SHPB device on the disc impact splitting. The X-axis constraint is applied from the prefabricated crack tip along the loading diameter to the corresponding loading platform surface, and the Y-axis is free. The Y-axis constraint is applied to the flat surface of the right end disc, and the X-axis is free of boundary conditions.

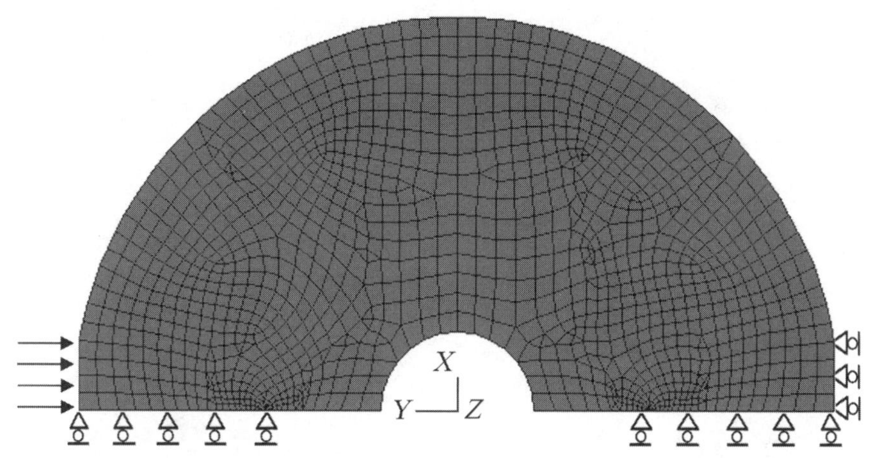

FIG. 2.43 Finite element model of the HCFBD specimen under dynamic impact (1/2 mode).

The HCFBD specimen model is divided by the Plane42 element, which is suitable for plane stress, plane strain, and axis symmetry. The finite element model is divided into 1251 four-node isoparametric elements.

The KSCON command is used to construct a singular element of cracks around the first layer of the crack tip. In order to reduce the influence of singular element meshing on the dynamic stress intensity factor, combined with the calculation speed and numerical precision, the crack tip unit angle α is 22.5 degrees, and the crack tip singular unit length l is 0.1 mm. The finite element model is divided into an adaptive free mesh with an overall defined size of 2 mm. The dynamic load is stored in the ".txt" document and loaded by reading the load corresponding to the time. The time step is 1 μs and the load history is from 0 to 315 μs.

2.6.4.3 *Dynamic stress intensity factor*

A partial enlarged view of the crack tip on the left side of the finite element model is shown in Fig. 2.44. It can be seen that the unit constructed by the crack tip is substantially octave, and the radius of the first layer unit is one quarter of the second layer unit. The nodes 50#, 72#, and 73# are defined along the path shown in Fig. 2.44 along the crack surface by the "path" command. The displacement-time history of the different nodes can be obtained from the calculation results.

The dynamic stress intensity factor history $K_I(t)$ determined by the displacement method of the crack tip node (r is equal to 0) combined with the linear interpolation extrapolation method is as follows in Formula (2.25).

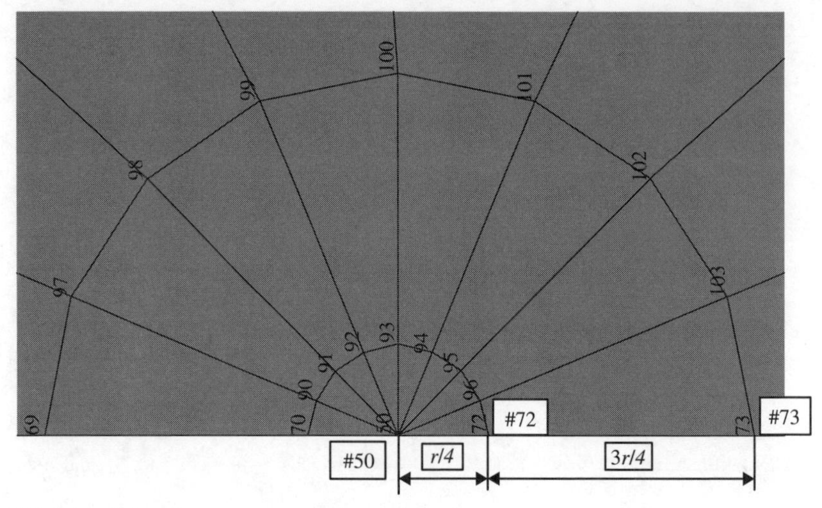

FIG. 2.44 Enlarged elements around the crack tip.

$$K_{\mathrm{I}}(t)=\sqrt{2\pi}\frac{2G}{1+\kappa}\frac{8u(t)_{\mathrm{B}}-u(t)_{\mathrm{A}}}{3\sqrt{r}} \tag{2.25}$$

In Formula (2.25), $\kappa=\begin{cases} 3-4\mu \Leftrightarrow \text{Plane strain} \\ (3-\mu)/(1+\mu) \Leftrightarrow \text{Plane stress} \end{cases}$

Where G and μ are the shear modulus and Poisson's ratio of the rock material, respectively. $u(t)_{\mathrm{A}}$ and $u(t)_{\mathrm{B}}$ are divided into the displacement-time history of 73# and 72# nodes along the vertical crack plane.

A plot of the displacement of the 72# and 73# nodes in the X-direction with time along the left end of the numerical model is shown in Fig. 2.45. It can be seen from Fig. 2.45 that the displacement of both nodes is small in the early stages of loading. However, as the loading time increases, the displacement of the 73# node is relatively far from the crack tip (50# node), and larger than the displacement value of the 72# node closer to the crack tip.

2.6.5 Results analysis and discussion

2.6.5.1 Central aperture influence on test values

The dynamic load history and the dynamic stress intensity factor $K_{\mathrm{I}}(t)$ curve at the crack tip applied to the N_1 specimen are shown in Fig. 2.46. According to the experimental-numerical method, the dynamic stress intensity factor $K_{\mathrm{I}}(t_{\mathrm{f}})$ corresponding to the fracture initiation time t_{f} determined by the strain gauge is equal to the test value K_{Id} of the rock dynamic fracture toughness.

The basic parameters and dynamic test results of 12 marble disc specimens with different circular apertures are shown in Table 2.9. It can be seen from

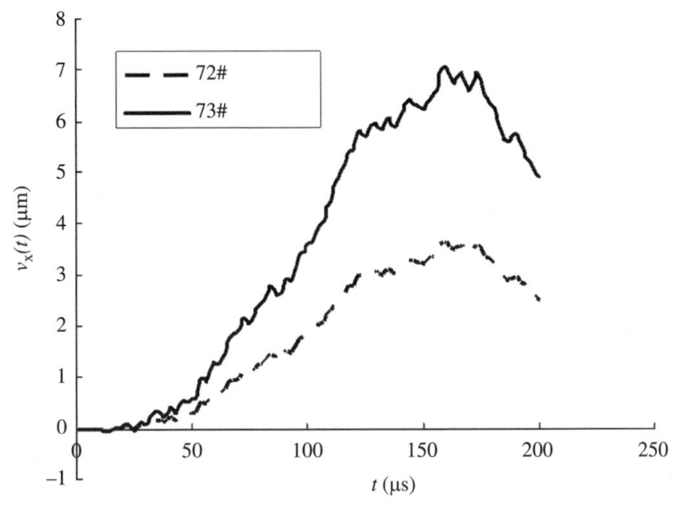

FIG. 2.45 Displacement curves of two nodes.

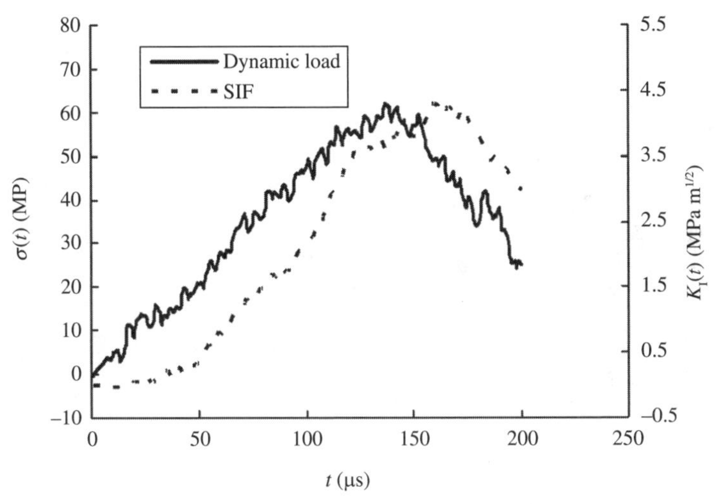

FIG. 2.46 Time histories of impact load and dynamic stress intensity factors.

Table 2.9 that the fracture initiation time of the HCFBD specimen is from 85 to 105 μs. For the HCFBD specimen with a diameter of 80 mm, the other conditions being the same, when the ratio of the diameter of the center circular hole to the diameter of the disc γ is between 0.10 and 0.30, and the average loading rate is 4.62×10^4 MPa m$^{1/2}$ s^{-1}, the average dynamic fracture toughness test value of marble is 4.57 MPa m$^{1/2}$.

All the dynamic fracture toughness test values obtained by testing different central circular aperture HCFBD specimens are shown in Fig. 2.47. Comparing

TABLE 2.9 Results of dynamic experiments of marble

Specimen no.	Disc diameter (mm)	r_0/R	Impact pressure (MPa)	Fracture initiation time (μs)	K_{Id}/t_f ($GPa\ m^{1/2}\ s^{-1}$)	K_{Id} ($MPa\ m^{1/2}$)
N1	80.23	0.10	0.32	101.0	4.47	4.52
N2	80.25	0.13	0.34	94.0	–	–
N3	80.27	0.13	0.35	98.0	4.20	4.12
N4	80.18	0.13	0.33	103.0	4.83	4.98
N5	80.33	0.15	0.35	101.0	4.80	4.85
N6	80.21	0.18	0.35	96.0	4.73	4.54
N7	80.28	0.19	0.32	88.0	4.92	4.33
N8	80.15	0.23	0.35	99.0	4.47	4.43
N9	80.11	0.24	0.34	100.0	4.61	4.61
N10	80.20	0.27	0.35	99.0	4.58	4.53
N11	80.19	0.27	0.33	97.0	4.53	4.39
N12	80.31	0.29	0.33	105.0	4.71	4.95
Mean	80.22	0.20	0.34	98.0	4.62	4.57

Note: *Specimen N2 is an invalid specimen.*

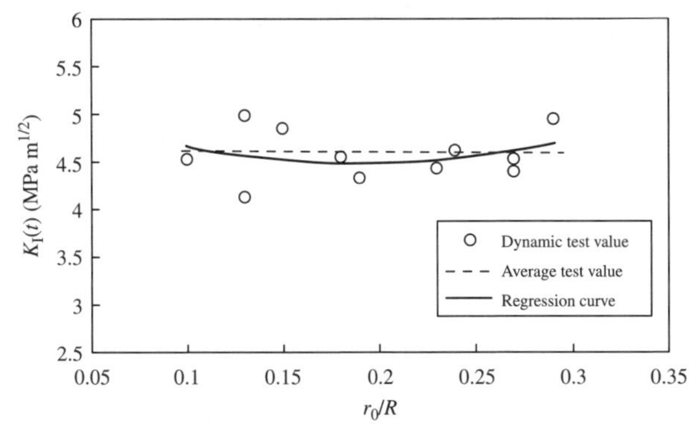

FIG. 2.47 Dynamic fracture toughness measured by the SHPB experiment.

the test results, it is found that the test value of the dynamic fracture toughness K_{Id} of marble determined by different specimens has a certain dispersion. The configuration of the disc specimen with different central circular holes will cause the reflection coefficient and transmission coefficient between the incident bar and the specimen, the specimen and the transmission bar, and thus affect the dynamic loading load of the specimen by the pressure signal of the compression bar. Different disc configurations will cause the difference of dynamic stress intensity factors of HCFBD specimens. The fracture initiation times are also different, but the test values of dynamic fracture toughness are not affected by the central hole diameter.

There are some effects on the test value of rock dynamic fracture toughness by the configuration of the specimen. In addition to the influence of the central aperture, the disc diameter, the platform loading angle, the disc thickness, and the prefabricated length and width will have some influence. Considering the difficulty of rock prefabrication cracks, the reliability of test results, and the limitation of the 75 mm diameter of the SHPB experimental equipment, it is recommended to select the nominal size of HCFBD specimens for general testing. The values are provided as the following: the diameter $2R$ is 80 mm, the center aperture ratio r_0/R is 0.2, the disc thickness ratio $b/(2R)$ is 0.4, the platform loading angle is 20 degrees, the prefabricated crack length ratio a_0/R is 0.5, and the prefabricated crack width is less than 0.8 mm.

In addition, the discreteness of the dynamic fracture toughness test value is a problem that must be noted. The mineral particle composition of the prefabricated crack tip, the processing accuracy of the specimen, and the environment surrounding the test system all affect the test value of the dynamic fracture toughness K_{Id}. In order to reduce the test error, on the one hand, the rock with small mineral grain size and good mean value should be selected, and the processing precision of the prefabricated crack should be ensured as much as possible while ensuring the parallelism of the platform. On the other hand, the accuracy of the dynamic test is increased as much as possible by increasing the number of specimens that are dynamically tested.

2.6.5.2 Final fracture mode of specimen

Under dynamic impact, the final fracture modes of HCFBD specimens with different center circular holes are shown in Figs. 2.48 and 2.49.

Generally, the fracture mode can be divided into a primary crack and a secondary crack according to the crack initiation position and the propagation direction. The primary crack refers to a crack that starts from the tip of the prefabricated crack and propagates in the loading direction. The secondary crack refers to a crack that cracks from the edge of the specimen and spreads toward the tip end of the crack. A specimen that does not have a macroscopic main crack and has significant secondary cracks before the primary crack is produced can be considered an invalid experiment (Zhang and Wang, 2007).

It can be seen from Fig. 2.48 that the specimen N_1 with a small aperture has many secondary cracks around the radial disc specimen along the central circular hole during the process of initiating, cracking, and propagating the primary crack. The smaller the central aperture, the more complex the fracture mode of the HCFBD specimen. From the point of view of energy dissipation, HCFBD specimens with smaller central apertures are more likely to induce more secondary cracks under the impact load to consume the energy carried by the increasing incident pulse and some energy converted into kinetic energy carried by

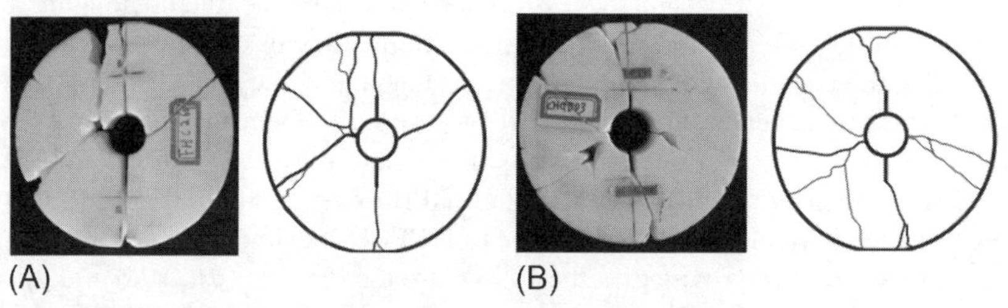

FIG. 2.48 Dynamic fracture modes and fracture sketches of specimen N_1 (A) and specimen N_2 (B).

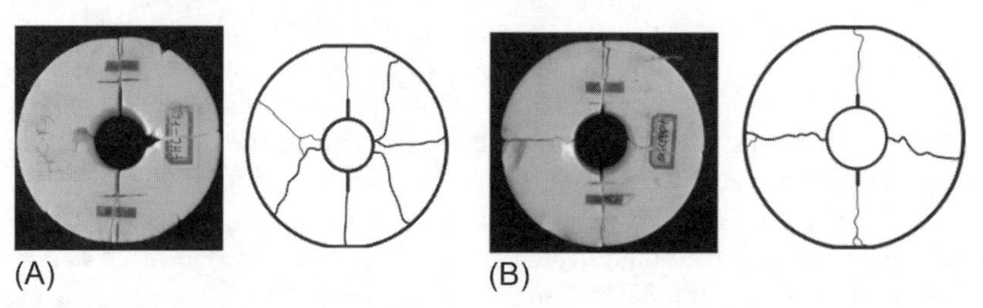

FIG. 2.49 Dynamic fracture modes and the fracture sketches of specimen N_9 (A) and specimen N_{10} (B).

flying debris. With the increase of the diameter of the central hole, HCFBD specimens show more macroscopic pull-through failure modes, the primary crack is very obvious, the number of secondary cracks is reduced, and the phenomenon of caving and fragmentation is weakened. In the process of crack evolution, the final fracture modes of HCFBD specimens are obviously divided between four and six parts.

The expansion of the secondary crack in Figs. 2.48 and 2.49 does not have significant symmetry or regularity, which is mainly due to the heterogeneity of the rock material. The distribution of microvoids, crystal particles, and impurities is random. Under dynamic load, the difference in local strength and defects caused by the marble cause changes in stress and displacement fields, resulting in the asymmetry of the secondary crack propagation path. It is worth pointing out that when the initiation of secondary cracks is produced after the crack initiation of the primary crack, it will not affect the effectiveness of the dynamic experiment.

2.6.6 Conclusions

As the fracture initiation time of the specimen is determined by using a strain gauge method, the crack initiation position must be determined first. The influence of the crack initiation point, the strain gauge attachment position, and the specimen thickness should be considered.

For specimens with a diameter of 80 mm, at an average loading rate of 4.62×104 MPa $m^{1/2}$ s^{-1}, the average dynamic fracture toughness of marble is 4.57 MPa $m^{1/2}$ when the ratios of the diameter of the center hole to the diameter of the disc r_0/R are between 0.10 and 0.30. The test value of rock dynamic fracture toughness has no obvious relationship with the change of the central hole diameter.

When an HCFBD specimen with a smaller central aperture is used, during the primary crack growth, cracking, and propagation, many secondary cracks are formed around the periphery of the radial disc specimen along the center circular hole, and the phenomenon of chip collapse occurred. When an HCFBD specimen with a larger central aperture is used, it is more likely to exhibit a macroscopic pull-through failure mode. Whether there is a significant macroscopic crack on the diameter of the disc specimen can be used as one of the criteria for judging the effectiveness of the experiment.

2.7 Dynamic fracture toughness of holed-cracked discs

2.7.1 Introduction

Dynamic fracture has been an area of sustained research for studying the behavior of engineering materials and structures under dynamic loading conditions (Freund, 1998). It is often encountered in events such as damage in the ground and buildings by earthquake, collision and impact of space vehicles with heavenly bodies, warhead penetration into military and civilian targets, etc. As compared with static fracture, dynamic fracture is much more complicated, considering the inertia effect or the stress wave propagation in structures. Therefore, the study on dynamic fracture requires expensive experimental setups and sophisticated computational resources. Dynamic fracture problems are usually divided into two categories: crack initiation of a static crack subjected to dynamic loading and fast crack propagation and/or the arrest of a running crack. The dynamic fracture initiation toughness investigated in this paper belongs to the first category; this mechanical property is very useful for evaluation of the dynamic fracture performance of materials and structures.

In early studies, the mode-I (opening mode) dynamic fracture toughness was tested by drop-weight loading of the three-point bend (3 PB) specimen; the test was improved by Yokoyama and Kishida (1989) and Popelar et al. (2000). They achieved a high loading rate by using the split Hopkinson pressure bar (SHPB) (Yokoyama and Kishida, 1989; Popelar et al., 2000). However, the phenomena of "loss of contact" may be a shortcoming of the 3PB test, as pointed out by Jiang and Vecchio (2007). Considering that core-based specimen configuration has a merit in specimen preparation for certain materials such as rock, concrete, and ceramics, Nakano et al. recently did the mixed mode dynamic fracture test for ceramics and glass using the cracked straight through Brazilian disc specimen loaded by the single pressure bar (Nakano et al., 1994), Lambert and Ross (2000) diametrically impacted the holed-cracked Brazilian disc for testing concrete and rock with the SHPB setup.

It has been observed that the fracture parameters of materials obtained from laboratory testing are affected by the size of specimens; this phenomenon is called the size effect or scaling effect. The size effect is especially complicated for quasibrittle materials such as rock, concrete, sea ice, wood, ceramics, etc. One of the reasons for the size effect is the fracture process zone developed at the front of the crack during the loading process (Carpinteri et al., 2006). Up to now, most of the studies on size effect for fracture toughness or strength are concerned with the static situation. It is more difficult to study the size effect in the dynamical loading condition, as the dynamic size effect is intervened with the loading rate effect (Petrov et al., 2003; Ruiz et al., 2000). The time factor, such as the fracture incubation time (Ruiz et al., 2000; Pugno, 2006), also plays a role in addition to the length factor, affecting the test values. The dynamic size effect hinders the application of material parameters tested in the laboratory with specimens to reliable engineering designs in the above-mentioned events. It should be elucidated based on the fundamental properties of test materials that this problem may be tackled by the cooperative efforts of scientists in the field of materials science or solid mechanics. Applying the recently proposed dynamic quantized fracture mechanics, Pugno (2006) studied dynamic crack propagation in a sense of the root mean square of stress intensity factors, considering the effect of both space and time. Elfahal et al. studied the size effect on the dynamic compression strength of concrete using the drop-weight method. They noted the influence of loading rate to the size effect (Elfahal et al., 2005; Krauthammer et al., 2003). Targeting Elfahal's work, Bindiganavile and Banthia (2004) proposed a dynamic impact factor that was claimed to "eliminate any rate effects and thus allow for a true determination of the size effect" (Bindiganavile and Banthia, 2004) However, Elfahal et al. disagreed with Bindiganavile's claim. It can be seen that the related references (Elfahal and Krauthammer, 2005; Krauthammer et al., 2003; Elfahal et al., 2005; Bindiganavile and Banthia, 2004; Elfahal and Krauthammer, 2004) only contributed to the dynamic compression strength. To the best of our knowledge, the size effect of dynamic tensile strength or fracture toughness has very rarely been reported in the literature.

In the present study, two types of the holed-cracked flattened Brazilian disc (HCFBD) specimens of marble were prepared. Of those, one type is geometrically similar with a different outside diameter (42, 80, 122, and 155 mm, respectively, the largest size ratio being 3.7 = 155/42) and with the crack length being constantly half the diameter;. The other type is with an identical outside diameter (80 mm) as the specimens only vary in crack length, so they are not geometrically similar. The HCFBD specimens were diametrically impacted by the SHPB. To cope with the size effect shown by the results derived by using the regular procedure, a method to determine the unique rock dynamic fracture toughness is proposed. It takes the average of the integration of the dynamic stress intensity factors in the spatial-temporal domain, which is defined by the fracture process zone length l and the incubation time τ jointly.

2.7.2 Dynamic fracture toughness test

2.7.2.1 Test specimens

The HCFBD specimens (Fig. 2.50) were prepared from white marble taken from Ya'an, Sichuan province. The marble has a Young's modulus of 16.3 GPa, a

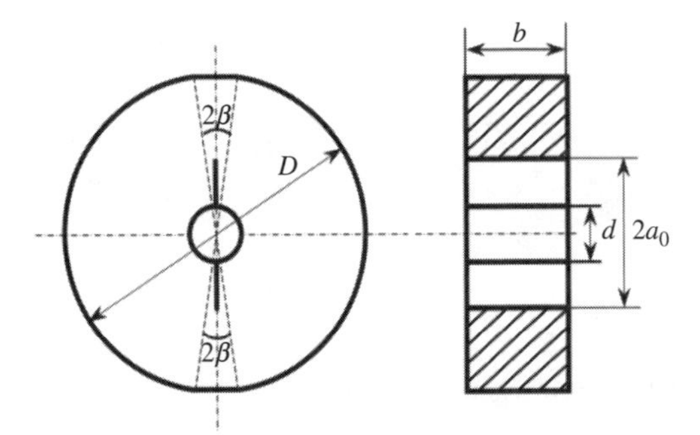

FIG. 2.50 Sketch of HCFBD specimen.

Poisson's ratio of 0.3, and a density of 2730 kg/m^3. The notch width is kept within 1 mm so that the prepared initial notch may be considered as a crack. Some consideration is given for the design of the specimen configuration. The central circular hole is beneficial for producing the initial notch at the hole circumference. The flat ends make it convenient for the alignment of loading the disc specimen in an accurate and favorable condition. They also provide advantages for the stress waves to travel through the bars and the specimen, and for reducing the stress concentration at the impact point. The crack length is $2a_0$, the inner hole diameter d, the outside diameter D, the flat end loading angle $2\beta = 20$ degrees, and the thickness b. There were 25 specimens in total; among them, 12 specimens were geometrically similar ($2a_0/D = 0.5$, unchanged, $D = 42$, 80, 122, 155 mm, respectively) and 13 specimens had a one-size outside diameter and different crack lengths ($D = 80$ mm, 25 mm $\le 2a_0 \le 41$ mm).

These two types of specimens are shown in Fig. 2.51A and B, respectively. The concrete values of geometric parameters for geometrically similar specimens are given in Table 2.10.

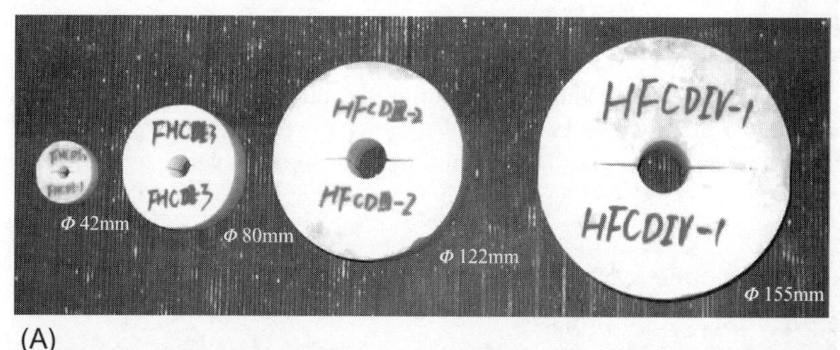

(A)

(B)

FIG. 2.51 Photos of two types of HCFBD specimens of marble. (A) Geometrically similar HCFBD specimens (size ratio of D is 1.0:1.9:2.9:3.7) and (B) one-size HCFBD specimens with different crack length ($D = 80$mm, 25 mm $\le 2a_0 \le$ 41 mm).

TABLE 2.10 Geometric parameters of geometrically similar HCFBD specimens

Size type	D (mm)	b (mm)	2β (degrees)	d (mm)	$2a_0$ (mm)	Size ratio
I	42	16.8	20	8.4	21.0	1.0
II	80	32.0	20	16.0	40.0	1.9
III	122	48.8	20	24.4	61.0	2.9
IV	155	62.0	20	31.0	77.5	3.7

2.7.2.2 Test setup

The SHPB system was specially designed to test larger specimens so all bars are 100 mm diameter. The length of the incident bar is 4500 mm and the length of the transmission bar is 2500 mm. The bars are made of an alloy structure steel of type 42CrMo (From the Chinese standard GB/T3077-1999, the main compositions in percentage: 0.38–0.45 C, 0.90–1.20 Cr, 0.15–0.25 Mo; the material's mechanical parameters are elastic modulus 210 GPa, tensile strength 1080 MPa, Poison's ratio 0.3, density 7850 kg/m^3). The incident pulse created in the Hopkinson pressure bar can be well controlled in amplitude, duration, and rise time by changing the length and velocity of the projectile as well as the suitable choice of the wave form shaper. The wave form shaper was used to obtain the ideal loading wave. It was a thin circular plate of paper glued at the end surface of the incident bar; this end with the shaper was stricken by the projectile. Strain gauges were stuck on the incident bar at a distance of 1000 mm and on the transmission bar at a distance of 800 mm, to the end contacting the flat end of the specimen, respectively. The strain signals recorded for the strain gauges on the bars are used for deriving the history of the load. The smallest specimen of 42 mm diameter and the largest specimen of 155 mm diameter, placed between the bars before the impact test, are shown in Fig. 2.52A and B, respectively. The strain gauges glued at the front of the crack tip of the specimen are used to detect the crack initiation time.

2.7.3 Experimental recordings and results

2.7.3.1 Strain signals on bars

Fig. 2.53 shows the representative experimental recordings of an 80 mm diameter specimen: the incident stress wave signal ε_i and the reflected stress wave signal ε_r recorded on the incident bar, and the transmitted stress wave signal ε_t

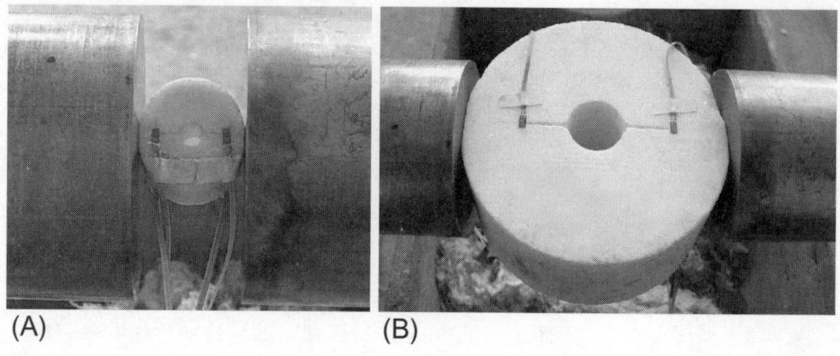

FIG. 2.52 HCFBD specimens positioned before impact by SHPB, good contact and accurate alignment achieved via the flat ends of the disc. (A) φ42 mm specimen and (B) φ155 mm specimen.

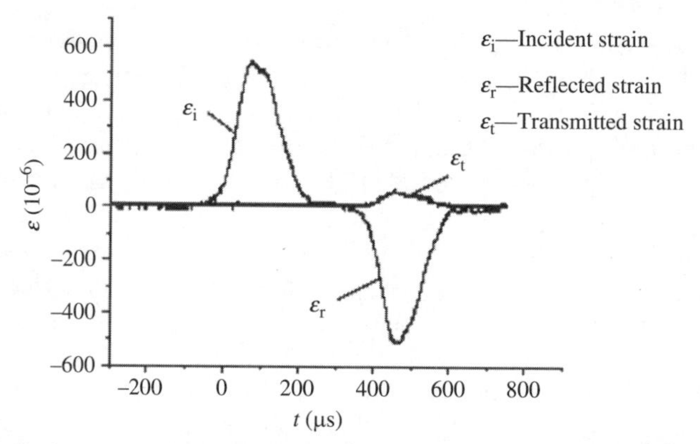

FIG. 2.53 Strain gauge signals on the bars.

on the transmission bar, as usual the compressive strain is defined as positive strain (Wang et al., 2009). It can be seen that the incident wave and reflected wave are distinctly separated, and there are no large trembles in the wave forms owing to the effective filtering function of the paper wave shaper. Fast Fourier transformation was also adopted to further smooth the wave form, as shown in Fig. 2.54.

In the test, the pressure exerted at the projectile is 0.15 MPa, the average projectile speed is 4.59 m/s, and other experimental details for the geometrically similar HCFBD specimens are given in Table 2.11.

It can be noted from Table 2.2 that the peak value of the incident stress wave ε_i is basically similar for specimens of different sizes, as the projectile speed is almost the same. The reflection coefficient c_r is the ratio of the reflected peak value over the incident peak value, and the transmission coefficient c_t is the ratio of the transmitted peak value over the incident peak value. The reflection coefficient c_r is also marginally influenced by the size of the specimen. The transmission coefficient c_t is obviously affected by the size of the specimen, as the

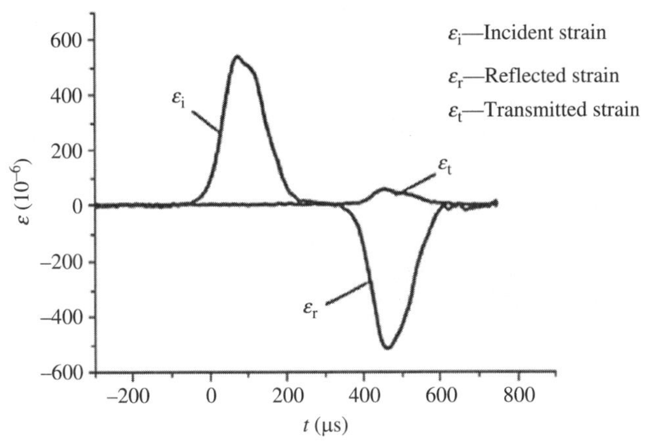

FIG. 2.54 Filtered strain gauge signals.

peak value of c_t for the largest specimen is almost an order of magnitude larger than that of the smallest specimen. It may be explained with a view of energy conservation and the passage of the wave in the specimen, as the energy is proportional to the square of the stress wave amplitude. The largest specimen only

TABLE 2.11 Experimental data for the geometrically similar HCFBD specimens

Specimen	Disc diameter D (mm)	Projectile velocity (m s^{-1})	Peak value of ε_i (με)	Reflected coefficient c_r	Transmitted coefficient c_t
M I-01	42	4.57	542.84	0.82	0.03
M I-02	42	4.86	533.25	0.91	0.03
M I-03	42	4.56	519.17	0.91	0.04
M II-01	80	4.63	540.98	0.85	0.11
M II-02	80	4.65	524.99	0.90	0.12
M II-03	80	4.64	534.51	0.91	0.15
M III-01	122	4.60	517.45	0.85	0.23
M III-02	122	4.37	516.45	0.84	0.20
M III-03	122	4.39	545.22	0.82	0.23
M IV-01	155	4.62	525.42	0.81	0.29
M IV-02	155	4.48	537.31	0.78	0.30
M IV-03	155	4.67	519.68	0.84	0.29

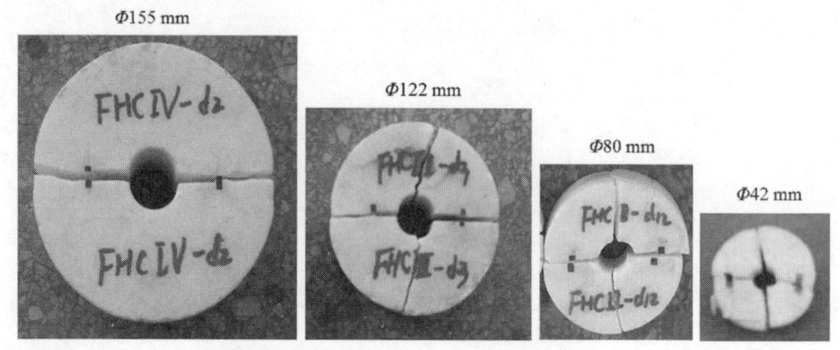

FIG. 2.55 Fracture patterns of geometrically similar HCFBD specimen.

broke into two halves along the loading diameter while the smallest specimens broke into up to six pieces flying away at high speed (see Fig. 2.6). Therefore, it may be more favorable for the largest specimen to allow the wave to pass through the specimen, hence more energy is transmitted to the transmission bar.

2.7.3.2 Fracture patterns of specimens

Geometrically similar and differently sized specimens were fractured in a way shown representatively in Fig. 2.55. All disc specimens were split along the loading diameter, and the strain gauges stuck at the front of the crack tip were cut through the middle of the gauge. No crush zone near the flat ends was observed, indicating that an ideal contact condition was realized because of the beneficial role of the flat ends. With the projectile of the same speed, more broken pieces were turned out for smaller specimens. For example, up to six broken pieces were produced for 42 mm diameter specimens, and in the test they flied away, consuming some kinetic energy. It is observed that, while 155 mm diameter specimens broke into two halves along the loading diameter, all other discs ruptured into at least four pieces. Notably, there were fractures along another diameter perpendicular to the loading direction. This phenomenon was not found in the test for dynamic tensile strength using a flattened Brazilian disc without a hole and initial crack, nor was it found in the counterpart static test using HCFBD specimens.

2.7.3.3 Test results analysis

The left end of the specimen was impacted by the incident bar (Fig. 2.52), and the time of this impact was taken as the time starting point. The load $P(t)$ was the average load derived from the loads applied on two flat ends of the HCFBD specimen; it was calculated by the incident stress wave, the reflected stress wave, and the transmission stress wave jointly (Wang et al., 2009). This load $P(t)$ was taken as input for the dynamic finite element analysis to calculate the time history of the dynamic stress intensity factor $K_{\mathrm{I}}(t)$. The commercial software ANSYS was used for the finite element computation, and six-node plane isoparametric elements (plane2 in ANSYS) were used to mesh half the disc specimen considering the symmetry, a typical mesh had 1382 elements and 2883 nodes, quarter-point elements were placed around the crack tip to capture the stress singularity, the dynamic stress intensity factor was calculated based on the displacement of the crack face. Then the dynamic fracture initiation toughness K_{Id} was determined by the crack initiation time t_f vertically intersecting the history curve of the dynamic stress intensity factor $K_I(t)$, as shown in Fig. 2.56; t_f is also called the time to fracture.

Table 12 presents test results for 22 HCFBD specimens, excluding 3 specimens of 80 mm diameter that were not tested successfully. The dynamic loading rate is defined as $\dot{K} = K_{Id}/t_f$. As shown in Table 2.12, the dynamic loading rate $\dot{K}$ is in the range of 2.00 GPa m$^{1/2}$ s^{-1} to 4.05 GPa m$^{1/2}$ s^{-1}.

Fig. 2.57A shows the mean value of the rock dynamic fracture toughness for specimens of the smallest diameter of 42 mm is 1.98 MPa m$^{1/2}$ and the mean value of the rock dynamic fracture toughness for specimens of the largest diameter of 155 mm is 3.12 MPa m$^{1/2}$; the latter is approximately 1.6 times the former. The size effect can be seen vividly, the dynamic fracture toughness increases with increment of specimen size, the law is similar to the static situation (Bazant and Kazemi, 1991). Fig. 2.57B shows for the specimens with identical diameters of 80 mm and varied relative crack lengths in the range of $2a_0/D \in [0.31, 0.51]$, the dynamic fracture toughness changes like a convex bow, the measured rock dynamic fracture toughness is basically not seriously influenced by the crack length, its mean value is 2.41 MPa m$^{1/2}$; however, this value does not reflect the size influence of outside diameter.

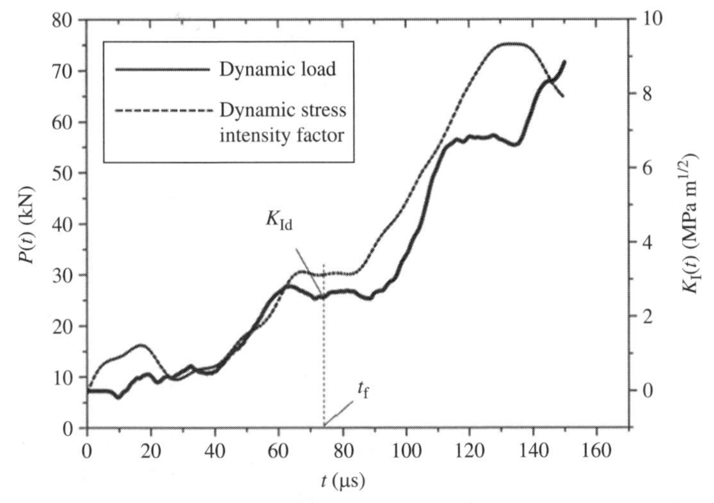

FIG. 2.56 History of dynamic load $P(t)$ and dynamic stress intensity factor $K_I(t)$ and the regular determination of dynamic fracture toughness K_{Id} using an HCFBD specimen.

TABLE 2.12 Test results for all the HCFBD specimens

Specimen	Disc diameter D (mm)	$2a_0/D$	Fracture time t_f (μs)	Loading rate $\dot{K}$ (GPa $m^{1/2}s^{-1}$)	Dynamic fracture toughness K_{Id} (MPa $m^{1/2}$)
M I-01	42	0.50	52.0	3.33	1.73
M I-02	42	0.50	55.0	4.05	2.23
M I-03	42	0.50	–	–	–
MII-01	80	0.50	76.0	3.88	2.95
MII-02	80	0.50	72.0	3.44	2.48
MII-03	80	0.50	72.0	3.07	2.21
MII-04	80	0.49	80.0	3.28	2.62
MII-05	80	0.39	85.0	3.44	2.92
MII-06	80	0.35	94.5	2.81	2.66
MII-07	80	0.36	93.0	2.23	2.07
MII-08	80	0.31	91.0	2.46	2.24
MII-09	80	0.34	81.0	2.31	1.87
MII -10	80	0.37	93.0	2.76	2.57
MII-11	80	0.39	76.0	3.11	2.36
MII-12	80	0.50	72.0	3.04	2.19
MII-13	80	0.51	85.0	2.73	2.32
M III -01	122	0.50	92.0	2.88	2.65
M III -02	122	0.50	86.0	3.51	3.02
M III -03	122	0.50	95.0	2.00	2.85
M IV -01	155	0.50	103.5	2.83	2.93
M IV -02	155	0.50	113.5	2.78	3.15
M IV -03	155	0.50	107.0	3.06	3.27

2.7.4 Dynamic stress intensity factor in spatial-temporal domain

The fracture incubation time also affects the dynamic fracture toughness in addition to the fracture process zone, and it is difficult to separate the two influencing factors. Much earlier, Kalthoff and Shockey (1977) pointed out that, "A minimum time for unstable crack growth is required because crack velocity

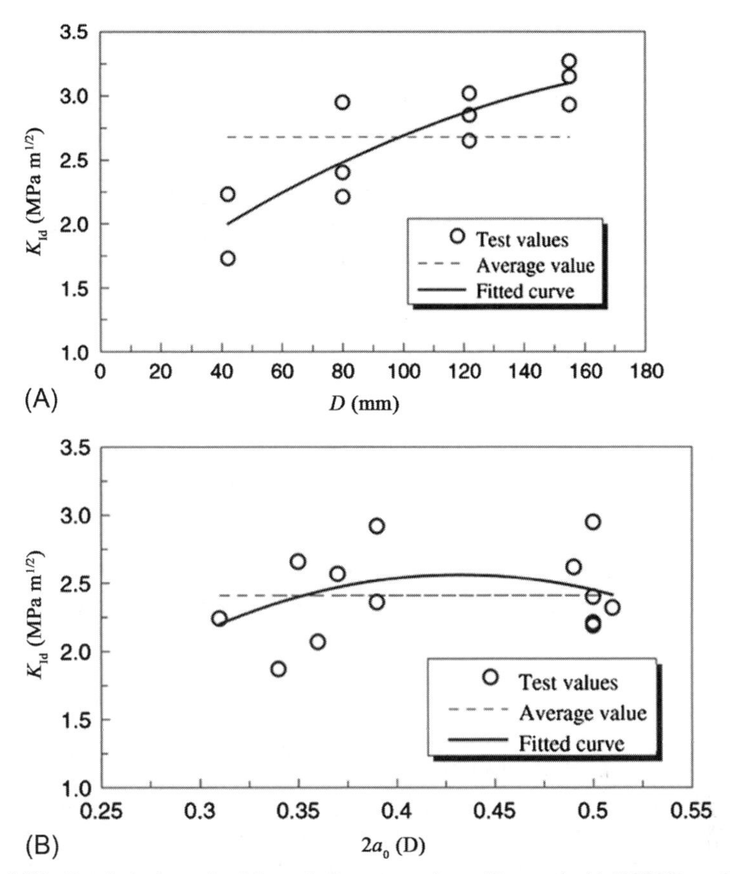

FIG. 2.57 Regularly determined dynamic fracture toughness K_{Id} tested with HCFBD specimens. (A) Geometrically similar specimens ($D = 42$, 80, 122, 155 mm) and (B) one-size specimens ($D = 80$mm, 25 mm $\leq 2a_0 \leq$ 41 mm).

does not change instantaneously from zero to the unstable value…time may be required for the process zone to build up to a critical state." Recently, Petrov et al. (2003) proposed the incubation time criterion for dynamic fracture problems, which "allows one to manage without the a priori given rate dependences of dynamic strength and fracture toughness." They assumed that "(dynamic) fracture occurs if the force momentum acting for a time τ (i.e., incubation time) reaches the critical value." However, they performed integration for the dynamic stress intensity factor in time scale only, not concurrently in the space scale (Petrov et al., 2003). Pugno used a "quantum" in space scale and a time scale to do the corresponding integration in a sense of the root mean square (Pugno, 2006).

Our understanding is that both fracture process zone (FPZ) length l and incubation time τ are material properties, and the fracture process zone length l

needs the time τ to build up. In this way, we also explain the meaning of "incubation." Although it is difficult to determine τ, we can obtain a suggested value of it, i.e., τ is assumed to be the time for the wave to travel the distance l, then $\tau = l/v$, with $v = \mathrm{d}a/\mathrm{d}t$. However, for simplicity we let $\tau = l/c$, where c is the speed of the longitudinal wave. Actually, the longitudinal wave travels the fastest among all waves, and the traveling route is zigzagged so the distance covered is longer than l. The introduction of the incubation time τ makes up the deficiency of the traditional method for the determination of dynamic fracture toughness, where only the space scale is handled, now time scale is added for the consideration. Especially worth mentioning is that τ cannot be neglected as compared to the time to fracture t_f, and t_f is affected by loading rate and specimen size. Although it is more difficult to add a time factor in the analysis for the problems of dynamic fracture, it seems to be more reasonable.

According to the idea of Pugno and Petrov on dynamic fracture, both the space quantum and the time quantum are considered as characteristic quantities (Ruiz et al., 2000; Pugno, 2006). Therefore, we define the space characteristic quantity and the time characteristic quantity more specifically, and use them in an approach we call averaging the dynamic stress intensity factor in the spatial-temporal domain. Thus, the dynamic fracture toughness for material is defined as K_{Id}^{m}, which is determined by averaging the double integration:

$$\frac{1}{l}\int \frac{1}{\tau}\int_{0}^{\tau} K_{\mathrm{I}}(\lambda, r)\mathrm{d}\lambda \mathrm{d}r = K_{\mathrm{Id}}^{m}(\tau, l) \tag{2.26}$$

where $K_{\mathrm{I}}(\lambda,r)$ is the local distribution and time history function of the dynamic stress intensity factor in the spatial-temporal domain; λ represents the time scale, the original point of λ, i.e., $\lambda = 0$ corresponds to the time $t = t_f$, the positive direction of λ is the direction that the time t decreases, this means that before crack initiation at $t = t_f$ there is a stage of incubation, characterized by the incubation time τ. Please note that t is used for experimental recordings in Figs. 2.53, 2.54, and 2.56 while λ is used in the following Fig. 2.58. Although t and λ both represent time, they are different in the starting point and positive direction, and they are also related. r is the distance to the crack tip. It is the space scale where the length of the fracture process zone l is considered, and the positive direction of r is toward the extension line of the crack. $K_{\mathrm{Id}}^{m}(\tau, l)$ is denoted as the dynamic fracture toughness that belongs to the test material and it is expected to have a marginal size effect.

Eq. (2.26) is used in the spatial-temporal domain, and it implies that fracture initiation occurs only after the FPZ length l is saturated and the incubation time τ is passed. Eq. (2.26) presents an approach of averaging integration in dual directions, one in space, the other in time. Especially worth mentioning is that the time scale is considered in the averaging integration process, which represents a marked difference from the static counterpart. In doing so, we think that the size effect of the dynamic fracture toughness may be alleviated or reduced.

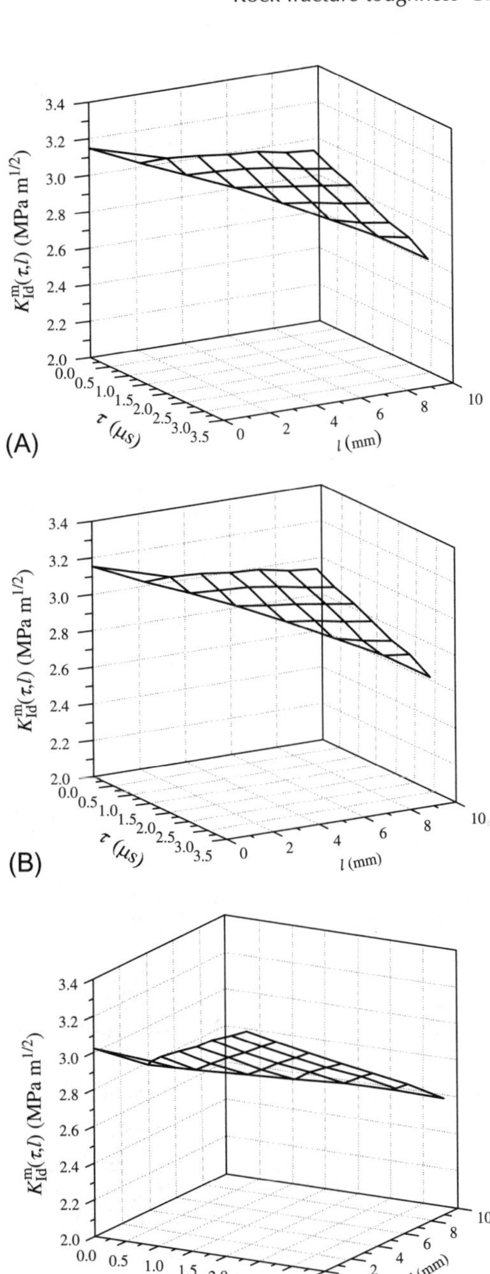

FIG. 2.58 Dynamic fracture toughness $K_{Id}^{m}(\tau, l)$ in the spatial-temporal domain. (A) Specimen MII-02 ($D = 80$ mm), (B) Specimen MIV-02 ($D = 155$ mm), and (C) Specimen MIII-02 ($D = 122$ mm).

However, it is crucial to determine the characteristic quantities. Take the FPZ length l, for example. Wecharatana and Shah (1980) pointed out that l was determined by the length of the uncracked ligament. Alexander and Blight (1986) showed through experiment that l was approximately 0.368–0.684 times the ligament length. Bazant et al. considered that l was the length characterizing the microstructure of the material. There has been no general agreement for this quantity. More than that, so far most discussions were restricted to the static situation. A similar problem exists for the determination of incubation time τ. Petrov et al. gave out $\tau = 7$ μs for ASTM 4340-steel, and $\tau = 9$ μs for Homalite-100 (Petrov et al., 2003).

We can describe in the spatial-temporal domain the variation of dynamic fracture toughness K_{Id}^{m} if FPZ length l and the incubation time τ are considered smaller or even neglected; in this way, we show the effect of l and τ. Taking HCFBD specimens, for example, divide l into n equal parts: $l = n \cdot \Delta l$, calculate the dynamic stress intensity factor $K_{\mathrm{I}}(t)$ corresponding to the crack length a_{i}, where $a_{\mathrm{i}} = a_0 + l_{\mathrm{i}}$, and $l_{\mathrm{i}} = i \cdot \Delta l (i = 0, 1, 2, \ldots n)$. For the incubation time τ, assume $\tau = l/c$, then divide τ into m equal intervals: $\tau = m \cdot \Delta\tau$, and $\tau_{\mathrm{j}} = j \cdot \Delta\tau (j = 0, 1, 2, \ldots m)$. Thus the dynamic fracture toughness $K_{\mathrm{Id}}^{m}(\tau, l_{\mathrm{i}})$ for the same incubation time τ and different FPZ length l_{i} can be determined using the following equation:

$$\frac{1}{\tau}\int_{0}^{\tau} K_{\mathrm{I}}(\lambda, l_{\mathrm{i}})\mathrm{d}\lambda = K_{\mathrm{Id}}^{m}(\tau, l_{\mathrm{i}}) \quad (i = 0, 1, 2, \ldots n) \tag{2.27}$$

Use Eq. (2.28) to get the dynamic fracture toughness for the same FPZ length l and a different incubation time τ_j:

$$\frac{1}{l}\int_{0}^{l} K_{\mathrm{I}}(\tau_{\mathrm{j}}, r)\mathrm{d}r = K_{\mathrm{Id}}^{m}(\tau_{\mathrm{j}}, l) \quad (j = 0, 1, 2, \ldots m) \tag{2.28}$$

We use Eqs. (2.27), (2.28) to illustrate how K_{Id}^{m} is calculated if either FPZ length or incubation time is considered to be fixed. If both are changing, then a curved surface will be generated, as shown in Fig. 2.58. Fig. 2.58 is a demonstration of the calculation results. The first two examples are from specimen MII-02 and MIV-02, respectively, using $l = 9.65$ mm, which is obtained from a static test for marble; no data seems to be available for dynamic FPZ length. Using $\tau = l/c$, then $\tau = 3.8$ μs, because $\Delta\tau = 0.5$ μs is used in recording the signals during experiment, so actually $\tau = 3.5$ μs is taken for the incubation time considering the discretization interval for the time scale. From Fig. 2.58A and B, we can see that if $l = 0$ and $\tau = 0$, then the dynamic fracture toughness for specimen MII-02 is 2.48 MPa m$^{1/2}$ and for MIV-02 it is 3.15 MPa m$^{1/2}$. However, if $l = 9.65$ mm and $\tau = 3.5$ μs, then the dynamic fracture toughness for specimen MII-02 is 2.66 MPa m$^{1/2}$ and 2.62 MPa m$^{1/2}$ for MIV-02. The third example is from MII-03, as shown in Fig. 2.9C. If $l = 0$ and $\tau = 0$, then the dynamic fracture toughness is 3.02 MPa m$^{1/2}$; however, if $l = 8.96$ or

10.24 mm and $\tau = 3.5$ μs, the dynamic fracture toughnesses are 2.63 MPa m$^{1/2}$ and 2.58 MPa m$^{1/2}$, respectively, and the last two values are consistent with the corresponding values obtained for MII-02 and MIV-02.

It can be seen from Fig. 2.58 that dynamic fracture toughness K_{Id}^{m} is a function of FPZ length l and the incubation time τ. If taking $l = 0$ and $\tau = 0$, then the regular and ordinary quantity K_{Id} is obtained, and K_{Id} shows a marked size effect. If l and τ can be determined for every specimen, in addition to its dynamic stress intensity factor history, then its corresponding value of K_{Id}^{m} can be obtained using Eq. (2.26), and the size effect is marginal.

2.7.5 Conclusions

Two types of the holed-cracked flattened Brazilian disc (HCFBD) specimens were impacted diametrically by the split Hopkinson pressure bar to test the dynamic fracture initiation toughness K_{Id} of marble and study its size effect. We obtained the following conclusions:

For one type of HCFBD specimen, which is geometrically similar with the different outside diameters of 42 mm, 80 mm, 122 mm, and 155 mm, respectively, corresponding to the size ratio of 1.0:1.9:2.9:3.7, while the crack length is kept to be half the diameter, K_{Id} determined with the regular method increases with increment of the specimen size.

For another type of HCFBD specimen, which is identical to the outside diameter of 80 mm and with a different crack length $2a_0$ in the range of 25 mm $\leq 2a_0 \leq$ 41 mm, the influence of crack length to K_{Id} is not obvious.

Considering that the fracture process zone length l and the fracture incubation time τ bring about the size effect for K_{Id}, we propose an approach of averaging the stress intensity factor distribution and history in the spatial-temporal domain to get a unique value for the dynamic fracture toughness, K_{Id}^{m}. In this way, the size effect of the dynamic fracture toughness is minimized.

Because both l and τ are taken into consideration, the dynamic fracture toughness K_{Id}^{m}, derived in the averaging process, is more reasonable than just taking an arithmetic mean for the specimens of different sizes. Another advantage of the approach is that the averaging process is manipulated for a single specimen but not for multiple specimens, so the required number of specimens is reduced. However, the averaging procedure is much more complicated than taking an arithmetic mean, and the determination of l and τ remains a difficult problem.

2.8 Dynamic fracture propagation toughness of P-CCNBD

2.8.1 Introduction

Dynamic fracture has been an area of sustained research for studying the behavior of engineering materials and structures under dynamic loading conditions

(Freund, 1998). It is often encountered in events such as damage to grounds and buildings by earthquakes; collision and impact of space vehicles with heavenly bodies; warhead penetration into military and civilian targets, etc. As compared with static fracture, dynamic fracture is much more complicated, considering the inertia effect or the stress wave propagation in structures. Therefore, the study on dynamic fracture requires expensive experimental setups and sophisticated computational resources. Dynamic fracture problems are usually divided into two categories: crack initiation of a static crack subjected to dynamic loading and fast crack propagation and/or arrest of a running crack. The dynamic fracture initiation toughness investigated in this paper belongs to the first category. This mechanical property is very useful for the evaluation of dynamic fracture performance of materials and structures.

In underground projects such as deep exploitation of solid resources, development of oil shale gas, and bedrock blasting of nuclear power plants, rocks are broken by the action of dynamic loads, which are related to dynamic rock fracture (Fan, 2006; She and Lin, 2014). There are two main types of rock mechanics parameters, namely dynamic crack initiation toughness and dynamic propagation toughness (Ravichandar, 2004; Zhang et al., 2014). These respectively characterize the ability of rock materials to resist crack initiation and dynamic propagation. Although research has been done, it is not yet mature, and there are fewer studies on rock dynamic crack arrest (Li et al., 2016).

Zhang et al. (2013a, b) carried out a semidisc three-point bending dynamic test using a split Hopkinson pressure bar test system to determine the dynamic cracking and propagation toughness of marble. Used a crack extension meter to monitor the velocity history of crack propagation in the Brazilian disk dynamic test of the central straight crack platform, and obtained the dynamic cracking and propagation toughness of the sandstone. Gao et al. (2015) obtained the dynamic crack initiation and extended toughness of the scuttled semidisc specimen marble in combination with the digital image correlation method. Yang et al. (2015) measured the dynamic cracking and propagation toughness of sandstone by a single-crack compression circular orifice specimen test combined with a numerical-analytical method. Dai et al. (2013) measured the dynamic fracture toughness of granite using a slotted semidisc specimen. The dynamic initiation toughness only involves the influence of the crack initiation time and the material inertia effect during the fracture process while the latter also needs to consider the crack propagation process after crack initiation and the effect of crack propagation. Therefore, the test of dynamic propagation toughness of rock materials is more difficult than dynamic crack initiation toughness.

The fracture toughness of rock, concrete, and other heterogeneous materials (Li et al., 2009) is determined by using large specimens, so a large-diameter split Hopkinson pressure bar (SHPB) test device is used. The device has mature research in the shape of the projectile (Lok et al., 2002), the loading waveform, the energy dissipation analysis (Ping et al., 2013), and the selection of the waveform shaper (Li et al., 2009; Man and Zhou, 2010).

The key factors in the study of dynamic fracture toughness are crack initiation time, crack growth rate, and time intensity of the stress intensity factor (SIF). According to the concept of the basic solution of the Green function, the universal function is derived from the correlation theory of the time-independent load on the semiinfinite crack surface of the infinitely large linear elastic body. It is then extended to the case where the crack propagates at any rate and is subjected to a general load (Ravichandar, 2004). The universal function embodies the effect of crack propagation speed on the dynamic stress intensity factor. Yang et al. (2015), Zhang et al. (2016), and Zhang et al. (2014) have applied the universal function to the study of rock dynamic propagation toughness.

Based on past achievements (Yang et al., 2015; Dai et al., 2013; Man and Zhou, 2010; Zhang et al., 2016; 2014), this paper promotes the work of the literature (Dai et al., 2013): On the one hand, a precracked chevron notched Brazilian disc (P-CCNBD) specimen was used for experimental research, overcoming the shortcomings of CCNBD's complex dynamic cracking and propagation under impact loading (Man and Zhou, 2010). On the other hand, only the dynamic crack initiation toughness has been studied in the literature (Dai et al., 2013; Man and Zhou, 2010), and the measurement of dynamic propagation toughness and the study of the crack arrest of P-CCNBD specimens have not been studied. In this paper, a dynamic impact test of a 160 mm diameter P-CCNBD specimen is carried out using a SHPB dynamic test device with a diameter of 100 mm. A crack propagation gauge (CPG) and a strain gauge (SG) are used to monitor the crack initiation time, propagation speed, and crack arrest of the specimen. The experimental-numerical-analytical method, including the application of universal function, was used to determine the dynamic fracture toughness and dynamic propagation toughness of Nanyang marble, and the possibility of crack arrest of P-CCNBD specimens was discussed.

2.8.2 Experimental preparation

2.8.2.1 P-CCNBD specimen

The material of the specimen is selected from the Nanyang marble of Henan Province. The rock is a fine-grained crystal structure with crystal grains between 0.1 and 0.3 mm. The main mineral components are calcite, dolomite, and wollastonite. The material density is 2.762 g/cm^3. The pine ratio is 0.26 and the modulus of elasticity is 69.04 GPa while the longitudinal wave velocity $c_d = 5681.1$ m/s, the transverse wave velocity $c_s = 3235.3$ m/s, and the Rayleigh wave speed $c_R = 2979.8$ m/s. On the RMT-150B rock mechanics test machine of Henan Polytechnic University, the average static fracture toughness of the marble was measured to be 1.57 MPa m$^{1/2}$.

The production process of the P-CCNBD specimen is shown in Fig. 2.59.

First, the homogeneous stone is made into a cylinder, and then four steps of disc cutting, end grinding, mechanical slitting (herringbone grooving), and artificial precracking (straight crack) are carried out.

(1) Using an RLS-100 automatic cutter manufactured by GCTS, a cylinder having a diameter of 160 mm was cut into a Brazilian disc rock specimen having a thickness of 50 mm.

(2) Use an SHM-200B double-face grinding machine to flatten both ends of the disc specimen to ensure that the parallelism deviation of the two ends of the specimen is not more than 0.02 mm.

(3) Fix discs fixed by self-manufacturing of polymer nylon rod materials to ensure that the grooving surfaces formed by the two cutting faces coincide. The cutting machine with the RSG-200 grinder is used to cut the center of the disc to form a herringbone groove, and the circular cutter with a grooving diameter is 100 mm.

(4) A sharp crack is formed on the crack tip of the specimen by using a polished thin saw blade (saw blade thickness of 0.6 mm).

Fig. 2.60 shows the geometry of the P-CCNBD specimen. The specimen was obtained by grinding a slotted tip into a straight crack on the basis of a cracked chevron notched Brazilian disc (CCNBD). B is the thickness of the disc, R is the

FIG. 2.59 Preparation process of P-CCNBD specimens. (A) Cutting disc, (B) grinding contact surface, (C) mechanical slit, (D) artificial prefabricated straight crack.

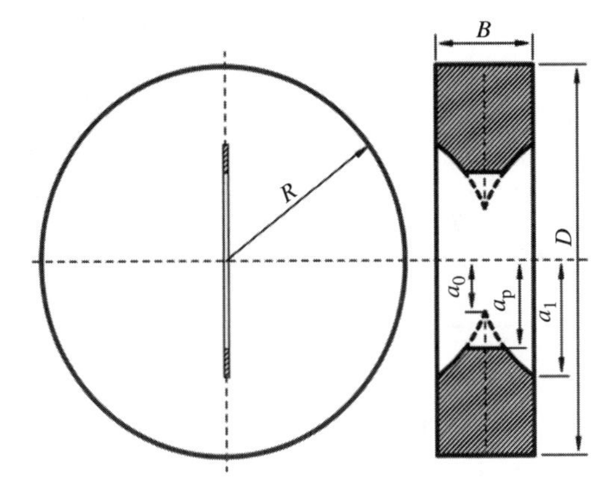

FIG. 2.60 Design of the P-CCNBD specimen.

TABLE 2.13 Geometric parameters of P-CCNBD specimen

D (mm)	B (mm)	a_0 (mm)	a_1 (mm)	a_p (mm)
160	50	25	46.8	36.5

radius of the disc, D is the diameter of the disc, and the slotted round tool has a diameter of 100 mm. a is the initial grooving length, a_1 is the maximum grooving length, and a_p is the precrack length. The geometric parameters of the specimens are shown in Table 2.13.

Table 2.13. Geometric parameters of the P-CCNBD specimen.

2.8.2.2 SHPB loading device

The dynamic test was carried out on an SHPB loading device with a plunger diameter of 100 mm in the Laboratory for Penetration of Explosive Effects of the Third Institute of Engineering Research, Luoyang Engineering Corps. The SHPB elastic bar material is 42 CrMo, the density is 7.85 g/cm^3, the elastic modulus is 210 GPa, the Poisson's ratio is 0.3, and the measured velocity of the elastic bar is 5244 m/s. The SHPB pressure bar has a diameter of 100 mm, the incident bar length is $l_i = 4500$ mm, and the transmission bar length is $l_t = 2500$ mm. The SHPB loading device is shown in Fig. 2.61. The distance between the strain gauge on the incident bar and the contact end of the specimen and the incident bar is $l_1 = 1500$ mm. The strain gauge on the transmission bar is at a distance of $l_2 = 1000$ mm from the contact end of the specimen to the transmission bar.

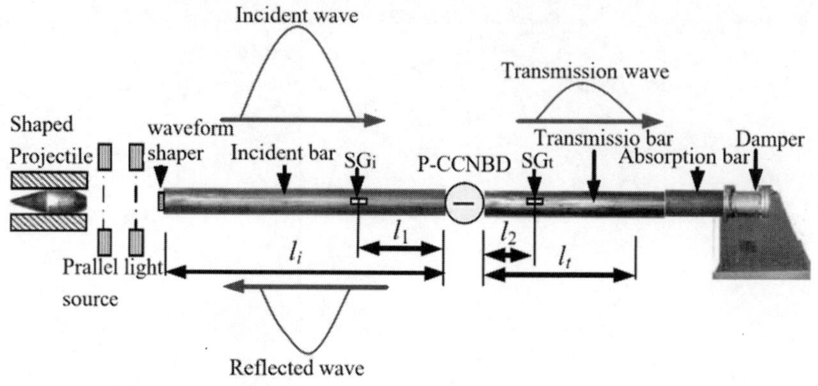

FIG. 2.61 Schematic of the SHPB setup.

2.8.2.3 Strain gauges and crack extension meters

As shown in Fig. 2.62, SG is used to monitor the crack initiation time of the crack in the P-CCNBD specimen, and CPG is used to monitor the crack initiation time and propagation law of the external crack of P-CCNBD. The initial total resistance of the CPG is about 2 Ω, which is made up of 10 Kama copper sheets with different resistances. The total length of CPG is $l = 10$ mm, the width is $h = 5$ mm, and the spacing between the two adjacent wires is $l_0 = 1.11$ mm.

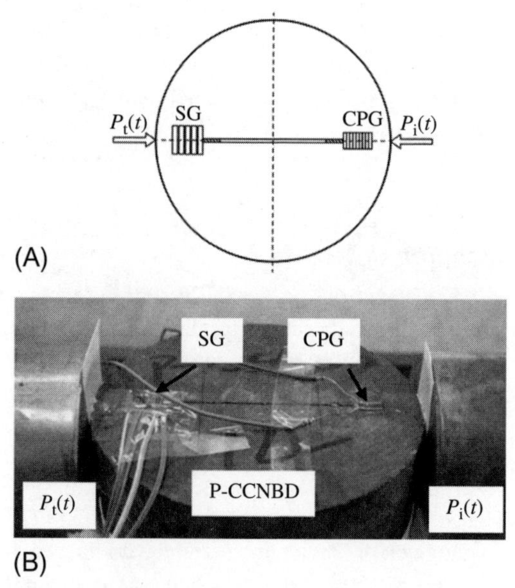

FIG. 2.62 Positions of SG and CPG on P-CCNBD. (A) Schematic diagram of SG and CPG adhesion on P-CCNBD and (B) real graphics of P-CCNBD.

The CPG circuit uses a constant voltage source to provide 20 V constant voltage. The resistor $R_{C2} = 50\ \Omega$ is connected in parallel with the CPG and then connected in series with the large resistor $R_{C1} = 1076\ \Omega$. When the crack of the specimen passes vertically through the positions of the resistive sheets of the CPG and causes the corresponding resistance wires to break, the parallel total resistance of CPG and R_{C2} will be sequentially mutated accordingly. In the test, the position reached by the crack tip can be judged based on the change of the voltage signal across the CPG. CPG has the advantages of simplicity, sensitivity, etc., in testing crack initiation and propagation (Zhang et al., 2013a, b, Yang et al., 2015; Zhang et al., 2016).

2.8.3 Experimental recording and data processing

2.8.3.1 Load determination

The signals of the incident wave and the reflected wave are measured by SG_i, and the transmitted wave signal is measured by SG_r. The signal on the incident and transmission bars during the test of M06 is shown in Fig. 2.63.

After the projectile hits the incident rod, an incident wave $\varepsilon_i(t)$ is generated in the incident bar. The incident wave passes through the strain gauge SG_i on the incident bar at time t_i, and reaches the contact end of the specimen and the incident bar at time t_i. Then, part of the incident wave is reflected by the end face to form a reflected wave $\varepsilon_r(t)$ and passes through the strain gauge SG_i on the incident rod at t_r. Part of the incident wave propagates in the specimen for a period of time Δt and reaches the contact end of the specimen and the transmission rod at t_1 to form a transmitted wave $\varepsilon_t(t)$, and passes the strain gauge SG_t on the transmission bar at time t_t.

The 1/5 method (Liu and Li, 1999) was used to determine the wave heads of the incident and transmitted waves t_i and t_t. The wave head of the reflected wave is $t_r = t_i + 2\,l_1/C_b$. The time at which the incident end of the specimen is

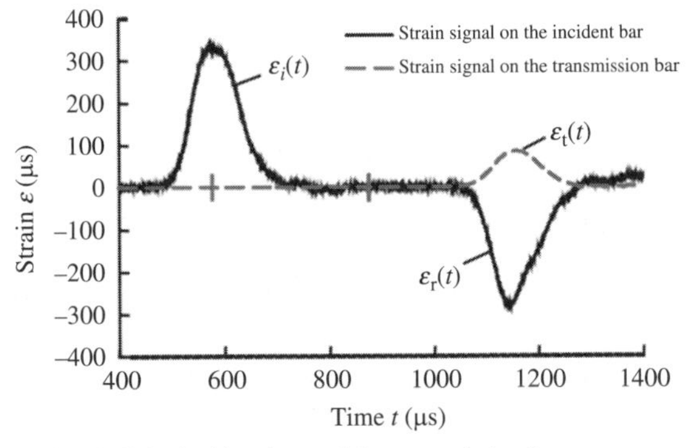

FIG. 2.63 Strain signal of the incident bar and the transmission bar.

subjected to the load is $t_0 = t_i + l_1/C_b$. The propagation time of the stress wave in the specimen is $\Delta t = t_t - l_1/C_b$.

The SHPB pressure bar satisfies the one-dimensional elastic stress wave assumption. The action of the incident rod on the incident end of the specimen $P_i(t)$ can be obtained by the superposition of the incident wave and the reflected wave. The force of the transmission rod on the transmission end of the specimen $P_t(t)$ can be calculated from the transmitted wave. Their expressions are:

$$\left.\begin{aligned} \mathrm{P_i}(t) &= E_b A_b[\varepsilon_i(t) + \varepsilon_r(t)] \\ \mathrm{P_i}(t) &= E_b A_b \varepsilon_t(t) \end{aligned}\right\} \tag{2.29}$$

where A_b and E_b are the cross-sectional area and elastic modulus of the SHPB pressure bar, respectively. $\varepsilon_i(t)$, $\varepsilon_r(t)$,and $\varepsilon_t(t)$ are the incident strain, the reflected strain, and the transmission strain, respectively. During the test, the time period that affects the specimen is after the stress wave is first transmitted to the incident end of the specimen. Therefore, the time t_0 at which the stress wave reaches the incident end face of the specimen is defined as 0, that is, $t_0 = 0$.

The load time history curve of the incident end and the transmitting end of specimen M06 can be obtained according to Formula (2.29), as shown in Fig. 2.64.

It can be seen from Fig. 2.64 that the load history of the incident end and the transmissive end of the specimen is greatly different. After the stress wave passes through the specimen, the peak value decreases and the wavelength is elongated. This phenomenon is caused by the stress wave transmitting part of the energy to the specimen during the action on the specimen. This part of the energy includes the elastic strain energy of the specimen, the fracture energy of the specimen fracture, the kinetic energy of the debris after the specimen is destroyed, and other forms of energy such as acoustic energy, thermal energy, and electromagnetic radiation (Li et al., 2009). A significant difference in the

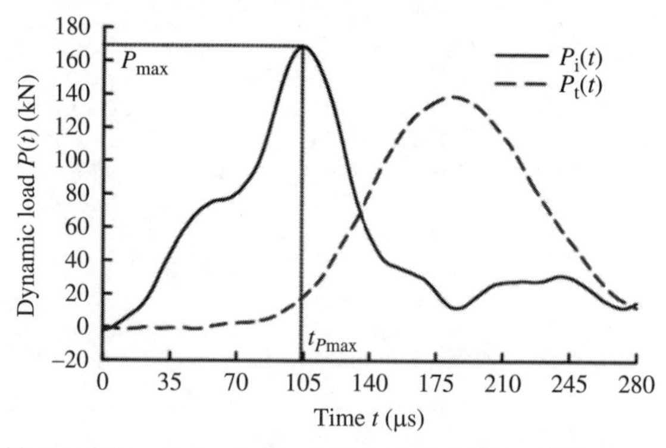

FIG. 2.64 Incident and transmission loading waves of the M06 specimen.

load at both ends of the specimen means that the specimen does not reach the stress balance during the fracture process.

2.8.3.2 Determination of cracking time

Because the P-CCNBD specimen has two crack tips, the crack initiation moments of the two crack tips need to be determined separately (Fig. 2.65). It can be seen from Fig. 2.65 that the crack initiation time of CPG1 of the M06 specimen is 18.7 s earlier than the crack initiation time of CPG2, and the crack propagation on the path of CPG1 has ended before the cracking of CPG2. This indicates that the cracking and propagation of the large-sized P-CCNBD specimen is asymmetric. The crack initiation time and propagation velocity on the crack initiation and propagation cracks should be used as the time to test the fracture toughness and the crack propagation velocity.

Because the three-dimensional crack tip of the P-CCNBD specimen is not easily measured directly inside the specimen, the crack initiation signal is monitored by using the SG1–SG4 strain gauge on the outer surface of the specimen crack. The voltage signal changes of SG3 and CPG1 are shown in Fig. 2.66.

The principle of strain gauges for monitoring internal cracks is that the tensile strain at the strain gauges on the outer surface of the crack tip rises sharply as the pressure bar loads the specimen. When the crack tip is cracked, unloading will occur, and the strain monitored by the strain gauge will decrease, resulting in a distinct peak (Jiang et al., 2004). The crack propagation speed of this paper is very fast, and the strain signal strain lags behind the crack propagation. The strain signal of SG has not reached the peak after the crack of the outer crack of the CPG1 path. Therefore, the SG is not sensitive enough to monitor the cracking of internal cracks, and the monitoring signal lags behind the true cracking moment of the internal crack. Therefore, this paper uses the average velocity of crack propagation to calculate the crack initiation moment of internal cracks: $t_f = t_1 - (a_1 - a_p)/v_a$. Where t_f, t_1, and v_a represent the crack initiation time of

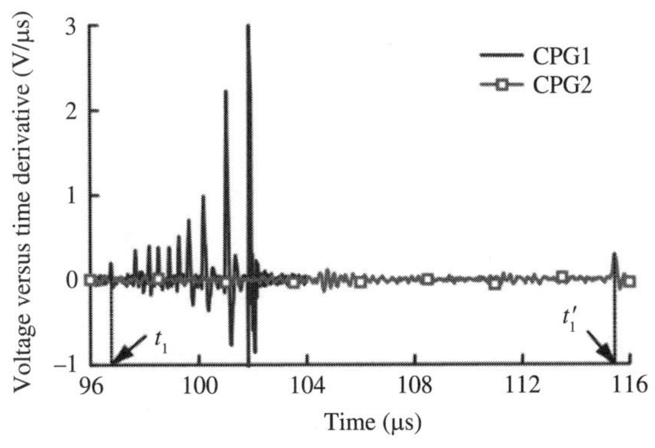

FIG. 2.65 Voltage vs time derivatives of CPG1 and CPG2.

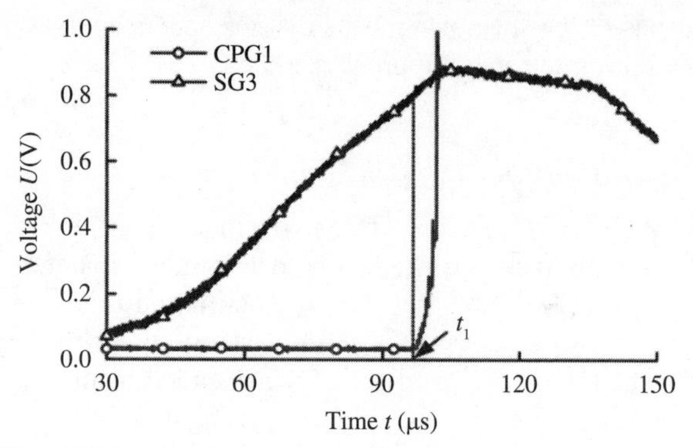

FIG. 2.66 SG3 and CPG voltage signal of specimen M06.

the specimen, the breaking moment of the first filament of CPG, and the average propagation speed of the crack, respectively.

2.8.3.3 Determination of crack propagation speed

The CPG voltage signal of specimen M06 and the derivative of voltage vs time are shown in Fig. 2.67. It can be seen from the figure that the voltage on the CPG rises stepwise, and the abrupt moments of the 10 steps correspond to the breaking moments of the corresponding resistive sheets on the CPG. The time of the break of the 10 resistance wires on the CPG is determined by the time corresponding to the derivative of the voltage vs time, t_1 to t_{10}, and the difference in the break time between the two adjacent resistor sheets is Δt_1 to Δt_9. The relationship between the position of the crack tip of specimen M06 and time is shown in Fig. 2.68. The velocity value of the crack in the CPG measurement range can be obtained from the distance between the two grids of the CPG and the difference in the break time between the corresponding two adjacent resistor sheets: $v_t = l_0/\Delta t_i$ ($i = 1$–9). As can be seen from Fig. 2.68, the maximum

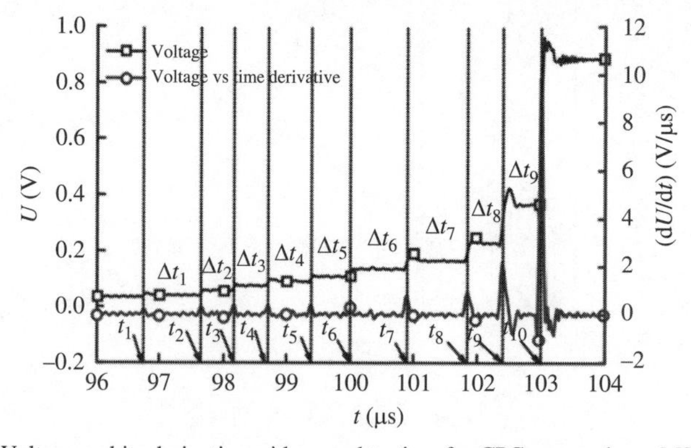

FIG. 2.67 Voltage and its derivative with regard to time for CPG on specimen M06.

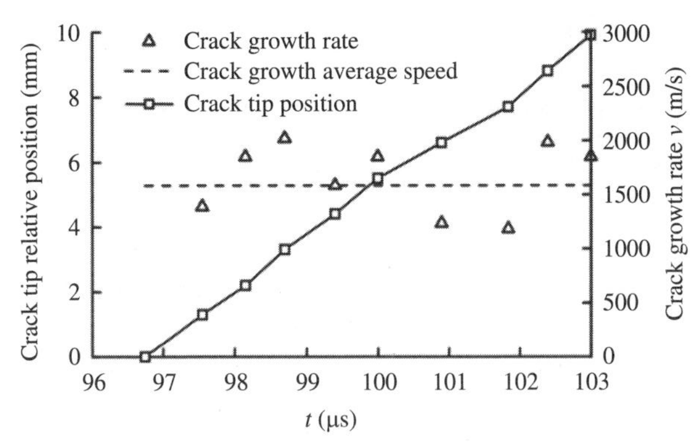

FIG. 2.68 Crack-tip position and crack propagation velocity of specimen M06.

propagation speed of the M06 specimen is $v_{max} = 2018.2$ m/s, minimum speed $v_{min} = 1180.9$ m/s, and average speed $v_a = 1584.0$ m/s.

Fig. 2.69 is an extended path diagram of crack propagation at high speed. It can be seen from Figs. 2.68 and 2.69 that the velocity of the crack propagates to a certain degree of up and down turbulence, and the propagation path of the crack propagation is unstable. This phenomenon can be explained in two ways. The first is the commonality of quasibrittle materials. After the brittle material and the quasibrittle material break, the crack propagation speed is very fast (Zhang et al., 2013a, b; Rabczuk et al., 2009). When the crack propagation velocity reaches a certain critical value, the crack propagation velocity begins to oscillate. There is a parabolic trench on the extended section, and the value on the crack propagation path is also indeterminate. Fineberg et al. (1992) believed that the Rayleigh wave velocity with a crack propagation speed exceeding 0.36 times will have a crack propagation velocity oscillation. However, the average velocity v_a of the crack propagation in this paper is 0.53 times that of the specimen

FIG. 2.69 Path of crack high-speed propagation.

Rayleigh wave velocity R_c, so crack propagation velocity oscillation and extended path tortuosity occur. The second is the characteristics of rock materials. As a typical heterogeneous pore material, rock contains a large number of randomly distributed micropores, microcracks, and weak interlayers in addition to the mineral particles that make up the rock material. When the crack propagates to the micropores and the weak interlayer, it will cause the crack propagation speed to increase. When the crack propagates to the mineral particles, the propagation of the mineral particles will cause the propagation speed to decrease, and expand along the mineral particle boundary or microjoin plane. This will cause the extension path to change. This is also an important reason for the rock material crack propagation velocity oscillation and the extension path tortuosity.

In order to reduce the influence of the propagation speed oscillation on the test results, the average speed v_a is extended as the crack propagation speed representative value at the midpoint of the CPG at the time of the crack $t_p = (t_1 + t_{10})/2$. t_p represents the moment at which the crack propagates to the midpoint of the CPG. Table 2.14 shows the crack initiation time t_f, the average propagation velocity v_a, and the time t_p at which the crack propagates to the midpoint of the CPG.

2.8.4 Numerical calculation of dynamic stress intensity factor

2.8.4.1 Loading of model

There are generally three ways to select the dynamic load at the incident end of the specimen: the one-wave method $P_1(t)$, the two-wave method $P_2(t)$, and the three-wave method $P_3(t)$ (Jiang and Vecchio, 2009). $P_1(t)$ is the result of the transmission end load $P_t(t)$ forward translation Δt and the $P_i(t)$ wave head, $P_2(t)$ is the incident end load $P_i(t)$, and $P_3(t)$ is the average of $P_1(t)$ and $P_2(t)$. The dynamic load is chosen differently depending on the configuration of the specimen and the purpose of the test. The specimen size of this test is large, and it is difficult to achieve stress balance during the test. The number of times the stress wave reflected back and forth in the specimen before the cracking of the P-CCNBD specimen was $(c_d \times t_f/(2D)) \approx 1.64$ times. Studies have shown that the stress wave can be considered to reach the stress balance after being reflected back and forth within the specimen 3–5 times (Jiang and Vecchio, 2009). Therefore, the internal stress is not balanced when the specimen is destroyed, and the load-time history at both ends of the specimen is large. The difference is that a wave method cannot be used as the applied load. In addition, the research method in this paper is the experimental-numerical-analytic method, which does not need to satisfy the quasistatic stress uniformity hypothesis, and only needs to satisfy the real load of the numerical model as close as possible to the incident end. However, the transmitted wave is the waveform of

TABLE 2.14 Experimental data of P-CCNBD specimens

P-CCNBD specimen	t_i (μs)	t_i (μs)	t_i (μs)	P_{max} (Kn)	t_{Pmax} (μs)	t_f (μs)	t_1 (μs)	t_{10} (μs)	t_p (μs)	v_a (m/s)
M01	490.3	1062.3	776.3	138.94	117.68	111.24	118.63	125.80	122.22	1393.3
M02	482.1	1054.1	768.1	130.72	132.48	107.17	114.33	121.28	117.80	1437.9
M03	498.5	1070.5	784.5	147.16	102.88	89.93	97.30	104.44	100.87	1398.4
M05	490.0	1062.0	776.0	146.51	109.10	101.67	108.50	115.12	111.81	1508.0
M06	499.4	1071.4	785.4	168.70	104.75	90.25	96.75	103.06	99.85	1584.0
M07	490.0	1062.0	776.0	133.92	104.10	94.44	101.80	108.94	105.37	1398.8
M08	492.0	1064.0	778.0	168.47	97.78	97.31	103.75	110.00	106.88	1598.4
M09	494.3	1066.3	780.3	146.84	105.99	95.96	102.75	109.34	106.04	1516.5

the stress wave reaching the transmission rod after being reflected and scattered by the inner and outer surfaces of the specimen and the crack surface during the propagation process, and the specimen has a certain degree of waveform shaping effect on the transmitted wave (Zhang et al., 2013a, b). Although the two-wave method has some human interference when the incident wave and the reflected wave are superimposed, the propagation law of the incident wave and the reflected wave in the SHPB pressure bar is one-dimensional elastic (Lu, 2013), so the two-wave method can better reflect the true load at the incident end of the specimen when the superposition calculation is reasonable. Therefore, this study uses the two-wave method, that is, the load on the left end of the specimen as the applied load of the dynamic test.

2.8.4.2 P-CCNBD numerical model

To ensure the accuracy of the ANSYS calculation results in the paper, the finite element analysis of the classical Chen problem (Chen and Wilkins, 1975) was carried out. Compared with Chen's results, the dynamic stress intensity factor time history curves obtained by the two are in good agreement. On this basis, according to the symmetry of the P-CCNBD specimen, the 1/4 model was established by the finite element software ANSYS, as shown in Fig. 2.70.

The model entity uses Solid 95 solid elements to mesh, and the singularity of the crack tip stress field and strain field is represented by 1/4 node singular

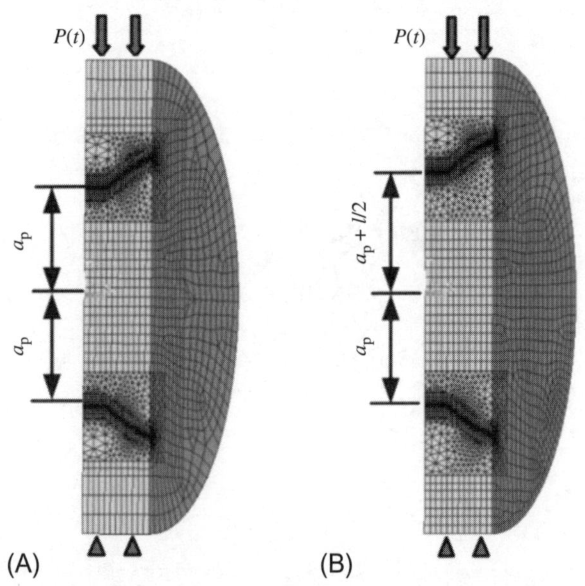

FIG. 2.70 1/4 finite element model of P-CCNBD. (A) Cracking finite element model and (B) propagation finite element model.

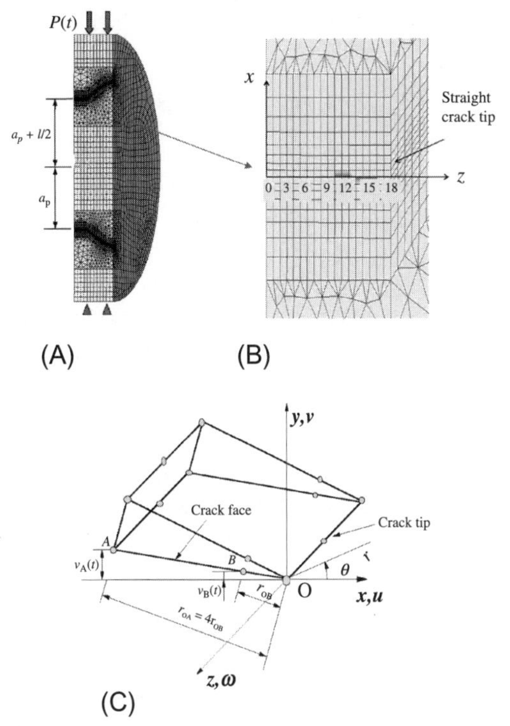

FIG. 2.71 Straight crack coordinate system and crack-tip coordinate system. (A) P-CCNBD straight crack coordinate system and (B) crack tip coordinate system.

elements. The three-dimensional singular unit division of the crack tip region is shown in Fig. 2.71A. The three-dimensional 1/4 node singular element coordinate system is shown in Fig. 2.71B. The model has a total of 33,422 units and 90,847 nodes. The incident end face dynamic loading history P_i(t) calculated by the first calculation of the Formula (2.29) is taken as the dynamic loading load of the specimen. The dynamic stress intensity factor of the static crack of the P-CCNBD specimen is determined by the displacement-time history of the crack tip node according to Eq. (2.30) (Chen and Sih, 1977), and the dynamic stress intensity factor obtained by the test-value-analysis method is represented by $K_I^0(t)$

$$K_I^0(t) = \frac{\sqrt{2\pi}E}{24(1-\mu^2)} \frac{8v_B(t) - v_A(t)}{\sqrt{r_{OB}}} \tag{2.30}$$

where E and μ are the elastic modulus and Poisson's ratio of the rock material, respectively; r_{OB} is 1/4 of the side length r_{OA} of the singular element; $v_A(t)$ is the time history of the displacement of the A node in the y direction in Fig. 2.13B;

and $v_B(t)$ is the time history of the displacement of the B node in the y direction in Fig. 2.71B.

2.8.4.3 Dynamic stress intensity factor

A local coordinate system is established at the center of the specimen thickness $Z = 0$ as shown in Fig. 2.71A. The Z coordinate in Fig. 2.71A indicates that the straight crack section of the three-dimensional crack is equally divided into 18 parts. The time history curve of the dynamic stress intensity factor of each layer of the straight crack is calculated by using the finite element model shown in Fig. 2.70A. It can be seen from Fig. 2.72 that the dynamic stress intensity factor of each layer node increases with the increase of Z in the local coordinate system; however, the final three points are significantly larger due to the influence of the edge arc grooving. Therefore, when calculating the dynamic fracture toughness of the specimen, the last three points affected by the edge arc grooving should be removed, and the average value of the previous data points is taken (Wang et al., 2013).

When the finite element model is shown in Fig. 2.70A, the crack length is pa, and the obtained dynamic stress intensity factor is recorded as $K_I^0(t, a_p)$, which is used to determine the dynamic fracture toughness of marble. When the finite element model is Fig. 2.70B, the crack length is $a_p' = a_p + l/2$ and the measured dynamic stress intensity factor is $K_I^0(t, a_p')$, which is used to determine the dynamic propagation toughness of marble. At $Z = 0$, the time-history curve of the dynamic stress intensity factor of the static crack at the two crack lengths of the M06 specimen is shown in Fig. 2.73.

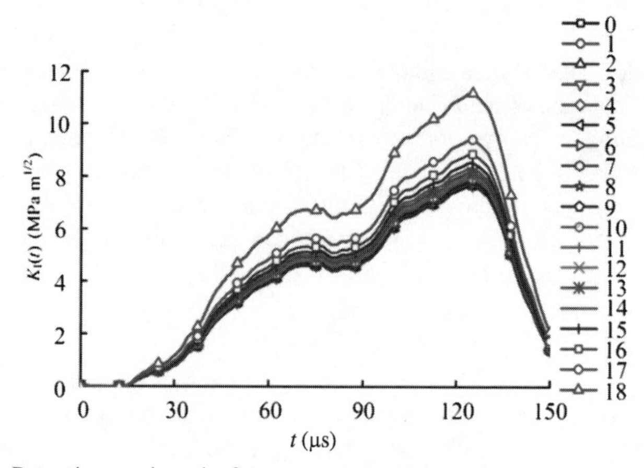

FIG. 2.72 Dynamic stress intensity factor curves on straight crack.

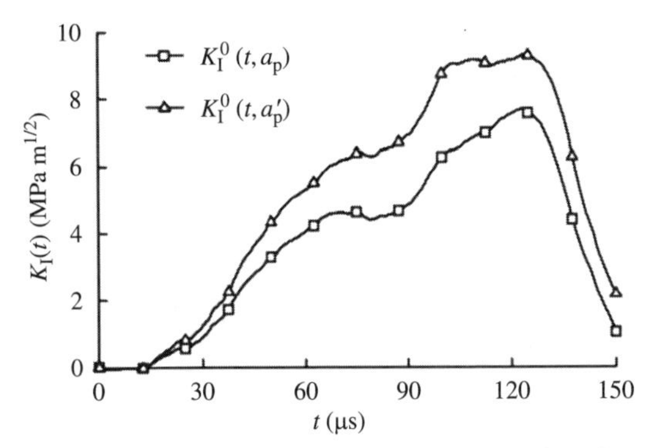

FIG. 2.73 Dynamic SIF of two types of crack lengths at the point of $Z=0$.

2.8.5 Determine dynamic fracture toughness

2.8.5.1 Universal function

Under the general load, the dynamic stress intensity factor of the open crack propagated at any speed is equal to the product of the stress intensity factor of the nonexpanded crack and the universal function value of the instantaneous crack velocity (Ravichandar, 2004), The universal function embodies the effect of the speed of the motion crack propagation on the dynamic stress intensity factor:

$$K_I{}^d(t) = k(v)\,K_I{}^0\left(t, a_p'\right) \tag{2.31}$$

where $K_I^d(t)$ is the dynamic stress intensity factor of the crack propagated at velocity v at time t; $K_I^0(t, a_p')$ is the dynamic stress intensity factor of the crack propagation to the static crack at a_p' under the same dynamic load at time t; and $k(v)$ is a universal function related to the crack velocity v. The approximate calculation formula of the universal function is

$$k(v) \approx (1 - v/c_R)/\sqrt{(1 - v/c_d)} \tag{2.32}$$

where c_d is the material propagation wave velocity; c_s is the shear wave velocity of the material; and c_R is the material Rayleigh wave velocity. It can be known from Eq. (2.32) that when the crack does not propagate, that is, $v=0$, $k(0)=1$. When the crack propagation speed $v=c_R$, the $k(c_R)=0$ crack rate monotonically decreases from 0 to c_R.

2.8.5.2 Dynamic cracking and propagated toughness

Before the P-CCNBD specimen is subjected to dynamic loading but not cracked, the crack propagation velocity is $v=0$, and the corresponding universal function has a value of $K(v=0)=1$. It can be seen from Eq. (2.33) that the dynamic stress intensity factor at this time is

$$K_{IC}^{D}\left(\dot{K}_{I}^{D}\right)=K_{I}^{d}\left(t, a_{p}, \quad v=0\right) \tag{2.33}$$

where K_{IC}^{D} is the dynamic crack initiation toughness, $\dot{K}_{I}^{D}=K_{IC}^{D}/t_{f}$ is the dynamic loading rate, and the dynamic cracking toughness is only related to the dynamic loading rate without considering the influence of other factors (FAN Tian-you et al., 2006). The dynamic fracture toughness of the P-CCNBD marble specimen is shown in Fig. 2.74.

It can be seen from Fig. 2.74 that the dynamic cracking toughness of the measured marble of the M06 specimen is 4.84 MPa m$^{1/2}$, corresponding to the loading rate $\dot{K}_{I}=5.28\,\text{GPa m}^{1/2}\text{ s}^{-1}$.

When the M06 specimen is under dynamic load and the crack initiation is propagating at a constant speed v_a, the corresponding universal function value is $k(v_a)$. It can be known from Eq. (2.31) that the dynamic stress intensity factor of the extended crack tip needs to be corrected by the universal function at the corresponding velocity to the dynamic stress intensity factor $K_{I}^{d}(t, a_p')$ in Fig. 2.70B, that is, $K_{I}^{d}(t)=k(v_a)K_{I}^{0}(t, a_p')$. If the influence of temperature is not considered, the dynamic propagation toughness of marble can be determined according to the dynamic propagation criterion.

$$K_{IC}^{d}(v_a)=K_{I}^{d}\left(t_p, a_p', v=v_a\right) \tag{2.34}$$

where K_{IC}^{d} is the dynamic propagation toughness. If the influence of temperature is not considered, the dynamic propagation toughness of the specimen is only related to the crack propagation speed (Fan, 2006; She and Lin, 2014; Ravichandar, 2004). As can be seen from Table 2.14, the average propagation speed of the M06 specimen $v_a=1584.0$ m/s; therefore, the dynamic propagation toughness at this time is 4.53 MPa m$^{1/2}$ (Fig. 2.74).

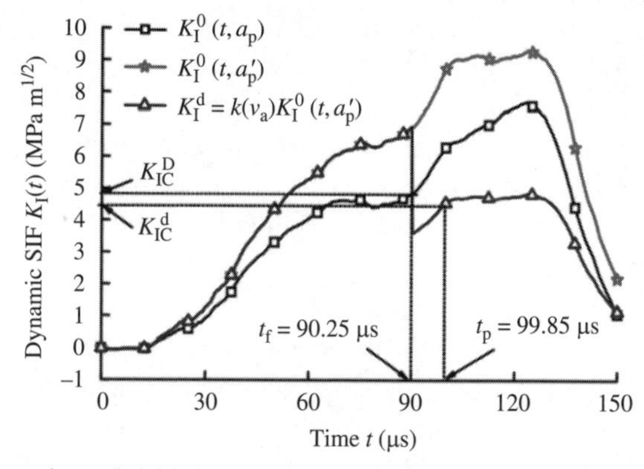

FIG. 2.74 Dynamic crack initiation toughness and dynamic crack propagation toughness of specimen M06.

It can be seen from the figure that the dynamic propagation toughness of the P-CCNBD specimen is slightly lower than the dynamic fracture toughness. This is because when the crack spreads rapidly, the motion crack generates kinetic energy. Combined with the universal function theory, when the propagation speed of the motion crack is large, the dynamic propagation toughness of the specimen will be smaller than the dynamic cracking toughness after the universal function correction. The calculation results of each P-CCNBD specimen obtained by the test-numerical-analytical method are shown in Table 2.15.

2.8.5.3 *Loading rate effect on dynamic cracking toughness*

The dynamic loading test must take into account the loading rate effect of the dynamic load and the inertial effect of the material. Therefore, during dynamic loading, due to the short loading time and high loading rate, the crack tip is not as good as the tiny cracks, which results in the dynamic cracking toughness being greater than the static cracking toughness under the same load amplitude. In the P-CCNBD dynamic test, the loading rate increased from 4.63×104 MPa $m^{1/2}$ s^{-1} to 7.17×10^4 MPa $m^{1/2}$ s^{-1}, causing the corresponding dynamic initiation toughness to increase from 4.68 MPa $m^{1/2}$ to 6.96 MPa $m^{1/2}$. The results obtained are compared with similar research results at home and abroad, as shown in Fig. 2.75.

It can be seen from Fig. 2.75 that the dynamic fracture toughness of marble has a significant upward trend with the increase of loading rate, which is consistent with the research results of other scholars (Zhang et al., 2014).

2.8.5.4 *Crack propagation speed on dynamic expansion toughness*

Once the crack of the rock material rapidly expands, a high stress concentration zone passes through the rock, producing a high strain rate or loading rate near the crack tip. This results in an enhanced plastic effect at the crack tip. At the same time, according to the microcrack expansion mode, there are two modes of intergranular fracture and transgranular fracture. Transgranular fracture consumes more energy than intergranular fracture. When the crack propagates in quasistatic or at low speed, the microdestruction mode is mainly along the crystal fracture. When the crack propagates at high speed, the microdestruction mode is an intergranular and transgranular coupling fracture or a transgranular fracture. The relationship between the dynamic expansion toughness of the P-CCNBD marble specimen and the crack propagation velocity is shown in Fig. 2.76.

It can be seen from Fig. 2.76 that the crack propagation speed range of this test is $(0.47 - 0.57)c_R$. In this range, the dynamic expansion toughness of marble increases with the increase of crack propagation speed, which is due to the fact that the crack propagation speed is fast and relatively oscillating while the extended path is curved, and the total path of crack propagation is not considered.

TABLE 2.15 Dynamic fracture toughness obtained by the experimental-numerical method

Specimen no.	$K_I^0(t_f)$ (MPa m$^{1/2}$)	$K_I^0(t_f)$ (MPa m$^{1/2}$)	$K_I^0(t_f)$ (MPa m$^{1/2}$)	$K_I^0(t_f)$ (MPa m$^{1/2}$)	v/c_R	$k(v)$	$K_I^0(t_f)$ (MPa m$^{1/2}$)
M01	4.97	4.97	4.96	6.91	0.47	0.61	4.23
M02	5.09	5.09	4.63	6.94	0.48	0.60	4.15
M03	5.34	5.34	5.91	7.69	0.47	0.61	4.70
M05	6.54	6.54	6.31	7.83	0.51	0.58	4.51
M06	4.99	4.99	5.45	8.66	0.57	0.52	4.49
M07	4.68	4.68	4.95	6.69	0.47	0.61	4.09
M08	6.96	6.96	7.17	9.25	0.54	0.55	5.06
M09	6.05	6.05	6.24	8.20	0.51	0.57	4.70

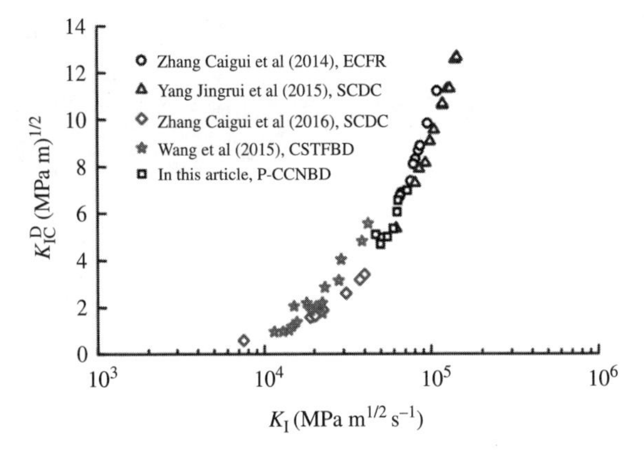

FIG. 2.75 Relation of dynamic crack initiation toughness and dynamic loading rate.

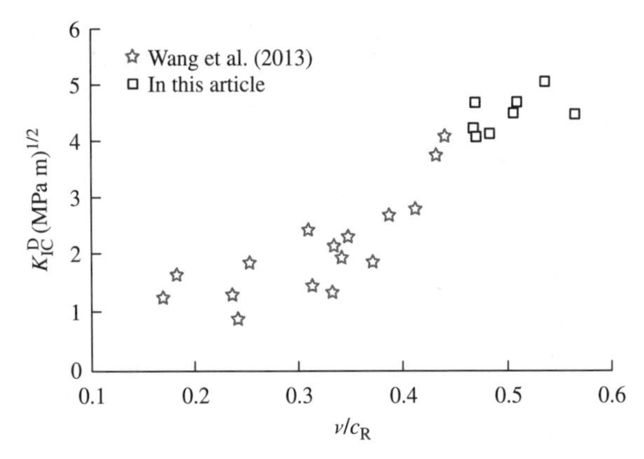

FIG. 2.76 Relation of dynamic crack propagation toughness and crack velocity.

Zhang et al. (2014) summarized the research results of dynamic expansion toughness and obtained that the dynamic expansion toughness of several rock materials increases proportionally with the crack growth rate, and the larger the crack velocity, the more discrete the data.

2.8.5.5 Dynamic crack arrest and DFT rationality

In this test, two specimens were selected for an impact test at an air gun pressure of 0.13 MPa in order to explore the possibility of crack arrest in the P-CCNBD specimen. The position where the strain gauges SG1 to SG5 and CPG of the specimen M-I are attached is shown in Fig. 2.77.

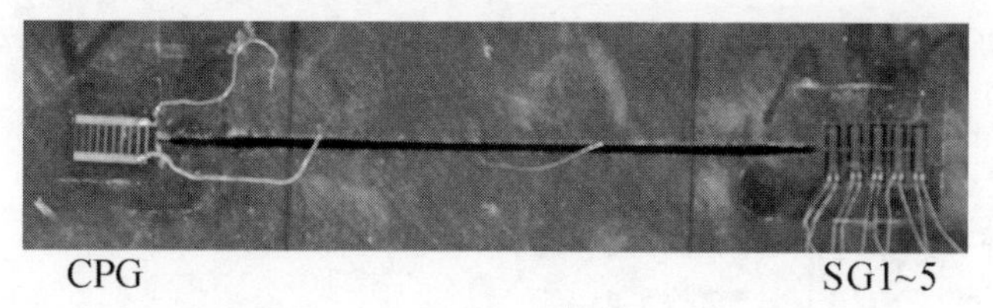

FIG. 2.77 Positions of SG and CPG on specimen M-I.

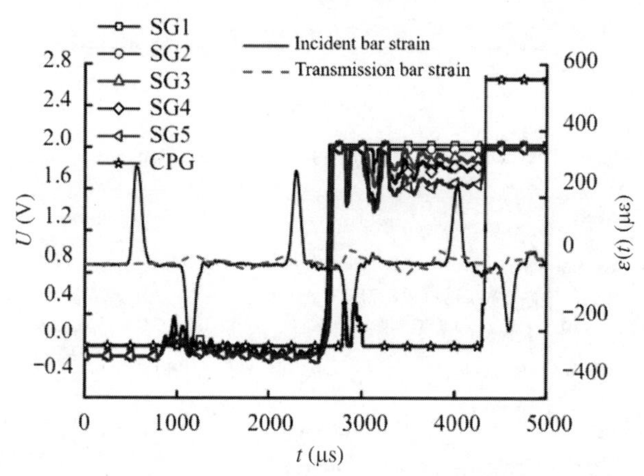

FIG. 2.78 Voltage signals of SG and CPG from specimen M-I and SHPB.

The strain signal of SG_i on the SHPB pressure bar under loading conditions and the voltage signals of the S1–S5 strain gauges and CPG on the specimen are shown in Fig. 2.78.

In order to see the voltage signal of CPG in Fig. 2.78, the CPG voltage value of the M-I specimen is increased by three times. It can be seen from the figure that the SHPB device has a phenomenon in which the specimen is loaded multiple times, and the peak value of the loading waveform gradually decreases as the stress wave is reflected back and forth in the loading system. Strain gauges SG1–SG5 When the stress wave is first loaded, the voltage signal only rises slightly, and the CPG signal does not change, indicating that neither SG nor CPG is broken. When the stress wave is loaded for the second time, the voltage signals of SG1–SG5 rise sharply, and SG1–SG3 are broken. Also, some voltages are restored in SG4–SG5, indicating that the specimen is not completely broken, and CPG is broken at 2788.8, 2799.03 μs stop cracking, during the period of 10.25 μs, CPG has seven filament breaks. After the interval of 1526.425 μs, the CPG cracked again and the strain gauges SG4–SG5 also broke. It is shown that the P-CCNBD specimen can achieve the crack arrest of the dynamic crack of the rock. The crack does not decelerate between the crack arrest, and it is suddenly stopped (Ravichandar, 2004; Zhang et al., 2014). It can be seen from Fig. 2.78 that CPG captures the whole process of crack initiation, propagation and crack arrest in 2788.8–2799.03 μs, which shows that it is possible to measure the rock crack arrest toughness with a P-CCNBD specimen under a reasonable selection of dynamic load. CPG can sensitively capture the stoppage moment of a sudden stop of a high-speed crack, which is a simple and efficient research method.

Due to the different particle size of rock and concrete and the existence of defects such as pores and fine cracks inside, it is a typical heterogeneous quasi-brittle material. Therefore, in order to reduce the influence of internal defects, the use of large-scale specimens for experimental research can better reflect the real dynamic performance of materials (Li et al., 2009; Zhang et al., 2009). However, the increase in the size of the specimen will make it difficult for the specimen to reach the state of stress equilibrium before the failure, and the quasistatic assumption is not satisfied. In the large-size P-CCNBD specimen used in this test, it is obvious that the time history of the dynamic load at the incident end and the transmissive end of the specimen are very different. Therefore, the quasistatic method is not suitable for the determination of rock dynamic fracture toughness in large disc specimens. The test-numerical-analytical method comprehensively considers the size of the specimen, the nature of the material, the influence of the inertial effect of the material, and the actual fracture time of the specimen during the test, which is more in line with the actual dynamic response of the specimen. In addition, the method also considers the influence of the crack propagation speed of the specimen on the dynamic stress intensity factor of the moving crack tip. It is an ideal method for the dynamic fracture toughness test using large rock specimens.

2.8.6 Conclusions

The P-CCNBD specimen overcomes the complex problem of the dynamic crack initiation of the herringbone groove in the CCNBD specimen, which makes it easier to determine the dynamic fracture toughness and extended toughness of the rock.

Large specimens are not easy to achieve stress balance. Therefore, using the test-numerical-analytic method, the method does not need to satisfy the stress uniformity assumption of the quasistatic method. At the same time, the influence of crack propagation speed on the dynamic stress intensity factor of the moving crack tip can be considered.

Under the premise of reasonable selection of impact load and specimen size, the P-CCNBD specimen can realize the dynamic crack arrest of rock material, and CPG can capture the crack arrest phenomenon of a sudden stop of a high-speed expansion crack.

When the loading rate is between 4.63 and 7.17 GPa $m^{1/2}$ s^{-1}, the dynamic cracking toughness of Nanyang marble is 4.68–6.96 MPa $m^{1/2}$, which basically increases with increasing dynamic loading rate. In the range of crack propagation speed (0.47–0.57)c_R, the dynamic propagation toughness increases with the increase of crack propagation speed.

References

Alexander, M.G., Blight, G.E., 1986. The use of small and large beams for evaluating concrete fracture characteristics. In: Wittmann, F.H. (Ed.), Fracture Toughness and Fracture Energy. Elsevier, Amsterdam, pp. 323–332.

ASTM, 1997. American society for testing and materials. In: ASTM Standard E399-90, Standard test method for plane-strain fracture toughness of metallic materials. 19428–2959. ASTM International, 100 Barr Harbor Drive, West Conhohocken, PA.

Bazant, Z.P., 2000. Size effect. Int. J. Solids Struct. 37 (1), 69–80.

Bazant, Z.P., Chen, E.P., 1997. Scaling of structure failure. Appl. Mech. Rev. 50, 593–627.

Bazant, Z.P., Kazemi, M.T., 1991. Size dependence of concrete fracture energy determined by RILEM work-of-fracture method. Int. J. Fract. 51 (2), 121–138.

Bindiganavile, V., Banthia, N., 2004. A comment on the paper, "Size effect for high-strength concrete cylinders subjected to axial impact" by T. Krauthammer et al. Int. J. Impact Eng. 30, 873–875.

Bouiadjra, B.B., Ouinas, D., Belhouari, M., Ziadia, A., 2007. Numerical analysis of the notch effect and the behavior of notch crack in adhesively bonded composite laminates. Comput. Mater. Sci. 38 (4), 759–764.

Carpinteri, A., Cornetti, P., Puzzi, S., 2006. Scaling laws and multiscale approach in the mechanics of heterogeneous and disordered material. Appl. Mech. Rev. 59, 283–304.

Chen, E.P., Sih, G.C., 1977. Transient response of cracks to impact loads, in mechanics of fracture Leyden. Noordhoff International Publishing, pp. 1–58.

Chen, Y.M., Wilkins, M.L., 1975. Numerical analysis of dynamic crack problem. Eng. Fract. Mech. 4 (7), 635–660.

Chen, M., Zhang, G.Q., 2004. Laboratory measurement and interpretation of the fracture toughness of formation rocks at great depth. J. Pet. Sci. Eng. 41, 221–231.

China Aviation Research Institute, 1993. Stress Intensity Factor Handbook (Updated Version). Higher Education Press, Beijing.

Chong, K.P., Kuruppu, M.D., 1984. New specimen for fracture toughness determination for rock and other materials. Int. J. Fract. 26 (2), 59–62.

Cui, Z.D., Liu, D.A., An, G.M., 2009. Research progress in mode-I fracture toughness testing methods for rocks. J. Test Measur. Technol. (3), 189–196.

Cui, Z.L., Gong, N.P., Jing, L.W., 2015. Experiment and finite element analysis of rock dynamic fracture toughness test on non-ideal crack disc specimens. Rock Soil Mech. 36 (3), 694–702.

Dai, F., Wang, Q.Z., 2004. Effects of finite notch width on stress intensity factor for CCNBD specimen. Rock Soil Mech. 25 (3), 427–431.

Dai, F., Xia, K.W., Zhang, H., Wang, Y.X., 2011. Determination of dynamic rock Mode-I fracture parameters using cracked chevron notched semi-circular bend specimen. Eng. Fract. Mech. 78 (15), 2633–2644.

Dai, S.H., Ma, S.L., Pan, Y.S., 2013. Evaluation of mixed-mode I-II stress intensity factors of rock utilizing digital image correlation method. Chin. J. Geotech. Eng. (7), 1362–1368.

Deng, H.F., Zhu, M., Li, J.L., Wang, Y., Luo, Q., Yuan, X.F., 2012. Study of mode-I fracture toughness and its correlation with strength parameters of sandstone. Rock Soil Mech. 33 (12), 3585–3591.

Dong, M.Y., Xia, Y.M., 2004. Effect of crack shape and size on stress intensity factors of central cracked circular disk. J. Univ. Technol. China 34 (4), 449–455.

Dong, S.M., Wang, Y., Xia, Y.M., 2006. A finite element analysis for using Brazilian disk in split Hopkinson pressure bar to investigate dynamic fracture behavior of brittle polymer materials. Polym. Test. 25 (7), 943–952.

Elfahal, M.M., Krauthammer, T., 2004. Authors' reply to comment on: size effect for high-strength concrete cylinders subjected to axial impact. Int. J. Impact Eng. 30, 877–880.

Elfahal, M.M., Krauthammer, T., 2005. Dynamic size effect in normal- and high-strength concrete cylinders. ACI Mater. J. 102 (2), 77–85.

Elfahal, M.M., Krauthammer, T., Ohno, T., Beppu, M., Mindess, S., 2005. Size effect for normal strength concrete cylinders subjected to axial impact. Int. J. Impact Eng. 31, 461–481.

Fan, T.Y., 2006. Principle and Application of Fracture Dynamics. Beijing Institute of Technology Press, Beijing.

Fan, H., Zhang, S., Gou, X.P., Wang, Q.Z., 2011. A completely new numerical calibration of the critical stress intensity factor for the rock fracture toughness specimen CCNBD. Chin. J. Appl. Mech. 28 (4), 416–422.

Feng, F., Wei, C.G., Wang, Q.Z., 2009. Size effect for rock dynamic fracture toughness tested with cracked straight through flattened brazilian disc. Eng. Mech. 26 (4), 167–173.

Fineberg, J., Gross, S.P., Marder, M., 1992. Instability in the propagation of fast cracks. Phys. Rev. B 45 (4), 5146–5154.

Fischer, M.P., Elsworth, D., Alley, R.B., Engelder, T., 1996. Finite element analysis of the modified ring rest for determining mode I fracture toughness. Int. J. Rock Mech. Min. Sci. 33 (1), 1–15.

Fowell, R.J., 1995. Suggested method for determining mode I fracture toughness using cracked Chevron notched Brazilian disc(CCNBD) specimens. Int. J. Rock Mech. Mining Sci. Geomech. Abstr. 32 (1), 57–64.

Fowell, R.J., Xu, C., Dowd, P.A., 2006. An update on the fracture toughness testing methods related to the cracked Chevron-notched Brazilian disk(CCNBD) specimen. Pure Appl. Geophys. 163 (5–6), 1047–1057.

Freund, L.B., 1998. Dynamic Fracture Mechanics. Cambridge University Press, Cambridge.

Gao, G., Huang, S., Xia, K., 2015. Application of digital image correlation (DIC) in dynamic notched semi-circular bend (NSCB) tests. Exp. Mech. 55 (1), 95–104.

Gong, F.Q., Lu, D.H., Li, X.B., Rao, Q.H., Fu, Z.T., 2016. Toughness increasing or decreasing effect of hard rock fracture with pre-static loading under dynamic disturbance. Chin. J. Rock Mech. Eng. (9), 1905–1915.

Guo, H., Aziz, N.I., Schmidt, L.C., 1993. Rock fracture toughness determination by the Brazilian test. Eng. Geol. 33 (2), 177–188.

Hao, X.F., Lu, D.H., Sun, J., 2016. Study of hard rock fracture properties with pre-static load under micro disturbance. Gold 37 (9), 34–38.

Hua, W., Dong, S.M., Xu, J.G., 2016. Experimental research on fracture toughness of rust stone under mixed mode loading conditions. Rock Soil Mech. 37 (03), 753–758.

ISRM Testing Commission, 1988. Suggested methods for determining the fracture toughness of rock. Int. J. Rock Mech. Mining Sci. Geomech. Abstr. 25, 71–96.

ISRM Testing Commission, 1995. Suggested method for determining mode I fracture toughness using cracked chevron notched Brazilian disc (CCNBD) specimens. Int. J. Rock Mech. Mining Sci. Geomech. Abstr. 32, 57–64.

Jia, X.M., Wang, Q.Z., 2003. Wide range calibration of the stress intensity factor for the fracture toughness specimen CCNBD. Rock Soil Mech. 21 (6), 907–912.

Jiang, F.C., Vecchio, K.S., 2007. Experimental investigation of dynamic effects in a two-bar/three-point bend fracture test. Rev. Sci. Instrum. 78 (6).

Jiang, F.C., Vecchio, K.S., 2009. Hopkinson bar loaded fracture experimental technique: a critical review of dynamic fracture toughness tests. Appl. Mech. Rev. 62 (6), 1–39.

Jiang, F.C., Liu, R.T., Zhang, X.X., 2004. Evaluation of dynamic fracture toughness K_{Id} by Hopkinson pressure bar loaded instrumented Charpy impact test. Eng. Fract. Mech. 71 (3), 279–287.

Kalthoff, J.F., Shockey, D.A., 1977. Instability of cracks under impulse loads. J. Appl. Phys. 48, 986–993.

Kolhe, R., Hui, C.Y., Zehnder, A.Z., 1998. Effects of finite notch width on the fracture ofchevron-notched specimens. Int. J. Fract. 94 (2), 189–198.

Krauthammer, T., Elfahal, M.M., Lim, J.H., Ohno, T., Beppu, M., Markeset, G., 2003. Size effect in high-strength concrete cylinders subjected to axial impact. Int. J. Impact Eng. 28, 1001–1016.

Kuruppu, M.D., Chong, K.P., 2012. Fracture toughness testing of brittle materials using semi-circularbend(SCB) specimen. Eng. Fract. Mech. 91, 133–150.

Kuruppu, M.D., Obara, Y., Ayatollahi, M.R., Chong, K.P., Funatsu, T., 2014. IRSM-suggested method for determining the modeIstatic fracture toughness using semi-circular bend specimen. Rock Mech. Rock Eng. 47 (1), 267–274.

Lambert, D.E., Ross, C.A., 2000. Strain rate effects on dynamic fracture and strength. Int. J. Impact Eng. 24, 985–998.

Li, Y.S., 1995. Experimental analysis on the mechanical effects of loading rates on red sandstone. J. Tongji Univ. 2 (3), 265–269.

Li, C., 2013. A study of the effect of temperature on the fracture properties of rock materials. J. Nanjing Institute of Technol. 11 (4), 39–46.

Li, M.L., Johnston, W., Choi, S.K., 1993. Stress intensity factors for semi-circular specimens under three-point bending. Eng. Fract. Mech. 44 (3), 363–382.

Li, Z.C., Zhang, J.Y., Liu, D.A., 2003. Fracture Mechanics and Fracture Physics. Huazhong University of Science and Technology Press, Wuhan.

Li, Y.P., Wang, Y.H., Chen, L.Z., Yu, F., Li, L.H., 2004. Experimental research on pre-existing cracks in marble under compression. Chin. J. Geotech. Eng. 26 (1), 1120–1124.

Li, Z.L., Wang, Q.Z., Li, W., 2006. Experiment on dynamic rock facture toughness using edge-notched disc specimen. J. Yangtze River Sci. Res. Inst. 23 (2), 54–57.

Li, Q.M., Lu, Y.B., Meng, H., 2009. Further investigation on the dynamic compressive strength enhancement of concrete-like materials based on split Hopkinson pressure bar tests. Part II: numerical simulations. Int. J. Impact Eng. 36 (12), 1335–1345.

Li, L., Yang, L.P., Cao, F., 2016. Complete dynamic fracture process of sandstone under impact loading: experiment and analysis. J. China Coal Soc. 41 (8), 1912–1922.

Liu, D.S., Li, X.B., 1999. Impact Mechanism System Dynamics. Science Press, Beijing, pp. 71–83.

Liu, S.Q., Li, H.B., Li, J.R., 2007. Mechanical properties of rock material under dynamic uniaxial tension. Chin. J. Geotech. Eng. 29 (12), 1904–1907.

Lok, T.S., Li, X.B., Liu, D., 2002. Testing and response of large diameter brittle materials subjected to high strain rate. J. Mater. Civ. Eng. 14 (3), 262–269.

Lu, F.Y., 2013. Hopkinson Bar Test Technique. Science Press, Beijing.

Man, K., Zhou, H.W., 2010. Research on dynamic fracture toughness and tensile strength of rock at different depths. Chin. J. Rock Mech. Eng. 29 (8), 1657–1663.

Meng, Q.B., Han, L.J., Pu, H., Li, H., Wen, S.Y., 2016. Effect of the size and strain rate on the mechanical behavior of rock specimens. J. China Univ. Min. Technol. 45 (2), 233–243.

Nakano, M., Kishida, K., Yamauchi, Y., Sogabe, Y., 1994. Dynamic fracture initiation in brittle materials under combined mode I/II loading. J. De Phys. IV Collogue C8, 695–700.

Ouinas, D., Sahnoune, M., Benderdouche, N., BachirBouiadjra, B., 2009. Stress intensity factor analysis for notched cracked structure repaired by composite patching. Mater. Des. 30 (7), 2302–2308.

Petrov, Y.V., Morozov, N.Y., Smirnov, V.I., 2003. Structural macro mechanics approach in dynamics of fracture. Fatigue Fract. Eng. Mater. Struct. 26, 363–372.

Ping, Q., Ma, Q.Y., Yuan, P., 2013. Energy dissipation analysis of stone specimens in SHPB tensile test. J. Mining Safety Eng. 30 (3), 401–407.

Popelar, C.H., AndersonJr, C.E., Nagy, A., 2000. An experimental method for determining dynamic fracture toughness. Exp. Mech. 40 (4), 99–104.

Pugno, N.M., 2006. Dynamic quantized fracture mechanics. Int. J. Fract. 140, 159–168.

Rabczuk, T., Song, J., Belytschko, T., 2009. Simulations of instability in dynamic fracture by the cracking particles method. Eng. Fract. Mech. 76 (6), 730–741.

Ravichandar, K., 2004. Dynamic Fracture. Elsevier, pp. 49–69.

Ruiz, G., Ortiz, M., Pandolfi, A., 2000. Three-dimensional finiteelement simulation of the dynamic Brazilian tests on concrete cylinders. Int. J. Numer. Methods Eng. 48, 963–994.

She, S.G., Lin, P., 2014. Some developments and challenging issues in rock engineering field in China. Chin. J. Rock Mech. Eng. 33 (3), 433–457.

Su, H.J., Jing, H.W., Zhao, H.H., 2014. Experimental investigation on loading rate effect of sandstone after high temperature under uniaxial compression. Chin. J. Geotech. Eng. 36 (6), 1046–1071.

Tang, H.M., 1993. Rock Fracture Mechanics. China University of Geosciences Press, Wuhan (in Chinese).

Tang, T.X., Bazant, Z.P., Yang, S.C., Zollinger, D., 1996. Variable-notch one-size test method for fracture energy and process zone length. Eng. Fract. Mech. 55 (3), 383–404.

Wang, Q.Z., Xing, L., 1999. Determination of fracture toughness K_{Ic} by using the flattened Brazilian disc specimen for rocks. Eng. Fract. Mech. 64 (2), 193–201.

Wang, Q.Z., Li, W., Xie, H.P., 2009. Dynamic split tensile test of flattened Brazilian disc of rock with SHPB setup. Mech. Mater. 41, 252–260.

Wang, H.L., Fan, P.X., Wang, M.Y., Li,W.P., Qian, Y.H. 2011. Influence of strain rate on progressive failure process and characteristic stresses of red sandstone. Rock Soil Mech., 32(5), 1340-1346.

Wang, Q.Z., Fan, H., Gou, X.P., 2013. Recalibration and clarification of the formula applied to the ISRM-suggested CCNBD specimens for testing rock fracture toughness. Rock Mech. Rock. Eng. 46 (2), 303–313.

Wecharatana, M., Shah, S.P., 1980. Slow crack growth in cement composites. J. Struct. ASCE 106 (2), 42–55.

Wu, L.Z., Jia, X.M., Wang, Q.Z., 2004a. A new stress intensity factor formula of cracked chevron notched Brazilian disc (CCNBD) and its application to analyzing size effect. Rock Soil Mech. 25 (2), 233–237.

Wu, L.Z., Wang, Q.Z., Jia, X.M., 2004b. Determination of mode-I rock fracture toughness with cracked chevron notched Brazilian disc (CCNBD) and application of size effect law. Chin. J. Rock Mech. Eng. (3), 383–390.

Xie, H.P., Chen, Z.H., 2004. Rock Mechanics. Science Press, Beijing.

Yang, S.C., Tang, T.X., Zollinger, D.G., Gurjar, A., 1997. Splitting tension tests to determine concrete fracture parameters by peak-load method. Adv. Cem. Based Mater. 5, 18–28.

Yang, J.R., Zhang, C.G., Zhou, Y., 2015. A new method for determining dynamic fracture toughness of rock using SCDC specimens. Chin. J. Rock Mech. Eng. 34 (2), 279–292.

Yin, Q., Jing, H.W., Su, H.J., Zhu, T., 2015. Loading rate effect on fracture properties of granite after high temperature. J. China Univ. Min. Technol. 44 (4), 597–603.

Yokoyama, T., Kishida, K., 1989. Novel impact three-point bend test method for determining dynamic fracture-initiation toughness. Exp. Mech. 6, 188–194.

Yu, X.Z., 1991. Fracture Mechanics of Rock and Concrete. Central South University of Technology Press, Changsha.

Yu, J.L., 1992. Dynamic fracture theory and experiment research progress. Mech. Eng. 14 (5), 7–14.

Zhang, Z.Z., Gao, F., 2012. Experimental research on energy evolution of red sandstone samples under uniaxial compression. Chin. J. Rock Mech. Eng. 31 (5), 953–962.

Zhang, S., Liang, Y.L., 2013. Influence of prefabricated crack width on determining rock fracture toughness for hole-cracked flattened Brazilian disc. J. Exp. Mech. (4), 517–523.
Zhang, S., Wang, Q.Z., 2006. Method for determination of dynamic fracture toughness of rock using holed-cracked flattened disc specimen. Chin. J. Geotech. Eng. 28 (6), 723–728 (in Chinese).
Zhang, S., Wang, Q.Z., 2007. Determination of marble fracture toughness by using variable crack one-size specimens. Eng. Mech. 24 (6), 31–35.
Zhang, S., Wang, Q.Z., 2009. Determination of rock fracture toughness by split test using five types of disc specimens. Rock Soil Mech. 30 (1), 12–18.
Zhang, Z.X., Yu, J., 1995. Microscopic characteristics of static and dynamic rock fracture surface. Trans. Nonferrous Metals Soc. China 5 (4), 21–24.
Zhang, Z.X., Yu, J., Zhao, Q., 1996. Effects of loading rates on rock materials. Nonferrous Metals 48 (1), 1–4.
Zhang, G.Q., Chen, M., Yang, X.Y., 2002. Influence of fracture width on rock toughness measurement. J. Univ. Petrol. China (6), 42–45.
Zhang, S., Wang, Q.Z., Xie, H.P., 2008. Size effect of rock dynamic fracture toughness. Explos. Shock Waves 28 (6), 544–551.
Zhang, S., Wang, Q.Z., Liang, Y.L., 2009. Research on influence of crack length on test values of rock dynamic fracture toughness. Chin. J. Rock Mech. Eng. 28 (8), 1691–1696.
Zhang, S., Li, X.J., Li, D.W., 2009a. A review of test technique and theory of mode-I rock fracture toughness. J. Henan Polytechnic Univ. Nat. Sci. 28 (1), 33–38.
Zhang, M., Wu, H.J., Li, Q.M., 2009b. Further investigation on the dynamic compressive strength enhancement of concrete-like materials based on split Hopkinson pressure bar tests. Part I: experiments. Int. J. Impact Eng. 36 (12), 1327–1334.
Zhang, Z.Y., Duan, Z., Zhou, F.H., 2013a. Experimental and theoretical investigations on the velocity oscillations of dynamic crack propagating in brittle material tension. Chin. J. Theor. Appl. Mech. 45 (5), 729–738.
Zhang, S., Li, X.W., Yang, X.H., 2013b. Influence of different dynamic load calculating methods on rock dynamic fracture toughness test. Rock Soil Mech. 34 (9), 2721–2726.
Zhang, C.G., Zhou, Y., Yang, J.R., Wang, Q.Z., 2014. Determination of model I dynamic fracture toughness of rock using edge cracked flattened ring specimen. J. Hydraul. Eng. 45 (6), 691–700.
Zhang, C.G., Cao, F., Li, L., 2016. Determination of dynamic fracture initiation, propagation, and arrest toughness of rock using SCDC specimen. Chin. J. Theor. Appl. Mech. 48 (3), 624–635.
Zhou, J., Wang, Y., Dong, S.M., 2005. Effect of specimen geometry on mode I fracture toughness of PMMA. Polym. Mater. Sci. Eng. 21 (6), 11–14.
Zhou, Y.X., Xia, K., Li, X.B., 2012. Suggested methods for determining the dynamic strength parameters and mode-I fracture toughness of rock materials. Int. J. Rock Mech. Min. Sci. 49, 105–112.
Zhou, H., Yang, Y.S., Xiao, H.B., 2013. Research on loading rate effect of tensile strength of hard brittle marble-test characteristics and mechanism. Chin. J. Rock Mech. Eng. 32 (9), 1868–1875.

Further reading

Bobet, A., 2000. The initiation of incidental cracks in compression. Eng. Fract. Mech. (66), 187–219.
Dai, F., Xia, K.W., 2013. Laboratory measurements of the rate dependence of the fracture toughness anisotropy of barregranite. Int. J. Rock Mech. Mining Sci. 60 (1), 57–65.
Pan, P.Z., Ding, W.X., Feng, X.T., Yao, H.Y., Zhou, H., 2008. Research on influence of pre-existing crack geometrical and material properties on crack propagation in rocks. Chin. J. Rock Mech. Eng. 27 (9), 1882–1889.
Zhang, Q.B., Zhao, J., 2014. A review of dynamic experimental techniques and mechanical behavior of rock materials. Rock Mech. Rock. Eng. 47 (4), 1411–1478.

Chapter 3

Scale effect of the rock joint

Joung Oh
School of Minerals and Energy Resources Engineering, The University of New South Wales, Sydney, NSW, Australia

Chapter outline

3.1 Fractal scale effect of opened joints

3.1.1 Introduction

The mechanical behavior of a rock joint can be affected significantly by the joint opening resulting from underground excavation (Oh, 2005; Li et al., 2014). The loss of asperity interlocking engenders an appreciable reduction in shear resistance and joint dilatancy (Ladanyi and Archambault, 1969; Li et al., 2015; Oh and Kim, 2010). Most of the developed formulations tacitly assumed that rock joints are closely mated at the onset of shearing. Therefore, an accurate account of the opening effect requires an adequate model for the frictional behavior of the initially mismatched joints.

On the other hand, due to geological processes leading to the creation of joints, rock surfaces have a roughness that radically alters the shear behavior along the joint walls at different sizes. The mechanical properties, especially for clean and rough joints, can vary with scale. Bandis (1980) observed that the peak shear stress decreased and the corresponding displacement increased as the size of the specimen was increased, which is termed the positive scale effect (Bandis et al., 1981). Similar results have been reported by Muralha and Cunha (1990), Ohnishi and Yoshinaka (1995), and Pratt et al. (1974). Small and steep asperities have a controlling effect on short joints, whereas a shallower asperity with a larger wavelength regulates the sliding behavior, resulting in a reduction of the peak shear stress (Bahaaddini et al., 2014; Bandis et al., 1981; Oh et al., 2015). However, others have shown the existence of conflicting results, that is, a negative scale effect (Kutter and Otto, 1990; Leal-Gomes, 2003) and no scale effect (Hencher et al., 1993a, b, Johansson, 2016). Dueto these contradictory findings, the nature of how scale affects the surface roughness needs further study.

The term "roughness" is a geometric measure of the inherent waviness and unevenness of a joint surface relative to its mean plane (ISRM, 1978). Joint roughness characterization has been studied by different approaches. Barton and Choubey (1977) introduced the joint roughness coefficient (*JRC*) to rank the roughness degree of the surface irregularities, which was subsequently adopted by ISRM (1978). Joint profiles with *JRC* values from 0 to 20 possess wavy and uneven asperities in a large range of heights and slopes. Several traditional statistical descriptors that represent the amplitude features have been reported, including the center line average (*CLA*) and the root mean square (*RMS*)(Krahn and Morgenstern, 1979). On the other hand, Tse and Cruden (1979), Maerz et al. (1990), and Grasselli (2001) suggested a textural characteristic, that is, an asperity inclination, is strongly correlated with the surface roughness. Hong et al. (2008) and Zhang et al. (2014) indicated that the joint roughness could be described properly by a combination of the inclination angle and the amplitude or wavelength. Recently, Li et al. (2016a) decomposed an irregular joint profile measured in the laboratory into wavy and uneven components. The interaction of a critical waviness and a critical unevenness controls the frictional behavior in the shear direction.

The seminal work of Mandelbrot (1967) and Mandelbrot (1985) provides the possibility to assess the joint roughness through the fractal method (Lee et al., 1990 Turk et al., 1987 Xie and Pariseau, 1992). The attraction of a fractal model lies in its ability to predict scaling behavior, that is, the relationship between surface geometry observed at various scales. A fractal is a natural phenomenon or a mathematical set that exhibits a repeating pattern at every scale (Mandelbrot, 1967). This replication can be self-similar or self-affine. As illustrated in Fig. 3.1, a self-similar fractal is a geometric object that retains its statistical properties through various magnifications of viewing, whereas a self-affine fractal remains statistically similar only if it is scaled differently in different directions (Mandelbrot, 1985). Joint profiles rarely satisfy the strict requirements of self-similarity (Kulatilake et al., 1998; Um, 1997). However, extensive studies have indicated that the self-affine method has great potential to describe the surface roughness of joints appropriately (Fardin et al., 2001; Jiang et al., 2006; Kulatilake et al., 1998; Kwašniewski and Wang, 1993; Lanaro, 2001; Um, 1997; Wei et al., 2013).

This study presents a constitutive model for the shear behavior of opened joints at the field scale. The opening state is quantified by the degree of interlocking that represents the true involved portion of joint walls with multiorder asperities. The roughness variance with joint size is investigated through fractal consideration. Morphological analysis of several large-scale profiles of typical rocks has demonstrated the capability of the developed scaling relationships in this study. The proposed model considers the shear resistance loss caused by the opening between joint walls in the field scale, which improves the reliability of the rock mass stability evaluation for underground excavations.

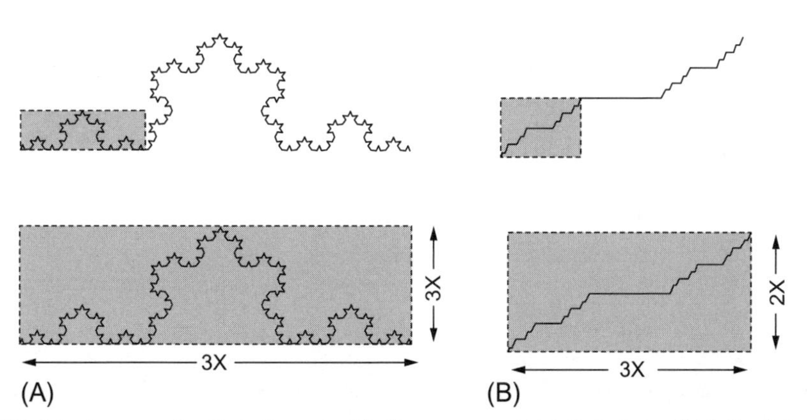

FIG. 3.1 Features of a self-similar and a self-affine geometry. (A) Self-similarity. (B) Self-affinity. *(After Um, J., 1997. Accurate Quantification of Rock Joint Roughness and Development of a New Peak Shear Strength Criterion for Joints. PhD Thesis, The University of Arizona.)*

3.1.2 Scale effect based on fractal method

3.1.2.1 Scale dependence of joint roughness

For a joint profile in the form of self-affinity, there is a power law relating $S_H(w)$, which is the standard deviation of the asperity height, to the window length (w) of the profile (Malinverno, 1990):

$$S_H(w) = A \cdot w^H \tag{3.1}$$

that is:

$$\ln[S_H(w)] = \ln(A) + H\ \ln(w) \tag{3.2}$$

where A is an amplitude parameter and H is the Hurst exponent, which can be readily estimated from the slope and intercept, respectively, of the plot between $ln[S_H(w)]$and$ln(w)$. When $w = 1$, $S_H(w) = A$. Thus, A captures the amplitude exaggeration of the profile at a specific scale. The Hurst exponent (H) or fractal dimension (D) indicates the rate at which the profile smooths with increasing sizes. The value of the Hurst exponent (H) is related to fractal dimension (D) (Voss, 1988) by:

$$H = E - D \tag{3.3}$$

where E is the Euclidean dimension (2 for a profile and 3 for a plane). Apparently, a joint profile in a high fractal dimension has a low value of the Hurst exponent.

As shown by Fig. 3.2, $S_H(w)$ in Eq. (3.2) is calculated as the root mean square (RMS) of the profile height residuals on a linear trend fitted to the measurement points in a window of length (w) through the following equation (Fardin et al., 2001; Kulatilake and Um, 1999; Malinverno, 1990):

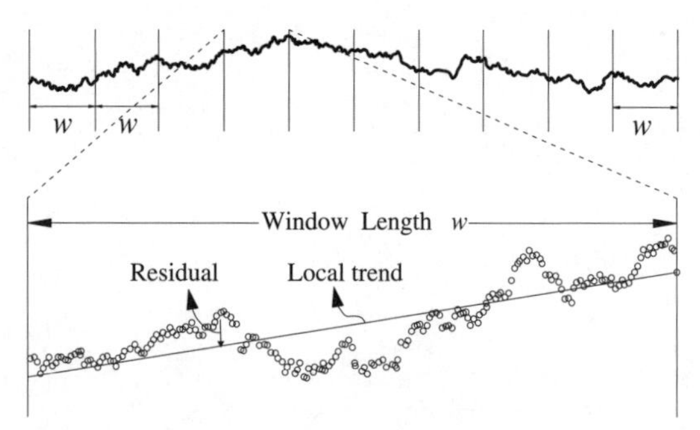

FIG. 3.2 Illustration of the local trend line and residual value in a window sized *(After Kulatilake, P., Um, J., 1999. Requirements for accurate quantification of self-affine roughness using the roughness-length method. Int. J. Rock Mech. Mining Sci. 36(1), 5-18.)*

$$S_H(w) = RMS(w) = \frac{1}{n_w}\sum_{i=1}^{n_w}\sqrt{\frac{1}{m_i - 2}\sum_{j\in w_i}\left(z_j - \overline{z}\right)^2} \tag{3.4}$$

where n_w is the total number of windows, m_i is the number of points contained in a window, z_j are the residuals on the trend, and $\overline{z}$ is the mean residual in windoww_i.

Because the amplitude of asperity is highly related to $S_H(w)$, Johansson (2009) suggested that the asperity height (h_{asp}) of a self-affine profile correlates with the asperity base length (L_{asp}) in a power form:

$$h_{asp} \propto \left(L_{asp}\right)^H \tag{3.5}$$

If $H = 1$, the amplitude of the asperity increases proportionally to the wavelength of the asperity. Nevertheless, a natural rock profile commonly possesses a value of H lower than 1 (Odling, 1994; Yang and Chen, 1999), implying that the increase of the asperity height is lower than the increase in the asperity wavelength. That is to say, the slope of the asperity decreases exponentially with the asperity base length, as the decline angle of a triangle-shaped asperity is denoted by:

$$i_{asp} = \arctan\left(\frac{2h_{asp}}{L_{asp}}\right) \tag{3.6}$$

The asperity angle (i_{asp}) shall be a function of the joint length due to the strong dependence of the asperity base length (L_{asp}) on the joint size (L) (Wei et al., 2013):

$$\tan i_{asp} \propto L^{H-1} \tag{3.7}$$

Li et al. (2016a) proposed a new constitutive law for the shear behavior of rock joints with a critical waviness and a critical unevenness. As depicted in Fig. 3.3, the critical waviness is the wavy asperity with highest amplitude among all the joint asperities. Along the critical waviness facing the shear direction, the critical unevenness is identified by selecting the uneven asperity whose wavelength is the largest. Fig. 3.3 illustrates that the critical waviness and critical unevenness display along the joint profiles at every scale. The occurrence of critical waviness could obey a self-affine scaling law:

$$A_w \propto (\lambda_w)^H \tag{3.8}$$

where A_w and λ_w are the amplitude and wavelength of the critical waviness, respectively. The slope of the critical waviness is:

$$\tan i = a \cdot L^{H-1} \tag{3.9}$$

that is:

$$\frac{\tan i_n}{\tan i_0} = \left(\frac{L_n}{L_0}\right)^{H-1} \tag{3.10}$$

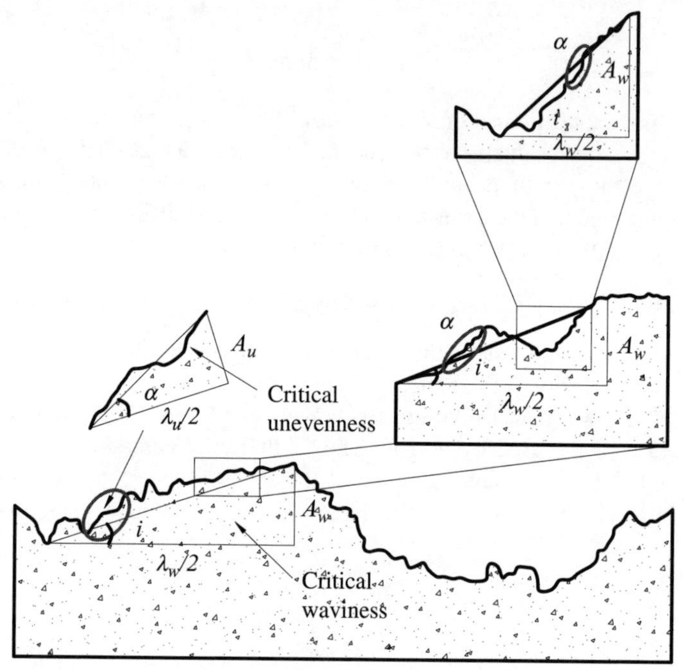

FIG. 3.3 Roughness decomposition of a natural joint profile at different scales.

where a is a proportionality constant relative to the amplitude parameter (A), and i_0 and i_n are the angles of critical waviness in the laboratory scale (L_0) and field scale (L_n), respectively.

Measurement of the amplitude of critical waviness (A_w) at different sizes is provided by:

$$\frac{A_w^n}{A_w^0} = \left(\frac{L_n}{L_0}\right)^H \tag{3.11}$$

where A_w^0, L_0 and A_w^n, L_n are the amplitude of the critical waviness and joint length at the laboratory and field, separately.

Sakellariou et al. (1991) suggested that the following relationship is warranted for a joint profile approximating a self-affine fractal curve:

$$\frac{\tan\theta_n}{\tan\theta_0} = \left(\frac{L_n}{L_0}\right)^{1-D} = \left(\frac{L_n}{L_0}\right)^{H-1} \tag{3.12}$$

where θ_n and θ_0 are the sliding angles obtained from the tilt test corresponding to a joint in the field size (L_n) and the laboratory size (L_0).

That the critical waviness and critical unevenness together govern the shear behavior implies:

$$\frac{\tan(i_n+\alpha_n)}{\tan(i_0+\alpha_0)}=\frac{\tan\theta_n}{\tan\theta_0}=\left(\frac{L_n}{L_0}\right)^{H-1} \tag{3.13}$$

where α_n and α_0 are the angles of critical unevenness in the field scale and laboratory scale, correspondingly.

Equate Eq. (3.13) to Eq. (3.10); the relationship below can be acquired:

$$\frac{\tan(i_n+\alpha_n)}{\tan(i_0+\alpha_0)}=\frac{\tan i_n}{\tan i_0} \tag{3.14}$$

Hence, the slope of the critical unevenness is:

$$\tan\alpha_n=\frac{(k-1)\tan i_n}{\left[1+k(\tan i_n)^2\right]} \tag{3.15}$$

where $k=\dfrac{\tan(i_0+\alpha_0)}{\tan i_0}$.

Likewise, the amplitude of the critical unevenness can be related to the wavelength of the critical waviness by the following equations:

$$A_u \propto (\lambda_w)^H \tag{3.16}$$

$$\frac{A_u^n}{A_u^0}=\left(\frac{\lambda_w^n}{\lambda_w^0}\right)^H \tag{3.17}$$

where A_u^n and A_u^0 are the amplitudes of critical unevenness in the field scale and laboratory scale, respectively.

3.1.2.2 Peak shear displacement for field-scale rock joints

Barton and Bandis (1982) reviewed a large number of shear tests for peak shear displacements reported in the literature (about 650 data points). Prediction of the peak shear displacements at various scales was suggested using an empirical equation:

$$\delta_p=\frac{L}{500}\left(\frac{JRC}{L}\right)^{0.33} \tag{3.18}$$

where δ_p is the shear displacement required to mobilize the shear strength of the rock joint with length (L). *JRC* is the joint roughness coefficient.

Oh et al. (2015) modified the above equation by replacing the qualitative *JRC* with the inclination angle of the large-scale waviness (i_n):

$$\frac{\delta_{p,n}}{\delta_{p,0}}=\left(\frac{L_n}{L_0}\right)^{0.33-0.67}\cdot(i_n)^{0.33} \tag{3.19}$$

where $\delta_{p,0}$ and L_0 are the laboratory-scale peak shear displacement and joint length and $\delta_{p,n}$ and L_n are the field-scale peak shear displacement and joint length, respectively.

The appropriate roughness quantification of an irregular joint profile requires a full description for the wavy and uneven asperities. As has been stated, self-affine fractals can reasonably capture the roughness features of natural joints through fractal descriptors. A simple formulation that involves peak shear displacement and joint size was proposed by taking into account the fractal attribute of a joint surface:

$$\frac{\delta_{p,n}}{\delta_{p,0}} = \left(\frac{L_n}{L_0}\right)^{(2-H)/2} \tag{3.20}$$

where $\delta_{p,\,0}$ and $\delta_{p,\,n}$ are the peak shear displacements of a laboratory-sized and a field-sized joint, respectively.

3.1.3 Constitutive model for opened rock joints

For large-scale rock joints, the mobilized shear stress (τ_{mob}) of joint surfaces under shear loading is generalized as:

$$\tau_{mob} = \sigma_n \cdot \tan\left(\varphi_r + i_n^{mob} + \alpha_n^{mob}\right) \tag{3.21}$$

where σ_n is the applied normal stress, $i_n^{mob}$and $\alpha_n^{mob}$are the mobilized asperity angles for critical waviness and critical unevenness in the field, respectively, and φ_r is the residual friction angle.

In the model of Li et al. (2016a), the degradation of laboratory-scale asperities is described by the wear process between sliding surfaces. Inclination angles of the critical waviness and critical unevenness decay exponentially as shear proceeds, the damage degree of which is dictated separately by two wear constants. By considering the geometric properties of field-scale joints, the wear constants can be expressed as:

$$c_w^n = K_w \frac{\sigma_n i_n}{\sigma_c \sqrt{\left(\lambda_w^n A_w^n\right)/2}} \tag{3.22}$$

$$c_u^n = K_u \frac{\sigma_n \alpha_n}{\sigma_c \sqrt{\left(\lambda_u^n A_u^n\right)/2}} \tag{3.23}$$

where c_w^n and c_u^n are wear constants for the critical waviness and critical unevenness in the field, correspondingly, K_w and K_u are dimensionless constants being 1.0 for most cases, and σ_c is the rock strength of the joint.

Once joint walls are opened, the asperity area in the shear between the upper and lower surface decreases. This areal reduction of triangulate asperities could be described by the degree of interlocking (η) (Ladanyi and Archambault, 1969). Shown clearly in Fig. 3.4, the degree of interlocking represents the true proportion in contact after an initial shear displacement ($\triangle x$), given by:

$$\eta = 1 - \frac{2\triangle x}{\lambda_w^n} \tag{3.24}$$

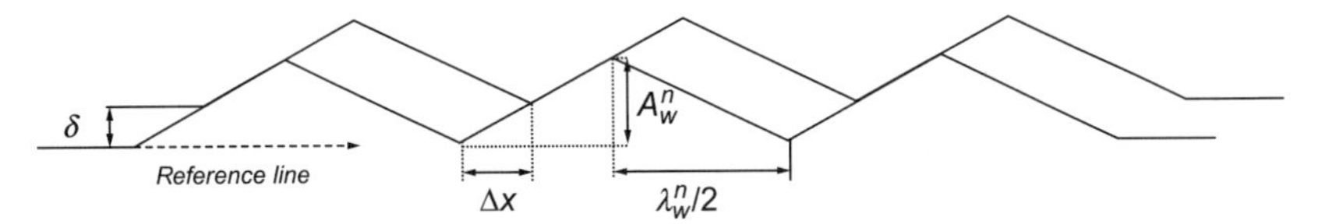

FIG. 3.4 The degree of interlocking for opened rock joints in the field. *(After Ladanyi, B., Archambault, G., 1969. Simulation of shear behavior of a jointed rock mass. In: The 11th US Symposium on Rock Mechanics (USRMS), Berkeley, California, pp. 105–125.)*

For saw-toothed joint profiles, there exists a simple relationship between the joint opening (δ) with respect to the reference line (Fig. 3.4) and the degree of interlocking (Li et al., 2014; Li et al., 2015; Oh and Kim, 2010):

$$\delta = (1-\eta)A_w^n \tag{3.25}$$

Eq. (3.25) indicates that the joint opening (δ) provides an alternative to quantify the degree of interlocking by:

$$\eta = 1 - \frac{\delta}{A_w^n} \tag{3.26}$$

Li et al. (2016b) proposed an algorithm to calculate the magnitude of joint opening (δ) with increasing values of the initial shear displacement (Δx), which permits the quantification of the degree of interlocking for a joint with an irregular surface. Utilization of Eq. (3.26) enables estimation of the geometric change of the critical waviness of large-scale opened joints. Fig. 3.5 illustrates the decrease in the amplitude of the critical waviness after the upper profile has been mismatched with an opening value (δ):

$$\frac{A_w^{o,n}}{A_w^n} = \frac{A_w^n - \delta}{A_w^n} = 1 - \frac{\delta}{A_w^n} = \eta \tag{3.27}$$

where $A_w^{o,n}$ is the amplitude of the critical waviness of opened joints in the field.

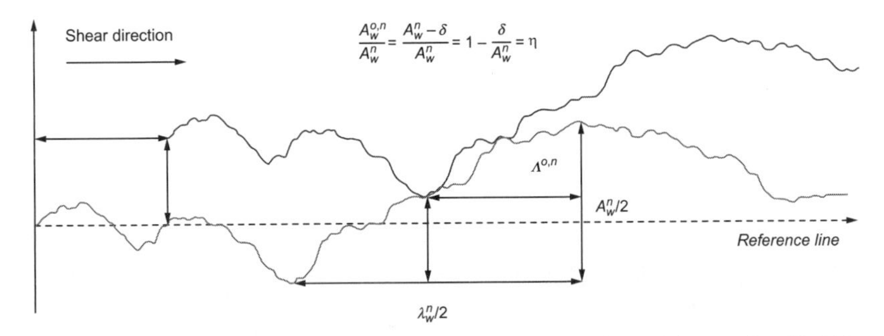

FIG. 3.5 Geometric change of the field-scale critical waviness due to joint opening (δ). *(After Li, Y., Oh, J., Mitra, R., Hebblewhite, B., 2016b. A constitutive model for a laboratory rock joint with multi-scale asperity degradation. Comput. Geotech. 72, 143-151.)*

Similarly, the wavelength of the critical waviness of the field-scale opened joints ($\lambda_w^{o,n}$) is given by:

$$\lambda_w^{o,n} = \eta\lambda_w^n \tag{3.28}$$

Therefore, the wear constant for the critical waviness of rock joints in the field with a degree of interlocking η is expressed as:

$$c_w^{o,n} = K_w \frac{\sigma_n i_n}{\sigma_c \sqrt{(\eta\lambda_w^n) \cdot (\eta A_w^n)]/2}} = \frac{\sigma_n i_n}{\eta\sigma_c \sqrt{(\lambda_w^n A_w^n)/2}} \tag{3.29}$$

Determination of the critical unevenness depends on the geometric distribution of asperities along the critical waviness after joints are opened. The corresponding wear constant ($c_u^{o,n}$) is expressed by:

$$c_u^{o,n} = K_u \frac{\sigma_n \alpha_n^o}{\sigma_c \sqrt{(\lambda_u^{o,n} A_u^{o,n})/2}} \tag{3.30}$$

where α_n^o, $\lambda_u^{o,n}$, and $A_u^{o,n}$ are the asperity angle, wavelength, and amplitude for the critical unevenness of opened rock joints at the field scale, respectively.

The geometric features of the field-scale rock joints involved in the above equations, including asperity angle, amplitude, and wavelength, can be estimated readily by the proposed scale-dependent formulations.

3.1.4 Validation of proposed scaling relationships

3.1.4.1 Validation of scale dependence of joint roughness

Four 400-mm long joint profiles were analyzed to validate the roughness scaling relations, Eqs. (3.9), (3.15). The digitization of two tensile joint surfaces of marble was carried out by the laser-scanning method. The other two are provided by Bahaaddini (2014), who reconstructed the rough sandstone surfaces using the photogrammetry technique. Fig. 3.6 shows the four profiles at identical sampling intervals of 0.2 mm. Following the approach by Bandis (1980), each 400 mm profile was divided into two 200 mm, four 100 mm, and eight 50 mm profiles. Slopes of the critical waviness and critical unevenness were measured in accordance with the roughness decomposition method proposed in Li et al. (2016a). Averaged values were used for 50-, 100-, and 200-mm profiles. Fig. 3.7 shows the natural-logarithmic relationship involving the standard deviation of asperity height, $S_H(w)$, and the corresponding window size (w). The length of the measured window (w) is prescribed as 5, 10, 20, 40, and 80 mm, respectively, based on the suggestion by Malinverno (1990) that the window size (w) should vary between 20% of the full joint length and the shortest span containing at least 10 measuring points. The Hurst exponent (H) of a joint

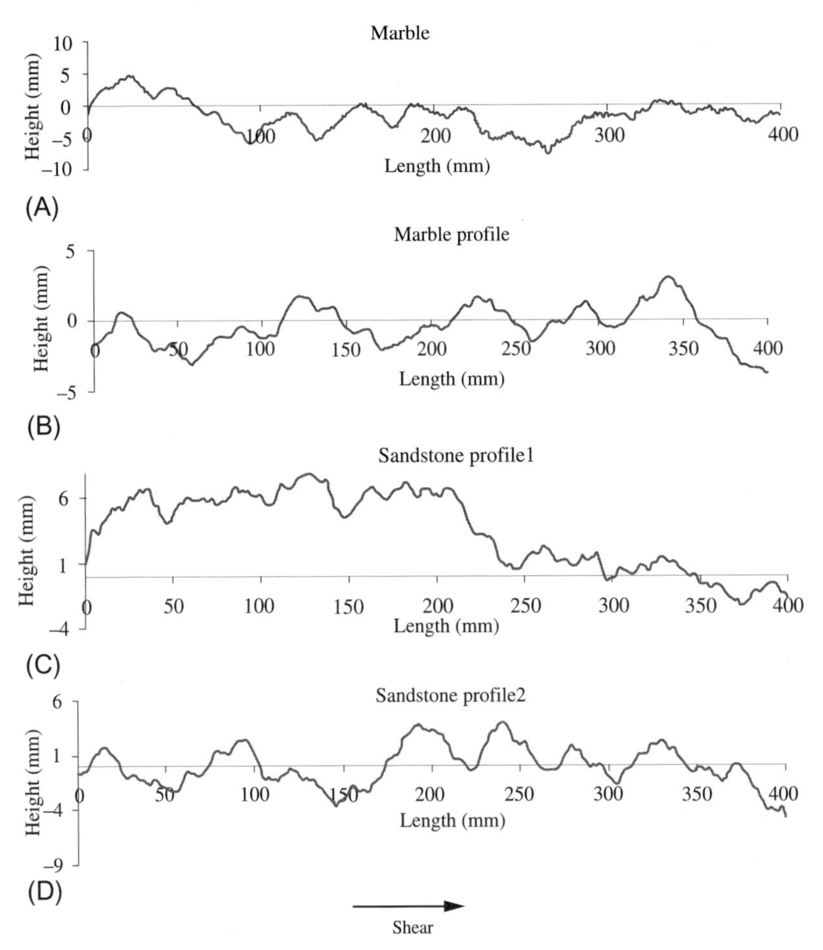

FIG. 3.6 400-mm long profiles of marble and sandstone joints.

profile equals the slope of the regression line. Fitting Eq. (3.9) to the experimental data yields the value of the proportionality constant (a).

Table 3.1 lists fractal parameters of the four profiles, including the Hurst exponent (H) and the proportionality constant (a). Fig. 3.8 clearly shows that the predicted results agree closely with the actual measurements.

3.1.4.2 *Predictive equation for peak shear displacement*

Verification of Eq. (3.20) is illustrated by providing its correlation with experimental results taken from the literature. Bandis (1980) conducted a comprehensive study regarding the scale effect of shear behavior. Four different-sized joint

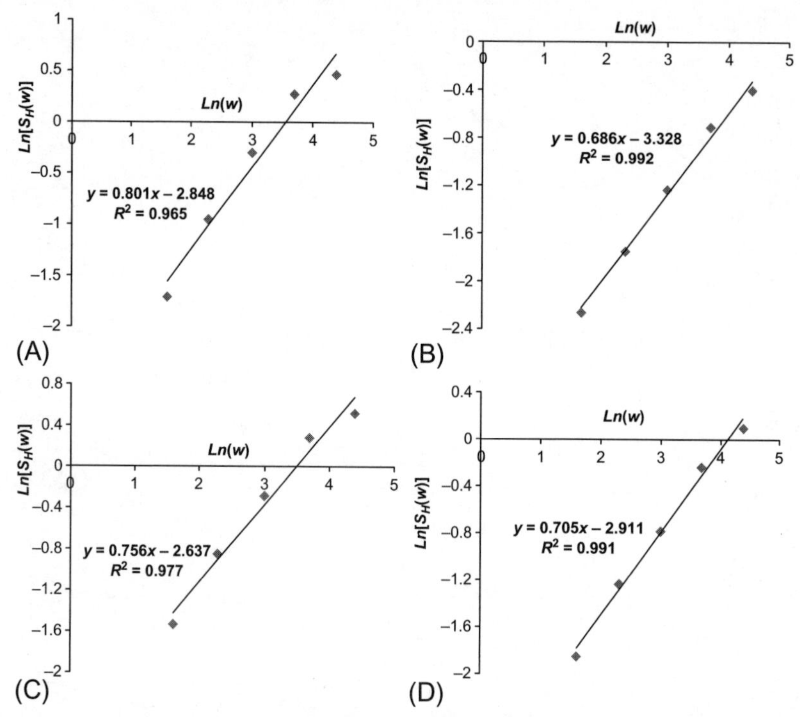

FIG. 3.7 The ln-ln plot of the standard deviation of asperity height, $S_H(w)$, against the window size (w). (A) Marble profile 1, (B) marble profile 2, (C) sandstone profile 1, (D) sandstone profile 2.

TABLE 3.1 Fractal description of the 400 mm long joint profiles.

Sample type	Marble 1	Marble 2	Sandstone 1	Sandstone 2
Proportionality constant (a)	34.07	48.34	31.36	53.5
Hurst exponent (H)	0.8019	0.6867	0.7568	0.7056

replicas, namely 6, 12, 18, and 36 cm, were subjected to direct shearing. Fig. 3.9 correlates the predictions of Eq. (3.9) to measured inclination angles of critical waviness for these joint profiles with distinct roughness degrees (Models No. 1 to No .4). Direct assessment of the Hurst exponent (H) is unavailable for joint profiles used in Bandis (1980). The fitted values of the proportionality constant (a) and the Hurst exponent (H) are included in Table 3.2.

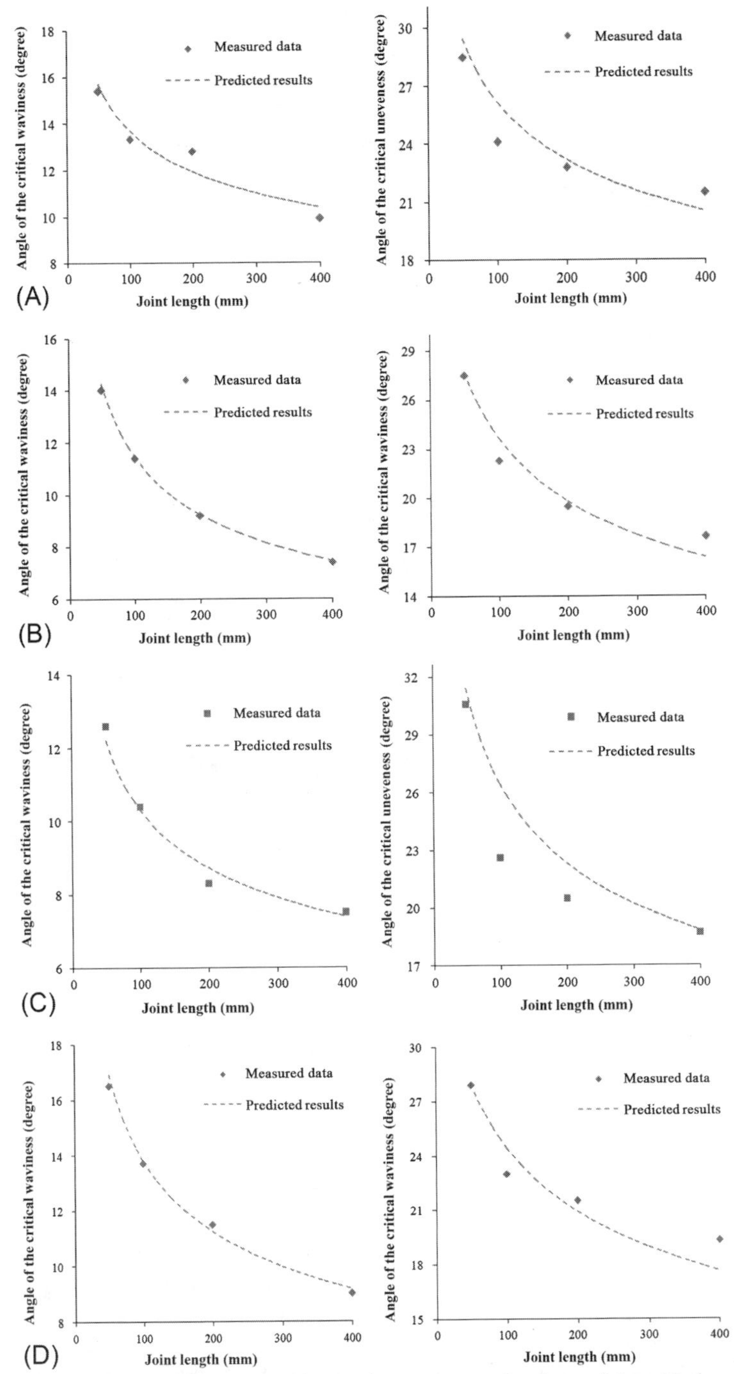

FIG. 3.8 Comparison between predictions and measurements for slopes of the critical waviness and the critical unevenness at different scales. (A) Inclination angles of the critical waviness and the critical unevenness (Marble 1). (B) Inclination angles of the critical waviness and the critical unevenness (Marble 2). (C) Inclination angles of the critical waviness and the critical unevenness (Sandstone 1). (D) Inclination angles of the critical waviness and the critical unevenness (Sandstone 2).

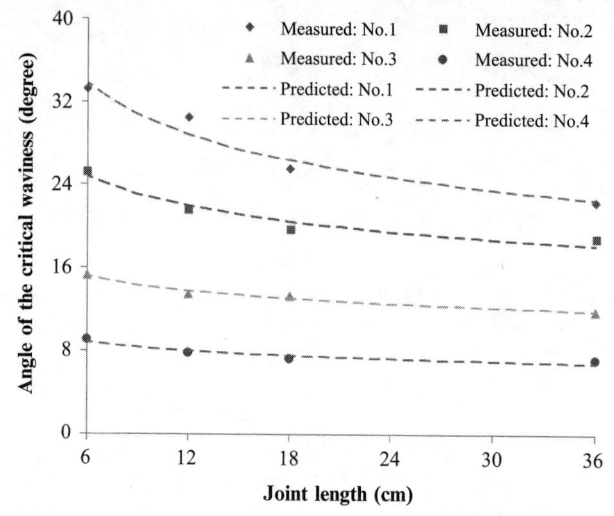

FIG. 3.9 Comparison between predicted and measured angles of the critical waviness for joints in Bandis (1980).

TABLE 3.2 Fractal description of the joint profiles used in Bandis (1980).

Joint sample	No.1	No.2	No.3	No.4
Proportionality constant (*a*)	50.62	33.68	19.39	11.41
Hurst exponent (*H*)	0.7746	0.8280	0.8638	0.8703

Peak shear displacements of the four joint replicas were compared to the estimations by Eq. (3.20). As shown in Fig. 3.10, the agreement between the experimental and analytical results is good.

3.1.5 Conclusions

The study in this paper investigates the scale effect of the surface roughness where a natural profile follows the properties of self-affine sets. However, it should be noted that a joint surface probably could not be considered fractal at all ranges. Brown and Scholz (1985) stated that the topography of real rock surfaces might be fractal within limited scales. Similar results have been observed by Fardin et al. (2001) and Lanaro (2001), who claimed that the applicability of self-affine models would be unwarranted beyond 1–3 m where a stationary threshold would be reached for the roughness. Despite this dissention, the fractal approach provides a useful framework for interpreting the surface

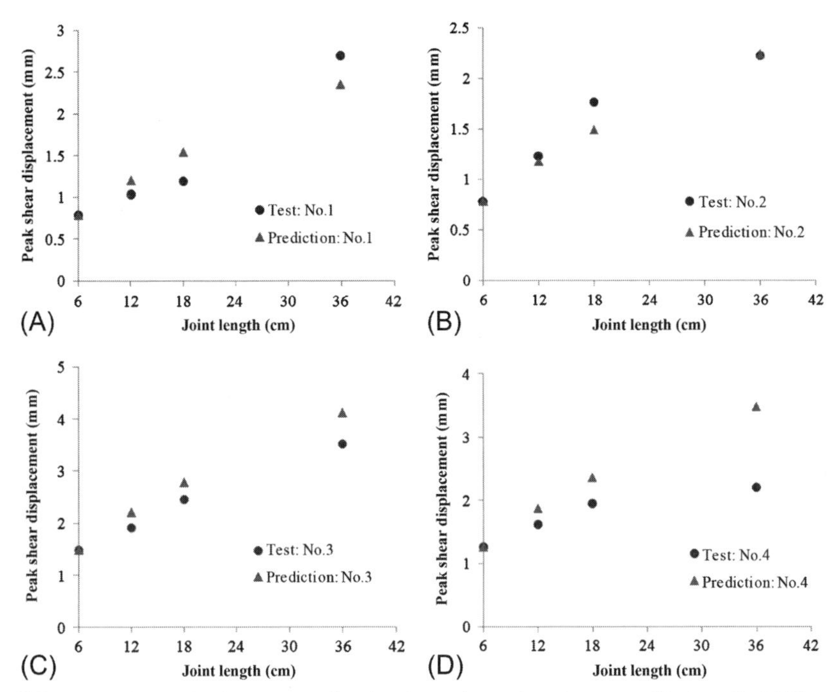

FIG. 3.10 Correlation between predicted and experimental results reported by Bandis (1980).

morphology. Unlike other conventional descriptions of the joint geometry (Barton and Choubey, 1977; Tse and Cruden, 1979), the fractal concept possesses scaling behavior as a basic principle. Because the fractal models recognize the existence of roughness at all sizes, they are uniquely capable of describing the variation of natural joint surfaces that occurs over a broad range in scale.

This study shows a shear model for an opened joint that incorporates the scale-dependent feature of the surface roughness. On the basis of fractal consideration, the roughness of a natural profile is described appropriately by a combination of fractal dimension and amplitude parameter. The critical waviness and critical unevenness featuring a typical joint surface are predicted to become shallower as the joint length increases. The capability of the developed formulations has been illustrated by the morphological examination on two types of 400 mm long rock surfaces. Moreover, the magnitude of peak shear displacement is estimated to vary with the size and fractal dimension of the sampled joint. Experimental data taken from the literature fit well with the proposed relationship.

The formulations developed for scale effect in this study enable a practical assessment of the roughness of joints with multiorder irregularities in the field based on the surface inspection of laboratory samples. With the advanced measurement techniques available today, joint exposures can be scanned with adequate accuracy, which reproduces the geometric features of rock surfaces up to 3 m in size (Fardin et al., 2001; Johansson, 2016; Tatone and Grasselli, 2013). This replication of the joint topography allows acquisition of the roughness and fractal parameters that are required for the proposed scaling relationships. Joints near the excavation are opened due to stress relief, which undermines the stability of rock masses. The presented shear model has the ability to analyze the failure state of joints with varying openings at the natural block size once it is incorporated into numerical codes such as discrete element or finite element programs.

3.2 Joint constitutive model for multiscale asperity degradation

3.2.1 Introduction

The response of a rock joint to shear loading depends largely on the rock strength as well as the surface morphology. Irregularities generate dilatancy between joint walls, which significantly enhances the resistance of a joint to further slip. On the other hand, the surface roughness degrades during the course of shearing. The degree of asperity damage is affected by a variety of factors, such as the magnitude of the applied normal loading, the material strength, and the surface morphology itself. Therefore, the frictional behavior of rock joints and the consequent shearing resistance vary greatly with the surface degradation.

Evaluation of the joint shear strength necessitates a quantitative estimation of the surface roughness. Barton and Choubey (1977) proposed a morphological parameter known as the joint roughness coefficient (JRC) to estimate the degree of joint surface roughness. Because this is the roughness parameter most commonly used in practice, many researchers have continued to improve the determination of JRC by proposing various approaches, such as statistical analysis (e.g., Tse and Cruden, 1979; Reeves, 1985; Yang et al., 2001a, b) and fractal consideration (e.g., Lee et al., 1990; Xie et al., 1997; Kulatilake et al., 2006). Most of the aforementioned methods for pure topography quantification tend to underestimate the joint roughness because the mechanical involvement of asperities in shearing is overlooked (Hong et al., 2008). Alternatively, many researchers (e.g. Patton, 1966a, b; Ladanyi and Archambault, 1969; Schneider, 1976; Jing et al., 1993; Wibowo, 1994) have employed an initial asperity angle to represent the surface roughness for nonplanar joints. Analyses of artificial profiles using statistical roughness parameters by Hong et al. (2008) indicated that at least two components of roughness parameters, such as the amplitude and inclination angle of joint asperities, should be included in the prediction model in order to represent the joint roughness properly.

It is plain that a joint surface inherently includes multiscale asperities. A recent study (Li et al., 2015) suggested that the structural composition of a joint surface substantially influences the tribological behavior of the rock joint, and therefore the role of waviness and unevenness involved in asperity shearing should be distinguished for estimation of the joint shear strength. This study presents a joint shear constitutive model that considers a wavy component as well as an uneven component scaled from laboratory samples. The degradation of asperities is modeled based on the principle of wear. Through a dimensional analysis, factors affecting roughness deterioration are taken into consideration. The proposed model's performance is assessed by conducting experimental tests and providing its correlation with the test results. Some experimental data taken from the literature are also compared with the model's simulations to make further verification.

3.2.2 Quantification of irregular joint profile

Surfaces of rock joints are irregular in nature. ISRM (1978) described said, "the roughness of discontinuity walls is characterized by a waviness (large-scale undulations that, if interlocked and in contact, cause dilation during shear displacement because they are too large to be sheared off) and by an unevenness (small-scale roughness that tends to be damaged during shear displacement unless the discontinuity walls are of high strength and/or the stress levels are low, so that dilation can also occur on these small-scale features)." Although the scale is different, the surface of a laboratory sample also shows similar structural components as described by ISRM (1978) and thus each component makes its own contribution to the joint shear behavior. Fig. 3.11 shows a pair of joint specimens casted with a standard JRC profile (JRC = 10–12). The sample length is 100 mm. This JRC profile is depicted by Barton and Choubey (1977) as being "rough and undulating." From a morphological viewpoint, the undulating component is equivalent to the lab-scale waviness (i_0) whose wavelength (λ_w) is typically on the order of tens of millimeters. The rough component, on the other hand, is equivalent to the lab-scale unevenness (α_0) with a wavelength (λ_u) in the range of one to a few millimeters. The quantification of lab-scale waviness and lab-scale unevenness in order to evaluate the shear strength of a rock joint requires consideration of how the asperity is involved in shearing. Seidel and Haberfield (1995a, b) investigated the failure modes of a tooth-shaped asperity (lab-scale waviness only). They reported that the steepest asperity predominated the shear behavior while the shallower asperities were involved later after a large amount of shear displacements in the postpeak stage. Direct shear tests for a composite joint profile with inclination angles of 30 and 15 degree (lab-scale waviness only) were conducted by Yang and Chiang (2000). The results showed that the shear resistance and dilation in both opposite directions were mainly contributed by the high-angle asperity (30-degree tooth). Consequently, the critical lab-scale waviness that determines the shearing characteristics can be identified as the steepest waviness facing the shear direction.

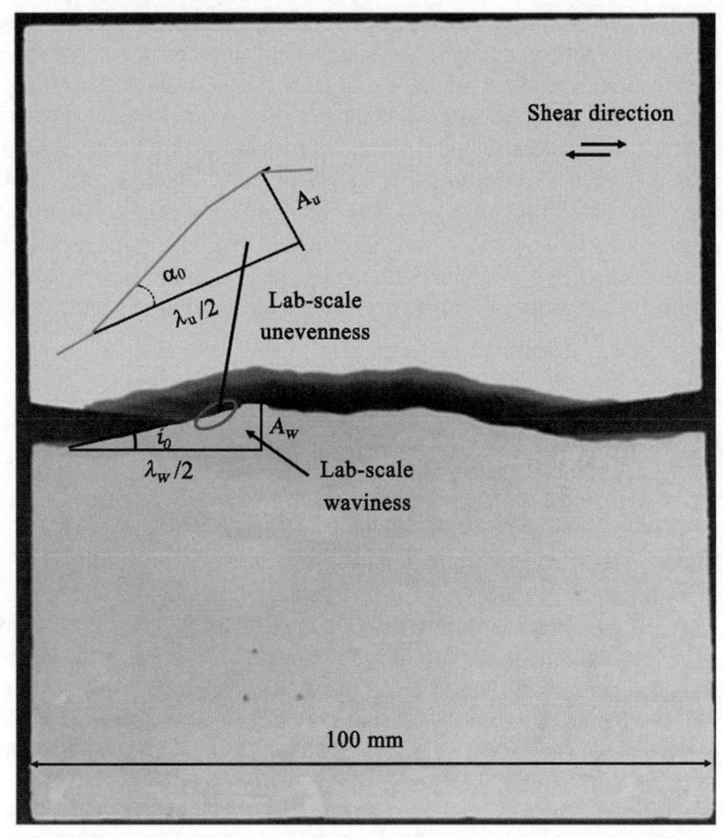

Notations: i_0 is the inclination angle of the critical lab-scale waviness, α_0 is the inclination angle of the critical lab-scale unevenness; A_w is the amplitude of the lab-scale waviness, λ_w is the wavelength of the lab-scale waviness; A_u is the amplitude of the lab-scale unevenness , λ_u is the wavelength of the lab-scale unevenness.

FIG. 3.11 Irregular joints used in the laboratory.

It should be noted that lab-scale waviness is dependent on the shear direction because a natural joint profile is rarely symmetric.

During shear, the damage starts from the steepest small-scale unevenness along the critical lab-scale waviness, which has been frequently observed by experiments (Gentier et al., 2000; Yang et al., 2001a, b; Grasselli et al., 2002). However, taking the steepest unevenness as the critical lab-scale unevenness is irrational for the following reasons. The inclination angles of some unevenness can be as high as 70 degree (Grasselli, 2001; Park et al., 2013; Park and Song, 2013). These tiny-waved asperities are sheared off quickly before shear stress reaches the peak value, even under low normal stress levels. This implies that the contribution of the steepest but small unevenness to the shear strength and dilation of the rock joint can be reasonably neglected. In other words, the unevenness with a relatively large wavelength plays a key role

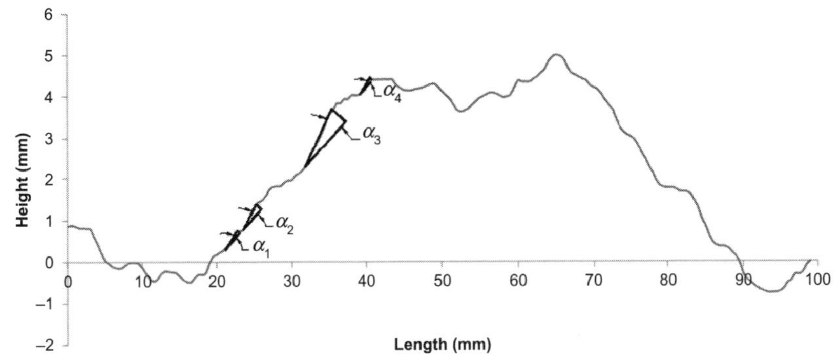

FIG. 3.12 The critical lab-scale unevenness ($\alpha_0=\alpha_3$) along the critical lab-scale waviness.

in the shearing behavior among these small-scale features. Thus, selecting the unevenness with a largest base length to represent the lab-scale unevenness is convincing. Fig. 3.12 illustrates which is the critical lab-scale unevenness α_0 along the critical lab-scale waviness.

The surface of a rock joint (JRC = 10–12) after a direct shear test under relatively low normal stress is shown in Fig. 3.13. After small shear displacements, severe damage occurs to the protruding tiny tips (unevenness) along the critical waviness. The lab-scale unevenness identified is partially sheared, indicating that this critical asperity provides major shear resistance and dilation among all the uneven irregularities.

3.2.3 Description of proposed model

In general, the shearing process of rough rock joints can be divided into five phases, as shown in Fig. 3.14. The proposed model represents the complete five

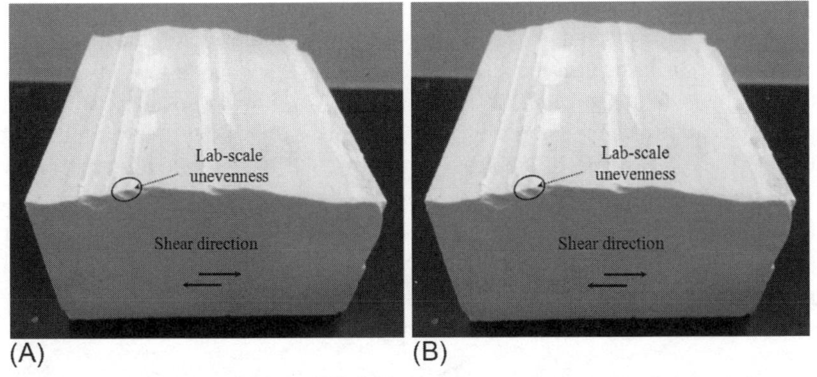

FIG. 3.13 Joint surfaces (JRC = 10–12) before and after shearing. (A) Joint surface before test. (B) Joint surface after test.

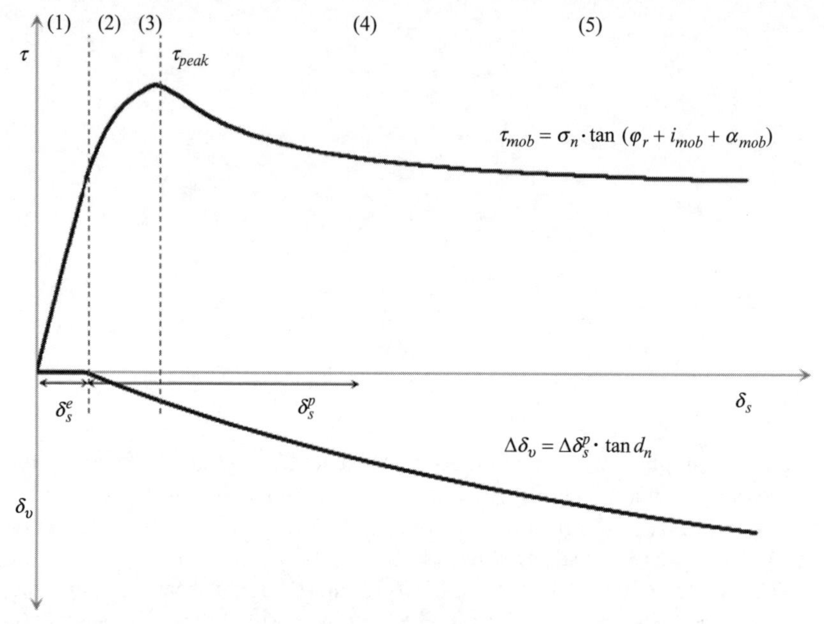

FIG. 3.14 Five phases in shear stress-shear displacement-dilation curves. *(After Oh, J., Cording, E.J., Moon, T., 2015. A joint shear model incorporating small-scale and large-scale irregularities. Int. J. Rock Mech. Min. Sci. 76, 78-87.)*

phases: (1) Elastic phase: shear stress increases linearly with increasing elastic shear displacements (δ_s^e) until it reaches the level of residual strength (Leong and Randolph, 1992; Lee, 2003; Oh et al., 2015). The asperities in contact deform elastically in this phase. (2) Prepeak softening phase: the joint slips or plastic shear displacement (δ_s^p) commences. In this nonlinear zone, the lab-scale unevenness suffers substantial degradation. Part of the lab-scale waviness is abraded, depending on the normal stress magnitude. Reduction of shear stiffness and dilation angle is commonly observed as the asperities are in gradual removal. (3) Peak shear stress: shear stress is mobilized to its peak value at the peak shear displacement (δ_p). The proposed model postulates the magnitude of the peak shear displacement (δ_p) as three times the value of the elastic shear displacement (δ_s^e) (Oh et al., 2015 Barton, 1982a, b). (4) Postpeak softening phase: shear resistance decreases progressively as the asperity wearing continues. The relatively large lab-scale waviness contributes to the postpeak shear strength because most unevenness has already been sheared off before this phase. (5) Residual phase: shear stress keeps stable after a large amount of shear displacement. Both waviness and unevenness have been almost smoothed, providing little dilatancy between joint walls.

The mobilized shear stress, in a general form, is given as:

$$\tau_{mob} = \sigma_n \cdot \tan\left(\varphi_r + i_{mob} + \alpha_{mob}\right) \tag{3.31}$$

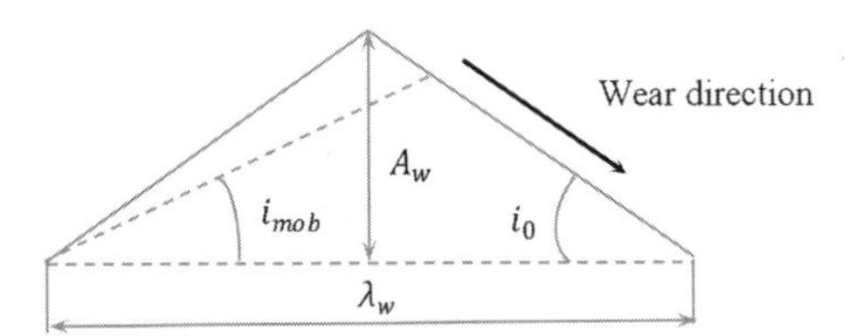

FIG. 3.15 Triangular asperity wear model.

where τ_{mob} is the mobilized shear stress, σ_n is the normal stress, i_{mob} and α_{mob} are the mobilized asperity angles for lab-scale waviness and lab-scale unevenness, respectively, and φ_r is the residual friction angle.

The mobilized asperity components i_{mob} and α_{mob} can be obtained from the theory of wear. For the lab-scale waviness shown in Fig. 3.15, the area increment of wear (dS_r)with respect to the increment of plastic shear displacement (δ_s^p) is considered proportional to the remaining roughness area S_r, that is:

$$\frac{dS_r}{d\delta_s^p} = -c_w S_r \tag{3.32}$$

where c_w is the wear constant for the lab-scale waviness.

Then the asperity area remains as:

$$S_r = S_0 e^{-c_w \delta_s^p} \tag{3.33}$$

where S_0 is the asperity area initially taking part in wear.

Considering the geometric relationship with the joint wavelength λ_w, the joint amplitude A_w, and the initial asperity angle i_0, as shown in Fig. 3.15, the remaining joint asperity area S_r:

$$S_r = \frac{\lambda_w{}^2}{2(\cot i_{mob} + \cot i_0)} \tag{3.34}$$

Combining Eq. (3.3) and Eq. (3.34), the mobilized asperity angle i_{mob} for the waviness is given as:

$$\tan i_{mob} = \frac{e^{-c_w \delta_s^p}}{\left(2 - e^{-c_w \delta_s^p}\right)} \cdot \tan i_0 \tag{3.35}$$

Similarly, for the lab-scale unevenness part, the mobilized asperity angle (α_{mob}) is expressed by:

$$\tan \alpha_{mob} = \frac{e^{-c_u \delta_s^p}}{\left(2 - e^{-c_u \delta_s^p}\right)} \cdot \tan \alpha_0 \tag{3.36}$$

where c_u is the wear constant for the lab-scale unevenness component.

Many researchers, such as Barton and Choubey (1977), Barton (1982a, b), and Oh et al. (2015), have indicated that the ratio of dilation angle to the mobilized asperity angle falls into the range from 0.4 to 1.0. The dilation angle approximates the mobilized asperity angle where the asperity damage is slight due to high joint wall strength or low applied normal stress. With the increase in the normal stress, the dilatancy drops gradually because the asperity degradation becomes more severe. It would be deducible from the description of irregularities by ISRM (1978) that the dilation angles of lab-scale waviness and lab-scale unevenness should be evaluated separately. As the unevenness degrades quickly in shear, the dilation angle (d_n^u) is reasonably assumed as half the mobilized asperity angle (α_{mob}), which is a lower bound suggested by Barton and Choubey (1977):

$$d_n^u = 0.5\alpha_{mob} \tag{3.37}$$

On the other hand, the ratio 0.7, higher than 0.5, is used to calculate the dilation angle for the lab-scale waviness:

$$d_n^w = 0.7 i_{mob} \tag{3.38}$$

The total dilation component (d_n) contributed by waviness and unevenness is calculated by:

$$d_n = d_n^w + d_n^u \tag{3.39}$$

Integrating Eq. (3.39) by plastic shear displacement (δ_s^p), the vertical displacement or joint dilation (δ_v) is expressed as:

$$\delta_v = \frac{7\tan i_0}{10c_w}\ln\left(2 - e^{-c_w\delta_s^p}\right) + \frac{\tan\alpha_0}{2c_u}\ln\left(2 - e^{-c_u\delta_s^p}\right) \tag{3.40}$$

Eq. (3.40) indicates that the waviness and unevenness together control the joint dilation behavior. The superimposed dilation angle is predicted to decrease exponentially. The rate of decrease in dilation is governed by the wear constants for the lab-scale waviness and the lab-scale unevenness.

The magnitude of the wear constant depends on a number of parameters such as the degree of surface roughness, the applied normal load, the material strength, etc. (Barwell, 1958; Queener et al., 1965; Leong and Randolph, 1992). Li et al. (2015) performed dimensional analysis as well as direct shear tests for specimens with triangular asperities (waviness only) and proposed a wear constant as:

$$c_w = K_w\frac{\sigma_n i_0}{\sigma_c\sqrt{(\lambda_w A_w)/2}} \tag{3.41}$$

where K_w is a dimensionless constant equal to 1.0 for most cases, i_0 is the inclination angle of lab-scale waviness, σ_n is the normal stress, λ_w is the wavelength for the waviness, and A_w is the amplitude for the waviness. Likewise, the wear constant for the lab-scale unevenness is calculated from the relationship given by:

$$c_u = K_u\frac{\sigma_n \alpha_0}{\sigma_c\sqrt{(\lambda_u A_u)/2}} \tag{3.42}$$

where K_u is a dimensionless constant equal to 1.0 for most cases, α_0 is the inclination angle of the lab-scale unevenness, and λ_u and A_u are the wavelength and amplitude for the unevenness, respectively.

Eqs. (3.41), (3.42) suggest the degree of surface degradation is exacerbated by increasing normal stress (σ_n) and roughness, whereas asperities experience relatively gentle damage for rock joints with high material strength (σ_c) and a large irregularity area profiled by wavelength and amplitude. Apparently, the value of the wear constant determined by Eq. (3.42) is much higher than that by Eq. (3.41) because the wavelength and amplitude of the lab-scale waviness are considerably larger than those of the lab-scale unevenness. As a result, the unevenness is damaged seriously compared to the waviness. This implication conforms to the experimental results reported by Yang et al. (2001a, b), who observed that the small-area unevenness was worn off noticeably at the end of shearing while nearly no damage occurred to the large undulating asperity, given that the joint samples were loaded under a relatively low normal stress level.

3.2.4 Joint model validation

3.2.4.1 Model implementation

The developed constitutive model is incorporated into the two-dimensional distinct element code UDEC (Itasca Consulting Group Inc, 1980) using the built-in programming language FISH. Shown in Fig. 3.16 is the model of a direct shear test. The model consists of a single horizontal joint with two elastic blocks. The interface or joint is defined by a contact surface composed of three point contacts. The lower block is fixed and the upper block, after being compressed by a normal pressure, is permitted to move horizontally.

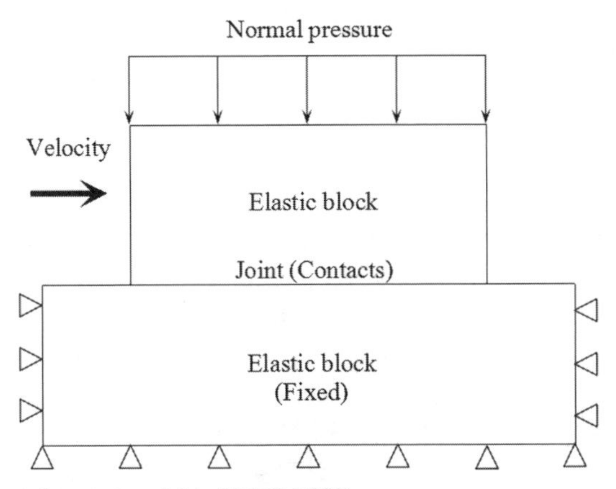

FIG. 3.16 Direct shear test model in UDEC (1980).

In the elastic phase, shear stress increases linearly with accumulating shear displacements to one-third of the peak shear displacement ($\delta_p/3$).

$$\Delta\tau_e = K_s \cdot \Delta\delta_s^e \tag{3.43}$$

where $\Delta\tau_e$ is the shear stress increment, $\Delta\delta_s^e$ is the incremental shear displacement in the elastic stage, and K_s is the elastic shear stiffness.

In the prepeak plasticity phase, shear stiffness softens progressively until peak shear stress is mobilized:

$$\tau_e + \int K_p d\delta_s^p = \sigma_n \tan\left(\varphi_r + i_{peak} + \alpha_{peak}\right) \tag{3.44}$$

where τ_e is the elastic shear resistance, K_p is the shear stiffness in the prepeak plastic region, σ_n is the normal stress, i_{peak} and α_{peak} are the mobilized peak asperity components for lab-scale waviness and unevenness, respectively, and φ_r is the residual friction angle. It is postulated that shear stiffness K_p decreases in a similar fashion as the degradation of asperities, that is:

$$K_p = K_s \frac{e^{-a\delta_s^p}}{2 - e^{-a\delta_s^p}} \tag{3.45}$$

where K_s is the elastic shear stiffness and a is the parameter governing the decreasing rate of the stiffness K_p.

Combining Eq. (3.44) and Eq. (3.45), the value of a is determined by:

$$e^{-\frac{2}{3}\delta_p \cdot a} + e^{\frac{\tau_p - \tau_e}{K_s} \cdot a} = 2 \tag{3.46}$$

where τ_p is the mobilized peak shear strength.

3.2.4.2 *Model validation*

A number of examples illustrating the proposed model's performance are presented in this section. Each example consists of numerically simulated direct shear tests for rock joints with various surface features and its correlation with experimental results. For numerical simulations, shear stiffness (K_s) value and peak shear displacement (δ_p) were acquired directly from laboratory test data. For all cases, K is set to be 1.0.

Correlation with JRC-profiled rock joints

Joint samples with three standard JRC profiles were prepared and subjected to direct shear tests. As shown in Fig. 3.17, the JRC value for each profile is 4–6, 10–12, and 14–16, respectively, and they represent typical natural profiles from slightly rough, planar to rough, and undulating. In order to make identical joint specimens, casting molds were designed and made of stainless steel. Fig. 3.18 shows the joint samples ready for test after 14 days of oven drying at a constant temperature of 40°C. The material used for casting joint replicas consisted of a

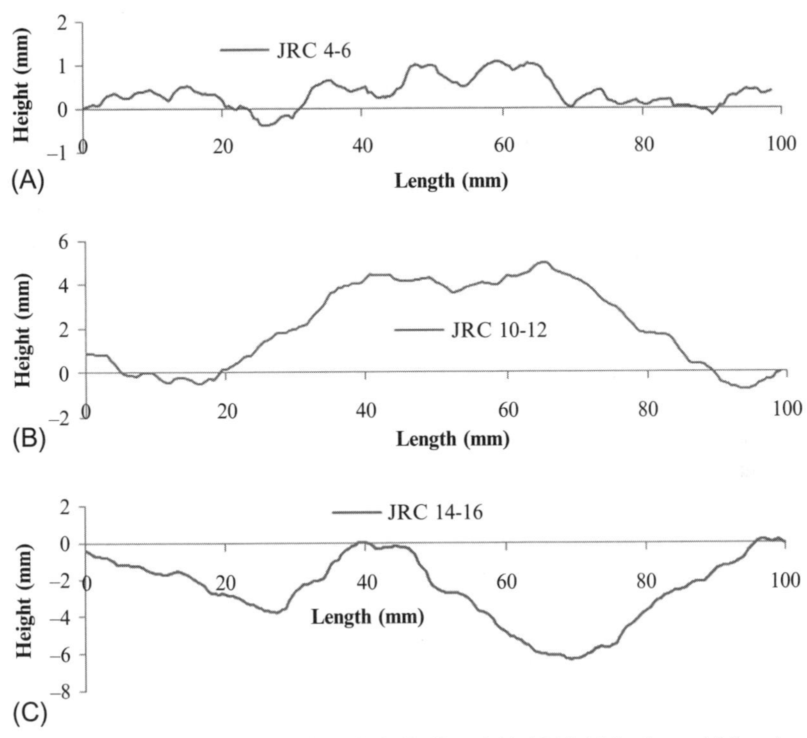

FIG. 3.17 Three standard JRC profiles: 4–6, 10–12, and 14–16 (digitizing interval 0.5 mm).

FIG. 3.18 Joint samples for direct shear test.

mixture of Hydro-stone TB Gypsum Cement and water, combined in the proportion 1:0.35 by weight. The uniaxial compression strength (σ_c) of the casting material was measured on cylindrical specimens with a 42-mm diameter and 100-mm height. It was found that the average compressive strength was 46 MPa. The indirect tensile strength (σ_t) of the same mixture was determined by the Brazilian method. The strength is averaged at 2.5 MPa. The residual friction angle (φ_r) of the specimens was measured by conducting direct shear tests on perfectly flat joint surfaces. The value of the basic friction angle was found to be 42 degree.

Shear experiments were carried out using a servo-controlled hydraulic testing machine RDS-300 manufactured by Geotechnical Consulting and Testing Systems. The shear box consists of upper and lower sections. The upper box section moves vertically and the lower section moves horizontally. The shear and normal forces are applied using two normal and shear actuators. Two linear rail bearings are used to guide the lower box and control the linear displacement of the lower box. Normal and shear loads are measured by load cells. The normal and shear displacements are measured by linear variable differential transducers (LVDTs). Four LVDTs are placed in a square pattern for measurement of the normal displacement of the upper block while one LVDT measures the relative displacement of the lower box to the upper box in the horizontal direction. The machine has the ability to undertake a shear test under constant normal stress mode by automatically decreasing the normal force proportionally to the reduction of the shear surface. During the test, the shear rate of the horizontal actuator was invariably set at 0.5 mm/min.

Fig. 3.19 shows comparisons between numerical simulations and experimental results for the joints with JRC=14–16 under different normal stress levels. Table 3.3 lists all the geometric parameters used in the numerical simulations. They are obtained from the critical waviness and unevenness of each profile, which are identified by the proposed approach described in Section 3.2. As shown in Fig. 3.19, the agreement between experimental results and the simulations from the proposed model is very good. Fig. 3.20 compares the numerical results to experimental curves of three JRC-profiled joints under the same normal stress. It is shown in Fig. 3.10 that the proposed model predicts shear stress and dilation well for different roughness profiles of rock joints.

3.2.4.3 Correlation with experimental data

Simulation of Bandis' direct shear test

In this example, comparisons are made between the simulations from the proposed model and the experiments performed by Bandis (1980). He carried out direct shear tests on identical replicas of size 90 × 50 mm with three surface profiles depicted in Fig. 3.21, for which joint roughness coefficients of 7.5, 10.6, and 16.6 were reported. The three profiles represent almost the planar surface, the smooth undulating surface, and the rough undulating surface, respectively (Bandis, 1980). The uniaxial compression strength (σ_c) of the molding material was 2.0 MPa. The tensile strength (σ_t) was averaged at 0.33 MPa and the elastic modulus (E) was about 800 MPa. The residual friction angle (φ_r) of the joint surfaces was 32 degree. The experiments employed constant normal stress at 90 KPa.

Details of joint amplitude and wavelength for lab-scale waviness and unevenness were acquired by digitizing the surface profile; they are shown in Table 3.4. As presented in Fig. 3.22, the degree of agreement between the model's simulations and the laboratory data is very high.

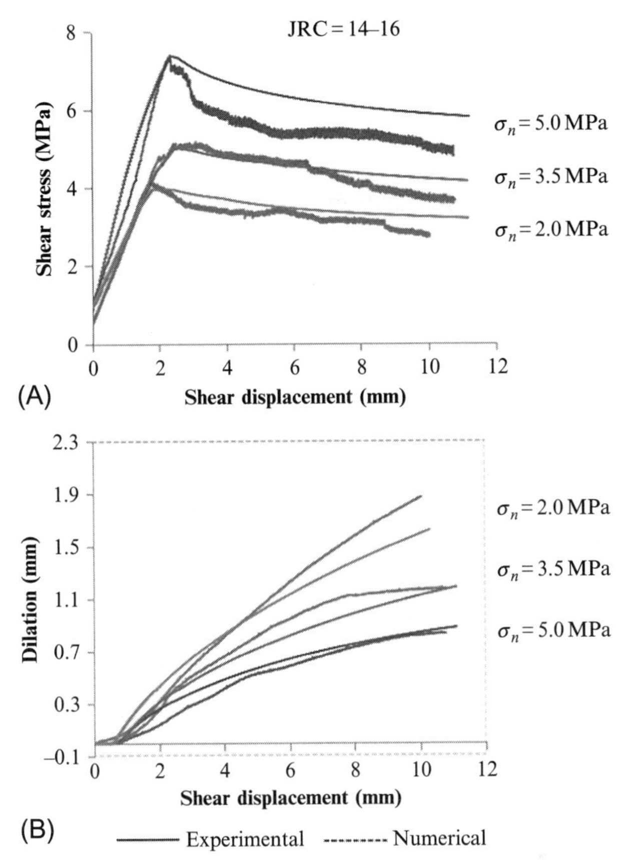

FIG. 3.19 Comparison between numerical simulations and experimental results under different normal stresses. (A) Shear stress vs shear displacement curves. (B) Dilation vs shear displacement curves.

TABLE 3.3 Geometric parameters used in simulations with the standard JRC profiles.

Joint sample	Lab-scale waviness component			Lab-scale unevenness component			
	i_0 (degree)	$\lambda_w/2$ (mm)	A_w (mm)	α_0 (degree)	$\lambda_u/2$ (mm)	A_u (mm)	$i_0+\alpha_0$ (degree)
JRC 4-6 (5.8)	2.6	33.0	1.47	7.7	5.50	0.75	10.3
JRC10-12 (10.8)	11.4	28.5	5.75	17.0	4.40	1.35	28.4
JRC 14-16 (14.5)	13.1	28.0	6.49	19.7	2.67	0.95	32.8

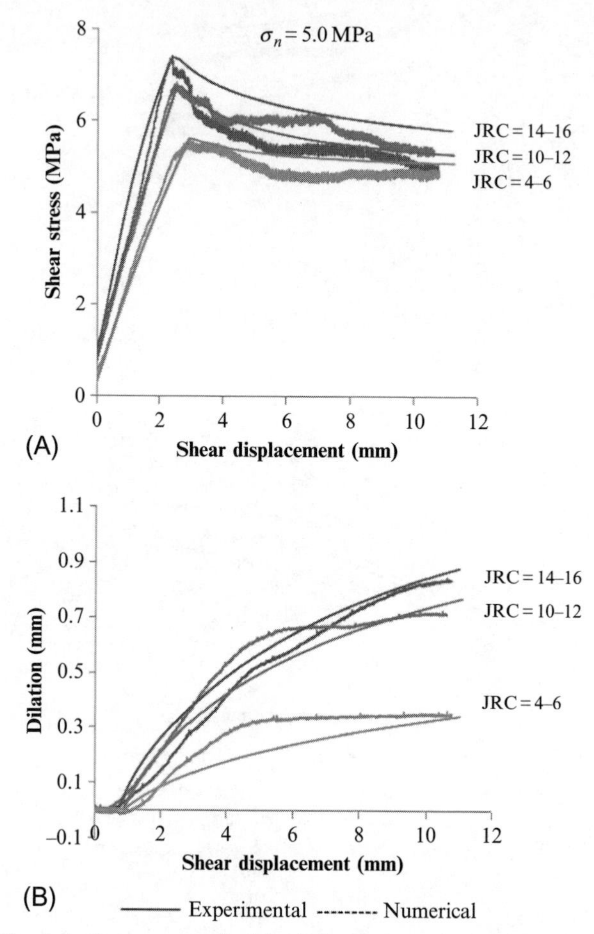

FIG. 3.20 Correlation between numerical simulations and experimental results on different joint profiles. (A) Shear stress vs shear displacement curves. (B) Dilation vs shear displacement curves.

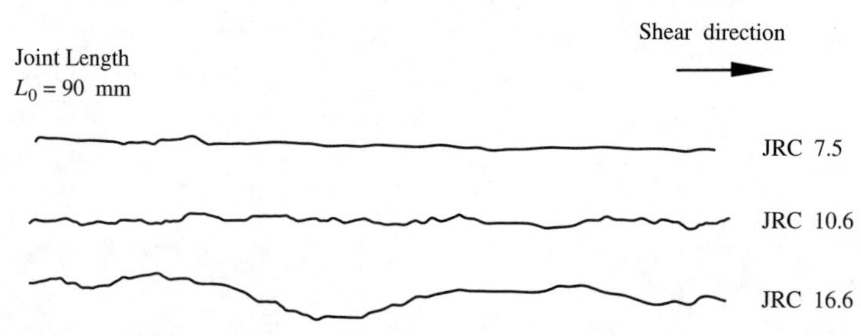

FIG. 3.21 Joint profiles used by Bandis (1980).

TABLE 3.4 Geometric parameters used in simulations with the joint profiles tested by Bandis (1980).

	Lab-scale waviness component			Lab-scale unevenness component			
Joint sample	i_0 (degree)	$\frac{\lambda_w}{2}$ (mm)	A_w (mm)	α_0 (degree)	$\frac{\lambda_u}{2}$ (mm)	A_u (mm)	$i_0+\alpha_0$ (degree)
JRC 7.5	2.4	31	1.3	7.6	1.5	0.2	10
JRC 10.6	4.3	28	2.1	14	1.6	0.4	18.3
JRC 16.6	8.2	33.2	4.8	20	2.2	0.8	28.2

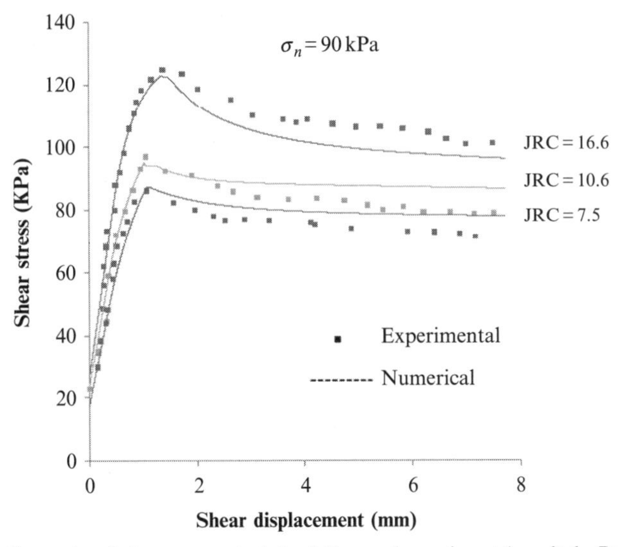

FIG. 3.22 Comparison between numerical simulations and experimental results by Bandis (1980).

Simulation of Flamand et al.'s direct shear test

The tests were conducted by Flamand et al. (1994) on identical replicas of a natural fracture in granite, molded from the original sample of a fracture in the Guéret Granite (France), drilled perpendicular to the fracture plane. As reported by Flamand et al. (1994), the casting material was nonshrinking mortar and had the mechanical properties; $\sigma_c=82$ MPa, $\sigma_t=6.6$ MPa, $E=32{,}200$ MPa, and $\varphi_r=37$ degree. Joint replicas were sheared monotonically under three normal stress magnitudes, namely, 7, 14, and 21 MPa.

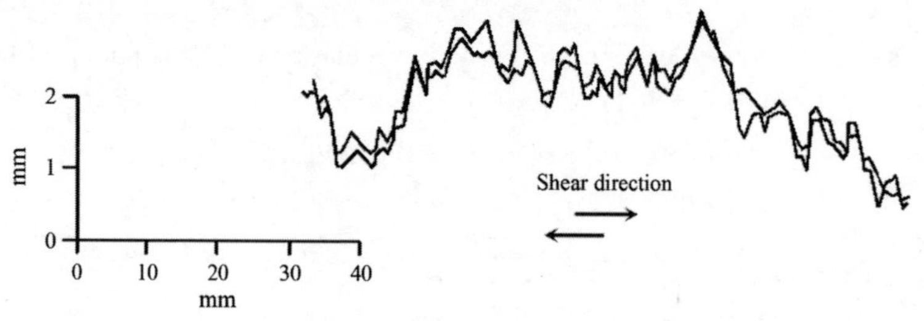

FIG. 3.23 Joint profile used in Flamand et al. (1994).

TABLE 3.5 Geometric parameters used in simulations with the joint profiles tested by Flamand et al. (1994).

	Lab-scale waviness component			Lab-scale unevenness component			
Joint sample	i_0 (degree)	$\frac{\lambda_w}{2}$ (mm)	A_w (mm)	α_0 (degree)	$\frac{\lambda_u}{2}$ (mm)	A_u (mm)	$i_0 + \alpha_0$ (degree)
Natural fracture	16.7	11.0	3.3	14.9	3.0	0.8	31.6

Fig. 3.23 shows the surface profile on which the geometric parameters were obtained. Table 3.5 lists geometric details of the critical asperity components, which were used in the numerical simulations. As illustrated in Fig. 3.24, the proposed model generates the shear stress-shear displacement-dilation curves that are well matched with experimental results.

3.2.5 Conclusions

The study in this paper addresses the shear strength of rock joints in the laboratory sample. It has been reported that the shear strength of rock joints should be evaluated assessing large-scale irregularities not present in the laboratory sample as well. In situ tests definitely increase the accuracy and reliability of estimations of shear strength of rock joints in the field. In practice, however, most shear strength is evaluated on the basis of the laboratory tests and then the scale effect is employed to extrapolate to field-scale properties (Bandis, 1980; Flamand et al., 1994; Hencher and Richards, 2014). Therefore, precise assessment of the joint shear strength from the laboratory sample is of great importance as the most fundamental element of the strength evaluation.

The shear process of rough rock joints involves multiscale asperities at the same time. Thus, the quantification of roughness is complex when certain

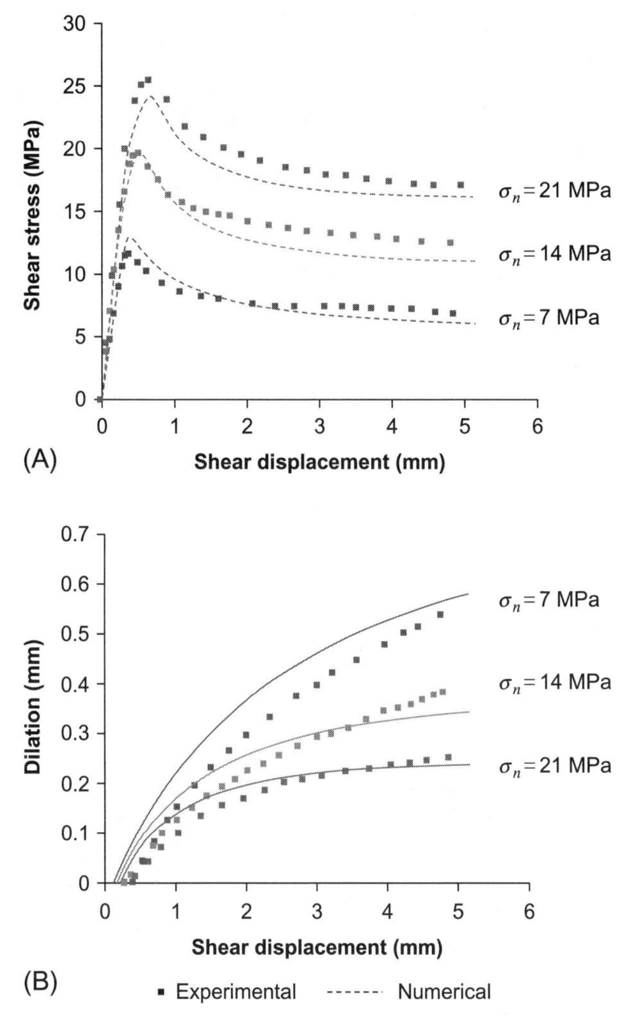

FIG. 3.24 Comparison between simulation results and experimental data in Flamand et al. (1994). (A) Shear stress vs shear displacement curves. (B) Dilation vs shear displacement curves.

values are required as input parameters to represent the shear behavior of those joints. The approach proposed in the present study to determine the critical asperities, that is, the lab-scale waviness and the lab-scale unevenness, is practical to implement. It is logically sound and consistent with the observations in experimental studies.

This study describes a joint asperity degradation model based on the wear process. The model is capable of describing the lab-scale waviness and the lab-scale unevenness separately and develops the different degradation constant for each asperity through dimensional analysis. The asperity degradation is formulated by reducing the area of the surface irregularities available for wear. The model features commonly held notions regarding the asperity degradation; the damage of the waviness is much slower than that of the unevenness, thereby dominating the overall shear behavior of the joint. When the joint asperity is of low strength under a high confining pressure, the contribution of the unevenness to shear resistance and dilation will be minimal. Several examples are provided to compare the constitutive law to experimental measurements. It needs attention that the dimensionless constant, K_w and K_u in Eqs. (3.41), (3.42) are set to be 1.0 for all simulations, which are introduced to fit the direct shear tests results. Nevertheless, the proposed model shows good overall agreement in all cases.

3.3 Shear model incorporating small- and large-scale irregularities

3.3.1 Introduction

The behavior of rock joints dominates the behavior of rock masses by providing planes of weakness along which shear and dilation can occur. A number of experimental studies have been conducted to understand the behavior of rock joints, and many joint constitutive models have been proposed to predict their mechanical behavior. Barton and Choubey, 1977, one of the earliest and most fundamental models for peak shear strength, was developed from the basic mechanics of sliding up the asperity with the inclination angle or shearing through the asperity depending on the normal stress level. Later, Ladanyi and Archambault (1969) proposed a semiempirical model that featured the curved failure envelopes. Barton (1982a, b) developed a useful empirical model by introducing a morphological parameter known as the joint roughness coefficient (JRC) and using the concept of roughness mobilization. A significant advanced theoretical model was developed by Plesha (1987), in which asperity degradation is a function of the plastic work during shear. More recently, approaches such as fractal (Jiang et al., 2006) and geostatistical analysis (Misra, 2002; Lopez et al., 2003) have been proposed to evaluate the mechanical behavior of rock joints under shear.

A large body of literature (e.g., Patton, 1966a, b; Pratt et al. (1974); Cording, 1976; Barton and Choubey, 1977; Bandis, 1980; Bandis et al., 1981; Cording and Barton and Bandis, 1982; McMahon, 1985; Hencher et al., 1993a, b; Fardin et al., 2001) indicates that strength and shear behavior of rock joints vary both qualitatively and quantitatively as a result of a change in a sample or an in situ block size. Ignoring scale effect may lead to overestimation or underestimation of the field shear strength of joints if the peak strength obtained from the laboratory joint shear test is used. The shear behavior of rock joints in the field should be evaluated by considering the dilation and strength along both the small-scale joint roughness scaled from laboratory data and the large-scale waviness determined from geologic observations. However, most rock joint constitutive models proposed in the literature have been developed on the basis of data obtained from laboratory tests on natural or model rock joints. Thus, they do not fully represent the behavior of rock joints in the field. This paper describes a rock joint constitutive model that can generate shear stress-displacement-dilation curves for both small-scale and large-scale joints. The model is incorporated in 3DEC (2003) using the built-in programming language FISH, and is correlated with the experimental results of direct shear tests taken from the literature.

3.3.2 Constitutive model for small-scale joints

3.3.2.1 Mobilized shear strength

The shear stress-displacement-dilation curves generated by the proposed joint model can be characteristically divided into five stages: (1) elastic region, (2) prepeak softening, (3) mobilized peak strength, (4) postpeak softening, and (5) residual strength. The schematic curves in Fig. 3.25 show that there are five stages in the shear stress-displacement-dilation curves. It is frequently observed from the experimental results of direct shear tests in the literature that the shear stress-displacement curves show almost linear elastic behavior to a stress approximately equivalent to the residual strength of the joint. The proposed joint model accounts for these observations. Therefore, during the elastic region, the shear stress is mobilized as a function of joint shear stiffness (K_s) and elastic shear displacement (δ_s^e). The shear stress increment ($\Delta\tau^e$) is calculated as

$$\Delta\tau^e = K_s \Delta\delta_s{}^e \tag{3.47}$$

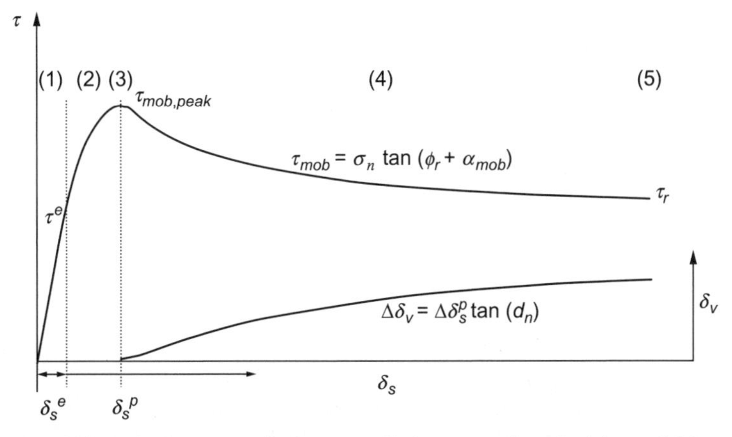

FIG. 3.25 Schematic shear stress-displacement-dilation curves from the joint model (curves of simulation results using the same material properties as used by Flamand et al. (1994) for direct shear tests as illustrated in Fig. 3.30. Note that superscript e and p indicate elastic and plastic, respectively).

After the elastic region, the joint starts to slide and dilation takes place, which means the plastic shear displacement occurs from this point on. The degradation in asperity and also dilation is modeled as a function of the plastic work done in the shear. This is discussed in detail in the following sections.

Most rock joints show that peak shear strength is mobilized at a very small deformation. A peak shear displacement is sometimes considered to be a material constant not affected significantly by changes of normal stresses, which is supported by experimental results such as Leichnitz (1985) and Herdocia (1985). On the other hand, as shown in some direct shear test results, such as Jaeger (1971), Schneider (1976), and Flamand et al. (1994), this parameter is not always constant for a given joint. On the basis of results from shear tests on model tension fractures, Barton and Choubey (1977) suggested that a peak shear displacement is dependent on sample scale and occurs after a shear displacement equal to 1% of the joint sample length up to some limiting size. Later, Barton (1982a, b) assumed that elastic shear displacement (δ_s^e) is 0.3 times the peak shear displacement (δ_{peak}). By performing direct shear tests on different sized replicas cast from various natural joint surfaces, Bandis et al. (1981) concluded that for practical purposes, a peak shear displacement (δ_{peak}) can be taken as approximately equal to 1% of the joint length for a large range of block sizes and types of roughness. Because no general relationship for peak shear displacement (δ_{peak}) is yet available, as the preceding discussion indicates, a peak shear displacement (δ_{peak}) is treated as an input parameter in the proposed joint model, as "n" times elastic shear displacement (δ_s^e), where δ_s^e is determined by the combination of stress level and joint shear stiffness (K_s). For the initial analysis or with limited test data available, the value of n can be selected to be 3, based on direct shear test results frequently observed in the literature (e.g., Barton, 1982a, b, 1993).

After peak shear strength, a mobilized shear stress is gradually decreased until it reaches a residual value. The mobilized shear stress during the plastic region is calculated as

$$\tau_{mob} = \sigma_n \tan(\phi_r + \alpha_{mob}) \tag{3.48}$$

where σ_n is a normal stress, τ_{mob} is a mobilized shear stress, ϕ_r is a residual friction angle, and α_{mob} is a mobilized asperity angle that degrades as plastic work increases.

3.3.2.2 *Asperity degradation*

The model proposed here simulates the progressive degradation of a joint asperity under shear. It is modeled by assuming that degradation is a function of the plastic work, W_p and its relationship is given by Eq. (3.3), which was suggested by Plesha (1987).

$$\alpha_{mob} = \alpha_0 \exp(-cW_p) \tag{3.49}$$

where α_0 is an initial asperity angle, c is an asperity degradation constant, and W_p is expressed as $W_p = \sum \Delta\delta_s^p \tau$

Although Eq. (3.49) possessed good qualitative and quantitative agreements with experimental observations, it is difficult, as Plesha, 1987 mentioned, to relate the asperity degradation constant, c to other properties of rock joints. From the cyclic shear test results on some 30 real rock granite and limestone joints, Hutson and Dowding, 1990 proposed an advanced relationship for the asperity degradation constant, c given by

$$c = -0.141\,\alpha_0 N/\sigma_c \tag{3.50}$$

where α_0 is an initial asperity angle, N is a normal stress, and σ_c is an unconfined compressive strength of rock. This relation indicates that asperity degradation is a function of material strength as well as stress level, which has been mentioned by many researchers (e.g., Ladanyi and Archambault, 1969; Cording, 1976; Barton and Choubey, 1977).

While Hutson's asperity degradation constant, c, Eq. (3.50) describes the general behavior of a rock joint, it still has some questions to be considered. First, it is not a dimensionless relation. Hutson's constant, c has a unit of [cm^2/J] with a dimensional constant of 0.141. Thus, Eq. (3.50) is valid in only one consistent system of units. Second, Eq. (3.50) is valid for a limited range of scale of irregularities. Hutson's model was determined from tests on rock joints having wavelengths of 3.2 and 5.0 cm, and is not expected to apply to the typical laboratory samples. Hutson mentioned in his thesis (Hutson, 1987) that large laboratory models, such as the largest used by Bandis (1980), should produce similar results. Finally, the effect of the applied normal stress in Eq. (3.50) is duplicated because this constant is multiplied by the plastic work, W_p as shown in Eq. (3.49).

In order for the asperity degradation constant to apply to any scaled laboratory samples, Eq. (3.50) appears to be modified. A geometric parameter, that is, a wavelength is introduced and thus the wavelength, λ with the initial asperity angle, α_0 describes the joint shape of typical laboratory samples. By removing the normal stress, an asperity degradation constant, c, which has a dimensionless relation along with the plastic work, is proposed as

$$c = k\alpha_0/\lambda\sigma_c \tag{3.51}$$

where σ_c is an unconfined compressive strength of rock, λ is a wavelength of asperity, and k is a constant. It should be noted that the degradation constant, Eq. (3.51) is independent of the applied stress but the effect of stress on the asperity degradation is included in the plastic work, W_p as shown in Eq. (3.49). The constant k is determined, which fits the direct shear test results for both natural and model joints. It has a range of 0.4–0.9 in most cases, as illustrated in Section 3.4.

3.3.2.3 *Dilation*

The mobilized dilation is calculated from the mobilized asperity angle. A dilation increment, $\Delta\delta v$ is calculated, based on the plastic shear displacement increment, $\Delta\delta_s^P$, as

$$\Delta\delta_v = \Delta\delta_s{}^P \tan(d_n) \tag{3.52}$$

where $\Delta\delta_v$ is a normal displacement, $\Delta\delta_s^p$ is a plastic shear displacement, and d_n is a dilation angle. The dilation angle is calculated from the relationship given by

$$d_{n,mob} = D\alpha_0 \exp\left(-cW_p\right) \tag{3.53}$$

where D is a coefficient for the joint surface damage and Eq. (3.51) proposed here is used for the asperity degradation constant, c.

The coefficient $D=0.5$ in Eq. (3.53) is to be a reasonably conservative estimation of dilation angle with limited test data available, based on Barton and Choubey's test results (Barton and Choubey, 1977). Through the test data obtained from 130 model fractures, Barton concluded that the coefficient 1.0 could be used for joints that suffer relatively little damage during shear while the coefficient 0.5 is for further accumulated damage of asperity (Barton and Choubey, 1977; Barton, 1993). A summary of the experimental results obtained by Bandis (1980) indicates that for small samples (5 or 6 cm), the average value of the ratio of d_n to $(\phi_{p,mob} - \phi_r)$ is about 0.4 and for large samples (36 or 40 cm), the average value of the ratio is about 0.5. Therefore, it is suggested that the coefficient values in Eq. (3.53) can be in a range between 0.4 and 1.0, which provides good agreement with the experimental results as illustrated in Section 3.3.4.

3.3.3 Constitutive model for large-scale joints

3.3.3.1 *Evaluation of peak shear strength*

Patton (1966a, b) and Cording and Mahar (1974), Cording (1976) obtained an estimate of the peak shear strength in the field by using the residual angle of friction obtained from laboratory direct shear tests and adding a component, i of the large-scale waviness to the angle of friction. McMahon (1985) reported the correlation of back-calculated effective friction angles from eight rockslides with results from laboratory direct shear tests and field measurements of roughness. He showed that the effective friction angle of potential rockslide surfaces could be estimated as the sum of the mean laboratory ultimate friction angles and the mean of the large-scale roughness angles. These approaches (that is, Patton, 1966a, b; Cording and Mahar, 1974; Cording, 1976; McMahon, 1985), however, may result in conservative estimates of the peak shear strength, especially in the case of rough and tensional fractured joints. Peak shear strength in the field is generally composed of not only components of residual

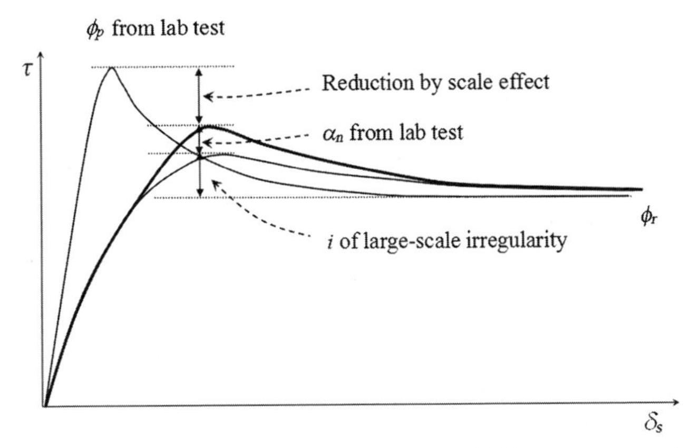

FIG. 3.26 Schematic shear stress-displacement curves. Bold represents a shear stress-displacement curve for a joint in the field.

(ϕ_r) and field-scale irregularity (i) angles, but also a component of field-scale small roughness, as illustrated in Fig. 3.26.

In this study, the primary reduction in shear strength with increasing block size in the field is assumed to be due to a reduction in a geometrical component (dilation angle) of shear strength. Shear strength in the field, therefore, can be expressed in the following equation as a general form:

$$\tau_{mob} = \sigma_n \tan(\phi_r + \alpha_{n,mob} + i_{mob}) \tag{3.54}$$

where ϕ_r and σ_n are obtained from a laboratory test and i can be determined from geologic observations in the field. The main question that remains to be addressed is: how could the field-scale small roughness (α_n) be determined from a laboratory test? During the shearing process of the joint in the field, large-scale irregularities generally lead to gradual dilation and thus the mobilized shear strength due to large-scale irregularities is also gradual. At the same time, shearing resistance contributed from small-scale irregularities is mobilized over a small displacement at the contact portions. Before the peak shear strength of discontinuity in the field is mobilized, shearing resistance due to small irregularity interlocking has already been fully mobilized and will degrade rapidly. This component of shear strength (α_n) can be practically obtained from a laboratory direct shear test for a typical size of a sample as a fraction of small-scale roughness (e.g., $0.6(\phi_{p,mob} - \phi_r)$), which would provide the value similar to the shearing-through component (s_n) of a small-scale joint. Bandis's test results show the average value of the dilation component (d_n) is 0.37 ($\phi_{p,mob} - \phi_r$). The field-scale small roughness (α_n) is, however, not solely a shearing-through component (s_n) because α_n would contribute to some dilatancy for the strength of small-scale roughness. $\alpha_{n,mob}$ at the peak in the field will be less than 0.6 ($\phi_{p,mob} - \phi_r$) because of the degradation before peak.

Increasing peak shear displacement (δ_{peak}) will cause more degradation before peak and lower $\alpha_{n,mob}$. The true scale reduction in the field, therefore, will depend on the amount of degradation that occurs prior to peak displacement. As shown in the examples later, a simulation of direct shear tests with 0.6 ($\phi_{p,mob} - \phi_r$) for the initial value of α_n shows a reasonably good correlation with experimental results taken from the literature.

A similar approach was used by Hencher and Richards (1982) to evaluate the field shear strength of discontinuity. They performed laboratory direct shear tests in order to get dilation-corrected friction angles; the stress measured in the horizontal and vertical plane is resolved tangentially and normally to the plane along which shearing is actually taking place. These corrected stresses may then be plotted to give a strength envelope reflecting the shear strength of nondilating, naturally textural surfaces. After determining a dilation-corrected friction angle, the angle, i of the large-scale irregularity can be added to it to represent field shear strength. Richards and Cowland (1982) successfully applied this procedure to slope stability design in Hong Kong.

3.3.3.2 Evaluation of peak shear displacement

Barton and Bandis (1982), after reviewing a large number of shear tests for peak shear displacements reported in the literature (about 650 data points), proposed the following empirical relation for estimation of the peak shear displacement in the field.

$$\delta_p = \frac{L}{500}\left(\frac{JRC}{L}\right)^{0.33} \tag{3.55}$$

where δ_p is a slip magnitude required to mobilize peak strength, or that occurring during unloading in an earthquake, and L is a length of a joint or a faulted block in meters.

Removing the difficulty in selecting the appropriate value of JRC for the in situ block, Eq. (3.55) is modified in the form, expressed with the ratio of block size and the angle of inclination of the large-scale waviness, i.

$$\frac{\delta_{p,n}}{\delta_{p,0}} = \left(\frac{L_n}{L_0}\right)^{0.11} i^{0.33} \tag{3.56}$$

$$\frac{\delta_{p,n}}{\delta_{p,0}} = \left(\frac{L_n}{L_0}\right)^{0.33\sim0.67} i^{0.33} \tag{3.57}$$

where $\delta_{p,0}$ and L_0 are the laboratory-scale peak shear displacement and joint length and $\delta_{p,n}$ and L_n are the field-scale peak shear displacement and joint length, respectively.

Eq. (3.56) is derived by correlating with Bandis's test results (Bandis, 1980), as shown in Fig. 3.27. The values of δ_p in the figure are for the largest samples (36 or 40 cm). Analyses are performed by considering 36 or 40 cm samples as

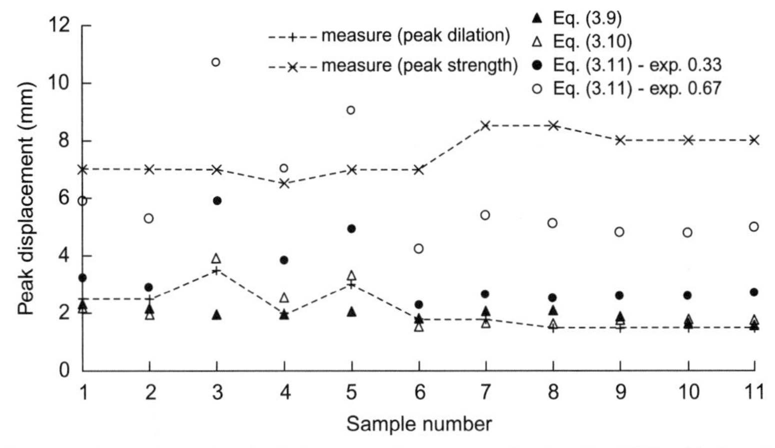

FIG. 3.27 Comparison of peak displacements from test results (Bandis, 1980) with the values derived from equations. Sample 1–5: strongly undulating; 6–8: moderately undulating; 9–11: almost planar joints.

models of the field scale; on the other hand, the smallest samples (5 or 6 cm) are models of the lab scale. At first sight, Eq. (3.56) appears sound. It provides a good agreement with the peak displacements estimated by Bandis (1980) and results in the similar estimation with Eq. (3.55), as shown in Fig. 3.27. However, the value of the exponent, 0.11 in Eq. (3.56) seems to be small. In Barton and Bandis's analyses (Bandis, 1980; Cording, 1976), the peak displacements (δ_p) for the largest sample or the in situ block were determined from the displacement at peak dilation, not at peak strength. Peak strength is often mobilized later than peak dilation for large blocks, as clearly shown in Bandis's test results (Bandis, 1980). Therefore, Eq. (3.56) could result in a smaller value of the peak displacement for rock joints in the field (increase in δ_p by a factor of 1.66 for the sample ratio of 100). Eq. (3.57) thus uses a larger value of exponent, 0.33–0.67, for the size ratio and provides a more reasonable value of the peak displacement that corresponds to the peak strength of the in situ block.

3.3.3.3 Degradation in dilation and postpeak strength

Field observations indicate that the large-scale irregularities are little sheared through and the strength component of the large-scale irregularities will not be lost with small shearing displacements. Also, they have low angles of inclination (usually less than 10 degree) and sinusoidal shapes. Based on these observations, the following assumptions were made to model the degradation in dilation and postpeak strength along a large-scale irregularity. First, the angle of inclination, i will contribute to dilation as well as the strength component; second, the degradation of a dilation angle is determined from the sine curve; and third, a dilation angle will be zero (that is, residual strength) after a shear displacement is half the wavelength of the large-scale irregularity. The sine curve is expressed in the following equation (Lee, 2003) and is called the Sinusoidal Function (f^s):

$$f^s = \left[\frac{\pi}{4} + \tan^{-1}\left(-\sin\left(\pi\left(\frac{\delta_s^p}{0.5 \cdot \lambda_{large}}\right) - \frac{\pi}{2}\right)\right)\right] \Big/ \frac{\pi}{2} \quad (3.58)$$

where λ_{large} is the wavelength of the large-scale irregularity observed in the field.

Fig. 3.28 depicts the normalized Sinusoidal Function, and the Sinusoidal Function (f^s) is used to obtain the mobilized dilation angle from a large-scale irregularity, as in Eq. (3.59):

$$d_{n,mob} = i_0 f^s \quad (3.59)$$

where i_0 is an initial angle of inclination of a large-scale irregularity.

3.3.3.4 Summary of proposed joint models

The proposed joint model has been developed by modifying some parameters of the existing models or adding new features to the existing models. This section thus describes what has been adopted in the proposed model from the previous work first and then what is proposed to the model in the current study. The complete form of the proposed joint model for typical laboratory-scale samples is expressed in the following equation:

$$\tau = \sigma_n \tan\left[\phi_r + \left\{\alpha_0 \exp\left(-cW_p\right)\right\}\right] \quad (3.60)$$

This model was originally suggested by Plesha (1987), as mentioned in Section 3.3.2.2. The mobilized dilation is also calculated from the mobilized

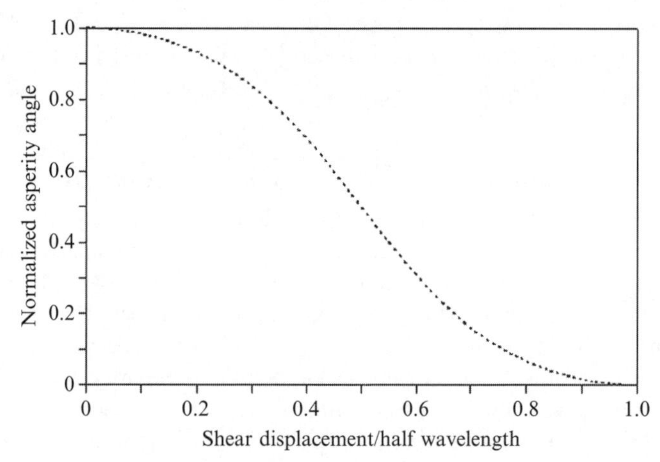

FIG. 3.28 Normalized sinusoidal function from Eq. (3.58).

asperities of the small scale. By combining Eqs. (3.52), (3.53), the formulation is given by

$$\Delta\delta_v = \Delta\delta_s^p \tan\left\{D\ \alpha_0 \exp\left(-cW_p\right)\right\} \tag{3.61}$$

This equation results from a combination of the asperity degradation model by Plesha (1987) and the asperity damage coefficient by Barton and Choubey (1977).

In the current work, the asperity degradation constant, $c = k\frac{\alpha_0}{\lambda\sigma_c}$, Eq. (3.51) is proposed. As shown in Eqs. (3.60), (3.61), by introducing the new asperity degradation constant, the dimensionless product of plastic work, rock strength, and wavelength of irregularities has been developed. The complete form of the joint model for field-scale asperities is given by

$$\tau = \sigma_n \tan\left[\phi_r + \left\{\alpha_n \exp\left(-cW_p\right) + i_0 f^s\right\}\right] \tag{3.62}$$

The mobilized dilation is calculated from the mobilized asperities of both small-scale and large-scale irregularities as

$$\Delta\delta_v = \Delta\delta_s^p \tan\ \left\{D\alpha_n \exp\left(-cW_p\right) + i_0 f^s\right\} \tag{3.63}$$

In this study, two new parameters are proposed to represent the shear strength and dilation behavior of a rock joint in the field. α_n is a field-scale small roughness obtained by a laboratory shear test (e.g., $0.6 \times (\phi_{p,\,mob} - \phi_r)$) and f^s is a sinusoidal function, expressed in Eq. (3.58). Finally, the study proposes an approach to scaling the peak shear displacement from the laboratory to the field scale, as in Eq. (3.57).

3.3.4 Correlation with experimental data

In this section, the verification of the model is illustrated by providing its correlation with experimental results taken from the literature. The simulation of a direct shear test, which consists of a single horizontal joint that is first subjected to a normal confining stress and then to a unidirectional shear displacement, is performed using the three-dimensional distinct element code, 3DEC (Itasca Consulting Group Inc, 2003). The model is incorporated in 3DEC using the built-in programming language, FISH. The average normal and shear stresses and normal and shear displacements along the joint are measured using FISH. The joint is defined by one contact that is composed of 10 subcontacts. Fig. 3.29 shows the direct shear test model and boundary conditions.

The first example shows the model's performance on the typical laboratory sample. In the second example, correlation is given for an artificial joint with a single tooth-shaped asperity. Finally, examples show the model's performance on both the small-scale and large-scale joints.

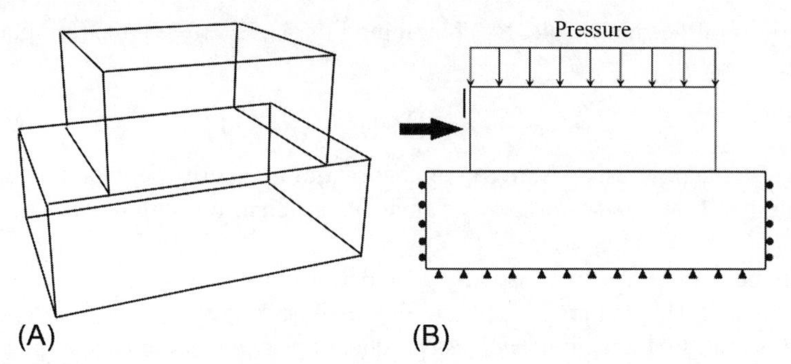

FIG. 3.29 Direct shear test models and boundary conditions. (A) 3D model for direct shear test. (B) Boundary conditions.

3.3.4.1 *Simulation of Flamand et al.'s test*

Flamand et al. (1994) conducted direct shear tests with identical replicas of a natural fracture in granite, modeled from the original sample of a fracture in the Gueret Granite (France). It was drilled perpendicular to the fracture plane under three different normal stresses (σ_n) applied on the shear plane. The samples have a circular section with a shear surface of 90 mm in diameter. The characteristics of the mortar used are: σ_c=82 MPa, E=32,200 MPa, and the basic friction angle (ϕ_b) is 37 degree. Joint profiles recorded parallel to the shear direction with a constant step, 0.5 mm are presented in Fig. 3.30.

Fig. 3.30 shows a comparison between simulation and experimental test results under a constant normal stress condition. Based on the joint profile shown in Fig. 3.30C, the average value of the wavelength (λ) is determined. Table 3.6 presents input parameters used in simulations. Because all material parameters were not provided in their study, some of them are estimated on the basis of experimental results. Based on the slopes of a straight line from the origin to the end of the elastic region of the experimental results shown in Fig. 3.30A, a joint shear stiffness (K_s) is estimated to be K_s=63,000 MPa/m. Without sufficient information, a joint normal stiffness (K_n) is assumed to be K_n=315,000 MPa/m. Based on the joint profile shown in Fig. 3.30C and the back-calculation from test results, an initial asperity angle (α_0) is estimated to be 22 degree (Flamand et al. in their study (Flamand et al., 1994), estimated initial asperity angle (α_0) to be 15–30 degree). As shown in Fig. 3.30, the agreement between simulation and experimental results is very good.

3.3.4.2 *Simulation of Yang and Chiang's test*

An experimental study was done by Yang and Chiang (2003) on the progressive shear behavior of rock joints with tooth-shaped asperities. For a basic understanding of shear behavior, direct shear tests were performed for the artificial joint with a single tooth-shaped asperity. The model material consisted of plaster and water mixed by weight ratios of 1:0.6. Its unconfined compressive

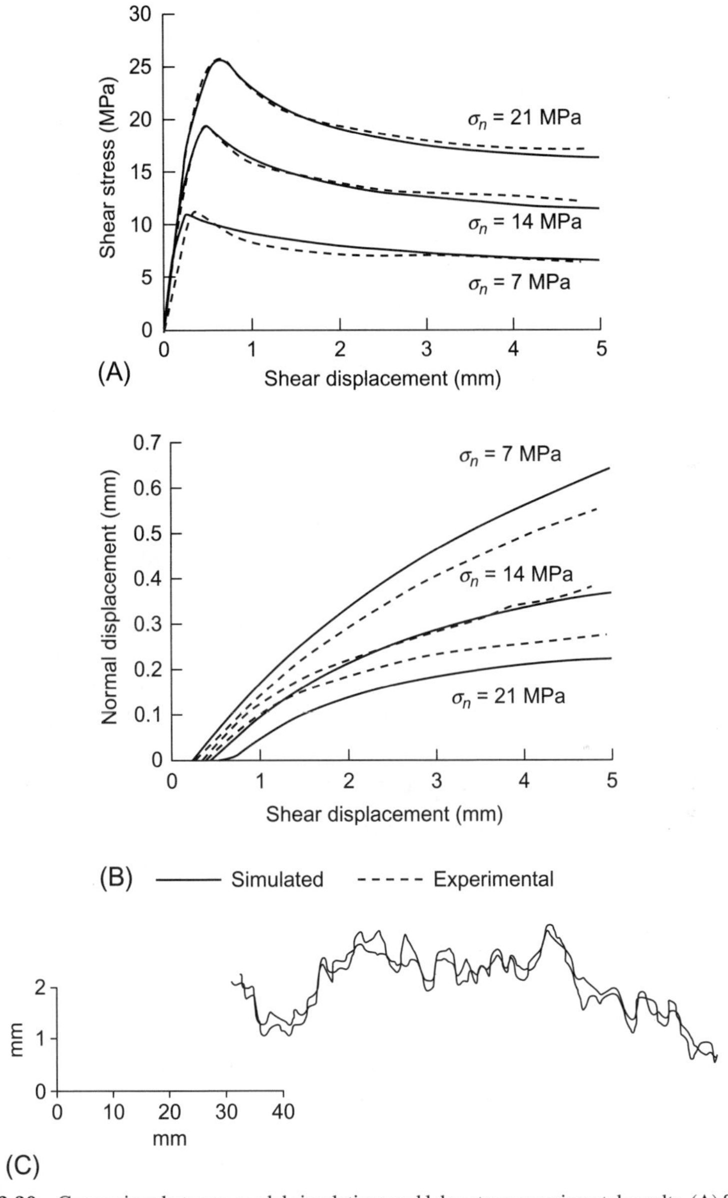

FIG. 3.30 Comparison between model simulations and laboratory experimental results. (A) Shear stress-displacement curves. (B) Normal-shear displacement curves. (C) Profile of surface geometries of joint. *(Experimental data from Flamand, R., Archambault, G., Gentier, S., Riss, J., Rouleau, A., 1994. An experimental study of the shear behavior of irregular joints based on angularities and progressive degradation of the surfaces. In: The 47th Canadian Geotechnical Conference, Halifax, Nova Scotia, Canadian Geotechnical Society, Halifax, Nova Scotia, pp. 253-262.)*

TABLE 3.6 Input parameters used in simulation.

E (MPa)	K_n (MPa/m)	K_s (MPa/m)	σ_c (MPa)	ϕ_b (degree)	α_0 (degree)	λ (mm)	k
32,200	315,000	63,000	82.0	37.0	22.0	3.0	0.45

JRC=8.5 to 10 estimated by Flamand et al.

strength (σ_c) was 7.36 MPa and the basic friction angle (ϕ_b) was about 35 degree. The single tooth-shaped asperity has a base length of 1.7 cm and an inclined angle of 30 degree.

With no data available, the elastic modulus (E) was assumed to be 2208 MPa, based on Deere's classification; for most rocks, the ratio E/σ_c lies in the range from 200 to 500 and averages 300 (Deere, 1968). Joint shear stiffness (K_s) was approximated to be 1300 MPa/m from test results by taking the slope from the origin to the end of the elastic region and a joint normal stiffness (K_n) was again assumed to be 21,000 MPa/m. Fig. 3.31 shows a comparison between model simulation and experimental results. Table 3.7 presents input parameters used in simulations. Unlike a natural joint surface, the test specimen has only one single triangular tooth-shaped asperity with a large wavelength (λ=1.7 cm) compared to typical laboratory samples; the model still provides a reasonable correlation.

3.3.4.3 *Simulation of Bandis's test*

Bandis (1980) conducted systematic studies for the effect of scale on the shear behavior of rock joints by performing direct shear tests on different sized replicas cast from various natural joint surfaces. Direct shear tests were performed under the constant normal stress of 24.5 kPa. The model joint compressive strength ($JCS=\sigma_c$) was set at 2 MPa for all types of surfaces tested. The E/σ_c ratio value was about 400 for all joint block sizes. The basic friction angle (ϕ_b) of the joint surfaces was about 32 degree. The description of the joint surface according to visual appearance can be classified as belonging to one of three categories: category *I*—strongly undulating, rough to moderately rough; category *II*—moderately undulating, very rough; and category *III*—moderately undulating to almost planar, moderately rough to almost smooth. Typical examples of the joint profiles of each category will be shown with experimental and simulation results.

Figs. 3.32–3.34 show the performance of the joint model for scale effect on the shear behavior of rock joints as well as the comparison with experimental results for each category. One of the main purposes of this study is, as previously mentioned, to propose a method of evaluating joint shear behavior in

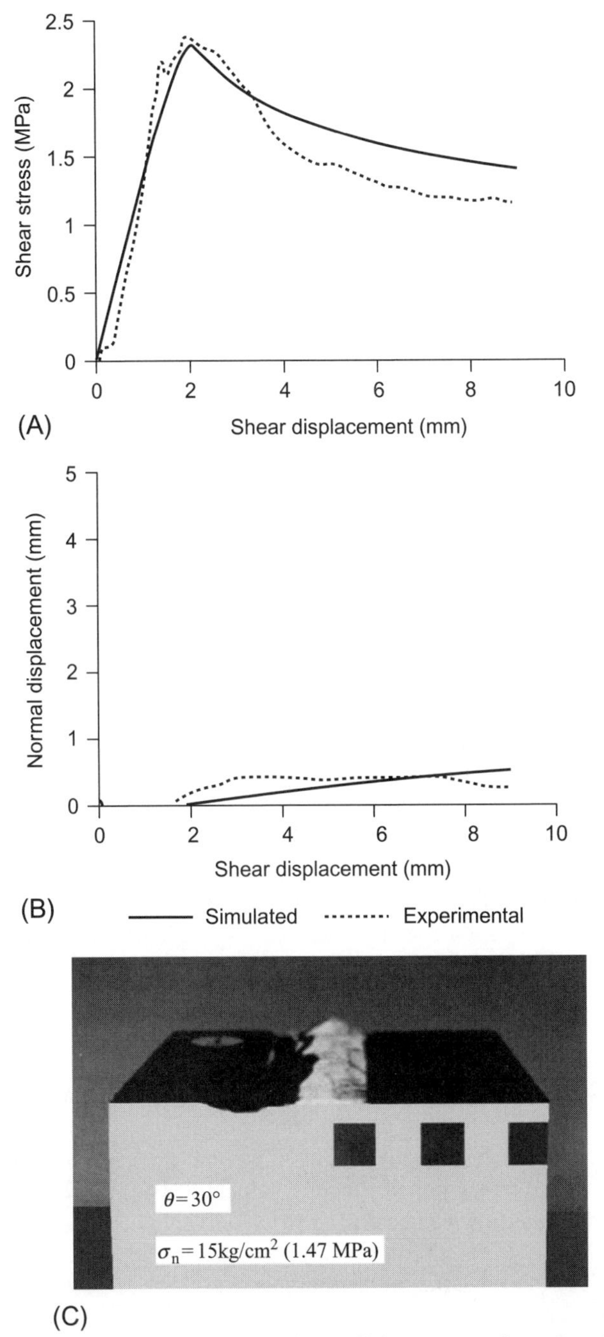

FIG. 3.31 Comparison between model simulations and laboratory experimental results. (A) Shear stress-displacement curves. (B) Normal-shear displacement curves. (C) Profile of surface geometries after shear test. *(Experimental data from Yang, Z.Y., Chiang, D.Y., 2000. An experimental study on the progressive shear behavior of rock joints with tooth-shaped asperities. Int. J. Rock Mech. Min. Sci. 37(8), 1247-1259.)*

TABLE 3.7 Input parameters used in simulation.

E (MPa)	K_n (MPa/m)	K_s (MPa/m)	σ_c (MPa)	ϕ_r (degree)	α_0 (degree)	λ (mm)	k
2208	21,000	1300	7.36	35.0	30.0	17.0	0.4

the field from a laboratory test. To evaluate this, the largest (36 or 40 cm) samples are considered as models of the field scale and the smallest (5 or 6 cm) samples are considered as models of the lab scale (Tables 3.8–3.10).

The input parameters used in simulations and joint profiles for each category are shown in Figs. 3.32–3.34. Each joint profile was constructed on the basis of a 0.7 mm sampling interval. Irregularities (α_0 and i_0) and wavelength (λ and λ_{large}) for both small (5 or 6 cm) and large (36 or 40 cm) samples were estimated by visual measurements and by obtaining the average values from a set of three surface profiles for each model. Figs. 3.32–3.34 also show the characteristics of joint asperities and wavelengths at different scales. The joint shear stiffness (K_s) of the small samples was back-calculated from the secant line of the elastic region of the shear stress-displacement curve obtained from tests. For lack of information on a joint normal stiffness (K_n), it is assumed to be about $15 \times K_s$. Because both experimental tests and simulations are performed under the constant-normal-stress boundary condition, the results of shear behavior are independent of the particular value chosen for K_n. The peak displacement (δ_p) for the lab scale was determined from fitting to the lab test. Using that value (δ_p for lab scale) and Eq. (3.57) with an exponent, 0.33, the peak displacement (δ_p) for the field scale was determined. The value of $0.6 \times (\phi_{p,\ mob} - \phi_r)$ obtained from the laboratory test of the lab-scale sample was used to obtain the value of the field-scale small roughness (α_n). Dilation of both the lab-scale and field-scale samples is modeled using Eqs. (3.61), (3.63), respectively, where $D=0.4$ is used.

As shown in the correlation with experimental results for both small and large samples, there are very good agreements between simulation and experimental results regarding the shear stress-shear displacement relation of the three different cases, but less agreement for dilatancy relation. This is partly due to the difficulty in estimating material parameters, especially the joint normal stiffness (K_n) as mentioned above. The main reason lies in the limitation of numerical modeling and can be explained as follows. The built-in programming language FISH has been used to incorporate the developed model into 3DEC. For this, FISH modifies the built-in joint model, Mohr-Coulomb, which is an elasto-perfectly plastic constitutive model. This joint model allows the dilation to take place at the peak shear displacement, which is quite a different characteristic of dilation behavior from the developed model, as explained in the

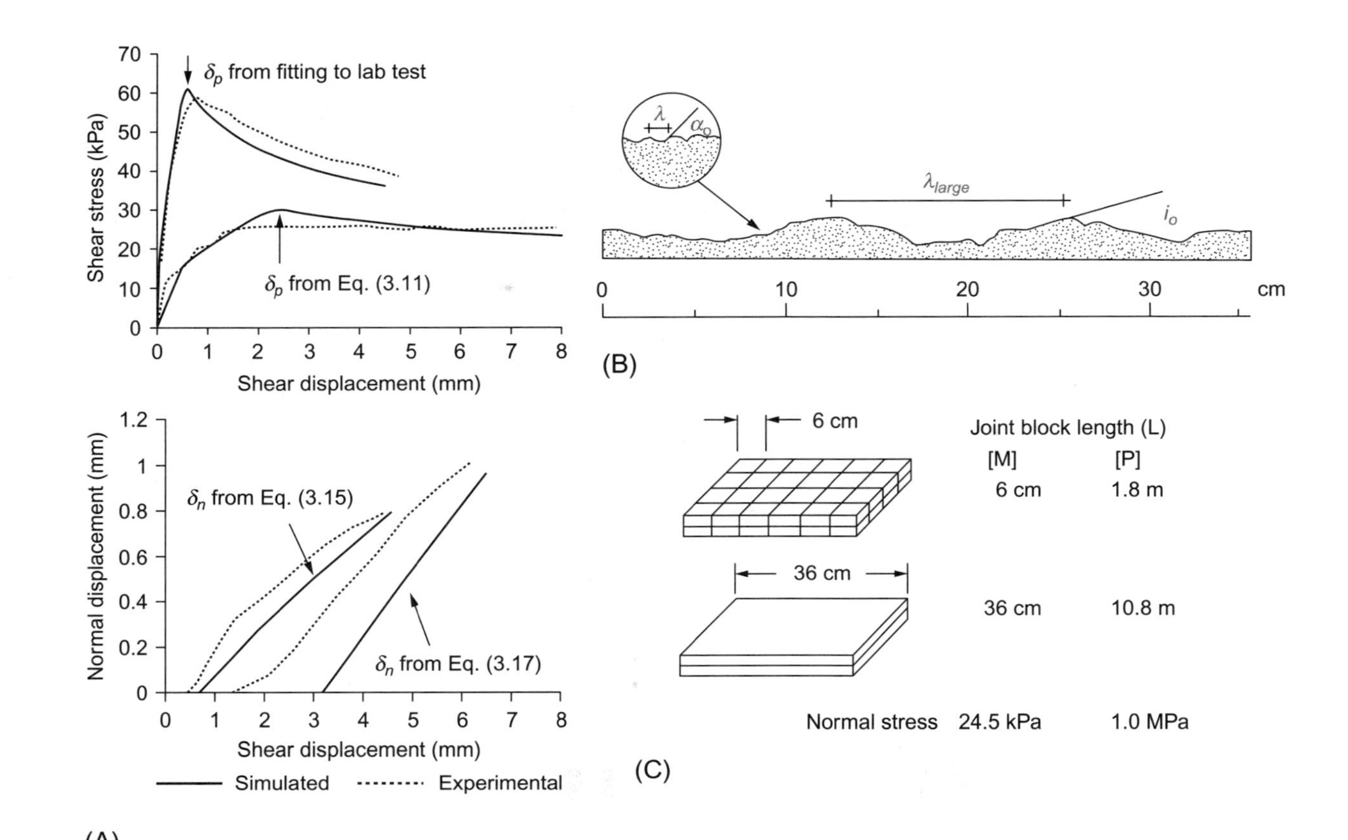

FIG. 3.32 Comparison between model 1 simulations and laboratory experimental results—Category I. (A) Shear stress-displacement curves and normal-shear displacement curves. (B) Profiles of surface geometries of joint. (C) Sample size; [M] = model, [P] = prototype. *(Experimental data from Bandis, S.C., 1980. Experimental Studies of Scale Effects on Shear Strengh, and Deformation of Rock Joints. PhD Thesis, The University of Leeds.)*

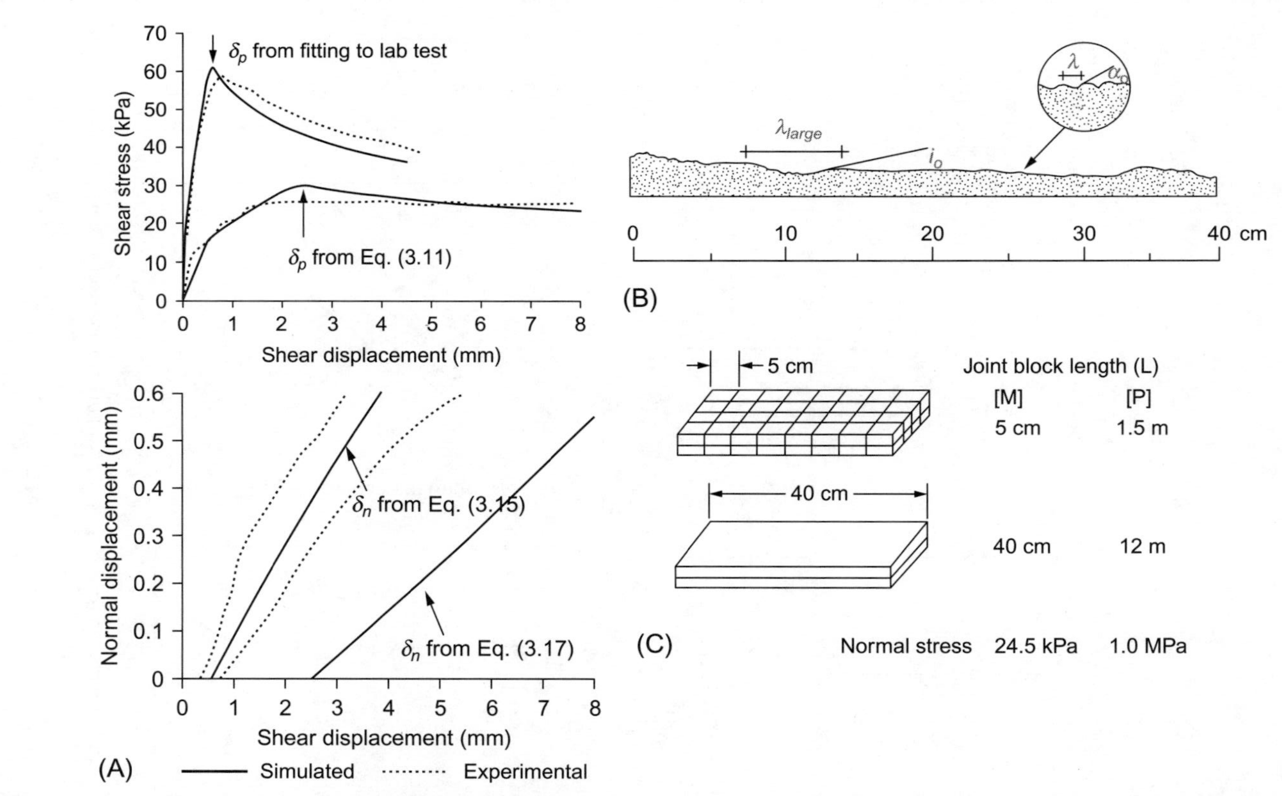

FIG. 3.33 Comparison between model 7 simulations and laboratory experimental results—Category II. (A) Shear stress-displacement curves and normal-shear displacement curves. (B) Profiles of surface geometries of joint. (C) Sample size; [M] = model, [P] = prototype. *(Experimental data from Bandis, S.C., 1980. Experimental Studies of Scale Effects on Shear Strengh, and Deformation of Rock Joints. PhD Thesis, The University of Leeds.)*

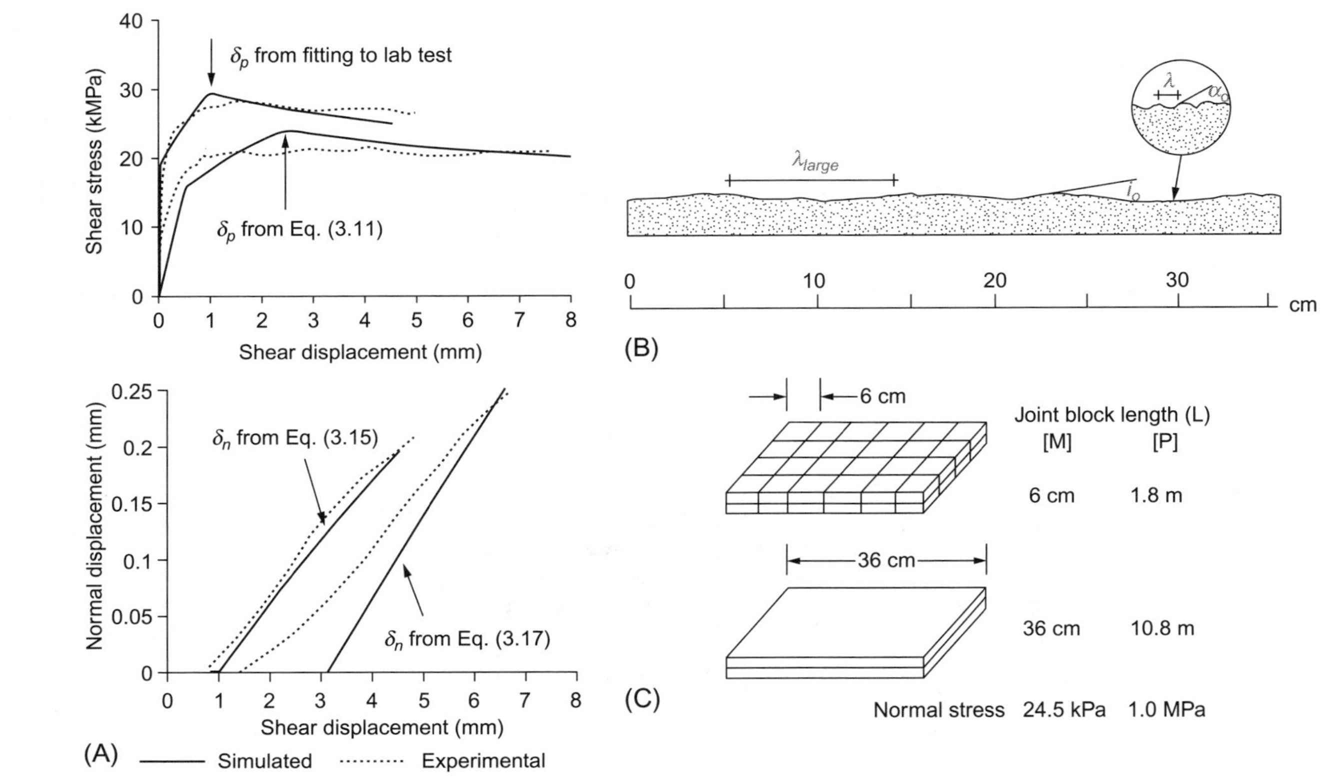

FIG. 3.34 Comparison between model 10 simulations and laboratory experimental results—Category III. (A) Shear stress-displacement curves and normal-shear displacement curves. (B) Profiles of surface geometries of joint. (C) Sample size; [M] = model, [P] = prototype. *(Experimental data from Bandis, S.C., 1980. Experimental Studies of Scale Effects on Shear Strengh, and Deformation of Rock Joints. PhD Thesis, The University of Leeds.)*

TABLE 3.8 Input parameters used in simulation.

Sample size (cm)	E (MPa)	σ_c (MPa)	K_s (MPa/m)	α_0/α_n (degree)	i_o (degree)	ϕ_b (degree)	λ (cm)	k
6	850	2	3000	40.0	–	32.0	0.3	0.4
36	850	2	23.9	20.8	12.0	32.0	10.0	0.8

JRC=18.1 for 6 cm and 12.5 for 36 cm estimated by Bandis

TABLE 3.9 Input parameters used in simulation.

Sample size (cm)	E (MPa)	σ_c (MPa)	K_s (MPa/m)	α_0/α_n (degree)	i_o (degree)	ϕ_b (degree)	λ (cm)	k
5	850	2	300	40.0	–	32.0	0.4	0.45
40	850	2	29.4	21.0	6.0	32.0	5.0	0.9

JRC=18.5 for 6 cm and 7.3 for 40 cm estimated by Bandis

TABLE 3.10 Input parameters used in simulation.

Sample size (cm)	E (MPa)	σ_c (MPa)	K_s (MPa/m)	α_0/α_n (degree)	i_o (degree)	ϕ_b (degree)	λ (cm)	k
6	850	2	300	20.0	–	32.0	0.15	0.5
36	850	2	29.4	9.6	3.0	32.0	8.0	0.9

JRC=8.4 for 6 cm and 4.5 for 36 cm estimated by Bandis

previous chapter. Therefore, the discrepancy results from when the joint starts to dilate. Nevertheless, the joint model reasonably well represents the shear stress-displacement-dilation relation of various natural joint surfaces and scales. The simulation results also support that the proposed approach to evaluating rock joint behavior in the field from laboratory data and field observations is reasonably reliable.

3.3.5 Conclusions

The most important feature of the model is that it considers the dilation and strength along both small-scale and large-scale irregularities. The study also includes how the scale effect is employed in the model. In the literature, there is considerable discussion on the scale effect of rock joint behavior and the extrapolation of the shear strength of lab-scale joints to large-scale joints in the field. Many studies show that there is a reduction in the joint shear strength with increasing length of the joint (that is, Bandis, 1980; Bandis et al., 1981; Pratt et al., 1974), but the opposite behavior has also been observed (that is, Locher and Rieder, 1970 Swan and Zongui, 1985). Another investigation reported that the scale dependency is limited to a certain size and for larger than the stationarity limit, the joint strength remains almost constant (e.g., Fardin et al., 2001). Although the problem of scale effect is very complex and under debate, it seems to be true that changing the scale involves a variation of the asperity size, which contributes to the change of its mechanical behavior. It is also true that the evaluation of the strength of the large-scale joint requires assessment of field-scale irregularity determined from geologic observations. It is often observed in the field that as the length of the joint increases, the size of the asperity controlling its strength increases. This controlling asperity has a relatively gentle slope and a large wavelength compared to that of a small-scale joint, thereby reducing its strength and brittle behavior.

This study presents a rock joint constitutive model that represents the behavior of both a small-scale joint in the laboratory and a large-scale joint in the field. A dimensionless product of plastic work, rock strength, and wavelength of irregularities has been developed to represent the degradation in strength and dilation along small-scale irregularities. On the other hand, a sinusoidal function is used to model the degradation in strength and dilation along large-scale irregularities. The comparisons of the model with experimental data taken from the literature show that the proposed model has very good overall agreement, although there is some discrepancy attributed to a lack of information on some parameters. The model can be incorporated into discrete element computer codes using a program language in order to solve boundary value problems. Although some simplifications are involved in the modeling, especially related to the scale effect of rock joints, the model predicts reasonably well the shear behavior of both lab-scale and field-scale joints.

3.4 Opening effect on joint shear behavior

3.4.1 Introduction

The presence of dilatancy can significantly enhance the stability of a rock mass, especially when rock blocks or wedges are placed between dilatant joints. Under the constrained normal displacement condition, dilatancy can lead to an increase in normal stress and thus the shear strength of the joint during shear. If the joint opens, however, the interlock between the joint is reduced. As the opening of the joint approaches the amplitude of the irregularities on the joint, then the dilatant component of strength is lost and the shear strength drops to the minimum or residual strength.

When an excavation is made in a rock mass, the stress that previously existed in the rock is changed and new stresses are induced in the rock in the vicinity of the

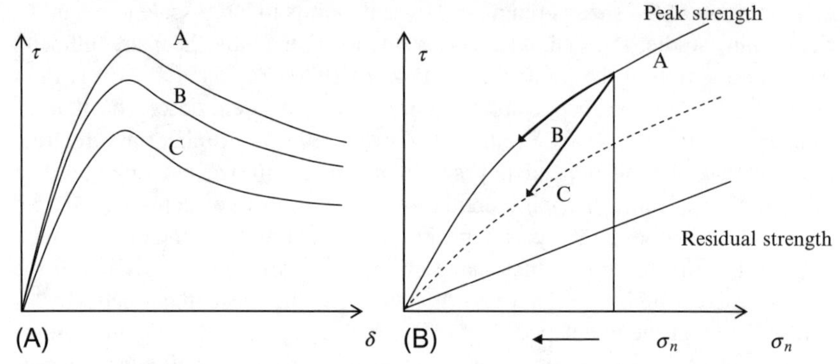

FIG. 3.35 Effect of the joint opening on shear strength due to the relief of confinement. (A) Shear stress-shear displacement relation. (B) Shear stress-normal stress relation.

opening. Many stability problems in the excavation of a jointed rock mass are triggered by such stress redistribution. Excavating the rock may result in a concentration of tangential stress in the rock immediately surrounding the opening. On the other hand, it may also allow relief of confinement normal to the excavated surface. The relief of confinement has two effects: first, a reduction in confining pressure along joints with a resulting reduction in shear strength; and second, a reduction in the dilatant component of shear strength due to the opening of the joint. The result is less of a dilatant component of strength and less shearing through of the smaller asperities. Fig. 3.1 illustrates this opening effect on the shear strength of a joint. Many numerical models consider the effect of confining pressure along joints (A→B in Fig. 3.35), but few have been able to model the effect of strength loss along joints as a result of joint opening (A→C in Fig. 3.35). This paper describes the effect of the opening of rough joints on the shear strength, incorporated in a joint constitutive model. Some hypothetical examples based on experimental results for direct shear tests taken from the literature are provided.

3.4.2 Constitutive model for joint opening effect

The effect of a previous decrease in the degree of interlocking of rock joints on slope stability problems was studied theoretically and experimentally by Ladanyi and Archambault (1969). For a regular joint system, such as in Fig. 3.36, Ladanyi and Archambault (1969) introduced the notion of true area, A_t and the degree of interlocking, η in order to consider the effect of a previous decrease in the degree of interlocking. According to Fig. 3.36, the true area, A_t can be expressed, after a shear displacement Δx

$$A_t = A\left(1 - \frac{\Delta x}{\Delta L}\right) = A\eta \tag{3.64}$$

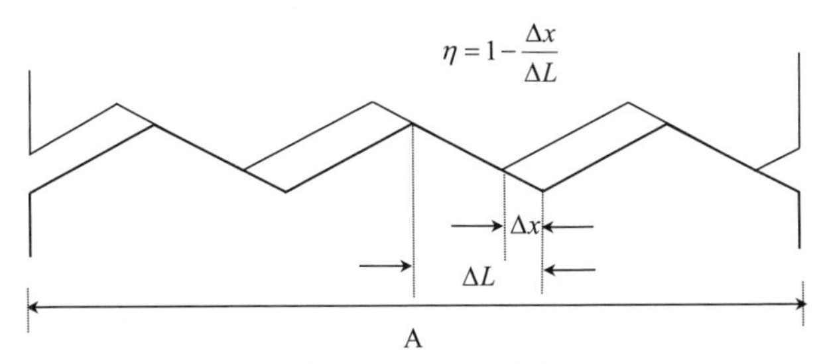

FIG. 3.36 Definition of the degree of interlocking, η. *(Modified after Ladanyi, B., Archambault, G., 1969. Simulation of shear behavior of a jointed rock mass. In: The 11th US Symposium on Rock Mechanics (USRMS), Berkeley, California, pp. 105-125.)*

From the concept of true area and the degree of interlocking, true stresses acting on the reduced shear area are derived (Ladanyi and Archambault, 1969):

$$\sigma_{n,t} = N/A\eta \tag{3.65}$$

$$\tau_t = S/A\eta \tag{3.66}$$

Eqs. (3.65), (3.66) are applied to joint constitutive models to consider the opening effect. In this paper, Barton's model, which is one of the most commonly used joint constitutive models proposed by Barton and Choubey (1977) and Barton (1982a, b), is used to describe how this opening effect is employed in the existing joint models. The joint model is expressed as

$$\tau = \sigma_n \tan\left(JRC \log\left(\frac{JCS}{\sigma_n}\right) + \varphi_r\right) \tag{3.67}$$

where *JRC* is the joint roughness coefficient, *JCS* is the joint wall compressive strength, and ϕ_r is the residual friction angle.

Applying Eqs. (3.65), (3.66) to the Barton's model, Eq. (3.67), the following expressions are obtained:

$$\frac{S}{A\eta} = \frac{N}{A\eta} \tan\left(JRC \log\left(\frac{JCS}{N/A\eta}\right) + \varphi_r\right) \tag{3.68}$$

$$\tau = \sigma_n \tan\left(JRC \log\left(\frac{JCS}{\sigma_n/\eta}\right) + \varphi_r\right) \tag{3.69}$$

From the geometric configuration shown in Fig. 3.37, it is possible to relate the degree of interlocking, η with a joint aperture, δ_n:

$$\delta_n = 2\Delta x \cdot \tan\alpha \tag{3.70}$$

Because the degree of interlocking, $\eta = 1 - \Delta x/\Delta L$, then,

$$\eta = 1 - \frac{\delta_n}{2\,\Delta L \cdot \tan\alpha} \tag{3.71}$$

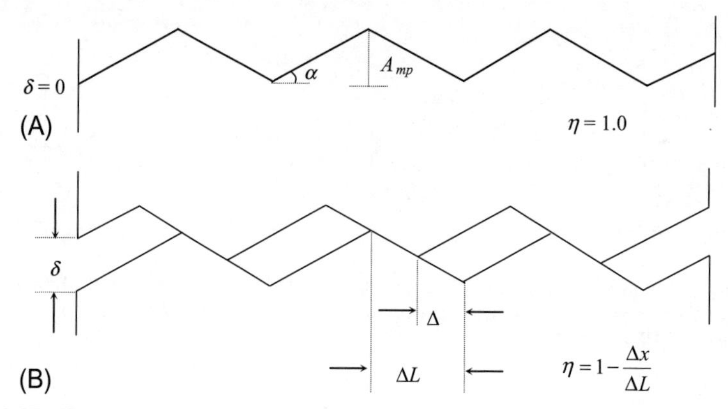

FIG. 3.37 Geometric configuration of joint opening. (A) a tightly closed joint, (B) an open joint.

$$\eta = 1 - \frac{\delta_n}{2A_{mp}} \tag{3.72}$$

By combining Eq. (3.72) into Eq. (3.69), the following equation, Eq. (3.73), is obtained for the shear strength of rough rock joints when they open.

$$\tau = \sigma_n \tan\left(JRC \log\left(\frac{JCS}{\sigma_n} \cdot \left[1 - \frac{\delta_n}{2A_{mp}} \right] \right) + \varphi_r \right) \tag{3.73}$$

The joint aperture (δ_n) in Eqs. (3.70)–(3.73) can be an initial joint opening (δ_O), that is, a previous decrease in the degree of interlocking as described by Ladanyi and Archambault (1969) or a joint normal displacement ($\Delta\delta_n$) induced by excavation.

3.4.3 Opening model performance

The performance of the joint opening model is illustrated by simulating a series of direct shear tests consisting of a single horizontal joint that is first subjected to a normal confining stress and then to a unidirectional shear displacement, using the three-dimensional distinct element code, 3DEC (Itasca Consulting Group Inc, 2003). The model is incorporated in 3DEC using the built-in programming language FISH. Depending on the boundary conditions, two types of tests are undertaken: a simulation under constant stress conditions and a simulation under constrained displacement conditions. For the first test, the normal pressure remains constant during shearing, which generally corresponds to the sliding of a rock block on a slope. The second test (under the constrained normal displacement condition) is intended to evaluate the effect of dilatancy in increasing the normal stress and the shear strength of the joint. Fig. 3.38 shows the direct shear test model and boundary conditions.

A series of direct shear tests on rock surfaces with various boundary conditions were simulated using the rock joint model proposed by the author-a

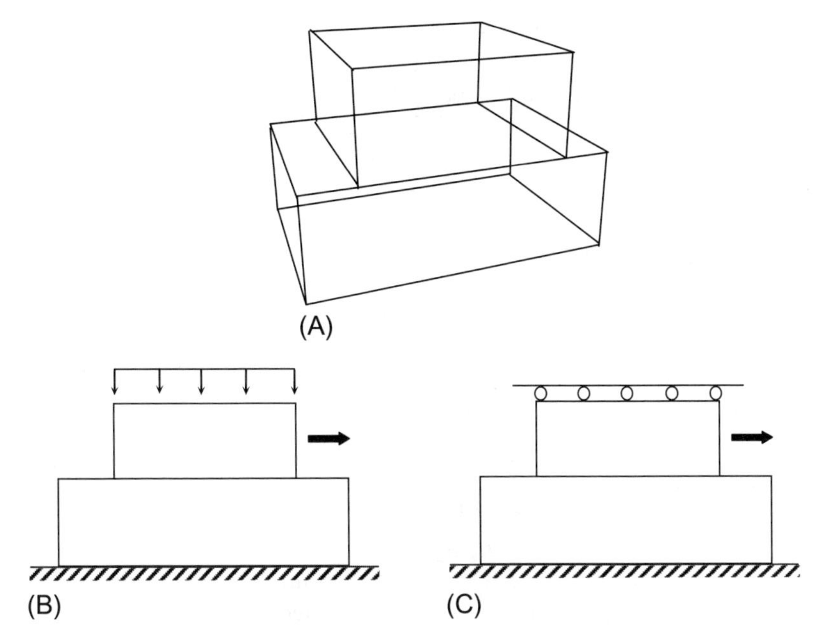

FIG. 3.38 Direct shear test model and boundary conditions. (A) 3D model for direct shear test. (B) Under stress condition. (C) Under constrained displacement condition.

dimensionless product of plastic work, rock strength, and wavelength of irregularities which fits the direct shear test results (Oh, 2005). The complete form of the joint model is expressed in the following equation:

$$\tau = \sigma_n \cdot \tan\left[\phi_r + \alpha_0 \times \exp\left(-c \cdot W_p\right)\right] \tag{3.74}$$

where α_0 is the initial asperity angle, c is the asperity degradation constant, and W_p is the plastic work, and c and W_p are expressed as (Oh, 2005; Plesha, 1987)

$$c = k \cdot \frac{\alpha_0}{\lambda \cdot \sigma_c} \tag{3.75}$$

$$W_p = \sum \Delta\delta_s^p \cdot \tau \tag{3.76}$$

where, k is constant, λ is the wavelength of asperity, σ_c is the unconfined compressive strength of rock, and $\Delta\delta_s^p$ is the plastic shear displacements. From the process described above, Eq. (3.76) can now be modified to employ the joint opening effect, given by

$$W_p = \sum \Delta\delta_s^p \cdot \tau \Big/ \left(1 - \frac{\delta_n}{2A_{mp}}\right) \tag{3.77}$$

The simulations of direct shear tests presented in this study thus use Eqs. (3.74), (3.75), and (3.77).

The material properties used in the simulations are the same as those used in the experimental tests performed by Flamand et al. (1994), who conducted direct shear tests with identical replicas of a natural fracture in granite, modeled from the original sample of a fracture in the Gueret Granite. The characteristics of the mortar used are $\sigma_c=82$ MPa, $E=32{,}200$ MPa, and $\phi_b=37$ degree. Based on the joint profile recorded parallel to the shear direction and back-calculation from test results, the average value of the wavelength (λ) and the asperity angle (α) as well as other properties not provided in their study were determined (Oh, 2005). All the parameters used in the simulations are shown in each figure, along with the simulation results. As no tests were performed with the joint open or under the constrained normal displacement condition by Flamand et al., these are hypothetical examples. Nevertheless, they provide a useful understanding of the characteristics of the joint opening behavior.

3.4.3.1 *Initial joint opening effect*

According to the literature (e.g., Bjerrum and Jorstad, 1963; Muller, 1963; Deere et al., 1967; Cording et al., 1971), most failures of a rock mass, particularly in rock slopes or shallow chambers, are related to specific geological features and usually preceded by a gradual loss of interlocking along the potential failure surface. The large displacements in a jointed rock mass generally indicate opening of discontinuities and loosening of rock blocks.

Direct shear tests are simulated with different initial opening (δ_0) values (Table 3.11). Fig. 3.39 plots the shear stress and dilatancy during shear under the constant normal stress condition. The conventional direct shear test simulation without opening effect and its correlation with the experimental results are also shown in the figure. As the value of the initial aperture is greater, not surprisingly the simulation results show the peak shear strength is lower and the dilatancy decreases due to the reduced interlock between the joints. Fig. 3.40 shows the simulation results for a direct shear test under the constrained displacement condition. As can be seen from Fig. 3.40C, there is an increase in normal stress due to the limited dilatancy during shear, which contributes to the increased shear strength. The increase in normal stress is dependent on the stiffness of the rock (E) as well as the joint (K_n), whose value is assumed in this example. The simulation results indicate that for some cases

TABLE 3.11 Input parameters used in simulation ($2A_{mp}=\lambda\tan\alpha_0=1.2$mm).

σ_n (MPa)	E (GPa)	K_n (GPa/m)	K_s (GPa/m)	ϕ_0 (degree)	λ (mm)	α_0 (degree)
21	32.2	31.5	63	37.0	3.0	22.0

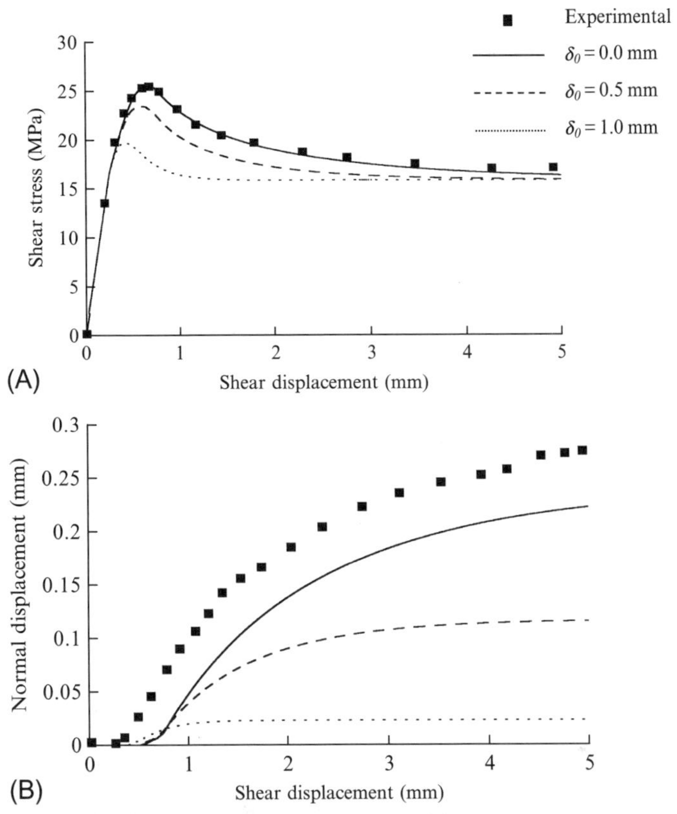

FIG. 3.39 Simulation of a direct shear test with a joint opening model under a constant normal stress condition. (A) Shear stress-shear displacement curves. (B) Normal displacement-shear displacement curves. *(Experimental results by Flamand, R., Archambault, G., Gentier, S., Riss, J., Rouleau, A., 1994. An experimental study of the shear behavior of irregular joints based on 6angularities and progressive degradation of the surfaces. In: The 47th Canadian Geotechnical Conference, Halifax, Nova Scotia, Canadian Geotechnical Society, Halifax, Nova Scotia, pp. 253-262.)*

in an underground opening, the stabilizing effect due to the additional shear strength acquired by the constrained dilation is reduced if the joint opens.

3.4.3.2 *Joint opening effect induced by excavation*

The relief of confinement due to excavation causes a reduction in normal stress to the excavated surface. The change of normal stress develops normal displacement as a function of the joint normal stiffness (K_n)–the rate of change of normal stress ($\Delta\sigma_n$) with respect to normal displacement ($\Delta\delta_n$). It is important to note that the normal stress-displacement relation is highly nonlinear, as shown in experimental results (Pratt et al., 1974; Goodman, 1976; Bandis et al., 1983). In this study,

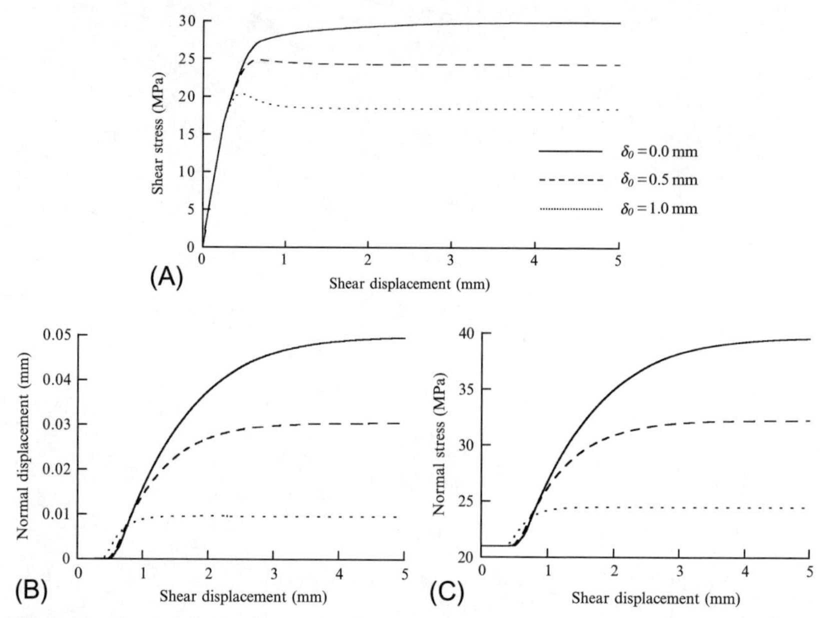

FIG. 3.40 Simulation of a direct shear test with a joint opening model under a constrained displacement condition. (A) Shear stress-shear displacement curves. (B) Normal displacement-shear displacement curves. (C) Normal stress-shear displacement curves.

however, for simplicity a linear stress-displacement relation (that is, constant K_n) is used to calculate the normal displacement due to the change of normal stress.

A series of direct shear tests, with the same material properties as used previously, were simulated under the constant normal stress condition with a constant confining pressure of 10 MPa. For this hypothetical example, the initial confining pressure is assumed to be 21 MPa. The value of joint normal displacement (joint opening) due to the change of normal stress is thus obtained from the linear normal stress-displacement relation and is used as an input parameter. In the full-scale, boundary-value problem, this numerical process can be implemented such that the value of joint normal displacement due to excavation is directly calculated and then used for the next calculation. Without sufficient information on the joint compressibility behavior, the value of joint normal stiffness is assumed to be equal to that of joint shear stiffness in this example. Fig. 3.41 shows the results of the simulation of a direct shear test under the constant normal stress condition (Table 3.12). It can be seen that with the opening effect due to relief of confinement, there is a reduction in shear strength and dilatancy, although the effect is not significant under the given conditions. It is interesting to mention again that the joint compressibility shows nonlinear behavior, as can be seen by the effect of joint opening being much greater than that from the linear stress-displacement relation, especially in the range of low normal stress levels encountered in near-surface excavations.

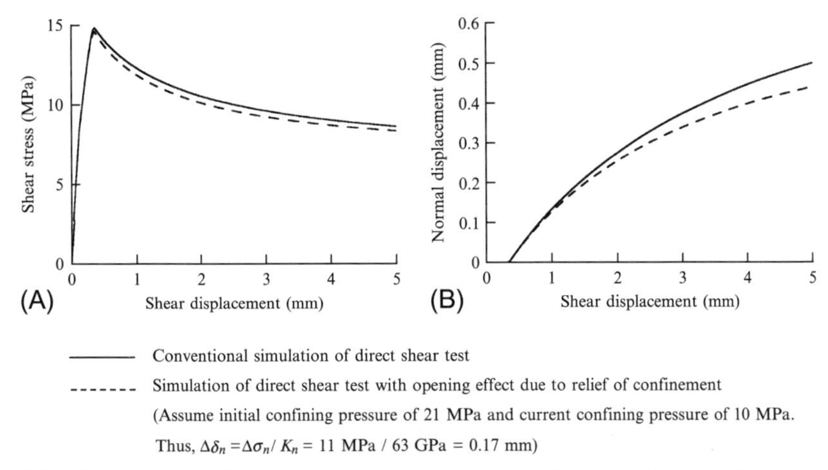

FIG. 3.41 Simulation of direct shear tests under constant normal stress condition—Effect of joint opening due to the relief of confinement. (A) Shear stress-shear displacement curves. (B) Dilatancy-shear displacement curves.

TABLE 3.12 **Input parameters used in simulation ($2A_{mp} = \lambda \tan \alpha_0 = 1.2$mm).**

σ_n (MPa)	E (GPa)	K_n (GPa/m)	K_s (GPa/m)	ϕ_0 (degree)	λ (mm)	α_0 (degree)
10	32.2	63	63	37.0	3.0	22.0

3.4.4 Discussion

In order to understand the engineering behavior of rock joints, it is necessary to evaluate the appropriate material properties of rock joints. Most rock surfaces in the field, however, are irregular, and this natural feature makes it more difficult to determine these values, especially geometric parameters such as the degree of interlocking (η). The value of the degree of interlocking (η) can reasonably be obtained, as mentioned by Ladanyi and Archambault (1969), if half the wavelength (ΔL) is replaced by an estimated average length of irregularities in the shear direction. The opening behavior of the rock joints can therefore be approximated sufficiently accurately, at least for rough joints for practical purposes.

In the examples presented here, the linear elastic model is assumed for the joint compressibility behavior in order to simplify evaluating the joint opening due to the relief of confinement. This linear elastic behavior is unrealistic, however. Instead, a highly nonlinear behavior (as shown in experimental studies of the joint compressibility behavior) has been represented by the hyperbolic equations (Goodman, 1976; Bandis et al., 1983). Although the hyperbolic equations were originally derived from the loading stress-joint closure relation, Bandis et al. (1983, 1985) showed that on the basis of the experimental results, the unloading stress-joint opening relation is also governed by the hyperbolic equations. Bandis et al. (1983) showed that the hyperbolic equation can be uniquely described by the values of two parameters, the initial normal stiffness (K_{ni}) and the maximum joint closure (V_{mc}), and suggested empirical relations to estimate these parameters from other joint properties. Based on a reinterpretation of approximately 60 datasets from published experiments, Alvarez et al. (1995) found that K_{ni} and V_{mc} are not independent and suggested the relation between these two parameters. A detailed description of the hyperbolic equation and its relation with joint properties is beyond the scope of this paper. However, it seems that a single equation cannot fully represent the joint compressibility behavior because of the complicating factors and uncertainties involved in reality. It is therefore necessary to have an appropriate compressibility law that is consistent with laboratory experiments as well as field measurements.

3.4.5 Conclusions

This study presents the development of a joint opening model that considers the effect of strength loss along the joint as a result of opening. The model, which takes account of geometric parameters and stress displacement, is easily incorporated into existing joint models. Hypothetical examples are presented to show the model is consistent with commonly held ideas about joint opening behavior. Although some simplifications are involved, the main mechanics of strength loss along the joint are well represented. Although the model is considered a useful tool for evaluating the shear strength of rock joints when they open, further studies are required to investigate its behavior, especially the joint compressibility behavior, using laboratory and field data.

3.5 Dilation of saw-toothed rock joint

3.5.1 Introduction

Joint surface irregularities cause dilation, which increases the shear resistance of rock joints. If a rock mass is restrained, constrained dilatancy due to boundary conditions leads to an increase in the normal stress on the joint. The magnitude of a normal stress increase under this circumstance depends on factors such as joint wall strength, asperity geometry, applied normal stress, etc. (Leichnitz, 1985; Saeb and Amadei, 1992). The increase in normal stress in turn leads to an increase in joint shear strength. Therefore, correctly estimating the dilation is important for rock mass stability analysis.

It has been well known after the study by Patton (1966a, b) that under low normal stress, sliding along asperities governs the failure mode leading to dilation of the joint, whereas most asperities are sheared through at high normal stress, preventing the tendency to dilate. In reality, dilation behavior is more complicated than what is described above. The two different patterns of failure occur simultaneously, particularly for natural joints. The asperity damage varies as a function of the properties of irregularities of the joint as well as the applied stress on the joint. However, in rock engineering practice, a dilation angle is often determined by simple assumption, especially when performing numerical modeling tasks. For example, the Mohr-Coulomb model, one of the most simplistic but widely used rock joint models, uses the dilation angle that is the same as the friction angle if an associated flow rule is employed to represent the plastic strain increments, which is unrealistic judging from experimental observations. On the other hand, in a nonassociated flow rule, the dilation angle is generally assumed to be some fraction of the asperity inclination angle. Barton and Choubey (1977) and Barton (1982a, b) showed, based on the experimental data from 136 joint samples, that the ratio of dilation angle to the asperity inclination angle could be in the range of 0.5–1.0, depending on the degree of asperity damage under given joint wall compressive strength and applied normal stress. Recent studies (Li et al. 2016; Oh et al., 2015) confirmed that dilation behavior is considerably affected by the geometric properties of joint asperities and that dilation angle should be estimated by considering those properties. Research in the literature has identified the critical parameters that affect the dilation behavior of a rock joint, but the approach to predicting the dilation angle still lacks some quantification.

The rapid development of modeling techniques has provided the opportunity for researchers to study the shear and dilation behavior by simulating the direct shear box. Rock joint behavior under monotonic shear using an explicit DEM code, PFC (Itasca, 2008), has been studied by several researchers (e.g. Park and Song, 2009; Asadi et al., 2012; Bahaaddini et al., 2013). As an alternative, Karami and Stead (2008) reproduced the shear behavior of rock joints with a hybrid FEM/DEM method, where the asperity damage in shearing was investigated using three standard JRC profiles provided by Barton and Choubey (1977). Through Voronoi tessellation in UDEC (Itasca, 2011), Kazerani et al. (2012) simulated joint asperities by a dense assemblage of irregular-sized blocks connected at the boundaries. The response of saw-toothed joints to shear force has been replicated, for which a new contact model was developed that required extra input parameters. Although each numerical method has limitations for representing shear and dilation behavior, they can be useful tools for performing detailed parametric studies after completing rigorous calibrations.

In this study, a DEM code with Voronoi tessellation was employed to investigate the dilation behavior of a saw-toothed rock joint. The important parameters such as joint strength, applied normal stress, and asperity geometry were studied to quantify how much they could affect the magnitude of rock joint dilation during shear. The applicability of the numerical model was assessed by calibrating and comparing the simulation results with experimental results. The study in this paper can be used as a practical guideline for predicting the dilation angle as an input parameter for rock mass stability analysis.

3.5.2 Constitutive law for contacts in DEM

In DEM, a rock mass is represented as an assembly of distinct blocks. Joints are regarded as interfaces between discrete blocks. The mechanical behavior at the interfaces of involved blocks is regulated by the contact constitutive law. The Coulomb-slip joint model is widely used in DEM. In the normal direction, the contact force varies linearly with the relative displacement, governed by the normal stiffness (k_n):

$$\Delta F_n = -k_n \Delta \delta_n \tag{3.78}$$

where ΔF_n is the normal force increment and $\Delta \delta_n$ is the incremental normal displacement. A contact has a limited tensile strength (F_n^{max}). Exceedance of F_n^{max}, that is, $F_n \leq -F_n^{max}$, causes F_n to drop to zero. On the contrary, blocks may overlap at the contacts under compression. The overlap amount is restricted to a maximum value that can be defined by the user. Similarly, the contact shear stiffness (k_s) dictates the shear response with a maximum shear force (F_s) limited by a combination of the cohesion (c^{cont}) and the friction coefficient (ϕ^{cont}), that is, if

$$|F_s| \leq c^{cont} A_c + F_n \tan \phi^{cont} = F_s^{max} \tag{3.79}$$

then

$$\Delta F_s = -k_s \Delta \delta_s^e \tag{3.80}$$

or if

$$|F_s| \geq F_s^{max} \tag{3.81}$$

then

$$F_s = \text{sign}(\Delta \delta_s) F_s^{max} \tag{3.82}$$

where A_c is the contact area, $\Delta \delta_s^e$ is the elastic component of the incremental shear displacement, and $\Delta \delta_s$ is the total incremental shear displacement. If a contact element is assigned with a residual attribute, the cohesion falls to zero at the onset of yielding. Fig. 3.42 illustrates the yielding mechanism of the contacts under normal and shear loadings.

3.5.3 Model calibration

Plaster mortar was used to replicate specimens in the laboratory and to calibrate numerical models. The mortar was made of Hydro-stone TB gypsum cement (>95% $CaSO_4 \cdot 1/2H_2O$, <5% Portland cement) and water mixed at a ratio of 1:0.35 by weight. All samples were cured at 40°C in a dry oven for 14 days.

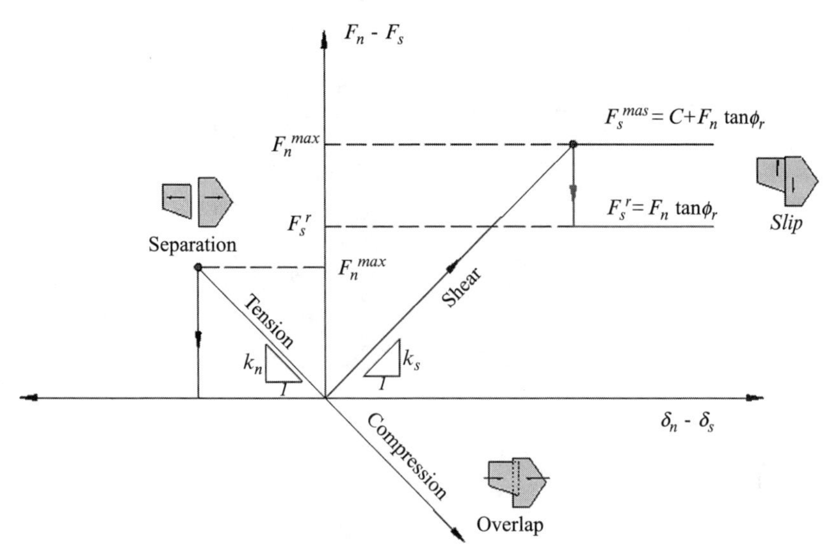

FIG. 3.42 Mechanical behavior of microcontacts in UDEC (Itasca, 2011).

TABLE 3.13 Macroproperties of plaster.

Type	σ_c(MPa)	E (GPa)	υ	σ_t (MPa)
Experimental results	46.3	14.9	0.20	2.40
Simulation results	46.8	14.9	0.21	2.42

The measured mechanical properties of the artificial material are shown in Table 3.13. The details can be found in Li et al. (2015).

Simulations of the uniaxial and Brazilian tests allow acquisition of the microparameters for Voronoi blocks and associated interfaces. The Coulomb-slip joint model with residual strength was employed for the fictitious contacts. The cylindrical sample for the compressive test measured 40 mm in diameter and 100 mm in height, and it consisted of 1797 blocks with an average edge size of 1.5 mm. The disk for the Brazilian test had a diameter at 40 mm with 570 blocks (Figs. 3.43–3.44). The rock sample was placed between two steel platens whose friction with the rock was assumed to be negligible. The upper platen moved downward with a certain velocity while the lower one was fixed. Continuous trace of the axial stress in the lower platen enabled estimation of the compressive and tensile strengths. Axial strain was calculated by dividing the normal separation between two platens by the sample height. The lateral movements of 10 symmetric side points in the middle region (30–70 mm) of the cylindrical specimen were averaged to obtain the diametric strain.

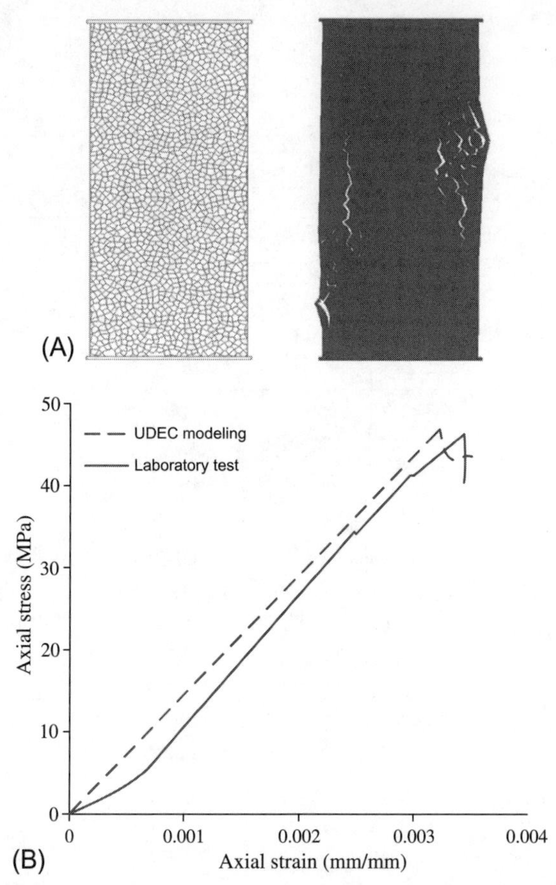

FIG. 3.43 Calibration of UDEC Voronoi model to the laboratory sample under the UCS test. (A) UCS test before (*left*) and after (*right*) test in simulation. (B) Comparison between experimental and numerical results.

A displacement-controlled loading rate at 0.01 m/s was initially set in both compressive and indirect tensile tests. A servo-control system was used to adjust the loading of each sample to limit the axial stress difference between the top and the bottom of the sample (Itasca, 2011). This ensured that the sample was not loaded faster than the speed through which stresses can be transferred numerically through the entire sample (Christianson et al., 2006). Procedures suggested by Ghazvinian et al. (2014) were adopted in the process of calibration. Table 3.14 shows the calibrated microproperties of the Voronoi blocks and contacts by simulating compression and tension tests (Fig. 3.43). A good agreement was achieved between the numerical and experimental results for the synthetic rock sample (Figs. 3.43 and 3.44).

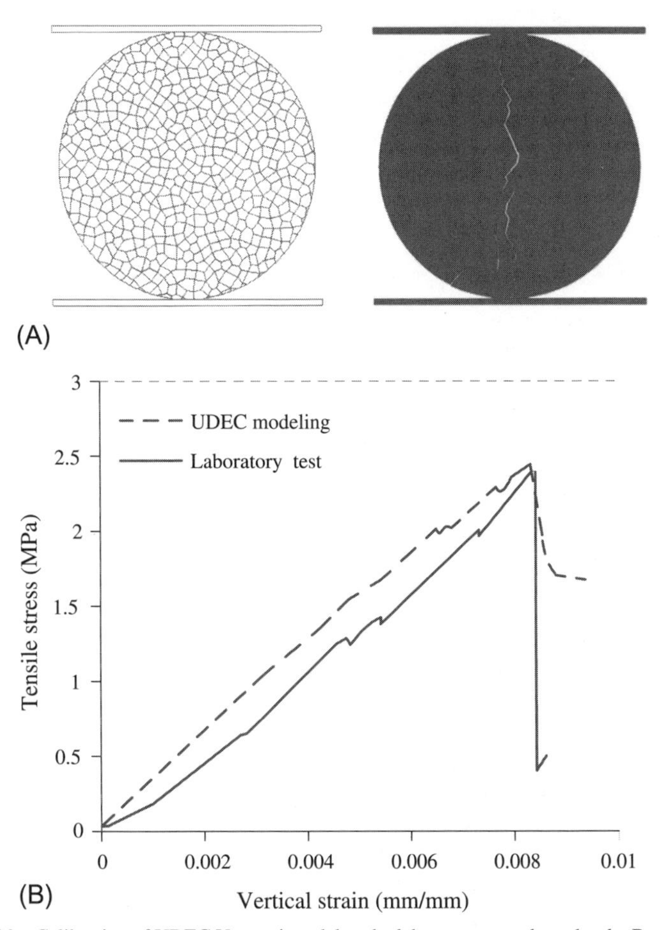

FIG. 3.44 Calibration of UDEC Voronoi model to the laboratory sample under the Brazilian test. (A) Tensile test before (*left*) and after (*right*) test in simulation. (B) Comparison between experimental and numerical results.

3.5.4 Direct shear test simulation

3.5.4.1 Joint surface calibration

Compression and direct shear tests were carried out on rock models containing planar joints to obtain suitable values of the physical parameters governing the mechanical behavior of simulated joint surfaces. The normal and shear stiffness were calibrated against the normal compression and direct shear tests, respectively.

The compression test involved the loading of single-jointed rectangular blocks and intact rock with identical side lengths of 100 mm, as illustrated in Fig. 3.45. The net closure of the flat joint is equal to the difference between the total deformation of the jointed rock and the corresponding value of the

TABLE 3.14 Calibrated microproperties used in the simulation.

Properties	Values
Grain size, D (mm)	1.5
Bulk modulus of blocks, K (GPa)	830
Shear modulus of blocks, G (GPa)	620
Normal stiffness, k_n (GPa/m)	13,000
Shear stiffness, k_s (GPa/m)	3700
Contact friction angle, $\varnothing^{cont}$ (degree)	43
Contact cohesion, c^{cont} (MPa)	12
Contact tensile strength, σ_t^{cont} (MPa)	3.9

solid rock under the same normal stress, similar to the approach described by Bandis et al. (1983) and Li et al. (2015). Fig. 3.45C compares numerical results, using the calibrated value of normal stiffness, to the joint deformation observed in the laboratory. The discrepancy between two deformability curves arises from the fact that a constant value of normal stiffness is used in the simulation, whereas the experimental stiffness increased from an initial value to a nearly constant value with growing normal compression. It should be noted that the simulation of joint compression was conducted under idealized conditions where the joint surface was in a perfectly closed state. Therefore, the simulation was not able to fully represent nonlinear joint closure behavior that results from being successively mobilized during closure. In this study, the calibration of the normal stiffness of the simulated joint surface was carried out by reproducing the slope of a straight-line portion of the experimental stress-closure curve.

Direct shear tests were simulated on flat surfaces for calibration of shear strength parameters (Fig. 3.46). Normal stress was applied and was held constant throughout the test. A constant shear velocity of 0.005 m/s was prescribed at the bottom part of the shear box so that it guaranteed the sample remained in quasistatic equilibrium (Kazerani and Zhao, 2010). The averaged shear stress of all contact points was recorded by a FISH function as suggested by Itasca (2011). The horizontal movement of the lower block was taken as the shear displacement. Fig. 3.46 illustrates the shear stress-shear displacement relationships using a calibrated shear stiffness of the contact surfaces under normal stresses from 1.0 to 3.0 MPa. The coefficient of friction was equal to the basic friction angle acquired experimentally from direct shear tests on flat surfaces. Table 3.15 lists the values of the mechanical properties of the surface contacts after calibration.

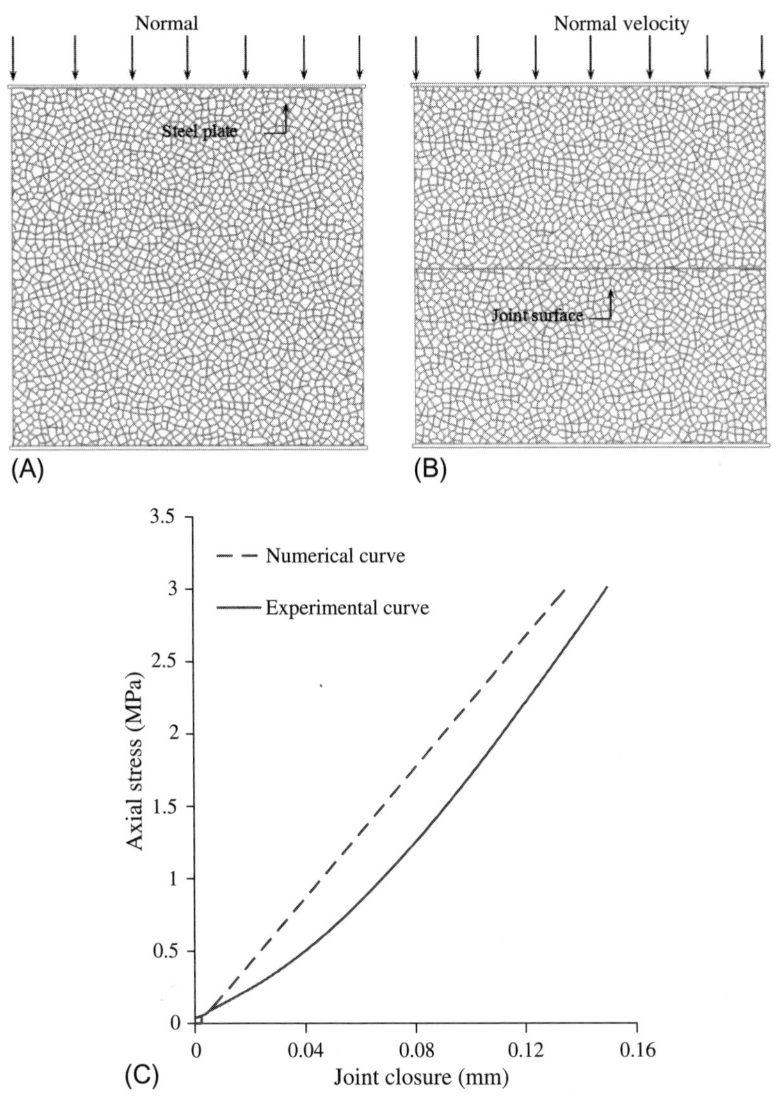

FIG. 3.45 Calibration of the normal stiffness of the simulated joint surface. (a) Compressive test on intact rock. (B) Compressive test on rocks with a planar joint. (C) Comparison of the joint closure between experimental and numerical results.

3.5.4.2 Parametric study on dilation of rock joints

Relative confining pressure effect

The dilation of a rock joint depends strongly on the magnitude of the applied normal stress in that the dilation is the difference between the normal displacements of the upper and lower block of the joint under confining pressure during shear.

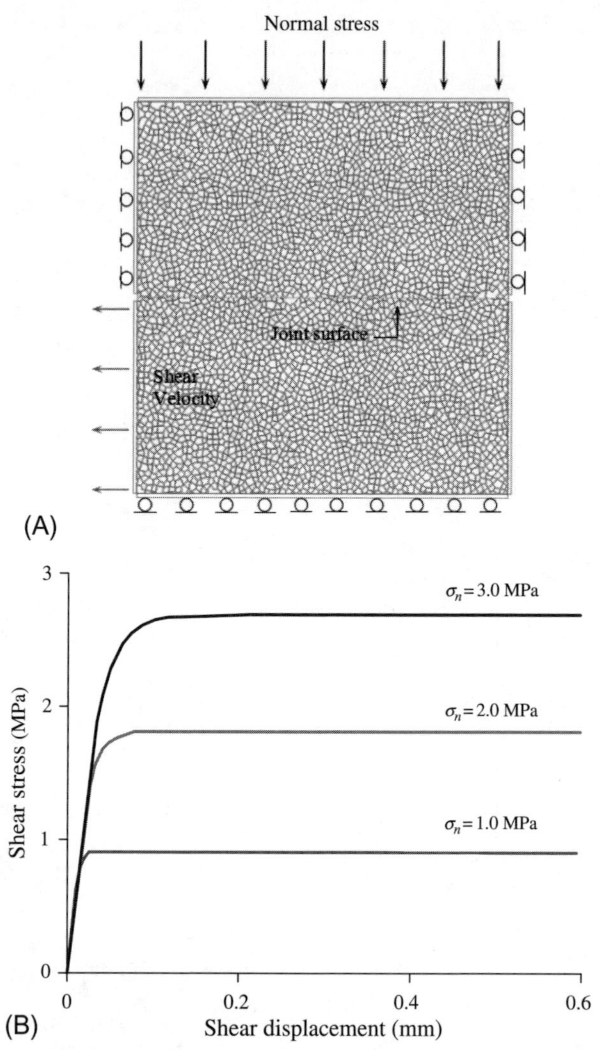

FIG. 3.46 Simulation of direct shear test on a planar surface with UDEC (Itasca, 2011). (A) Numerical model and boundary conditions. (B) Shear stress vs shear displacement curves under different normal stresses.

A low normal stress is usually deemed where the ratio of the applied normal stress over the rock strength, that is, σ_n/σ_c is not greater than 0.01 (Barton, 1973; Barton and Choubey, 1977). On the other hand, Hong et al. (2008) classified three distinct levels of normal stress relative to the rock strength, namely, $\sigma_n/\sigma_c \leq 0.01$ as low, $0.01 < \sigma_n/\sigma_c < 0.1 \sim 0.2$ as intermediate, and $\sigma_n/\sigma_c > 0.1 \sim 0.2$ as high. Recently, Indraratna et al. (2012) considered $\sigma_n/\sigma_c \leq 0.05$ and $\sigma_n/\sigma_c \geq 0.1$ as low and high degrees of normal confinement, respectively. Similarly, in this

TABLE 3.15 **Mechanical properties of the simulated joint surface after calibration.**

Properties	Values
Normal stiffness, k_n (GPa/m)	1760
Shear stiffness, k_s (GPa/m)	200
Friction angle, ∅ (degree)	42

study, 0.5, 2.0, and 4.0 MPa, that is, $\sigma_n/\sigma_c \approx 1\ \%$, $5\ \%$, 10% have been applied as the normal stresses ranging from low to high.

Two types of saw-tooth shaped joint profiles having base angles $i = 20$ and 30 degree with equal wavelengths $\lambda = 25$ mm were first simulated. The geometric configurations of the joint profiles were identical to the replicas prepared in the laboratory (Fig. 3.47) so that additional comparisons could be made to make further verification of the numerical simulations. Direct shear tests under varying normal stress levels were carried out numerically. Fig. 3.48 presents an example of the simulation results, showing the shear stress-displacement-dilation curves

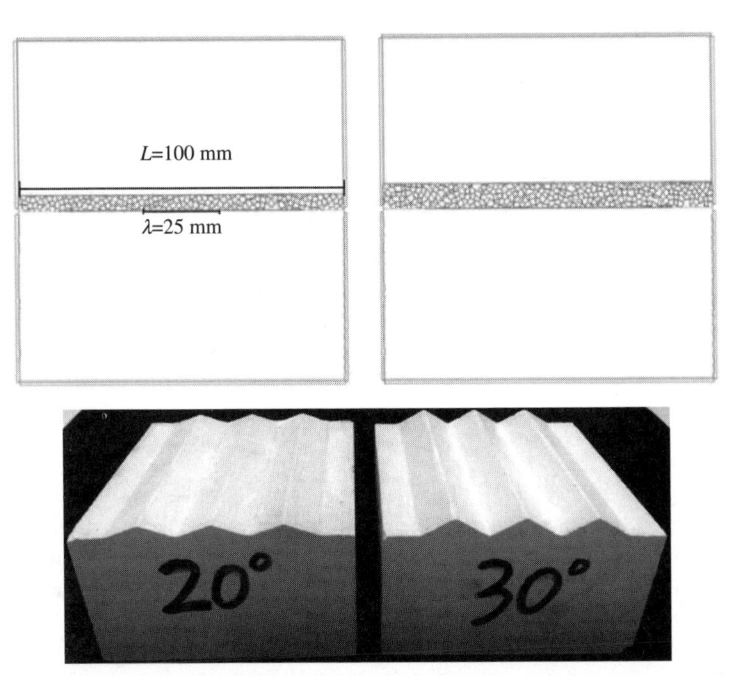

FIG. 3.47 Triangular joints with 25-mm wavelength in simulation (*upper*) and laboratory (*lower*).

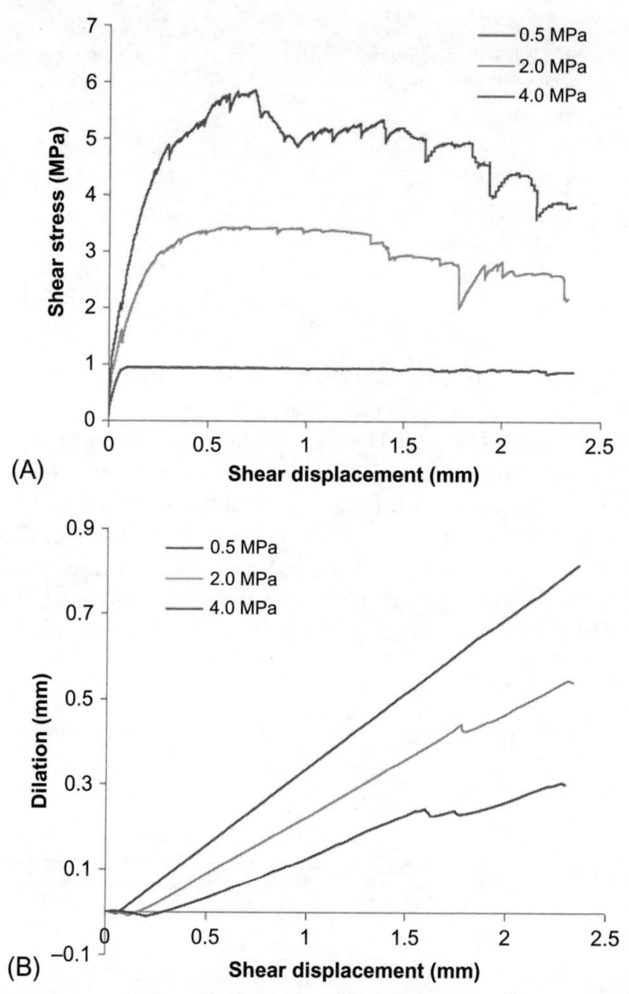

FIG. 3.48 Results of numerical direct shear tests on a 20-degree saw-toothed joint with a 25 mm asperity wavelength. (A) Shear stress vs shear displacement. (B) Normal displacement vs shear displacement.

for 20 degree-asperity rock joints. Under a low normal stress, the asperities remained nearly intact, where the saw-tooth roughness slid up the opposite one, thereby resulting in a few cracks of the contacts (Fig. 3.49A). As aforementioned, the simulation commenced in a perfectly closed state of the joint surface, thereby resulting in the linear dilation curves while nonlinear curves were frequently observed from experimental results of natural joint surfaces.

When the normal stress rose to 4.0 MPa, severe damage occurred to the triangular asperities, as illustrated in Fig. 3.49B, which reduced the shear resistance to

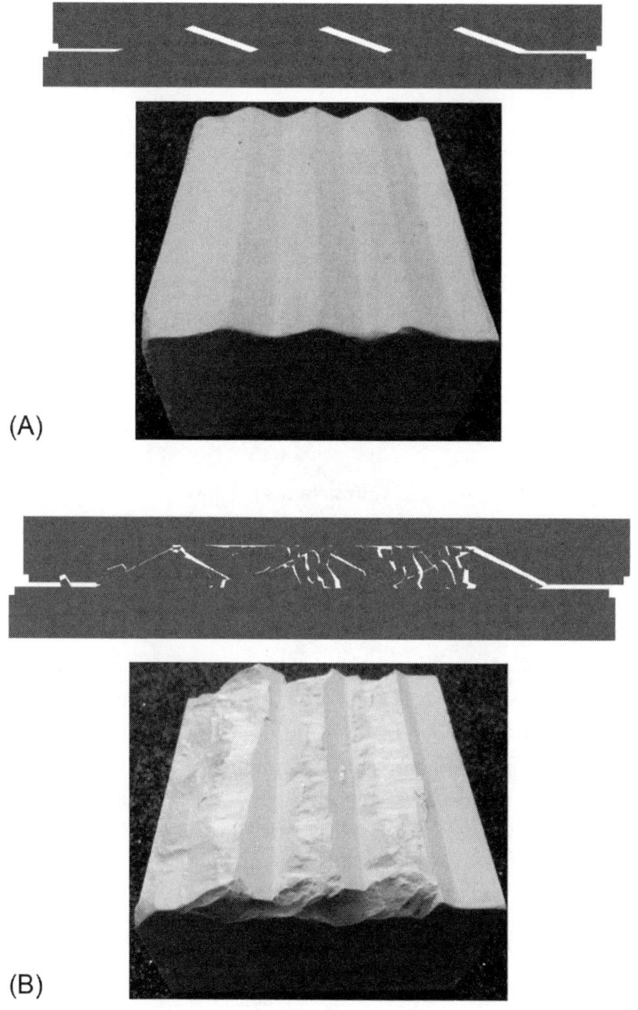

FIG. 3.49 Surface degradation of 20-degree saw-toothed joint with 25 mm wavelength under different normal stresses. (A) Surface damage of a sample after shear under 0.5 Pa in simulation (*upper*) and experiment (*lower*). (B) Surface damage of a sample after shear under 4.0 MPa in simulation (*upper*) and experiment (*lower*).

nearly the residual value. Fig. 3.49 demonstrates that the asperity failure modes are similar for the simulation and the experiment. As seen clearly in Fig. 3.50, the increase in normal stress substantially decreases the peak dilation, for example, to about 11 degree and 13 degree for 20 degree- and 30 degree-asperity rock joints, respectively. A comparison between numerical and experimental results in terms of peak dilation angles for these two idealized rock joints (Fig. 3.50) indicates that the DEM with Voronoi logic after strict calibration has the capability to reproduce the dilation behavior of saw-toothed rock joints.

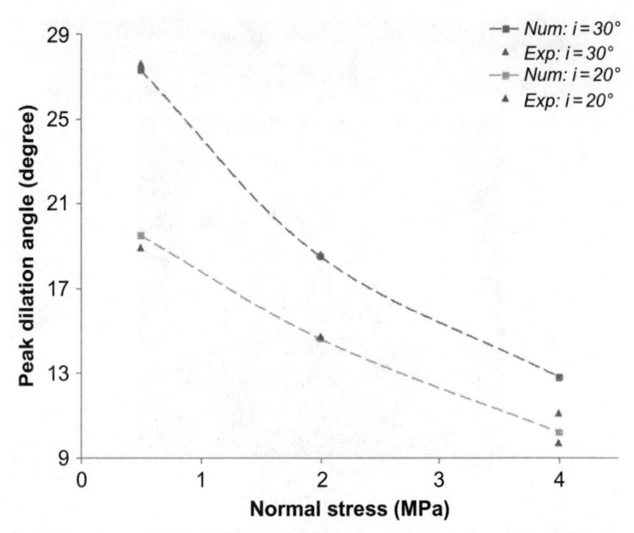

FIG. 3.50 Numerical and experimental comparison for peak dilations of saw-toothed joints with 25 mm wavelength.

Asperity wavelength effect

The dilation behavior can be affected by the asperity wavelength of a rock joint with the same inclination angle (Hong et al., 2008; Leal-Gomes and Dinis-da-Gama, 2014). Fig. 3.51 shows the simulation result of the relationship between peak dilation angle and asperity wavelength for idealized rock joints under intermediate normal stress. In both cases, that is, 20-degree and 30-degree inclined asperities, the decrease of asperity wavelength reduces the peak dilatancy rate. This dilation reduction can be attributed to the fact that the joint with a smaller asperity wavelength required less energy to cause sliding failure, thereby resulting in more severe asperity damage, as discussed by Leal-Gomes and Dinis-da-Gama (2014).

Effect of multifaceted factors

In reality, the dilation behavior of a rock joint is affected simultaneously by the aforementioned factors. In order for this study to have substantial practical significance, those factors should be considered together and represented in such a way that the users of numerical analysis tools are able to determine an appropriate dilation angle. As depicted in Fig. 3.52, triangular asperities with inclination angles $i = 10$, 20, and 30 degree and wavelengths $\lambda = 25$, 10, and 4 mm, are subjected to monotonic shearing under various normal stress levels. Fig. 3.53 illustrates the relationships of peak dilation angles (d_n/i) to normal stress (σ_n/σ_c) for different wavelengths and to the asperity wavelength (λ/L) for different normal stress levels, respectively. In Fig. 3.12A, a decrease in normalized asperity base length (λ/L) leads to a reduction of normalized peak

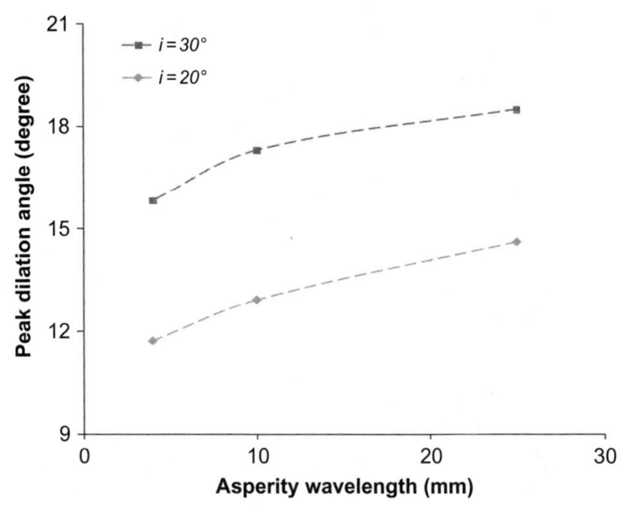

FIG. 3.51 The effect of asperity wavelength on the dilation of simulated joints under normal stress at 2.0 MPa.

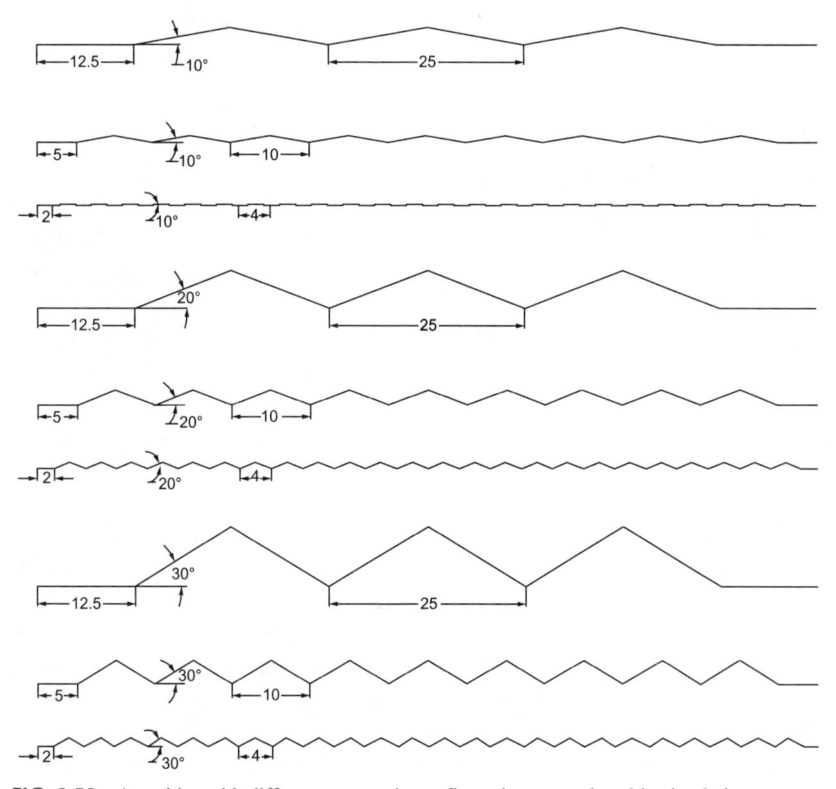

FIG. 3.52 Asperities with different geometric configurations reproduced in simulation.

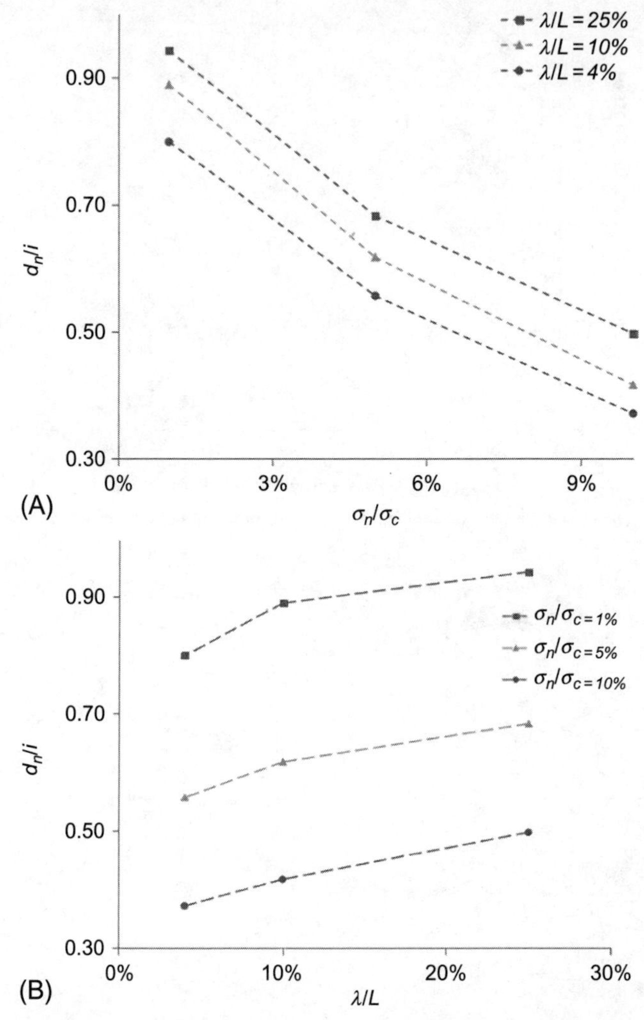

FIG. 3.53 The effect of wavelength and normal stress on peak dilation. (A) Normalized peak dilation angle vs normalized normal stress. (B) Normalized dilation angle vs normalized wavelength.

dilatancy rate (d_n/i). Fig. 3.53B shows the influence of relative normal stress (σ_n/σ_c) on the peak dilatancy of simulated joints (d_n/i) with varied wavelength ratios (λ/L). As can be expected, an increase in σ_n/σ_c considerably reduces the normalized peak dilation angles (d_n/i). For instance, for a simulated joint with the smallest wavelength $\lambda/L = 4\%$, the normalized dilation (d_n/i) is 0.8 while the ratio drops to nearly 0.35 when a high level of normal stress $\sigma_n/\sigma_c = 10\%$ is applied.

Fig. 3.54 provides an empirical graph to estimate peak dilation angles of rock joints based on numerical/experimental results from this study and

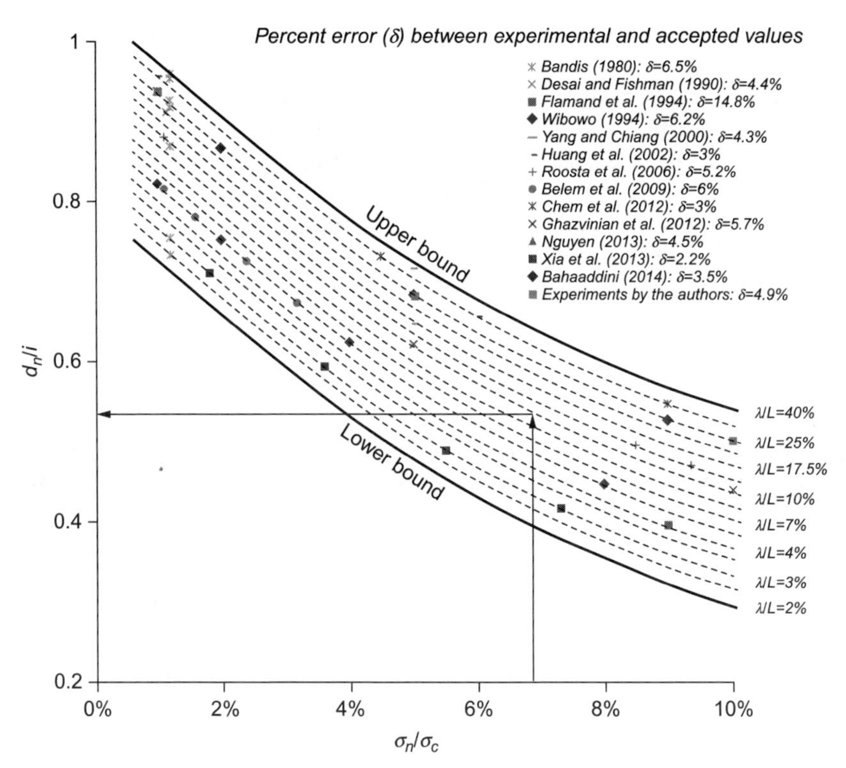

FIG. 3.54 A chart for predicting peak dilation angles of rock joints.

extensive experimental data taken from the literature (Bahaaddini, 2014; Bandis, 1980; Belem et al., 2009; Chern et al., 2012; Desai and Fishman, 1991; Flamand et al., 1994; Ghazvinian et al., 2012; Huang et al., 2002; Nguyen, 2013; Roosta et al., 2006; Wibowo, 1994; Yang and Chiang, 2000). The chart considers the peak dilations of rock joints in a wide range of asperity wavelength ratios (λ/L), from a lower bound ($\lambda/L = 2\%$) to an upper bound ($\lambda/L = 40\%$). As indicated by the values of percent error (δ), which are less than 10% for most cases, the predicted values in Fig. 3.54 agree well with the experimental results. Therefore, this chart can serve as a practical guideline to assess joint dilation angles with adequate accuracy. As demonstrated in Fig. 3.54, for a rock joint with $\lambda/L = 10\%$ under a moderate normal stress $\sigma_n/\sigma_c = 7\%$, the peak dilation angle can be conveniently predicted as 0.53 of the initial asperity angle.

3.5.5 Conclusions

This study presents an investigation of the dilatation behavior of an idealized saw-tooth shaped joint under direct shear. The magnitude of dilation is concurrently dominated by several parameters such as surface geometry, normal stress, strength of the joint wall, and sample length. Due to the complexity involved, with dissimilar scales of factors, the process of normalization used in this study allows variables to be comparable to each other through making the parameters dimensionless. In other words, the effects of dominant factors on joint dilatancy have been interpreted independently. The simulation results indicate that the magnitude of the relative normal stress plays the most important role in affecting dilation. Furthermore, the normalized relationship involving critical factors in dilation behavior enables the users of numerical analysis tools to readily estimate a dilation angle for various ranges of parameter values.

The dilative features of rock joints have been investigated where the relative normal stress (σ_n/σ_c) grows from 1% to 10%. This range covers typical levels of normal stress acting across the joints in many rock engineering problems (Barton, 1976). In addition, research indicated that the base length of critical asperities governing the dilatancy is about 4% as a rough estimate of length for large-scale (block-size) joints (Bandis et al., 1981; Barton, 2016). Thus, the values of parameters used in this study are not arbitrary but can be applicable to preliminary rock engineering studies. It is important to note that the shear behavior simulated in this study is more brittle than that of a natural joint surface due to the distinct characteristics of sample profiles used in the simulation (Yang and Chiang, 2000). The study also does not include the effect of the joint initial opening on the reduction of dilation (Li et al., 2015; Oh and Kim, 2010).

This study presents a systematic parametric study on the dilation of saw-toothed rock joints under direct shear using a distinct element code, UDEC (Itasca, 2011). Joints with saw-tooth shaped asperities were replicated in the direct shear simulation. This simplification is made to avoid the complication caused by the effect of varying roughness during shear, but a consideration of natural joint surfaces has been recognized as a focus for further study. The simulation results prove that the magnitude of dilation decreases with increasing normal stresses and decreasing wavelength values, and quantify how much those parameters affect the dilation behavior during shear. After the limitations concomitant with the idealized saw-toothed joints are recognized, the findings of this study can be employed practically when dilation values need to be determined for stability analysis or support design of rock masses with dilative joints.

3.6 Joint mechanical behavior with opening values

3.6.1 Introduction

Investigations on the mechanical response of rock joints have drawn increasing attention since the discontinuous character of a rock became evident. Particularly in near-surface construction works, the mechanical behavior of the rock mass is often dominated by rough joints rather than by intact rocks. Factors governing the joint normal and shear deformation include rock type, joint wall strength, surface roughness, aperture distribution, and contact area percentage or joint matching state. On the other hand, joint opening due to excavation has a significant influence on the stability of a rock mass, especially when rock blocks are bounded by irregular joints. When an excavation is made in a rock mass, it can possibly result in the relief of confinement normal to the excavated surface, thereby reducing the interlock between the rock joints (Li et al., 2014; Oh, 2005; Oh and Kim, 2010). The shear strength of rock joints, therefore, is closely related with the joint normal behavior in this view.

Observations on the joint normal deformation suggested that the relation between joint closure ΔV_j and the normal loading stress σ_n is typically nonlinear. Approximation by some functions fitting the curves has been made in empirical fashion. A hyperbolic curve was proposed by Goodman (1976), later modified by Bandis et al. (1983), to improve the prediction performance. Two free parameters, maximum joint closure V_m and initial normal stiffness K_{ni}, were used to shape the hyperbola. Goodman (1976) also tried a semilogarithmic function to represent the nonlinear curve using the best fitting method. Evans et al. (1992) and Zhao and Brown (1992) applied the semilogarithmic law to both shear and tensile fractures. Zangerl et al. (2008) showed that the agreement between a semilogarithmic function and the collected data from the laboratory as well as in situ tests was acceptable for numerical modeling purposes. Two parameters are involved in the semilogarithmic law, that is, stiffness characteristic A and reference normal stress σ_n^{ref}. Their values can be estimated from normal compression tests conducted over a low normal stress range in the laboratory or the field.

Alternative models based on Hertzian contact theory have also been used to describe the nonlinear normal stress-normal deformation behavior (Cook, 1992; Greenwood and Tripp, 1970; Greenwood and Williamson, 1966; Matsuki et al., 2001; Swan, 1983; Xia et al., 2003). Their work suggested that the increasing contact areas as the normal stress increases led to the nonlinear behavior. That is to say, the asperities are assumed to undergo elastic deformation, which is contrary to the claim by Goodman (1976) that the nonlinearity arises mainly from crushing. It seems that both factors contribute to the nonlinearity. The asperity crush accounts more for hysteresis and irrecoverable fracture deformation and the elastic deformation represents more the recoverable deformation observed in cyclic tests (Malama and Kulatilake, 2003).

It has been repeatedly observed that unmated or opened rock joints feature much smaller stiffness than closed ones (Bandis, 1980; Bandis et al., 1983; Barton, 1982a, b; Goodman, 1976; Tang et al., 2013). Bandis et al. (1983) reported that an empirical semilogarithmic function is suitable to interpret the normal deformation of an opened rock joint. However, various surface dislocations had not been considered until Saeb and Amadei (1992), who suggested that the normal stiffness was dependent on the degree of joint mating. The analytical equation proposed by Saeb and Amadei (1992) predicted the ultimate joint closure of the opened joint, which was the sum of the initial aperture caused by mismatching and the maximum joint closure V_m of the closed joint. The asperities of the opened joint will be totally crushed and will be converted into a closed joint eventually, conflicting with the implication by Bandis et al. (1983) that mismatched joints would never reach the maximum closure state. Xia et al. (2003) and Tang et al. (2013) decomposed the irregular topography into a large-scale waviness component and a small-scale unevenness component, and classified three cases of contact state by considering the composition of the waviness and the unevenness component. They showed that joint dislocation altering the contact state substantially affected the joint closure behavior. The mathematical model based on Hertzian contact theory proposed by those authors was used to describe the normal closure of a mismatched joint. Although acceptable agreement was achieved between the model and experimental results, the model is so complex with many parameters involved that it can hardly make the model practical.

Shear behavior of closed joints has been studied by many researchers. Patton (1966a, b) and Ladanyi and Archambault (1969) are among those who first obtained different failure modes between the joint shear resistance and the normal confinement. Afterward, Jaeger (1971), Barton (1973), Goodman (1976), Schneider (1976), Barton and Choubey (1977), Hungr and Coates (1978), Saeb and Amadei (1992), Jing et al. (1993), Wibowo (1994), Kulatilake et al. (1995), Maksimović (1996), Haque and Indrarata (2000), Grasselli and Egger (2003), Asadollahi et al. (2010), Asadollahi et al. (2010), Asadollahi and Tonon (2010), Ghazvinian et al. (2012), Park et al. (2013), and Xia et al. (2013) derived empirical models to describe the shear strength along a rock joint by considering joint properties such as material mineralogical prosperities, surface roughness, and compressive strength of the joint wall. Alternatively, the shear behavior of rock joints has been studied based on plasticity theory by, for example, Plesha (1987), Desai and Fishman (1991), Huang et al. (1993), Qiu et al. (1993), Dong and Pan (1996), Roosta et al. (2006). Nevertheless, little work has been carried out on the shear strength reduction as a result of joint opening (Ladanyi and Archambault, 1969; Li et al., 2014; Oh, 2005; Oh and Kim, 2010; Zhao, 1997a, b).

This study presents the experimental results of the deformation characteristics of opened rock joints with triangular-shaped asperities under normal and shear loading. The study proposes a new semilogarithmic equation to better represent the normal deformation behavior of the rock joint with various initial openings. The model has been developed by modifying existing equations in the literature and its performance is verified by providing a correlation with experimental results. Evidence to date shows that no laboratory studies have been undertaken to investigate the shear behavior of opened rock joints. In this paper, a systematic experimental study on the effects of initial opening on the shear behavior under constant normal loading is presented.

3.6.2 Normal deformation of opened joints

3.6.2.1 Semilogarithmic model

Among fitting functions attempted to describe the joint deformability, the semilogarithmic closure law, the power law, and hyperbolic relation are widely used. The power equation given by Swan (1983) has the following form:

$$\Delta V_j = \alpha \sigma_n{}^{\beta} \tag{3.83}$$

where α and β are coefficients determined empirically, with $\beta < 1$. It should be noted that initial normal stiffness $K_{ni} = 0$ is suggested in Eq. (3.83), which is not the case for most rock joints. In other words, the power law tends to underestimate the joint closure in low normal stress ranges.

Barton (1982a, b) and Bandis et al. (1983) proposed a hyperbolic formula for joint normal deformation:

$$\Delta V_j = \frac{V_m \sigma_n}{V_m K_{ni} + \sigma_n} \tag{3.84}$$

where V_m is the maximum joint closure and K_{ni} is the initial normal stiffness.

The initial normal stiffness K_{ni} and ΔV_j can be derived empirically by considering the joint roughness coefficient *JRC*, the joint wall compressive strength *JCS*, and the joint wall aperture a_j. The initial normal stiffness K_{ni} for the first loading cycle is:

$$K_{ni} = 0.0178 \left(\frac{JCS}{a_j} \right) + 1.748 JRC - 7.155 \tag{3.85}$$

The maximum joint closure V_m for the first loading cycle is:

$$V_m = -0.296 - 0.0056 JRC + 2.241 \left(\frac{JCS}{a_j} \right)^{-0.245} \tag{3.86}$$

The initial joint aperture a_j is estimated empirically as:

$$a_j = \frac{JRC}{5} \left(0.2 \frac{\sigma_c}{JCS} - 0.1 \right) \tag{3.87}$$

where σ_c is the uniaxial compressive strength.

Although the hyperbolic model proposed by Barton (1982a, b) and Bandis et al. (1983) is used popularly to model the deformation behavior of the closed joint, there are several difficulties in practice. First, the quantitative estimation of *JRC* is difficult. Visual comparison is prone to be subjective (Beer et al., 2002). Objective *JRC* measuring methods including statistical approaches (Gao and Wong, 2013; Jang et al., 2014; Reeves, 1985; Tse and Cruden, 1979; Yang et al., 2001a, b) and fractal approaches (Kulatilake et al., 2006; Lee et al., 1990; Seidel and Haberfield, 1995a, b; Xie and Pariseau, 1992; Xie et al., 1997) were attempted by considering the geometric features of joint profiles. The value of statistical and fractal roughness parameters, however, can be heavily influenced by measurement quality (Bryan, 2009). Moreover, few approaches have fully captured the microroughness features such as amplitude, slope, and wavelength (Hong et al., 2008). Second, in terms of the initial normal stiffness K_{ni} and the maximum joint closure V_m, the method of fitting experimental data is preferred to obtain them rather than using the above empirical equations (Malama and Kulatilake, 2003). The main reason presumably is that the initial normal stiffness K_{ni} and the maximum joint closure V_m are significantly sample-dependent. Besides joint roughness, joint wall strength, and aperture thickness, roughness amplitude, asperity size, and spatial distributions also affect the normal deformation behavior (Belem et al., 2000; Re and Scavia, 1999).

The joint normal stiffness K_n is the instantaneous slope of the normal stress versus joint closure increment. For the semilogarithmic closure law suggested in Evans et al. (1992) and Zangerl et al. (2008), the predicted change in joint deformation ΔV_j results from the change in normal stress from the reference value σ_n^{ref} to a value σ_n:

$$\Delta V_j = \frac{1}{A} \ln\left(\frac{\sigma_n}{\sigma_n^{ref}}\right) \tag{3.88}$$

where $A = \frac{dK_n}{d\sigma_n}$ = constant, defined as the stiffness characteristic, equals the slope of the linear log-stress versus closure curve; and σ_n^{ref} is the effective normal stress at the start of the test when $\Delta V_j = 0$. Constant $\frac{dK_n}{d\sigma_n}$ indicates that the relationship between normal stiffness K_n and normal stress σ_n is linear and passes through the origin, that is, zero stiffness at zero normal stress:

$$K_n = \left(\frac{dK_n}{d\sigma_n}\right) = A\sigma_n \tag{3.89}$$

It follows that the normal stiffness K_n at any normal stress level σ_n can be predicted by multiplying the value of the stiffness characteristic by the normal stress value. As indicated by Zangerl et al. (2008), however, the value of stiffness characteristic A varies within an unpredictable range, which is strongly dependent on the joint morphology, weathering state, surface profile, joint matching state, and testing conditions. Commonly, it is assumed that the joint closure is zero and the normal stiffness is the initial normal stiffness K_{ni} at starting normal stress (Bandis et al., 1983; Malama and Kulatilake, 2003; Swan, 1983), and thus Eq. (3.88) can be modified as:

$$\Delta V_j = \frac{1}{A'} \ln\left(\frac{\sigma_n}{B} + 1\right) \tag{3.90}$$

where $A' = K_{ni}/B$ and has a unit of mm^{-1}; K_{ni} is the initial normal stiffness; and B is a free parameter determined from the normal deformation test, and has a unit of MPa. The normal stiffness is also proportional to the normal stress:

$$K_n = A'(\sigma_n + B) = K_{ni}(B\sigma_n + 1) \tag{3.91}$$

The above equation indicates that the normal stiffness K_n at zero normal stress is the initial normal stiffness K_{ni} and increases linearly along with increasing normal stress σ_n.

Compared to the normal deformability of closed rock joints, much less research work has been done regarding the joint closure of the opened rock joint. Goodman (1976), Barton (1982a, b), and Bandis et al. (1983) showed that opened rock joints are easier to be deformed and the behavior is representable by a semilogarithmic scale. Saeb and Amadei (1992) suggested that the normal stiffness decreases as the joint opening increases. To describe the deformation behavior of opened rock joints, first, the degree of interlocking presenting the mating state of rock joints proposed by Ladanyi and Archambault (1969) was introduced, as shown in Fig. 3.55. The degree of interlocking of the rock joint is estimated as:

$$\eta = 1 - \frac{\Delta x}{\Delta L} \tag{3.92}$$

where Δx is the initial shear displacement, and ΔL denotes half the joint wavelength.

Oh (2005), considering geometric relationship, related the degree of interlocking η to the measurable parameters joint opening δ_0 and joint amplitude A_{mp} in Fig. 3.55:

$$\delta_0 = 2\Delta x \cdot \tan i_0 \tag{3.93}$$

$$A_{mp} = \Delta L \cdot \tan i_0 \tag{3.94}$$

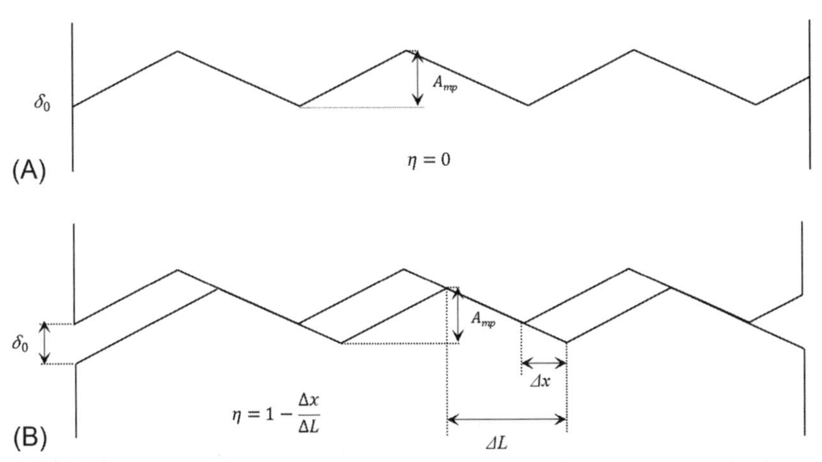

FIG. 3.55 Geometric configuration of the joint opening. (A) A tightly closed joint. (B) An opened joint. *(After Ladanyi, B., Archambault, G., 1969. Simulation of shear behavior of a jointed rock mass. In: The 11th US Symposium on Rock Mechanics (USRMS), Berkeley, California, pp. 105-125.)*

Then the degree of interlocking is given as:

$$\eta = 1 - \frac{\delta_0}{2\Delta L \cdot \tan i_0} = 1 - \frac{\delta_0}{2A_{mp}} \tag{3.95}$$

The following shows how the degree of interlocking is incorporated into the proposed semilogarithmic equation and how the model can be used to predict the normal deformation of opened rock joints.

The initial normal stiffness of closed rock joints K_{ni}, which is the outset slope of the normal stress-joint closure curve, can be given as:

$$K_{ni} = \lim_{\sigma_n \to 0} \frac{\sigma_n}{\Delta V_j} = \left(\frac{d\sigma_n}{\Delta V_j} \right)_{\sigma_n = 0} \tag{3.96}$$

If the same joint opens, the initial normal stress required to compress the joint sample to the same closure as the closed one should be $\eta \times \sigma_n$. The initial normal stiffness of the opened rock joint K_{ni}^o is expressed as:

$$K_{ni}^o = \lim_{\sigma_n \to 0} \frac{\eta \sigma_n}{\Delta V_j} = \eta K_{ni} \tag{3.97}$$

According to Ladanyi and Archambault (1969), the true normal stress σ_n^o acting on the opened joint should be substituted as:

$$\sigma_n^o = \frac{\sigma_n}{\eta} \tag{3.98}$$

Substituting Eqs. (3.97), (3.98) into Eq. (3.90), we have:

$$\Delta V_j = \left(\frac{1}{\eta A'} \right) \ln \left(\frac{\sigma_n}{\eta B} + 1 \right) \tag{3.99}$$

Eq. (3.99) shows that the deformability of the rock joint is affected evidently by the degree of interlocking. Under the same compressive stress, the normal closure of opened rock joints could be increased to be much higher than closed ones. This significant increase in the normal deformation has been rarely studied as the rock joints are tacitly assumed closed. The quantitative performance of Eq. (3.99) describing the deformability of the rock joint under varying degrees of interlocking will be discussed in the next section.

3.6.2.2 Experiments and correlation

Compression tests

Normal compression tests were conducted on artificial intact rocks and rocks with a triangular joint profile at 20 degree and 30 degree. Hydro-stone TB Gypsum Cement ($CaSO_4 \cdot 1/2H_2O > 95\%$, Portland cement $< 5\%$) was used to cast joint samples, mainly because the material is easily accessible and inexpensive. When mixed with water, it can be molded into any shape and the long-term strength is constant when the chemical hydration ends. Hydro-stone and water

TABLE 3.16 Mechanical properties of the casting material.

Material	Uniaxial compressive strength σ_c (MPa)	Tensile strength σ_t (MPa)	Young's Modulus E (GPa)	Poisson's ratio υ	Basic friction angle φ_b
Hydrostone mixed with water	46.33	2.45	14.93	0.20	42.41

were mixed in the ratio of 1:0.35 by weight. All the test samples were cured at a constant 40° in a curing oven for 14 days. To determine the mechanical characteristics of the samples, 10 cylindrical specimens of 41 mm in diameter and 102 mm in height were tested to obtain the uniaxial compressive strength, and 10 cylinder specimens with a diameter-to-thickness ratio at around 2.0 were split under diametral compression (Brazilian test). The basic friction angle of joint mineralogy is determined from a direct shear test on flat joint surfaces and was found to be 42.41 degree. The mechanical properties are presented in Table 3.16.

Casted artificial joints had two triangular profiles: 20-degree and 30-degree inclination angles. Details of the triangular profiles are shown in Fig. 3.56. The full-length of the profiles is 100 mm. The joint wavelength of the two inclination angles is 25 mm. The joint amplitudes are 4.55 mm and 7.22 mm for 20-degree and 30-degree profiles, respectively.

An MTS 815 Rock Mechanics Test System (Fig. 3.57A) was used to conduct normal compression tests. This testing system is ideal for uniaxial and triaxial rock tests in rock mechanics research. The highly stiff load frames offer high axial force capacity, with compression ratings up to 4600 kN and tension ratings up to 2300 kN. A monotonic force control configuration with maximum normal stress of 10 MPa was applied in the tests. The loading speed was 0.01 MPa per second. The test setup is also shown in Fig. 3.57. A normal deformation test on intact blocks was repeated three times. The final normal stress-axial deformation is the average value of the three tests. The normal deformation of close jointed blocks was tested five times. Ideally, the deformation curve should be the same for replicas made of the same mixture and joint molds. The joint closure curves, however, showed slight differences. Sample variation (mainly in porosity) during casting may be responsible; however, manual setup before starting the test affects more. Because the samples were fresh and unloaded, the initial normal stiffness is affected significantly by the starting point of the machine. The difference in initial normal stiffness gives differing curves, though the trend is similar (Nassir et al., 2010). To eliminate this discrepancy, the mean value of the three closest curves in five tests was selected as the normal

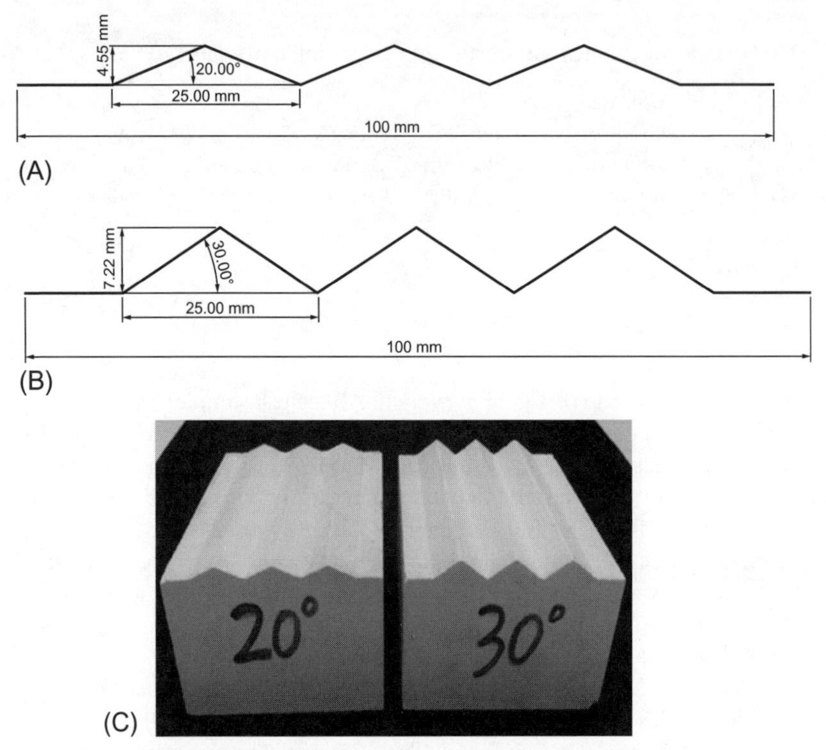

FIG. 3.56 Triangular joint profiles of samples. (A) 20-degree joint profile. (B) 30-degree joint profile. (C) Rock joint samples.

deformation of the close jointed block. As for the tests of opened rock joints, a series of mismatching displacements was created before starting the tests. Each test was conducted twice. The averaged normal displacement versus normal stress curves was used for interpretation. Table 3.17 shows the details of all the tests conducted.

Goodman (1976) measured joint closure in the laboratory as a function of normal stress on artificial tensile fractures. Joint closure is determined by measuring the displacement across a gauge length of an intact cylindrical specimen, as a function of normal stress. The measurement across a tensile fracture is repeated within the same length. The difference between these two displacements at every value of the normal stress is the joint closure at that stress (Fig. 3.58). The method was also adopted by Bandis et al. (1983). The means to measure joint normal displacement under compressive load used by Sun et al. (1985) was distinct. An apparatus was designed with two linear displacement transducers (LVDTs), both of which were allowed to record the incremental change directly from the contacting surface of the joint samples under test. The joint closure is the average value of the normal increase recorded by the two LVDTs.

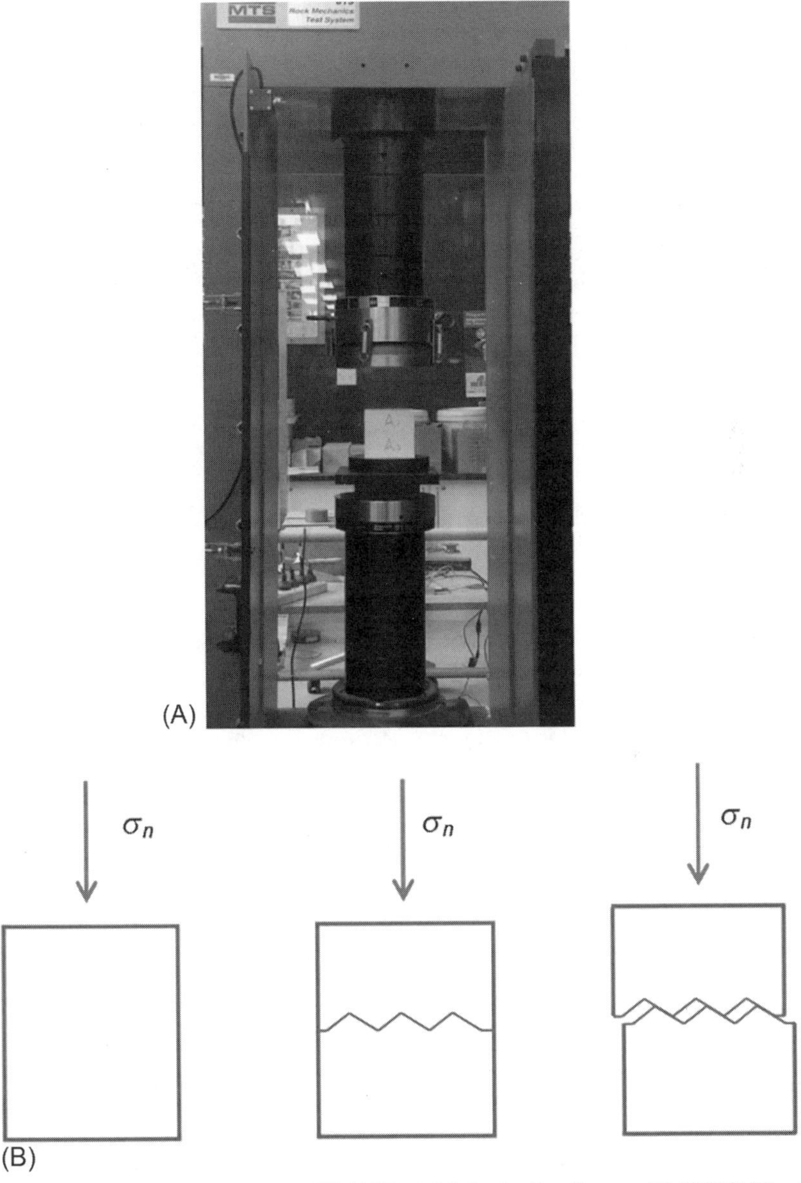

FIG. 3.57 Compression test using MTS 815 Rock Mechanics Test System. (A) MTS 815 Rock Mechanics Test System. (B) Experimental setup of compression tests for intact and jointed rock specimens.

In this study, the joint closure ΔV_j was obtained from the difference between the total deformation of the jointed rock under normal stress σ_n and the corresponding deformation of the solid rock, as done by Goodman (1976) and Bandis et al. (1983):

TABLE 3.17 Normal deformation test details.

Sample type	Average sample dimensions $L \times W \times H$ (mm)	Test type	Tested quantities
Intact rock	101.47 × 101.32 × 108.02		3
20-degree jointed rock	100.34 × 99.77 × 108.37	Closed	5
30-degree jointed rock	100.21 × 99.83 × 108.36	Closed	5
20-degree jointed rock	100.45 × 99.87 × 108.09	Opened	11
30-degree jointed rock	100.30 × 99.95 × 108.40	Opened	12

$$\Delta V_j = \Delta V_t - \Delta V_r \tag{3.100}$$

where ΔV_j is the net deformation or joint closure, ΔV_t is the total deformation of the jointed block, and ΔV_r is the deformation of the solid block.

Results analysis

Fig. 3.59 shows the normal stress versus joint closure of two closed rock joints. The closure curves possess a high degree of nonlinearity. As the normal stress increases, the normal stiffness increases quickly. These characteristics have been described by many researchers, such as Goodman (1976) and Bandis

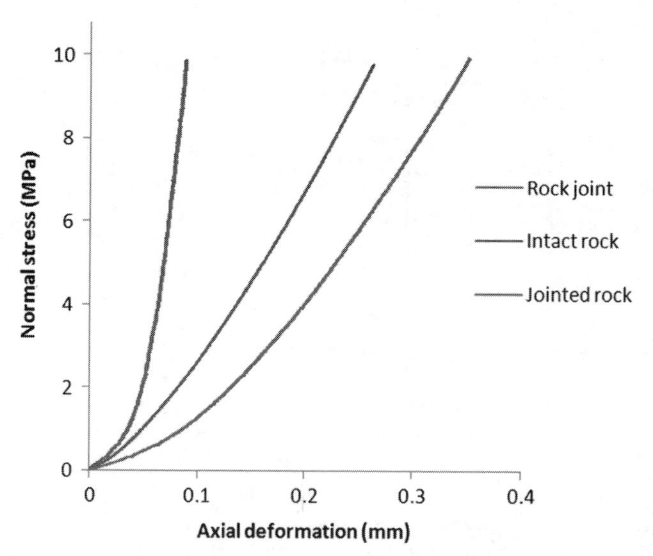

FIG. 3.58 Normal stress versus axial deformation relationship of intact and jointed rock specimens.

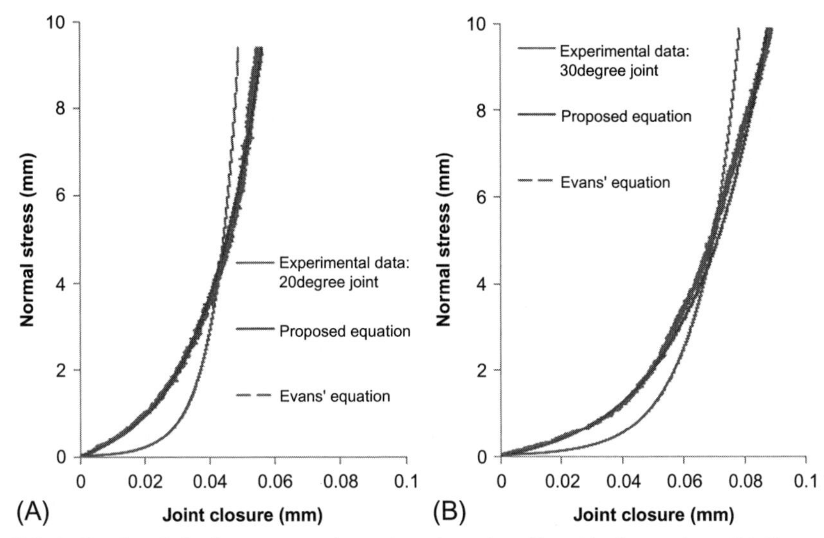

FIG. 3.59 Correlation between experimental results and semilogarithmic equations. (A) Closure curve of 20-degree triangular joint. (B) Closure curve of 30-degree triangular joint.

et al. (1983). Evans' equation and the proposed equation are correlated with the experimental data. The input values of parameters involved in the two semilogarithmic expressions are listed in the figure. As shown in Fig. 3.59 and Table 3.18, it is evident that the proposed model presented higher correlation than Evans' equation. This can be explained as follows. The lower determination coefficient for Evans' equation is attributed to the fixed value of σ_n^{ref}, which is the normal stress at the first positive joint deformation value, while the free parameter B in the proposed equation is obtained directly by testing data. Zangerl et al. (2008) collected 115 normal closure experiments mostly on natural joints after several loading cycles. Regression analysis using Evans'

TABLE 3.18 Parameter values used in semilogarithmic equations.

	Evans' equation		Proposed equation	
Rock joint sample	$\frac{1}{A}$ (mm)	σ_n^{ref} *(MPa)*	$\frac{1}{A'}$ (mm)	B *(MPa)*
20-degree joint	0.00775	0.0161	0.01892	0.4965
30-degree joint	0.01336	0.027294	0.02460	0.3294
R^2	0.7744		0.9966	
	0.8691		0.9964	

function showed that the average determination coefficient was around 0.96, higher than those shown in this study. Both Evans et al. (1992) and Zangerl et al. (2008) indicated that the Evans' equation may be more suitable to predict the deformation behavior of rock joints for the second and following loading cycles than for the first cycle while the joint samples used in this study were loaded for the first time and did not have any loading history. However, it is the first loading cycle that displays the strongest nonlinearity and the largest joint deformation that demands the highest prediction accuracy.

The proposed semilogarithmic law was also correlated with the normal deformation curve of the natural rock joint. Swan and Sun (1985) carried out compression tests on grey granite rock joints before and after shear testing. Fig. 3.60 shows the rock material properties (Tables 3.19 and 3.20). The data of the two deformation tests before shearing were randomly selected and compared with the proposed semilogarithmic equation. Parameters used in the proposed equation and the determination coefficients are shown in the figure. It shows that the proposed equation fits well with the experimental data on two grey granite rock joints, with high determination coefficients R^2 of 0.9946 and 0.9980.

Fig. 3.61 shows the deformation curves of rock joints with various degrees of interlocking (Tables 3.21 and 3.22). All curves obtained from experimental tests are compared with the proposed Eq. (3.99). The parameters used in the proposed equation and the equivalent initial opening values corresponding to

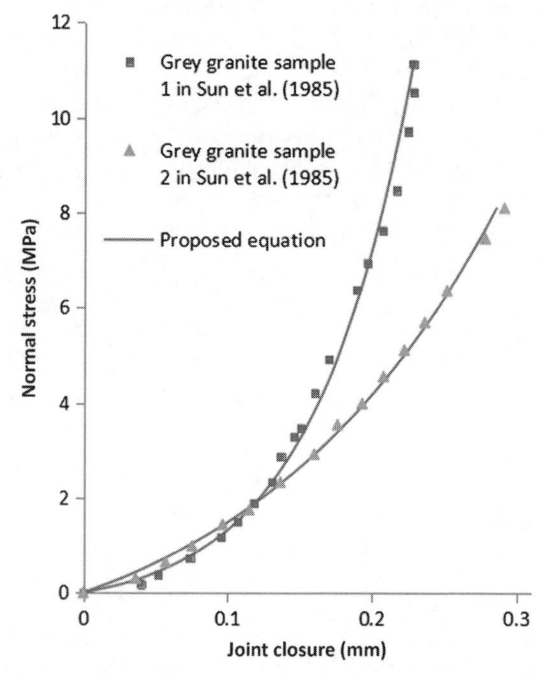

FIG. 3.60 Normal deformation of the grey granite samples in Sun et al. (1985).

TABLE 3.19 Grey granite properties in Sun et al. (1985).

Rock type	Uniaxial compressive strength σ_c (MPa)	Young's modulus E (GPa)	Poisson's ratio υ	Sample shape	Surface area (cm^2)
Grey granite	207.5	60.2	0.22	Cylinder	1009

TABLE 3.20 Parameter values used in the proposed semilogarithmic equation.

Rock joint sample	Maximum applied normal stress (MPa)	$\frac{1}{A'}$ (mm)	B (MPa)
Grey granite sample 1	11.3	0.06703	0.3871
Grey granite sample 2	8.2	0.1705	1.875
R^2	–	0.9946	0.998

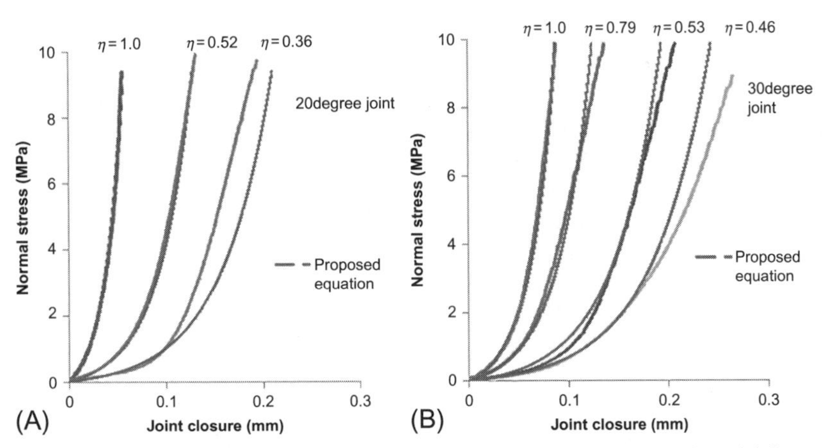

FIG. 3.61 Comparison between proposed equation and experimental data. (A) Normal deformation curves of 20-degree triangular joint. (B) Normal deformation curves of 30-degree triangular joint.

each degree of interlocking are shown in the figure. As would be expected and can be seen in Fig. 3.61, under the same normal stress level, the larger the initial opening, the more the normal closure. That is to say, the normal stiffness decreases with the increase in joint opening, as already referenced by Saeb and Amadei (1992). The results indicate that the new equation agrees fairly well

TABLE 3.21 Parameter values used in proposed equation for different degrees of interlocking.

Rock joint sample	Proposed equation		
	$\frac{1}{A'}$ (mm)	B (MPa)	η
20-degree joint	0.01892	0.4965	1.0, 0.52, 0.36
30-degree joint	0.02460	0.3294	1.0, 0.79, 0.53, 0.45

TABLE 3.22 Initial opening value equivalent to each degree of interlocking.

Rock joint	Joint amplitude A_{mp} (mm)	Joint wavelength λ (mm)	Degree of interlocking η	Initial opening value δ_0 (mm)
20-degree joint	4.55	25	1.0	0.0
			0.52	4.368
			0.36	5.824
30-degree joint	7.22	25	1.0	0.0
			0.79	3.032
			0.53	6.787
			0.45	7.942

with the laboratory results. Particularly, under the low normal stress level, a remarkable agreement was achieved using the suggested model.

3.6.2.3 *Shear deformation of opened joints*

Zhao (1997a, b) introduced the joint matching coefficient (*JMC*) into Barton's model in order to consider the reduction of rock joint shear strength due to a joint surface mismatch:

$$\tau = \sigma_n \tan\left[JMC \cdot JRC \log\left(\frac{JCS}{\sigma_n}\right) + \varphi_r\right] \tag{3.101}$$

where *JMC* is an independent geometrical parameter to represent the degree of joint surfaces matching and estimated by visual comparison with five standard profiles.

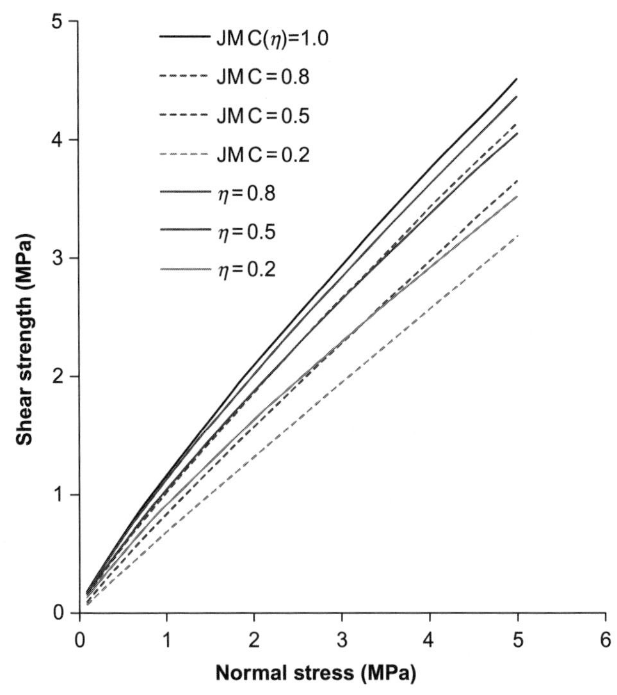

FIG. 3.62 Shear strength reduction due to initial joint openings. Illustration of the effect of joint initial opening.

By considering the true stresses acting on the joint profiles, Oh (2005) incorporated the degree of interlocking η proposed by Ladanyi and Archambault (1969) into Barton's model:

$$\tau = \sigma_n \tan\left[JRC \cdot \log\left(\frac{\eta \cdot JCS}{\sigma_n}\right) + \varphi_r\right] \tag{3.102}$$

where η is expressed as in Eq. (3.95) for regular-shaped joint profiles. Fig. 3.62 shows the peak shear strength obtained from these two models to illustrate the effect of joint initial opening (Table 3.23). Although hypothetical input parameters are used in this example, it shows clearly that the peak shear strength decreases significantly with the increase in the values of the joint initial

TABLE 3.23 Input parameters used in two shear strength models for opened rock joints.

JRC	*JCS* (MPa)	*JMC* or η	φ_r (degree)
10	80	0.2, 0.5, and 0.8	30

opening. In the following section, experiments are conducted to describe the effect of joint opening on the joint shear strength quantitatively.

3.6.3 Direct shear tests

For investigation of the effect of initial opening on the shear behavior of rock joints, direct shear tests were conducted on the same artificial joint specimens used in compression tests. The tests were conducted under constant normal stress conditions.

A Geotechnical Consulting and Testing Systems (GCTS) servo-hydraulic testing machine RDS-300 (Fig. 3.63) was used to carry out the experiments.

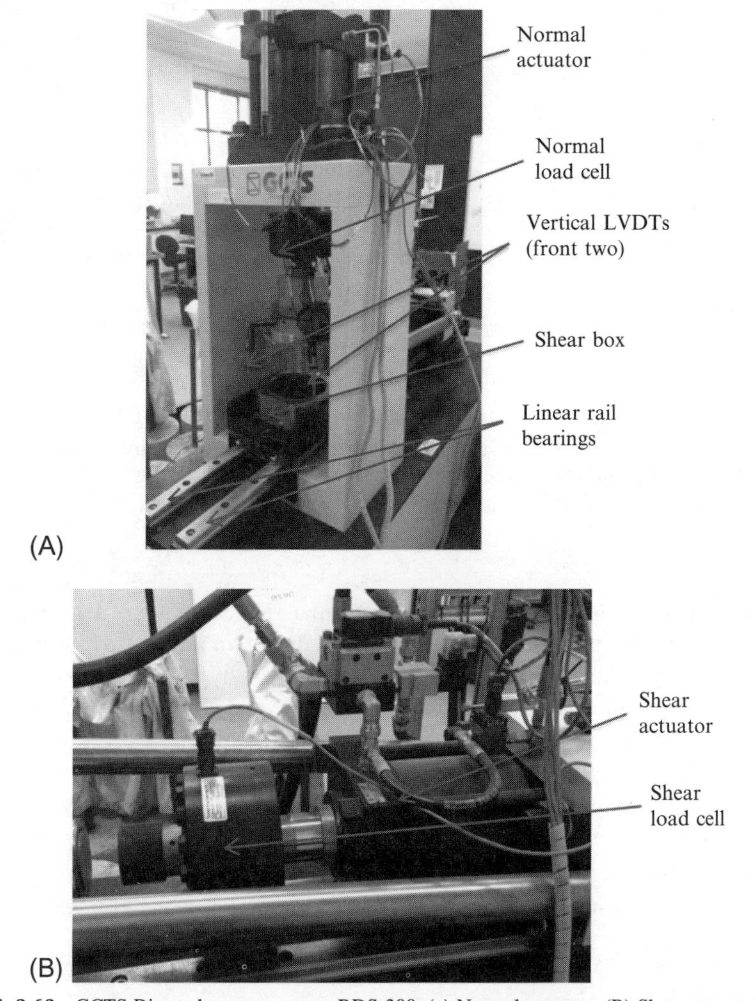

FIG. 3.63 GCTS Direct shear test system RDS-300. (a) Normal actuator. (B) Shear actuator.

The servo-hydraulic testing machine is composed of a 500 kN compression frame, a direct shear apparatus, and electro-hydraulic shear and normal load actuators with 300 kN and 500 kN load capacity, respectively. The maximum stroke is 100 mm in the vertical direction and ±50 mm in the horizontal direction. The normal and shear displacements are measured by linear variable differential transducers (LVDTs).The vertical displacement between two shear box parts is measured by four LVDTs, positioned in a square pattern around the sample, one in each corner (Fig. 3.63A). Each of the LVDTs has a measurement range of 12 mm. The normal displacement is presented by the average value of the four LVDTs. The relative displacement of the two shear box parts in the horizontal direction is measured by one LVDT, which has a range of 100 mm. The sensitivities of the LVDTs are 0.025 mm for shear displacement and 0.0025 mm for normal displacement. Test output, including normal and shear stresses, shear displacement, and normal displacement, are collected automatically by the GCTS software.

Direct shear tests were performed on both closed and opened rock joints with 20-degree and 30- degree asperity angles. During tests, the shear rate of the horizontal actuator was set at 0.5 mm/min. The maximum shear displacement was set at 10 mm. For closed rock joints, the degree of interlocking η before the test was 1.0. Two initial shear displacements were set in the opened rock joint tests, that is, $\Delta x = 4.0$ mm and 8.0 mm. The corresponding degrees of interlocking were 0.68 and 0.36. In tests, the normal stress level ranged from 0.05 MPa ($\sigma_n \approx 0.001\sigma_c$) to 5.0 MPa ($\sigma_n \approx 0.1\sigma_c$). Fig. 3.64 illustrates the experimental setup of the joint samples and shows the equivalent initial opening value corresponding to each degree of interlocking (Table 3.24).

3.6.4 Results analysis and discussion

Fig. 3.65 illustrates the shear strength of closed rock joints with 20-degree and 30-degree asperities under various constant normal stress values (Table 3.25).

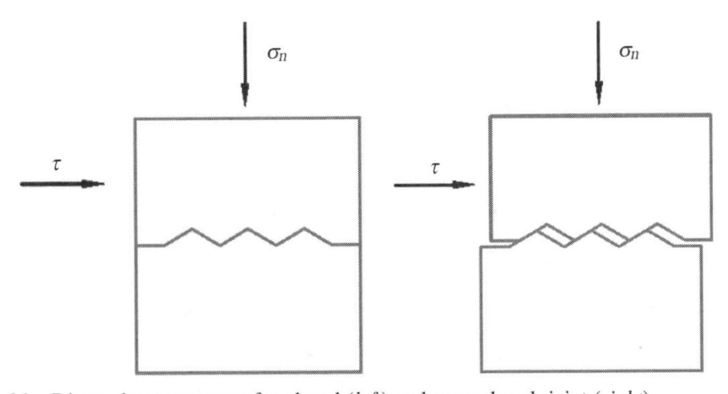

FIG. 3.64 Direct shear test setup for closed (*left*) and opened rock joint (*right*).

TABLE 3.24 Direct shear test details under constant normal stress conditions.

Sample type	Normal stress range σ_n (MPa)	Joint amplitude A_{mp} (mm)	Initial shear displacement Δx (mm)	Initial joint opening δ_0 (mm)	Equivalent degree of interlocking η
20-degree joint	0.05 to 5.0		0.0	0.0	1.0
		4.55	4.0	2.912	0.68
			8.0	5.824	0.36
30-degree joint	0.05 to 5.0		0.0	0.0	1.0
		7.22	4.0	4.621	0.68
			8.0	9.242	0.36

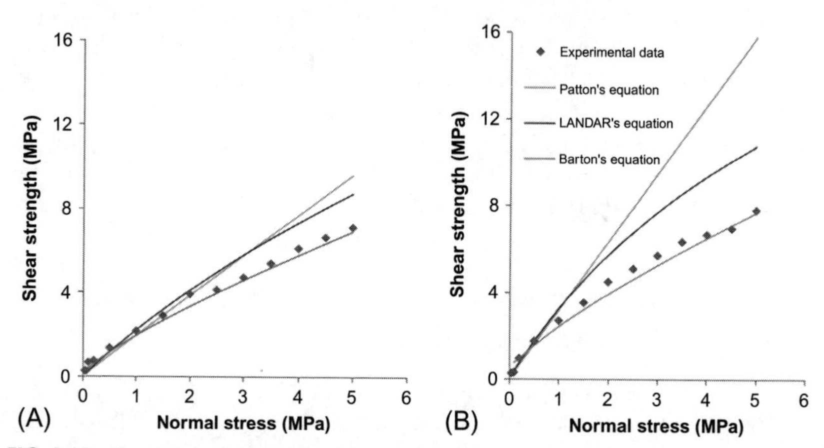

FIG. 3.65 Correlation of shear strength between experimental data and existing joint models for closed rock joints. (A) Shear strength of 20-degree closed rock joints. (B) Shear strength of 30-degree closed rock joints.

TABLE 3.25 Input parameters for used models.

Sample type	JRC	JCS/σ_c(MPa)	σ_t(MPa)	φ_r (degree)	i_0(degree)
20-degree joint	10	46.33	2.45	42.41	30
30-degree joint	15	46.33	2.45	42.41	20

The results are correlated with Patton's model (Patton, 1966a, b), LANDAR's model (Ladanyi and Archambault, 1969), and Barton's model (Barton, 1982a, b) in order to confirm that the test results are within the range of possible shear strength of rock joints. The material parameters used in the three models are also shown in the figure. It should be noted that *JRC* values of 20-degree and 30-degree joint profiles are estimated to be 10 and 15, respectively, on the basis of the study by Maksimović (1996), where he suggested that the *JRC* value was reasonably assumed to be half the initial asperity angle for triangle-shaped joint profiles. Fig. 3.12 shows the shear strength of rock joints with 20-degree and 30-degree asperities for three degrees of interlocking, namely $\eta = 1.0, 0.68, 0.36$. As can be expected, the opened rock joints present lower frictional resistance than the closed ones. The increase in initial opening leads to the decrease in peak shear strength. In other words, the peak shear strength of a rock joint is approximated by the degree of asperity interlocking if the peak shear strength of a closed joint is known. Fig. 3.66 shows a correlation of experimental data of both 20-degree and 30-degree asperity joints with Patton's equation, LANDAR's equation, and Eq. (3.102). Clearly, Patton's model and LANDAR's relation overestimate the peak shear stress of opened rock joints, due to the ignorance of the opening effect. Taking the degree of interlocking into account, Eq. (3.102) results in better agreement with the shear strength of joints with various initial openings.

Fig. 3.67 is an example of experimental results, which shows the shear stress-shear displacement-dilation relations (Table 3.26). It also shows the asperity geometric parameters for the tested joint. As would be expected, shear strength and dilation decrease as a joint initially opens. This can be explained as follows. When a joint is sheared in a closed state, the asperity with a full

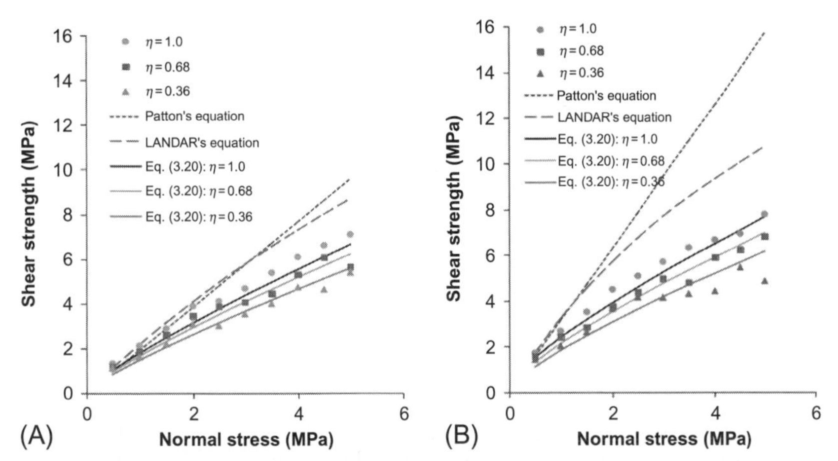

FIG. 3.66 Direct shear test results of triangular rock joints with different degrees of interlocking. (A) Shear strength of 20-degree opened rock joints. (B) Shear strength of 30-degree opened rock joints.

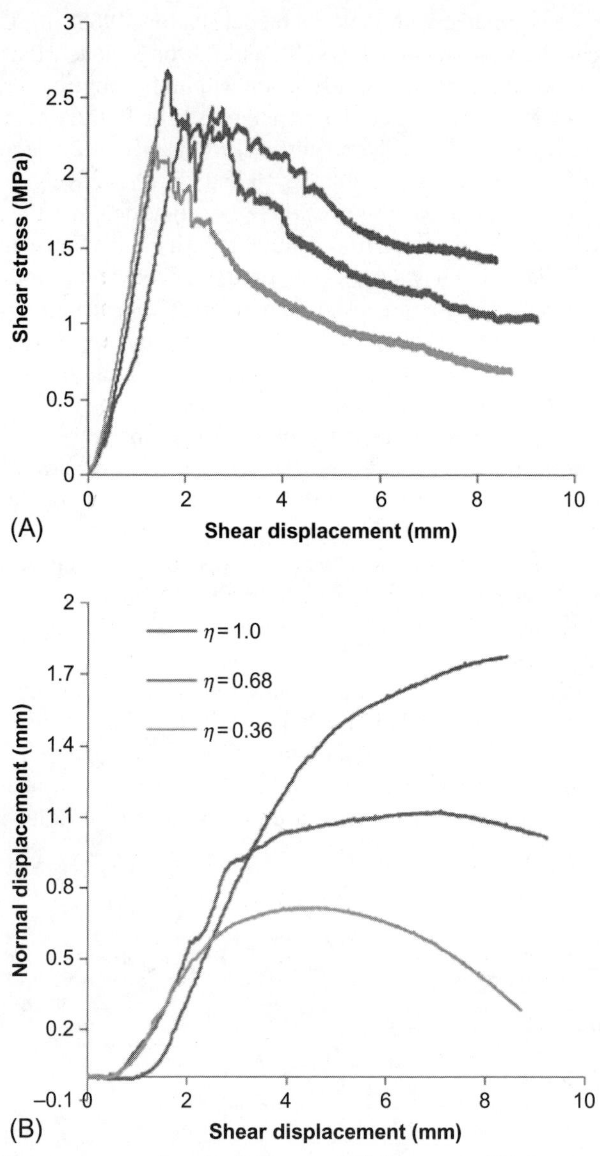

FIG. 3.67 Shear stress versus shear displacement and dilation curves of 30-degree rock joints with different initial openings. (A) Shear stress vs shear displacement curves. (B) Normal displacement vs shear displacement curves.

wavelength may be mobilized during shear. On the other hand, when a joint opens, it reduces the asperity interlocking and thus a partial asperity is involved directly in shear. This process results in the increase in actual stresses on the smaller contact areas, thereby degrading the smaller asperity more easily and

TABLE 3.26 Direct shear test details on opened rock joints.

Sample type	Normal stress σ_n (MPa)	Joint amplitude A_{mp} (mm)	Wavelength λ/λ_0 (mm)	Initial joint opening δ_0 (mm)	Equivalent degree of interlocking η
30-degree joint	1.0	7.22	25	0.0	1.0
			17	4.621	0.68
			9	9.242	0.36

reducing shear strength and dilation compared to a closed joint. It is also interesting to note that when a joint opens, a negative dilation (contraction) occurs quickly because the wavelength of asperity involved in the shear is reduced. The figure shows that the open joint with $\eta = 0.36$ starts contracting after shear displacements of about 4.5 mm, which is the value of the half-wavelength. Fig. 3.68 shows the appearance of specimens after shearing and illustrates the involvement of asperities during shear for both closed and opened joints.

The hyperbolic law proposed by Barton (1982a, b) and Bandis et al. (1983) involves estimation of the initial normal stiffness K_{ni} and the maximum joint closure V_m. Equations developed by Bandis et al. (1983) to determine the values of these two parameters are empirical. Evans' semilogarithmic formulation was reported to represent the normal closure curve with good agreement (Evans et al., 1992; Zhao and Brown, 1992). However, according to the compilation in Zangerl et al. (2008), the value of the involved parameter stiffness characteristic A varies enormously from 3 to 700 mm^{-1}. This huge range makes Evans' law much less practical, even in the empirical level. Additionally, in Fig. 3.5, Evans' equation tends to underestimate the joint closure magnitude at low normal stress levels where the normal stiffness is most stress-dependent. The experimental study in this paper shows that the normal deformation behavior

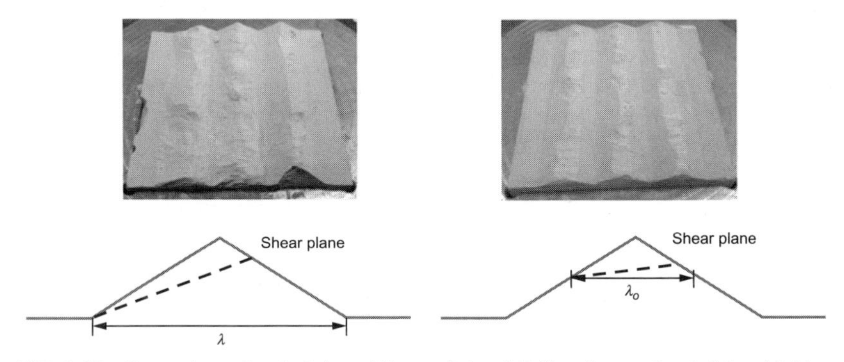

FIG. 3.68 Comparison of typical shear failures of closed (*left*) and opened rock joints (*right*).

can be represented by the proposed equation with high accuracy. Instead of empirical estimation, free parameters, that is, the initial normal stiffness K_{ni} and coefficient B, can be obtained from normal compression tests, which significantly improves the precision of the prediction of joint normal deformation behavior when implimented in numerical modeling.

In order to illustrate the effect of joint opening on the joint shear strength, Barton's model, which is mainly used in practice, has been employed by incorporating the degree of joint opening. It should be noted that the strength loss due to the reduction in asperity interlocking can be modeled along with not only Barton's model, but also other joint constitutive models. The degree of interlocking proposed by Ladanyi and Archambault (1969) has been used in this paper to describe the opening state of rock joints. The value of the degree of interlocking is calculated exactly using mismatched shear displacement and joint waviness if the surface profile is an idealized triangular or sinusoidal shape (Oh, 2005). However, natural joints are irregular, composed of first-order and second-order roughness, and the evaluation of the degree of interlocking for the real joint is very difficult, as the contact area changes discontinuously from an interlocked state (Swan and Sun, 1985). Quantitative estimation of the joint opening degree is necessary to describe both the normal and shear deformation of opened joints. Joint aperture is the perpendicular distance between upper and lower joint walls. The aperture value is assumed zero for perfectly matched rock joints. When the joint is completely opened, the aperture magnitude equals the double joint amplitude. Xia et al. (2003) and Tang et al. (2013) indicated that the joint aperture increases continuously following increasing joint mismatching. Thus, the ratio between joint aperture and joint amplitude is able to quantify the degree of joint opening. The variation of joint aperture ratio is nonlinear and topography-dependent. An algorithm computing the joint opening degree for natural joint surfaces is required to improve the stability analysis associated with underground excavation. This has been recognized as a focus for further study.

3.6.5 Conclusions

The results of the compression tests in this study show that the normal stiffness of rock joints decreases significantly as the initial joint opening increases. A semilogarithmic equation is proposed to represent the normal deformation behavior of opened rock joints. The formulation takes into account the effect of joint opening by considering the true normal stress acting on the joint surfaces. The agreement between experimental results and the prediction from the proposed equation is very good.

The results of direct shear tests clearly show that shear strength and dilation are reduced because smaller asperities are involved in the shear process for opened rock joints. Although experimental studies employ triangular-shaped asperities, they illustrate commonly held notions regarding the opened rock joint behavior quantitatively. They also indicate joint constitutive models considering that the true asperity contacting area and the actual stress on it can be used to predict the shear strength of opened rock joints.

Underground excavation loosens the surrounding rock mass and the joints become opened. As presented in this study, both the normal and shear deformation behavior of opened rock joints is different from that of closed ones. By predicting more reliable joint normal displacements and considering the strength loss due to less asperity interlock, an advanced joint constitutive model can be developed and implemented in finite element or discrete element computer codes. It could render the stability prediction of rock masses under excavation more accurate and reliable.

3.7 Joint constitutive model correlation with field observations

3.7.1 Introduction

The presence of joints substantially reduces the capability of a rock mass to support loadings in both normal and shear directions. Slip and separation along joint walls occur during the excavation of jointed rock masses. The dislocation or opening of joint surfaces decreases the normal stiffness and shear resistance of rock joints by a large value (Li et al., 2014), thereby undermining the stability of rock structures.

The distinct element method, for example, the UDEC (Universal Distinct Element Code) (Itasca, 2014) and 3DEC (3D Distinct Element Code) (Itasca, 2013) provide a powerful discontinuum modeling approach to simulate the behavior of jointed rock masses under quasistatic or dynamic loadings. In DEM, a rock mass is represented as an assembly of rigid or deformable blocks interacting at joints that are viewed as interfaces between discrete bodies (Itasca, 2014). Because DEM allows large displacements and rotations of the blocks, the effect of joints on the response of excavations in jointed rocks can be considered appropriately.

Over the last several decades, numerous constitutive models have been developed to describe the mechanical response of rock joints to normal compression and shearing (Goodman, 1976; Barton, 1982a, b; Plesha, 1987; Jing et al., 1993; Qiu et al., 1993; Grasselli and Egger, 2003; Oh et al., 2015). Most of the models assumed that joint walls are initially tightly matched, whereas the mechanical softening due to joint opening has not been fully addressed. Ladanyi and Archambault (1969), Zhao (1997a, b), and Oh and Kim (2010) indicated that shear resistance can be remarkably reduced due to the initial opening of joint walls. Tang et al. (2016) and Li et al. (2016a, c) observed that joint shear stiffness also decreases as the magnitude of the opening increases. On the other hand, it has been frequently reported that opened joints exhibit much lower normal stiffness than that of closed ones (Goodman, 1976; Li et al., 2016a; Bandis et al., 1983; Xia et al., 2003; Tang et al., 2013). That is to say, joint opening softens both the normal and shear stiffness, the magnitudes of which depend on the degree of opening. Recently, Li et al. (2016a) and Li et al. (2016c) proposed a constitutive model that is capable of predicting the stiffness variation of joints with different initial opening states. The scaling dependence of the joint surface roughness in the study of Li et al. (2016b) permits the proposed model to estimate the mechanical behavior of rock joints in the field. Although the performance of the proposed model's features (Li et al., 2016a, b, c, d) has been illustrated through experimental studies, their applications to field case studies are required to make further verification.

This study initially shows an example that verifies the implementation of the proposed model in UDEC (2014). Three case studies, including a 10-meter high rock slope and two underground excavations, were studied numerically. In the modeling, rock joints were represented by the developed constitutive model that took into account the effect of joint opening on the mechanical response to both compressive and shear loadings. Comparisons between numerical analyses and in situ measurements demonstrated the model's capability for field-scale rock mass stability analysis.

3.7.2 Model description and implementation

The constitutive law for the mechanical behavior of opened joints proposed by Li et al. (2016a), Li et al. (2016b), Li et al. (2016c), and Li et al. (2016d) has been incorporated into UDEC via the built-in programming language, FISH. The joint model characterizes a semilog relationship between the joint closure and the normal stress acting on the joints with varying initial opening values (Li et al., 2016a, c):

$$\Delta V_j = \left(\frac{K}{\eta A'}\right) \ln\left(\frac{\sigma_n}{\eta B} + 1\right) \tag{3.103}$$

where ΔV_j is the joint closure, σ_n is the normal stress, η is the degree of interlocking (Ladanyi and Archambault, 1969; Li et al., 2016c), and K, A', and B are coefficients.

In the shear direction, Li et al. (2016b) accounted for the influence of opening on the joint friction by incorporating the degree of interlocking. The mobilized shear stress (τ_{mob}) is generalized as:

$$\tau_{mob} = \sigma_n \cdot \tan\left(\phi_r + i_n^{mob} + \alpha_n^{mob}\right) \tag{3.104}$$

where σ_n is the applied normal stress, i_n^{mob} and α_n^{mob} are the mobilized asperity angles for the critical waviness and critical unevenness at the field scale, respectively; and ϕ_r is the residual friction angle.

Joint roughness is scale-dependent and its descriptors vary with the scale of the investigation. Li et al. (2016b) proposed scaling laws for the joint roughness at varying sizes, provided that a natural joint profile obeys self-affinity, that is:

$$\frac{\tan i_n}{\tan i_0} = \left(\frac{L_n}{L_0}\right)^{H-1} \tag{3.105}$$

$$\tan \alpha_n = \frac{(k-1)\tan i_n}{\left[1 + k(\tan i_n)^2\right]} \tag{3.106}$$

where $k = \frac{\tan(i_0 + \alpha_0)}{\tan i_0}$, i_0 and i_n and α_0 and α_n are the inclination angles of the critical waviness and critical unevenness in the laboratory scale (L_0) and field scale (L_n), correspondingly; and H is the Hurst exponent.

Moreover, a fractal-based correlation is proposed to assess the peak shear displacements of rock joints at differing sizes (Li et al., 2016b):

$$\frac{\delta_{p,n}}{\delta_{p,0}} = \left(\frac{L_n}{L_0}\right)^{(2-H)/2} \tag{3.107}$$

where $\delta_{p,\,0}$ and $\delta_{p,\,n}$ are the peak shear displacements of a laboratory-sized and a field-sized joint, respectively.

To illustrate the model's behavior, a normal deformability test and a direct shear test were simulated. Different values of the degree of interlocking (Ladanyi and Archambault, 1969; Li et al., 2016c) were prescribed. Fig. 3.69 shows the closure curves of rock joints in varying opening states under the same monotonic loading path in the normal direction. Table 3.27 lists the input parameters for the semilog formulation relating the normal stress to the normal deformation of joints with different degrees of interlocking. The joint normal stiffness remarkably decreases as the joint opening increases (Fig. 3.69).

Simulations of direct shear tests on joints with varied initial openings under constant normal stress ($\sigma_n = 1.0$ MPa) are shown in Fig. 3.70. Input parameters used in this example are included in Table 3.28. For a rock joint undergoing a monotonic shear process under constant normal stress, the implemented model exhibits a shear curve featuring a nonlinear prepeak softening followed by a gradual softening in the postpeak stage. Joint dilatancy increases gradually at

TABLE 3.27 **Input parameters for the semilog formulation describing the joint normal deformation.**

Semilog equation			
1/A′ (mm)	*B*(MPa)	*K*	*η*
0.01	2.0	1.0	1.0/0.7/0.4

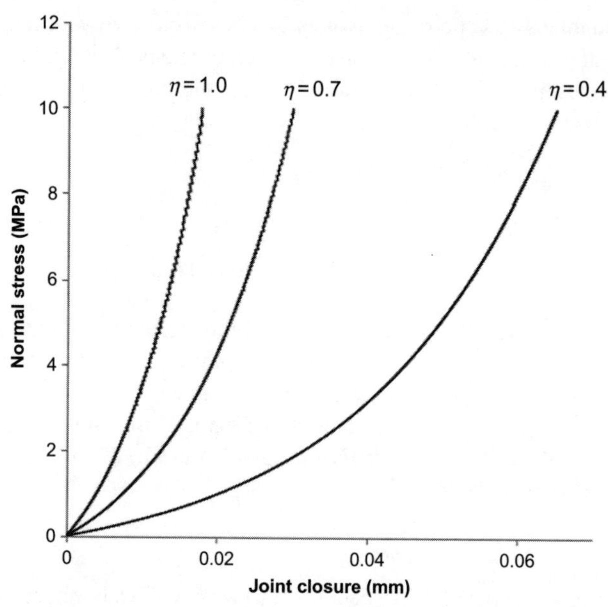

FIG. 3.69 The semilog relationship between normal stress and the closure of joints with different interlocking degrees.

a decreasing dilation rate with an increase of the shear displacement in the plastic region. The effect of joint opening on the shear behavior is illustrated in Fig. 3.28, where both the shear resistance and the dilatancy of a joint are reduced progressively as the opening value increases.

An example in UDEC is used to verify the implementation of the proposed model. The model shown in Fig. 3.71 is a two-dimensional, plane-strain representation of a 4×4 m square opening. Two continuous sets of joints are present in the model, one with an orientation of 70 degree and the other with 310 degree (Fig. 3.71 and Table 3.29). A single fault dipping at 70 degree with a 5-degree friction angle cuts through the entire model. The mechanical behavior of a single joint (the bold dashed line in Fig. 3.71) is represented by the Mohr-Coulomb model and the proposed model. Table 3.30 lists the equivalent input parameters for the Mohr-Coulomb joint model. In the implementation of the proposed model, the calculation of the degree of interlocking was updated via a FISH function that continuously traced the variation in the opening of contacts.

Fig. 3.72 shows that, at point A, the shear displacement predicted by the proposed joint model is higher than that by the Mohr-Coulomb joint model. The difference can be explained as follows. The Mohr-Coulomb slip law features an elastic-perfectly plastic relationship between the shear stress and shear displacement, indicating that the shear resistance retains the peak value after failure. The joint dilation grows linearly without any degradation, which maximizes the possibility of stabilization of the jointed rocks. The proposed

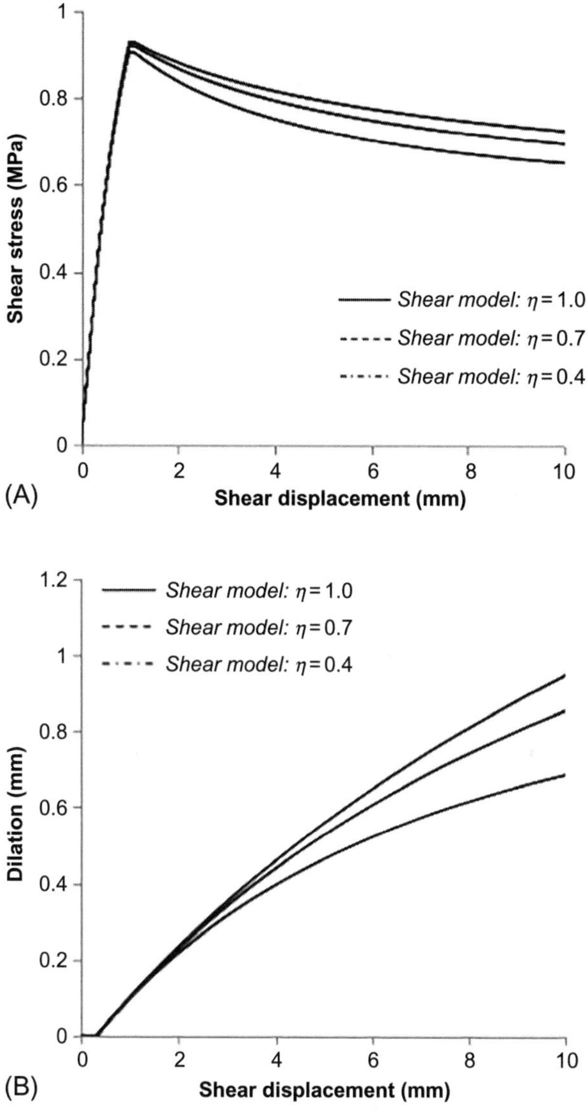

FIG. 3.70 Shear behavior of rock joints with different interlocking degrees ($\sigma_n = 1.0$ MPa). (A) Shear stress-shear displacement curves. (B) Dilation-shear displacement curves.

model, which characterizes a continuous deterioration of the surface roughness, yields a progressive decrease in shear stress and dilatancy rate in the postpeak stage. Moreover, in the simulation with the proposed joint model, the joint opening reduced the sliding resistance of the jointed rock masses. Consequently, the shear displacement at point A is larger compared to the numerical evaluation with the Mohr-Coulomb joint model.

TABLE 3.28 Input parameters for the shear behavior of rock joints.

Elastic shear stiffness K_s(MPa/m)	UCS σ_c (MPa)	Critical waviness			Critical unevenness			Residual friction angle ϕ_r (degree)	Peak shear displacement δ_p (mm)
		i (degree)	$\lambda_w/2$ (mm)	A_w (mm)	α (degree)	$\lambda_u/2$ (mm)	A_u (mm)		
1200	30	10	20	3.5	4	2	0.14	30	1.0

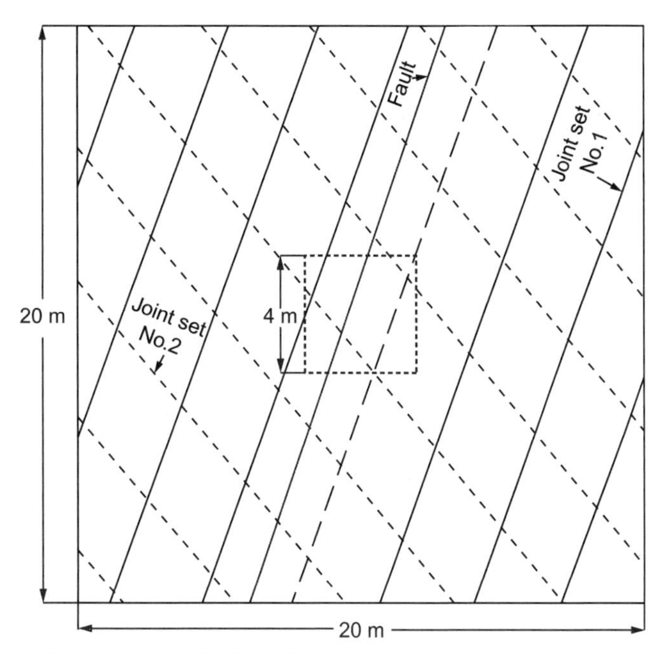

FIG. 3.71 Joint orientation in the simulation.

TABLE 3.29 Joint sets in the model.

Joint set	Dip (degree)	Spacing (m)	Origin
No.1	70	3	(−2,0)
No.2	310	3	(1,2)
No.3 (Single fault)	70	-	(3,10)

TABLE 3.30 Equivalent parameters for the Mohr-Coulomb model for the interested joint.

Constitutive type	K_n(MPa/m)	K_s (MPa/m)	ϕ (degree)	d (degree)
Mohr-Coulomb model	2000	1200	44	14

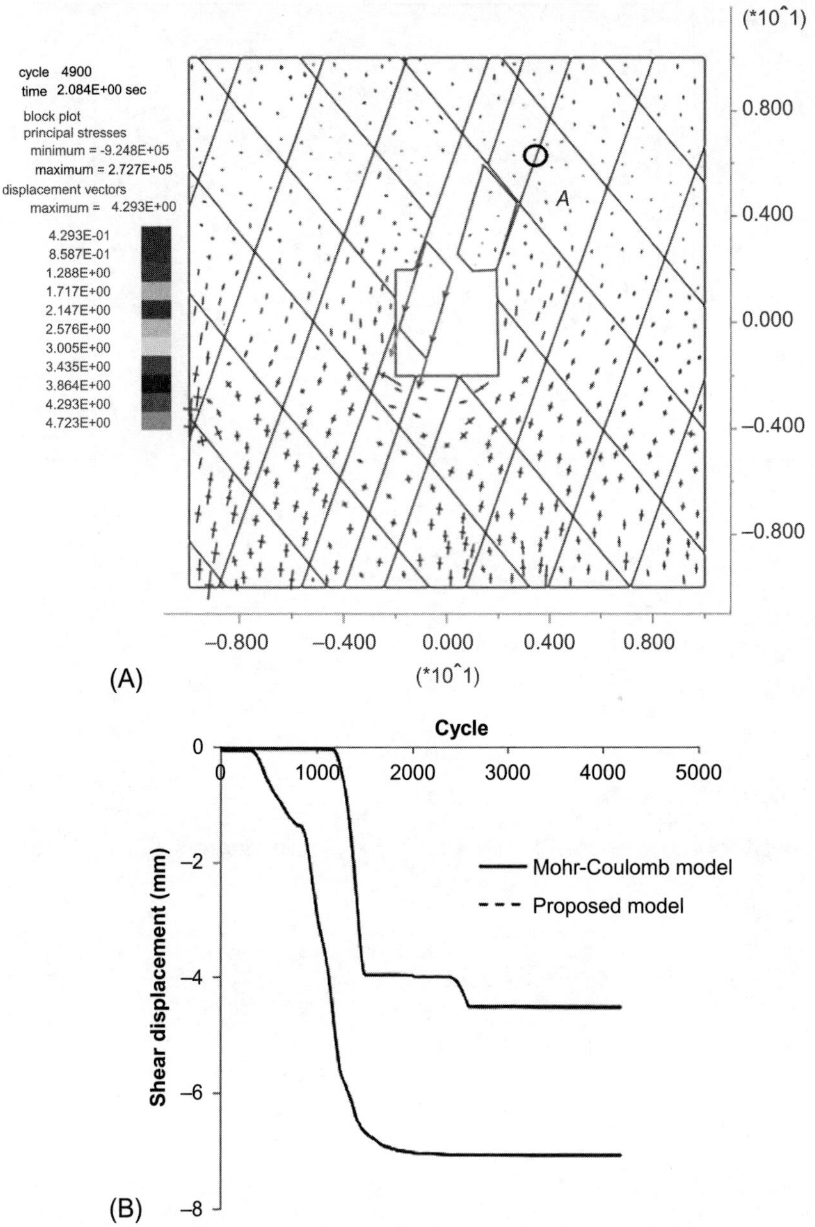

FIG. 3.72 Simulation results using the proposed joint model and the comparison of shear displacements at point A. (A) Simulation results using the proposed joint model. (B) Comparison of the shear displacements at point A.

3.7.3 Stability analysis of large-scale rock structures

3.7.3.1 Rock slope case

Site description

Kim et al. (2013) investigated the stability of a large rock slope located in the Tambourine Mountain area, Gold Coast, Australia (Fig. 3.73A). The length of this slope was about 200 m with a height varying from 8 to 10 m (Fig. 3.73B). This slope had experienced slope stability problems in the past few years, especially during long periods of rain (Kim et al., 2013).The surface of a 302 mm long joint sampled from the site (Fig. 3.73C) had been scanned. Estimation of the joint roughness using a statistical approach in Tse and Cruden (Tse and Cruden, 1979) yielded JRC =8.0. Fractal analysis through the roughness-length method (Malinverno, 1990; Kulatilake and Um, 1999) gave a value of the Hurst exponent (H) at 0.703.

Properties of rock mass

Table 3.31 lists the input parameters used in the modeling performed by Kim et al. (2013), in which the intact rock behaved elastically and the mechanical behavior of the rock joints followed the Mohr-Coulomb criterion.

Utilization of the proposed constitutive law to represent the mechanical behavior of rock joints requires specification of the input parameters, as exemplified in Tables 3.27 and 3.28. Parameter values for the semilog deformation law are desirably attained by correlating with joint closure curves that are determined experimentally. Because no normal deformability test was reported by Kim et al. (2013), it is assumed that the hyperbolic equation in Bandis et al. (1981) could describe the normal stress-joint closure relationship for closed joints ($\eta = 1.0$). Input parameters for the semilog equation were acquired by calibrating with the hyperbolic curve. Fig. 3.74 shows the correlation between the joint closure curves of hyperbolic and semilog formulations. Table 3.32 lists the parameter values used in the semilog relation after calibration.

Table 3.33 presents the input parameters for the shear behavior of the proposed model. Geometric features of the critical waviness and critical unevenness are attained by inspecting the joint profile in Fig. 3.73 through the roughness quantification approach in Li et al. (2016d). It should be noted that the predictive equation in Barton and Bandis (1982) and Barton (1982a, b) was utilized to estimate the peak shear displacement of the sampled joint because no relevant experimental data has been reported by Kim et al. (2013). The residual friction angle of the joint sample can be found in the study by Kim et al. (2015).

Comparison numerical results with site investigation

The simulation result using the proposed model is illustrated in Fig. 3.75A, where the surface blocks slide down toward the toe of the slope and a rock block

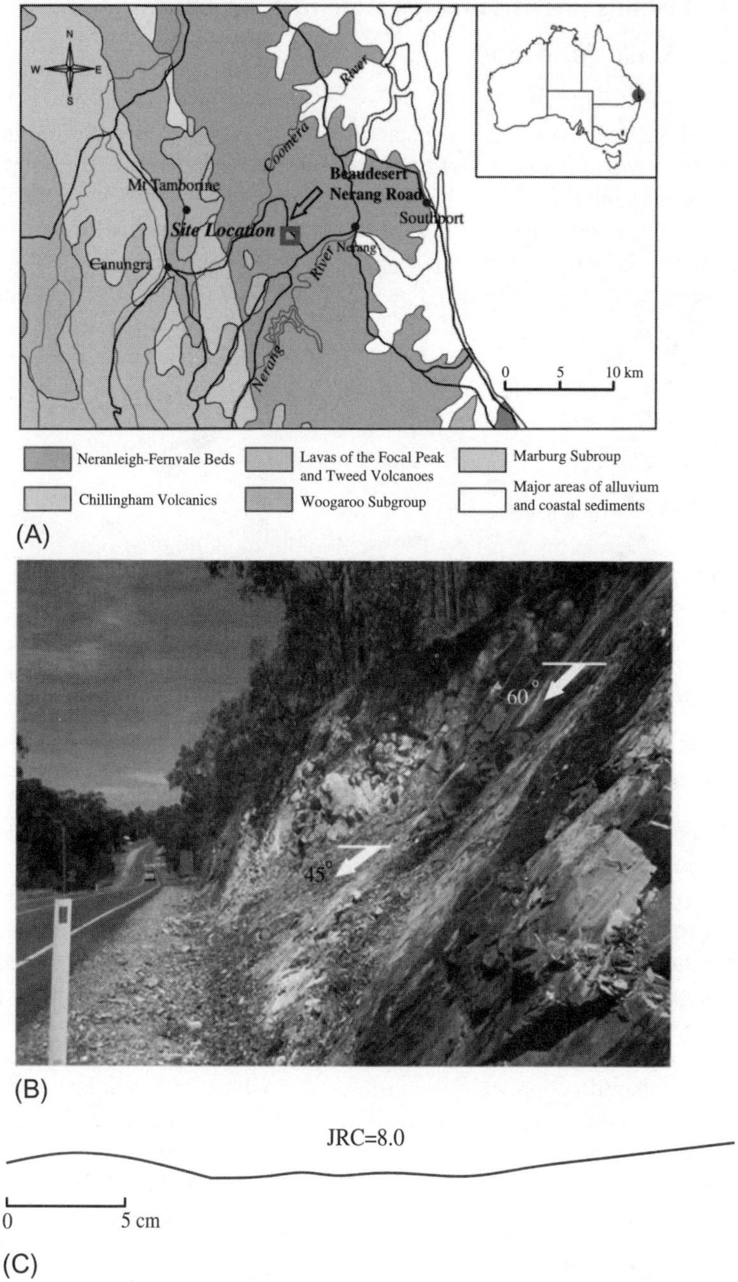

(A)

(B)

(C)

FIG. 3.73 Geological survey of the rock slope. (A) Geological map. (B) Photo of the studied area. (C) Profile of the joint sample collected from the site. *(After Kim, D.H., Gratchev, I., Balasubramaniam, A., 2013. Determination of joint roughness coefficient (JRC) for slope stability analysis: a case study from the Gold Coast area, Australia. Landslides, 10(5), 657-664.)*

TABLE 3.31 Input parameters used for the simulation in Kim et al. (2013).

Rock type	L_0 (mm)	JRC	σ_c (MPa)	γ kN/m^3	ϕ (degree)	K_s (MPa/mm)	K_n (MPa/mm)
Sandstone	302	8.0	13.1	25.7	42	5.68	17.04

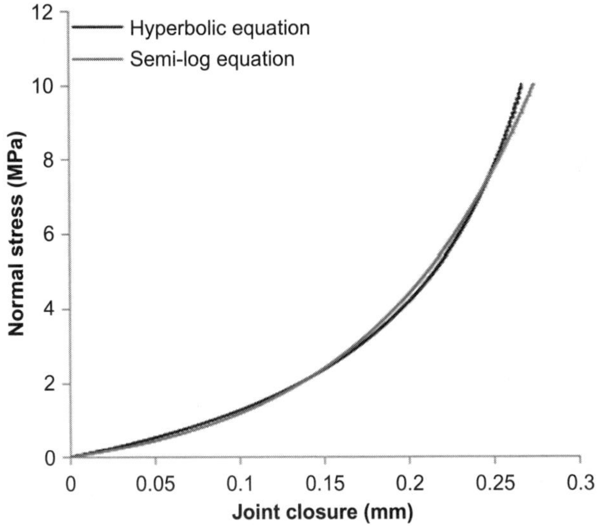

FIG. 3.74 Calibration of the semilog equation for the joint closure.

TABLE 3.32 Input parameters for the hyperbolic and semilog equations.

Joint roughness	Hyperbolic equation		Semilog equation		
JRC	K_{ni} (MPa/mm)	V_m(mm)	$1/A'$ (mm)	B (MPa)	K
8.0	8.976	0.35	0.0988	0.6652	0.75

fell. The numerical evaluation is consistent with the site observation shown in Fig. 3.75B.

Because no onsite quantitative measurements were available in terms of the rock slope movements, numerical estimations from UDEC with the proposed joint model were compared to those predicted by Kim et al. (2013). Three measuring points are placed along the face of the rock slope (Fig. 3.76). Vertical displacements of the points were recorded during the simulation. In Fig. 3.76, the displacements predicted by the proposed model were substantially higher than those predicted by Kim et al. (2013). Results of slope stability

TABLE 3.33 Input parameters for the shear behavior of the proposed model.

	Critical waviness			Critical unevenness						
L_0 (mm)	i_0 (degree)	$\lambda_w^0/2$ (mm)	A_w^0 (mm)	α_0 (degree)	$\lambda_u^0/2$ (mm)	A_u^0 (mm)	σ_c/JCS (MPa)	ϕ_r (degree)	$\delta_{p,0}$ (mm)	H
302	7	55	6.5	4	14	1.0	13.1	30	1.78	0.703

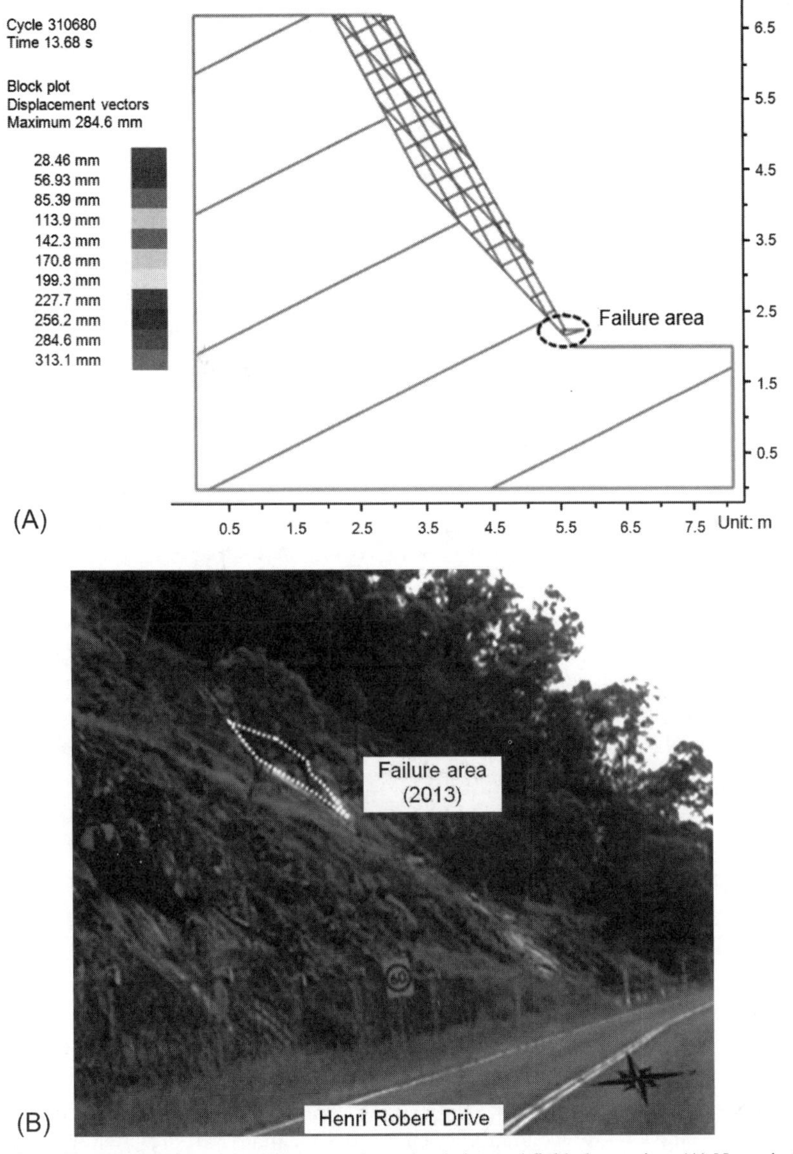

FIG. 3.75 Comparison of the failure area from simulation and field observation. (A) Numerical results. (B) Site observation.

analysis in Kim et al. (2013) showed that the joint roughness, that is, the asperity interlocking, had a significant effect on the safety factor; when the degree of roughness decreased, the safety factor of the slope decreased. The opening of the joint walls appreciably reduces the interlocking between surface roughness.

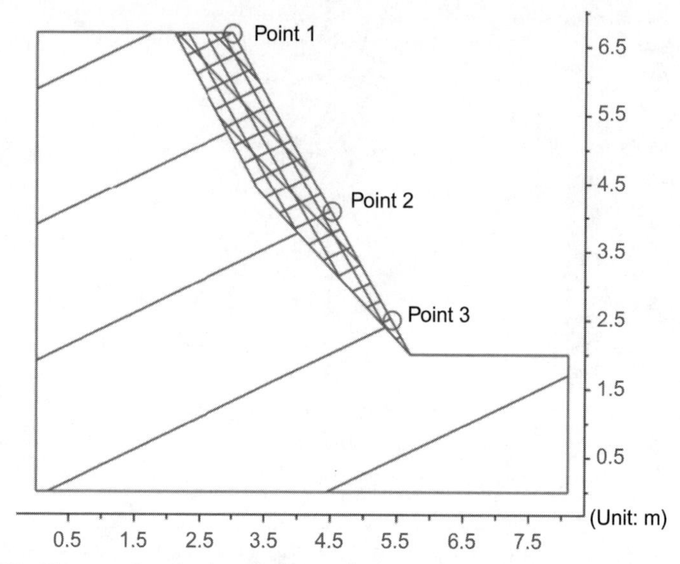

FIG. 3.76 Three monitored points along the rock slope in the modeling.

Thus, the simulation with the proposed model gave much larger displacements of measured points. Moreover, the proposed model captures the progressive damage of the joint roughness under shear, which is considered more realistic than the joint model employed by Kim et al. (2013). Kim et al. (2015) reported that rock failure had occurred several times in the last few years, implying that the slope was in a severely unsteady condition. This is in good agreement with the simulation results using the proposed joint model (Fig. 3.77).

3.7.3.2 The underground powerhouse case

Site description

Li (2009) modeled the sequential excavations of a pumped storage power plant and compared the numerical results with in situ measurements. The underground powerhouse is 425 m deep from the ground surface, with a warhead shaped cross-sectional profile (Fig. 3.78). The height, width, and length of the powerhouse are 48 m, 24 m, and 187 m, respectively. The opening is situated in the slightly weathered granodiorite, the main part of which was classified as class C_H (*the rock strength is nearly unaffected although the rock-forming minerals and grains (except quartz) are slightly weathered. Due to the intrusion of limonite, the cohesion between rock joints was slightly decreased. When hit heavily by a hammer, the rock block exfoliated along the joint where a thin layer of clay could be observed. Light noise was heard if the rock was hit by a hammer*) with a few discrete parts classified as C_M (*the rock strength is decreased as the rock-forming minerals and grains*

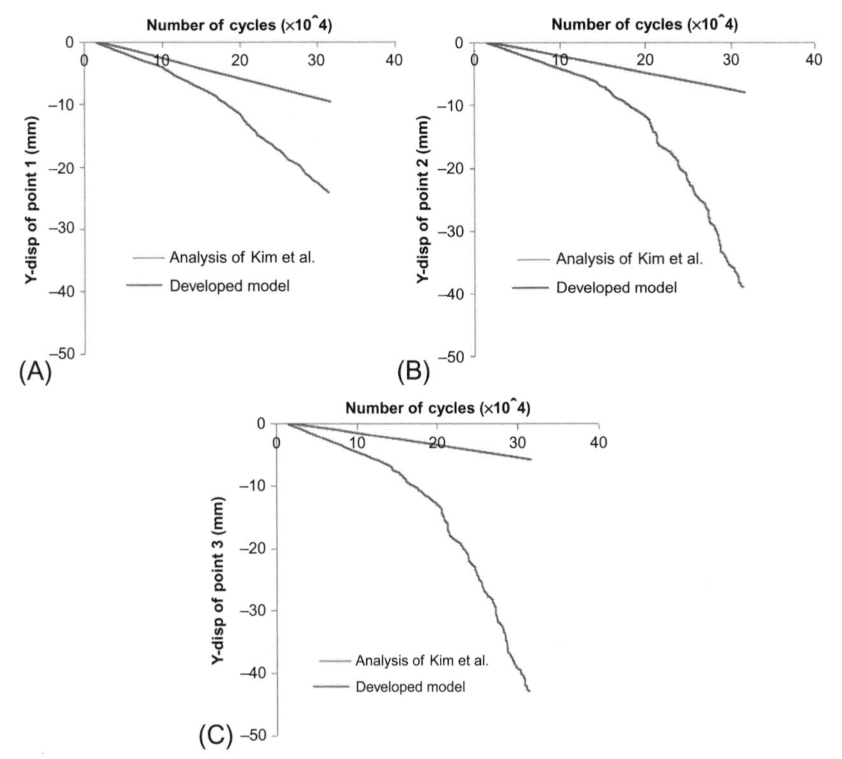

FIG. 3.77 Comparison of the numerical results using the developed model and the analysis of Kim et al. (A) Point 1. (B) Point 2. (C) Point 3.

(except quartz) are moderately weathered. The cohesion between rock joints is decreased to a certain extent. When hit by a hammer, the rock block exfoliated along the joint where a thin layer of clay could be observed. Light noise was heard if the rock was hit by a hammer.) according to the rock mass classification standard of Japan (Nippon Expressway Company, 1991; Li, 2009). In situ stress monitoring in the marked point indicated that the coefficient of lateral pressure K_0 was 0.6 (Fig. 3.78). Fig. 3.78B shows the boundary conditions of the two-dimensional model.

Properties of rock and joints

Table 3.34 lists the mechanical properties of the rock in the construction site. The intact rock in the model was considered an elastic-perfectly plastic material that obeyed the Mohr-Coulomb failure criterion.

Three joint specimens (Fig. 3.79) were collected from the field during different survey stages (Fig. 3.80) (Li, 2009). Estimations of the *JRC*, Hurst exponent, and peak shear displacement of the laboratory-scale samples can be found in Li (2009). Parameter values used in the simulation of the joint shear behavior

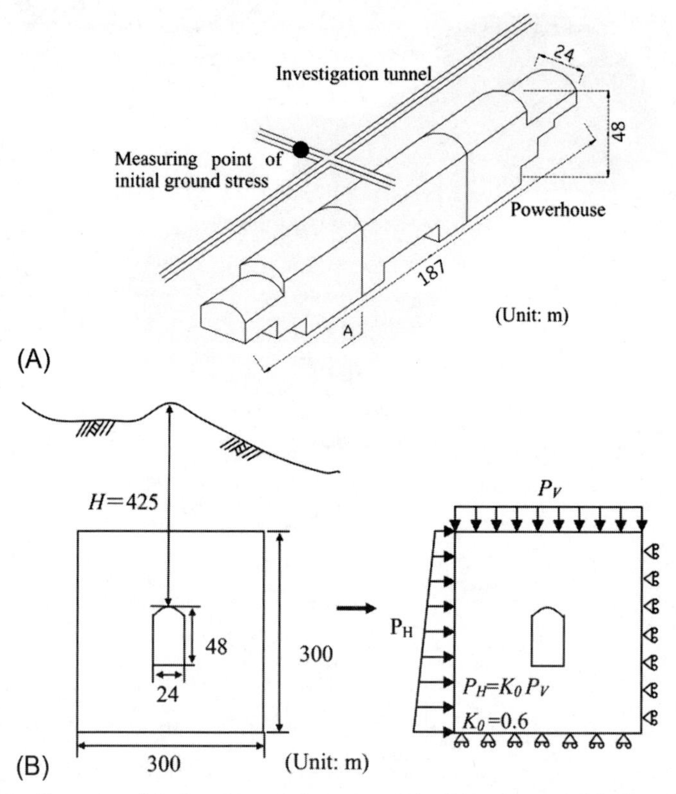

FIG. 3.78 Illustration of the location and dimension of the object cavern and the boundary conditions with the initial ground stresses for numerical modeling. (A) Schematic view of the powerhouse. Section A is chosen to generate numerical model. (B) Location of the study area and boundary conditions of the numerical model, where P_H and P_V are horizontal and vertical pressures; K_0 is the coefficient of lateral pressure.

TABLE 3.34 Mechanical properties of the rock (Li, 2009).

Item	Value
Rock type	Granodiorite
Unit volume weight (γ), kN/m^3	27.1
Elastic modulus (E), MPa	20,000
Poisson's ratio(ν)	0.23
Cohesion (c), MPa	1.6
Friction angle (ϕ), degree	60
Tensile strength (σ_t), MPa	0.22

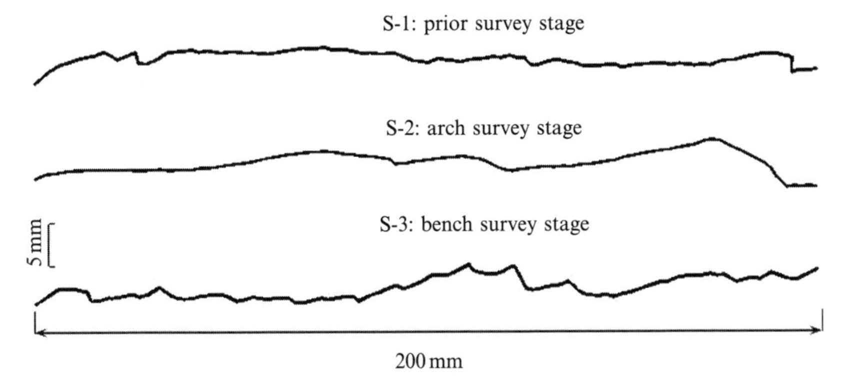

FIG. 3.79 Two-dimensional profiles of three joint samples.

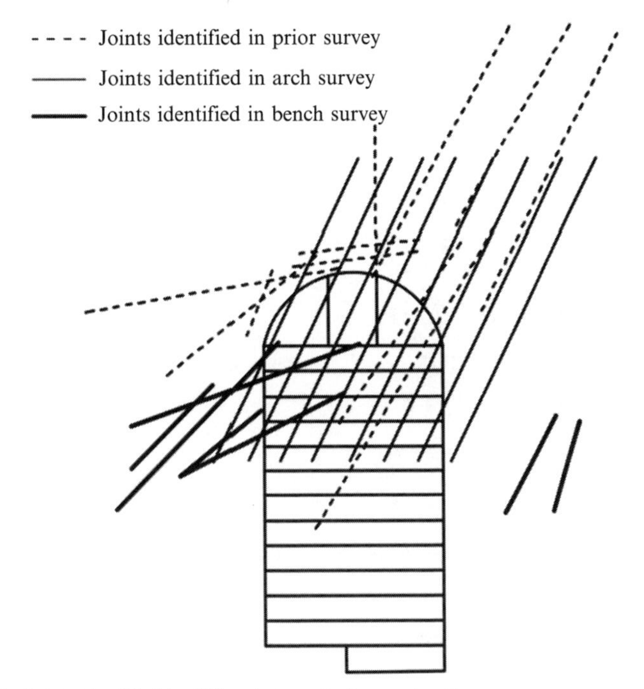

FIG. 3.80 Joints identified in different stages in the construction.

are listed in Table 3.9. Table 3.10 shows the input parameters for the joint normal deformability, the values of which were acquired by correlating the semilog formulation with the hyperbolic formulation (Fig. 3.81) (Tables 3.35 and 3.36).

The joints depicted in Fig. 3.80 were reproduced in the numerical model. Similar to the approach adopted by Li (2009), the generation of nonpersistent joints in UDEC was fulfilled by adding fictitious joints that connected the heads

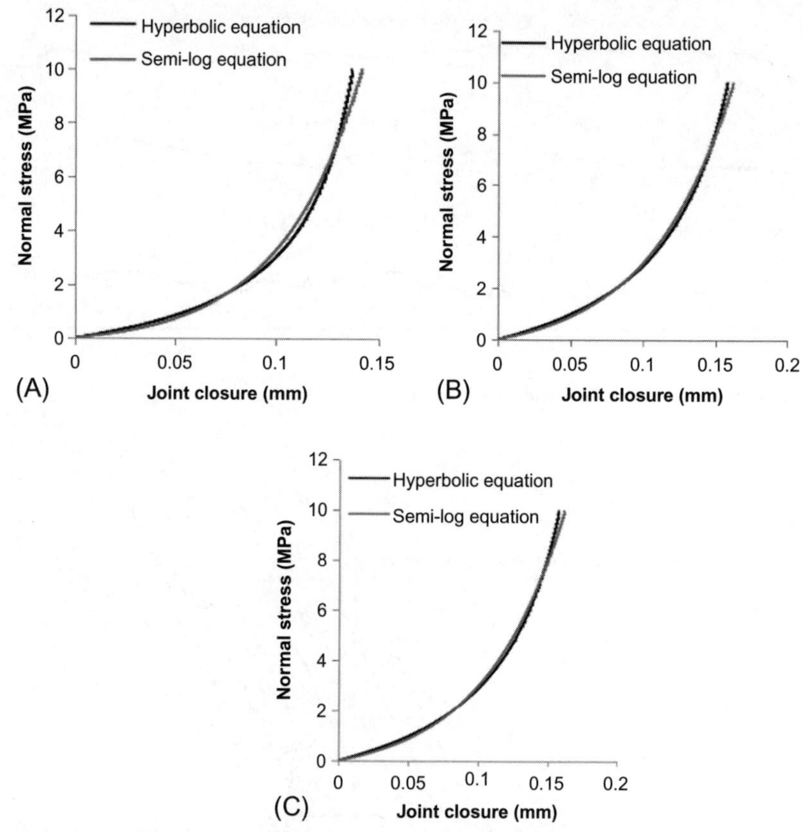

FIG. 3.81 Calibration of the semilog equation for the joint closure. (A) Joint No. 1. (B) Joint No. 2. (C) Joint No. 3.

of nonpersistent joints to the model boundary. The physical parameters were selected as follows so that the fictitious joints could behave like the intact rock (Kulatilake et al., 1992):

(a) Identical strength parameter values should be chosen for both the intact rock and the fictitious joints.
(b) The shear stiffness (K_s) of the fictitious joints should be chosen to produce a G/K_s ratio between 0.008 and 0.012 m, where G is the shear modulus of the rock.
(c) A value between 2 and 3 should be chosen for the normal to shear stiffness ratio K_n/K_s. The most appropriate value to choose in this range may be the value of E/G for the particular rock.

Accordingly, the properties of the fictitious joints were determined, as shown in Table 3.37.

TABLE 3.35 Input parameters for the shear behavior of three rock joints.

Joint sample	JRC	Critical waviness			Critical unevenness			H	$\delta_{p,0}$ (mm)	σ_c (MPa)	ϕ_r (degree)	L_0 (mm)
		i_0 (degree)	$\lambda_w^0/2$ (mm)	A_w^0 (mm)	α_0 (degree)	$\lambda_u^0/2$ (mm)	A_u^0 (mm)					
S-1	5.8	13.5	17.8	4.3	11.9	4.1	0.8	0.80	1.22	170	29	200
S-2	9.5	4.7	51.3	4.3	2.4	7.0	0.3	0.74	1.13			
S-3	12.8	9.3	28	4.6	4.3	7.9	0.6	0.69	1.35			

TABLE 3.36 Input parameters for the hyperbolic and semilog equations.

Joint roughness	Hyperbolic equation		Semilog equation		
JRC	K_{ni} (MPa/mm)	V_m(mm)	$1/A'$ (mm)	B (MPa)	K
5.8	11.67	0.163	0.03943	0.282	0.71
9.5	14.76	0.206	0.0571	0.624	0.77
12.8	19.16	0.229	0.0705	1.111	0.82

TABLE 3.37 Mechanical properties of the fictitious joints.

Item	Value
Normal stiffness (K_n), GPa/m	2000
Shear stiffness (K_s), GPa/m	813
Cohesion (c), MPa	1.6
Tensile strength(σ_t), MPa	0.22

Excavation process and reinforcements

Fig. 3.82 illustrates the excavation process. Rock bolts and prestressed anchors were used as reinforcements; the rock bolts were installed in a pattern of a 2×2 m grid over the entire arch and a 3×2 m grid on the two sidewalls. Reinforcement in each stage was conducted in the following sequence: (1) install rock bolts, (2) embed prestressed anchors, and (3) place shotcrete. The rock bolt and prestressed anchor were both treated as cable elements with different properties in the numerical model; the shotcrete was modeled as a structural element with a thickness of 0.32 m. A FISH function was written to perform automatic excavation and reinforcement installation.

Table 3.38 lists the input parameters for the structural elements and cable elements in the simulation (Li, 2009). Blasting dramatically affects the rock mass stability. In the simulation, the rock masses within 2 m from the wall of the excavation have been treated as a blast-induced damaged zone with a decreased deformation modulus that is 36% of the original value (Li, 2009).

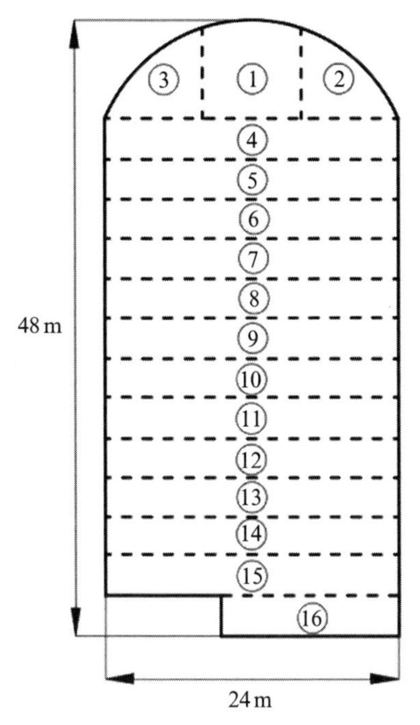

FIG. 3.82 Illustration of the excavation sequence.

Site monitoring and result analysis

A number of measurement instruments were placed in the investigation tunnels and drifts at various locations around the excavation (Li, 2009). Fig. 3.83 shows the layout of measuring lines in one section. The displacements of identical measuring points in the numerical model were traced. Fig. 3.16 illustrates the distribution of plastic zones after excavation stages 3, 10, and 16. Fictitious joints were hidden to make a clear demonstration of the simulation results. The effect of the presence of rock joints on the plastic zone was not obvious until step 3, after which yielded zones commenced to develop from the near-wall region toward the highly jointed area along the joint orientations. The most remarkable progress of the plastic zones occurred in the left side of the excavation where a high density of rock joints existed (Fig. 3.84).

Joints in the opening state were traced in the simulation (Fig. 3.84). The number of opened joints increased as the excavation progressed, which was mainly contributed to by the joints in the left side of the excavation. Fig. 3.85 compares the predicted and measured performance of the excavation. In all cases, the displacements decreased along with the increase in the distance from the wall side, whereas the magnitude of deformation was exacerbated by further excavation. For instance, the displacement at the arch was around

TABLE 3.38 Input parameters for the reinforcements (Li, 2009).

Shotcrete				Rock bolt				PS anchor			
Structural element	Value	Interface	Value	Cable element	Value	Grout	Value	Cable element	Value	Grout	Value
Density (kg/m^3)	2300	Normal stiffness (MPa)	2400	Cross-section area (mm^2)	446	Shear stiffness (MN/m/m)	16,906	Cross-section area (mm^2)	446	Shear stiffness (MN/m/m)	13,409
Elastic modulus (MPa)	20,000	Shear stiffness (MPa)	260	Density (kg/m^3)	7700	Shear strength (MN/m)	0.225	Density (kg/mm^2)	7700	Shear strength (MN/m)	0.351
Poisson's ratio	0.20	Friction (degree)	31	Elastic modulus (MPa)	210,000			Elastic modulus (MPa)	190,000		
Tensile yield strength (MPa)	18	Dilation (degree)	0	Compressive yield force (MN)	0.228			Compressive yield force (MN)	0		
Residual tensile yield strength (MPa)	18	Cohesion (MPa)	0	Tensile yield force (MN)	0.228			Tensile yield force (MN)	0.318		
Compressive yield strength (MPa)	18	Tensile strength (MPa)	0	Extensional failure strain	0.00335			Extensional failure strain	0.00923		

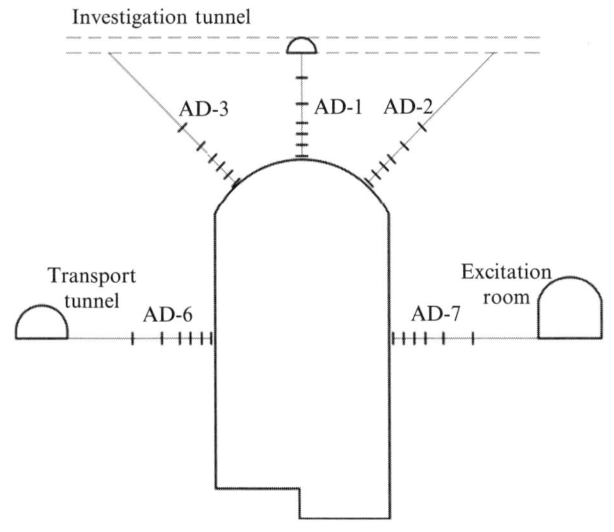

FIG. 3.83 Layout of the monitoring lines.

30 mm at 1 m away from the roof and decreased gradually to around 10 mm at 15 m away from the roof. The displacement of AD-7 at 1 m to the right side wall grew from around 1 mm after step 3 to 18 mm after stage 16. In Fig. 3.85, the numerical evaluation slightly underestimates the displacement at AD-1 but agrees well with the measured values at AD-2, AD-3, and AD-7. The largest displacement occurred at the measuring line AD-6 after stage 16 due to the intensive distribution of joints near the left side wall. Moreover, the increase in the number of opened joints degraded the strength of the rock masses, thereby leading to more damage zones (Fig. 3.84). Overall, simulation results provided close agreement with field measurements.

3.7.3.3 The gold mine case

Site description

This section presents a numerical analysis for the deformation of an upper-level backfill tunnel that is used for backfilling the lower-level goaf in a goldmine located at Jiaojia, Laizhou, Shandong province, China. The strike of the orebody at the Jiaojia goldmine is NE10-50 degree and the dip is NW25-45 degree. The thickness of the orebody ranges from 0.5 to 30 m and its average value is 16.4 m. Slightly weathered beresitization granite constitutes the main part of the orebody. Rock mass classification using *RMR* (Bieniawski, 1993) and the *Q*-system (Barton et al., 1974) suggested that the rock mass quality is fair to poor. Fig. 3.86 illustrates the layout of the roadways, including the heading tunnel and the backfill tunnel, the level of which is –480 m with reference to the ground surface.

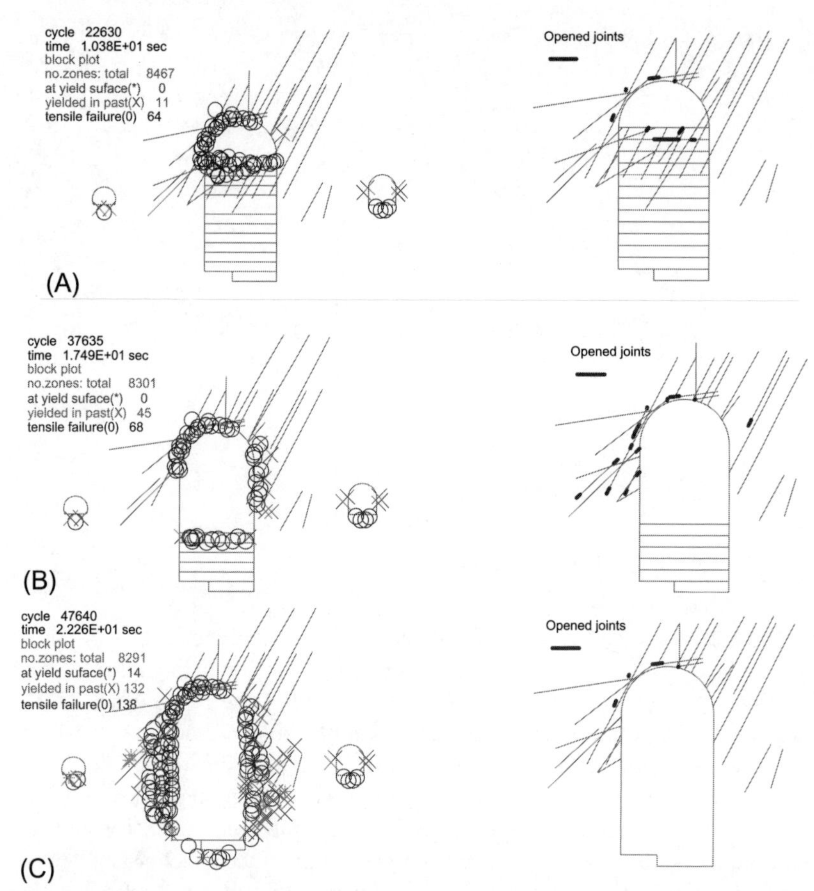

FIG. 3.84 Plastic state (*left*) and opened joints (*right*) at different excavation stages. (A) Step 3. (B) Step 10. (C) Step 16.

An overcoring method was used to measure the local ground stresses. Analysis of the field stress of 20 points in Table 3.39 showed that the maximum principal stress (σ_h) was nearly horizontal and its magnitude was approximately twice the vertical stress (σ_v). Thus, the in situ stress field was initialized as $\sigma_h = 2\sigma_v$ in the simulation.

Rock mass properties

Table 3.40 shows the mechanical properties of the rock, which were obtained from laboratory tests conducted in the Key Laboratory of the Ministry of Education on Safe Mining of Deep Metal Mines, China Northeastern University. Three main joint sets (Fig. 3.87) were identified around the roadway region. Fig. 3.88 illustrates the two-dimensional joint profiles. Table 3.41 shows the input parameters involved in representing the response of rock joints to shear.

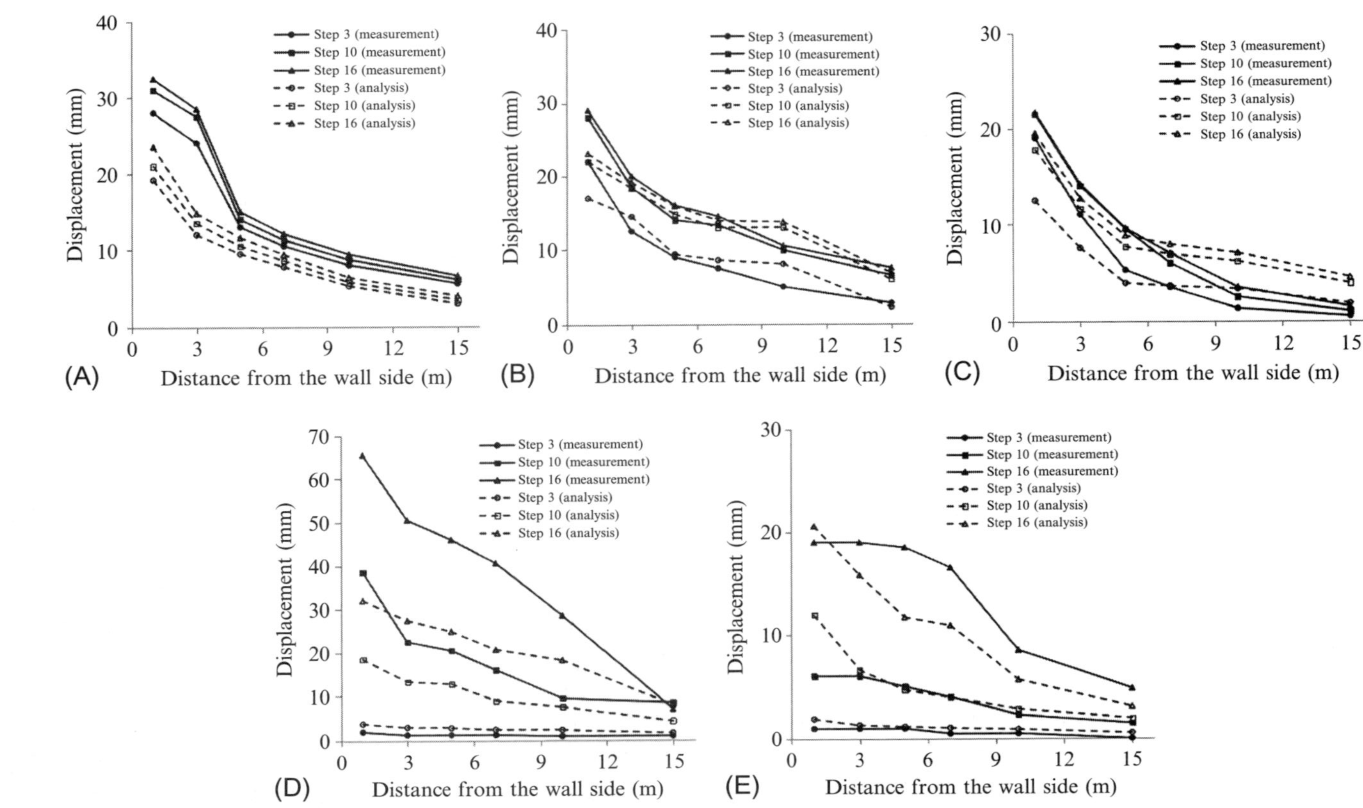

FIG. 3.85 Comparison between measured and predicted deformation. (A) AD-1. (B) AD-2. (C) AD-3. (D) AD-6. (E) AD-7.

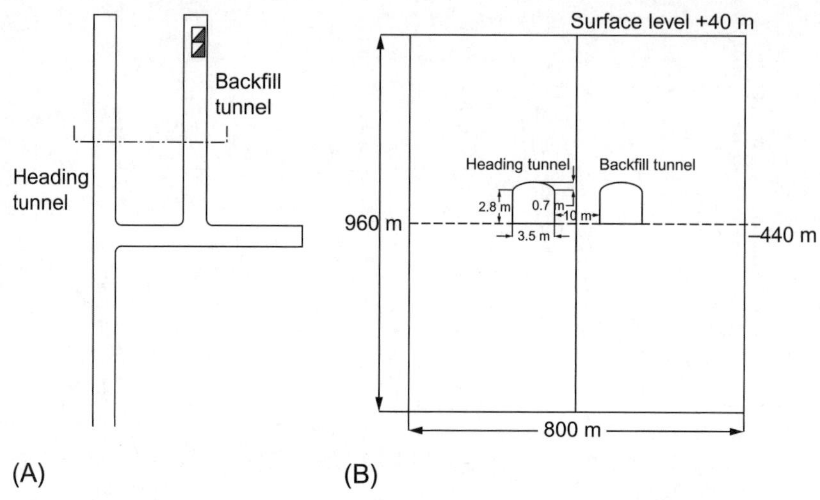

FIG. 3.86 Schematic layout of the tunnels in the Jiaojia goldmine. (A) Top view. (B) Side view.

The values of *JRC*, the Hurst exponent, and peak shear displacement were estimated by the Z_2 approach (Tse and Cruden, 1979), the roughness-length method (Malinverno, 1990), and Barton's model (Barton, 1982a, b), respectively. The critical waviness components of the three joints were acquired through examining the joint profiles in Fig. 3.20. Because the geometric details of the unevenness were not available, the critical unevenness components of these joints were represented by the morphologic properties of the critical unevenness of the corresponding standard *JRC* profiles (*JRC* 10–12, *JRC* 8–10, and *JRC* 6–8). It should be noted that the residual friction angle of the joint material was evaluated as the typical value of granite (Barton and Choubey, 1977) because no direct shear tests have been conducted on joint samples with flat surfaces. Fig. 3.89 and Table 3.42 show the semilog curves for the normal deformation of the three joint sets and the corresponding values of the input parameters, respectively.

Comparison between predicted and measured performance

During the excavation, 2.5 m long resin-grouted rock bolts were installed in a pattern with 0.7 m spacing along the tunnel. Table 3.17 shows the input parameters used in the modeling. Following the approach suggested by Itasca (2014), the parameter values were estimated based on the steel properties and in situ pull-out tests (Table 3.43).

Relative horizontal movements of the tunnel sides and corners and the vertical settlement of the roof center relative to the floor were measured by convergence meters (Fig. 3.90). Fig. 3.91 shows the numerical predictions in comparison with the field measurements. The numerically recorded values

TABLE 3.39 In situ stress measurement results.

Number	Depth (m)	σ_1			σ_2			σ_3		
		Value (MPa)	Strike (degree)	Dip (degree)	Value (MPa)	Strike (degree)	Dip (degree)	Value (MPa)	Strike (degree)	Dip (degree)
1	205	11.45	307.1	−17.6	5.69	286.3	71.3	4.03	35.1	6.2
2	205	11.54	270.0	4.3	6.77	181.5	−19.0	5.72	347.8	−70.04
3	205	11.27	218.9	10.2	5.68	220.2	−79.8	3.98	129	−0.2
4	235	14.62	237.6	9.2	10.17	329.9	13.9	5.63	295.1	−73.2
5	235	13.69	128.7	−7.8	6.83	131.3	82.2	5.06	38.8	0.3
6	235	12.99	301.9	−0.6	6.14	208.2	−81.3	5.00	212.0	8.7
7	235	13.60	311.0	−1.4	8.93	220.7	−10.4	6.85	228.8	79.5
8	235	12.58	280.0	−13.2	7.85	187.3	−11.1	6.92	238.5	72.6
9	235	12.80	127.1	−7.2	7.41	35.9	−9.7	5.89	72.4	78.0
10	310	18.39	123.1	−1.6	11.65	213.2	−3.3	10.73	187.7	86.4
11	310	18.50	285.5	−17.7	8.89	80.8	−70.6	7.05	13.0	7.6
12	310	20.73	109.9	−0.4	9.00	201.9	−79.1	7.01	199.8	10.9
13	310	16.32	82.9	3.2	9.19	13	−80.7	7.99	172.4	−8.7
14	410	29.62	308.9	−5.3	13.77	193.2	−78.0	11.98	219.9	10.7
15	410	31.49	148.4	−6.9	14.13	267.7	−76.0	11.08	236.9	12.0
16	410	31.55	327.2	11.77	13.89	219	−79.1	11.77	237.8	10.3
17	410	25.98	90.7	−4.5	11.54	106.7	85.3	5.78	0.8	1.3
18	610	30.21	300	−22	17.12	357	−72	13.09	28	15
19	660	33.35	280	−9	21.46	322	−78	20.08	11	8
20	660	31.68	276	−10	19.04	252	−80	16.55	9	6

TABLE 3.40 Mechanical properties of the rock.

Item	Value
Rock type	Beresitization granite
UCS (σ_c), MPa	72
Unit volume weight (γ), kN/m^3	28.2
Elastic modulus (E), MPa	35,000
Poisson's ratio(ν)	0.22
Cohesion (c), MPa	12
Basic friction angle (ϕ), degree	41
Tensile strength (σ_t), MPa	6.8

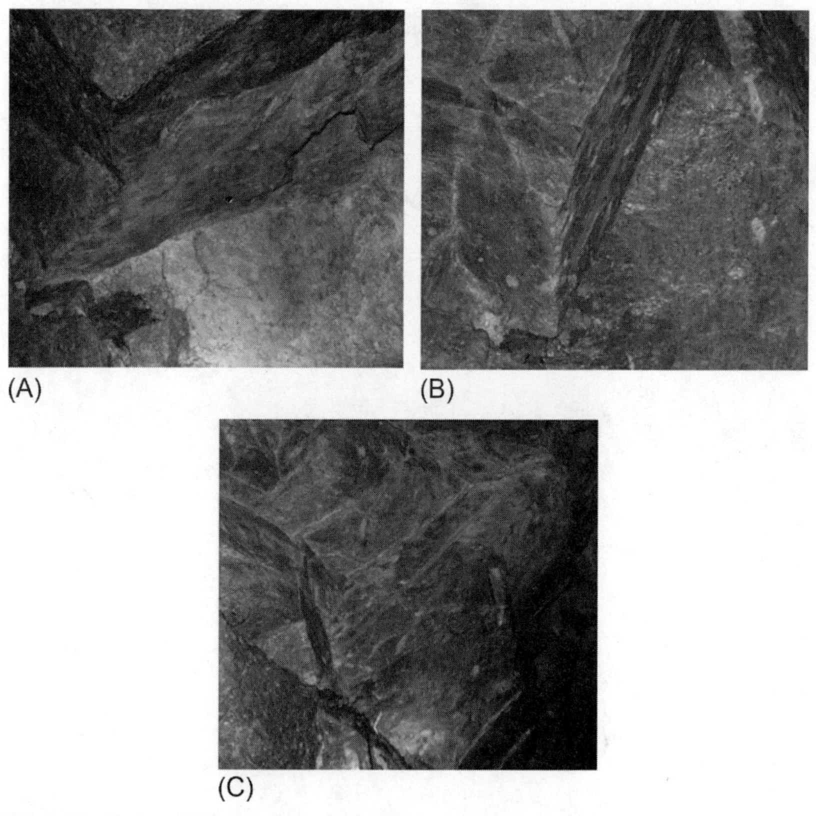

FIG. 3.87 Photos of joint sets in the field. (A) No. 1. (B) No. 2. (C) No. 3.

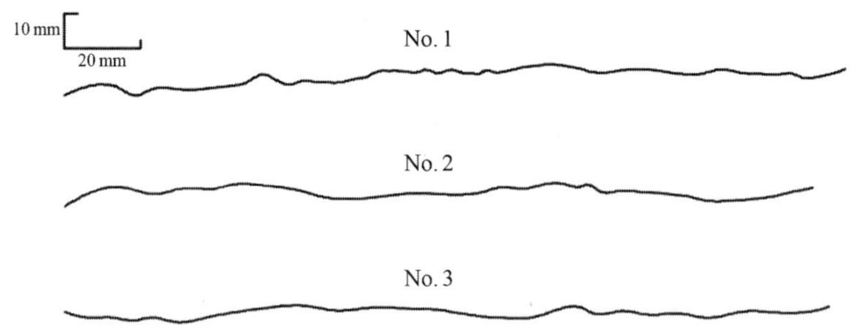

FIG. 3.88 Profiles of three joint sets.

slightly underestimated the measured convergence along the periphery of the backfill tunnel. At measuring points with equal distance of 2.0 m to the cavern sidewalls, numerical estimations were insignificantly higher than the measurements by the convergence meters.

Fig. 3.92 illustrates the rock bolting and opened joints around the tunnel. As predicted by the simulation, few joints were opened and the tunnel was stable. The high stability can be mainly attributed to the strong reinforcements and small-sized excavation. The prediction closely matched the magnitude of deformations from in situ investigations (Fig. 3.91).

3.7.4 Conclusions

Accurate representation of the mechanical behavior of rock joints improves the reliability of rock mass stability analysis. Previous studies indicated that the joint opening can reduce the normal stiffness and shear resistance of rock joints (Bandis et al., 1983; Zhao, 1997a, b; Oh and Kim, 2010; Li et al., 2014). The constitutive models in Li et al. (2016a), Li et al. (2016c), and Li et al. (2016d) provided realistic simulations for the mechanical behavior of laboratory-sized rock joints under different contact states. The scale dependence of joint roughness in Li et al. (2016b) enabled the modeling of the behavior of opened joints in the field. However, the performance of the proposed model for evaluating the stability of field-scale jointed rock masses has not been demonstrated. In this paper, the deformation of three large-scale rock structures was studied numerically, where the mechanical behavior of the joints was modeled by the proposed constitutive law. The capability of the proposed model was illustrated by the comparison of predictive and actual measurements.

In this study, parameter values for the joint model were acquired by inspecting the geometric properties of two-dimensional profiles of the joint samples collected from the field. The two-dimensional simplification has led to the discrepancies between in situ measurements and numerically determined estimations. During the process of numerical investigation, reproduction of the joints existing in the field was desired. However, this was hardly

TABLE 3.41 Input parameters for the shear behavior of rock joints.

Joint set	JRC	L_0 (mm)	Dip (degree)	Spacing (m)	Critical waviness			Critical unevenness			$\delta_{p,0}$ (mm)	H	ϕ_r (degree)
					i_0 (degree)	$\lambda_w^0/2$ (mm)	A_w^0 (mm)	α_0 (degree)	$\lambda_u^0/2$ (mm)	A_u^0 (mm)			
No.1	11.1	204	42	1.9	23.3	6.5	2.8	17.0	4.4	1.3	1.53	0.75	32
No.2	9.7	195	70	2.5	20.5	15.0	5.6	13.6	3.3	0.8	1.42	0.83	
No.3	6.8	200	55	1.9	16.4	11.2	3.3	21.8	2.5	1.0	1.28	0.87	

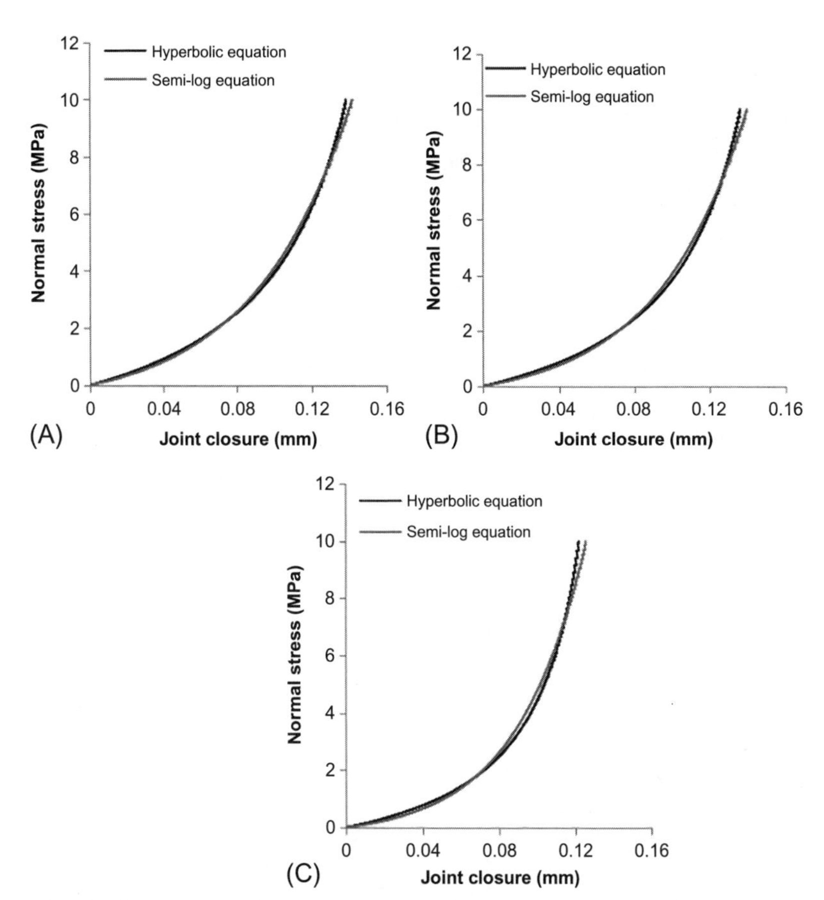

FIG. 3.89 Joint closure of three rock joints. (A) Joint No. 1. (B) Joint No. 2. (C) Joint No. 3.

TABLE 3.42 Input parameters for the hyperbolic and semilog equations.

	Hyperbolic equation		Semilog equation		
Joint	K_{ni} (MPa/mm)	V_m (mm)	$1/A'$ (mm)	B(MPa)	K
No.1	18.02	0.185	0.05291	0.7279	0.80
No.2	16.407	0.176	0.04802	0.573	0.77
No.3	14.155	0.148	0.03681	0.3336	0.73

possible for large-scale rock structure problems due to technical and economic difficulties. Despite these limitations, in the study of the two underground excavation cases, the numerical results generally matched well with the site measurements.

TABLE 3.43 Input parameters for the rock bolts.

Rock bolt			
Cable element	*Value*	*Grout*	*Value*
Cross-section area (mm^2)	380	Shear stiffness (MN/m/m)	16,000
Density (kg/m^3)	8000	Shear strength (MN/m)	0.20
Elastic modulus (MPa)	220,000		
Compressive yield force (MN)	0.1900		
Tensile yield force (MN)	0.1900		
Extensional failure strain	0.018		

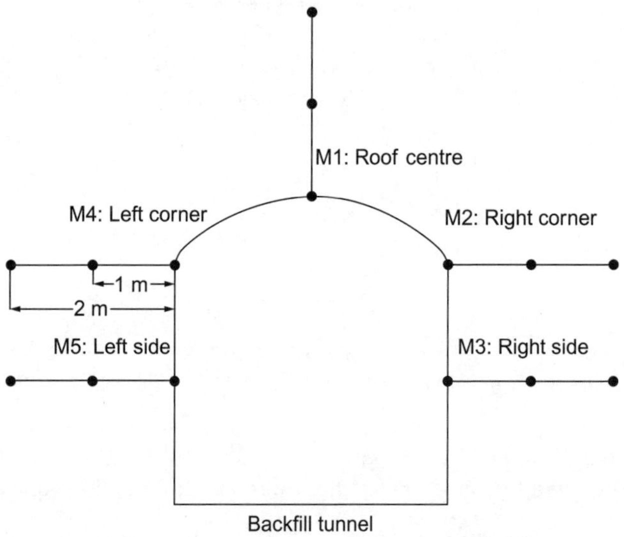

FIG. 3.90 Site measurement points along the tunnel.

This study examined the performance of a new joint constitutive model proposed for the stability analysis of large-scale jointed rock masses. The model features that joint stiffness under both normal and shear loads decreased progressively as the degree of opening increased. The stiffness-joint opening relationship is absent in most of the previous studies. Numerical analyses of three rock structures indicated that the opening of joints degraded the resistance of rock masses to loadings, thereby resulting in more deformations along rock slopes and around underground caverns. The results of the deformation

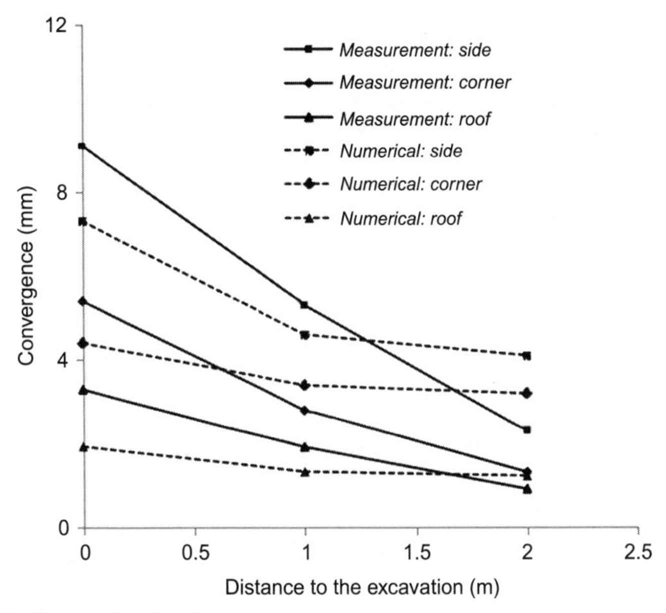

FIG. 3.91 Numerical and in situ measurements.

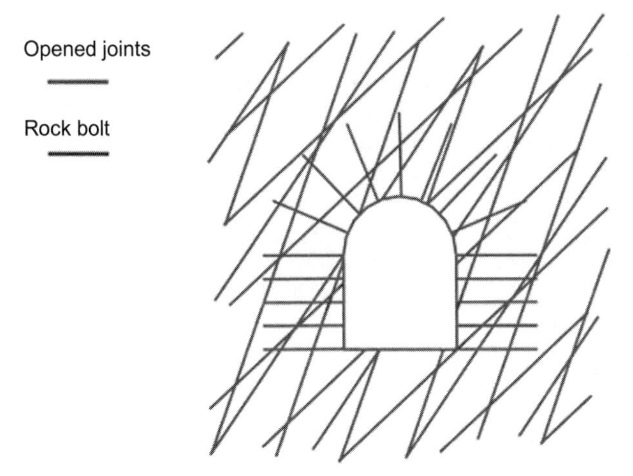

FIG. 3.92 Opened joints and rock bolting around the tunnel.

measurements around the two underground excavations have been compared to those predicted through numerical approaches. Good agreement between numerical estimations and field observations suggested that the proposed joint model could be employed to assess rock mass stability with adequate accuracy.

References

Alvarez, T.A., Cording, E.J., Mikhail, R.A., 1995. Hydromechanical behavior of rock joints: a re-interpretation of published experiments. In: Proceeding of the 35th US Symposium on Rock Mechanics, Lake Tahoe, Nevada, Balkema, pp. 665–671.

Asadi, M.S., Rasouli, V., Barla, G., 2012. A bonded particle model simulation of shear strength and asperity degradation for rough rock fractures. Rock Mech. Rock. Eng. 45 (5), 649–675.

Asadollahi, P., Tonon, F., 2010. Constitutive model for rock fractures: revisiting Barton's empirical model. Eng. Geol. 113 (1), 11–32.

Asadollahi, P., Invernizzi, M.C.A., Addotto, S., Tonon, F., 2010. Experimental validation of modified Barton's model for rock fractures. Rock Mech. Rock. Eng. 43 (5), 597–613.

Bahaaddini, M., 2014. Numerical Study of the Mechanical Behaviour of Rock Joints and Non-Persistent Jointed Rock Masses. PhD Thesis, University of New South Wales, Australia.

Bahaaddini, M., Sharrock, G., Hebblewhite, B., 2013. Numerical direct shear tests to model the shear behaviour of rock joints. Comput. Geotech. 51, 101–115.

Bahaaddini, M., Hagan, P., Mitra, R., Hebblewhite, B., 2014. Scale effect on the shear behaviour of rock joints based on a numerical study. Eng. Geol. 181, 212–223.

Bandis, S.C., 1980. Experimental Studies of Scale Effects on Shear Strengh, and Deformation of Rock Joints. PhD Thesis, The University of Leeds.

Bandis, S.C., Lumsden, A.C., Barton, N.R., 1981. Experimental studies of scale effects on the shear behaviour of rock joints. Int. J. Rock Mech. Min. Sci. Geomech. Abstr. 18 (1), 1–21.

Bandis, S.C., Lumsden, A.C., Barton, N.R., 1983. Fundamentals of rock joint deformation. Int. J. Rock Mech. Min. Sci. Geomech. Abstr. 20 (6), 249–268.

Bandis, S.C., Barton, N., Christianson, M., 1985. Application of a new numerical model of joint behavior to rock mechanics problems. In: Proceeding of International Symposium on Fundamentals of Rock Joints, Bjorkliden, Lulea, Sweden, pp. 345–356.

Barton, N.R., 1973. Review of a new shear-strength criterion for rock joints. Eng. Geol. 7 (4), 287–332.

Barton, N.R., 1976. The shear strength of rock and rock joints. Int. J. Rock Mech. Mining Sci. Geomech. Abstr. 13 (9), 255–279.

Barton, N.R., Choubey, V., 1977. The shear strength of rock joints in theory and practice. Rock Mech. 10, 1–54.

Barton, N.R., 1982a. Modelling Rock Joint Behavior From In Situ Block Tests: Implications for Nuclear Waste Repository Design. Terra Tek Inc, Salt Lake City, UT.

Barton, N.R., 1982b. Modeling Rock Joint Behavior From In Situ Block Tests: Implications for Nuclear Waste Repository Design. Office of Nuclear Waste Isolation, Columbus, Ohio, p. 96. ONWI-308.

Barton, N.R., 1993. Physical and discrete element models of excavation and failure in jointed rock. In: Pasamehmetoglu, et al., (Ed.), Assessment and Prevention of Failure Phenomena in Rock Engineering. Balkema, Rotterdam, pp. 35–46.

Barton, N.R. 2016. Personal communication.

Barton, N.R., Bandis, S., 1982. Effects of block size on the shear behavior of jointed rock. In: 23rd U.S. Symposium on Rock Mechanics. Berkeley, California, pp. 739–760.

Barton, N.R., Choubey, V., 1977. The shear strength of rock joints in theory and practice. Rock Mech. 10 (1-2), 1–54.

Barton, N.R., Lien, R., Lunde, J., 1974. Engineering classification of rock masses for the design of tunnel support. Rock Mech. 6 (4), 189–236.

Barwell, F., 1958. Wear of metals. Wear 1 (4), 317–332.

Beer, A.J., Stead, D., Coggan, J.S., 2002. Estimation of the joint roughness coefficient (JRC) by visual comparison. Rock Mech. Rock. Eng. 35 (1), 65–74.

Belem, T., Homand-Etienne, F., Souley, M., 2000. Quantitative parameters for rock joint surface roughness. Rock Mech. Rock. Eng. 33 (4), 217–242.

Belem, T., Souley, M., Homand, F., 2009. Method for quantification of wear of sheared joint walls based on surface morphology. Rock Mech. Rock. Eng. 42 (6), 883–910.

Bieniawski, Z., 1993. Classification of rock masses for engineering: the RMR system and future trends. In: Comprehensive Rock Engineering. vol. 3. Pergamon, Oxford, pp. 553–573.

Bjerrum, L., Jorstad, F., 1963. Discussion of paper by K. John. Proc. Am. Soc. Civil Eng. 89, 300–302.

Brown, S.R., Scholz, C.H., 1985. Broad bandwidth study of the topography of natural rock surfaces. J. Geophys. Res. Solid Earth 90 (B14), 12575–12582.

Bryan, S.A.T., 2009. Quantitative Characterization of Natural Rock Discontinuity Roughness In-Situ and in the Laboratory. M.S. Dissertation, University of Toronto.

Chern, S.G., Cheng, T.C., Chen, W.Y., 2012. Behavior of regular triangular joints under cyclic shearing. J. Mar. Sci. Technol. 20 (5), 508–513.

Christianson, M., Board, M., Rigby, D., 2006. UDEC simulation of triaxial testing of lithophysal tuff. In: Golden Rocks 2006, The 41st US Symposium on Rock Mechanics (USRMS). American Rock Mechanics Association, pp. 1–8.

Cook, N.G.W., 1992. Natural joints in rock: mechanical, hydraulic and seismic behaviour and properties under normal stress. Int. J. Rock Mech. Min. Sci. Geomech. Abstr. 29 (3), 198–223.

Cording, E.J., 1976. Shear strength of bedding and foliation surface. In: Proceedings of ASCE Specialty Conference, Boulder, Coloradopp. 172–192.

Cording, E.J., Mahar, J.W., 1974. The effect of natural geologic discontinuities on behavior of rock in tunnels. In: Proceedings of 1974 Rapid Excavation and Tunneling Conference, San Francisco, Canada, pp. 107–138.

Cording, E.J., Hendron, A.J., Deere, D.U., 1971. Rock engineering for underground caverns. In: Proceedings of ASCE Symposium on Underground Rock Champers, Phoenix, pp. 567–601.

Deere, D.U., 1968. Geological considerations. In: Stagg, K.G., Zienkiewicz, O.C. (Eds.), Rock Mechanics in Engineering Practice. Wiley, New York.

Deere, D.U., Hendron, A.J., Patton, F.D., Cording, E.J., 1967. Design of surface and near-surface construction in rock. In: Fairhurst, C. (Ed.), Failure and Breakage of Rock. AIME, New York, pp. 237–302.

Desai, C.S., Fishman, K.L., 1991. Plasticity-based constitutive model with associated testing for joints. Int. J. Rock Mech. Min. Sci. Geomech. Abstr. 28 (1), 15–26.

Dong, J.J., Pan, Y.W., 1996. A hierarchical model of rough rock joints based on micromechanics. Int. J. Rock Mech. Min. Sci. Geomech. Abstr. 33 (2), 111–123.

Evans, K.F., Kohl, T., Rybach, L., Hopkirk, R.J., 1992. The effects of fracture normal compliance on the long term circulation behavior of a hot dry rock reservoir: a parameter study using the new fully-coupled code 'Fracture'. Geotherm. Resourc. Counc. Trans. 16, 449–456.

Fardin, N., Stephansson, O., Jing, L., 2001. The scale dependence of rock joint surface roughness. Int. J. Rock Mech. Min. Sci. 38 (5), 659–669.

Flamand, R., Archambault, G., Gentier, S., Riss, J., Rouleau, A., 1994. An experimental study of the shear behavior of irregular joints based on angularities and progressive degradation of the surfaces. In: The 47th Canadian Geotechnical Conference, Halifax, Nova Scotia. Canadian Geotechnical Society, Halifax, Nova Scotia, pp. 253–262.

Gao, Y., Wong, L.N.Y., 2013. A modified correlation between roughness parameter Z2 and the JRC. Rock Mech. Rock Eng., 1–10.

Gentier, S., Riss, J., Archambault, G., Flamand, R., Hopkins, D., 2000. Influence of fracture geometry on shear behavior. Int. J. Rock Mech. Min. Sci. 37 (1), 161–174.

Ghazvinian, A.H., Azinfar, M.J., Vaneghi, R.G., 2012. Importance of tensile strength on the shear behavior of discontinuities. Rock Mech. Rock. Eng. 45 (3), 349–359.

Ghazvinian, E., Diederichs, M.S., Quey, R., 2014. 3D random Voronoi grain-based models for simulation of brittle rock damage and fabric-guided micro-fracturing. J. Rock Mech. Geotech. Eng. 6 (6), 506–521.

Goodman, R.E., 1976. Methods of Geological Engineering in Discontinuous Rock. West, New York, p. 472.

Grasselli, G., 2001. Shear Strength of Rock Joints Based on Quantified Surface Description. PhD Thesis, The Swiss Federal Institute of Technology (EPFL).

Grasselli, G., Egger, P., 2003. Constitutive law for the shear strength of rock joints based on three-dimensional surface parameters. Int. J. Rock Mech. Min. Sci. 40 (1), 25–40.

Grasselli, G., Wirth, J., Egger, P., 2002. Quantitative three-dimensional description of a rough surface and parameter evolution with shearing. Int. J. Rock Mech. Min. Sci. 39 (6), 789–800.

Greenwood, J.A., Tripp, J.H., 1970. The contact of two nominally flat rough surfaces. Proc. Inst. Mech. Eng. 185, 625–633.

Greenwood, J.A., Williamson, J.B.P., 1966. Contact of nominally flat surfaces. Proc. R. Soc. Lond. Ser. A. Math. Phys. Sci. 295, 300–319.

Haque, A., Indrarata, B., 2000. Shear Behaviour of Rock Joints. CRC Press, Rotterdam, Netherlands.

Hencher, S.R., Richards, L.R., 1982. The basic frictional resistance of sheeting joints in Hong Kong granite. Hong Kong Eng. 11, 21–25.

Hencher, S.R., Richards, L.R., 2014. Assessing the shear strength of rock discontinuities at laboratory and field scales. Rock Mech. Rock. Eng. 48 (3), 1–23.

Hencher, S.R., Toy, J.P., Lumsden, A.C., 1993a. Scale dependent shear strength of rock joints. In: da Cunha, A.P. (Ed.), Scale Effects in Rock Masses. Balkema, Rotterdam, pp. 233–240.

Hencher, S., Toy, J., Lumsden, A., 1993b. Scale-dependent shear-strength of rock joints. In: The 2nd International Workshop on Scale Effects in Rock Masses, Lisbon, Portugal. A.A. Balkema, Rotterdam, pp. 233–240.

Herdocia, A., 1985. Direct shear tests of artificial joints. In: Proceedings of International Symposium on Fundamental of Rock Joints, Björkliden, pp. 123–132.

Hong, E.S., Lee, J.S., Lee, I.M., 2008. Underestimation of roughness in rough rock joints. Int. J. Numer. Anal. Methods Geomech. 32 (11), 1385–1403.

Huang, X., Haimson, B.C., Plesha, M.E., Qiu, X., 1993. An investigation of the mechanics of rock joints—part I. Laboratory investigation. Int. J. Rock Mech. Min. Sci. 30 (3), 257–269.

Huang, T., Chang, C., Chao, C., 2002. Experimental and mathematical modeling for fracture of rock joint with regular asperities. Int. J. Rock Mech. Min. Sci. 69 (17), 1977–1996.

Hungr, O., Coates, D.F., 1978. Deformability of joints and its relation to rock foundation settlements. Can. Geotech. J. 15 (2), 239–249.

Hutson, R.W., 1987. Preparation of Duplicate Rock Joints and Their Changing Dilatancy Under Cyclic Shear. PhD Thesis. Northwestern University.

Hutson, R.W., Dowding, C.H., 1990. Joint asperity degradation during cyclic shear. Int. J. Rock Mech. Min. Sci. Geomech. Abstr. 27 (2), 109–119.

Indraratna, B., Mirzaghorbanali, A., Oliveira, D., Premadasa, W.N., 2012. Shear behaviour of rock joints under cyclic loading. In: The 11th Australia–New Zealand Conference on Geomechanics: Ground Engineering in a Changing World, Melbourne. Melbourne, Engineers Australia, pp. 1256–1261.

ISRM, 1978. Commission on standardization of laboratory and field tests of the international society for rock mechanics: "Suggested methods for the quantitative description of discontinuities" Int. J. Rock Mech. Min. Sci. Geomech. Abstr. 15 (6), 320–368.

Itasca, 2008. PFC (Particle Flow Code) Version 4.0. Minneapolis, USA.

Itasca, 2011. UDEC (Universal Distinct Element Code) Version 5.0. Minneapolis, USA.

Itasca, 2013. 3DEC (Three-dimensional Distinct Element Code) Version 5.0. Minneapolis, USA.

Itasca, 2014. UDEC (Universal Distinct Element Code) Version 6.0. Minneapolis, USA.

Itasca Consulting Group Inc, 2003. 3DEC 3-dimensional Distinct Element Code. Minneapolis, USA.

Jaeger, J.C., 1971. Friction of rocks and stability of rock slopes. Geotechnique 21 (2), 97–134.

Jang, H.S., Kang, S.S., Jang, B.A., 2014. Determination of joint roughness coefficients using roughness parameters. Rock Mech. Rock. Eng., 1–13.

Jiang, Y., Li, B., Tanabashi, Y., 2006. Estimating the relation between surface roughness and mechanical properties of rock joints. Int. J. Rock Mech. Min. Sci. 43 (6), 837–846.

Jing, L., Stephansson, O., Nordlund, E., 1993. Study of rock joints under cyclic loading conditions. Rock Mech. Rock. Eng. 26 (3), 215–232.

Johansson, F., 2009. Shear Strength of Unfilled and Rough Rock Joints in Sliding Stability Analyses of Concrete Dams. PhD Thesis, The Royal Institute of Technology.

Johansson, F., 2016. Influence of scale and matedness on the peak shear strength of fresh, unweathered rock joints. Int. J. Rock Mech. Min. Sci. 82, 36–47.

Karami, A., Stead, D., 2008. Asperity degradation and damage in the direct shear test: a hybrid FEM/DEM approach. Rock Mech. Rock. Eng. 41 (2), 229–266.

Kazerani, T., Zhao, J., 2010. Micromechanical parameters in bonded particle method for modelling of brittle material failure. Int. J. Numer. Anal. Methods Geomech. 34 (18), 1877–1895.

Kazerani, T., Yang, Z.Y., Zhao, J., 2012. A discrete element model for predicting shear strength and degradation of rock joint by using compressive and tensile test data. Rock Mech. Rock. Eng. 45 (5), 695–709.

Kim, D.H., Gratchev, I., Balasubramaniam, A., 2013. Determination of joint roughness coefficient (JRC) for slope stability analysis: a case study from the Gold Coast area, Australia. Landslides 10 (5), 657–664.

Kim, D.H., Gratchev, I., Balasubramaniam, A., 2015. Back analysis of a natural jointed rock slope based on the photogrammetry method. Landslides 12 (1), 147–154.

Krahn, J., Morgenstern, N., 1979. The ultimate frictional resistance of rock discontinuities. Int. J. Rock Mech. Min. Sci. 16 (2), 127–133.

Kulatilake, P., Um, J., 1999. Requirements for accurate quantification of self-affine roughness using the roughness-length method. Int. J. Rock Mech. Min. Sci. 36 (1), 5–18.

Kulatilake, P., Ucpirti, H., Wang, S., Radberg, G., Stephansson, O., 1992. Use of the distinct element method to perform stress analysis in rock with non-persistent joints and to study the effect of joint geometry parameters on the strength and deformability of rock masses. Rock Mech. Rock. Eng. 25 (4), 253–274.

Kulatilake, P.H.S.W., Shou, G., Huang, T.H., Morgan, R.M., 1995. New peak shear strength criteria for anisotropic rock joints. Int. J. Rock Mech. Min. Sci. Geomech. Abstr. 32 (7), 673–697.

Kulatilake, P., Um, J., Pan, G., 1998. Requirements for accurate quantification of self-affine roughness using the variogram method. Int. J. Solids Struct. 35 (31), 4167–4189.

Kulatilake, P.H.S.W., Balasingam, P., Park, J., Morgan, R.M., 2006. Natural rock joint roughness quantification through fractal techniques. Geotech. Geol. Eng. 24 (5), 1181–1202.

Kutter, H.K., Otto, F., 1990. Influence of parallel and cross joints on shear behaviour of rock discontinuities. In: Rock Joint: A Regional Conference of the International Society for Rock Mechanics, Loen, Norway, 4-6 June, pp. 243–250.

Kwašniewski, M., Wang, J., 1993. Application of laser profilometry and fractal analysis to measurement and characterization of morphological features of rock fracture surfaces. In: Géotechnique et Environnement, Colloque Franco-Polonais, pp. 163–176.

Ladanyi, B., Archambault, G., 1969. Simulation of shear behavior of a jointed rock mass. In: The 11th US Symposium on Rock Mechanics (USRMS), Berkeley, California, pp. 105–125.

Lanaro, F. 2001. Geometry, Mechanics and Transmissivity of Rock Fractures. PhD Thesis, The Royal Institute of Technology.

Leal-Gomes, M., 2003. Some new essential questions about scale effects on the mechanics of rock mass joints. In: The 10th ISRM Congress: Technology Roadmap for Rock Mechanics, Vila Real, pp. 721–728.

Leal-Gomes, M.J.A., Dinis-da-Gama, C.A.J.V., 2014. Limit equilibrium model for rock joints based on strain energies. Int. J. Geomech. 14 (4), 1–5.

Lee, S.W., 2003.Stability Around Underground Openings in Rock With Dilative, Non-Persistent and Multi-Scale Wavy Joints Using a Discrete Element Method. PhD Thesis. University of Illinois, Urbana-Champaign, Illinois.

Lee, Y.H., Carr, J.R., Barr, D.J., Haas, C.J., 1990. The fractal dimension as a measure of the roughness of rock discontinuity profiles. Int. J. Rock Mech. Min. Sci. Geomech. Abstr. 27 (6), 453–464.

Leichnitz, W., 1985. Mechanical properties of rock joints. Int. J. Rock Mech. Min. Sci. Geomech. Abstr. 22 (5), 313–321.

Leong, E., Randolph, M., 1992. A model for rock interfacial behaviour. Rock Mech. Rock. Eng. 25 (3), 187–206.

Li, B., 2009. Coupled Shear-Flow and Deformation Properties of Fractured Rock Mass. PhD Thesis, Nagasaki University.

Li, Y., Oh, J., Mitra, R., Hebblewhite, B., 2014. A study of a joint constitutive model incorporating excavation-induced opening effect. In: EUROCK 2014, ISRM European Regional Symposium, Vigo, Spain. CRC Press/Balkema, Vigo, Spain, pp. 625–629.

Li, Y., Oh, J., Mitra, R., Hebblewhite, B., 2015. A joint asperity degradation model based on the wear process (Accepted). In: The 49th US Rock Mechanics and Geomechanics Symposium, San Francisco, USA.

Li, Y., Oh, J., Mitra, R., Canbulat, I., 2016a. A fractal model for the shear behaviour of large-scale opened joints. Rock Mech. Rock. Eng. https://doi.org/10.1007/s00603-016-1088-8.

Li, Y., Oh, J., Mitra, R., Hebblewhite, B., 2016b. A constitutive model for a laboratory rock joint with multi-scale asperity degradation. Comput. Geotech. 72, 143–151.

Li, Y., Oh, J., Mitra, R., Hebblewhite, B., 2016c. Experimental studies on the mechanical behaviour of rock joints with various openings. Rock Mech. Rock. Eng. 49 (3), 837–853.

Li, Y., Oh, J., Mitra, R., Hebblewhite, B., 2016d. Modelling of the mechanical behaviour of an opened rock joint. In: EUROCK 2016, ISRM International Symposium, Cappadocia, Turkey, pp. 1–7.

Locher, H.G., Rieder, U.G., 1970. Shear tests on layered Jurassic Limestone. In: Proceedings of 2nd Congress International Society for Rock Mechanics, Belgrade. vol. 2. pp. 1–3.

Lopez, P., Riss, J., Archambault, G., 2003. An experimental method to link morphological properties of rock fracture surfaces to their mechanical properties. Int. J. Rock Mech. Min. Sci. 40, 947–954.

Maerz, N., Franklin, J., Bennett, C., 1990. Joint roughness measurement using shadow profilometry. Int. J. Rock Mech. Min. Sci. 27 (5), 329–343.

Maksimović, M., 1996. The shear strength components of a rough rock joint. Int. J. Rock Mech. Min. Sci. Geomech. Abstr. 33 (8), 769–783.

Malama, B., Kulatilake, P., 2003. Models for normal fracture deformation under compressive loading. Int. J. Rock Mech. Min. Sci. Geomech. Abstr. 40 (6), 893–901.
Malinverno, A., 1990. A simple method to estimate the fractal dimension of a self-affine series. Geophys. Res. Lett. 17 (11), 1953–1956.
Mandelbrot, B.B., 1967. How long is the coast of Britain. Science 156 (3775), 636–638.
Mandelbrot, B.B., 1985. Self-affine fractals and fractal dimension. Phys. Scr. 32 (4), 257.
Matsuki, K., Wang, E.Q., Sakaguchi, K., Okumura, K., 2001. Time-dependent closure of a fracture with rough surfaces under constant normal stress. Int. J. Rock Mech. Min. Sci. Geomech. Abstr. 38 (5), 607–619.
McMahon, B.K., 1985. Some practical considerations for the estimation of shear strength of joints and other discontinuities. In: Proceedings of International Symposium on Fundamentals of Rock joints, Bjorkliden, Sweden, pp. 475–485.
Misra, A., 2002. Effect of asperity damage on shear behavior of single fracture. Eng. Fract. Mech. 69, 1997–2014.
Muller, L., 1963. Die Standfestigkeit von Felsboschungen als spezifisch geomechanische Aufgabe. Rock Mech. Eng. Geol. 1 (1), 50–71.
Muralha, J., Cunha, P., 1990. About LNEC experience on scale effects in the mechanical behaviour of joints. In: The 1st International Workshop on Scale Effects in Rock Masses, Luen, Norway. A.A. Balkema, Rotterdam, pp. 131–148.
Nassir, M., Settari, A., Wan, R., 2010. Joint stiffness and deformation behaviour of discontinuous rock. J. Can. Pet. Technol. 49 (9), 78–86.
Nguyen, V.M., 2013. Static and Dynamic Behaviour of Joints in Schistose Rock. Freiberg University of Mining and Technology.
Nippon Expressway Company, 1991. Design Guidelines: Bridge Constructions. vol. 2.
Odling, N., 1994. Natural fracture profiles, fractal dimension and joint roughness coefficients. Rock Mech. Rock. Eng. 27 (3), 135–153.
Oh, J, 2005. 3-D Numerical Modeling of Excavation in Rock With Dilatant Joints. PhD Thesis, Department of Civil and Environmental Engineering, University of Illinois at Urbana-Champaign, Urbana, IL 61801, USA.
Oh, J., Kim, G.W., 2010. Effect of opening on the shear behavior of a rock joint. Bull. Eng. Geol. Environ. 69 (3), 389–395.
Oh, J., Cording, E.J., Moon, T., 2015. A joint shear model incorporating small-scale and large-scale irregularities. Int. J. Rock Mech. Min. Sci. 76, 78–87.
Ohnishi, Y., Yoshinaka, R., 1995. Laboratory investigation of scale effect in mechanical behavior of rock joint. In: The 2nd International Conference on the Mechanics of Jointed and Faulted Rock–MJFR-2, Vienna, Austria. A.A. Balkema, Rotterdam, pp. 465–470.
Park, J.W., Song, J.J., 2009. Numerical simulation of a direct shear test on a rock joint using a bonded-particle model. Int. J. Rock Mech. Min. Sci. 46 (8), 1315–1328.
Park, J.W., Song, J.J., 2013. Numerical method for the determination of contact areas of a rock joint under normal and shear loads. Int. J. Rock Mech. Min. Sci. 58, 8–22.
Park, J.W., Lee, Y.K., Song, J.J., Choi, B.H., 2013. A constitutive model for shear behavior of rock joints based on three-dimensional quantification of joint roughness. Rock Mech. Rock. Eng. 46 (6), 1513–1537.
Patton, F.D., 1966a. Multiple Modes of Shear Failure in Rock and Related Material. PhD Thesis. University of Illinois, Urbana-Champaign, Illinois.
Patton, F.D., 1966b. Multiple modes of shear failure in rock. In: The 1st Congress of International Society for Rock Mechanics, Lisbon, Portugal, pp. 509–513.

Plesha, M.E., 1987. Constitutive models for rock discontinuities with dilatancy and surface degradation. Int. J. Numer. Anal. Methods Geomech. 11 (4), 345–362.

Pratt, H.R., Black, A.D., Brace, W.F., 1974. Friction and deformation of jointed quartz diorite. In: Advances in Rock Mechanics: The 3rd Congress of the International Society for Rock Mechanics, Denver, pp. 306–310.

Qiu, X., Plesha, M.E., Huang, X., Haimson, B.C., 1993. An investigation of the mechanics of rock joints—part II. Analytical investigation. Int. J. Rock Mech. Min. Sci. Geomech. Abstr. 30 (3), 271–287.

Queener, C., Smith, T., Mitchell, W., 1965. Transient wear of machine parts. Wear 8 (5), 391–400.

Re, F., Scavia, C., 1999. Determination of contact areas in rock joints by X-ray computer tomography. Int. J. Rock Mech. Min. Sci. 36 (7), 883–890.

Reeves, M.J., 1985. Rock surface roughness and frictional strength. Int. J. Rock Mech. Min. Sci. Geomech. Abstr. 22 (6), 429–442.

Richards, L.R., Cowland, J.W., 1982. The effect of surface roughness on the field shear strength of sheeting joints in Hong Kong granite. Hong Kong Eng. 11, 39–43.

Roosta, R.M., Sadaghiani, M.H., Pak, A., Saleh, Y., 2006. Rock joint modeling using a visco-plastic multilaminate model at constant normal load condition. Geotech. Geol. Eng., 24 (5), 1449–1468.

Saeb, S., Amadei, B., 1992. Modelling rock joints under shear and normal loading. Int. J. Rock Mech. Min. Sci. Geomech. Abstr. 29 (3), 267–278.

Sakellariou, M., Nakos, B., Mitsakaki, C., 1991. On the fractal character of rock surfaces. Int. J. Rock Mech. Min. Sci. 28 (6), 527–533.

Schneider, H.J., 1976. The friction and deformation behaviour of rock joints. Rock Mech. 8 (3), 169–184.

Seidel, J.P., Haberfield, C.M., 1995a. The application of energy principles to the determination of the sliding resistance of rock joints. Rock Mech. Rock. Eng. 28 (4), 211–226.

Seidel, J.P., Haberfield, C.M., 1995b. Towards an understanding of joint roughness. Rock Mech. Rock. Eng. 28 (2), 69–92.

Sun, Z.Q., Gerrard, C., Stephansson, O., 1985. Rock joint compliance tests for compression and shear loads. Int. J. Rock Mech. Min. Sci. Geomech. Abstr. 22 (4), 197–213.

Swan, G., 1983. Determination of stiffness and other joint properties from roughness measurements. Rock Mech. Rock. Eng. 16 (1), 19–38.

Swan, G., Zongui, S., 1985. Prediction of shear behaviour of joints using profiles. Rock Mech. Rock. Eng. 18, 183–212.

Tang, Z.C., Liu, Q.S., Xia, C.C., Song, Y.L., Huang, J.H., Wang, C.B., 2013. Mechanical model for predicting closure behavior of rock joints under normal stress. Rock Mech. Rock. Eng., 1–12.

Tang, Z.C., Huang, R.Q., Liu, Q.S., Wong, L.N.Y., 2016. Effect of contact state on the shear behavior of artificial rock joint. Bull. Eng. Geol. Environ., 1–9.

Tatone, B.S., Grasselli, G., 2013. An investigation of discontinuity roughness scale dependency using high-resolution surface measurements. Rock Mech. Rock. Eng. 46 (4), 657–681.

Tse, R., Cruden, D.M., 1979. Estimating joint roughness coefficients. Int. J. Rock Mech. Min. Sci. Geomech. Abstr. 16 (5), 303–307.

Turk, N., Greig, M., Dearman, W., Amin, F., 1987. Characterization of rock joint surfaces by fractal dimension. In: The 28thUS Symposium on Rock Mechanics (USRMS), 1987. American Rock Mechanics Association.

Um, J., 1997. Accurate Quantification of Rock Joint Roughness and Development of a New Peak Shear Strength Criterion for Joints. PhD Thesis, The University of Arizona.

Voss, R.F., 1988. Fractals in Nature: From Characterization to Simulation. Springer, New York.

Wei, M., Liu, H., Li, L., Wang, E., 2013. A fractal-based model for fracture deformation under shearing and compression. Rock Mech. Rock. Eng. 46 (6), 1539–1549.

Wibowo, J.L., 1994. Effect of Boundary Conditions and Surface Damage on the Shear Behaviour of Rock Joints: Tests and Analytical Predictions. PhD Thesis, The University of Colorado.

Xia, C.C., Yue, Z.Q., Tham, L.G., Lee, C.F., Sun, Z.Q., 2003. Quantifying topography and closure deformation of rock joints. Int. J. Rock Mech. Min. Sci. 40 (2), 197–220.

Xia, C.C., Tang, Z.C., Xiao, W.M., Song, Y.L., 2013. New peak shear strength criterion of rock joints based on quantified surface description. Rock Mech. Rock. Eng. 47 (2), 1–14.

Xie, H.P., Pariseau, W.G., Myer, L.R., Tsang, C., 1992. Fractal estimation of joint roughness coefficient. In: Cook, N.G.W., Goodman, R.E. (Eds.), International Conference on Fractured and Jointed Rock Masses, Lake Tahoe, California, USA. A.A. Balkema, Rotterdam, Lake Tahoe, California, USA, pp. 125–131.

Xie, H.P., Wang, J.A., Xie, W.H., 1997. Fractal effects of surface roughness on the mechanical behavior of rock joints. Chaos, Solitons Fractals 8 (2), 221–252.

Yang, Z., Chen, G., 1999. Application of the self-affinity concept to the scale effect of joint roughness. Rock Mech. Rock. Eng. 32 (3), 221–229.

Yang, Z.Y., Chiang, D.Y., 2000. An experimental study on the progressive shear behavior of rock joints with tooth-shaped asperities. Int. J. Rock Mech. Min. Sci. 37 (8), 1247–1259.

Yang, Z., Di, C., Yen, K., 2001a. The effect of asperity order on the roughness of rock joints. Int. J. Rock Mech. Min. Sci. 38 (5), 745–752.

Yang, Z.Y., Lo, S.C., Di, C.C., 2001b. Reassessing the joint roughness coefficient (JRC) estimation using Z2. Rock Mech. Rock. Eng. 34 (3), 243–251.

Yang, Z.Y., Chiang, D.Y., 2003. An experimental study on the progressive shear behavior of rock joints with tooth-shaped asperities. Int. J. Rock Mech. Min. Sci. 37, 1247–1259.

Zangerl, C., Evans, K., Eberhardt, E., Loew, S., 2008. Normal stiffness of fractures in granitic rock: a compilation of laboratory and in-situ experiments. Int. J. Rock Mech. Min. Sci. 45 (8), 1500–1507.

Zhang, G., Karakus, M., Tang, H., Ge, Y., Zhang, L., 2014. A new method estimating the 2D joint roughness coefficient for discontinuity surfaces in rock masses. Int. J. Rock Mech. Min. Sci. 72, 191–198.

Zhao, J., 1997a. Joint surface matching and shear strength part A: joint matching coefficient (JMC). Int. J. Rock Mech. Min. Sci. 34 (2), 173–178.

Zhao, J., 1997b. Joint surface matching and shear strength part B: JRC-JMC shear strength criterion. Int. J. Rock Mech. Min. Sci. 34 (2), 179–185.

Zhao, J., Brown, E.T., 1992. Hydro-thermo-mechanical properties of joints in the Carnmenellis granite. Q. J. Eng. Geol. Hydrogeol. 25 (4), 279–290.

Further reading

Itasca Consulting Group Inc, 2011. UDEC (Universal Distinct Element Code). Minneapolis, USA.

Pease, K.A., Kulhawy, F.H., 1984. Load Transfer Mechanisms in Rock Sockets and Anchors. Report EL-3777, Electric Power Research Institute, Palo Alto.

Rengers, N., 1970. Influence of surface roughness on the friction properties of rock planes. International Society of Rock Mechanics, Proceedings 1, 229–234.

Chapter 4

Microseismic monitoring and application

Shuren Wang[a] and Xiangxin Liu[b]

[a]*School of Civil Engineering, Henan Polytechnic University, Jiaozuo, China,* [b]*School of Mining Engineering, North China University of Science and Technology, Tangshan, China*

Chapter outline

4.1 Acoustic emission of rock plate instability

4.1.1 Introduction

Because rock failure is always accompanied by the acoustic emission (AE) phenomenon, it can facilitate the understanding of the mechanical properties and failure rules of rocks to research the rock failure mechanism and evolution characteristics of acoustic emission.

Currently, the study methods of AE mainly contain the means of laboratory tests, numerical simulation, or both. In the experimental research field, both domestic and foreign scholars have conducted plenty of studies. For example, Li et al. (2004) investigated the rock mechanics and AE characteristics in the whole failure process for rock samples and pointed out that the test results could be used to explain the AE phenomenon in an engineering project. Fu (2005) compared the similarities and differences of these AE performances for different rocks in the uniaxial compression tests. Based on the locations of AE events, Zhao et al. (2006) found that the AE distribution could reflect the shape and development of the cracks in rocks, which is meaningful to study the deformation and failure laws of rocks. Liu et al. (2012a, b) observed the changes of the AE number in a loading test, discovering that the accelerating release of energy was meaningful to predict the instability or failure of the rock mass. Zhang (2013) inferred the AE magnitude by the ratio of the whole released energy and AE number in a particular time, discovering that the higher the ratio, the fewer AE events had produced the energy. Jia et al. (2013) monitored the process of the uniaxial compression test for limestone by acoustic emission techniques, and elaborated the crack development law in the rock sample after scanning the destroyed samples by a computerized tomography (CT) machine. Huang and Liu (2013) analyzed the stress-strain curves and AE characteristics for coal-bearing rock samples under different stress paths. Zhao et al. (2015) conducted the research on AE characteristics of phyllite specimens under uniaxial compression tests. Karakus and Perez (2014) analyzed the AE signal in the rock drilling process, which would provide the guidance to improve the drilling effect. Aker et al. (2014) analyzed the differences of AE between the shear failure and tensile failure in the triaxial compression tests of sandstone and predicted the failure mechanism using the proportion of the isotropic and anisotropic moment tensors. Frash et al. (2015) conducted the granite tests using the AE technique and simulated the evolution of the crack in the geothermal development, providing useful information for the engineering application. The numerical simulation can show some information that the laboratory tests cannot provide. For example, some scholars simulated AE by using the particle flow code (PFC) and the rock failure process analysis system (RFPA), which provided great help to understand AE phenomenon (Liu et al., 2012a, b; Ren et al., 2012). Additionally, based on the correspondence between AE and cell rupture in the FLAC/FLAC3D code, Wang (2008) and Han and Zhang (2014) respectively simulated the AE in both laboratory and engineering scales, and they all obtained good results. And other related works (Poerbandono and Suprijo, 2013; Garcia-Barruetabena et al., 2014; Miljenko et al., 2015).

Based on the laboratory test, further research following the above-mentioned results will be conducted by using the FLAC3D technique under the criterion that the rupture of a cell or several adjacent cells is regarded as an AE event. The temporal and spatial distribution characteristics of the AE events with large magnitudes will be analyzed and the relationships between AE and natural earthquakes will be discussed.

4.1.2 Materials and methods

4.1.2.1 Samples of rock plates

The rock plate samples in the tests were Hawkesbury sandstones, which were obtained from the Gosford Quarry in Sydney, Australia. The quartz sandstones were formed in the marine sedimentary basin of the Mid-Triassic and located on the top of the coal-bearing strata, which contained a small quantity of feldspars, siderite, and clay minerals. According to the definition of the thick plate in elastic mechanics, the specimen sizes of the thick plate were designed as 190 mm × 75 mm × 24 mm (length, width, and thickness).

4.1.2.2 Equipment and AE acquisition system

The MTS-851 rock mechanics testing machine was selected as the loading equipment, the load was controlled by vertical displacement, and the loading rate was set at 1×10^{-2} mm/s. The vertical force and displacement in the process of the test were automatically recorded in real time by the data acquisition system.

As shown in Fig. 4.1, the concentrated loading tests were designed to mainly consist of three parts. The top was a point loading for the concentrated loading. The middle was a loading framework that included four bolts with nuts connecting the steel plates on both sides, and the lateral pressure cell was placed

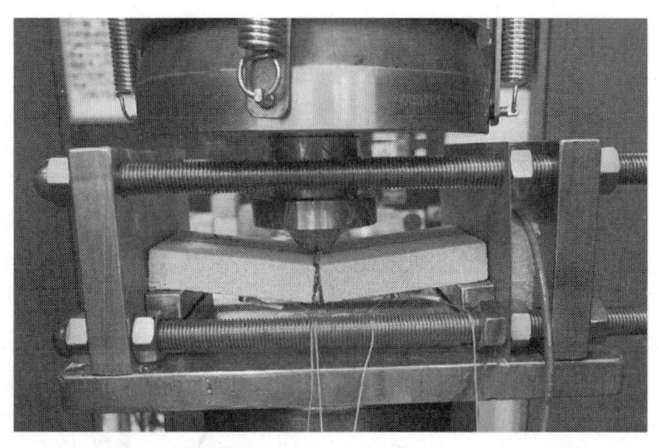

FIG. 4.1 The concentrated loading test for the rock plate.

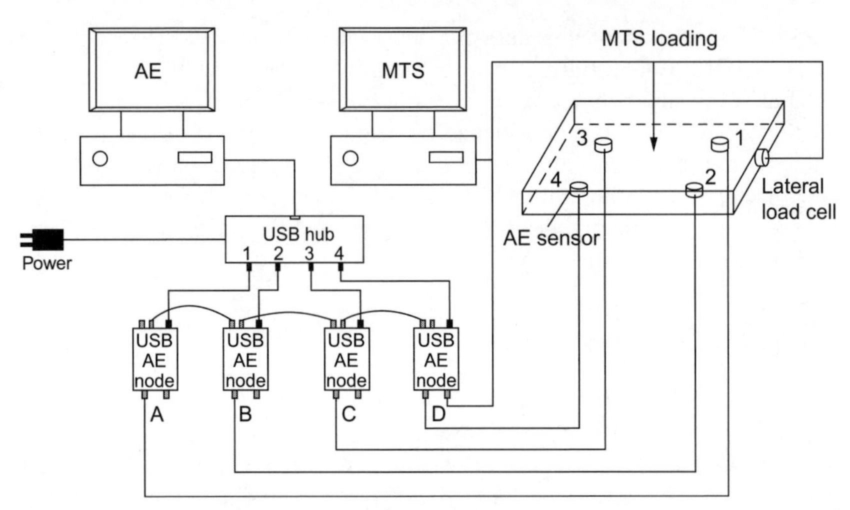

FIG. 4.2 MTS and AE monitoring system diagram.

between the deformable steel plate and the thick steel plate so as to monitor the horizontal force. The capacity of the lateral pressure cell LPX was 1000 kg. The bottom was a rectangle steel foundation, and the rotatable hinge support was set on both sides of the loading framework to maintain the connections with the steel plates.

To monitor the cracks initiating and to identify the failure location of the rock plate, the USB AE nodes were used in the test. The USB AE node is a single-channel AE digital signal processor with a full AE hit and real-time features. In the test, there were four USB AE nodes connected to a USB hub for multichannel operation (Fig. 4.2). All these AE nodes were made by the MISTRAS Group (USA).

4.1.2.3 Numerical simulation scheme

AE is due to the internal microfracture by tension, shearing, and compression stress in the rock plate, and this process is accompanied by the release of elastic waves. The rupture of a cell in FLAC^{3D} is also accompanied by the release of elastic energy, so it can be used to simulate an AE event (Han and Zhang, 2014). In fact, if several adjacent cells rupture in the rock plate in a calculating cycle (step), it should also be regarded as an AE event, the advantage of which is that we can record the number of ruptured cells reflecting the magnitude of the energy in an AE event. Then, we can analyze the evolution characteristics of the temporal and spatial of AE events, which may facilitate the understanding of the mechanical properties and failure laws of rocks.

We assumed that if an AE event during the simulation test corresponded to only one ruptured cell, the center of this cell was defined as the AE event location, or if an AE event corresponded to several adjacent cells, the center of the

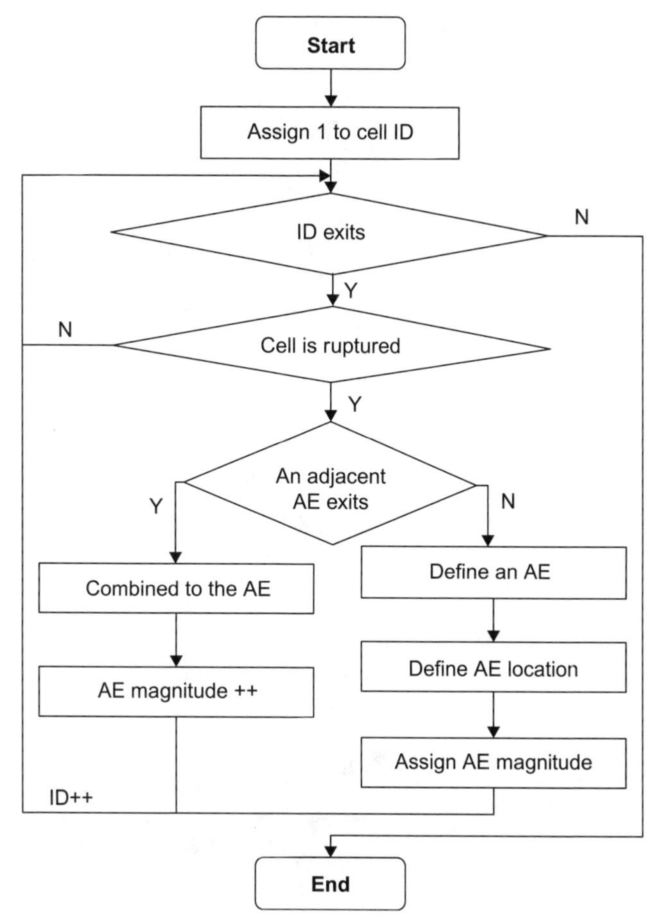

FIG. 4.3 The flow chart of recording AE events.

cells near the nuclear of the block was defined as the AE event location. Then, a recording function was written by the FISH language embedded in the FLAC^{3D} code to record the information of AE events such as the number of ruptured cells as well as the locations and magnitudes of the AE events. The realization process of this function was shown in Fig. 4.3. The state of each cell was defined as ruptured or not and all the cells were judged during each calculating step. In addition, the cells were regarded as "adjacent cells" if the distance between two cells was less than 3.16 mm. Finally, the large magnitude AE events, namely the events corresponding to two or more ruptured cells, were abstracted for being analyzed. So these AE characteristics in the test were analyzed through observing the temporal and spatial distribution of the large magnitude AE events.

4.1.2.4 Computational model and parameters

The computational model was the same size of the sandstone sample as shown in Fig. 4.4, which combined 68,992 cells totally and each cell was a cube element measuring 1.7 mm on each side. To simulate the defects in the rock plate, about 14,000 defective cells were generated by a FISH function. These defects were randomly distributed in the main cells to form a block group. In addition, a new block group containing these weak cells was defined to reflect the tension strength decreasing in the middle of the rock plate bottom after the initial crack being developed in the test. Therefore, all the cells belonged to three groups: the main cells, the defective cells, and the weak cells (Fig. 4.4).

The Mohr-Coulomb strength criterion was applied and the physical and mechanical parameters of the model are shown in Table 4.1.

4.1.2.5 Loading and boundary conditions

As shown in Fig. 4.5, the rock plate was hinged to both ends with the fixed bearing in the vertical direction and the spring bearing with a stiffness 6.0×10^4 N/m in the horizontal direction. The concentrated loading was applied in a circular zone with a diameter of 10 mm in the center of the upper surface of the model. To avoid a drastic disturbance of the calculating system, the compressive stress was increased linearly from 0 to a final stress of 12.0 MPa.

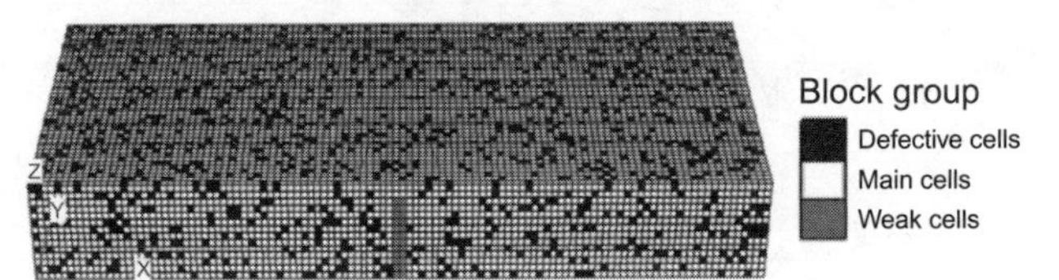

FIG. 4.4 The computational model and its grids.

TABLE 4.1 Physical and mechanical parameters of the model

Name	Density (kg/m^3)	Bulk modulus (GPa)	Shear modulus (GPa)	Cohesion (MPa)	Friction angle (degree)	Tension (MPa)
Main cells	2650	15	11	2.8	45	0.60
Defective cells	2650	15	11	2.8	45	0.55
Weak cells	2650	15	11	2.8	45	0.14

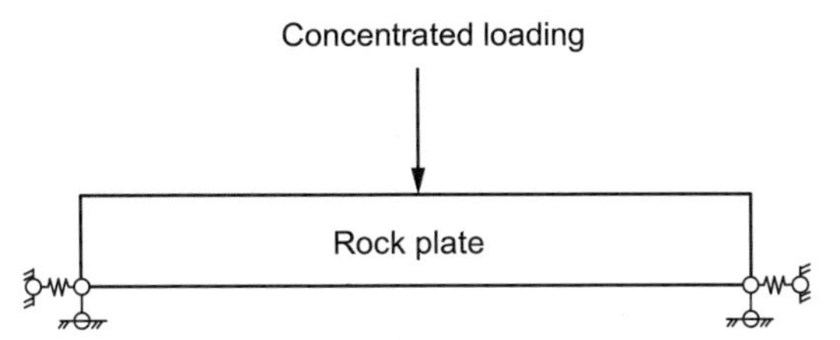

FIG. 4.5 Loading and boundary conditions.

4.1.3 Results analysis

4.1.3.1 AE in the failure process of the rock plate

As shown in AE location maps (Fig. 4.6), the results showed the obvious distribution differences between the initial cracks and the ultimate cracks of the rock plate in the test. Also, the AE hits-time curve could be divided into four stages in the process of bearing load to the instability of the rock plate under the concentrated loading condition (Fig. 4.7).

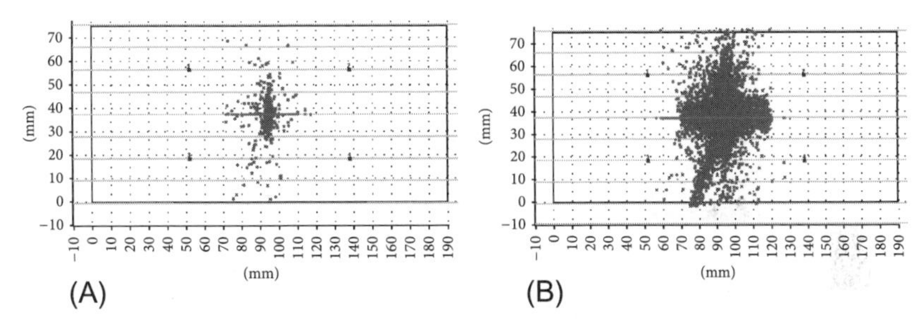

FIG. 4.6 AE locations on the rock plate. (A) Initial stage and (B) ultimate stage.

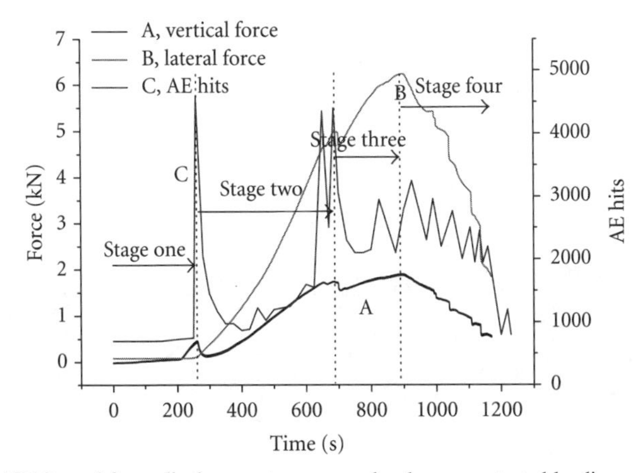

FIG. 4.7 AE hits and force-displacement curves under the concentrated loading.

4.1.3.2 AE characteristics in numerical simulation test

The results were saved in each particular time interval to observe the AE states of cells, as shown in Fig. 4.8. The recording function was applied to record the AE amount (Fig. 4.9) and AE locations (Fig. 4.10). During the loading process, the stress in the rock plate transmitted in every step and the cell ruptured when the stress in a particular cell reached the shearing or the tension strength. The ruptured cells began in the center of the rock plate bottom, then extended along the center line of the rock plate, resulting in a fracture in the middle, and finally

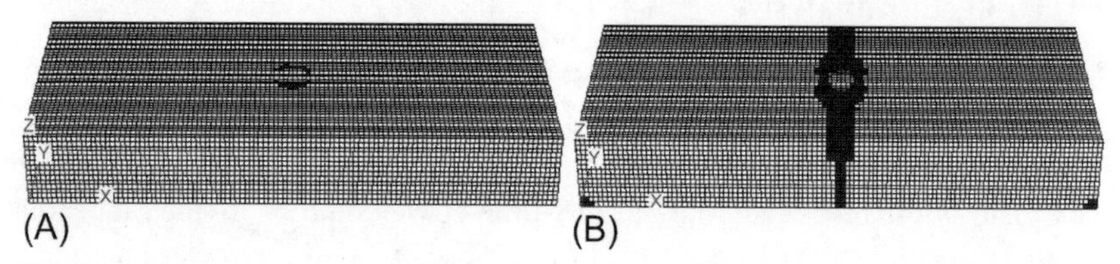

FIG. 4.8 AE locations on the top of the rock plate. (A) Initial stage (3000 steps) and (B) ultimate stage (12,000 steps).

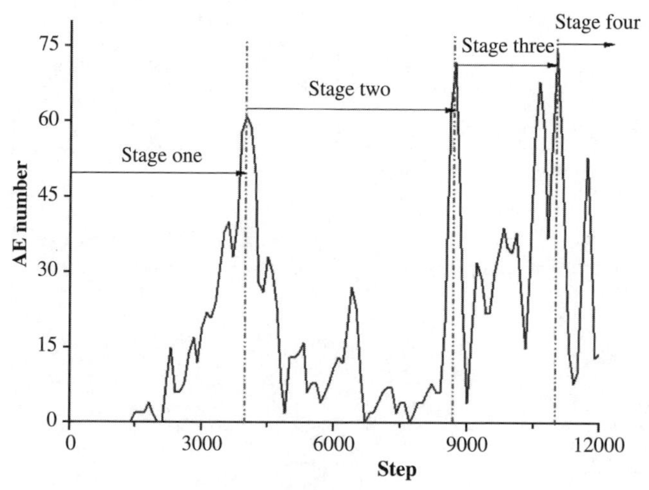

FIG. 4.9 AE hits-step curve in the numerical simulation.

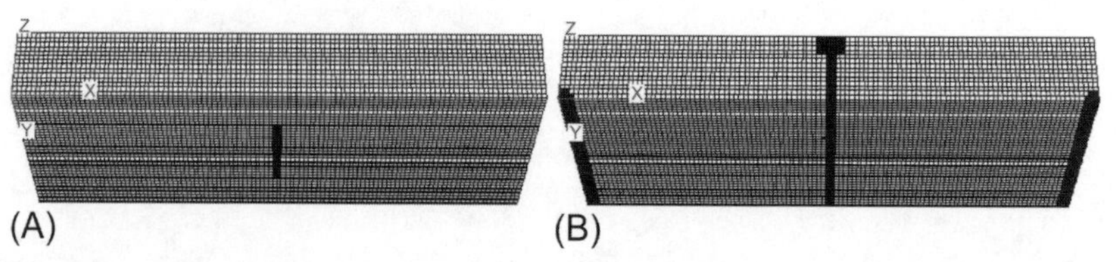

FIG. 4.10 AE locations on the bottom of the rock plate. (A) Initial stage (3000 steps) and (B) ultimate stage (12,000 steps).

formed a rock-arch structure. Meanwhile, the AE events mainly gathered on the hinged lines of the rock-arch structure in the rock plate.

From the above-mentioned results, we can see that the numerical simulation showed nearly the same AE characteristics with those of the laboratory test. As shown in Fig. 4.9, the AE hits-step curve can be divided into four stages:

Stage 1 (stress adjusting stage): The stress has not yet reached the tensile strength of the sandstone sample in the early time, so a large number of AE events were avoided. And then an initial crack paralleling with the short sides formed in the center of the rock plate boring the largest tensile stress. The stresses transmitted in FLAC^{3D} by the neighboring nodes in the calculating process, redistributing in each step. For the stress in the rock plate increased gradually under the concentrated loading, the sample generated the displacements on its bottom of the both ends, which induced the horizontal constraining force. Few AE events were recorded at first and then they increased sharply to the first peak. The AE events induced by the tensile rupture mainly gathered near the first crack.

Stage 2 (the brittle fracture stage): In this stage, the extended crack resulted in the rock plate fracturing into two halves and a hinged rock-arch structure being formed. Both the laboratory test and the numerical simulation showed that the number of AE events decreased rapidly and stabilized in a low level. This indicated that the rock plate produced a brittle fracture induced by the extent of the first crack. Both the laboratory test and the simulation showed the same characteristics in the spatial distribution, namely AE events spread from the center to the ends with the extent of the initial crack (Fig. 4.10).

Stage 3 (rock-arch structure bearing loading): The hinged rock-arch structure boring the loading and the horizontal force continued to increase with the loading increasing. The AE characteristics in the laboratory test and the numerical simulation were nearly the same, namely the AE number performed a sharp increase to reach the second peak and then it reduced and stabilized. The AE spatial distribution focused on the hinge lines of the rock-arch structure, namely the center line and the ends of the rock plate (Figs. 4.8 and 4.10).

Stage 4 (instability and failure of the rock-arch structure): Once the concentrated load exceeded the bearing capacity of the rock-arch structure, it would lead to the instability and failure of the rock plate. The AE events reduced unsteadily and distributed in the middle and ends of the rock plate, both in the laboratory test and the numerical simulation (Figs. 4.8 and 4.10).

4.1.4 Discussion of the magnitudes of AE events

As we can see, all these 156 large magnitude events revealed the similar spatial distribution to the AE events, and were located on the main crack of the sample, namely the middle hinge of the rock-arch structure (Fig. 4.10).

There are 1145 AE events and the number of the ruptured cells corresponded to one AE event from 1 to 5 in the numerical simulation, which represented the different energy scale released in one AE event. In the fracture and instability

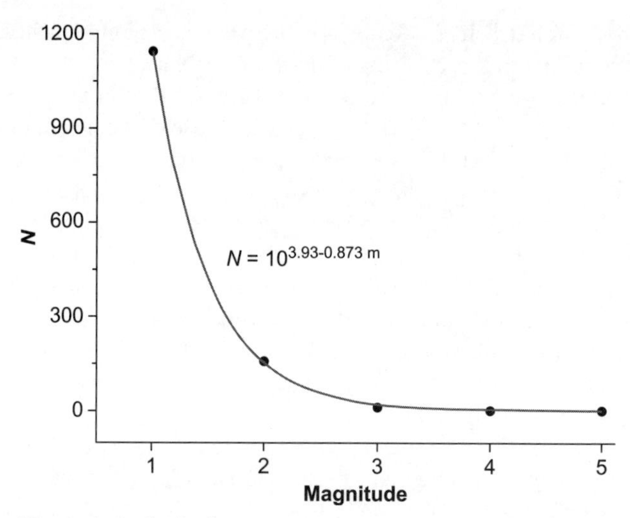

FIG. 4.11 AE magnitude distribution curve.

process of the rock plate, the number of AE events reduced with the AE magnitude increasing. As shown in Fig. 4.11, we can obtain the negative exponential formula by fitting the curve by using Origin software, wherein N indicates the number of AE with magnitudes greater than or equal to m while m indicates the AE magnitude.

In seismology, the earthquake with a larger magnitude is relatively rare while some small magnitude earthquakes occur frequently. The relationship between the magnitude and frequency is generally described by a probability distribution, which is derived based on the statistics of the observed seismic activity. The most widely used relationship is the following one (Yoder et al., 2012; Yucemen and Akkaya, 2012).

$$N = 10^{a-bm} \tag{4.1}$$

where N indicates the number of earthquakes with magnitudes greater than or equal to m; m indicates the magnitude; and a and b are regional parameters. For example, El-Isa and Eaton (2014) compiled the seismicity data for all earthquakes with magnitudes $m \geq 4.5$ that occurred globally from January 1990 to December 2012. The fitting result is presented in Fig. 4.12.

The AE phenomena and tectonic earthquakes both release energy processes induced by the slippage or breakage of the rock sample or stratum, and have a substantial connection in the failure mechanism. In addition, the similarity of distribution in frequency corroborates this law indirectly.

4.1.5 Conclusions

Based on the laboratory test and simulation results, the process of the fracture and instability for the rock plate could be divided into four stages: the stress

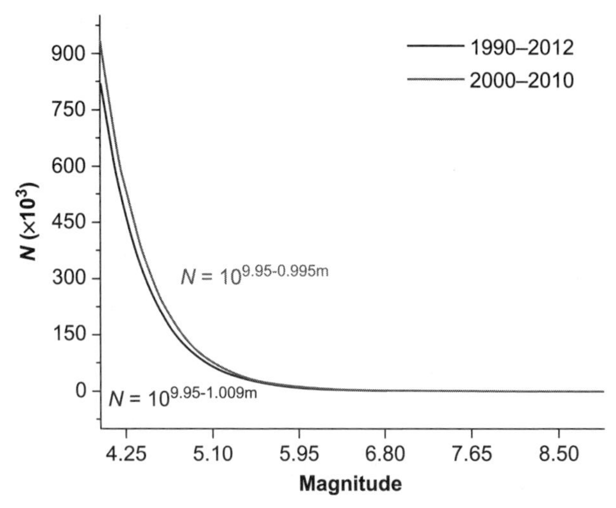

FIG. 4.12 Magnitude frequency distribution of the seismicity.

adjusting stage, the brittle fracture stage, the rock-arch bearing load stage, and the rock-arch instability stage. The acoustic emission exhibited the different characteristics in each stage.

By a self-programming recording function in $FLAC^{3D}$, these AE events with large magnitudes were abstracted. We found similar temporal and spatial characteristics between the large magnitude AE events and the whole AE events, which reflected the feature of the instability process of the rock plate.

The AE distribution with the magnitude showed that the AE events reduced with the AE magnitudes followed a negative exponential function. This distribution was similar to the tectonic earthquakes, which reflected the intrinsic link between the AE events and the AE magnitudes.

4.2 Prediction method of rockburst

4.2.1 Introduction

A rockburst is one of the most serious geological disasters in worldwide mining engineering (Jiang et al., 2014a, b). However, it is difficult to be predicted timely because it often occurs without obvious precursors. In recent years, microseismic monitoring has gradually become an important early warning method for preventing rockbursts. Compared with other monitoring methods, it can provide detailed information on potential risks in both the space and time domains.

Microseismic monitoring has been tested in forecasting rockbursts in various mines. For example, Mansurov (2001) proposed a prediction technique by analyzing microseismic data. This technique showed high efficiency when it was applied retrospectively for the induced seismicity database in a Bauxite coal mine. Driad-Lebeau et al. (2005) delineated other high-risk areas by analyzing the cause and process of a strong rockburst that occurred in the Frieda 5 coal mine in eastern France. Lesniak and Isakow (2009) analyzed the spatial and temporal characteristics of two-month microseismic events in the Zabrze-Bielszowice coal mine in Poland, and pointed out the relations between the occurrence of a rockburst while establishing a hazard assessment function. Xia et al. (2010) selected five indicators to predict rockbursts and discussed their prediction efficiencies, respectively. Yuan et al. (2012) and Zhao et al. (2012) found that some significant changes in the microseismic waveforms happened before rockbursts, and they thought these signal characteristics would be useful for rockburst prediction. Based on the induced seismicity data in the excavation process of the Base Gotthard tunnel, Husen et al. (2013) discussed the induced effect of rockbursts caused by multiple fractures in fault zones and stress redistribution. Yu et al. (2014) explained the preparation process of rockbursts by the crack extension theory, and revealed the relations between the microseisms and rockbursts. Pastén et al. (2015) analyzed more than 50,000 microseismic events recorded in the Creighton mine by the fractal geometry method, and they tried to verify the relations between the fractal dimension and the occurrence of large magnitude events. By means of the so-called 3S principle in seismology, Ma et al. (2016) proposed four rockburst criteria based on the distribution of the microseismic events in the process of rock damage.

Because the mechanism of a rockburst is very complex and each mine has its own characteristics, rockburst prediction is a challenging problem that still lacks a general prediction method applicable in various cases (Li et al., 2015; Lu et al., 2015; Wang et al., 2015a, b). In this paper, with the 80-day continuous microseismic monitoring data recorded in the Qixing Coal Mine in China, we analyzed the distribution characteristics of the destructive seismic events with high released energy (rockbursts). By analyzing the evolution characteristics of microseismic events prior to the occurrence of rockbursts, a prediction model including four indicators was built. Then the conditional probability in probability theory was introduced as the assessment indicators to evaluate the prediction efficiency.

4.2.2 Microseismic monitoring system

The Qixing Coal Mine belongs to the Shuangyashan Mine Bureau of the Longmei Group in China, and it was put into operation in 1973. After the technological transformation, the annual designed production capacity was 2.4 million tons. There were 16 layers of coal in the field, and the cumulative thickness was 21.5 m, of which four, six, and eight coal seams were in the whole region while the rest was locally recoverable. The Qixing Coal Mine was divided into two levels of mining, respectively at −100 m and −450 m.

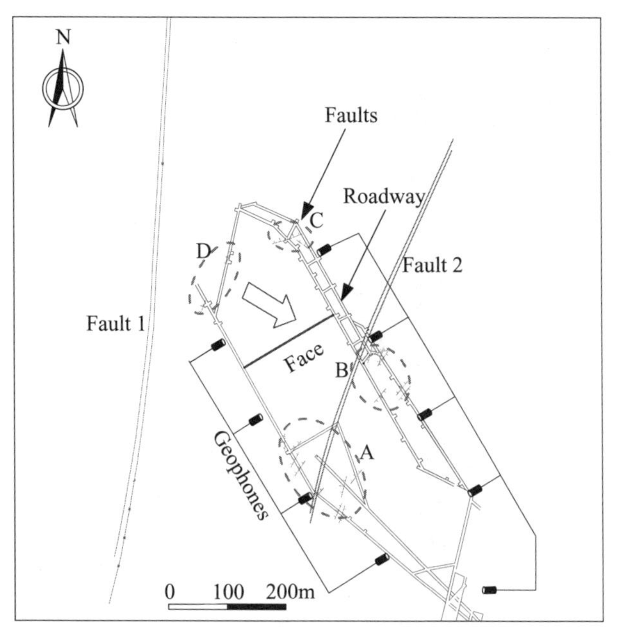

FIG. 4.13 The working face and the locations of seismic sensors.

The high-grade ordinary mining method was used and the annual mining velocity was about 600 m in the Qixing Coal Mine. The buried depth of the Dongsan working face was about 550 m, the coal thickness was about 2.45–3.60 m, and the mining direction of the working face is shown in Fig. 4.13. There were two large north-south trend faults in the mining areas, and there were several small faults found in the excavation process of the roadways. These small faults are mainly distributed in the four regions marked by A, B, C, and D in Fig. 4.13, respectively. The roof and floor of the coal seam are both sandstone, with a fine integrity in general except for the local broken zones affected by faults.

As shown in Fig. 4.13, real-time continuous microseismic monitoring was conducted during the mining process. There were nine high-sensitivity sensors buried in the roadway to collect seismic waveforms. The sketch of the data acquisition system was shown in Fig. 4.14. The data acquisition modules (Paladin, Fig. 4.15A) were connected to the sensors (Fig. 4.15B), which converted the waveform signals into the high-resolution digital signals. These signals were transmitted to the data processing center through the optical fiber cables in real time. Then, the microseismic events were detected and located in the processing center.

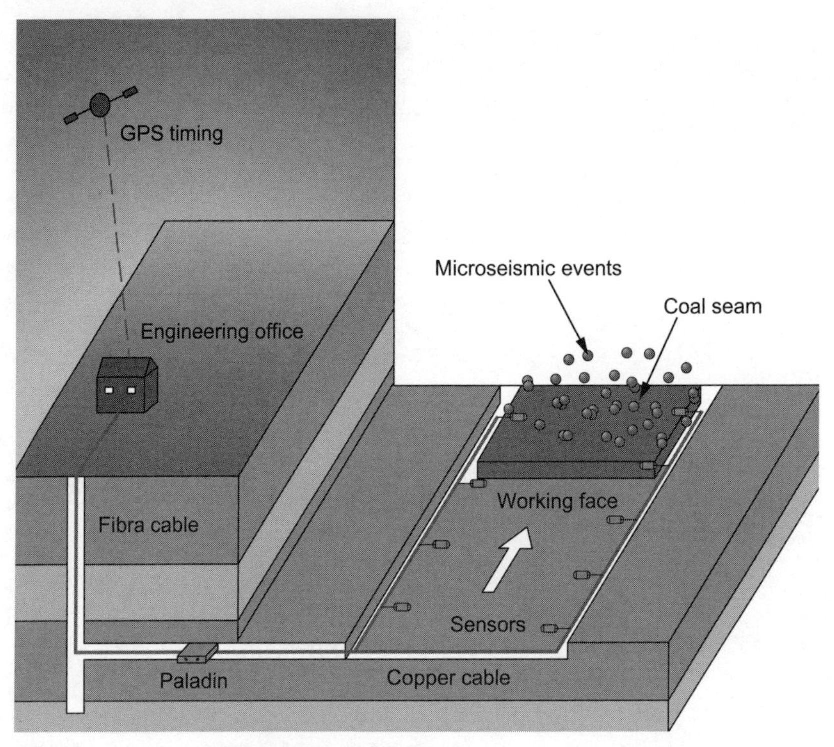

FIG. 4.14 The sketch of the data acquisition system.

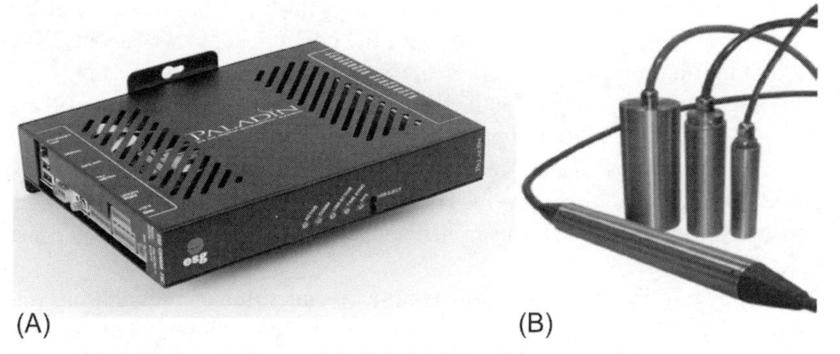

FIG. 4.15 The data acquisition module (Paladin) (A) and the sensors (B).

4.2.3 Active microseismicity and faults

The geological structures and stress distributions were not uniform during the mining process, which resulted in the uneven distribution of the rockburst risks in space. Here, we rely on analyzing microseismic event distribution in the mining process of rockbursts to delineate the high-risk zones for early warning in mining.

4.2.3.1 Microseismic event distribution

The spatial distribution of microseismic events detected and located by the monitoring system for about 80 days during the mining process can be categorized into four stages (Fig. 4.16). With the working face advancing, these microseismic events were mainly concentrated in three zones.

As shown in Fig. 4.16, for the first 40 days, the microseismic events were mostly concentrated in zone 1. Then they were mostly clustered around zone 2 in the next 20 days while some microseismic events were clustered around zone 3 during the same time period. Finally, all the microseismic events in the last 20 days were located in zone 3. Zone 1 is close to zone 2 while there is big gap between zones 2 and 3. The seismic events with high released energy (i.e., rockbursts) were mainly located in these three zones. In each zone, the seismic events often occurred before the rockbursts, indicating the induced

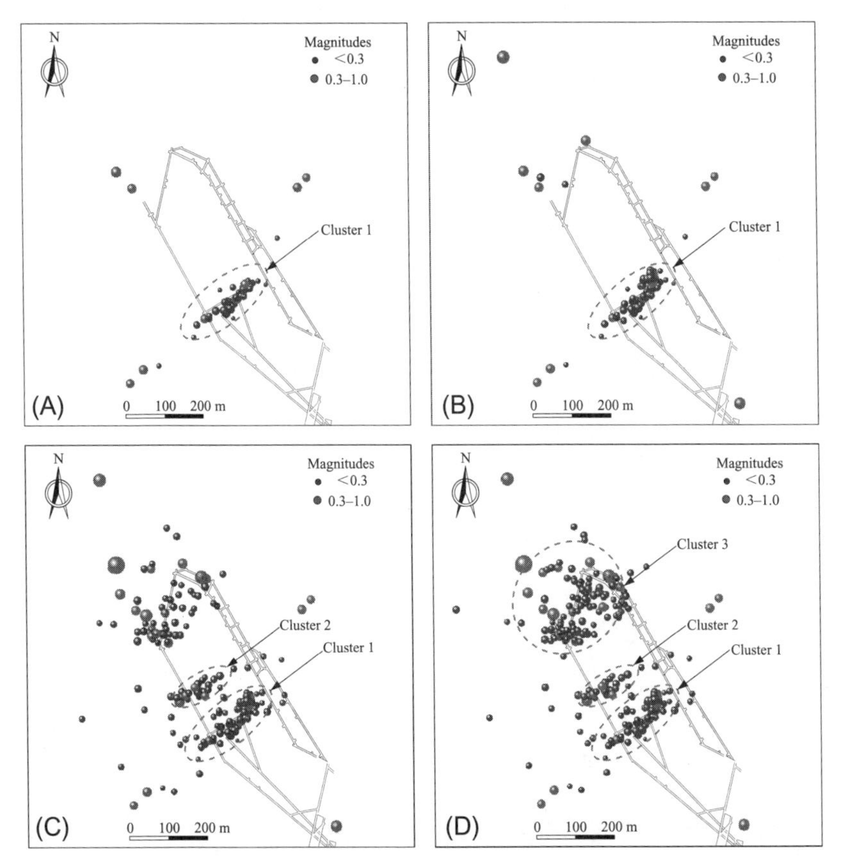

FIG. 4.16 Distribution of microseismic events for different times during the 80-day monitoring period. (A) 20 days, (B) 40 days, (C) 60 days, and (D) 80 days.

effect of the microseismic events by the rockbursts. Based on the relationships between microseismicity and rockbursts, we can hence delineate the rockburst high-risk zones by identifying active microseismic zones.

4.2.3.2 Fault structures on rockburst distribution

There are many complex factors that could induce rockbursts, among which the existence of faults has been proved to be a crucial one. Large strain energy can be accumulated around the faults due to the horizontal tectonic stress (Li et al., 2014). In the process of mining, the stress redistribution can easily activate faults (McKinnon, 2006), thus the fault structures have a great influence on the spatial distribution of rockbursts. As shown in Fig. 4.17, some rockbursts concentrated in three zones are clearly linked to the existing small faults. In zone 1, there were a number of small faults (A, B in Fig. 4.17) detected in the adjacent roadways, implying that there were some faults distributed between A and B. In addition, there were a number of rockbursts (zones 2 and 3 in Fig. 4.17) and the small faults overlapped (C, D in Fig. 4.17), which implied that some rockbursts were caused by the fault activation.

In addition to microseismic locations, we can calculate source mechanisms by using the seismic waveforms (Šílený and Milev, 2008). As shown in Fig. 4.17, the focal mechanism of an event located on fault F_1 suggested that the strike of the slipping surface followed the faulted structure trend, indicating that this event was induced by the activation of F_1. It can be seen that the stress

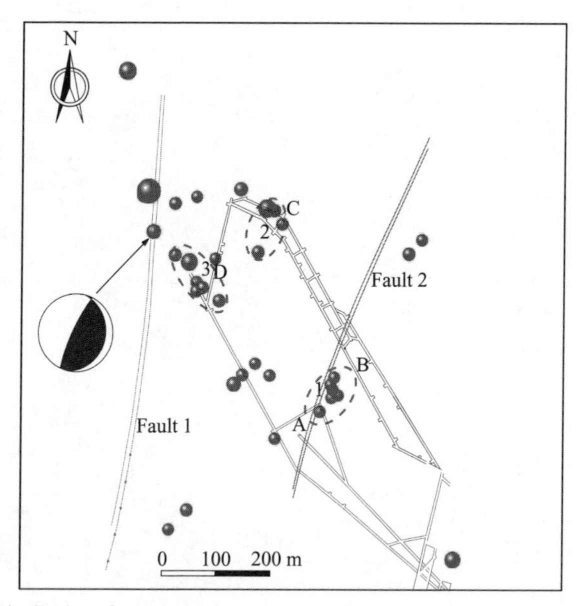

FIG. 4.17 Distribution of microseismic events (*red circles*) and the larger faults (*red lines*).

redistribution in the mining process may induce the activation of the faults accompanied by rockbursts. Thus, the regions with faults often bear high risks of rockbursts. Identifying the active microseismic zones and the existing faults can help delineate the high-risk zones of rockbursts, which are often associated with intensive microseismicity and are near existing faults. After identifying the high-risk zones, the supporting measures in these areas can be strengthened to avoid or reduce accidents. Beyond what has been stated, during the mining process, only a small shear microseismic event occurred in fault F_1, and a large amount of strain microseismic events occurred in the mining of the surrounding rock. Compared with the latter, the occurrence probability of microseismic events in the fault F_1 is small; therefore, this study mainly focused on the strain rockburst forecast in the mining of the surrounding rock.

4.2.4 Rockburst prediction indicators

4.2.4.1 Constructing prediction indicators

The timely prediction of rockbursts is hard but essential, hence it is necessary to comprehensively analyze the microseismic monitoring data to identify the precursors of rockbursts.

Average number *N* and average released energy *E*

Rockbursts are the release of elastic waves in the process of deformation and failure of rock mass in a high-stress state while the stress evolution directly affects the microseismic activity. As shown in Fig. 4.6, the curves of the average number N and the average released energy E of microseismic events displayed a similar trend with time, which reflecting the microseismic activity at different stages.

It can be seen from Fig. 4.18A that the whole monitoring curve could be divided into two periods while each period could be divided into three stages sequentially: the quiet stage (S_1), the active stage (S_2), and the transition stage of the microseismic activity (S_3). All three stages (S_1, S_2, and S_3) corresponded to that of the stress evolution process respectively in the engineering rock mass, namely the stress concentration stage, the stress weakening stage, and the stress transition stage. In the stress concentration stage, the accumulated stress did not yet reach the maximum strength of the rock mass, with few microseismic events occurring. However, the stress accumulation had prepared for the increased microseismicity and released energy in the next stage. In the stress weakening stage, the local stress increased markedly and the rock mass produced deformation and failure while the number of microseismic events increased sharply and remained at a high level during a certain time. After releasing the local concentrated stress, the curve entered the stress transition stage, that is, the local high stress transferred to the near rock mass with a medium activity of microseismicity. A new period including the above-mentioned three stages started after the energy fully released.

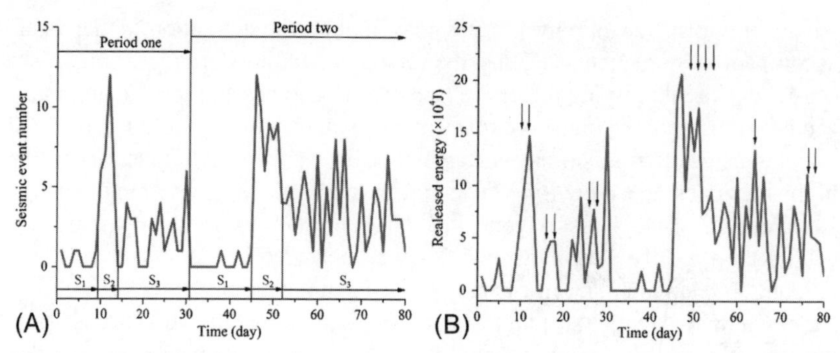

FIG. 4.18 The stages of the monitoring curves and the arrows in (B) mark the start of rockbursts. (A) The curve of seismic event number and (B) the curve of the released energy.

In the stress weakening stage and the beginning of the stress transition stage, the local stress was so high that it tended to reach the maximum strength of the rock mass, often accompanied by high energy microseismic events that pointed toward rockburst prediction. The occurrence times of the rockbursts were marked by arrows in Fig. 4.18B, which suggests that the day when the released energy reached the peak value and the subsequent 3 days had a high risk of rockbursts. Therefore, the sharp increase of both the number of microseismic events and the released energy can be regarded as the precursors of rockbursts, that is, the average seismic number N and the average released energy E can be chosen as the prediction indicators of rockbursts.

Seismological parameter b and its decrease Δb

In seismology, the frequency and magnitude relations of the seismic events in a particular area often follow the formula:

$$\lg N = a - bM \tag{4.2}$$

where M indicates the magnitude of the seismic event, N indicates the number of earthquakes whose magnitude is M, and a and b are regional parameters (Alvarez-Ramirez et al., 2012). This relationship has been widely applied in the prediction of earthquakes. The parameter b indicates the proportion of the seismic events with different magnitudes, reflecting the stress level and rupture scale of the medium, is the capacity of hindering releasing energy of the rock mass (Yin et al. 1987).

As shown in Fig. 4.19, the b value tends to be decreasing in the whole process of microseismic monitoring, which reflected the increased trend of large magnitude events. This also corresponded to the increase of the released energy, as seen in Fig. 4.6B. It is noted that the rockbursts often occurred after a sharp decrease of the b value. Therefore, the decrease of the b value compared to the previous day's Δb can be chosen as one of the prediction indicators.

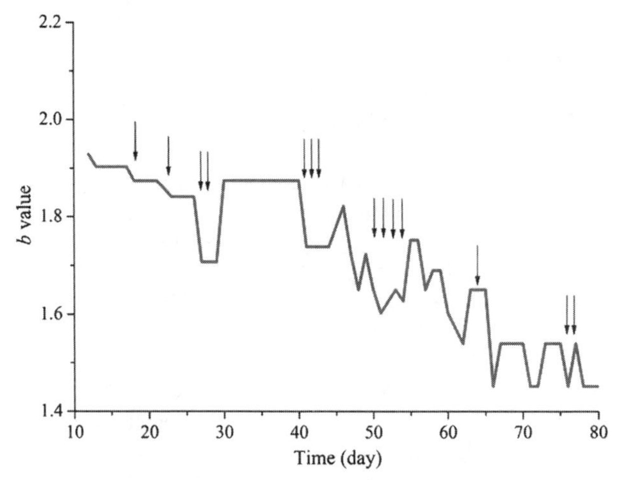

FIG. 4.19 The curve of the b value during the microseismic monitoring period.

Potential maximum magnitude M_m

According to the seismicity absence theory, if the seismicity is always active in a particular area but a large magnitude earthquake did not occur in a certain time period, an earthquake may occur in the near future in this area. This theory is often used in the prediction of tectonic earthquakes, and it is also applicable in the microseismic prediction in the engineering scale.

An application of the theory in the working face in the Qixing Coal Mine is shown in Fig. 4.20. During days 41–60, the microseismicity was active in the whole area A while the accumulated strain energy released through the seismic activity. From day 61–72, the seismic events occurred in the whole area except region B, in which the strain energy kept accumulating, indicating that the

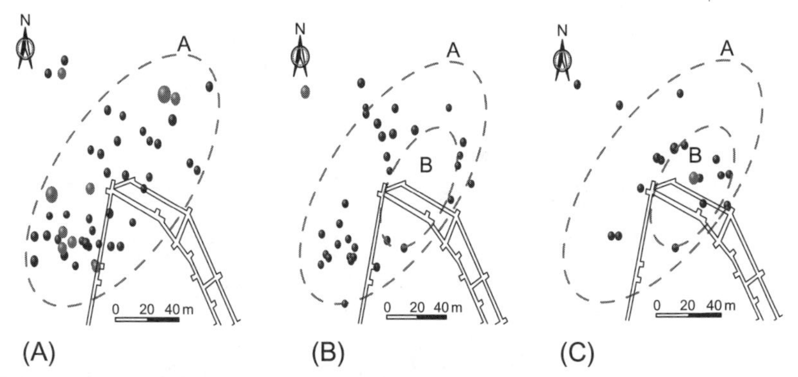

FIG. 4.20 Prediction of rockbursts using the seismicity absence theory. (A) 41–60 days, (B) 61–72 days, and (C) 73–76 days. (Microseismicity is marked as *blue circles* and rockbursts are marked as *red stars*.)

rockbursts might occur in this region. Entering days 73–76, the microseismic events are mainly located in or near region B, besides some rockbursts that occurred in the center of region B. Therefore, the local absence of seismicity can be regarded as a precursor of rockbursts.

In Formula (4.2), when $N=1$, $M_m = a/b$ indicates the potential maximum magnitude in the current magnitude distribution law. In the M-y coordinate system, it represents the intercept on the horizontal axis of a linear function $y = a - bM$, reflecting the capacity of creating a rockburst. Therefore, it is reasonable to choose M_m as a prediction indicator.

4.2.4.2 Assessing prediction indicators

Based on the recorded microseismic data, after selecting the appropriate alarm threshold, the alarm results of the four indicators were calculated in Fig. 4.21. It can be seen from Fig. 4.21 that the successful prediction, that is, the indicator, exceeded the threshold value and the rockburst occurred in the same time had been identified. The graphs showed that each indicator could predict some of the rockbursts while missing other ones. To quantitatively evaluate the successful prediction rate of various predictors, probability theory is introduced here. The conditional probability was chosen to assess the prediction efficiency of the

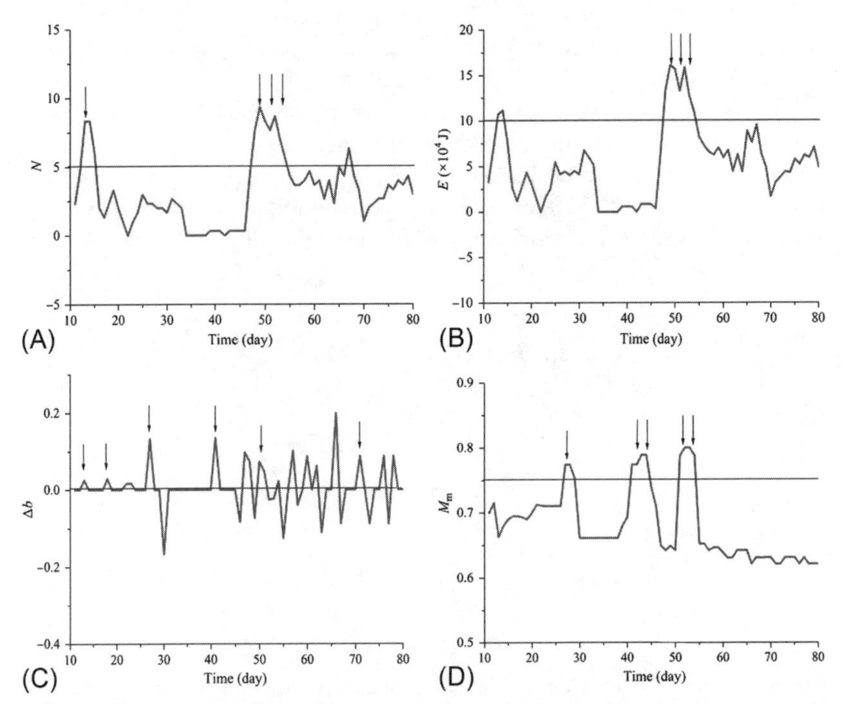

FIG. 4.21 Prediction efficiency of the four indicators. (A) The curve of seismic event number with time, (B) the curve of the released energy with time, (C) the curve of Δb value during the monitoring period, and (D) the maximum magnitude M_m during the monitoring period.

indicators. Event A indicated that a seismic event larger than magnitude 0.3 (rockburst) occurred and B indicated that the indicator exceeded the threshold and gave alarms while $\overline{A}$ and $\overline{B}$ indicated their opposite states, respectively. There were four conditions in total:

(1) *Positive successful prediction* (AB)*:* the prediction parameter exceeded the preset threshold and then gave the alarm before a rockburst. This was the case of an ideal prediction.
(2) *Missing alarm* ($A\overline{B}$)*:* the prediction indicator did not reach the threshold but a rockburst occurred in the absence of an alert. This situation showed a sudden rockburst without precursors, which was the main source of accidents.
(3) *False alarm* ($\overline{A}B$)*:* the prediction model alarms did not find rockbursts. In this case, taking measures relying on the alarm result may lead to production delays and the waste of resources.
(4) *Negative successful prediction* ($\overline{A}\overline{B}$)*:* there was no warning or occurrence of rockbursts. In this case, no special measures should be taken.

The probabilities of the four mentioned conditions reflected the efficiency of the prediction results, which could be quantitatively assessed by the conditional probability as following:

$P(B/A)$: it indicated the rate of forecasting the rockbursts successfully so the model should try to improve this value. $P(A/B)$: it indicated the rate of effective warning; a model showed good efficiency when this value was high. $P(\overline{B}/A)$:$P(\overline{B}/A) = 1 - P(B/A)$, the missing alarm rate significantly affected the prediction effect. When it was high, a lot of rockbursts failed to be forecasted, including the risk of serious accidents. $P(\overline{A}/B)$:$P(\overline{A}/B) = 1 - P(A/B)$, the false alarm rate indicated the valid warning, having an important impact on the prediction efficiency. $P(A/B + \overline{B}/\overline{A})$: it indicated that the predicted results were consistent with the actual results. The prediction efficiencies of the above-mentioned four indicators are listed in Table 4.2.

As seen from Table 4.2, we think that the probabilities of the prediction indicators are not less than 50% are good, so the probability 50% can be a threshold value to evaluate the prediction efficiency of the prediction indicators. All four

TABLE 4.2 Prediction efficiency of the four indicators

Name	$P(B/A)$	$P(A/B)$	$P(\overline{B}/A)$	$P(\overline{A}/B)$	$P(A/B+\overline{B}/\overline{A})$
N	0.37	0.50	0.63	0.50	0.73
E	0.32	0.55	0.68	0.45	0.74
Δb	0.63	0.39	0.37	0.61	0.63
M_m	0.53	0.71	0.47	0.29	0.81

indicators had a prediction effect in different degrees, in which M_m performed best while Δb performed poor due to its high false alarm rate.

4.2.5 Conclusions

Sufficient and accurate information on geological, stress, and microseismic conditions is required for the development of reliable rockburst prediction models. In this study, microseismic monitoring is used for the prediction of rockburst events and their location. It is shown that the accuracy of seismic events is one of the key issues, which depends primarily on the quality of the processing software and hardware of the monitoring system as well as the skills of the data processing analyst.

Based on collected microseismic data, a prediction model is developed, which includes the following main steps: analyzing the precursors, extracting the indicators, and determining the alarm thresholds. Microseismicity often shows similar characteristics before a rockburst. To quantitatively describe the precursory characteristics, it is essential to extract accurate prediction indicators that have a solid physical basis. For example, the "b" value chosen in this paper is widely used in seismology to describe the seismic magnitude distribution and its physical meaning engineering scale is thoroughly discussed, making itself a reasonable indicator. Second, these quantitative indicators must be calculated automatically because the monitoring is continuous and lasts for a long time. Finally, the chosen thresholds significantly affect the model efficiency, hence it is necessary to continuously train and optimize the threshold values in an established model.

Four indicators are extracted from the microseismic data to predict rockburst risks, including the average number of microseismic events N, the average energy released E, the decrease Δb of the seismological parameter b, and the potential maximum seismic magnitude M_m. To quantitatively compare the performance of these four prediction indicators, a number of rockbursts are assessed using conditional probabilities. The results show that all four indicators well predict rockbursts and the prediction indicator using the potential maximum seismic magnitude M_m appears to be the best.

The prediction model in this study relies only on microseismic monitoring data. By using other types of data such as mining-induced stress, it is possible to derive better prediction models. However, this would increase the complexity of the prediction model.

4.3 Near-fault mining-induced microseismic

4.3.1 Introduction

When there are fault structures in the mining field, near-fault mining will lead to the concentration of the mining stress and the elastic strain energy, which often induces the sudden slip and instability of the fault structure and causes a serious rockburst or a mine earthquake (Zhao et al., 2008). For example, coal mines in China such as the Dongtan, the Baodian, the Huafeng, the Muchengjian, and the

Yima have experienced rockbursts. In the Czech and Polish underground hard coal mines of the Upper Silesian Coal Basin as well as the Merlebach mine in East France and the Kopanang gold mine in South Africa, high-energy seismic phenomena are periodically recorded. With the mining depth increasing, the risk and frequency of rockbursts or earthquakes will become increasingly serious. Due to this kind of disaster with instantaneous characteristics, high energy, and magnitude, the damage caused by a rockburst under near-fault mining is more serious. Thus, it is of great significance for mining safety to forecast and prevent rockbursts.

It is extremely complicated and difficult to predict the time, place, region, and source of a rockburst or earthquake, problems that remain unsolved. This is because near-fault mining usually has a potent effect on the stress field near the fault. So it is essential to monitor the surrounding rock in the vicinity of the fault during the mining or underground excavation for early warning and to prevent potential geological hazards (Jiang et al., 2011; Yang et al., 2011). As one of the monitoring methods, the microseismic monitoring technique can record seismic waves released from the rock mass (microseismic events) to obtain comprehensive information about the disturbed rock mass, which is widely used in engineering practices (Sun et al., 2011; Riemer and Durrheim, 2012; Wang et al., 2015a, b). At present, for earlier microearthquake monitoring, South Africa, Poland, the Czech Republic, Canada, and other countries have formed the national mine microseismic monitoring network. In China, great progress has been made in this respect as the Mentougou Coal Mine began using microearthquake monitoring in 1959. So it is important to research the distribution and evolution characteristics of the microseismic events induced by near-fault mining based on the formers' achievements.

With microseismic monitoring and other technical means, a great deal of research has been conducted on the mining engineering, which was affected by the fault structure. For example, Driad-Lebeau et al. (2005) proposed that a fault structure in the floor under high-stress conditions resulted in a serious rockburst accident in the Merlebach Coal Mine in France based on microseismic monitoring data. Jiang et al. (2010) divided the fault-control rockbursts into two types: increasing and decreasing pressure modes, and pointed out the difference between the two types for risk assessment. Using the Integrated Seismicity System (ISS), Zhang et al. (2012) evaluated the rock mass stability of a fault-control area near the Dagangshan Hydropower Station. Snelling et al. (2013) explored the relationships in geologic structures, stress fields, and microseismic events based on the background of the Creighton Mine in Canada. They found that the microseismic events with large energy often occurred in the shear zones near faults. Li et al. (2014) summarized the key parameters inducing dynamic disasters in rock masses and studied their impacts on the reverse faults with different dip angles. Jiang et al. (2014a, b) found that the fault structure had a significant influence on the stress field by analyzing the stress evolution process in the working face approaching a normal fault with a thick overburdened rock mass.

Throughout the previous work, the research results were mainly focused on the microseismic events and the stress field distributions near the fault structures. However, few studies addressed the comparative analysis of the microseismic event distribution between the fault-control area and the engineering disturbance area. Moreover, the sensitive factors affecting the microseismic events during near-fault mining were rarely reported. So for early warnings and geological disaster prevention, this paper has tried to research the different characteristics of the microseismic event distribution and the stress field evolution in the above-mentioned two different areas based on the microseismic monitoring data in the East-3 mining area of the Qixing Coal Mine in China, and then to study the sensitive factors affecting the microseismic activity by using FLAC^{3D}.

4.3.2 Engineering situations

The Qixing Coal Mine located in the Heilongjiang Province of China was put into operation in 1973, and the annual output was 2.4 Mt. The two mining levels were −100 m and −450 m, and the thickness of the main coal was 2.45–3.60 m with a dip angle of about 10 degrees. The roof and floor of the coal seam were mainly sandstone and siltstone, with poor integrity in some areas. The longwall high-grade general mining method was adopted in the Qixing Coal Mine.

As shown in Fig. 4.22, the depth of the working face in the East-3 mining area was about 550 m, the mining area was about 330 m long and 160 m wide,

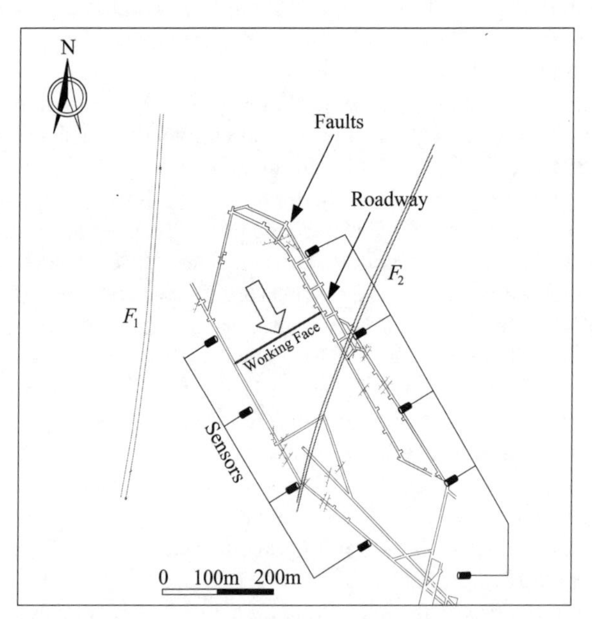

FIG. 4.22 Sketch map of East-3 mining area and the distribution of sensors.

and the stop mining position was about 200 m away from the open-off cut. The working face was arranged between two normal faults, F_1 and F_2. The inclination of each fault was about 70 degrees. The distance of F_1 was 0–25 m and that of F_2 was 15–50 m. Several small faults had been detected in the roadway excavation as well.

In order to obtain dynamic information about the mining stress field and fault activity during the mining process, the Canadian ESG microseismic monitoring system was introduced in the East-3 mining area of the Qixing Coal Mine in September 2013. The hardware part of the system mainly consisted of microseismic sensors, a digital acquisition instrument, a data transmission network, a GPS time synchronization source, a ground data-processing center, and a wireless transmission system.

The microseismic event acquisition process: First, the waveform signals of the microseismic events were detected by the microseismic sensors and were transmitted to the digital acquisition instrument. Second, the instrument converted the analog signals to digital signals and packaged them into a packet. Third, these signals were sent to the network switch through the LAN or remote wireless network. Finally, these signals were imported into the microearthquake monitoring system and stored in the database after data analysis and positioning.

As shown in Fig. 4.23, the general processing schemes for the collected microseismic data were to detect, locate, and interpret them. At first, the microseismic events were detected automatically by an advanced algorithm, and then artificial participation was involved to control the data quality. Finally, the microseismic events were precisely located and the source mechanism was determined before a comprehensive interpretation of the results.

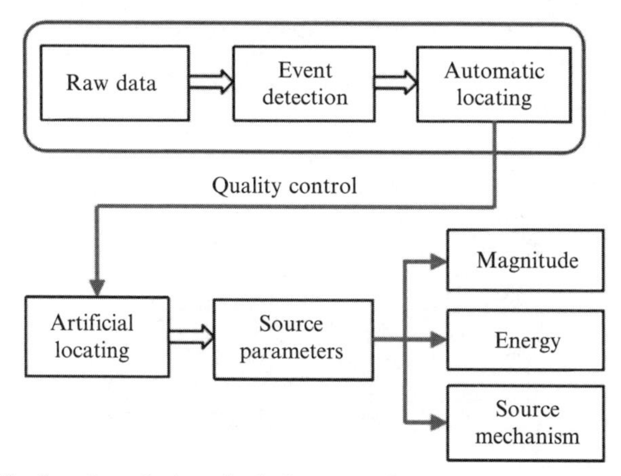

FIG. 4.23 The flow chart of microseismic data processing.

4.3.3 Computational model

As shown in Fig. 4.24, the numerical model was built by using FLAC^{3D} and the size of the model was 600 m long, 600 m wide, and 300 m high. The model was divided into 113,600 units and 123,533 nodes. Considering the physical and mechanical properties, the rock mass in the model was generalized into three rock groups: roof, coal seam, and floor. The bottom of the model was fixed and the lateral displacements were limited. However, the top of the model was applied the weight of the overlying strata. Besides, the model material strength was assumed to meet the Coulomb-Mohr criterion.

The initial stress field of the model was set at $\sigma_H = \sigma_{yy} = \sigma_v$, $\sigma_h = \sigma_{xx} = 0.5\sigma_v$. The displacement and velocity fields were cleared after the whole model was calculated to a balance state. Then, the four-step excavation was executed to simulate the mining process (Fig. 4.24B). The faults F_1 and F_2 were represented by two interfaces whose distribution patterns were observed to

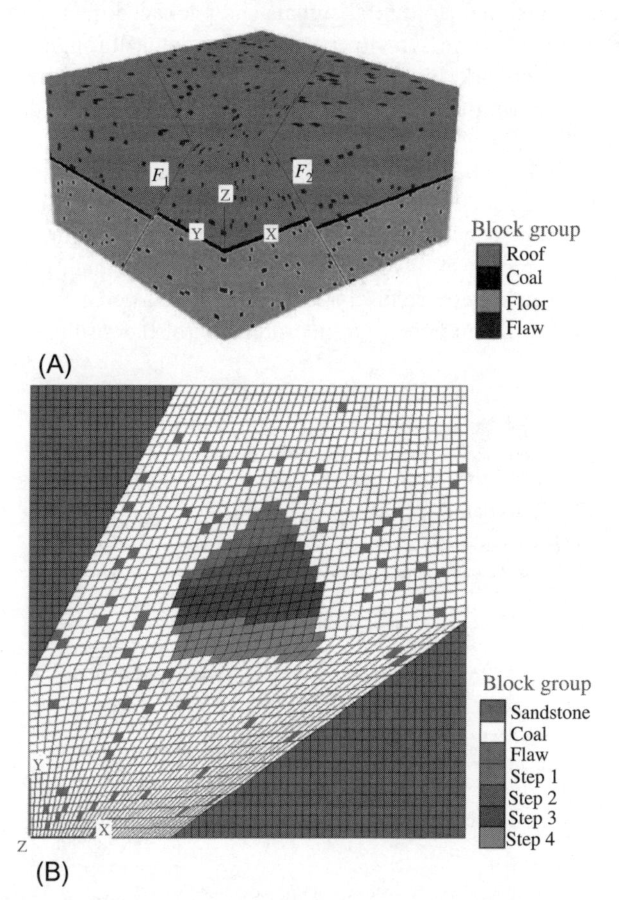

FIG. 4.24 The computational model and the excavation sequence in the mining area. (A) 3D numerical model and (B) four-step excavation in mining area on $z = 2$ m plane.

TABLE 4.3 Physical and mechanical parameters of the model

Groups	Density (kg/m³)	Bulk modulus (GPa)	Shear modulus (GPa)	Cohesion (MPa)	Friction angle (degree)	Tensile strength (MPa)
Roof	2500	4.50	1.50	3.50	45	0.10
Floor	2600	6.40	2.20	5.80	48	0.50
Coal	1300	4.20	1.20	1.60	30	0.05
Flaw	2500	4.20	1.30	0.80	21	0.05

TABLE 4.4 Physical and mechanical parameters of the faults

Fault name	Strike (degree)	Dip (degree)	Displacement of fault (m)	Normal stiffness (GN/m)	Shear stiffness (GN/m)	Cohesion (MPa)	Friction angle (degree)
F_1	7	70	10	6.00	6.00	0.50	15
F_2	22	70	20	3.00	3.00	0.50	15

the geological reports. The physical and mechanical parameters of the model and two faults are listed in Tables 4.1 and 4.2, respectively.

Due to the weak joints and faults, the engineering rock mass always showed strong nonuniformity and nonequal strength characteristics, which were mainly the sources of microseismic events. To reflect this feature, a FISH function was written in FLAC^{3D}; there were 14,000 defective units being set randomly in the computational model to represent the weak parts of the rock. In addition, the microseismic event was the microrupture in the rock mass accompanied by the release of elastic energy, and the rupture of a unit in the model also represented the release of energy. So, the rupture of a defective unit was taken as the criterion for the occurrence of a microseismic event (Wang, 2008; Han and Zhang, 2014) (Tables 4.3 and 4.4).

4.3.4 Result analysis and discussion

4.3.4.1 Average energy of microseismic events

As shown in Fig. 4.25, to reveal the influence of F_1 on the microseismic event distribution, the whole monitoring region was divided into two parts: the fault-control region R_1 and the mining disturbance area R_2. The East-3 mining area was initially mined on November 1, 2013, while the microseismic monitoring system began continuous monitoring for 150 days from the same day. There

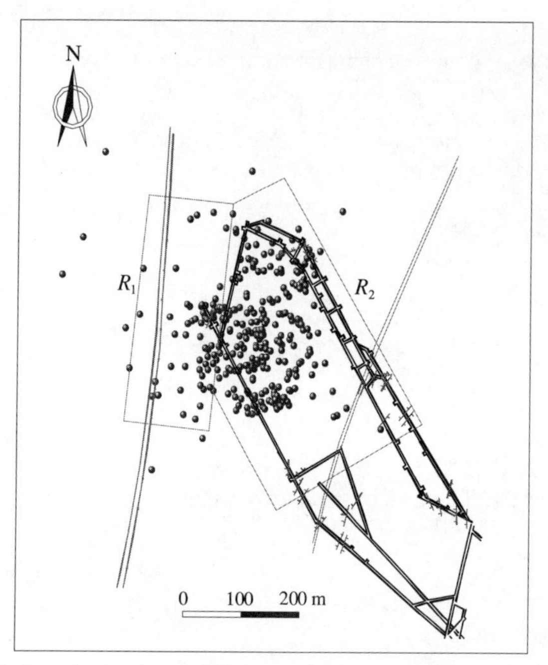

FIG. 4.25 The microseismic event distributions in different regions.

were a total of 133 microseismic events that occurred in R_1 while 1216 microseismic events occurred in R_2.

In the monitoring period, the moment magnitude and released energy of each microseismic event was calculated by the system instantly. The average energy values of the microseismic events in these two regions were 2.68×10^4 J in R_1 and 1.59×10^4 J in R_2, respectively. The former was 69% higher than the latter, indicating that there were more large-magnitude microseismic events in the fault-control area R_1 than in R_2.

4.3.4.2 Different characteristics of parameter b *value*

In seismicity, the earthquake magnitude and occurring frequency in a region were often regarded to observe to the empirical formula: $\lg N = a - bM$, where M was the magnitude of the earthquake, N was the number of the earthquakes whose magnitudes were greater than M, and a and b were the regional geological parameters (Varley et al., 2010). The formula was widely used worldwide. Many researchers were interested in the physical meaning of the b value and its influence on the temporal and spatial distribution of earthquakes (Alvarez-Ramirez et al., 2012; Grzegorz et al., 2016). In general, the b value reflected the stress change levels in the rock mass, and the releasing potential of energy was high if the b value was low.

As shown in Fig. 4.26, the data of microseismic events was fitted in both R_1 and R_2 using this formula, where the b values were 2.184 in R_1 and 2.782 in R_2.

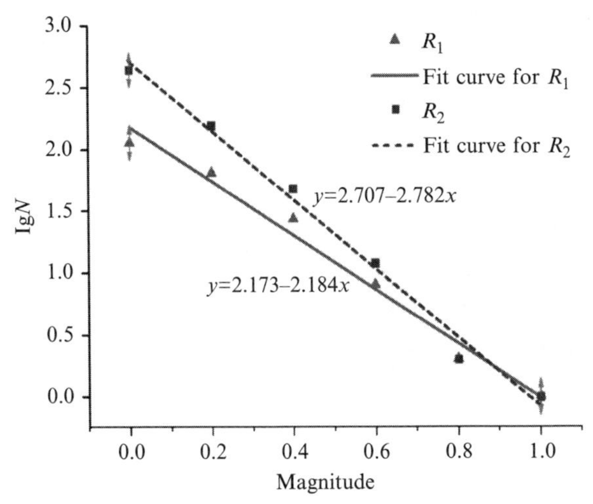

FIG. 4.26 Comparison curves of different b values in R_1 and R_2.

It can be seen that the b value in the fault-control area R_1 is lower than that in R_2 due to the F_1 fault structure, which indicated that the occurring possibility of the large-magnitude microseismic events in R_1 was more than that in R_2. This can be proved by the measured results that the average energy of the microseismic events in R_1 (2.68×10^4 J) was much higher than that in R_2 (1.59×10^4 J).

4.3.4.3 Local-mechanism solutions and fracture modes

The local-mechanism solution was widely used to describe the mechanical process in the source area during an earthquake, which was usually interpreted by the so-called double-couple model. In this theory, the source area was divided into four quadrants, including compression and expansion regions, through comparing the initial directions of P waves recorded by different sensors (Xie et al., 2015; Li et al., 2016). In order to analyze the failure characteristics of the disturbed rock mass in the fault-control area and reveal its fracture mechanism, the fracture modes of the disturbed rock mass were statistically analyzed (Fig. 4.27).

As can be seen from Fig. 4.27, the shearing failure occupied a dominant percentage, as high as 79.7% in R_1, followed by compression failure and tensile failure, with 13.5% and 6.8%, respectively. In R_2, 82.6% of microseismic events experienced shearing failures, accounting for an overwhelming rate as well. However, the proportion of the compression failure mode was obviously lower (9.1%) while the tensile failure was higher (8.3%) than the former ones. It seemed that the rock mass in R_1 had experienced a high potential of compression failure because there was a relatively high compression stress concentration near the fault F_1 than that in R_2.

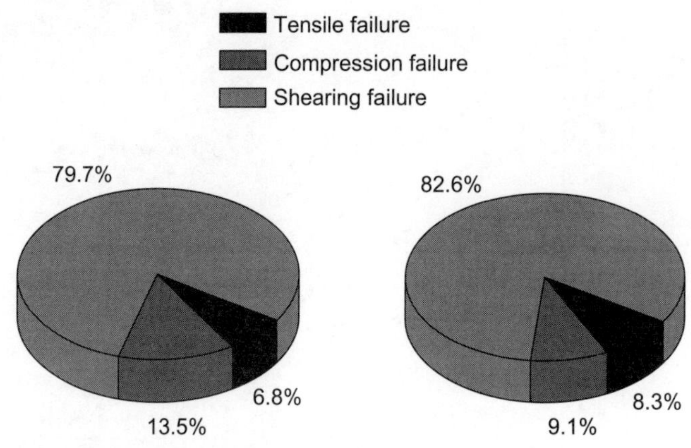

FIG. 4.27 Pie charts of focal-mechanism solutions in R_1 and R_2.

4.3.4.4 Distribution of microseismic events

During the numerical calculation process, the software automatically recorded the information of rupture units (i.e., microseismic events). After the first and second step excavations, the simulated microseismic events were mainly concentrated in the front of the working face and the surrounding rock of the roof while few microseismic events occurred near the F_1 or F_2 normal faults. As shown in Fig. 4.28, after the third excavation, apart from the ones in the working area like before, a small amount of microseismic events occurred between the left roadway and the F_1 fault (Fig. 4.28A). It was shown that the disturbance

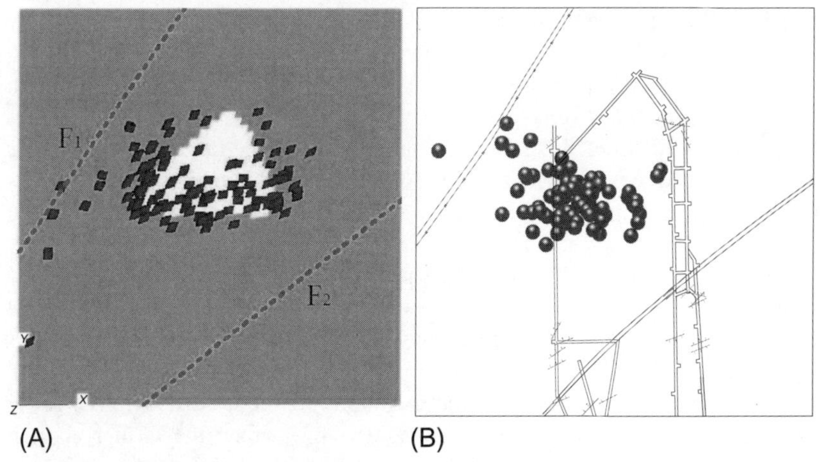

FIG. 4.28 Microseismic event distribution after the third excavation. (A) The calculation results and (B) the monitoring results.

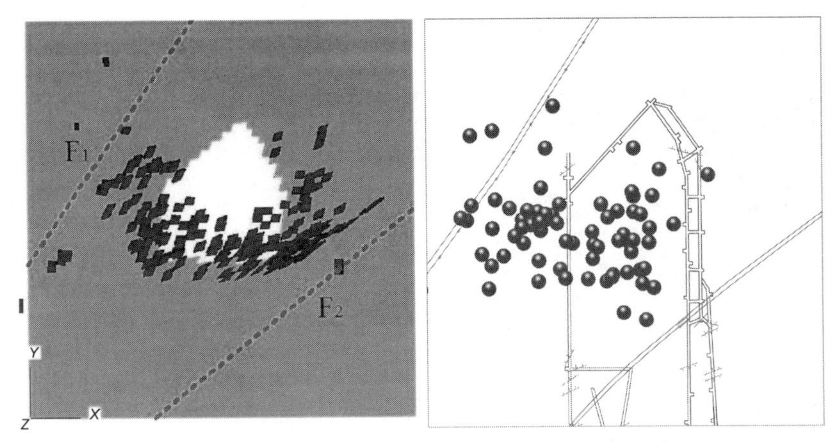

FIG. 4.29 Microseismic event distribution after the fourth excavation. (A) The calculation results and (B) the monitoring results.

area had influenced the F_1 control area R_1, which could be confirmed by the measured result (Fig. 4.28A). At the end of the last excavation, with the disturbance area further enlarging, both the calculated and the measured microseismic events accumulated in the F_1 control area R_1 (Fig. 4.29), indicating that the F_1 fault had begun to activate. The numerical and monitoring results coincided with each other, showing that the F_1 normal fault began to activate after 90 days due to the redistribution of the stress field.

4.3.4.5 Principal stress difference and elastic energy

In rock mechanics, the principal stress difference can better present the essence of rock deformation and failure (Xie et al., 2015) because it can reflect the vertical and horizontal stress synthetically, namely the level of the shearing stress under many complex loading conditions. As shown in Fig. 4.30, the distribution of the principal stress difference in the East-3 mining area after the fourth excavation showed as discontinuous near the two faults. The principal stress difference near F_1 was lower than that in the other surrounding rock mass, which was also obviously lower than the nearby F_2. The decrease of the principal stress difference near F_1 might be because of the active deformation of itself. In addition, the faults also had significant impacts on the elastic energy distribution (Fig. 4.31), and the calculation method of the elastic energy can be found in reference (Sun et al., 2007). The elastic energy distribution in the vicinity of the two faults also showed discontinuous features: the elastic energy near F_1 was lower than that in the other surrounding rock and the nearby F_2. These characteristics also reflected the loss of elastic energy due to the activation of the F_1 fault.

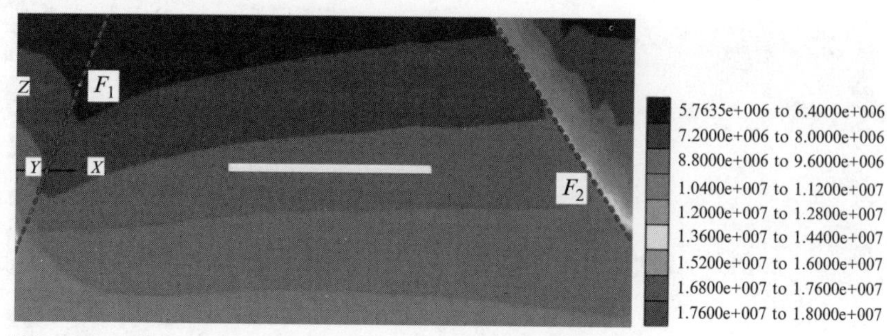

FIG. 4.30 Principal stress differences near the two faults (MPa).

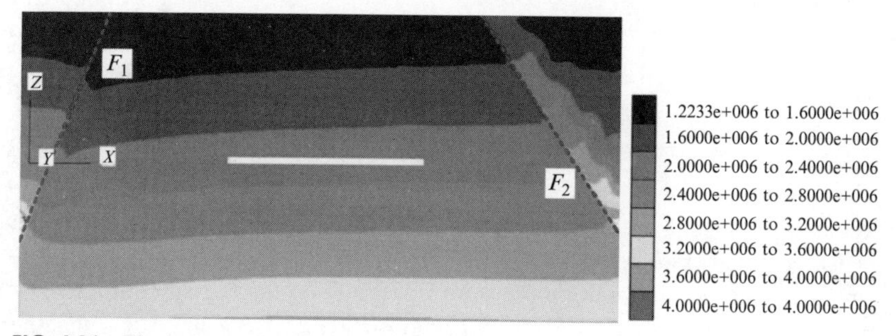

FIG. 4.31 Elastic energy distribution near the two faults (J).

4.3.4.6 Sensitive factors of microseismic events

To investigate the sensitive factors that may affect the microseismic events (ME), the number of microseismic events in different mining stages being the indicator, the curves were obtained under different working conditions by setting the different burial depths, the different friction angles of the faults, and the different stress fields.

It can be seen from Fig. 4.32 that the number of microseismic events displayed an increasing trend with the mining depth increasing: it went up slowly in the range of 450 m and 600 m, and rose sharply after 600 m.

As shown in Fig. 4.33, the number of microseismic events decreases sharply when the friction angle of the fault was less than 22 degrees, and then decreases slowly. It can be seen from Fig. 4.34 that with the stress ratio $R=\sigma_{yy}/\sigma_{xx}$ increasing by setting σ_{xx} stable and σ_{yy} changeable, the number of microseismic events increased gradually before 1.0 and rapidly after that value.

4.3.5 Conclusions

To reveal the microseismic event characteristics induced by near-fault mining, this study conducted the contrastive analysis of the microseismic events in the fault-control area R_1 and the mining disturbance area R_2. From the point of

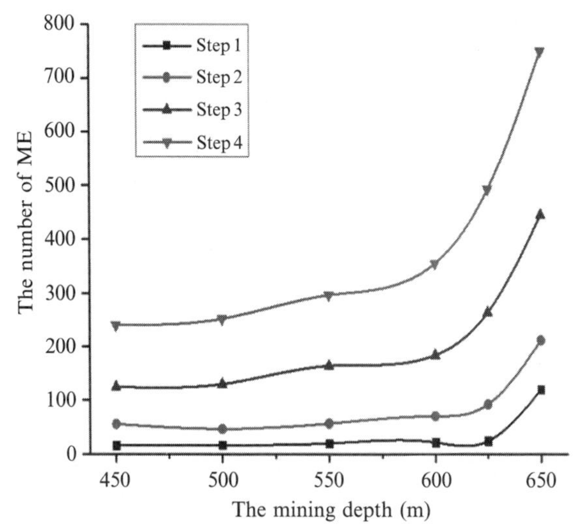

FIG. 4.32 Curves of ME numbers with the mining depth.

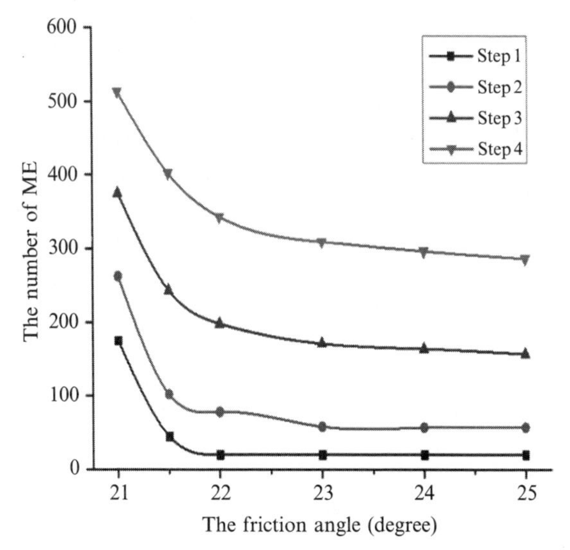

FIG. 4.33 Curves of ME numbers with the friction angle.

views of the seismic parameter b value and the focal-mechanism solution, we analyzed the microseismic event distribution characteristics and the fracture modes in those two areas. Moreover, we built the computational model to study the sensitive factors affecting the microseismic events during the near-fault mining. We finally achieved the conclusions as follows:

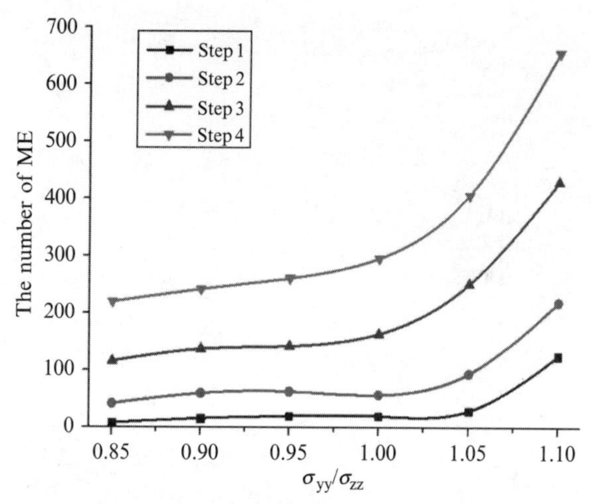

FIG. 4.34 Curves of ME numbers with the stress ratio.

(1) We found that the b value in R_1 was much lower than that in R_2, indicating that the occurring possibility of large-magnitude microseismic events in R_1 was higher than the latter. In terms of the focal-mechanism solution, it seemed that the rock mass in R_1 had experienced a high potential of compression failure because there was a relatively high compression stress concentration near the fault R_1 than that in R_1.

(2) The simulation results verified the evolution characteristics in R_1 with the mining area increasing and showed the discontinuity distributions of the principal stress difference and the elastic energy near the faults. Through the sensitive factor analysis, we found that there were some threshold values for the sharp increase of the microseismic event numbers with these factors changing, such as the mining depths, the friction angles of the faults, and the stress fields.

This study provides a new idea to analyze the microseismic event distributions under the near-fault mining condition, and the results can greatly promote microseismic monitoring in practice. However, further studies are needed to obtain more important information with the large amounts of data increasing by the long time continuously monitoring.

4.4 Acoustic emission recognition of different rocks

4.4.1 Introduction

Many kinds of rocks are fracturing due to human activity or geological processes. It is well known that elastic waves are emitted from rock mass during the fracturing process. Therefore, various symptoms related to the breakdown of rock mass can be detected by the acoustic emission (AE) technique (Arosio et al., 2009).

Rock mass is an environmental geological body. It is formed in a certain environment by the mineral composition and structural plane due to the movement of geological structures and a complex atmospheric environment. Different types of rocks contain different kinds of minerals and different geological structures, such as joints and microcracks. The fracture involves debonding and slipping between the grains, minerals, and geological structures in rocks. Each AE signal is an indication that some part of the released energy due to rock crack propagation is transformed into an elastic wave. Therefore, different types of rocks will generate different types of elastic waves. Laboratory experiments and field monitoring are often conducted to investigate the characteristics of rock AE signals. These AE signals are often mingled with electric signals and artificial noises. Therefore, how to distinguish these signals becomes a significant topic in AE investigations (Yang et al., 2002; Yi et al., 2002; Bhat et al., 2003).

Monitoring techniques and an artificial intelligence algorithm have been widely used in rock slopes and tunnels (Liang et al., 2013a, b; Lin et al., 2014), concrete dams (Lin et al., 2012, 2014), and high-rise structures (Yi et al., 2011; Yi et al., 2012). An artificial neural network (ANN) has some capability of learning from examples through iteration, without requiring prior knowledge of the relationships between the process parameters. The major benefits in using ANN are the excellent management of uncertainties, noisy data, and nonlinear relationships. Neural network modeling has become increasingly accepted and is an interesting method for application to the AE technique (Grabec and Kuljanić, 1994; Kwak and Song, 2001; Yi et al., 2002; Bhat et al., 2003; Kwak and Ha, 2004; Leone et al., 2006).

Many authors have conducted AE investigations using ANN. Kwak and Ha constructed a neural network to achieve an intelligent diagnosis for chattering vibration and burning phenomena on a grinding operation. The static power, the dynamic power, the peak of RMS, and the peak of FFT have been used as an input feature of the neural network to diagnose the grinding faults (Kwak and Song, 2001). Samanta and Al-Balushi (2003a) presented a procedure for fault diagnosis of rolling element bearings through ANN. The time-domain vibration signal of the rotating machinery with normal and defective bearings has been used as the input feature of the ANN. The results showed that the effectiveness of the ANN can diagnosed the physical condition of the machine. Samanta and Balushi compared the performance of bearing fault detection by ANNs and support vector machines (SMVs) (Samanta and Al-Balushi, 2003b). Kim and Yoon (2004) trained an ANN to recognize the stress intensity factor in the time interval of the microcrack to the fracture by an AE measurement. Hill et al. (1993) used the AE flaw growth activity to train a back-propagation neural network to predict the ultimate strengths in the remaining six specimens. However, few investigations can be found that use an ANN to distinguish rock AE signals and determine their characteristics for different rock types.

The objective of the current research was to develop a neural network for the prediction of rock types or other noises from their AE measurements. The AE signals were recorded in the rock failure process under uniaxial loading. The wavelet analysis helped to obtain the basic parameters of the AE signals of the rocks and the environmental noises. These parameters were used to establish input layers in the ANN. The trained ANN was applied to predict rock types and noise types. The predictions obtained from the ANN are in good agreement with the laboratory experiments. The ANN based on the AE measurement can be used to distinguish different rock AE signals and predict different rock specimens in rock engineering.

4.4.2 Experiment preparation and methods

4.4.2.1 Laboratory experiments

Four types of rocks were selected to conduct uniaxial compression. To guarantee the diversity of the rock types, these rock specimens were collected from four mines in China. The granulite specimens were obtained from a gold mine in Fujian province, the limestone specimens were from a tin ore mine in Guangxi province, the granite specimens were from an open pit in central Jiangxi province, and the siltstone specimens were obtained from a mine in the southern part of Jiangxi province.

As shown in Fig. 4.35, all the mechanical tests were conducted on a servo-controlled rock mechanical machine RMT-150C. This machine has a visualization operation platform based on Windows, and it can record the load, stress, and strain during the rock failure process.

An AE monitoring system SAEU2S with eight parallel detection channels was applied to collect AE events in the compression. Each channel with an AE sensor, a preamplifier, and an acquisition card can collect the parameters of AE events, such as the amplitude, energy, and counts.

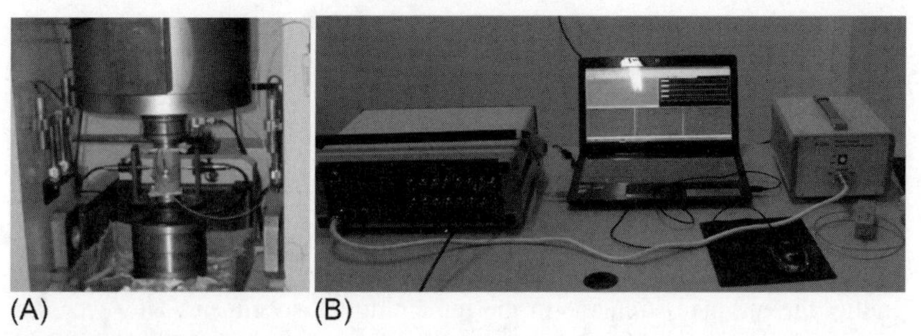

(A) (B)

FIG. 4.35 (A) The servo-controlled rock mechanical test machine RMT-150C and (B) the AE monitoring system SAEU2S.

4.4.2.2 AE signals

AE signals in the time domain

AE signals as well as electronic noises were recorded and processed during rock fracturing. The AE signals in the time domain and frequency domain were analyzed in this section. Different rocks generated different AE signals in the time domain under the same stress conditions. AE counts, cumulative AE counts, energy, rise time amplitudes, event rates, and energy rates are often used to describe the AE features in the time domain. Cumulative AE counts can reflect internal damage in the rock specimens.

According to the curves of the cumulative AE counts, axial stress, and time (Fig. 4.36), the curves can be divided into four periods: the prelinear period, the linear period, the postpeak and nonlinear period, and the residual strength period.

All the AE signals of the specimens show a sudden jump before their final failure, accompanied by a stress drop. The granulite specimens show a sudden increase before failure without precursors. The stress for the granite specimens lasted many times of stress buildup and stress release before final failure. The AE counts for the limestone specimens increased gradually before the peak strength points. The siltstone specimens demonstrated an obvious residual failure process, and the cumulative AE counts reached the peak point after the peak strength points.

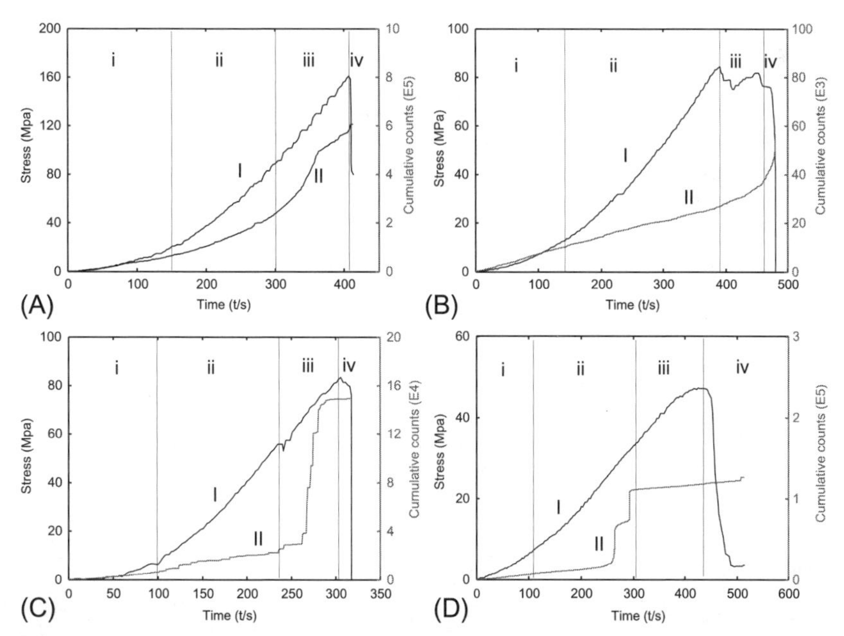

FIG. 4.36 Curves of cumulative counts, axial stress, and load time. (A) Granulite sample, (B) granite sample, (C) limestone sample, and (D) siltstone sample. I, Stress-time curve and II, cumulative counts-time curve.

The granulite, limestone, and granite specimens showed a hard and brittle failure mode and their curves for the postpeak period were not obtained. The soft siltstone showed a slight plastic failure mode.

AE signals in different domains

A continuous wavelet transform (CWT) is used to divide a continuous time function into wavelets. Unlike Fourier transform, the continuous wavelet transform possesses the ability to construct a time-frequency representation of a signal that offers very good time and frequency localization. CWT is very efficient in determining the damping ratio of oscillating signals. CWT is also very resistant to the noise in the signal.

According to the Mallat theory, the fast wavelet transform algorithm, the signal function can be decomposed into the low-frequency component and the high-frequency component under the scale j of the wavelet packets transform:

$$\begin{cases} f(n) = A_0 f(n) \\ A_0 f(n) = A_1 f(n) + D_1 f(n) \\ \cdots \\ A_{J-1} = A_J f(n) + D_J f(n) \end{cases} \tag{4.3}$$

where $f(n)$ is the signal function, A_i is the low-frequency component coefficient, and D_i is the high-frequency component coefficient. So, the signal function can be described as follows under the scale J:

$$f(n) = \mathrm{A}_J f(n) + \sum_{j=1}^{J} D_j f(n) \tag{4.4}$$

A topology structure of a wavelet transform with three layers is shown in Fig. 4.37.

Nowadays, wavelet analysis has been widely used to analyze nonstationary random signals (Wu et al., 2008). The wavelet analysis is a time-frequency localized analysis method in which the window size is fixed but its shape

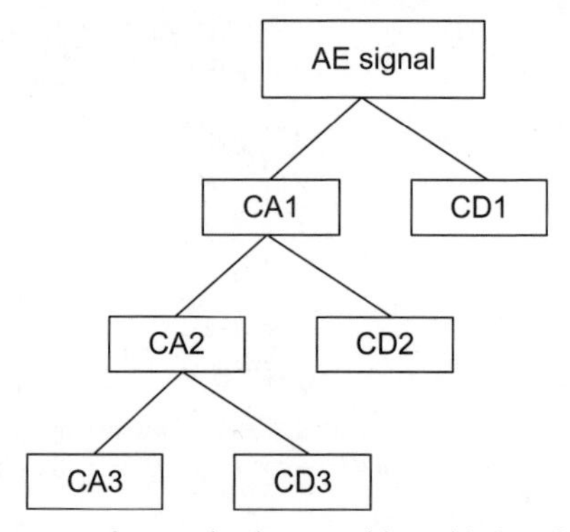

FIG. 4.37 Topology structure of a wavelet decomposition with three layers.

can be changed. The time and frequency window can also be changed, which means the low-frequency part with higher frequency resolution and lower time resolution, and the high-frequency part with a higher time resolution and lower frequency resolution (Wu et al., 2008). The wavelet analysis can decompose AE signals both in the frequency domain and the time domain. The wavelet analysis provides a kind of adaptive time and frequency domain localization analysis method.

The dynamical damage processes and characteristics of different rocks (granite, granulite, siltstone, and limestone) under the stress condition were obtained. It is found that the transformed AE signals for different rock types were different.

The frequency distribution reflected rock fracture and associated AE characteristics. As shown in Table 4.5, the frequency domain of the granulite specimens was mainly distributed in CD5 and CD4, taking up more than 90% of all the frequency bands. Most of the frequency of the granulite signals was located in the high-frequency bands [156, 625 kHz]. The frequency domain of the granite specimens was distributed in the low bands from CD1 to CD6, and CA6 took up more than 89% of all the signals, which ranged from [0, 78 kHz]. Most of the granite AE signals belonged to low-frequency bands while the AE signals of the limestone specimens mainly concentrated in the frequency of CA6 (59.835%) and CD6 (35.511%).

4.4.2.3 Artificial neural network

The artificial neural network can be seen as a set of parallel processing elements, and the suitable mathematical methods can be used to change the weights and thresholds to perform specific functions. The BP neural network can figure out each layer's error derivatives by using the back propagation algorithm, according to the generated weight matrices and threshold matrices. And then, the BP adjusts the corresponding matrices on the basis of error derivatives and square error sum to approach the mapping relation between the system input variables and the output variables step by step. The typical structure of a BP neural network is shown in Fig. 4.38. It has one input layer, one or more hidden layers, and one output layer, with each layer consisting of one or more neurons.

The number of neurons (m) in the input layer is the same as the number of mechanical parameters to be solved, and the number of neurons (n) in the output layer is the number of the measured displacements. Usually, only one hidden layer is needed. The number of neurons (p) in the hidden layer can be specified either manually or by an optimization method. The training specimens are often used to adjust the weight values by making the summed squared error between the displacements from numerical simulation and those from the BP network a minimum. For the training specimens, the input parameters can be prepared by the parameter experiment design method while the corresponding output parameters can be prepared by numerical simulation.

The calculating procedure of a three-layer BP neural network is shown in Fig. 4.39. W1 and b1 are the weight matrix and threshold matrix between the input layer and the hidden layer, respectively; W2 and b2 are the weight

TABLE 4.5 Energy ratio of the wavelet transformed AE signals in failure

Decomposition layers		CA6	CD6	CD5	CD4	CD3	CD2	CD1	Band (kHz)
Frequency layers (kHz)		*0–78*	*78–156*	*156–312.5*	*312.5–625*	*625–1250*	*1250–2500*	*2500–5000*	
Energy ratio (/%)	Granulite	0.61645	4.7345	15.846	74.723	4.0579	0.022976	0.0018851	[156, 625]
	Granite	89.339	9.9742	0.47787	0.14506	0.016137	0.16799	0.031258	[0, 78]
	Limestone	59.835	35.511	2.4313	1.926	0.26136	0.013166	0.022664	[0, 156]
	Siltstone	52.215	44.053	3.5185	0.17952	0.011493	0.0076762	0.015349	[0, 156]

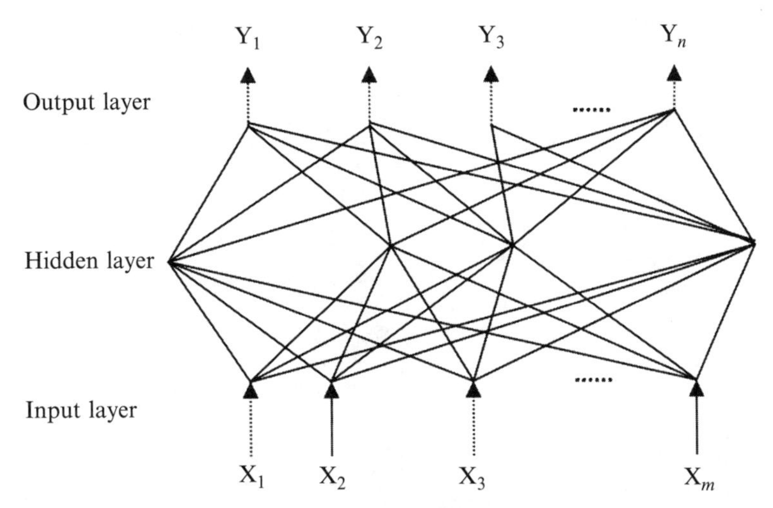

FIG. 4.38 Typical structure of a BP neural network (Kim and Yoon, 2004).

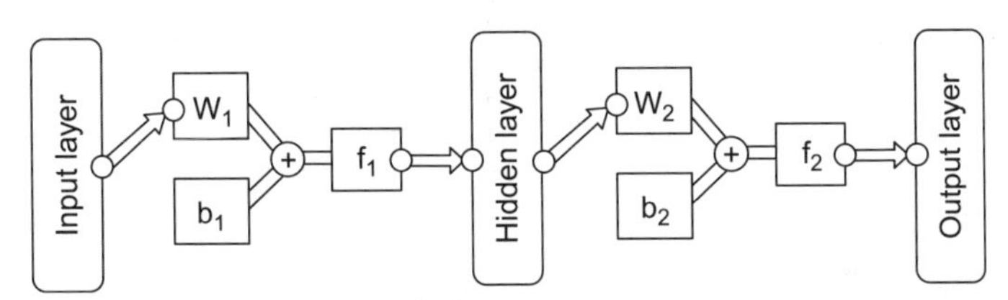

FIG. 4.39 Flow chart of a BP neural network.

matrix and threshold matrix between the hidden layer and the output layer, respectively; and Function *f* is the transfer function between two adjacent layers. Three transfer functions, including the Tan-Sigmoid transfer function (tansig), the Log-sigmoid transfer function (logsig), and the linear transfer function (purelin), are the most commonly used transfer functions for multilayer networks.

4.4.3 Results and discussion

4.4.3.1 Mechanical experiment results

Fig. 4.40 shows the typical fracture patterns for these four types of rock specimens. These rock specimens demonstrated different fracture patterns under uniaxial compression. A shear crack was formed in the granulite specimen, and the rupture is a typical single shear failure mode. Some thin flakes spalled from the granite specimen vertically, and the failure mode is a typical splitter failure. Several parallel cracks occurred in the limestone specimen that ran through the whole specimen from the bottom to the top, accompanied by some small cracks. It had the same tensile failure mode with the granite specimen. The siltstone presented a typical slightly plastic rock. The cracks in the siltstone

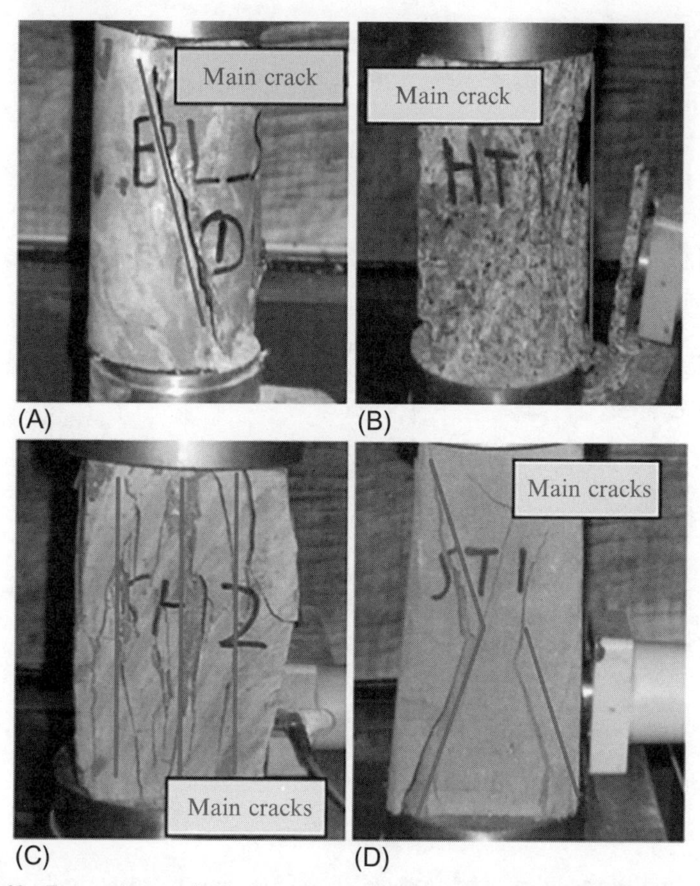

FIG. 4.40 Fracture patterns for the four types of rock specimens. (A) Specimen BL1 (granulite), (B) specimen HT1 (granite), (C) specimen SH2 (limestone), and (D) specimen ST1 (siltstone).

specimen developed relatively slow, and an X fracture pattern was observed on the surface, indicating a typical shear failure mode.

The four types of rock specimens have different failure modes. On one hand, different types of rocks have different distributions of mineral particle size, the hardness of mineral grains, and different microgeological structures. On the other hand, different rocks contain different properties and scales of weakness structural plane.

4.4.3.2 AE characteristics

Fig. 4.41 shows the curves of the accumulated AE counts as the load increased for the four typical specimens. Fig. 4.42 shows the curves of the AE rate and the load for the specimens. It can be observed that the limestone specimen and the granite specimen produced more AE events than the siltstone specimen and the granulite specimen in the beginning loading stage. However, the AE rates

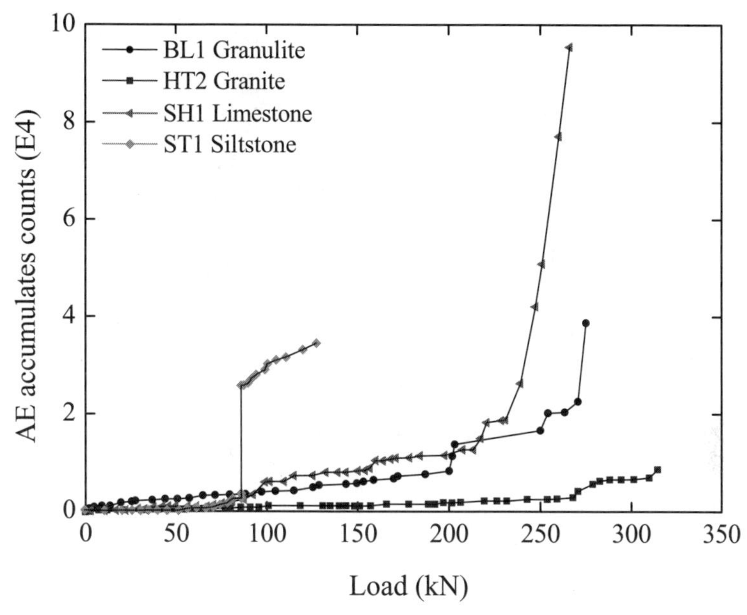

FIG. 4.41 The curves of the accumulated AE counts and load for the four specimens.

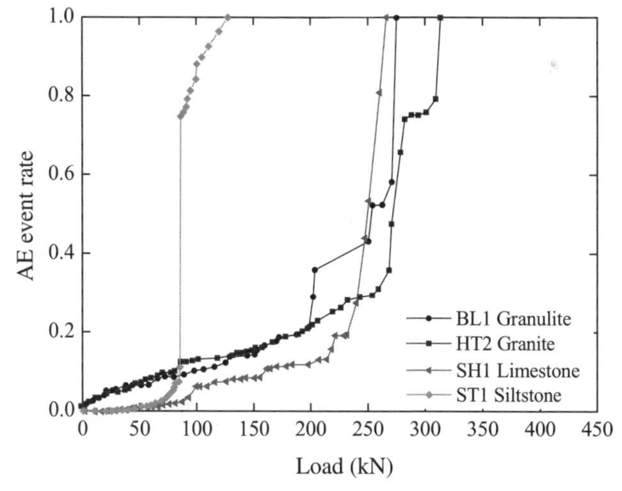

FIG. 4.42 The curves of the AE rate and load for the four specimens.

of the granite and granulite specimens were higher than the limestone and siltstone specimens. The granite specimen generated the smallest number of AE events, whereas it had the largest number of the AE rate. The siltstone did not generate a large number of AE events before the sudden burst failure.

Fig. 4.43 shows the wavelet transformed energy spectrum coefficient of the rock specimens. It can be observed that the spectrum coefficients of the

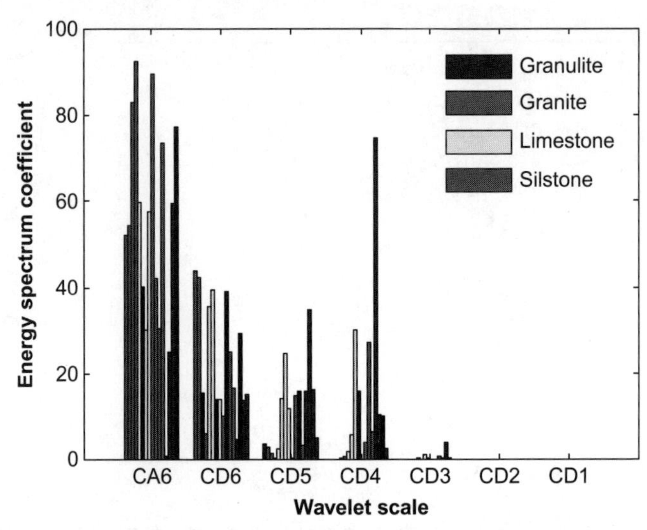

FIG. 4.43 The distribution of the energy spectrum after the wavelet transform.

TABLE 4.6 The distribution of the frequency band in each layer

	The layer of wavelet decomposition						
	CA6	*CD6*	*CD5*	*CD4*	*CD3*	*CD2*	*CD1*
Frequency range (kHz)	0–78	78–156	156–312.5	312.5–625	625–1250	1250–2500	2500–5000

granulite, granite, and limestone specimens were mainly distributed in the CA6, CD6, CD5, and CD4 bands. The distribution for the siltstone specimen was distributed in the CA6 and CD6 bands.

Table 4.6 lists the distribution of the frequency band in each layer. The frequency range of the AE signals could be determined by the wavelet transform. The frequency of the hard and brittle rock (the granulite, granite, and limestone specimens) ranged from 0 to 625 kHz. The siltstone belongs to a moderate strength and slightly plastic rock, and the frequency ranged from 0 to 312.5 kHz in a narrower band.

4.4.4 AE signal recognition using ANN

4.4.4.1 ANN structure

The number of neurons in the input layer, the output layer, and the hidden layer as well as a proper transfer function should be determined in a typical ANN structure.

Input layer vector: There were 11 input neurons in the ANN model, as shown in Table 4.7: the rise time (X1), ring count (X2), energy (X3), duration time (X4), amplitude (X5), peak frequency (X6), the CA6 value of the wavelet decomposition (X7), the CD6 value of the wavelet decomposition (X8), the CD5 value of the wavelet decomposition (X9), the CD4 value of the wavelet decomposition (X10), and the CD3 value of the wavelet decomposition (X11).

Output layer vector: There were three neurons in the output layer. The output parameters ($y1$, $y2$, and $y3$) should be either one or zero. Their combined value indicated the signal types (Table 4.8). For example, the output value 001 ($y1 = 0$, $y2 = 0$, and $y3 = 1$) predicted the signal generated by a granulite specimen.

According to the number of input neurons and output neurons, the number of neurons in the hidden layer can be obtained as follows (Hill et al., 1993):

$$n_1 = \sqrt{n+m} + a \tag{4.5}$$

where n_1, m, and n are the numbers of neurons in the hidden layer, the input layer, and the output layer, respectively, and a is a constant between 0 and 10. Because the input vector dimension was set to be 11 and the output vector dimension was set to be 3, n_1 could range from 4 to 14. The number of neurons in the hidden layer was set to be 14 in the ANN model.

Three transfer functions, including the Tan-Sigmoid transfer function (*tansig*), the Log-sigmoid transfer function (*logsig*), and the linear transfer function (*purelin*), are the most commonly used transfer functions for multilayer networks.

Two transfer functions were required in the ANN structure. The input parameters, such as the rise time, ring count, energy, duration time, amplitude, etc., had been normalized in the range $[-1, 1]$ before being input into the transfer function *tansig* as arguments. Compared with the transfer function *purelin*, *logsig* was better to link the hidden layer and the output layer. After 1 min training at the 360th iterative step, the mean squared error was less than 0.005 (Fig. 4.44). For the *purelin* function, the mean squared error did not reach 0.009 until 100,000 steps (Fig. 4.45).

4.4.4.2 BP network training

Six types of elastic wave signals were considered in the ANN model, including the four types of rocks, the electrical noise, and the artificial knock noise. A total of 120 sets of rock AE signals (30 sets of the granulite AE signals, 30 sets of the granite AE signals, 30 sets of the limestone signals, and 30 sets of the siltstone AE signals) were used for training. There were 30 sets of electrical noise signals and 15 sets of artificial noise signals for training. The RPROP algorithm was applied in the training, the mean squared error of the objective function was set to be 005, the maximum number of iterative steps was 100,000, and the number of independent training times was set to be more than 50. As shown in

TABLE 4.7 Input parameters and their signal types

Rise time /*a*1	Ring count/ *a*2	Energy/ *a*3	Duration time/*a*4	Amplitude/ *a*5	Peak frequency/ *a*6	CA6 layer of wavelet decomposition/ *a*7	CD6 layer of wavelet decomposition/ *a*8	CD5 layer of wavelet decomposition/ *a*9	CD4 layer of wavelet decomposition/ *a*10	CD3 layer of wavelet Decomposition/ *a*11
−0.9944	−0.9946	−0.9996	−0.9955	−0.2645	0.1404	0.4449	−0.2088	−0.4542	−0.6116	0.0746
−0.9917	−0.8019	−0.9658	−0.8734	0.4908	−0.7638	−0.5627	−0.3218	0.7551	−1.0000	0.0746
−0.9981	−0.9986	−0.9998	−0.9986	−0.3013	−0.7796	−0.6560	−0.7819	−0.8793	0.6216	0.0746
−0.9907	−0.9813	−0.9981	−0.9847	−0.0919	−0.0960	0.2005	−0.1534	−0.0943	−0.8261	0.2388
−0.9986	−0.9989	−0.9997	−0.9826	−0.3833	−0.1998	−0.0113	−0.2583	−0.5884	−0.2963	0.2388
−0.9954	−0.9774	−0.9968	−0.9811	0.0835	−0.8997	−0.9157	−0.8796	−0.8954	0.8161	0.0000
−0.9995	−0.9739	−0.9961	−0.9304	0.0580	−0.7373	−0.6852	−0.7980	−0.8384	0.6148	0.0746
−0.9801	−0.9766	−0.9971	−0.9245	−0.1372	−0.4244	−0.2463	−0.5013	−0.3729	−0.1488	0.0746
−0.9958	−0.9671	−0.9957	−0.9710	0.0580	−0.2490	−0.0722	−0.2650	−0.5771	−0.2764	−0.0896
−0.9991	−0.9644	−0.9954	−0.8638	−0.1796	−0.4675	−0.3698	−0.5240	−0.7964	0.2235	0.6866
−0.7740	−0.7066	−0.5099	0.2520	0.9661	−1.0000	−0.9925	−0.9809	−0.2206	0.4489	−0.6567
−0.9092	−0.8845	−0.8938	−0.4225	0.8218	−0.9947	−0.9993	−1.0000	−0.9561	1.0000	−0.8358
−0.8796	−0.8671	−0.7102	−0.1123	0.9123	−0.9911	−0.9945	−0.9418	−0.4143	0.5553	−0.8358
−0.9801	−0.9842	−0.9674	−0.9293	0.7624	−0.9947	−0.9882	−0.8576	0.6103	−0.2624	−0.7015
−0.7032	−0.6484	−0.4425	0.2268	0.9349	−0.9987	−1.0000	−0.9919	−0.9503	0.9891	−0.8507

TABLE 4.8 Output parameters and their signal types

Signal type	y1	y2	y3	Output value
Granulite	0	0	1	001
Granite	0	1	0	010
Limestone	0	1	1	011
Siltstone	1	0	0	100
Electrical noise	1	0	1	101
Knock noise	1	1	0	110

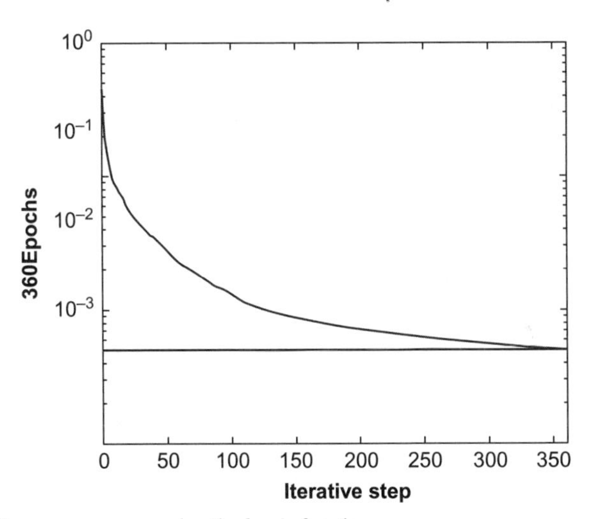

FIG. 4.44 Convergence curve using the *logsig* function.

Fig. 4.46, the mean squared error reached the specified minimum value after 374 training steps.

4.4.4.3 ANN recognition

Some basic parameters, such as the rise time, the ring count, the energy, the duration time, and the amplitude, that described the characteristics of the AE signals were combined as an input vector in the network. Moreover, the wavelet transform method was applied to decompose the signal waves to obtain the frequency spectrum. The decomposed energy spectrum at different layers was also treated as an input parameter. The trained neural network was used to predict the signal types by establishing the mapping function between the input parameters and the output parameters.

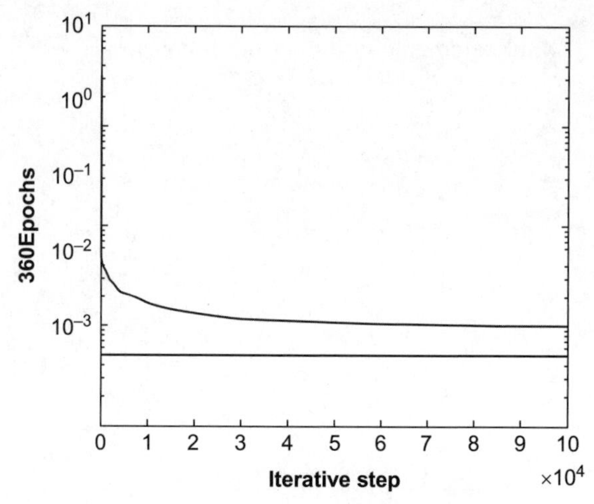

FIG. 4.45 Convergence curve using the *purelin* function.

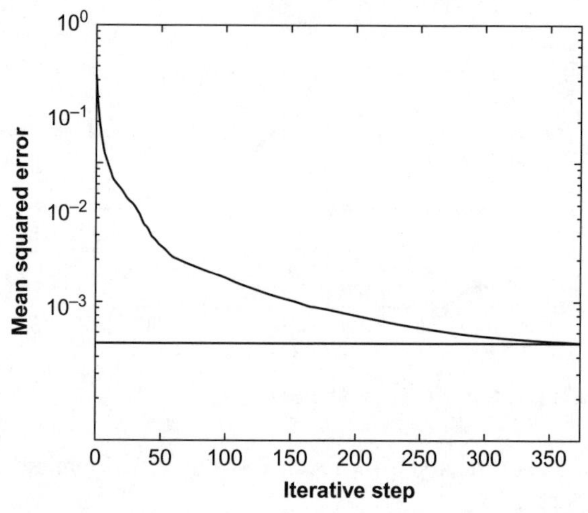

FIG. 4.46 The convergence curve in the training.

The predicted results proved that the BP neural network based on the wavelet transform analysis can achieve a high accurate ratio to recognize different rock AE signals. Table 4.9 listed the recognized AE signal types of the 110 sets of signals using the ANN model. There were 20 sets of the granulite signals, 20 sets of the granite signals, 20 sets of the limestone signals, and 20 sets of the siltstone signals. The average accuracy of signal prediction for the rocks was greater than 90%. It should be noted that all the signals for the granulite, siltstone, and electrical noises were predicted by the ANN model. Only one set of

TABLE 4.9 Signal type prediction using the ANN

Signal types	Signal set number	Predicted results						Accuracy
		Granulite	*Granite*	*Limestone*	*Siltstone*	*Electrical noise*	*Knock noise*	
Granulite	20	20	0	0	0	0	0	100%
Granite	20	0	17	0	2	0	1	85%
Limestone	20	1	0	19	0	0	0	95%
Siltstone	20	0	0	0	20	0	0	100%
Electrical noise	20	0	0	0	0	20	0	100%
Knock noise	10	0	4	0	0	0	6	60%

the limestone AE signals was predicted to be generated by the granulite specimens, and two sets of the granite AE signals were predicted to be generated by the siltstone specimens. Four sets of knock noise signals were recognized as the granite signals, indicating that the signal waves generated by the granite specimens were very similar to the artificial knock signals. Influenced by the random factors and the environmental factors, the accuracy for the artificial knock noise recognition was as low as 60%. To achieve good monitoring results, man-made noises should be reduced or gotten rid of in laboratory AE tests on rocks.

4.4.5 Conclusions

The wavelet transform and artificial neural network were applied to determine rock types from their AE characteristic parameters. The wavelet transform was used to decompose the AE signals, and the artificial neural network (ANN) was established to recognize the rock types and noises (artificial knock nose and electrical noise). The following conclusions can be drawn:

(1) Different rocks had different rupture features and AE characteristics. The wavelet transform provided a powerful method to acquire the basic characteristics of the rock AE and the environmental noises, such as the energy spectrum and the peak frequency. The signal parameters were input into the network, and the predicted results showed that the wavelet transform method was effective and accurate for AE signal decomposition.
(2) The ANN proved to be a good method to recognize AE signals from different types of rocks and environmental noises. The AE signal parameters decomposed by the wavelet transform composed the input layer, the *tansig* function was selected as the transfer function of the hidden layer, and the *logsig* function was selected as the output layer. The average recognition accuracy of the four kinds of rock AE signals was above 95% in the BP neural network.
(3) The signals generated by the granite specimens were very similar to the artificial knock signals. The electrical noises were easy to recognize by the BP neural network, but it had a low accuracy for the artificial knock noise recognition. To avoid adjacent-channel interference, man-made noises should be reduced as much as possible in laboratory AE tests on rocks.

4.5 Acoustic emission in tunnels

4.5.1 Introduction

The rockburst is a class of rock failure where the strain energy is suddenly released by an unstable rock fracture (Hoek and Brown, 1997; Rudajev et al., 2000; Beck and Brady, 2002; Weng et al., 2017). Typically occurring in deep underground mines, a rockburst is a common disaster. The opening of a tunnel relieves neighboring rocks of tremendous pressure, which can literally cause the rock to explode as it attempts to reestablish a stress and strain equilibrium (Lo and Yuen, 1981). The released energy leads to a violent fracture of the surrounding rock around the excavation and reduces the potential energy of the rock, which makes the surrounding rock reach another equilibrium state (Liang et al., 2013a, b, c).

In the past few years, many methods of forecasting rockbursts have been proposed, including the assessment of rock, stress and strain detection, and modern mathematical theories. However, a comprehensive understanding of rockbursts and associated damage mechanisms has remained elusive (Tang et al., 2010; Xu et al., 2011). Based on the agitated behavior of animals prior to earthquakes, a semiquantitative study has been carried out on the possibility of the emission of acoustic strain and fracture radiation prior to such events. Such emissions have been observed in the laboratory and in mines, and are termed the acoustic emission (AE) phenomenon (Armstrong, 1969). The AE phenomenon is defined as elastic mechanical waves associated with a rapid release of localized stress energy propagated within the material. AE monitoring is a useful tool for studying rock fracturing (Lavrov, 2003; Li et al., 2017). In order to understand the physical process generating seismicity within volcanic edifices, some studies have been based on monitoring an array of transducers around a rock sample, permitting the full-waveform capture, location, and analysis of microseismic events (Frid, 1997; Benson et al., 2008). In order to distinguish between the seismic events and blasts generate seismic waveforms, several characteristic parameters were extracted as discriminant indicators. The Fisher classifier, the naive Bayesian classifier, and the logistic regression were used to establish discriminators between them. Research results showed a reasonably good discriminating performance (Dong et al., 2016a, b). In order to solve the accuracy of localization methods based on the arrival time difference, the collaborative localization method using analytical and iterative solutions (CLMAI) was proposed. It was combined with the arrivals of multisensors and inversion of the real-time average wave velocity to seek the optimal locating results. This method highlights four advantages: without an iterative algorithm, without a premeasured velocity, without an initial value, and without square root operations (Dong et al., 2017, 2018). In in situ direct shear test studies, an initially intact region of rock bounded by joints and a seam is fractured, generating an AE. Large-scale inhomogeneous rock fracturing experiments, such as the in situ direct shear tests, may provide useful insights as analog models of seismogenic faulting (Ishida et al., 2010). The rockburst proneness index and the AE energy rise as the temperature increases. That is, the degree of rockburst increases rather than decreases with rising temperature, and this is helpful for explaining rockburst disasters in tunnels at a high ground temperature (Chen et al., 2014). A true triaxial unloading testing machine was utilized to perform rockburst tests on granite specimens with changing heights for size-effect investigations. A size effect exists during a rockburst simulation process, and affects the rock failure strength and fracture mode (He et al., 2010, 2015a, b; Zhao and He, 2016). Because of the high stiffness and strength in high-stress conditions, a rockburst in brittle fracturing can produce intensive AE activity and large amounts of energy. It is appropriate to locate the failure positions and determine the energy released in rock fracturing by using the AE technique (Cai and Kaiser, 2005; Dixon and Spriggs, 2007; Arosio et al., 2009).

It is generally accepted that a rockburst is dependent on the size and depth of the excavations, and we also know that the likelihood of rockbursts occurring increases as depth increases. In this study, a rockburst in a tunnel was monitored on granite rocks using the AE monitoring system, and different horizontal stresses were considered. The characteristics of AE signals were analyzed to predict the occurrence of a rockburst.

4.5.2 Rockburst experiments in a tunnel

4.5.2.1 Sample preparation

Granite rocks were chosen as the samples in this experiment, and the size of the rock samples was set at 150 mm × 150 mm × 150 mm. The specimen indexes were prepared in accordance with the Standard for Test Method of Engineering Rock Mass (GB/T50266-99).

As shown in Fig. 4.47, to simulate the tunnel model, a hole of $\Phi = 45$ mm was set in the front and rear center of the rock sample, and the parallelism error of two head faces was within 0.02 mm. The filling body is filled around the hole, and the filling body is made by the quartz sand and the special expansion cement at a mixture ratio of 1:1.

4.5.2.2 Laboratory equipment

The experimental system comprised a loading system and an AE system (Fig. 4.48).

Loading system (RLW-3000): The servo-controlled rock-testing machine is produced by Chaoyang Test Instrument Corporation, China. The deformation and applied vertical force can be monitored. The capacity of the axial load transducer is up to 3000 kN, and the capacity in the horizontal direction is up to 1000 kN.

AE monitoring system (PCI-2): The AE monitoring system is produced by Physical Acoustics Corporation in the United States. The AE activities of rock fracturing were recorded by an AE detector with eight channels. The multiparameter AE data, including waveform, hits, ring-down counts, and amplitudes, were obtained using the AE system. In addition, the air conditioner, which is used to guarantee a constant temperature in the laboratory, is produced by Gree Company, China.

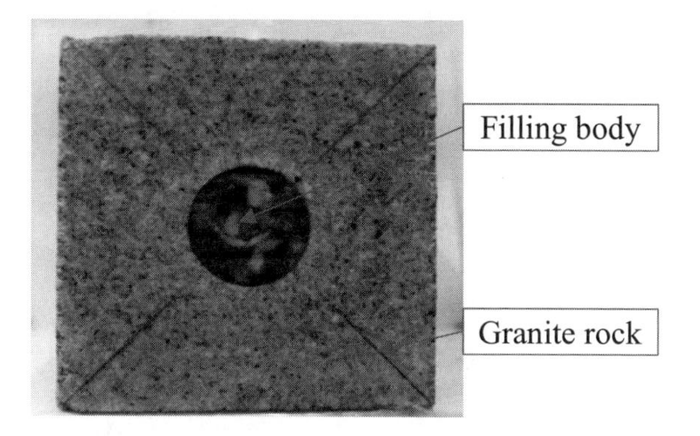

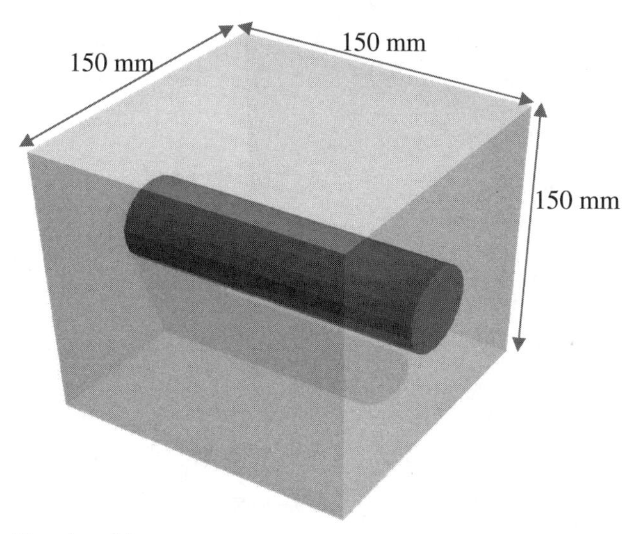

FIG. 4.47 Tunnel model.

In order to guarantee the conformity of the experimental data, the equipment setup should be kept consistent (Ishida et al., 2017). The AE devices were set as follows: The sampling time was set at 0.2 μs, and the memory length was set at 2 k (i.e., 2048 words). In this case, the recording time was approximately 0.4 ms (0.2 μs × 2048). The pretrigger was set at 1 k and the sampling rate was set at 1 MHz.

4.5.2.3 Loading condition

In general, the underground rock is mainly affected by vertical stress and horizontal stress, both of which form the initial stress field (Wang et al., 2015a, b). The vertical stress is affected by depth and upper overburden while the horizontal stress (also called the tectonic stress) is affected by diagenesis and tectonic

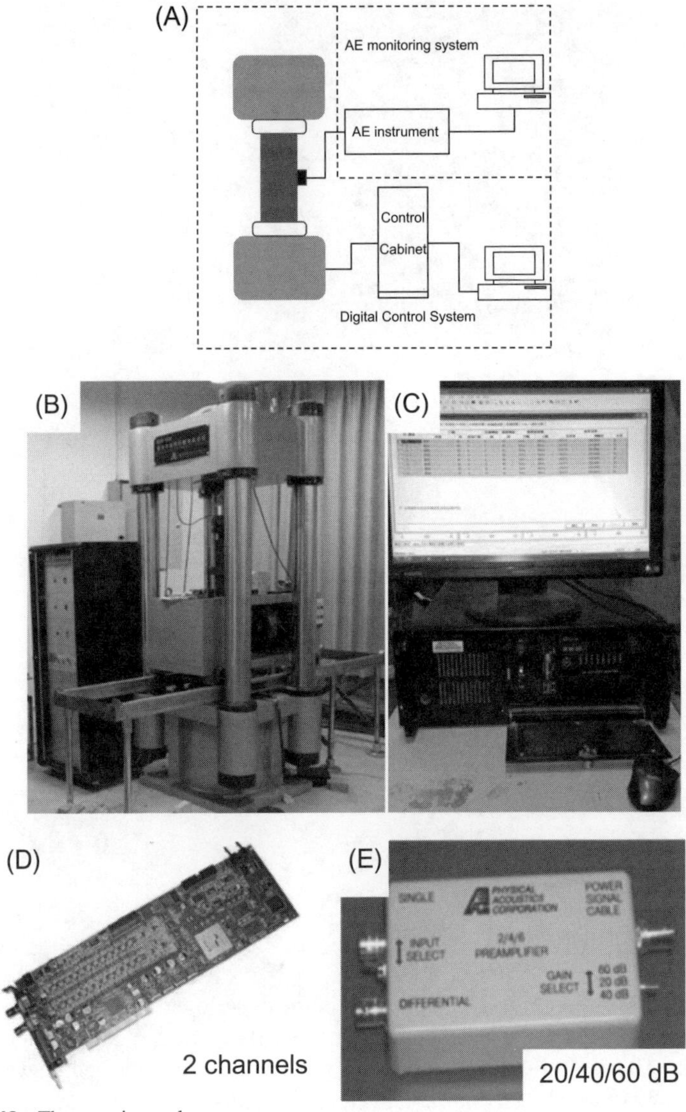

FIG. 4.48 The experimental system.

movement. Thus, in this experiment, the biaxial servo experimental testing machine is used, which applies pressure both horizontally and axially.

The operating steps of the biaxial servo experimental testing machine are as follows (Fig. 4.49):

(1) The first step is to preload 20 kN horizontally and axially before the experiment officially starts.

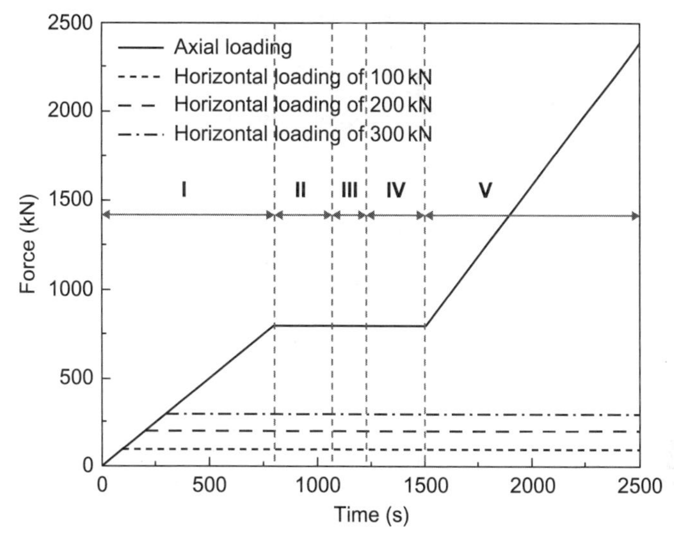

FIG. 4.49 Loading process. I, Initial loading; II, maintain the loading number in two directions; III, detach the filling body; IV, maintain the loading number in two directions; and V, load in the vertical direction again.

(2) Both directions are loaded at a rate of 1000 N/s. The horizontal directions are loaded to 100 kN, 200 kN, and 300 kN, respectively, and the axial directions are loaded to 800 kN.

(3) After the mechanical state completion (where the horizontal force is 100 kN, 200 kN, or 300 kN, and the axial force is 800 kN), the force boundary condition is maintained for 5 min.

(4) The filling body is detached by the extrusion device, and the current stress state is maintained for another 5 min to the stress adjustment.

(5) Finally, the tunnel model is loaded in the vertical direction at a rate of 0.3 mm/min until the rockburst appears around the tunnel wall.

4.5.3 Experimental results

4.5.3.1 Rockburst tendency

Mineral composition analysis

A number of engineering cases show that rockbursts often take place at rock masses that have hard texture, good brittleness, and elasticity (Graham et al., 1991). Granite rock has a high modulus of elasticity. This study uses granite rock from Laizhou in China. Rockbursts often occur in the deep mines located around this area.

The granite samples were observed under a polarizing microscope by transmitted light as well as detected by X-ray diffraction. The results are shown in Fig. 4.50.

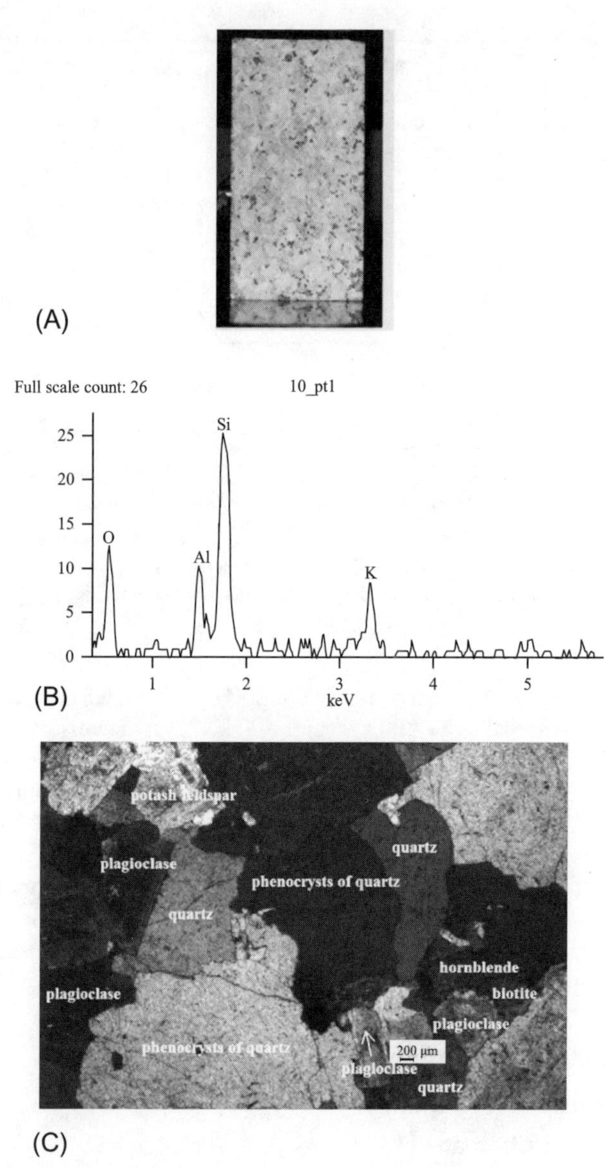

FIG. 4.50 Rock composition. (A) Granite sample, (B) X-ray diffraction results, and (C) image from the result of polarizing microscope.

The dominant components of the granite samples are plagioclase, potash feldspar, quartz, hornblende, and biotite. The specific mineral form and its components are accounted for as follows. The flaggy plagioclase content is approximately 35%–40%, and it has albite twins and a circle-type structure, obviously kaolinitization and sericitization. The flaggy potash feldspar content is approximately 25%–30%, and shows different degrees of kaolinitization and sericitization. The granular quartz content is approximately 20%–25%. The column's hornblende content is approximately 3%–5%, also showing chloritization. Some other minerals, such as biotite, titanite, monazite, and magnetite, also appear, with their contents being approximately 2%–3%.

Analysis of rockburst tendency

One of the necessary conditions for a rockburst to occur is that the rock should have the capacity for a large amount of elastic energy storage (Xu et al., 2017). Thus, we selected the energy criterion to calculate the rockburst tendency of granite rock. The energy criterion can be expressed as:

$$U = \frac{\sigma_c^2}{2E} \tag{4.6}$$

where σ_c is the uniaxial compressive strength and E is the elastic modulus.

When U is larger, the energy storage ability of the rock is higher, so that the rockburst tendency is greater. Based on these results, we can divide the rockburst tendency into four levels: (1) Grade I, the weak rockburst, $U < 40$ kJ/m^3; (2) Grade II, the medium rockburst, 40 kJ/m^3 $< U \leq$ 100 kJ/m^3; (3) Grade III, the strong rockburst,100 kJ/m^3 $< U \leq$ 200 kJ/m^3; and (4) Grade IV, the strongest rockburst, $U \geq$200 kJ/m^3.

The mechanical parameters of granite rock can be seen in Table 4.10. Based on the experimental results, the rockburst tendency of the granite samples is 74.63–94.03; it is Grade II, corresponding to a medium rockburst.

4.5.3.2 Destruction phenomenon of rockburst

Along with rockburst evolution, the evolution of a rockburst in a tunnel can be divided into four stages (He et al., 2010; Kusui et al., 2016): the quiet period stage, the particle ejection stage, the flaking and particle ejection stage, and

TABLE 4.10 Rockburst proneness of granite samples

No.	Elastic modulus (GPa)	Uniaxial compressive strength (MPa)	Energy criterion (kJ/m^3)	Level of rockburst tendency
HGY–1	61.1	72.8	74.63	II
HGY–2	51.8	83.6	94.03	II
HGY–3	56.2	75.8	93.84	II

FIG. 4.51 Process of a rockburst in a tunnel. (A) 669 s, (B) 889 s, (C) 1001 s, (D) 1042 s, (E) 1054 s, and (F) 1100 s.

the complete ejection stage. In this research, the evolution of the rockburst in a tunnel can be seen in Fig. 4.51. There is no significant significance with regard to the early loading (Fig. 4.51A). An increase of loading results in small particles of rock appearing up along the tunnel model (Fig. 4.51B), and then the particles become larger (Fig. 4.51C). Thereafter, misty rock powder can be observed in the tunnel model (Fig. 4.51D and E). As the loading continues to increase, a rockburst occurs and many rockburst pits can be found at approximately waist height in the surrounding rock in the tunnel.

The relationship between the horizontal force and the rockburst fracturing characteristics is seen in Table 4.11.

4.5.3.3 Horizontal stress and rockburst intensity

The rockburst intensity in a tunnel can be described by the morphological characteristics of the burst surface, the depth and width of fracturing, and the acoustic features (Zhou et al., 2015). The final fracturing morphology of different horizontal stresses can be seen in Fig. 4.52.

During the same time, there is a positive correlation between the horizontal stress and the accumulation of strain energy (D'Agostino et al., 2005). The horizontal stress has a close influence on the final fracturing morphology of a rockburst in a tunnel, such as (1) when the horizontal stress is 100 kN (Fig. 4.52A), the intensity of the rockburst in the tunnel is the lowest; (2) when the horizontal stress is 200 kN (Fig. 4.52B), the intensity of the rockburst in the tunnel is greater than 100 kN (the bottom of the tunnel has several rock fragments, and both sides of the tunnel have a cratering shape from the rockburst); and (3) while the horizontal stress is 300 kN (Fig. 4.52C), the rockburst intensity

TABLE 4.11 Experimental results of a rockburst in a tunnel

Horizontal force (kN)	Peak force ofaxially direction (kN)	Fracturing characteristics during rockburst evolution
100	2223	When the axial loading reaches 1927 kN, the rock granules catapult from the surrounding rock, and then flake peeling occurs in the surrounding rock. When the axial loading reaches 2128 kN, "V-shaped" rockburst pits appear
200	2492	When the axial loading reaches 2127 kN, the rock granules catapult from the surrounding rock, then flake peeling appears with a small amount of rock dust discharged as a mist. When the axial loading reaches 2437 kN, "V-shaped" rockburst pits appear
300	2423	When the axially loading reaches 1754 kN, the rock granules and flake peeling catapult from the surrounding rock. When the axially loading reaches 2400 kN, "V-shaped" rockburst pits appear

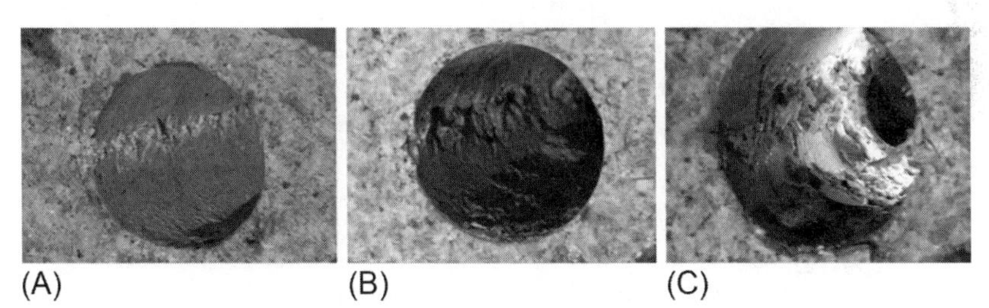

FIG. 4.52 Final fracturing morphology of the rockburst in the tunnel under different horizontal stresses. (A) Horizontal stress is 100 kN, (B) horizontal stress is 200 kN, and (C) horizontal stress is 300 kN.

causes the most serious damage, forming a clear "V-shape" of continuous rockburst pits along both sides of the tunnel.

4.5.3.4 Macroscopic morphology of rockbursts

In underground mines, the elastic strain energy release of rocks is accompanied by the formation of new free surfaces during the rock fragmentation by blasting; this is a transient process (Yang et al., 2012). In the different areas of the surrounding rock of a tunnel, the tensile and shear crack can be found from time

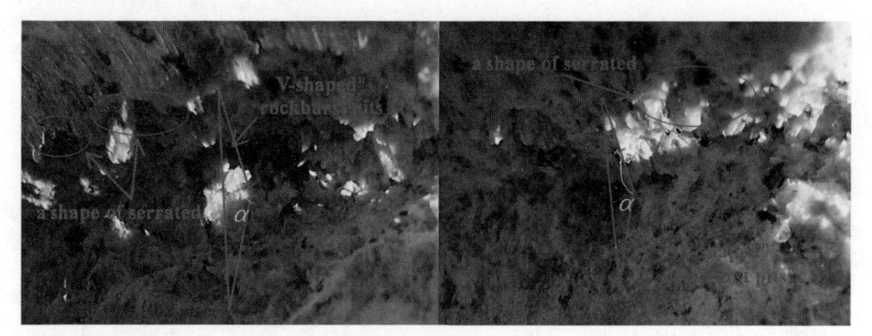

FIG. 4.53 Macroscopic morphology of a rockburst.

and space scales. For this problem of a rockburst in a tunnel, the following process occurs: split (tension stress) into a plate → shear (shearing stress) into a column.

From Fig. 4.53, we can analyze the geometric features of a rockburst. The overall fracturing surface of a rockburst is seen to have a V-shape. The edge of the fracturing surface is seen to have a serrated shape. Both sides of the rockburst pits are seen to have a stepped shape.

4.5.4 AE characteristics of rockburst

4.5.4.1 AE characteristics under different horizontal stresses

The phases of the evolutionary process of AE energy characterize the energy evolution during a rockburst process (Fig. 4.54). Based on the pattern of the axis loading curve, the schematic change of AE timing parameters (energy and cumulate energy) can be divided into four stages.

(1) *The stage of pregnant rockburst:* the amount of AE energy and cumulative AE energy is at a very low level. Some random fracturing occurs.
(2) *The stage of small particle ejection:* the curves of AE energy and cumulative AE energy have a slight upward trend, so that the strength of fracturing is greater.
(3) *The stage of flake peeling:* the AE energy curve displays several obvious surge events, and the cumulative AE energy curve displays a phenomenon of "step-like" rising. Some rockburst events with high intensity appear.
(4) *The stage of rockburst occurrence:* the AE energy curve starts to display the phenomenon of a sudden increase in frequently, and the cumulative AE energy curve is in a rising state. The high-intensity and high-energy release of rock fracturing appears in this.

The amount of cumulative AE energy referents the energy release during rockburst evolution. From the amount of energy released during rockburst evolution, the time of the rockburst occurrence and the precursor time for different

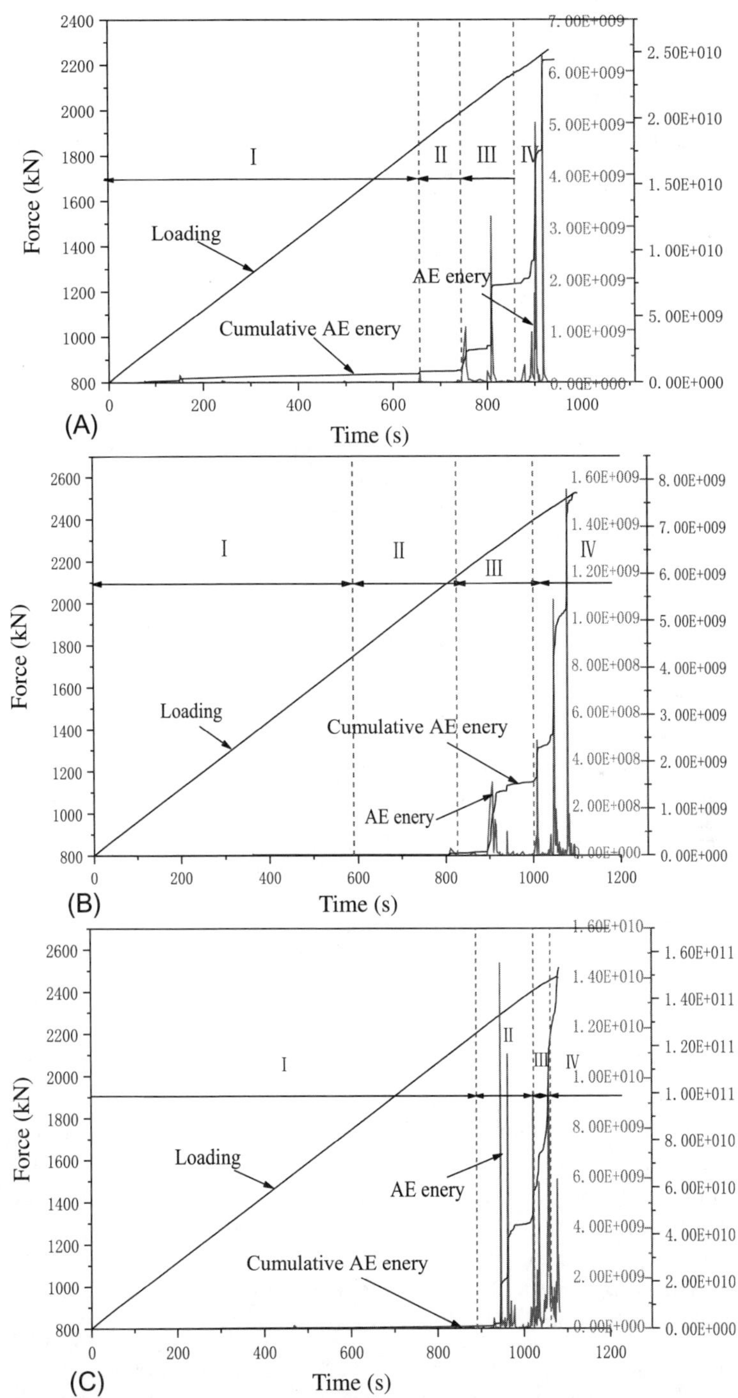

FIG. 4.54 Loading curves and AE parameters. (A) Horizontal stress of 100 kN, (B) horizontal stress of 200 kN, and (C) horizontal stress of 300 kN. I, Pregnant rockburst stage; II, small particle ejection stage; III, flake peeling stage; and IV, rockburst occurrence stage.

TABLE 4.12 Accumulation AE energy and precursor time for different horizontal stresses

Horizontal stress (kN)	Accumulation AE energy (E 10 × aJ)	Relative time of precursor
100	4.82	0.90
200	6.56	0.89
300	9.39	0.86

horizontal stresses also have some variances (Table 4.12). The relationship between horizontal stress and amount of energy released is positively correlated, and is the same as the horizontal stress and degree of intensity of a rockburst. By different conditions of horizontal stress, the cumulative AE energy curve displays step-like inflection points before the rockburst occurs (Fig. 4.54).

4.5.4.2 Rockburst fracturing model

The original stress of the surrounding rock of a tunnel will cause redistribution because of the unloading excavation (Zhou and Shou, 2013). It forms a disturbance stress distribution, and the disturbance stress is in an unstable state. Research shows that the relationship between RA and AF reflects the crack mode during rock fracturing (Ohtsu et al., 2007; Shiotani, 2008), where

$$\mathrm{RA} = \frac{\text{Rise Time}}{\text{Amplitude}} \tag{4.7}$$

$$\mathrm{AF} = \frac{\text{Counts}}{\text{Duration Time}} \tag{4.8}$$

This type of cracking can be classified by the following factors (Farhidzadeh et al., 2014): (1) Mode I, tensile crack, has a high AF value and a low RA value; and (2) Mode II, shear movement, has a low AF value and a high RA value (Fig. 4.55).

By analysis of the rockburst evolution, the process can be divided as follows: the stage of quiet period → the stage of particle ejection → the stage of flaking and particle ejection → the stage of complete ejection. From the view of crack analysis, the process of "V" formation can be divided as follows: split (tension stress) into a plate → shear (shear stress) into a column → fly out. As shown in Fig. 4.56, the crack distribution during rockburst evolution reflects and characterizes the effect of horizontal stress as follows:

(1) In general, there is a common feature in the three types of horizontal stress. The main fracturing mode is tensile in the quiet period stage, which

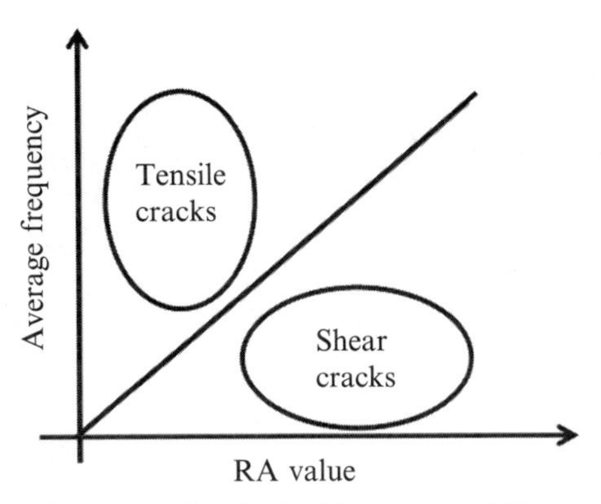

FIG. 4.55 Conventional crack classification (Farhidzadeh et al., 2014).

matches the process of split (tension stress) into a plate. In the complete ejection stage, the fracturing mode switches to shear (shear stress) into a column.

(2) There are also some personality characteristics between the three types of horizontal stress. When the horizontal stress is 100 kN, the fracturing mode from the quiet period stage to the flaking and particle ejection stage is tensile fracturing; shear fracturing appears in the final stage. When the horizontal stress is 200 kN, the fracturing mode in the particle ejection stage begins to appear as shear fracturing. When the horizontal stress is 300 kN, the quiet period stage and the particle ejection stage are a mixed mode of tensile and shear fracturing, but display a state of pure shear fracturing in the stage of flaking and particle ejection as well as the complete ejection stage.

The results are consistent with the relationship between the stress boundary conditions and the characteristics of rockburst occurrence. Thus, as the horizontal stress increases, the fracturing type in the early stage is displayed as a mixed mode of tensile and shear fracturing; pure shear fracturing occurs at a later stage. The strength of the rockburst occurrence is positively correlated with the horizontal stress.

4.5.5 Discussion

4.5.5.1 Tunneling model of excavation mechanics

This is supported by the excavation operation in the experimental setting (Fig. 4.57), where q and P have been constructed as the boundary conditions of the tunnel model. First, the axis and horizontal force should be kept constant. Then, the surrounding tunnel wall appears as the prima facie by the tunnel excavation. Lastly, the tensile stress can be found on both sides of the surrounding wall; fracturing will have taken place first in these areas.

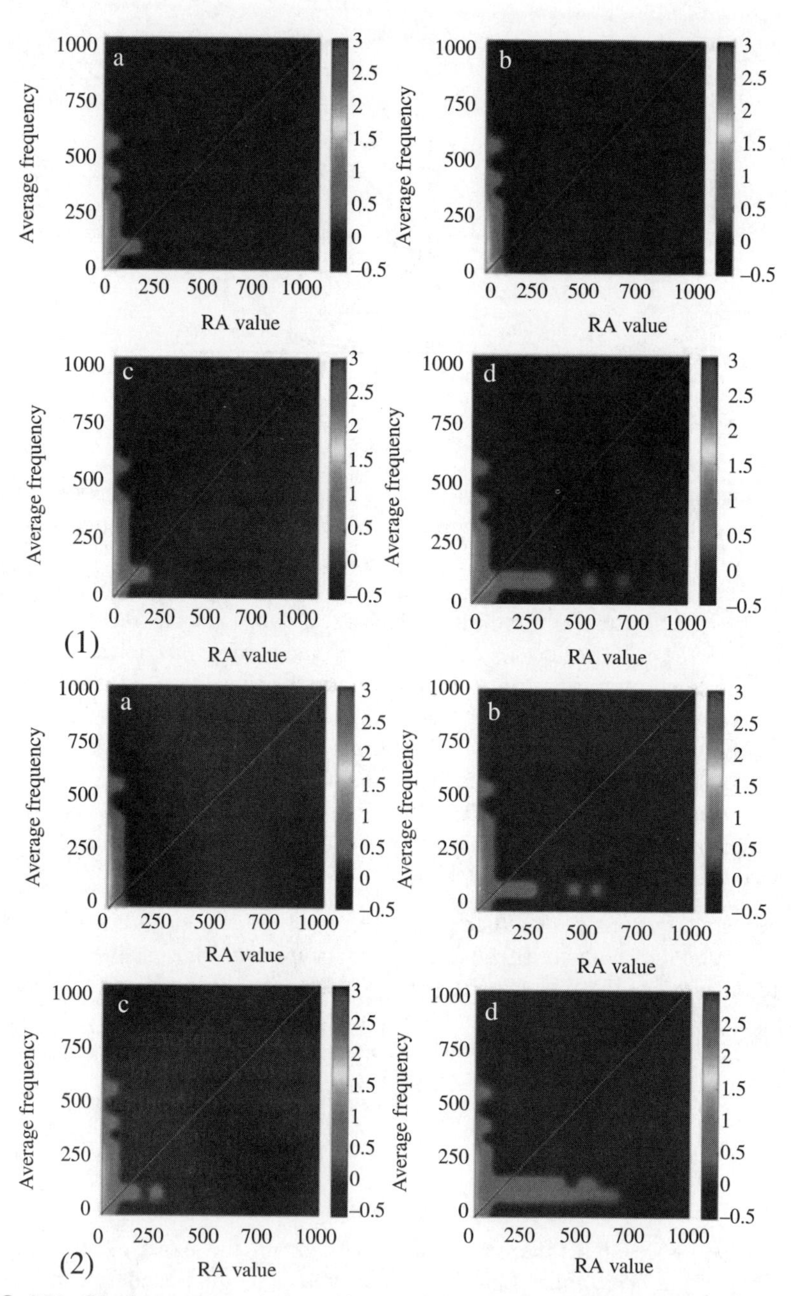

FIG. 4.56 Distribution patterns in the rockburst evolution process under different confining pressures. (1) Horizontal stress of 100 kN, (2) horizontal stress of 200 kN, (3) horizontal stress of 300 kN. a, the quiet period stage; b, the particle ejection stage, c, the flaking and particle ejection stage, and d, the complete ejection stage.

(Continued)

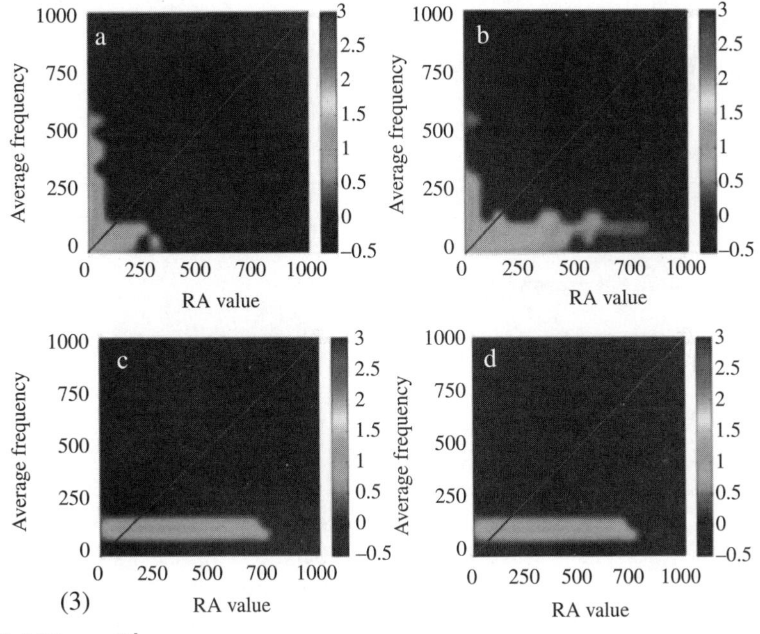

FIG. 4.56—cont'd

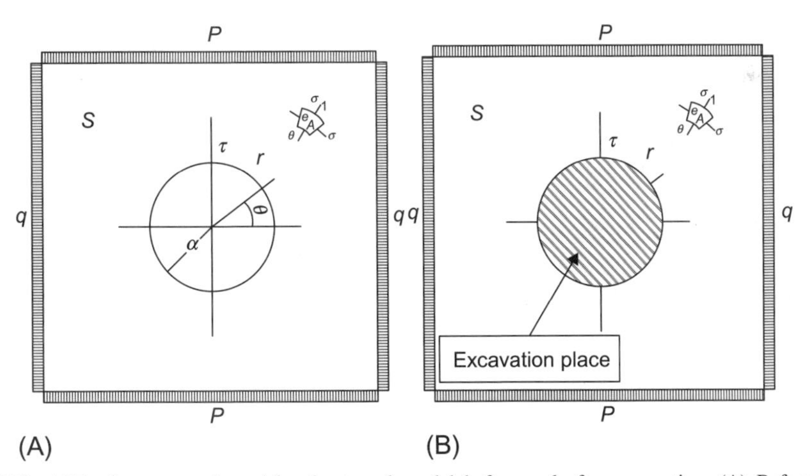

FIG. 4.57 Stress experienced by the tunnel model before and after excavation. (A) Before and (B) after.

The force of decomposition of the V-shaped block of rock is shown in Fig. 4.58. In accordance with the effects of force, the formation of V-shaped rockburst pits can be broken down into the following three characteristics: First, the effect of the normal stress of Q_P is to form a number of tensioning surfaces. Second, the effect of the shear stress of Q_S is to cut the block out of the mother rock and to put the block out of the surrounding rock. Last, under the action of Q_P and Q_S, the area of OAB forms the V-shaped rockburst pits.

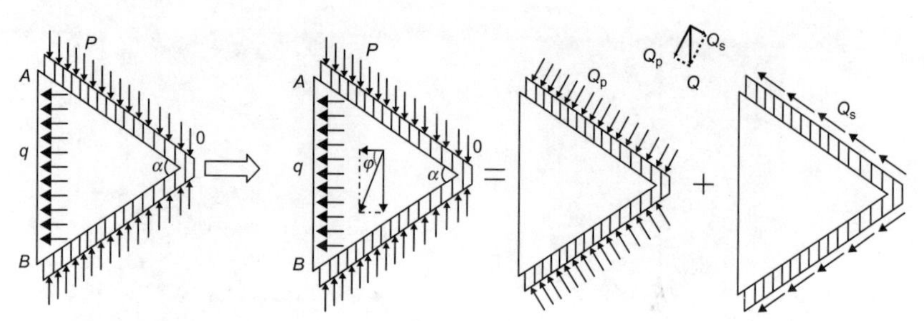

FIG. 4.58 Analyzing the stress in the surrounding rock.

The calculations for Q_S and Q_P can be expressed as

$$\begin{cases} Q_P = Q \times \cos\alpha \\ Q_S = Q \times \sin\alpha \end{cases} \tag{4.9}$$

where $Q = \sqrt{q^2 + P^2}$.

$Q = q^2 + P^2$ with the operation of excavation unloading, both the surrounding walls of a tunnel form a V-shaped rockburst pit; the shear stress of Q_S is parallel to the V-shaped block of rock, and the normal stress of Q_P is perpendicular to the V-shaped block of rock. The other stresses are shown in Table 4.13. We propose that the size of the V shape is determined by three parameters: Q_P, Q_S, and α (Fig. 4.58). The size of the V shape is determined by the magnitude of the destructive force of a rockburst occurrence. As described later, there is a positive correlation between the horizontal stress, Q, and Q_P. There is also a positive correlation between the horizontal stress and Q_P, Q_S, and α.

Along with the above analysis, the rockburst intensity is positively related to horizontal stress, which means that the greater the horizontal stress, the more serious the rockburst situation. Onsite, the horizontal stress usually refers to the tectonic stress. If it is observed that this area has high tectonic stress, we should pay more attention to the rockburst occurrence.

4.5.5.2 Key areas of rockbursts

The split fracturing caused by the unloading at the left and right sides of a tunnel is the formation mechanism of a single rockburst (Liu, 2016). Essentially, the tunnel enters into the stage of complete ejection and the inner wall of the tunnel produces a rockburst that has the following circulating process: rockburst → stress adjustment → stress adjustment failure → repeated rockburst. While the accumulated energy is released, the process of stress adjustment is over. Finally, a rockburst pit with a continuous V shape can be formed on both side walls of the tunnel. For example, by analyzing the advantages and disadvantages of qualitative classification methods for the rockburst intensity at the Jinping II Hydropower Station in Sichuan province, China, we found that the rockburst area along the axis of the hole

TABLE 4.13 Mechanical response characteristics of the rock blasting process.

Horizontal stress (MPa)	Vertical stress (MPa)	Average vertical stress (MPa)	φ (°)	Average φ (°)	Q (MPa)	Average Q (MPa)	α (°)	Q_P (MPa)	Q_S (MPa)
4.44	100.62	98.80	87.47	87.41	100.72	98.90	45.94	68.78	72.38
	86.31		87.05		86.43				
	101.20		87.48		101.30				
	107.07		87.62		107.16				
8.89	107.64	107.71	85.28	85.28	108.01	108.08	76.42	76.42	45.00
	109.78		85.37		110.14				
	111.47		85.44		111.82				
	101.96		85.02		102.35				
13.33	108.44	110.79	82.99	83.14	109.26	111.59	81.41	76.32	43.15
	109.78		83.07		110.58				
	111.47		83.18		112.26				
	101.96		82.55		102.82				

was either intermittent or continuously distributed. Some rockbursts in partial areas lasted for 7–10 days or even up to a month (Feng et al., 2015). The experimental conditions are very similar to those encountered at this site.

As can be seen from the above discussion, the middle part of the tunnel first has small particles of popped-up rock, then the rock cuttings begin to eject one after another, and finally the rockburst occurs in the same place. Thus, for this type of rockburst, we should focus on monitoring the space of the middle part of the tunnel.

4.5.6 Conclusions

In this research, we carried out the simulation test of a rockburst in a tunnel, and the process of rockburst evolution was analyzed under different horizontal stresses. In order to capture the characteristics of a rockburst in a tunnel in various stages (the quiet period stage, the particle ejection stage, the flaking and particle ejection stage, and the complete ejection stage), the meso-fracturing mechanical method and AE monitoring were used. We considered the relationship between horizontal stress and rockburst intensity. The results indicate the following:

(1) The rockburst tendency of granite samples is in Grade II, corresponding to a medium rockburst. The granite samples have a higher elastic modulus and compressive strength, and their capability for withstanding strain is stronger. They are suitable samples for studying a rockburst in a tunnel.
(2) According to the stress analysis of the rockburst pits, the horizontal stress and rockburst intensity are positively correlated. Thus, the greater the horizontal stress, the higher the impact of the rockburst.
(3) The horizontal stress and fracture type are related. The early stages of rockburst evolution are characterized by a tensile-shear mixed model. Thereafter, the ratio of shear fracturing is positively correlated to the horizontal stress.
(4) In the early stages of a rockburst, the curve of cumulative AE energy shows a "step-like" rising trend. The high-intensity and high-energy release from the rock-fracturing event appears at the rockburst occurrence stage. Thus, we can use this characteristic of a "step-like" rising trend to predict the occurrence of a rockburst in tunnel.

4.6 AE and infrared monitoring in tunnels

4.6.1 Introduction

Rich groundwater content can produce a complex geological environment for underground tunnels. Overall, the water (moisture) effects include weathering, erosion, freeze-thaw cycling, chemical and physicochemical degradation, and instigation of biological degradation (Verstrynge et al., 2014). Deep underground engineering sites typically include tunnels of various types (including drift tunnels, shafts, and inclined shafts) and the probability of rockburst development is quite high at such sites. A watery state is a common case in the rock mass, and the tunneling is easily to meet the rock mass in water-bearing state (Lisa et al., 2012). Achieving safety and economy in tunnel construction can be challenging because of complex geological conditions such as faults, fractures, and caverns as well as high in situ stress and the presence of water (Xu et al., 2016). Rockbursts have been defined in many ways (Brown, 1984).

Despite the fact that monitoring, preventing, and controlling rockbursts pose many issues, important achievements have been made in the monitoring and prediction of rockbursts in tunnels. Currently, acoustic emission (AE)/microseismic and infrared monitoring are widely used in the prevention of rockbursts (Cress et al., 1987; Saltas et al., 2014). By analyzing the characteristics of the AE time and frequency domains, the wavelet transform was used to decompose the AE signals. The artificial neural network (ANN) approach was used to recognize rock types and distinguish between AE and noise signals. The study on rock fracturing evolution has been carried out. The AE/microseismic monitoring systems are used to analyze the stability of rock samples. Low frequencies were associated with pore fluid decompression and found to be located in the damage zone of the fractured sample. Excavation unloading has been found to induce the formation and localization of microcracks, which eventually produced local rock failure. The differences and similarities between tunnel support designs can be briefly examined by their AE characteristics (William, 2001; Philip et al., 2008). It is important to collect full-wave AE data and capture rockburst characteristics by analyzing the frequency spectra of the AE signals. Real-time microseismic data can be used to establish a rockburst warning system and provide a dynamic warning of rockburst risk during tunnel excavations (He et al., 2010; Feng et al., 2015). Continuous microseismic monitoring was conducted to capture microseismic precursory information at the Jinping II Hydropower Station in China. The relationship between the spatial and temporal evolution of microseismic activities and rockbursts was thus revealed (Ma et al., 2015).

Far infrared (FIR) is a region in the infrared spectrum of electromagnetic radiation. FIR technology is often applied to monitor temperature variations in stressed rocks (Frid, 2000, 2001). studied the infrared radiation (IRR) of concrete in the process of loading and fracturing with thermal imaging technology. A series of rockbursts was studied using IRR at the laboratory scale. The relation between rock stress and IRR temperature was assessed using the thermo-mechanical coupling theory. The study showed that IRR image abnormalities are important precursors for rock fracturing and can be used in the forecast of rockbursts and tectonic earthquakes. FIR has been used to monitor progressive failure in tunnel models. The results showed that vertical stress enhances tunnel stability, and tunnels with higher confining pressure demonstrated more abrupt and strong rockbursts (Liang et al., 2013a, b, c).

The brittle mechanical characteristics of hard rock exposed by tunneling in a moist environment are of great significance. The rock in a moisture state can affect the brittle hard rock and will improve the design and construction of deep tunnels (Chen et al., 2017). In order to investigate the influence of moisture content of a rock mass on a rockburst, we conducted a series of laboratory rockburst experiments of sandstone under three different moisture contents by the modified true-triaxial apparatus (MTTA). The rockburst process, type, and intensity under different moisture contents were discussed. With the increase of moisture content, the rock strength was softened while the elastic and cumulative damage of the rock were reduced, resulting in a gradual decrease in AE cumulative counts and cumulative energy over the course of the rockburst (Sun et al., 2016). In the study of the Päijänne water-conveyance tunnel in southern Finland, many of the locations where water-conducting fracturing occurs, or where groundwater inflow has been measured at a larger scale, are associated with intersecting or individual topographically interpreted fracture zones (Lipponen and Airo, 2006). Based on detailed analysis of geological/hydrogeological data, it will provide an insight into the permeability distribution in granitic rocks affected by relevant brittle tectonic deformation and the consequences of water inflow during excavation (Perello et al., 2014). Groundwater can cause the problems of open pit slope stability, rock consolidation, and surface subsidence, and rockburst prevention can be prevented by using water injection (Mironenko and Strelsky, 1993; Frid, 2000; Song et al., 2014).

The mode of action and mechanisms of the water state can divided into three categories: weathering, chemical erosion, and stiffness degradation, it will changes the mechanical properties of rock (Wong et al., 2016). A considerable amount of research has focused on rock mechanical characteristics as a function of watery states. However, few experiments have discussed the acting mechanisms of moisture from the IR monitoring and AE monitoring, and the influence of water on rockbursts is rarely studied. Understanding the functional mechanisms of a rockburst in a hard and brittle rock with water effects entails additional research.

This research analyzes rockbursts in a tunnel model, assuming dry and saturated conditions. The AE and infrared thermal imaging systems are used to monitor the evolution of rockbursts during the tunneling process. The mechanism of water influencing the rockburst in tunnels is thus revealed. Finally, the premonitory information of a rockburst on the time and space scale should be analyzed and discussed.

4.6.2 Simulating rockbursts in a tunnel

4.6.2.1 Sample preparation

Granite rocks (150 mm × 150 mm × 150 mm) were chosen as samples. The specimens were prepared in accordance with the Standard for Test Method of Engineering Rock Mass (GB/T50266-99). A round hole of diameter

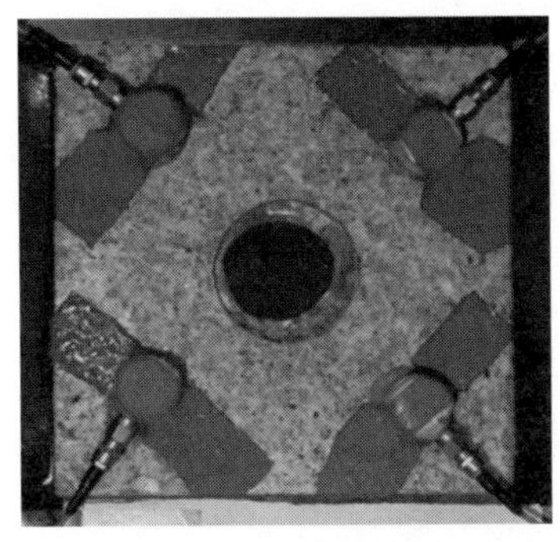

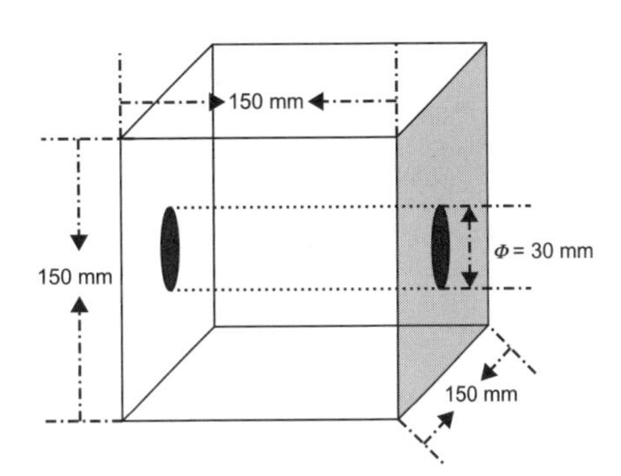

FIG. 4.59 A rock sample and its tunnel model.

30 mm was drilled centrally in the front and rear of the rock specimens, and the parallelism error of two head faces was within 0.02 mm (Fig. 4.59).

The samples were prepared as follows: (a) Dry samples: At least five rock samples were oven dried at a constant temperature of 105°C for 48 h. (b) Saturated samples: Another five rock samples were used as a saturated treatment. The free moisture absorption method was used. First, a quarter of the samples were immersed in a sink of water for 2 h. Second, one-half of the samples were immersed after another 2 h. This was followed by immersing three quarters after another 2 h. Then, all the samples were immersed after the last 2 h. Finally, the samples absorbed water for 48 h.

4.6.2.2 Laboratory equipment

The experimental system comprised the loading system, the AE system, the infrared monitoring system, and the air conditioner (Fig. 4.60).

Loading system: A servo-controlled rock testing machine was used (RLW-3000, Chaoyang Test Instrument Corporation, China). The deformation and applied vertical force can be monitored. The capacity of the axial load transducer is up to 3000 kN while that in the horizontal direction is up to 1000 kN.

AE monitoring system: The AE activities of rock fracturing were recorded by an AE detector with eight channels (PCI-2, Physical Acoustic Corporation, United States). Multiple parameters of AE data, including waveforms, hits, ring-down counts, and amplitudes, can be obtained with this AE system.

Infrared monitoring system: The spectral range of the infrared thermal imager system (InfraTec Image IR8325, Infra Technology Company, Germany) ranges from 3.7 to 4.8 μm, with a resolution of up to 640 × 512 pixels and a thermal sensitivity of less than 25 mK at 25°C. The temperature measurement accuracy is ±1 K. The sampling rate is up to 100 S/s (sampling/second).

A constant temperature of 25°C was maintained in the laboratory by using an air conditioner (Gree Company, China). In order to guarantee the conformity of the experimental data, the equipment setup should be consistent (Ishid, 2017). The various components of the system were set as follows: (1) The loading system: After the formation of the tunnel, the area of the surrounding rock

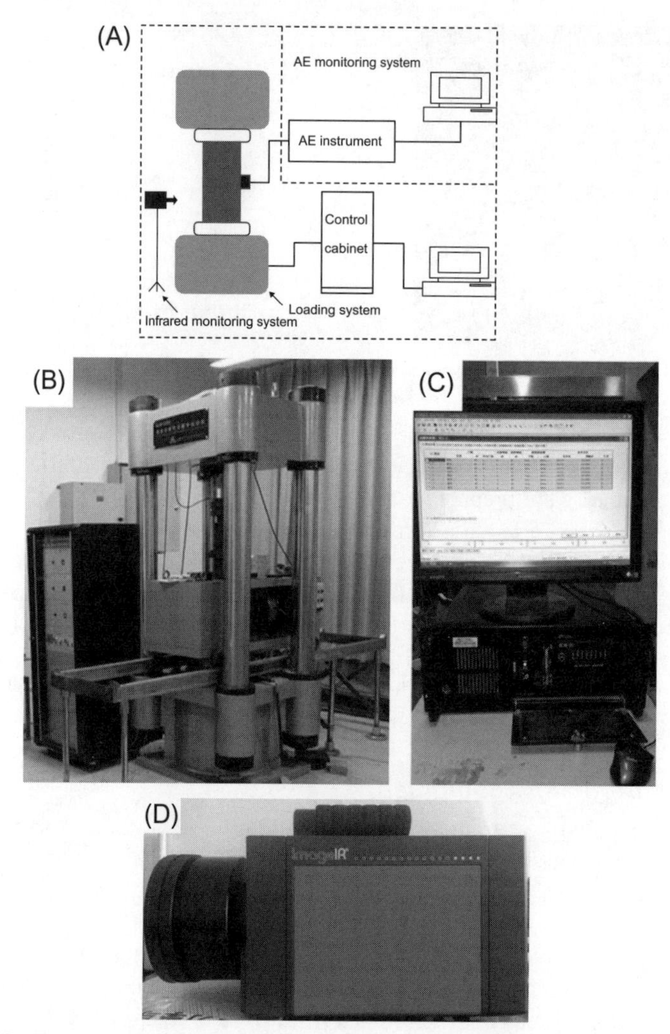

FIG. 4.60 Experimental equipment used in this study. (A) The experiment system, (B) the loading system (RLW-3000), (C) acoustic emission monitoring system (PCI-2), and (D) infrared monitoring system (InfraTec ImageIR8325).

surface approximated bidirectional loading conditions. The horizontal force was set to 30 kN, and the axial loading rate of 0.3 mm/min was applied until the rockburst occurred at the displacement control loading mode. (2) The AE system: The sampling time was set to 0.2 μs, with a memory length of 2 k (2048 words). In this case, the recording time period was set to 0.4 ms (0.2 μs × 2048). The pretrigger was set to 1 k, and the sampling rate to 1 MHz. (3) The infrared monitoring system: The resolution and sampling rate were set to 640 × 512 pixels and 80 S/s, respectively.

4.6.3 Experimental results

4.6.3.1 Rockburst evolution process

In this experiment, the inner holes of the tunnel model experienced particle ejections of rock blocks, and finally, the tunnel model collapsed. The evolution of the rockburst in the tunnel can be divided into four stages (He et al., 2010; Kusui et al., 2016): quiet period, particle ejection, flaking and particle ejection, and ejection completed.

Quiet period stage: There are no obvious fracturing phenomena around the tunnel model at the beginning of loading. It can be called the quiet period (Fig. 4.61A).

Particle ejection stage: When the loading is increased to a certain point, the rockburst dissipates the accumulated energy, and the holes begin to eject tiny particles (Fig. 4.61B).

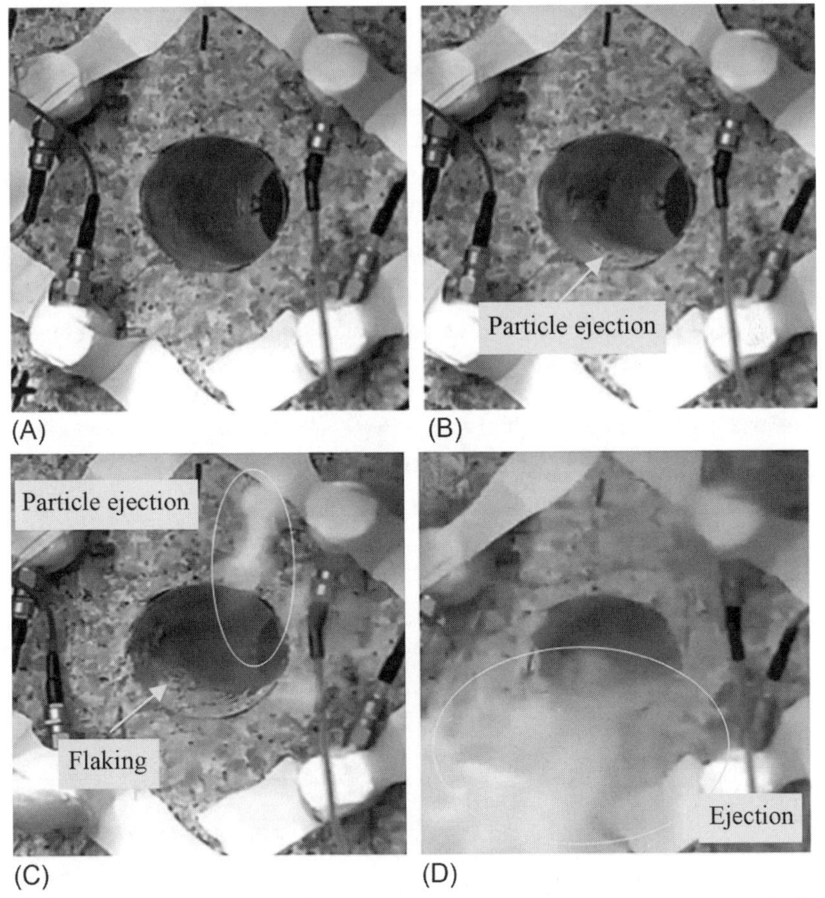

FIG. 4.61 The four stages of a rockburst process. (A) Quiet period stage, (B) particle ejection stage, (C) flaking and particle ejection stage, and (D) ejection completed.

Flaking and particle ejection stage: As the experiment proceeds, rock fragments are ejected from the holes and more rock flakes off. The internal holes go into a state of "smoking" because of these violent ejections. Gradually, the rock spalls, accompanied by ringing bursts (Fig. 4.61C).

Ejection completed stage: The area around the holes shows major spalling, a considerable amount of rock is ejected, and the hole becomes seriously deformed. The rock specimen suddenly loses its stability, and this phenomenon is accompanied by loud noises. Because of the violent ejections, smoke is observed around the specimen and visible cracks gradually appear on both sides of the holes (Fig. 4.61D).

Rockbursts are typically defined as damage to an excavation in a sudden or violent manner. The rockburst mechanism is the same as that of rock fracturing, and it exhibits tensile mode, splitting mode, and shearing mode (Zhou et al., 2015). The morphology of rock fracturing observed after a burst indicates the mechanism underlying the failure/rockburst process.

Some comparatively continuous V-shaped rockburst pits were observed at the left and right sides of the tunnel (Fig. 4.62). After the rockburst occurred in a saturated tunnel model, a considerable amount of flaked rock was suspended in the rockburst pits, which were distributed on the left and right sides of the tunnel (Fig. 4.62A). The large rock blocks were found to be scattered at the tunnel bottom, when the rockburst occurred in a dry tunnel model, most of the debris rock is ejected and finally scattered on the tunnel bottom (Fig. 4.62C).

In the saturated state, the surrounding rock is weaker than the dry one, and hence, it is easier to form a wider flake rock and the wall parallel to the cracked area bears an additional load. This increases the size of the V shape in the saturated tunnel model. As we can see from Table 4.14, obvious differences exist in width (L) and depth (D) between the saturated and dry states. The average L of the V shape for saturated samples is 15.056 mm while that for the dry one is 13.436 mm. The corresponding values of D are 4.584 mm and 3.734 mm, respectively. However, the opening angle (α) is quite similar: 128.746 degrees

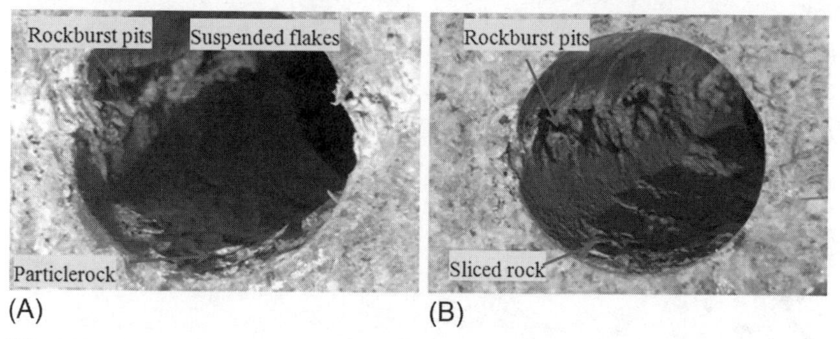

FIG. 4.62 Fissure damage morphology of the tunnel model under the (A) saturated and (B) dry states.

TABLE 4.14 Measurements of the "V" shape of the rockburst pits

State	Specimen number	Width (L) (mm)		Depth (D) (mm)		Opening angle (α) (degree)	
		Measured value	*Average*	*Measured value*	*Average*	*Measured value*	*Average*
Dry	ZRHG-1	13.48	13.436	3.49	3.734	125.28	128.066
	ZRHG-2	12.52		3.34		130.33	
	ZRHG-3	14.41		4.56		128.77	
	ZRHG-4	13.22		3.52		127.65	
	ZRHG-5	13.55		3.76		128.30	
Saturated	BHHG-1	15.88	15.056	4.76	4.584	129.56	128.746
	BHHG-2	16.12		5.33		130.71	
	BHHG-3	14.98		4.56		128.22	
	BHHG-4	13.23		3.67		126.35	
	BHHG-5	15.07		4.60		128.89	

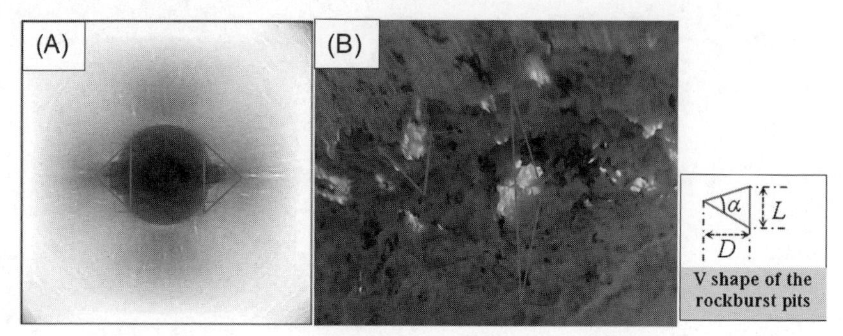

FIG. 4.63 The V shape of the rockburst pit. (A) Internal form of computed tomography (CT) diagram and (B) inner wall of the tunnel model.

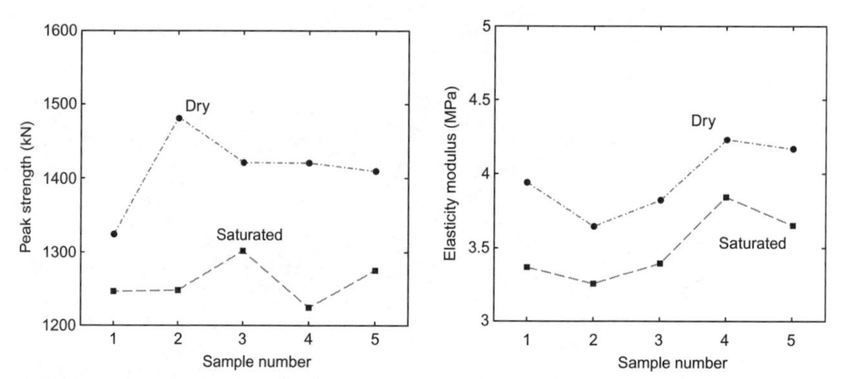

FIG. 4.64 Peak strength and elasticity modulus of the tunnel model under dry and saturated conditions.

and 128.066 degrees, respectively, for the saturated and dry states (Figs. 4.63 and 4.64; Table 4.15).

Table 4.14**.** Measurements of the V shape of the rockburst pits.

Table 4.15**.** Mechanical properties of the rock in the dry and saturated states.

When the tunnel model is saturated, the water affects the rockburst process. It softens the rock and decreases its mechanical parameters (such as the elastic modulus and peak strength). The average ejection speed of the rock blocks declines. In fact, some types of debris rock cannot be ejected. However, the V shape widens. Thus, in the saturated state, the tunnel model exhibits a lower dynamic failure rate, leading to a quicker static failure.

4.6.3.2 AE characteristics

AE activity is fairly common during such an experiment (Fig. 4.65). The tunnel model showed the same characteristics in dry and saturated states. The AE

TABLE 4.15 Mechanical properties of rock in dry and saturated states

State	Specimen number	Elasticity modulus (GPa)		Peak strength (kN)		ΔE (%)	ΔF (%)
		Measured value	*Average (E)*	*Measured value*	*Average (F)*		
Dry	ZRHG-1	4.34	4.28	1324	1411.2	18.3	10.8
	ZRHG-2	3.94		1481			
	ZRHG-3	4.23		1420			
	ZRHG-4	4.17		1421			
	ZRHG-5	4.73		1410			
Saturated	BHHG-1	3.36	3.50	1246	1258.4		
	BHHG-2	3.25		1247			
	BHHG-3	3.39		1301			
	BHHG-4	3.84		1224			
	BHHG-5	3.65		1274			

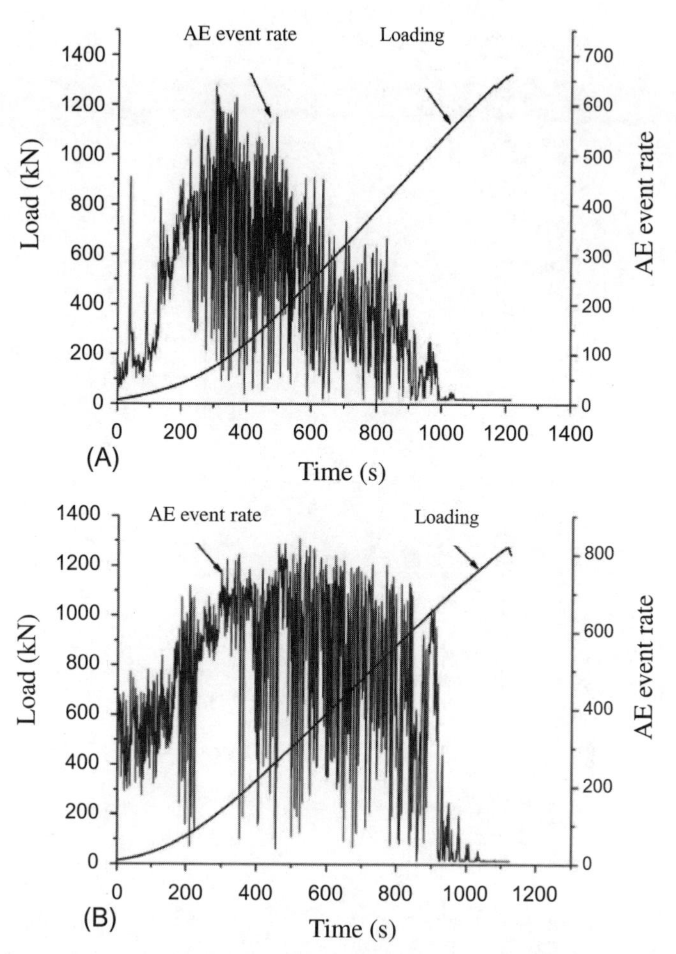

FIG. 4.65 Loading and AE event rate-time curves of granite rock. (A) Saturated sample and (B) dry samples.

activity is mainly concentrated when the loading number ranges from 10% to 80% of the peak number, and the values vary. These cluster characteristics show a relationship between material composition and internal structure. The specimens in this study were mainly composed of feldspar, quartz, and hornblende. The differences in the strength properties of various minerals, the inhomogeneity of the crystallization connection, and the dry flaws in the internal structure make the rocks sensitive to stress and crack propagation under loading.

As shown in Table 4.16, the quiet period lasted longer in the dry state than in the saturated state. The tunnel model in the saturated state underwent crush failure and formed a typical single bevel shearing fracture surface. The dry model first underwent tension failure, typically characterized by a splitting crack. When the axial loading reached 65%–80% of the peak number, the AE events entered a quiet period. The AE quiet period for the saturated samples was shorter than that for the dry ones, and the presence of water weakened the bonds

TABLE 4.16 **Quiet period of AE events for the tunnel model under dry and saturated states**

Dry samples	Duration times of AE quiet period (s)	Saturated samples	Duration times of AE quiet period (s)
ZRHG-1	308	BHHG-1	116
ZRHG-2	220	BHHG-2	120
ZRHG-3	210	BHHG-3	58
ZRHG-4	431	BHHG-4	80.6
ZRHG-5	266	BHHG-5	127
Average	287	Average	100.3

between the rock particles. Water is also associated with chemical corrosion, which reduces the elastic modulus and strength of rock. Cracks are more likely to extend under these conditions. The AE signals were severely attenuated before rupture. The AE quiet period lasted longer than that under dry conditions.

The results suggest that during the rockburst process, there is a reciprocal relationship between the AE event rate and the AE energy rate (Fig. 4.66). During the early stage of rockburst, the rock is mainly subject to the production and extension of microcracks, the AE event rate is higher, and the AE energy rate is low and stable. While the AE event rate decreases in the quiet period, the AE energy rate increases sharply. Rockbursts produce a large number of small AE events during inoculation, but a few large AE events must occur before the rock is destroyed and energy is released suddenly.

Thus, the quiet period of the AE event rate and the sharp increase of the AE energy rate should be the precursors to a rockburst on the time scale.

4.6.3.3 IR characteristics

When rock is under stress, the physical mechanism of thermal radiation includes thermoelastic and friction heat effects (Wu et al., 2006). According to the thermoelastic effect theory, temperature decreases when the rock is under tensile stress. Heat absorption temperature increases when substances are compressed.

Fig. 4.67 shows an infrared image during the process of a rockburst in a tunnel. In the initial stage (80–589 s), the tunnel model was warming up. The movement of small debris rock is recorded between 685 and 845 s. For instance, at low temperatures, particle rock ejection can be observed inside the tunnel at 685 s, slice rock flakes out, and sillar rock can be seen to project outward at 760 s. The larger size of the sillar rock and particle rock in the left and right

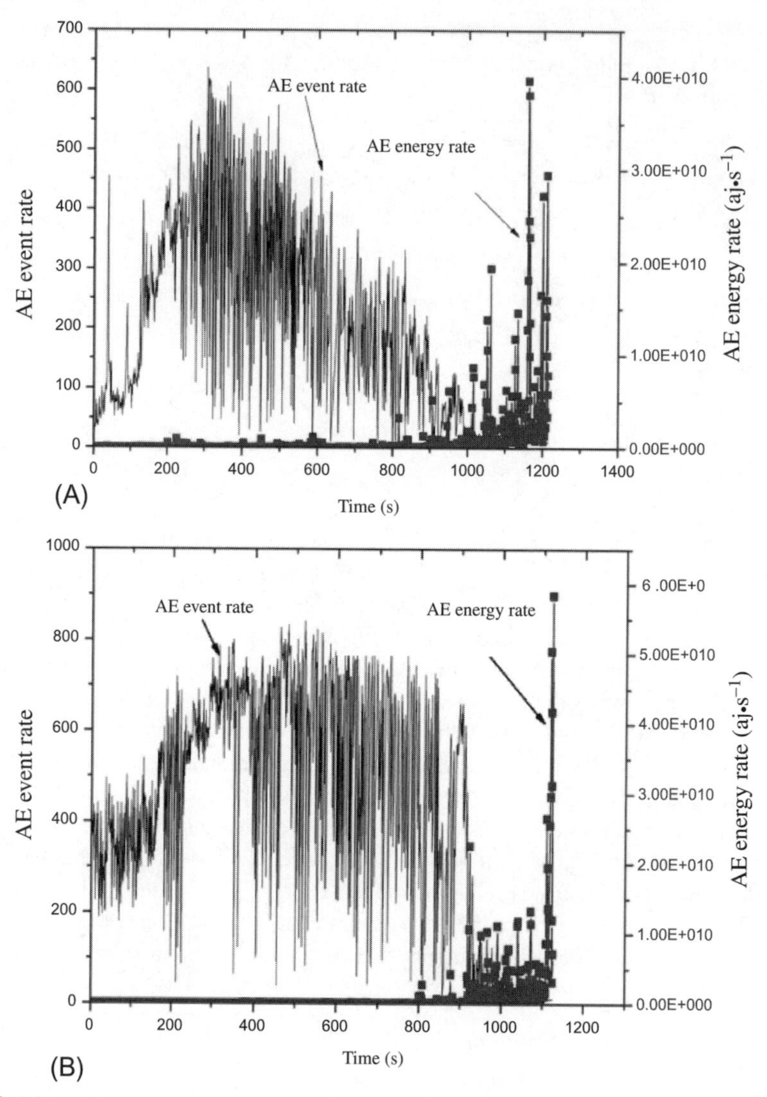

FIG. 4.66 AE event rate and energy rate-time curves for granite. (A) Saturated sample and (B) dry samples.

side walls begins to erupt outward at 845 s. At 1015 s, the rockburst occurs, showing considerable flaking and detachment of the low-temperature rock. The low-temperature dynamic area is in sharp contrast to the entire high-temperature field of the tunnel model. Finally, a large low-temperature area is observed at the tunnel bottom.

The infrared characteristics of tunnel models are marked by several differences between the saturated and dry states (Fig. 4.68). The cutting rock in the

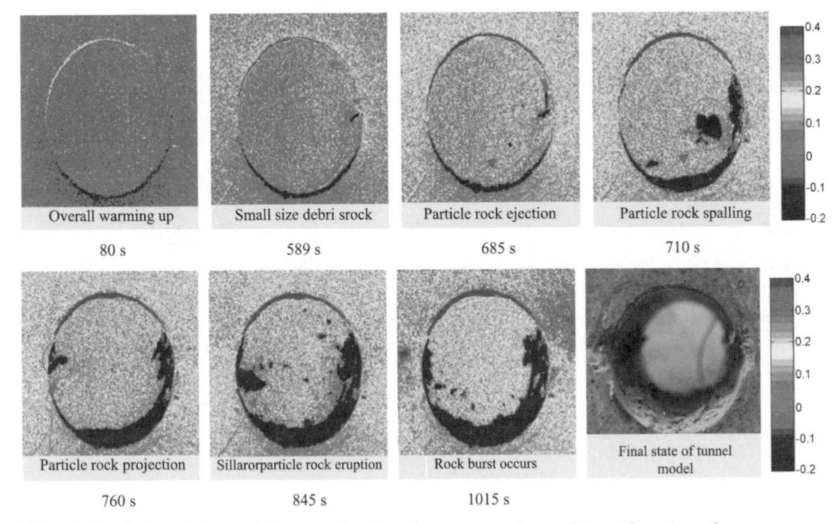

FIG. 4.67 Infrared thermal images showing the process of a rockburst in a tunnel.

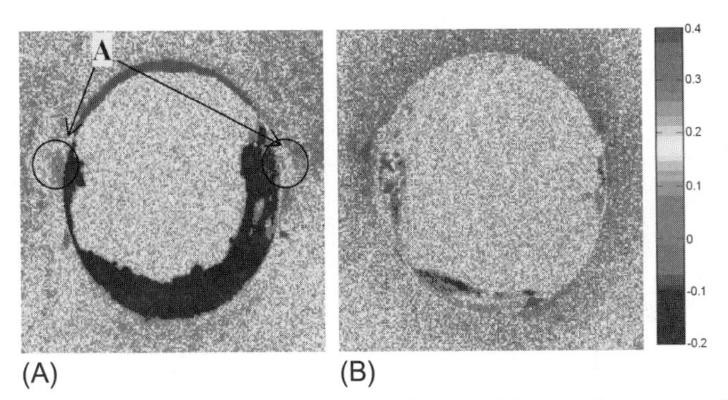

FIG. 4.68 Infrared thermal images of the rockburst in a tunnel for the (A) saturated and (B) dry samples.

saturated state is observed to be at a low temperature typically, but in the dry state, the temperature tends to be mixed. Furthermore, the saturated tunnel model is subject to more serious damage during the rockburst process. It forms a low-temperature zone (the black arrows of A in Fig. 4.68(A) in the high-temperature band around the left and right side walls.

As shown in Table 4.17, the AIRT of dry samples is higher than that of saturated samples. The AIRT of a dry sample is approximately 20°C, and that of a saturated sample is between 24°C and 25°C. During the rockburst evolution, the AIRT of the saturated tunnel model is 24.78°C, and that of the dry tunnel model

TABLE 4.17 AIRT during rock fracturing under dry and saturated states

Dry samples	AIRT (°C)	Saturated samples	AIRT (°C)
ZRHG-1	20.21	BHHG-1	24.67
ZRHG-2	20.06	BHHG-2	24.79
ZRHG-4	20.20	BHHG-3	25.13
ZRHG-5	20.36	BHHG-5	24.51
Average	20.21	Average	24.78

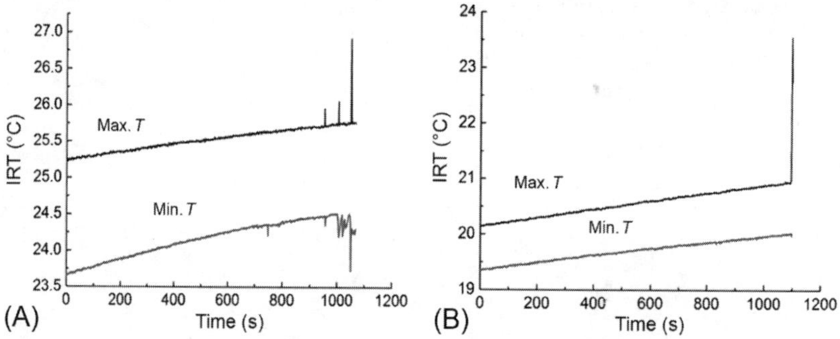

FIG. 4.69 The curves of max. T and min. T of IRT for the tunnel models. (A) Saturated state and (B) dry state.

is 20.21°C. The AIRT of the saturated tunnel model is thus higher than that of the dry one, the difference being 4.57°C.

The max. T and min. T curves are shown in Fig. 4.69. Water can affect the IR characteristics in the rockburst process. Before the rockburst occurs in the tunnel, the max. T and min. T curves for the saturated tunnel model change suddenly and are relatively visible (Fig. 4.69A). The precursor features of the burst in saturated samples were thus more easily captured than those of the dry samples. The max. T of the dry and saturated states increases sharply when the rockburst occurs. The min. T of the dry samples remains stable and uniform without any obvious precursors at any time, and it decreases sharply at the moment of rockburst for the saturated state (Fig. 4.69B).

According to the IR thermoplastic effect theory, the tensile fracturing reduces the IR temperature, and the shear fracturing increases it (Wu et al., 2002). The presence of water weakens the frictional effect between crack surfaces. It first reduces the rate of increase of the IR temperature between the cracks and then leads to further lowering of the IR temperature. In addition, because of the thermoelastic effect, the saturated tunnel model has a higher IR strength

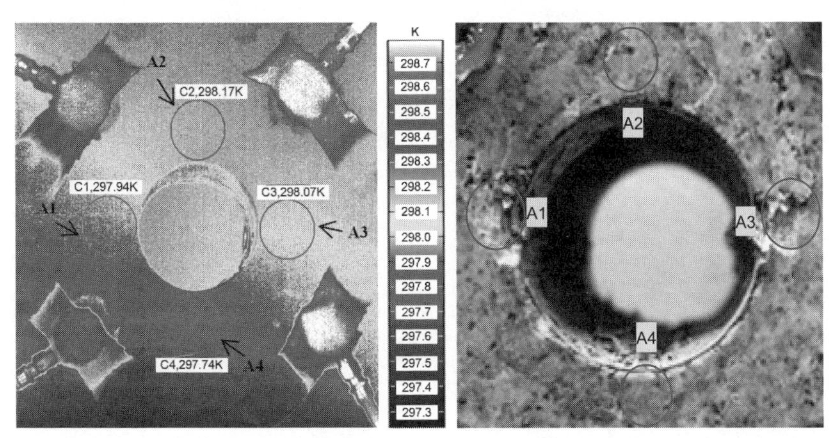

FIG. 4.70 Infrared thermal pictures and delineation of the four areas (A1–A4) around the holes.

under stress. Therefore, water promotes the effects of IR on rock deformation and fracturing.

The variations of average IR temperature (AIRT) in different stress areas were analyzed. Four areas were delineated around the holes of A1, A2, A3, and A4 (Fig. 4.70). The AIRTs of these four areas were monitored, and the following characteristics were observed.

The AIRT curves of the tunnel model fluctuated obviously in the dry state, and this phenomenon was increased by volatility (Fig. 4.71). The AIRT curves in this state were strongly influenced by rock fracturing morphology. The microcracks in the saturated state influence water movement. Under stress, the original cracks/pores become compressed and the water was squeezed out from the interior of the rock. It produced a heat reflection, which offset some of the releasing energy. This phenomenon leads to slight fluctuations in AIRT. At the moment the main fracturing occurs, the AIRT curves of the dry state model show a sudden increase at both holes sites in the tunnel. The AIRT of the saturated model declines, and the curves do not show changes in response to the rockburst explosion.

Thus, the AIRT curves in different areas have regional characteristics (Fig. 4.71). The top and bottom of the surrounding rock is the tensile-stress concentrated area. The tensile-stress concentrated area will produce tensile failure when the tensile-stress concentrates to a certain extent. The left and right of the surrounding rock is the compressional-stress concentrated area; the compressional-stress concentrated area will produce shearing failure when the compressional-stress concentrates to a certain extent. Under the influence of water content, the sudden changing (sudden increase by the saturated tunnel model and sudden decrease by the dry tunnel model) of the AIRT curves in the left and right sides should be used as the precursor to a rockburst on the space scale.

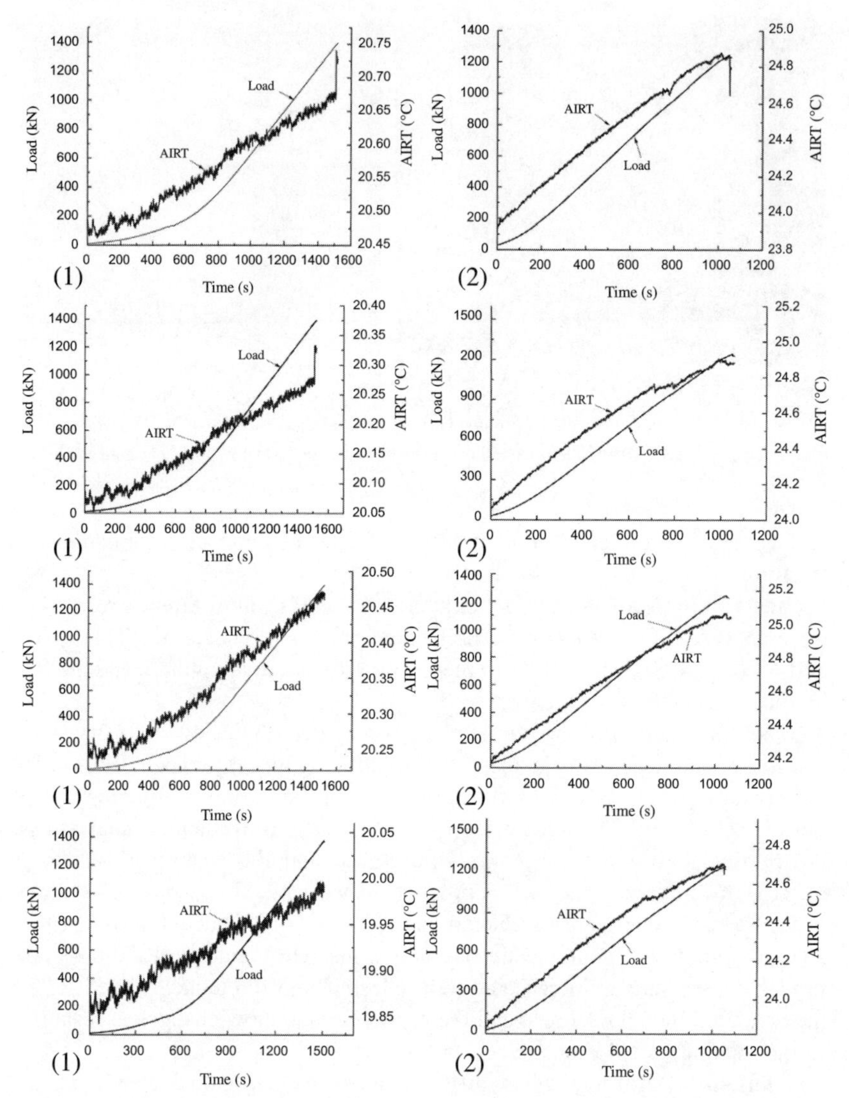

FIG. 4.71 AIRT curves of the tunnel model in saturated and dry states. (1) A1–A4 saturates states and (2) A1–A4 dry states.

4.6.4 Rockburst characteristics in tunnels

The rockburst in the tunnel in this study showed localization characteristics, not only for the simulation experiment at the laboratory scale (Fig. 4.4) but also at the engineering site (Fig. 4.72). An analysis of the evolution process of rockbursts in tunnels shows that the V shape in rockburst pits is a typical characteristic.

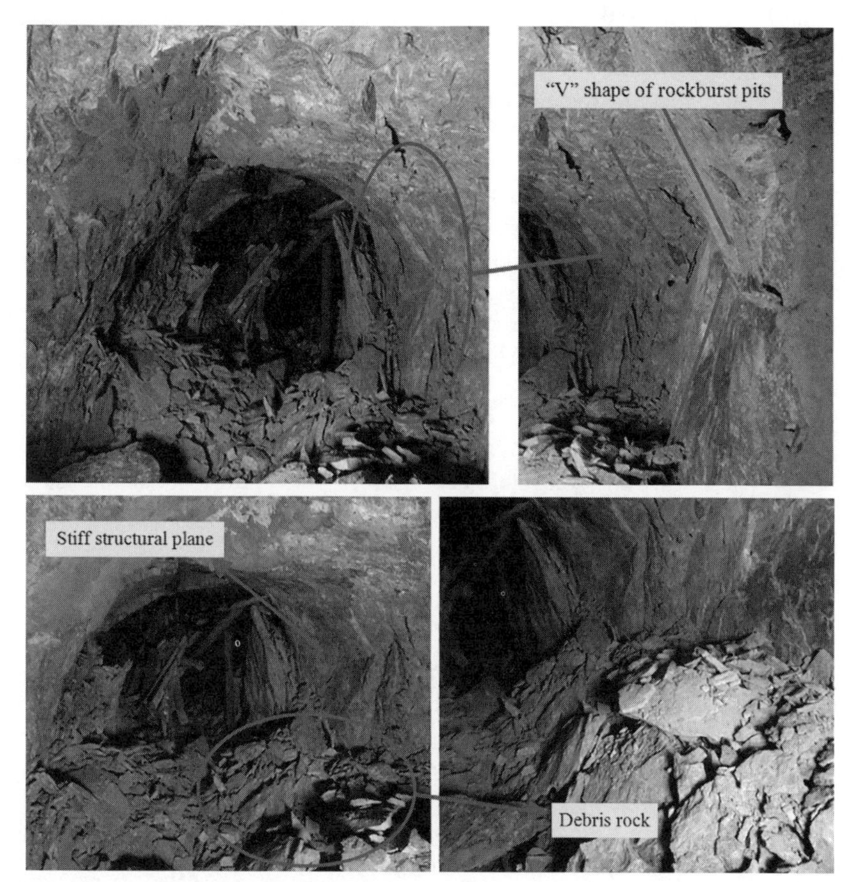

FIG. 4.72 Rockburst in the tunnel at a gold mine site.

This V shape is also called a stiff structural plane (He et al., 2015a, b; Xiao et al., 2016) and is formed by the cycle of stress rearrangement → rearrangement failure → rockburst occurrence → stress rearrangement → rearrangement failure → rockburst recurrence. It is a gradual process, wherein the destruction area formed is smaller and shallower near the tunnel surface at the left and right side walls, and then it constantly expands deep inside the tunnel, finally resulting in the V shape of the rockburst.

Combining the results of the model experiments (Figs. 4.62 and 4.63) and site monitoring observations (Fig. 4.72), we can sum the characteristics of rockbursts in tunnels as follows. (1) Because of the concentration of stress on both the waistlines of the tunnel, rockbursts generally occur in these areas. (2) Rockburst regions are typically characterized by the V shape, which extends along the tunnel to form a continuous crater (rockburst pit). (3) The debris rocks usually occur in the form of flakes or blocks, which often have sharper edges. This also explains, to some extent, the intensity of rock fracturing accompanied by rockburst occurrences.

Typically, rock is a fractured material whose porosity is mainly the result of microcracks (Hoek and Brown, 1997; Steen et al., 2005). When a rock is in the saturated condition, cracks have significant effects on effective thermal conductivity. Because of the considerable differences in the heat transfer characteristics of solids, liquids, and gases, the pores and microcracks between mineral grains before the invasion of liquid/fluid can increase thermal resistance (Liang et al., 2012).

The results of this research showed that when the tunnel model is saturated, the peak intensity number for AE energy is reduced by 11.47% while the elasticity modulus reduces by 8.16%. These findings may be explained as follows: (1) The weakening effect of water in the tunnel surrounding rock: It is difficult for saturated samples to accumulate large amounts of energy in short periods. Thus, particle ejection takes place later. Under the same boundary conditions, static fracturing (large-sized rock block spalling) is more obvious, but dynamic fracturing (ejection of rock cuttings) weakens the rock; (2) The immersion effect of water leads to a rockburst: The immersion effect reduces the frictional force of the shearing slip markedly, and the saturated sample easily forms slipping cracks. Water can soften rock, weaken its mechanical properties (e.g., elasticity modulus and intensity), and decrease its tendency to burst.

Granite is a brittle rock, its elastic modulus is large, and its ability to store energy elastically is large. This brittleness is closely associated with a strong tendency toward bursting. Because there are so few targeted measurements, rockbursts are often handled poorly, and their prediction is also difficult. As the AE phenomenon is accompanied by rock fracturing, it is closely related to the extent of cracking and other internal damage. The presence of water inhibits tensile cracking on the left and right sides of the tunnel model and causes splitting cracks. Pressure shear cracking occurs as water does not have shearing capacity. The pressure shear fracturing of rock requires more energy than tensile fracturing. The presence of water can reduce the threshold of subcritical crack growth, and thus saturated rocks display different internal crack propagation characteristics (Nikolaevskiy et al., 2006). The saturated tunnel model produced a certain amount of fracturing and cracking under lower loading conditions.

Microfractures under a dry state show different characteristics as well as thermal effects. If a tension fracture occurs, no friction is produced between the fractured surfaces and there is no friction heat effect. Fractures expand during tensile bursts, leading to a slight temperature decrease. If shear fracturing takes place, it is usually attributed to the disruption of surface friction and rupture and the increased temperature, causing thermal radiation. The two sides of the holes became the rupture zones during the experimental simulation of a rockburst. The saturated condition promotes the IR temperature variation at the lateral wall (rockburst area) of the tunnel. Thus, the presence of water can improve the IR effect during rock fracturing. This phenomenon can be attributed to the heat capacity of water being higher than that of the rock's mineralogical composition; a saturated tunnel model will absorb more thermal energy during the process of rockburst evolution. The IR temperature variation for the saturated tunnel model is higher compared to the dry state. That is, the IR effect of rock fracturing can be promoted by water. A rockburst involves both rock static and rock dynamic factors (Wang et al., 2011). Water invasion decreases rock strength, its elastic modulus, and the elastic energy storage ability. The peak strength and elasticity modulus of the saturated model were significantly lower than that of the dry samples. The elasticity modulus reflects the energy storage property of rock, and the peak strength reflects the breakdown strength of rock (Wang, 2018). For AE monitoring, the saturated model displays a higher AE event rate, shorter times of the quiet period, and a lower AE energy rate. For IR monitoring, the saturated model is higher than the dry state. Thus, the rockburst of the saturated tunnel model has the characteristics of static damage increases and weakens the dynamic fracturing.

During testing, some characteristics of AE and IR were found before the rockburst. To some extent, the AE and IR activities implied the stress state of the microfracture and reflected the damage evolution in the rock. But there must be some factors that caused the AE and IR intermittent existed in the whole process of rockburst (Wang, 2018). The strain energy of the tunnel model system transformed as deformation energy and dissipated energy. In the different stage of the rockburst process, the energy percent of deformation and dissipation was relatively changing, which could be due to rock inhomogeneity. Deformation energy could also be divided into volume energy and fracture energy, as the dissipated energy could be divided into the AE energy, thermal energy, etc. There exists some change character in before the rockburst occurrence. The AE event rate decreases in the quiet period and the AE energy rate increases sharply can be served as the precursor to rockburst on the time-scale. On the other hand, the AIRT curves of left and right of surround-rock take a sudden increasing by the saturated tunnel model and a sudden dropping by the dry tunnel model, these characteristics of AIRT should be used as the precursor to rockburst on the space-scale.

4.6.5 Conclusions

The moisture state can raise the devastation range of a rockburst, and the phenomenon of static damage increases and weakens the dynamic fracturing will appear when the rockburst occurs. The tunnel model in a saturated condition is more prone to shear fracturing, and it's easily formed atypical fracturing surface of single-cant shear.

When the tunnel model goes into the saturated state, the water can decrease its peak strength and elastic modulus. In this research, the elastic modulus is decreased by 18.3% and the peak intensity by 10.8%.

According to the AE activity in rockburst evolution, the tunnel model in the saturated state takes more colorful than dry model. The AE event rate and the AE energy rate are both effective parameters to monitor the rock failure process.

The rock in a watery situation can reduce thermal resistance, increase heat conduction coefficients, and accelerate the heating process. When the tunnel model is in the saturated state, during the rockburst in tunnels evolution, the infrared radiation (IR) effect is stronger, the changed rate of IR temperatures are largely under unit loading.

The AE event rate decreases in the quiet period and the AE energy rate increases sharply can be served as the precursor to rockburst on the time-scale, and the sudden changing of AIRT curves in left and right sides should be used as the precursor to rockburst on the space-scale.

References

Aker, E., Kühn, D., Vavryèuk, V., Soldal, M., Oye, V., 2014. Experimental investigation of acoustic emissions and their moment tensors in rock during failure. Int. J. Rock Mech. Min. Sci. 70, 286–295.

Alvarez-Ramirez, J., Echeverria, J.C., Ortiz-Cruz, A., Hernandez, E., 2012. Temporal and spatial variations of seismicity scaling behavior in Southern Mexico. J. Geodyn. 54, 1–12.

Armstrong, B.H., 1969. Acoustic emission prior to rockbursts and earthquakes. Bull. Seismol. Soc. Am. 3, 1259–1279.

Arosio, D., Longoni, L., Papini, M., Scaioni, M., 2009. Towards rockfall forecasting through observing deformations and listening to microseismic emissions. Nat. Hazards Earth Syst. Sci. 9 (4), 1119–1131.

Beck, D.A., Brady, B.H.G., 2002. Evaluation and application of controlling parameters for seismic events in hard-rock mines. Int. J. Rock Mech. Min. Sci. 39 (5), 633–642.

Benson, P.M., Vinciguerra, S., Meredith, P.G., Young, R.P., 2008. Laboratory simulation of volcano seismicity. Science 322 (10), 249–252.

Bhat, C., Bhat, M.R., Murthy, C.R.L., 2003. Acoustic emission characterization of failure modes in composites with ANN. Compos. Struct. 61 (3), 213–220.

Brown, E.T., 1984. Rockbursts: prediction and control. Tunnels Tunnell 84, 17–19.

Cai, M., Kaiser, P.K., 2005. Assessment of excavation damaged zone using a micromechanics mode. Tunn. Undergr. Space Technol. Incorp. Trenchless Technol. Res. 20 (4), 301–310.

Chen, G., Li, T., Zhang, G., Yin, H., Zhang, H., 2014. Temperature effect of rock burst for hard rock in deep-buried tunnel. Nat. Hazards 72 (2), 915–926.

Chen, G., Li, T., Wang, W., Zhu, Z., Chen, Z., Tang, O., 2019. Weakening effects of the presence of water on the brittleness of hard sandstone. Bull. Eng. Geol. Environ., 78 (3), 1471–1483.

Cress, G.O., Brady, B.T., Rowell, G.A., 1987. Sources of electromagnetic radiation from fracture of rock samples in the laboratory. Geophys. Res. Lett. 14, 331–334.

D'Agostino, N., Cheloni, D., Mantenuto, S., Selvaggi, G., Michelini, A., Zuliani, D., 2005. Strain accumulation in the southern alps (ne Italy) and deformation at the northeastern boundary of adria observed by cgps measurements. Geophys. Res. Lett. 32 (19), 477–485.

Dixon, N.D., Spriggs, M.S., 2007. Quantification of slope displacement rates using acoustic emission Mon. Can. Geotech. J. 44 (8), 966–976.

Dong, L.J., Wesseloo, J., Potvin, Y., Li, X.B., 2016a. Discriminant models of blasts and seismic events in mine seismology. Int. J. Rock Mech. Min. Sci. 86, 282–291.

Dong, L., Wesseloo, J., Potvin, Y., Li, X., 2016b. Discrimination of mine seismic events and blasts using the fisher classifier, naive bayesian classifier and logistic regression. Rock Mech. Rock. Eng. 49 (1), 183–211.

Dong, L., Shu, W., Li, X., Han, G., Zou, W., 2017. Three dimensional comprehensive analytical solutions for locating sources of sensor networks in unknown velocity mining system. IEEE Access 5 (99), 11337–11351.

Dong, L., Zou, W., Li, X., Shu, W.,Wang, Z.2018. Collaborative localization method using analytical and iterative solutions for microseismic/acoustic emission sources in the rockmass structure for underground mining. Eng. Fract. Mech.https://doi.org/10.1016/j.engfracmech.

Driad-Lebeau, L., Lahaie, F., Heib, M.A., Josien, J.P., Bigarré, P., Noirel, J.F., 2005. Seismic and geotechnical investigations following a rockburst in a complex French mining district. Int. J. Coal Geol. 64 (1), 66–78.

El-Isa, Z.H., Eaton, D.W., 2014. Spatiotemporal variations in the b-value of earthquake magnitude-frequency distributions: classification and causes. Tectonophysics 615–616, 1–11.

Farhidzadeh, A., Mpalaskas, A.C., Matikas, T.E., Farhidzadeh, H., Aggelis, D.G., 2014. Fracture mode identification in cementitious materials using supervised pattern recognition of acoustic emission features. Constr. Build. Mater. 67 (2), 129–138.

Feng, G.L., Feng, X.T., Chen, B.R., Xiao, Y.X., Yu, Y., 2015. A microseismic method for dynamic warning of rockburst development processes in tunnels. Rock Mech. Rock. Eng. 48 (5), 2061–2076.

Frash, L.P., Gutierrez, M., Hampton, J., Hood, J., 2015. Laboratory simulation of binary and triple well EGS in large granite blocks using AE events for drilling guidance. Geothermics 55, 1–15.

Frid, V., 1997. Rockburst hazard forecast by electromagnetic radiation excited by rock fracture. Rock Mech. Rock Eng. 30 (4), 229–236.

Frid, V., 2000. Electromagnetic radiationmethodwater-infusion control in rockburst-prone strata. J. Appl. Geophys. 43 (1), 5–13.

Frid, V., 2001. Calculation of electromagnetic radiation criterion for rockburts hazard forecast in coal mines. Pure Appl. Geophys. 158 (5–6), 931–944.

Fu, X.M., 2005. Experimental study on uniaxial compression deformation and acoustic emission property of typical rocks. J. Chengdu Univ. Technol. 32 (1), 17–21.

Garcia-Barruetabena, J., Cortes-Martinez, F., Isasa-Gabilondo, I., 2014. Study of the acoustic absorption properties of panels made from ground tire rubbers. DYNA 89 (2), 236–242.

Grabec, I., Kuljanić, E., 1994. Characterization of manufacturing processes based upon acoustic emission analysis by neural networks. CIRP Ann. Manuf. Technol. 43 (1), 77–80.

Graham, G., Crampin, S., Fernandez, L.M., 1991. Observations of shear-wave polarizations from rockbursts in a south african gold field: an analysis of acceleration and velocity recordings. Geophys. J. Int. 107 (3), 661–672.

Grzegorz, M., Aleksandra, P., Adam, B., 2016. B-value as a criterion for the evaluation of rockburst hazard in coal mines. In: 3rd international symposium on mine safety Science and engineering/ Montreal, pp. 13–19.

Han, T.C., Zhang, H.J., 2014. Numerical simulation of acoustic emission for defective rock. Chin. J. Rock Mech. Eng. 33 (S1)3198–3024.

He, M.C., Miao, J.L., Feng, J.L., 2010. Rock burst process of limestone and its acoustic emission characteristics under true-triaxial unloading conditions. Int. J. Rock Mech. Min. Sci. 47 (2), 286–298.

He, M.C., Zhao, F., Cai, M., Du, S., 2015a. A novel experimental technique to simulate pillar burst in laboratory. Rock Mech. Rock. Eng. 48 (5), 1833–1848.

He, M., Sousa, L.R.E., Miranda, T., Zhu, G., 2015b. Rockburst laboratory tests database—application of data mining techniques. Eng. Geol. 185, 116–130.

Hill, E.K., Israel, P.L., Knotts, G.L., 1993. Neural network prediction of aluminum-lithium weld strengths from acoustic emission amplitude data. Mater. Eval. 51 (9), 1040–1045.

Hoek, E., Brown, E.T., 1997. Practical estimates of rock mass strength. Int. J. Rock Mech. Min. Sci. 34 (8), 1165–1186.

Huang, B.X., Liu, J.W., 2013. The effect of loading rate on the behavior of samples composed of coal and rock. Int. J. Rock Mech. Min. Sci. 61, 23–30.

Husen, S., Kissling, E., Deschwanden, A.V., 2013. Induced seismicity during the construction of the Gotthard Base tunnel, Switzerland: hypocenter locations and source dimensions. J. Seismol. 17, 63–81.

Ishida, T., Kanagawa, T., Kanaori, Y., 2010. Source distribution of acoustic emissions during an in-situ direct shear test: implications for an analog model of seismogenic faulting in an inhomogeneous rock mass. Eng. Geol. 110 (3–4), 66–76.

Ishida, T., Labuz, J.F., Manthei, G., Meredith, P.G., Nasseri, M.H.B., Shin, K., Yokoyama, T., Zang, A., 2017. ISRM suggested method for laboratory acoustic emission monitoring. Rock Mech. Rock. Eng. 50 (3), 665–674.

Jia, L.C., Chen, M., Zhang, W., Xu, T., Zhou, Y., Hou, B., Jin, Y., 2013. Experimental study and numerical modeling of brittle fracture of carbonate rock under uniaxial compression. Mech. Res. Commun. 50, 58–62.

Jiang, F.X., Miao, X.H., Wang, C.W., Song, J.H., Deng, J.M., Meng, F., 2010. Predicting research and practice of tectonic-controlled coal burst by microseismic monitoring. J. China Coal Soc. 35 (6), 900–903.

Jiang, Y.D., Lu, Y.K., Zhao, Y.X., Gao, Z.X., 2011. Multi-parameter monitoring the stability of rock around roadway while fully mechanized coal face passing through fault. J. China Coal Soc. 36 (10), 1601–1602.

Jiang, Y.D., Pan, Y.S., Jiang, F.X., Dou, L.M., Ju, Y., 2014a. State of the art review on mechanism and prevention of coal bumps in China. J. China Coal Soc. 39 (2), 205–213.

Jiang, J.Q., Wu, Q.L., Qu, H., 2014b. Evolutionary characteristics of mining stress near the hard-thick overburden normal faults. J. Min. Saf. Eng. 31 (6), 881–887.

Karakus, M., Perez, S., 2014. Acoustic emission analysis for rock-bit interactions in impregnated diamond core drilling. Int. J. Rock Mech. Min. Sci. 68, 36–43.

Kim, K.B., Yoon, D.J., 2004. Determining the stress intensity factor of a material with an artificial neural network from acoustic emission measurements. NDT E Int. 37 (6), 423–429.

Kusui, A., Villaescusa, E., Funatsu, T., 2016. Mechanical behaviour of scaled-down unsupported tunnel walls in hard rock under high stress. Tunn. Undergr. Space Technol. 60 (11), 30–40.

Kwak, J.S., Ha, M.K., 2004. Neural network approach for diagnosis of grinding operation by acoustic emission and power signals. J. Mater. Process. Technol. 147 (1), 65–71.

Kwak, J.S., Song, J.B., 2001. Trouble diagnosis of the grinding process by using acoustic emission signals. Int. J. Mach. Tools Manuf. 41 (6), 899–913.

Lavrov, A., 2003. The Kaiser effect in rocks: principles and stress estimation techniques. Int. J. Rock Mech. Min. Sci. 40, 151–171.

Leone, C., Caprino, G., De Iorio, I., 2006. Interpreting acoustic emission signals by artificial neural networks to predict the residual strength of pre-fatigued GFRP laminates. Compos. Sci. Technol. 66 (2), 233–239.

Lesniak, A., Isakow, Z., 2009. Space-time clustering of seismic events and hazard assessment in the Zabrze-Bielszowice coal mine, Poland. Int. J. Rock Mech. Min. Sci. 46, 918–928.

Li, S.L., Yin, X.G., Wang, Y.J., Tang, H.Y., 2004. Studies on acoustic emission characteristics of uniaxial compressive rock failure. Chin. J. Rock Mech. Eng. 23 (15), 2499–2503.

Li, S.G., Lv, J.G., Jiang, Y.D., Jiang, W.Z., 2014. Coal bump inducing rule by dip angles of thrust fault. J. Min. Saf. Eng. 31 (6), 869–875.

Li, J.R., Song, B., Cui, J.Y., 2015. Seismic dynamic damage characteristics of vertical and batter pile-supported wharf structure systems. J. Eng. Sci. Technol. Rev. 8 (5), 180–189.

Li, S.L., Hu, J.Y., Zhou, A.M., Lin, F., Yu, Z.F., 2016. Comprehensive research on character of collapse and fracture of large thick overburden in caving mine. Chin. J. Rock Mech. Eng. 35 (9), 1729–1739.

Li, L.R., Deng, J.H., Zheng, L., Liu, J.F., 2017. Dominant frequency characteristics of acoustic emissions in white marble during direct tensile tests. Rock Mech. Rock. Eng., 50 (5), 1337–1346.

Liang, Z.Z., Xing, H., Wang, S.Y., Williams, D.J., Tang, C.A., 2012. A three-dimensional numerical investigation of the fracture of rock specimens containing a pre-existing surface flaw. Comp. Geotech. 45, 19–33.

Liang, Z.Z., Liu, X.X., Zhang, Y.B., Tang, C.A., 2013a. Analysis of precursors prior to rockburst in granite tunnel using acoustic emission and far infrared monitoring. Math. Prob. Eng. 2013.

Liang, Z., Liu, X., Zhang, Y., Tang, C., 2013b. Analysis of precursors prior to rock burst in granite tunnel using acoustic emission and far infrared monitoring. Math. Probl. Eng. 2013 (3), 1–10.

Liang, Z., Xu, N., Ma, K., Tang, S., Tang, C., 2013c. Microseismicmonitoring and numerical simulation of rock slope failure. Int. J. Distrib. Sensor Netw.

Lin, P., Li, Q.B., Hu, H., 2012. A flexible network structure for temperature monitoring of a super high arch dam. Int. J. Distrib. Sensor Netw. 2012.

Lin, P., Li, Q.B., Jia, P.Y., 2014. A real-time temperature data transmission approach for intelligent cooling control of mass concrete. Math. Prob. Eng. 2014.

Lipponen, A., Airo, M., 2006. Linking regional-scale lineaments to local-scale fracturing and groundwater inflow into the päijänne water-conveyance tunnel, Finland. Near Surf. Geophys 4 (2), 97–111.

Lisa, H., Christian, B., Asa, F., Gunnar, G., Johan, F., 2012. A hard rock tunnel case study: characterization of the water-bearing fracture system for tunnel grouting. Tunn. Undergr. Space Technol. 30 (4), 132–144.

Liu, X.R., Liu, J., 2016. Analysis on the mechanical characteristics and energy conversion of sandstone constituents under natural and saturated states. Adv. Mater. Sci. Eng.

Liu, J.P., Xu, S.D., Li, Y.H., Dong, L.B., Wei, J., 2012a. Studies of AE time-space evolution characteristics during failure process of rock specimens with prefabricated holes. Chin. J. Rock Mech. Eng. 31 (12), 2538–2547.

Liu, N., Zhang, C.S., Chu, W.J., Wu, X.M., 2012b. Microscopic characteristics analysis of brittle failure of deep buried marble. Chin. J. Rock Mech. Eng. 31 (S2), 3557–3565.

Lo, K.Y., Yuen, C.M.K., 1981. Design of tunnel lining in rock for long term time effects. Can. Geotech. J. 18 (1), 24–39.

Lu, J.T., Yang, N.D., Ye, J.F., Liu, X.G., Mahmood, N., 2015. Connectionism strategy for industrial accident-oriented emergency decision-making: a simulation study on PCS model. Int. J. Simul. Model. 14 (4), 633–646.

Ma, T.H., Tang, C.A., Tang, L.X., Zhang, W.D., Wang, L., 2015. Rockburst characteristics and microseismic monitoring of deep-buried tunnels for Jinping II Hydropower Station. Tunn. Undergr. Space Technol. 49, 345–368.

Ma, T.H., Tang, C.A., Tang, L.X., Zhang, W.D., Wang, L., 2016. Mechanism of rock burst forcasting based on micro-seismic monitoring technology. Chin. J. Rock Mech. Eng. 25 (3), 470–483 (in Chinese).

Mansurov, V.A., 2001. Prediction of rockbursts by analysis of induced seismicity data. Int. J. Rock Mech. Min. Sci. 38, 893–901.

McKinnon, S.D., 2006. Triggering of seismicity remote from active mining excavations. Rock Mech. Rock. Eng. 39 (3), 255–279.

Miljenko, K., Sinia, F., Marin, M., 2015. Principal component analysis in determining the best setup for acoustic measurements in sound control rooms. Tehnicki Vjesnik 22 (5), 1327–1335.

Mironenko, V., Strelsky, F., 1993. Hydrogeomechanical problems in mining. Mine Water Environ. 12 (1), 35–40.

Nikolaevskiy, V.N., Kapustyanskiy, S.M., Thiercelin, M., Zhilenkov, A.G., 2006. Explosion dynamics in saturated rocks and solids. Transp. Porous Media 65 (3), 485–504.

Ohtsu, M., Tomoda, Y., Suzuki, T., 2007. Damage evaluation and corrosion detection in concrete by acoustic emission. In: Carpinteri, et al. (Ed.), Fracture Mechanics of Concrete and Concrete Structure-Design. Assessment and Retrofitting of RC Structures, Taylor and Francis Group, London.

Pastén, D., Estay, R., Comte, D., Vallejos, J., 2015. Multifractal analysis in mining microseismicity and its application to seismic hazard in mine. Int. J. Rock Mech. Min. Sci. 78, 74–78.

Perello, P., Baietto, A., Burger, U., Skuk, S., 2014. Excavation of the aica-mules pilot tunnel for the brenner base tunnel: information gained on water inflows in tunnels in granitic massifs. Rock Mech. Rock. Eng. 47 (3), 1049–1071.

Philip, M.B., Sergio, V., Philip, G.M., Young, R.P., 2008. Laboratory simulation of volcano seismicity. Science 10 (322), 249–252.

Poerbandono, R., Suprijo, T., 2013. Modification of attenuation rate in range normalization of echo levels for obtaining frequency-dependent intensity data from 0.6 MHz and 1.0 MHz devices. J. Eng. Technol. Sci. 45B (2), 140–152.

Ren, X.H., Wang, H.J., Zhang, J.X., 2012. Numerical study of AE and DRA methods in sandstone and granite in orthogonal loading directions. Water Sci. Eng. 5 (1), 93–104.

Riemer, K.L., Durrheim, R.J., 2012. Mining seismicity in the Witwatersrand Basin: monitoring, mechanisms and mitigation strategies in perspective. J. Rock Mech. Geotech. Eng. 4 (3), 228–249.

Rudajev, V., Vilhelm, J., Lokajı´Ček, T., 2000. Laboratory studies of acoustic emission prior to uniaxial compressive rock failure. Int. J. Rock Mech. Min. Sci. 37 (4), 699–704.

Saltas, V., Fitilts, I., Vallianatos, F., 2014. A combined complex electrical impedance and acoustic emission study in limestone samples under uniaxial loading. Tectonophysics 637, 198–206.

Samanta, B., Al-Balushi, K.R., 2003a. Artificial neural networks and support vector machines with genetic algorithm for bearing fault detection. Eng. Appl. Artif. Intell. 16 (7–8), 657–665.

Samanta, B., Al-Balushi, K.R., 2003b. Artificial neural network based fault diagnostics of rolling element bearings using time-domain features. Mech. Syst. Signal Process. 17 (2), 317–328.

Shiotani, T., 2008. Parameter analysis. In: Grosse, C., Ohtsu, M. (Eds.), Acoustic Emission Testing. Springer, Berlin (Heidelberg), pp. 41–51.

Šílený, J., Milev, A., 2008. Source mechanism of mining induced seismic events—resolution of double couple and non double couple models. Tectonophysics 456, 3–15.

Snelling, P.E., Godin, L., Mckinnon, S.D., 2013. The role of geologic structure and stress in triggering remote seismicity in Creighton Mine, Sudbury, Canada. Int. J. Rock Mech. Min. Sci. 58 (1), 166–179.

Song, D.Z., Wang, E.Y., Liu, Z.T., Liu, X.F., Shen, R.X., 2014. Numerical simulation ofrockburst relief and prevention by water-jet cutting. Int. J. Rock Mech. Min. Sci. 70, 318–331.

Steen, B.V.D., Vervoort, A., Napier, J.A.L., 2005. Observed and simulated fracture pattern in diametrically loaded discs of rock material. Int. J. Fract. 131 (1), 35–52.

Sun, Z.W., Dai, J., Yang, C.M., Yang, J.F., 2007. Elastic energy criterion of rockburst in roadway and coal face of mine. J. China Coal Soc. 32 (8), 794–798.

Sun, J., Wang, L.G., Tang, F.R., Shen, Y.F., Gong, S.L., 2011. Microseismic monitoring failure characteristics of inclined coal seam floor. Rock Soil Mech. 32 (5), 1589–1595.

Sun, X.M., Xu, H.C., Zheng, L.G., He, M.C., Gong, W.L., 2016. An experimental investigation on acoustic emission characteristics of sandstone rockburst with different moisture contents. Sci. China Technol. Sci. 59 (10), 1–10.

Tang, C., Wang, J., Zhang, J., 2010. Preliminary engineering application of microseismic monitoring technique to rockburst prediction in tunneling of jinping ii project. J. Rock Mech. Geotech. Eng. 2 (3), 193–208.

Varley, N., Arámbula-Mendoza, R., Reyes-Dávila, G., Sanderson, R., Stevenson, J., 2010. Generation of vulcanian activity and long-period seismicity at Volcán de Colima, Mexico. J. Volcanol. Geotherm. Res. 198 (1–2), 45–56.

Verstrynge, E., Adriaens, R., Elsen, J., Van Balen, K., 2014. Multi-scale analysis onthe influence of moisture on the mechanical behaviorofferruginous sandstone. Constr. Build. Mater. 54, 8–90.

Wang, X.B., 2008. Numerical simulation of failure processes and acoustic emissions of rock specimens with different strengths. J. Univ. Sci. Technol. Beijing 30 (8), 837–843.

Wang, C., 2018. Experimental investigation on AE precursor information of rockburst. Evolution, Monitoring and Predicting Models of Rockburst. Springer, Singapore.

Wang, B., Zhao, F.J., Yin, T.B., 2011. Prevention of buckling rockburst with water based on statics and dynamics experiments on water-saturated rock. Chin. J. Geotech. Eng. 33 (12), 1863–1869.

Wang, S.R., Li, N., Li, C.L., Zou, Z.S., Chang, X., 2015a. Instability mechanism analysis of pressure-arch in coal mining field under different seam dip angles. DYNA 90 (3), 279–284.

Wang, H.N., Utili, S., Jiang, M.J., He, P., 2015b. Analytical solutions for tunnels of elliptical cross-section in rheological rock accounting for sequential excavation. Rock Mech. Rock. Eng. 48 (5), 1997–2029.

Weng, L., Huang, L., Taheri, A., Li, X., 2017. Rockburst characteristics and numerical simulation based on a strain energy density index: a case study of a roadway in linglong gold mine, China. Tunn. Undergr. Space Technol. 69, 223–232.

William, D.O., 2001. The behaviour of tunnels at great depth under large static and dynamic pressures. Tunn. Undergr. Space Technol. 16, 41–48.

Wong, L.N.Y., Maruvanchery, V., Liu, G., 2016. Water effects on rockstrength and stiffness degradation. Acta Geotech. 11, 713–737.

Wu, L., Liu, S., Wu, Y., Wu, H., 2002. Changes in infrared radiation with rock deformation. Int. J. Rock Mech. Min. Sci. 39 (6), 825–831.

Wu, L., Liu, S., Wu, Y., Wang, C., 2006. Precursors for rock fracturing and failure—Part I: Irr image abnormalities. Int. J. Rock Mech. Min. Sci. 43 (3), 473–482.

Wu, R., Zai, Y., Liao, L., Kong, X., 2008. Wavelets application in acoustic emission signal detection wire related events in pipeline. Can. Acoust. 36 (2), 96–103.

Xia, Y.X., Kang, L.J., Qi, Q.X., Mao, D.B., Ren, Y., Lan, H., Pan, J.F., 2010. Five indexes of microseismic and their applications in rock burst forecast. J. China Coal Soc. 35 (12), 2011–2016.

Xiao, Y.X., Feng, X.T., Li, S.J., Feng, G.L., Yu, Y., 2016. Rock mass failure mechanisms during the evolution process of rockbursts in tunnels. Int. J. Rock Mech. Min. Sci. 83, 174–181.

Xie, S.R., Li, S.J., Huang, X., Sun, Y.D., Yang, J.H., Qiao, S.X., 2015. Surrounding rock principal stress difference evolution law and control of gob-side entry driving in deep mine. J. China Coal Soc. 40 (10), 2355–2360.

Xu, N.W., Tang, C.A., Li, L.C., Zhou, Z., Sha, C., Liang, Z.Z., Yang, J.Y., 2011. Microseismic monitoring and stability analysis of the left bank slope in jinping first stage hydropower station in southwestern China. Int. J. Rock Mech. Min. Sci. 48 (6), 950–963.

Xu, N.W., Li, T.B., Dai, F., Zhang, R., Tang, C.A., Tang, L.X., 2016. Microseismic monitoring of strainburst activities in deep tunnels at the jinpingII hydropower station, China. Rock Mech. Rock. Eng. 49 (3), 981–1000.

Xu, J., Jiang, J., Xu, N., Liu, Q., Gao, Y., 2017. A new energy index for evaluating the tendency of rockburst and its engineering application. Eng. Geol. 230, 46–54.

Yang, Z.X., Feng, D.Q., Chen, T.J., Wan, H., 2002. Application of artificial neural networks to nondestructive testing. Nondestr. Testing 24 (6), 244–252.

Yang, J.L., Zuo, J.P., Sun, K., Meng, B.B., Lin, X., 2011. In-situ observation and numerical analysis of surface subsidence of high working face with multi-fault induced by full-mechanized mining activity. Chin. J. Rock Mech. Eng. 30 (6), 1216–1224.

Yang, J.H., Lu, W.B., Chen, M., Yan, P., Zhou, C.B., 2012. Mechanism and identification of triggered microseism by transient release of in-situ stress in deep rock mass. Acta Seismol. Sin. 34 (5), 581–592.

Yi, R.X., Liu, S.F., Geng, R.S., 2002. Application of artificial neural network to acoustic emission testing. Nondestr. Testing 24 (11), 488–491.

Yi, T.H., Li, H.N., Gu, M., 2011. Optimal sensor placement for structural health monitoring based on multiple optimization strategies. Struct. Design Tall Spec. Build. 20 (7), 881–900.

Yi, T.H., Li, H.N., Zhang, X.D., 2012. Sensor placement on Canton tower for health monitoring using asynchronous-climb monkey algorithm. Smart Mater. Struct. 21 (12), 1–12.

Yin, X.C., Li, S.Y., Li, H., Wang, M., 1987. On the physical essence of b value for AE of rock tests and natural earthquakes in terms of fracture mechanics. Acta Seismol. Sin. 9 (4), 364–374.

Yoder, M.R., Holliday, J.R., Turcotte, D.L., Rundle, J.B., 2012. A geometric frequency-magnitude scaling transition: Measuring b = 1.5 for large earthquakes. Tectonophysics 532–535, 167–174.

Yu, Q., Tang, C.A., Li, L.C., Li, H., Cheng, G.W., 2014. Nucleation process of rockbursts based on microseismic monitoring of deep-buried tunnels for Jinping II hydropower station. Chin. J. Geotech. Eng. 36 (12), 2315–2322 (in Chinese).

Yuan, R.F., Li, H.M., Li, H.Z., 2012. Distribution of microseismic signal and discrimination of portentous information of pillar type rockburst. Chin. J. Rock Mech. Eng. 31 (1), 80–85 (in Chinese).

Yucemen, M.S., Akkaya, A.D., 2012. Robust estimation of magnitude-frequency relationship parameters. Struct. Saf. 38, 32–39.

Zhang, Y.B., 2013. Spectral character analysis of sandstone under saturation condition in rupture procedure. Rock Soil Mech. 34 (6), 1574–1578.

Zhang, B.H., Deng, J.H., Zhou, Z.H., Lv, H.X., Wu, J.C., Wu, S.H., 2012. Analysis of monitoring microseism in areas controlled by faults near powerhouse in Dagangshan hydropower station. Rock Soil Mech. 33 (S2), 213–218.

Zhao, F., He, M.C., 2016. Size effects on granite behavior under unloading rockburst test. Bull. Eng. Geol. Environ., 1–15.

Zhao, X.D., Tang, C.A., Li, Y.H., Yuan, R.F., Zhang, J.Y., 2006. Study on AE activity characteristics under uniaxial compression loading. Chin. J. Rock Mech. Eng. 25 (S2), 3673–3678.

Zhao, H.J., Ma, F.S., Li, G.Q., Ding, D.M., Wen, Y.D., 2008. Fault effect due to underground excavation in hangingwalls and footwalls of faults. Chin. J. Geotech. Eng. 30 (9), 1372–1375.

Zhao, Y.X., Jiang, Y.D., Wang, T., Gao, F., Xie, S.T., 2012. Features of microseismic events and precursors of rock burst in underground coal mining with hard roof. J. China Coal Soc. 37 (12), 1960–1966 (in Chinese).

Zhao, K., Zhu, Z.C., Zeng, P., Cheng, S.J., 2015. Experimental study on acoustic emission characteristics of phyllite specimens under uniaxial compression. J. Eng. Sci. Technol. Rev. 8 (3), 53–60.

Zhou, X.P., Shou, Y.D., 2013. Excavation-induced zonal disintegration of the surrounding rock around a deep circular tunnel considering unloading effect. Int. J. Rock Mech. Min. Sci. 64 (64), 246–257.

Zhou, H., Meng, F., Zhang, C., Hu, D., Yang, F., Lu, J., 2015. Analysis of rockburst mechanisms induced by structural planes in deep tunnels. Bull. Eng. Geol. Environ. 74 (4), 1435–1451.

Further reading

Assi, A.H., 2011. Engineering Education and Research Using MATLAB. InTech Publications, Rijeka, Croatia, pp. 219–238.

Chen, B.R., Feng, X.T., Li, Q.P., Luo, R.Z., Li, S., 2015. Rock burst intensity classification based on the radiated energy with damage intensity at jinping ii hydropower station, China. Rock Mech. Rock. Eng. 48 (1), 289–303.

Duda, R.O., Hart, P.E., Stork, D.G., 2000. Pattern Classification, second ed. Wiley-Interscience Publications, New York.

Gong, Q.M., Yin, L.J., Wu, S.Y., Zhao, J., Ting, Y., 2012. Rock burst and slabbing failure and its influence on tbm excavation at headrace tunnels in jinping ii hydropower station. Eng. Geol. 124 (1), 98–108.

Liu, X.X., Liang, Z.Z., Zhang, Y.B., Yao, X.L., Li, H.Y., 2016. Unloading test for rockburst mechanism in tunnel model. J. Eng. Geol. 24 (5), 967–975.

Ma, H., Liu, Q., 2017. Prediction of the peak shear strength of sandstone and mudstone joints infilled with high water-cement ratio grouts. Rock Mech. Rock. Eng. 15, 1–17.

Wang, Y.L., Tang, J.X., Dai, Z.Y., Yi, R., 2018. Experimental study on mechanical properties and failure modes of low-strength rock samples containing different fissures under uniaxial compression. Eng. Fract. Mech. 197 (15), 1–20.

Zhuang, D.Y., Tang, C.A., Liang, Z.Z., Ma, K., Wang, S.Y., Liang, J.Z., 2017. Effects of excavation unloading on the energy-release patterns and stability of underground water-sealed oil storage caverns. Tunn. Undergr. Space Technol. 61, 122–133.

Chapter 5

Structural effect of rock blocks

Shuren Wang[a] **and Wenbing Guo**[b]
[a]*School of Civil Engineering, Henan Polytechnic University, Jiaozuo, China,* [b]*School of Energy Science and Engineering, Henan Polytechnic University, Jiaozuo, China*

Chapter outline

5.1 Cracked roof rock beams

5.1.1 Introduction

The breaking instability of the roof in shallow mining shows particularity that the mining pressure behaves abnormally, especially when the working face is too long. This often results in step sinking, bracket damage, etc. For high yield and efficient mining of a large coal field, it is a key problem to research the deformation and breaking law as well as the controlling countermeasures for roof stability under shallow mining. Considering the stratified depositing feature of coal-bearing strata, the stratified roof is often regarded as the elastic rock beam or block beam by scholars.

Hou et al. (2000) regarded the stratified cracked roof as a beam-type elastic component, and studied the static bifurcation behavior and the mechanism of bending catastrophe phenomena separately using the stability theory of elastic systems (Gao, 2004). Yang (2010) concluded that the mechanism of the main roof breaking at a shallow depth was caused by the bifurcation instability of the roof structure. Taking the unit-wide rock stratum structure of the middle part of the mining face as the study object, Pan et al. (2012) derived the related expressions of roof deflection, bending moment, and bending strain energy density distribution of the rock stratum ahead of the coal face before and after the initial fracturing. Li et al. (2014) analyzed a new mechanical model of the roof rock beam with both ends fixed in cemented filling mining, and deduced the calculating formulas of limit span based on tensile and shear strength. Wang et al. (2015c) obtained the limit position of the rotary instability of the roof rock beam according to the principle of minimum potential energy, and put forward a dynamic method for determining the supporting resistance according to the specific instability form of the shallow buried high-intensity mining face.

Under the shallow mining conditions, Zhao and Song (2016) thought that the small angle of rotation and the horizontal thrust of the block with different fragmentation increases slightly with the length of the mining face and the angle of rotation. Diederichs and Kaiser (1999) summarized the critical span-thickness-modulus relationship of the unsupported stability of the jointed roof rock beam. Nomikos et al. (2002) found that the multijointed roof rock beam had a relatively small deflection, which indicated an increase in the deflection and a decrease in the extreme strain, especially at the abutment. Alejano et al. (2008) thought it was very difficult to predict the occurrence of buckling instability of the roof rock beam because small changes produced significant stability variations from the practical engineering perspective. Marcak (2012) believed that the distribution of stresses, the average energy of seismic events and their frequencies, whose significant changing trend was caused by the bending of the roof layers over the exploited area in many Polish underground mines. Assuming the undermined sedimentary rock layers as multicracked hinged beams, Tsesarsky (2012) concluded that the thickness of the compressing arch at the abutments was inversely proportional to beam stiffness, and at the mid-span was positive proportional. Please et al. (2013) found that the roof layer formed a beam clamped by pillars of two ends, whose adjacent cracks would occur when the stresses exceeded the tensile strength of the beam.

In the mining process analysis, the cracked roof is usually simplified as a simple supported beam, which often leads to defects if the effect of the deflection and the horizontal thrust is ignored. In addition, current research rarely involved the bending instability process of the roof rock beam under the combined effect of the vertical and horizontal loads. It is difficult to judge the deflection effects on the structure stability of the beam pre- and postbreaking. In this paper, considering a mining face at the Shangwan Mine of Inner Mongolia in China as the background, a modified model of the Euler beam was used to analyze the bending instability process of the cracked roof rock beam under shallow mining conditions. Applying the stability theory of elastic systems, the relationship between the deflection of the roof rock beam and the state of the block rotary motion in the voussoir beam was established, which would provide the theoretical reference for roof management in mining practice.

5.1.2 Mechanical model of a cracked roof beam

5.1.2.1 Formation of cracked roof beam

Taking the main roof of a mining face at the Shangwan Mine as the investigated object, the comprehensive mechanized longwall mining method is used with a double-drum coal cutter of the type JOY 7LS7/lws630 and the ZY18000/32/70 shield type hydraulic support type. The mining face is advanced in an inclined back type as shown in Fig. 5.1A, the length of the mining face is 290 m, and the advanced distance of 2970 m. The dip angle of the coal seam is 1–3 degrees with a nearly horizontal coal seam at an average depth of 117 m. The designed mining height is 5 m, the full thickness of the coal seam is mined in one time, and the goaf area is handled by the caving method, as shown in Fig. 5.1B. Assuming the dimensions of the working face satisfied that the ratio of the advancing distance a to the working face length b is less than one ($a/b \leq 1$) before initial weighting of the main roof.

The main roof before being fractured can be regarded as a fixed beam at two ends, then the span L, thickness h, and the action of uniform load p are shown in Fig. 5.2A. According to the bending stress theory of materials mechanics, the maximum moment M_{max} acting at two ends of the roof rock beam in a constant section equals $-\frac{1}{12}pL^2$ while the maximum normal stress σ_{max} acting at the same section keeping the furthest distance from the neutral axis is $\sigma_{max} = M_{max}y_{max}/I_z$, where y_{max} and I_z can be regarded as constants. When σ_{max} reaches the tensile strength σ_t, the tensile cracks will occur at two ends of the roof rock beam, as shown in Fig. 5.2B.

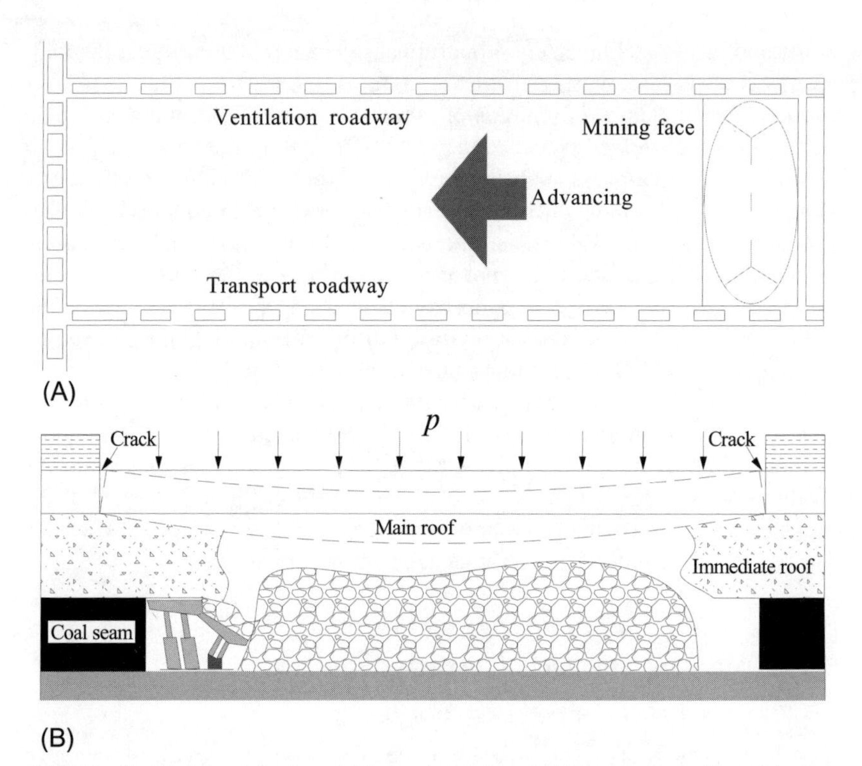

FIG. 5.1 Sketch of the mining face and the cracked rock roof. (A) Arrangement of the mining face and (B) cracked rock proof.

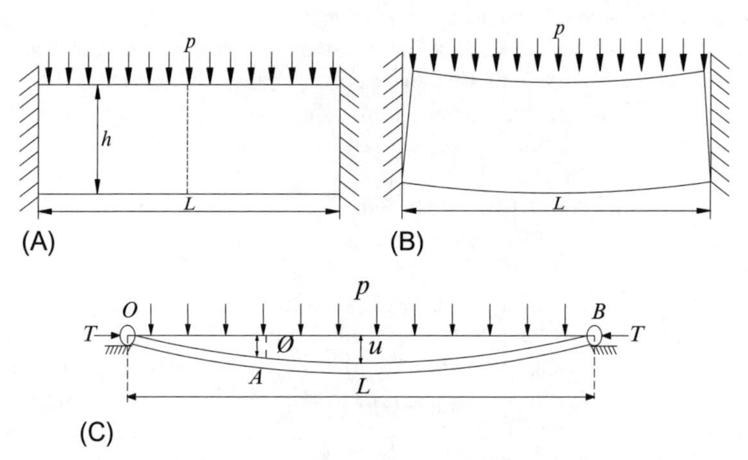

FIG. 5.2 Mechanical model for the basic roof of the mining face. (A) The fixed beam, (B) the simple-supported beam, and (C) Euler model for the cracked roof beam.

Based on the maximum shear stress value, $\tau_{max}=3F_s/2A$, where A is a constant sectional area and F_s is the shear force operating at the cross-section, when the maximum value of the shear stress acting at the beam ends equals $\frac{pL}{2}$, the cracks will begin to form at both ends of the roof rock beam. So, the fixed beam converts to the simple-supported structure as shown in Fig. 5.2, and the cracked roof rock beam can be described as this kind of restraint at both ends.

During loading, a vertical tension fracture develops at the mid-span of the cracked roof rock beam, and the roof beam is broken into two blocks. The voussoir beam is formed with the rotation of two blocks. The cracked roof rock beam of the simple-supported structure is the previous configuration of the voussoir beam. The vertical deflection and lateral thrust of the cracked roof affects the stability of the voussoir beam. So it is essential to analyze the mechanical behavior of the cracked roof rock beam converting to the voussoir beam.

5.1.2.2 Model of roof rock beam

With the working face advancing, the cracks formed at both ends of the main roof beam when the beam reached the limit span, as shown in Fig. 5.2A. Before the initial weighting of the main roof, the deflection of the roof is not enough to form a vertical fracture at the mid-span of the roof beam, the roof layer can be seen as continuous, the failure of the roof is related to the strain energy, focusing on the initial deflection and strain energy of the roof, the roof beam is treated as elastic when ignoring local plastic deformation in small scale (Yang, 2010; Pan et al., 2012; Wang et al., 2015c).

Because almost no tensile resistance exists in the cracks, the cracked roof in unit width can be regarded as a no-tension beam. As shown in Fig. 5.2C, an Euler beam model is established considering its boundary condition to simulate a cracked roof rock beam. The overlying strata weight and dead weight are simplified as uniform load p. For the roof rock beam, the horizontal thrust is T, the span is L, and its deflection ω can be expressed as follows:

$$\omega = u\sin\frac{\pi x}{L} \tag{5.1}$$

where x is the arch length between origin O and arbitrary point A, m; ω *is* the x point deflection, m; and u the midpoint deflection, m.

In general, rock failure is an instability phenomenon driven by the energy (Jiang et al., 2016a). So, the instability state of the roof rock beam can be analyzed through examining the total potential energy changes in the roof system.

The total potential energy U is composed of the structural strain energy U_1 and the external load potential energy (the work done by the vertical load p and the horizontal thrust T). Due to the beam bending, then the strain energy U_1 is accumulated as follows.

$$U_1 = \frac{EI}{2}\int_0^L k^2 dx \tag{5.2}$$

where I is the inertia moment of the beam cross-section, m^4; E is the elastic modulus, GPa; and k is the curvature at the arbitrary point A of the beam.

The work was done by q and T, that is,

$$W_q = \int_0^L pu \sin\frac{\pi x}{L} dx \tag{5.3}$$

$$W_T = \left(L - \int_0^L \sqrt{dx^2 - d\omega^2}\right) T \tag{5.4}$$

Hence, the total potential energy U is given by

$$U = \frac{EI}{2}\int_0^L \left(\frac{d^2\omega}{dx^2}\right)^2 \left(1 - \left(\frac{d\omega}{dx}\right)^2\right)^{-1} dx - \left(L - \int_0^L \sqrt{1 - \left(\frac{d\omega}{dx}\right)^2} dx\right) T - \int_0^L pu \sin\frac{\pi x}{L} dx \tag{5.5}$$

If Eq. (5.5) was developed as a Taylor expansion and omitting the higher-order trace, then the integral calculation is

$$U = \frac{EI\pi^6}{16L^5} u^4 + \frac{\pi^2}{4L}\left(\frac{EI\pi^2}{L^2} - T\right) u^2 - \frac{2pL}{\pi} u \tag{5.6}$$

The first and second partial derivatives of the total potential energy U to u separately are given by

$$\frac{\partial U}{\partial u} = \frac{EI\pi^6}{4L^5} u^3 + \frac{\pi^2}{2L}\left(\frac{EI\pi^2}{L^2} - T\right) u - \frac{2pL}{\pi} \tag{5.7}$$

$$\frac{\partial^2 U}{\partial u^2} = \frac{3EI\pi^6}{4L^5} u^2 + \frac{\pi^2}{2L}\left(\frac{EI\pi^2}{L^2} - T\right) \tag{5.8}$$

As a mechanic system of the roof rock beam, the external force p and thrust T function as the control variables, producing deflection u, that is, the state variable. The curved surface of the system balance path is deduced from $\frac{\partial U}{\partial u} = 0$, and any arbitrary point (p, T, u) on the surface represents a state of equilibrium. Because the second-order partial derivative $\frac{\partial^2 U}{\partial u^2}$ of the total potential energy U and the two-order variation $\delta^2 U(u)$ share the same positive and negative, the stability of the equilibrium state can be judged by the positive and negative of the formula $\frac{\partial^2 U}{\partial u^2}$ (Yu et al., 2015).

The bifurcation set formula of the system equilibrium state is obtained from Eqs. (5.7), (5.8), that is,

$$\Delta = \frac{54p^2 EI}{\pi^2} + \left(\frac{EI\pi^2}{L^2} - T\right)^3 = 0 \tag{5.9}$$

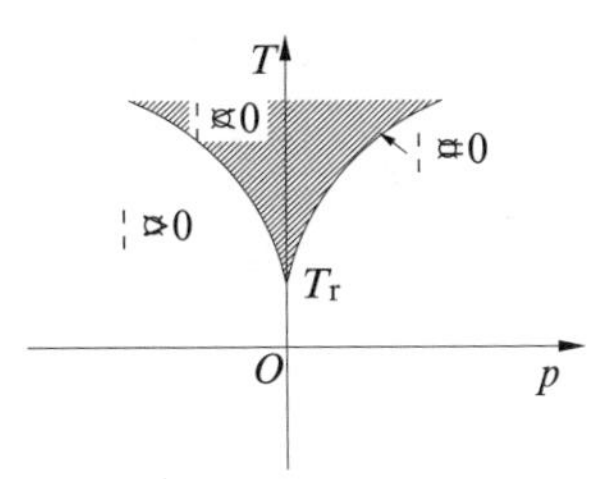

FIG. 5.3 Bifurcation curve of the equilibrium state.

With the gradual deflection of the roof rock beam, the increasing of the horizontal thrust T is getting faster, and the beam structure approaches the transient state. $T_r = \frac{EI\pi^2}{L^2}$ is a critical horizontal thrust derived from Eq. (5.9), which equals the critical pressure in the tiny bending equilibrium bar in elasticity theory. When $T < T_r$ and $\frac{\partial^2 U}{\partial u^2} > 0$, the total potential energy has minimum value and the structure reaches gradual equilibrium, as illustrated in Fig. 5.3.

Only when $T \geq T_r$, the equilibrium path appears bifurcation and the vertical displacement occurs a sudden-jump. If the horizontal thrust T satisfies Eq. (5.8), the system will be in the critical equilibrium state, $\frac{\partial^2 U}{\partial u^2} = 0$. Under the further disturbance of the mining action, the equilibrium point will appear in the shadow area (Fig. 5.4), which means $\frac{\partial^2 U}{\partial u^2} < 0$, that is,

$$u^2 \leq \frac{2L^4}{3EI\pi^4}\left(T - \frac{E\pi^2}{L^2}\right) \tag{5.10}$$

In this condition, the structure is unstable and the deflection u increases sharply. After the cracked roof rock beam is broken, then a new equilibrium state is formed. The necessary condition for the equilibrium path bifurcating is $T \geq \frac{EI\pi^2}{L^2}$.

Based on the nonlinear mechanic characteristic, the change of the horizontal thrust T is related to the load p and the deflection u, and the broken or buckling instability of the roof rock beam is determined by the combined action of T and p.

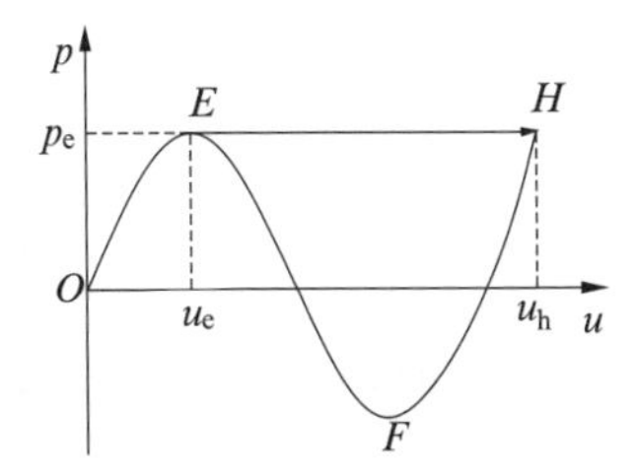

FIG. 5.4 Evolution curve of the equilibrium path.

5.1.2.3 Instability process of cracked roof beam

In the Euler model for the cracked roof rock beam, considering the actual situation of the load p and deflection u, when Eq. (5.7) equals zero, $\frac{\partial U}{\partial u}=0$, the equilibrium path of the beam system is deduced as follows:

$$p=\frac{EI\pi^7}{8L^6}(u-2.38u_e)^3+\frac{\pi^3}{4L^2}\left(\frac{EI\pi^2}{L^2}-T\right)(u-2.38u_e) \tag{5.11}$$

where u_e is the deflection of the beam at the mid-span when the beam is in the critical state, m.

Supposing the horizontal thrust $T>\frac{EI\pi^2}{L^2}$, the equilibrium path will bifurcate, the changing curve of the load p and deflection u, that is, the equilibrium path of the roof rock beam structure, can be obtained based on the practical stress and the vertical displacement of the roof rock beam as shown in Fig. 5.4. Each point of the curve corresponds to the equilibrium state of a different structural configuration.

In the curve OE segment, $\frac{\partial^2 U}{\partial u^2}>0$, and the structure is stable.

Point E is the bifurcation point of the equilibrium path as the structural configuration is in the critical equilibrium state, here $\frac{\partial^2 U}{\partial u^2}=0$. The corresponding deflection u_e is given by

$$u_e=0.61L\sqrt{\frac{TL^2}{EI}-9.9} \tag{5.12}$$

The load p_e at the extreme point is given as

$$p_e=0.85\left(T-\frac{9.9EI}{L^2}\right)\sqrt{\frac{TL^2}{EI}-9.9} \tag{5.13}$$

In the curve EF segment, $\frac{\partial^2 U}{\partial u^2}<0$, and the structure is unstable corresponding to the maximum total potential energy. When the structure maintains the equilibrium state in the OE segment until the bifurcation point E, the critical point jumps directly to point H, causing the sudden increase of vertical displacement Δu, given by

$$\Delta u=u_h-u_e=2.5L\sqrt{\frac{TL^2}{EI}-9.9} \tag{5.14}$$

The deflection u of the roof rock beam varies parabolically with the load p as shown in Fig. 5.4, under the action of the increasing overlying strata load p as well as the antihorizontal thrust of the adjacent strata T, the cracked beam appears flexural subsidence. Before the load reaches the extreme point E, the structure preserves a dynamic balance. When it reaches a critical value p_e, the roof rock beam loses stability while an incremental deflection Δu is induced suddenly. Then, the original dynamic balance is broken. The evolution process can be described as the crack formed at the mid-span of the roof rock beam and the voussoir beam structure arriving as a new structure due to the large deformation in the central section of the beam.

5.1.2.4 Voussoir beam of cracked roof beam

A voussoir beam model is considered after the instability of the cracked roof rock beam, with a span of L and thickness of h. The model is composed of two blocks and three crosscut cracks at the two abutments and the mid-span, as shown in Fig. 5.5A. The hinged arch is formed among the block by the vertical deflection and lateral trust, and it can be simplified as a hinged beam based on the abutment contact mechanism and geometry of the block, as shown in Fig. 5.5B. The thickness/span ratio i_1 of one block is given by

$$i_1 = \frac{2h}{L} \tag{5.15}$$

Hence, for the sake of simplicity, the included angle α_0 between the neighbor hinged end connection line and the horizontal line is approximately equal to $\arctan\frac{2h}{L} = \arctan i_1$ before rotation of the block. α_0 is reduced by β_0 after rotation. The variables such as horizontal thrust, friction, thickness/span ratio, and initial rotating angle are influencing factors of block moving.

The relation of the deflection u_e and the vertical displacement incremental Δu of the cracked roof beam, and the initial rotating angle β_0 of the hinged block are given by

$$\beta_0 = \arcsin\frac{2(u_e + \Delta u)}{L} \tag{5.16}$$

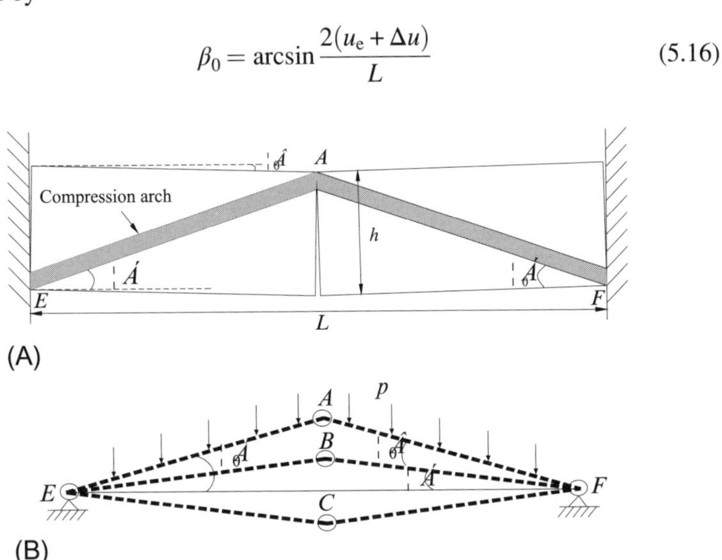

FIG. 5.5 Voussoir beam configuration. (A) Schematic of the voussoir beam and (B) three hinged points of the rock-arch structure.

The angle α between the truss and the horizontal direction in the initial state is expressed in terms of ratio i_1 and the initial rotating angle β_0, by

$$\alpha = \alpha_0 - \beta_0 = \arctan i_1 - \arcsin \frac{2(u_e + \Delta u)}{L} \tag{5.17}$$

The instability of the hinged beam structure with these determination variables α, α_0, and β_0 includes three possibilities:

(1) *Sliding instability*. Because the initial rotating angle β_0 is small, the friction force at the hinge abutment is not big enough to resist the shear force, so these blocks slide down along two end hinge points, as shown in Fig. 5.6A.

(2) *Rotating instability*. When the initial rotating angle β_0 is big enough but still $\beta_0 < \alpha_0$, the friction force at the hinge abutment can resist the shear force and then the hinged beam structure can be formed. The broken blocks rotate backward, and the truss in the mechanic model descends with decreasing of α. Until $\alpha = 0$ degree, that is, $\beta_0 = \alpha_0$, three plastic hinge points are collinear, the truss reverses and sags, and the broken blocks lose stability, as shown in Fig. 5.6B. This is the smallest influencing situation to the mining field stability.

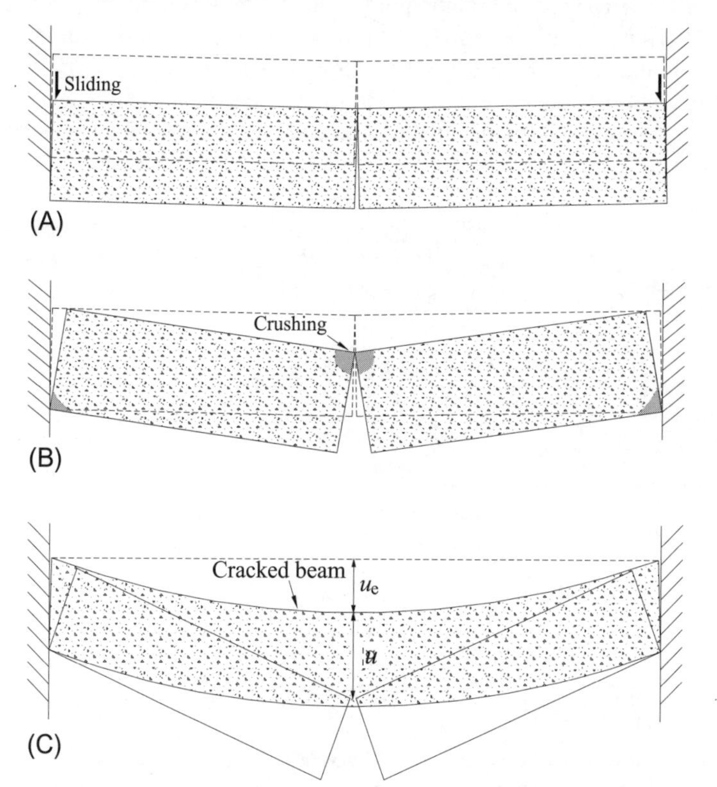

FIG. 5.6 Different instability types of the voussoir beam structure. (A) Sliding instability, (B) rotating instability, and (C) sudden-jump instability.

(3) *Sudden-jump instability*. In the initial stage, the cracked roof beam buckled and the vertical displacement increased sharply. Then $\beta_0 > \alpha_0$ and the broken blocks directly reversed down without forming a convex hinged beam structure, that is, they immediately lost stability, as shown in Fig. 5.6C. Thus it can be seen that the deflection and failure forms of the cracked beam directly influence the voussoir beam structure types and the stability of the broken blocks. The effect of the structural buckling cannot be ignored (Zhou et al., 2015). So the bending and breaking of the cracked roof rock beam and the block movement should be considered to analyze the stability of the roof movement more objectively.

5.1.3 Instability feature of cracked roof beams

5.1.3.1 Influence factors of critical deflection

The sudden-jump instability characteristics and the affecting factors of the cracked rock beam can be discussed by using Eqs. (5.11)–(5.14). The critical deflection u_e and the sudden-jump displacement value Δu are mainly affected by three factors: the thickness/span ratio i, the Young's modulus E, and the buried depth h_1 of the main roof. Regarding mining engineering in the Shangwan Mine of the Shendong Zone in China as an example, the sandstone main roof of 4 m thickness is located at a depth of 150–350 m. The main roof and the overlying rock strata unit weigh 25 kN/m^3, and Young's modulus E is 30 GPa. Considering the unit width of the cracked roof rock beam, the inertia moment of the beam cross-section is $I = bh^3/12$ ($b = 1.0$ m) and the thickness/span ratio is $i = \frac{h}{L}$. According to Eqs. (5.12), (5.13), the critical deflections with different buried depths and different sizes of the roof rock beam can be obtained, as shown in Fig. 5.7.

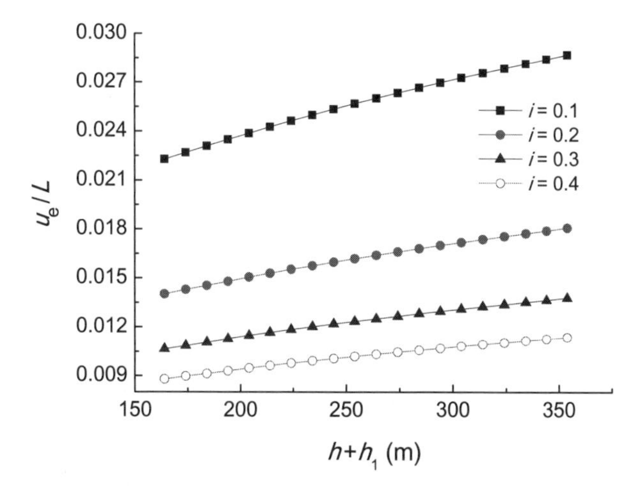

FIG. 5.7 Critical deflection variation curves with different buried depths.

Each point on the deflection curves in Fig. 5.7 indicates one critical stability state of the roof rock beam. The area over the curve means the unstable state while the area under the curve means the dynamic equilibrium state. With the roof rock beam bending and sinking, these curves can be used as references to analyze the stability, supposing the buried depth and the rock feature are known. The critical deflection u_e of different sizes of the roof rock beam increases linearly with increasing mining depth. The smaller the thickness/span ratio i, the bigger the value u_e/L, and the spacing between the curves is more obvious.

Considering an example of the mining face buried at a 200 m depth, the effects of thickness/span ratio i and Young's modulus E on the deflection u_e are reflected in Fig. 5.8. The deflection u_e decreases nonlinearly with the increasing value of i and E. The higher ratio i leads to u_e/L curve being flatter. For a certain depth h_1, the values of i and E directly affect the deflection u_e and Δu. The relational degree between correlation variables (I, E, and characteristic variable u_e, and Δu) is $R_i = 0.992$, $R_E = 0.978$. The deflection curve rises more obviously with lower ratio i. For example, with a change in i from 0.15 to 0.1 (by 33.3%), the deflection u_e increases on average by 31.2%, and with a change in i from 0.4 to 0.35, the deflection u_e increases on average by 9.5%.

It can be known that the deflection of the thin roof is obviously larger than that of the thick one. The thick and hard roof often produces the brittle broken during the mining process, that is, the instability type belongs to the strength failure. The thin and mid-hard or soft roof often produces large deflection and buckling broken, that is, the instability type belongs to the structure failure. For shallow mining, a sudden-jump instability of structural buckling tends to occur in the thin and soft roof.

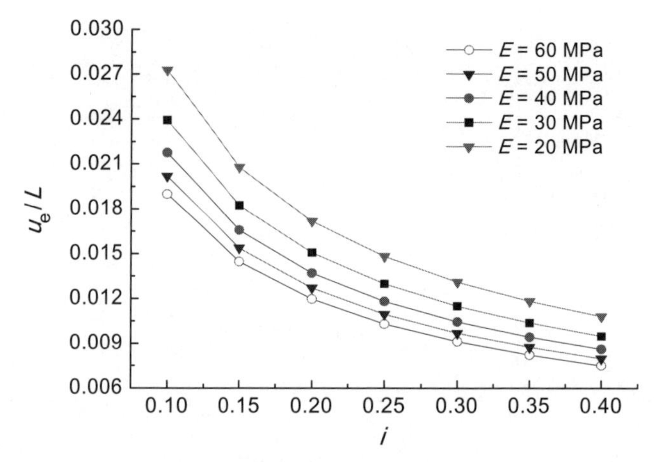

FIG. 5.8 Variation curves of critical deflection to thickness/span ratio.

5.1.3.2 Hinged blocks structure after roof instability

The movement of the broken block is affected by the thickness/span ratio i of the roof rock beam and the initial rotating angle β_0. The broken blocks tend to slip down and fall off in the initial formation stage, leading to the bench convergence. This is very common in the shallow mining field, and is a direct threat to mining safety (Shabanimashcool and Li, 2015; Wang et al., 2015c). The relationship between β_0 and i is illustrated in Fig. 5.9. The higher ratio i or smaller angle β_0 leads to the broken blocks easily slipping down. The angle β_0 is directly proportional to the mining depth $h + h_1$, and the broken blocks tend to slip down at a small mining depth and with a small angle β_0, which exactly corresponds to the common sliding instability facts.

Whether the three-hinged arch structure of broken blocks can be formed depends on the parameters β_0 and α_0. As shown in Fig. 5.5B, when $i = 0.15$, $\beta_0 < \alpha_0$, and the broken block rotates with angle $\alpha_0 - \beta_0$ before losing stability. When $i = 0.10$, $\beta_0 > \alpha_0$, and the initial rotating angle β_0 is larger than the included angle α_0 between the hinged arch end connection line and the horizontal line; therefore, there is no way to form the convex hinged rock-arch structure. Therefore, only under the condition that the broken blocks satisfy the criteria of Eq. (5.11) will the thin roof with low thickness/span ratio produce gradual movement of a three-hinged rock-arch structure.

5.1.4 Mechanical analysis of roof rock beams

5.1.4.1 Building computational model

The mechanical model for the roof rock beam can be simulated and verified by $FLAC^{3D}$ numerical analysis. Regarding mining engineering with a depth of 160–200 m at the Shangwan Mine as an example, based on the mining

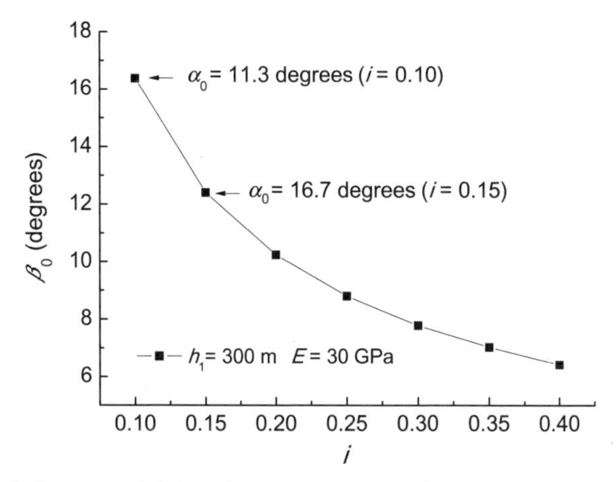

FIG. 5.9 Variation curve of the initial rotating angle with thickness/span ratio.

conditions of an immediate roof of 9 m thick sandy shale, a 5 m thick coal seam, and a 4–7 m thick sandstone main roof, the computational model was built as shown in Fig. 5.10A. The unit weight of the overlying rock is 25 kN/m^3. The computational model was assumed in the hydrostatic stress field and the Mohr-Coulomb criteria is used as the material strength criteria. Based on the initial caving distance of the roof, its span is assumed to be in the interval of 24–40 m. The model bottom was assumed to be fixed, and the horizontal movement was restricted. A uniformly distributed load is applied at the model top equivalent to the overlying rock weight. The plasticity state of the model in the initial loading stage is shown in Fig. 5.10B to indicate the cracks.

The physical and mechanical parameters of the model are available in Table 5.1.

5.1.4.2 Results analysis and discussion

The uniformly distributed load of 4.0–5.0 MPa was applied at the roof rock beam top to simulate its instability at a buried depth of 160–200 m. The buried depth and thickness of the simulated rock beam was 160 and 4 m, respectively.

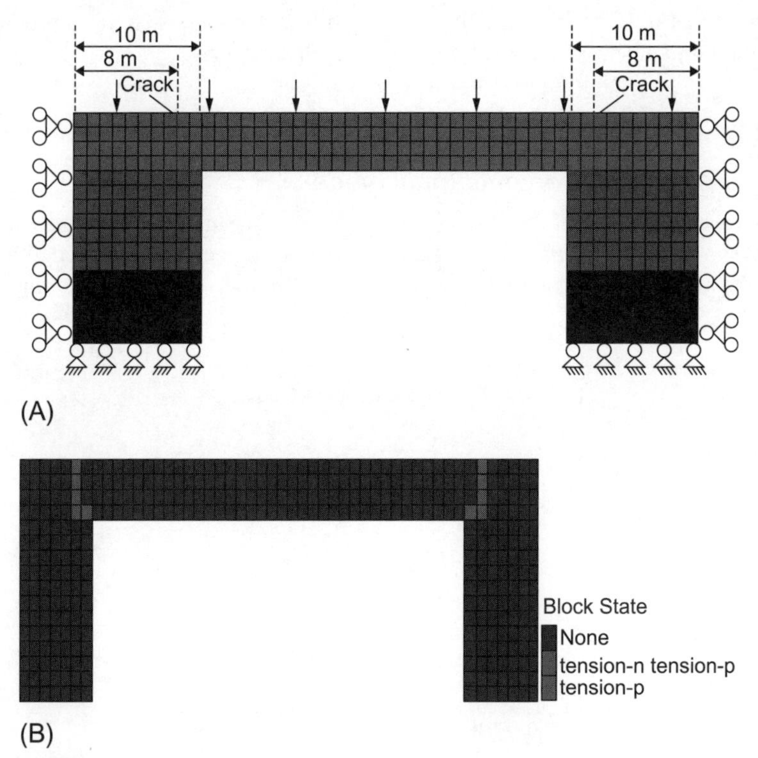

FIG. 5.10 Numerical model of the cracked roof rock beam. (A) The numerical model and its boundary conditions and (B) plasticity state of the model in the initial stage.

TABLE 5.1 **Physical and mechanical parameter of the model.**

Name	Density (kg/m^3)	Friction angle (degree)	Cohesion (MPa)	Elasticity modulus (GPa)	Poisson's ratio
Sandstone	2470	10.3	34	30	0.32
Sandy shale	2590	8.9	39	27	0.23
Coal	1310	4.0	30	5.7	0.26

The relationship curves of the horizontal thrust T and the vertical displacement with different spans are shown in Fig. 5.11. In the initial stage (O-A-B), the curve rises quickly, then the horizontal thrust T shows a gradient ascent with an increase of the vertical displacement. Before reaching the peak value of point P, the horizontal thrust T first decreased and then increased to point P. This mechanical changing corresponds to the process of the cracked roof beam being broken into two blocks. Finally, the postpeak curve tends to occur fluctuate falling, the broken blocks rotate in the hinged arch structure.

Obvious differences are observed among the curves for different beam spans. (1) With the decreasing of the beam span, the OA curve increases significantly and the peak points A, B, and P constantly improve. (2) With decreasing of the beam span, the fluctuation between points B and P reduces while the vertical displacement of point P becomes gradually smaller. (3) With decreasing beam span, the postpeak curve falls with a gradually fluctuating pattern.

As seen from Fig. 5.12, the peak value of horizontal thrust decreases with increasing beam span, and the smaller buried depth leads to the smaller horizontal thrust.

With fixed values of depth, span, and Young's modulus, the deflection of the roof increases more obviously with decreasing thickness of the roof. The deflection increases sharply when the roof thickness reduces from 5 to 4 m (Fig. 5.13).

The roof rock beam structure of the underground space makes it easy to produce buckling instability and a strong destructive disaster under the action of horizontal thrust (Zhou et al., 2015). Once the roof rock beam is unstable, a slight disturbance can lead to deformation growing rapidly, and the structure is destroyed suddenly. Considerable previous work based on elasticity mechanics was completed to explore the mechanical behavior of the main roof, before the initial weighting of main roof, the roof failure is related to the strain energy, the elastic rock beam was used to analyze the instability of the roof under shallow mining condition, this method is efficient in previous cases (Yang, 2010; Li et al., 2014; Wang et al., 2015a, b, c; Zhao and Song, 2016; Please et al., 2013).

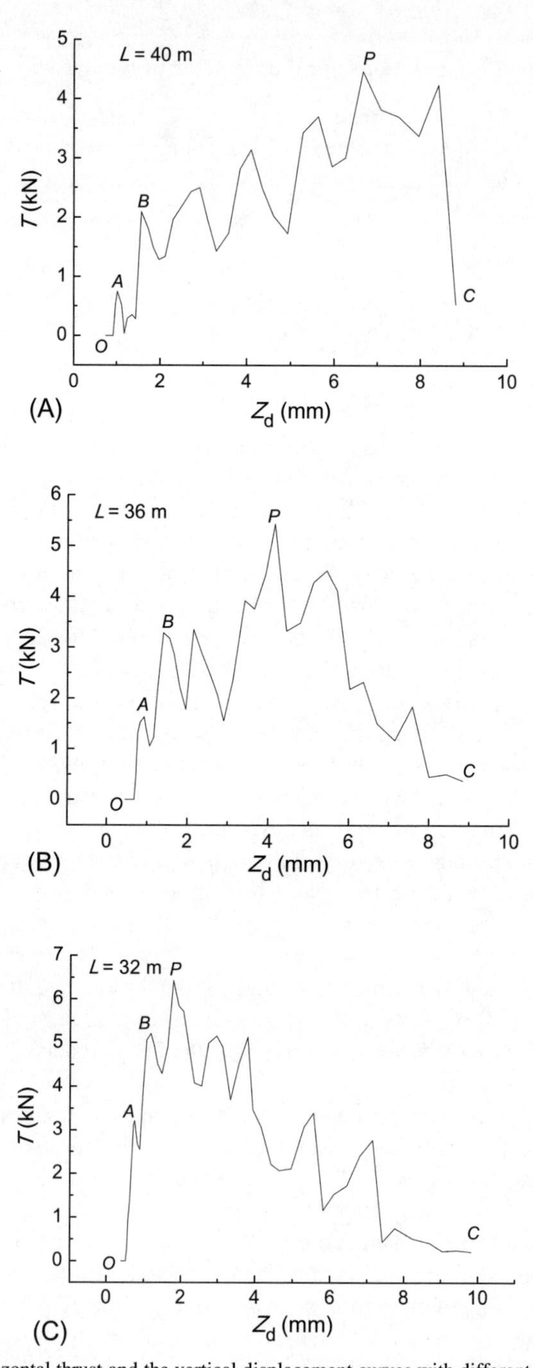

FIG. 5.11 Horizontal thrust and the vertical displacement curves with different beam spans. (A) 40 m, (B) 36 m, and (C) 32 m.

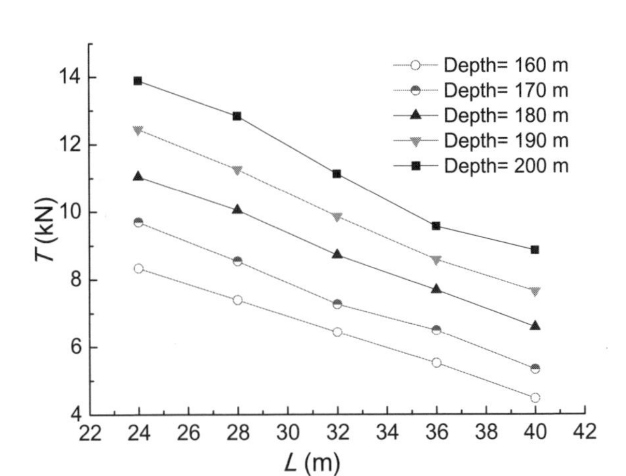

FIG. 5.12 Curves of horizontal thrust and spans of the beam with different buried depths.

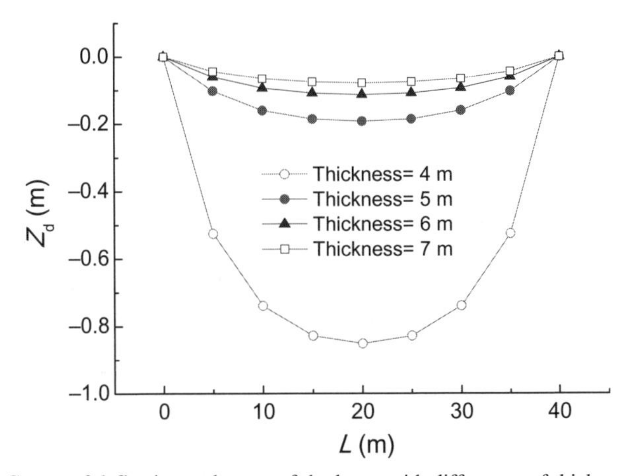

FIG. 5.13 Curves of deflection and spans of the beam with different roof thicknesses.

The structural model of the Euler beam for the cracked roof rock beam, satisfying the condition that the mining face length is far greater than the advancing distance, can be applied in the stability analysis of the main roof under shallow mining conditions. However, considering that the vertical force alone will lead to defects of the mechanical model, the horizontal thrust is added in the modified model to reflect the instability process of the roof rock beam. The roof structure determines the characteristics of ground pressure and mining safety, and the control of it was the purpose of the working face support (Jiang et al., 2016a, b; Li et al., 2017a). If the roof structure occurred instability under the critical force, the vertical displacement increased to final value immediately, the instability with sudden-jump characteristics was the failure type of the extreme value point (Gao, 2004; Yin et al., 2015).

It is insufficient to determine the potential energy of the cracked roof rock beam structure simply by considering the vertical concentrated force alone (Hou et al., 2000; Wang et al., 2015c). This paper set out a new mechanical model for the cracked roof rock beam under the interaction of the vertical load and the horizontal thrust based on the stability theory of the elastic system, which reflects the physical process from the gradual to sudden instability through the equilibrium path curve.

The structure is in the stable state before arriving at the critical point. The potential energy of the structure is constantly accumulated and its deflection is gradually changed. When arriving at the critical point, the structure occurs stability bifurcation, and vertical displacement appears sudden-jump, which results in the structure losing stability and being broken. The present roof control theory of the longwall mining face seldom relates to the cracking and subsidence process of the main roof while the deflection of the roof rock beam before cracking at the mid-span directly influences the movement of the broken blocks (Yang, 2010). The buckling instability of the roof rock beam cannot be ignored. When analyzing the instability process of the roof rock beam, the main roof deflection and the broken block movement should be considered as well as the thickness/span ratio and elastic modulus.

5.1.5 Conclusions

Based on the interaction of vertical load and horizontal thrust, the mechanical model of the cracked roof rock beam was built under shallow mining conditions. Using the stability theory of the elastic system, the equilibrium path, and the bifurcation condition, the critical load causing structural buckling instability, the deflection and the sudden-jump value of vertical displacement were gained.

By establishing the mechanical model and numerical model, the deflected process of the cracked roof rock beam was analyzed, and the evolution law of the horizontal thrust and the influencing factors of that were discussed. The thickness span ratio and elastic modulus are affecting factors of the roof deflection and the initial rotation of the broken block.

The instability of the broken block is affected by the deflection and deformation of the cracked roof rock beam, and the sudden-jump value of displacement cannot be ignored. The relation between the roof deformation and rotary movement of the block should be established under shallow mining conditions, thus the stability analysis method of the cracked roof rock beam is improved.

5.2 Evolution characteristics of fractured strata structures

5.2.1 Introduction

The large-scale mining of a shallow coal seam will destroy the thin bedrock. The mining disturbance often causes the fractured strata falling and strong roof weighting, so the ground subsidence, water loss, and soil erosion can seriously damage the ecological environment (Zhang and Peng, 2005; Zhang et al., 2011). The types and bearing capacities of the block structures will affect the strata movement, if grasping the structure evolution characteristics of the overlying strata, and predicting the mining damage by using the structure stability criterion, the roof weighting and the surface movement can be controllable. With the construction and development of large coal bases in China, the stability control of the overlying strata structure has become an important issue that needs to be solved urgently for safety and environmental protection under shallow coal mining conditions.

The strong roof weighting during shallow coal mining leads to frequent support crushing. This problem was caused by the load-bearing structure failure of the overlying strata; the basic roof block slid under the overburdened load and strong pressure was formed on the support (Xu et al., 2014). The stable strata structure can carry the upper overburdened load and protect the lower strata. The load-bearing capacity of the strata structure is the key factor affecting load transferring among the layers, and the efficient bearing load of various block structures deserves attention. The self-carrying capacity of the strata was derived from the arching stress that was delivered from the mid-span to both abutments. The effect of the stress arch should be taken into account when analyzing the load-bearing capacity of the overlying strata. The alternated variation of roof weighting was caused by the difference of the structure types of the strata. The instability of the main load-bearing structure would lead to the failure of other structures in advance, so the combined motion of the strata induced the stronger roof weighting (Wang et al., 2016a, b, c, d, e, f, g). But the failure situation and migration range of the strata vary in different mining stages. The pressure arch evolution has an important influence on the overall stability and fractured characteristics of the overlying strata, so the structure characteristics of the overlying strata under the influence of the pressure arch should be further studied.

Aiming at revealing the revolution characteristics of the strata structure under shallow coal mining, based on the observed weighting law of a mining face in the Shangwan Mine in the Shendong Mining Area in China, considering the influence of the horizontal stress and the arching stress in the overlying strata, the mechanical models of strata blocks were established in different mining stages and the corresponding formulas were deduced to analyze the load-bearing capacity of different structures and the influencing factors. To reveal the revolution rules of the pressure arch in the surrounding rock and the characteristics of the fractured strata structure by using UDEC, the reliability of the mechanical models and data analysis were verified. The research work provided a theoretical supplement to the issue of the strata structure under shallow coal mining, and had a guiding significance for the prediction of roof weighting and the subsidence control in engineering practice.

During shallow coal mining, the load of the hanging strata was transferred from the mid-span to the two abutments, and the voussoir beam structure was formed by the arching stress in the roof strata. Based on the practical support design for the roof of a deep metal mine, Li (2006) proposed that the pressure arch could carry a considerable load through delivering ground pressure to the surrounding rock. A numerical simulation with DDA was carried out by He and Zhang (2015) to investigate the roof stability in underground excavation in the laminated rock mass. The arching mechanism of compressive stress was observed to keep the voussoir beam of the roof stable. The overburden depth and the horizontal stress affected the stability of the roof structure, shallow buried depth and lower horizontal stress favored the sliding or snap-through failure of the voussoir beams. Shabanimashcool and Li (2015) conducted the analytical approaches for the stability of voussoir beams by considering the horizontal loading condition of the beams. The analysis showed that the maximum stable span of voussoir beams increased with the horizontal stress and the failure mode might be changed from buckling to crushing with increasing horizontal stress. Mazor et al. (2009) found that the pressure arch was essential to reach the efficient thickness to sustain the stability of the roof structure. The stable arch thickness and the instability types of the structure varied with the block size and stress condition (Tsesarsky, 2012). These studies indicated that the arching mechanism of stress kept the stability and load-bearing capacity of the fractured strata structure, and the stability of the stress arch was affected by horizontal stress, buried depth, and block size.

The previous studies about the load-bearing structure of the stress arch in the strata were concentrated on the rock excavation engineering. For the stability of the fractured roof block structure under shallow coal mining, the voussoir beam and the stepped rock beam were applied in analyzing the initial weighting mechanism of horizontal coal seam mining (Yang, 2010), the cause of strong weighting of the large height mining face (Ju and Xu, 2013), and determining the resistance of the support (Huang et al., 2016). However, the influences of the boundary horizontal stress and the arching stress in the block on the load-bearing capacity of the structure were not taken into account in the established models. According to the rock strength and strata thickness in the Shendong Mining Area, X. Liu et al. classified the overlying strata of the shallow coal seams. They pointed out that the weak and laminated strata was unable to form a load-bearing structure while the strong and massive strata carried the loads of weak strata. The splitting of the key stratum would lead to caving of the weak and laminated strata above and the weight would be applied to the strata below (Liu et al., 2015a, b; Hu et al., 2016). The large-scale mining of the shallow coal seam easily caused the bedrock broken totally. Most of the existing studies were focused on the basic roof and ignored the differences of the structure characteristics and the carrying capacity among multilayer bedrocks. The mechanisms of long/short and strong/weak weighting have not been reasonably revealed.

The overlying strata instability caused by shallow coal mining was common in many countries. Soni et al. (2007) conducted ground subsidence monitoring of shallow cover coal mining by resistance imaging and ground penetrating radar at Kamptee Colliery in India. However, the work was limited to monitoring image analysis and lacked a study on the movement law of overlying strata and the effect of strata structure. A shallow postmining subsidence caused the formation of several crown holes in Edinburgh in the United Kingdom. The sinkholes affected the East Coast Main Railway and led to enormous economic losses; this event attracted the attention of many scholars. Helm et al. (2013) performed the ground investigation and the numerical modeling to analyze the potential causes of instability, including the variations of the level of the groundwater table, the influence of the structure of the rock mass, and the geometry of the abandoned workings. Salmi et al. (2017) adopted the tributary area method and FLAC3D to investigate the effect of gradual deterioration on the stability of the shallow abandoned room and pillar, and pointed out that the long-term deterioration due to mechanical and chemical factors would decrease the strength of the pillar and reduce its effective bearing width. Gradual weakening due to weathering would result in the bearing capacity failure of the floor strata as well as pillar collapses and subsidence. Based on numerous sudden surface collapse cases in China's Datong coal field, Cui et al. (2014) investigated the mechanism of subsidence induced by shallow partial mining by using the similar material simulation and the UDEC technique. It was considered that the inner stress arches merged, accompanying the pillar rupture and gradually combined with the external stress arch, putting the contained rock mass in a stress state of tension and compression. Failure in the thick and hard overburden rock strata resulted in the loss of the complex stress arch and a sudden collapse at the surface. This work was concentrated on the stress revolution between the pillars, so the influence of the overlying strata structure evolution on subsidence was not involved. The inducing factors of the instability of the overlying strata, such as pillar stability and hydrogeological conditions, were analyzed in the above studies. The synchronous evolution characteristics of the global pressure arch and the fractured strata structure need to be further researched.

To sum up the above-mentioned literature, we found that many results had been achieved regarding the load-bearing structure of the stress arch in strata. However, it was not perfect enough regarding the pressure arching characteristics of the fractured strata during shallow coal mining. So, this study will conduct further research on this aspect.

5.2.2 Engineering background

Taking the No. 51104 working face in the west panel at the Shangwan Mine in the Shendong Mining Area as the engineering background, the working face was at an average depth of 115.4 m, with a dip angle of 0–5 degrees. The coal thickness was 5.03–7.90 m and the average thickness was 6.7 m. The average thickness of the ground overburden of the alluvial sand was 13.70 m, and that of the bedrock was 102.10 m. The immediate roof was mostly sandy mudstone or siltstone with a thickness of 0.50–7.10 m, 3.00 m in average. The basic roof was sandstone with a thickness of 6.00–15.10 m, 9.00 m in average, and the compressive strength was 16.60–33.80 MPa, 25 MPa in average.

The designed mining height was 5.20–5.50 m, the strike length was 3654 m, the dip length was 301 m, and the advancing length was 3360 m. The inclined longwall backward comprehensive mining method was adopted, matching the German Eickhoff SL-500 shearer. The total caving method was used to manage the roof and DBT double column shield electrohydraulic control supports were arranged to protect the working face. The initial resistance of the support was 6900 kN and the rated working resistance was 8638 kN. The top beam length was 4.08 m, and the center distance was 1.75 m. The initial weighting and 19 times periodic weighting occurred during the observation period. As shown in Fig. 5.14, the initial weighting interval was 27.20 m and the support resistance was 5921.33 kN. The periodic weighting interval varied at 9.40–32.3 m and the support resistance was 5571.69–8974.55 kN, the 11th time weighting exceeded the rated resistance of the support.

Four times periodic weighting occurred with the working face advancing 61.20 m after initial weighting. The weighting law was similar in this stage, and the stronger roof weighting appeared with the longer interval. Thereafter, the weighting interval and intensity appeared the alternated regularity of long/short and strong/weak periodically, The long interval corresponded to the weak weighting (L/W) while the short interval corresponded to the strong weighting (S/S). The sixth and eighth time weighting was the S/S type while the seventh and ninth time weighting was the L/W type. The 10th to 15th time weighting appeared as normal regularity, with the working face advancing. The 16th to

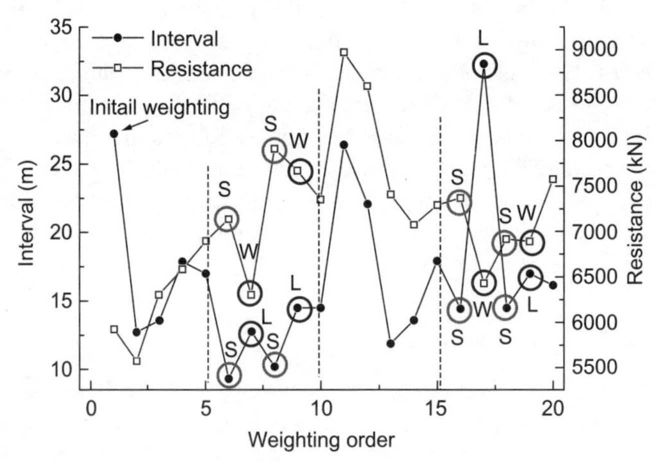

FIG. 5.14 The roof weighting law of the 51,104 working face.

20th weighting alternated with the long/short and strong/weak law again. There was a long interval between the normal weighting and the alternated long/short and strong/weak weighting.

5.2.3 Mechanical and computational model

5.2.3.1 Simplified mechanical model

In the initial stage of mining, the compressive stress deviated toward the mined out area under the disturbance of excavation and showed significant arching effect in the surrounding rock. The global pressure arch of the nonfractured zone was formed to carry the load from the overlying strata and self-weight. The arch foot was located in the surrounding rock at the work face and open-off cut side. There was a low-pressure area under the shield of the global pressure arch, and the face support was required to prevent the fall of detached basic roof blocks. The removable space and structure characteristics of the key roof blocks were affected by the filling of the caving waste of the immediate roof.

Taking the roof blocks in the initial mining stage as research objects, as shown in Fig. 5.15. The roof block K_A was the main key block above the support. The block K_B slid with the displacement Δ relative to the left vertex of the block K_A. Assuming that the thickness of the coal seam was M, the thickness of the immediate roof was $\sum h$, the rock bulking factor was K_p, and the displacement Δ depended on the calculation of $M - \sum h(K_p - 1)$. The hinged structure was formed by the sliding block and main key block in the small movable space. For the large height mining field, the step structure was formed by the sliding blocks with unequal displacement in the large movable space. The block K_A rotated with the angle α, and the half stress arch was formed in the block by the horizontal thrust T from the front rock wall and the rear sliding block. Using the distribution function of the horizontal stress σ_{xx} at the boundary of the block (Mazor et al., 2009):

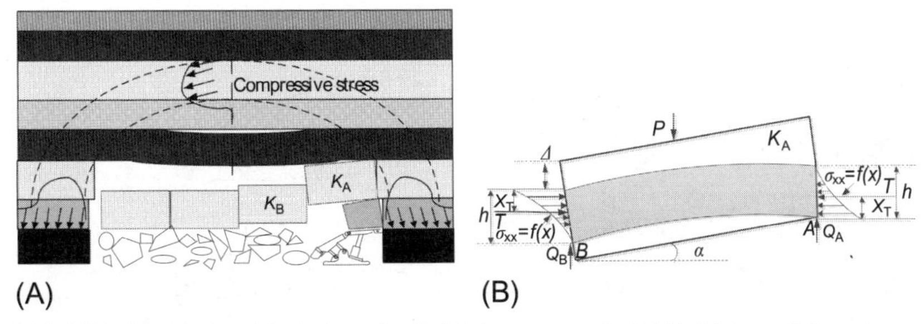

FIG. 5.15 Key blocks of the basic roof and global pressure arch. (A) Sketch map of the engineering model and (B) mechanical model of the main key block.

$$\sigma_{xx} = \sigma_m (x - h)^b \tag{5.18}$$

where h was the distributed height of the horizontal stress at the abutments, σ_m was the maximum horizontal stress, and x was the coordinate position.

The horizontal thrust T and its position X_T were respectively given by

$$T = \frac{\sigma_m h}{b+1} \tag{5.19}$$

$$X_T = \frac{h}{b+2} \tag{5.20}$$

The main key block K_A was taken as the study object, the block thickness was H, and the length was L, taking the moment at the right origin point A, $\sum M_A = 0$, then

$$Q_A = 0.5P + T\frac{H - L\sin\alpha - \Delta - 2X_T}{L} \tag{5.21}$$

When the shear force Q_A was larger than the contacting friction $T \tan \varphi$, the support was required to provide sufficient resistance to prevent sliding of the key block. According to Eqs. (5.19), (5.21), the working resistance F_S of the support was given by:

$$F_S = 0.5P + \frac{\sigma_m h}{b+1}(H - L\sin\alpha - \Delta - 2X_T - \tan\varphi) + \gamma L \sum h \tag{5.22}$$

If the stress in the global pressure arch exceeded the rock strength, the strata was fractured and gradually lost its carrying capacity, the global pressure arch of nonfractured zone developed toward the upper stable surrounding rock, the fractured roof sank and changed into near the mining field area, the pressure arch in roof blocks was formed in the initial and periodic roof weighting stage. As shown in Fig. 5.16A, the symmetrical stress arch was formed under the inner boundary of the global pressure arch of the nonfractured zone in the initial stage of strata separation. After the breaking of the strata, the symmetrical stress arch still could be maintained by the horizontal stress transferring between the blocks. The rotation and sliding motion of the block in the stress arch structure

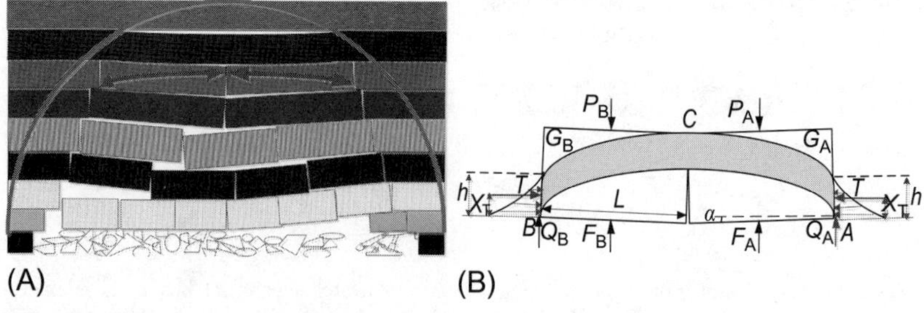

FIG. 5.16 The caving arch of the fractured area and the global pressure arch. (A) Sketch map of the engineering model and (B) mechanical model of the symmetrical stress arch.

would increase the load on the lower strata. The successive instability of the symmetrical stress arch in each strata was the main cause of roof weighting and this type of structure first appeared in the initial mining stage. Establishing the mechanical model as shown in Fig. 5.16B, let the thickness of the block be G_A, G_B was H, the length was L, and the rotating angle was α. Taking the moment at the left point B, $\sum M_B = 0$, then

$$0.5P_AL + 1.5P_BL - 1.5F_AL - 0.5F_BL - 2Q_AL = 0 \tag{5.23}$$

Taking the moment of the G_A at the middle point C, $\sum M_C = 0$, then

$$T(H - L\sin\alpha - 2X_T) + 0.5P_AL - 0.5F_AL - Q_AL = 0 \tag{5.24}$$

Considering the symmetry of the structure, the overburden load and the supporting force on the two blocks could be treated as equal, $P_A = P_B = P$, $F_A = F_B$. Combining Eqs. (5.23), (5.24), the supporting force on the block was given as:

$$F_A = P - \frac{2T(H - L\sin\alpha - 2X_T)}{L} \tag{5.25}$$

As the global pressure arch of the nonfractured zone developed into the loose layer of sand, because of the lower strength of the sand layer, the arching effect vanished gradually with the plastic failure of the loose layer, the ground subsidence appeared and the subsidence basin was formed in the fully mining stage. Thereafter, the hinged structure and the squeezed arch structure existed in the fractured roof under the global pressure arch of the nonfractured zone. As in Fig. 5.17B, the hinged blocks rotated relatively and delivered load to the lower

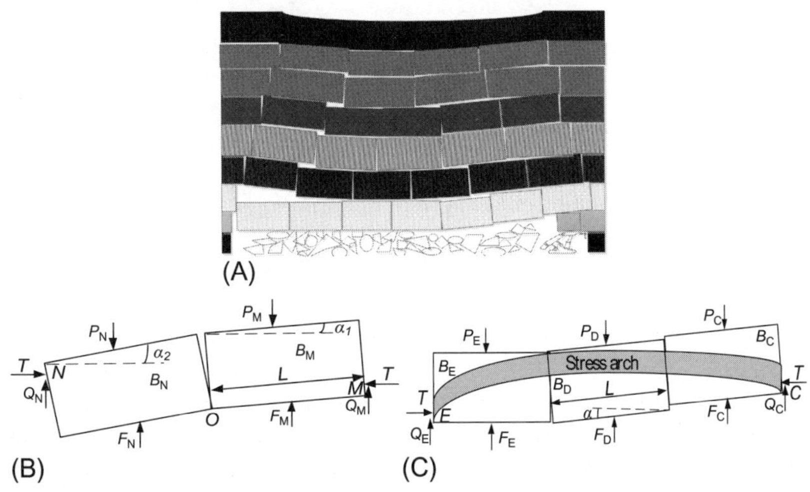

FIG. 5.17 The hinged and pressure arch structure in multilayer overlying strata. (A) Sketch map of the engineering model, (B) mechanical model of the hinged structure, and (C) mechanical model of the squeezed arch structure.

strata. Establishing the mechanical model of the roof blocks of the initial rotating position during periodic roof weighting, the shear force Q_N on the block B_N at the left hinged point N was applied by the rear compacted block, $Q_N = 0$, taking the moment at the left hinged point N, $\sum M_N = 0$, and the vertical resultant force $\sum F_y = 0$, then

$$T(H - L\sin\alpha_1 - L\sin\alpha_2 - 2X_T) + 0.5L(P_M - F_M) + 1.5L(P_N - F_N) - 2Q_M L = 0 \tag{5.26}$$

$$P_M + P_N - F_M - F_N - Q_M = 0 \tag{5.27}$$

Taking the moment of the front block B_M at the middle point O, $\sum M_B = 0$, then

$$0.5(P_M - F_M) - T\sin\alpha - Q_M = 0 \tag{5.28}$$

This model was the structure with one degree of indeterminacy, assuming that the carrying load of the two blocks was equal, the equation set could be solved, let $P_M = P_N = P$, combining Eqs. (5.26)–(5.28), the supporting force on the block was given:

$$F_M = P - T\left(\frac{8X_T}{L} + 4\sin\alpha_2 - 4\frac{H}{L} - 2\sin\alpha_1\right) \tag{5.29}$$

As shown in Fig. 5.17C, the squeezed arch structure was formed by the horizontal thrust between the blocks, and the latter arch foot was located in the compacted block B_E which contacting the waste, establishing the mechanical model of the two blocks in front half arch, for the compacted block B_E, $P_E \approx F_E$, $Q_E = \frac{T(H-2X_T)}{L}$, taking the moment at the left supported point E, $\sum M_E = 0$, and the vertical resultant force $\sum F_y = 0$, then

$$2TL\sin\alpha - 1.5P_D L - 2.5P_C L + 1.5F_D L + 2.5F_C L + 3Q_C L = 0 \tag{5.30}$$

$$P_C + P_D - Q_E - Q_C - F_C - F_D = 0 \tag{5.31}$$

Assuming that the overburden load and the supporting force on the two blocks are equal, $P_C = P_D = P$, $F_C = F_D$, combining Eqs. (5.30), (5.31), the supporting force on the block was given:

$$F_C = P - T\frac{3(H - 2X_T) - 2L\sin\alpha}{2L} \tag{5.32}$$

In the initial mining stage, the symmetrical stress arch occurred in the separated strata near the mining field and the global pressure arch was formed in the nonfractured strata zone far from the mining field. With the working face advancing in the periodic roof weighting stage, the strata near the mining field separated from the nonfractured strata zone and fractured. The hinged structure and the squeezed stress arch structure were formed in the broken blocks under the inner boundary of the global pressure arch of the nonfractured strata zone.

5.2.3.2 Building computational model

Based on the geological condition and the mining field in the Shanwan Mine, the discrete element software UDEC was used to analyze the evolution characteristics of the strata structure under the shallow mining condition. As shown in Fig. 5.18, the thickness of the coal seam was 7.0 m, and the total thickness was mined in one time with large height mining technology. The thickness of the floor was 10.0 m, the thickness of the immediate roof was 4.0 m, and the thickness of the basic roof was 9 m. The thickness of the strata over the basic roof was set as 95.0 m and the thickness of the ground sand layer was 12.0 m.

The model size was 400.0 m long and 137.0 m high. The upper boundary was free, the bottom boundary of the model was fixed, and the lateral boundaries of the model were fixed in the horizontal direction.

The position of the start of mining was 50 m away from the left boundary, and the advancing distance was 300 m. Each strata was arranged with three monitoring lines along the *Y* direction to obtain the evolution law of the stress field and displacement field during coal mining. The physical and mechanical parameters of the model were listed in Table 5.2, and the Mohr-Coulomb criteria was used in the numerical calculation.

5.2.4 Results and discussion

Based on the established mechanical models in different mining stages, the load-bearing capacity of the fractured strata structure could be analyzed. Assuming that the thickness of the overburden layer was H_Z, the average unit weight was γ. The pressure P on the block with length L was $\gamma H_z L$, the downward delivering load q of the block was equal to the support force in unit length F/L, then the efficient bearing load of the structure was $\gamma H_z - q$, substituting Eqs. (5.19), (5.20) into Eqs. (5.25), (5.29), (5.32), then the efficient bearing load

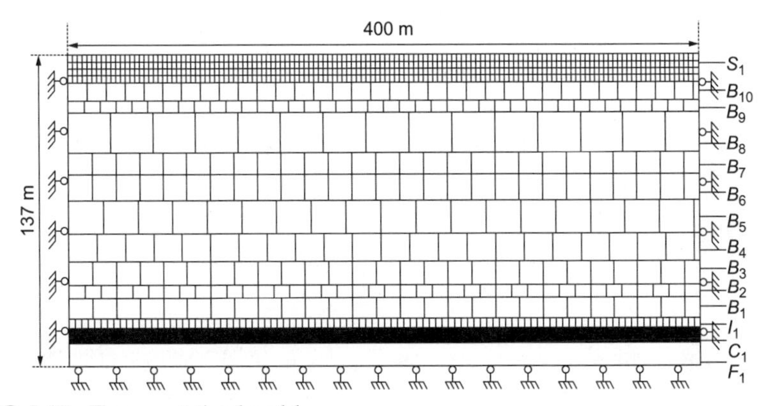

FIG. 5.18 The computational model.

TABLE 5.2 Physical and mechanical parameters of the materials.

Name	Unit weight (kN/m^3)	Elastic modulus (GPa)	Poisson's ratio	Cohesive strength (MPa)	Tensile strength (MPa)	Frictional angle (degree)
Soil	19.2	0.08	0.32	0.1	0.04	28
Sandy mudstone	24.0	29	0.23	3.7	2.1	35
Coarse sandstone	24.3	35	0.21	5.5	4.0	33
Sandstone	25.0	32	0.24	7.3	4.9	35
Mudstone	22.4	23	0.15	2.5	1.7	30
Coal	13.1	15	0.29	0.79	0.57	27
Siltstone	24.6	26	0.22	3.9	2.5	38

f of the symmetrical stress arch structure, the squeezed arch structure and the hinged structure was, respectively,

$$f_{\mathrm{A}}=\gamma H_{\mathrm{Z}}-\frac{F_{\mathrm{A}}}{L}=\sigma_{\mathrm{m}}\frac{2ti}{b+1}\left(i-\sin\alpha-\frac{2ti}{b+2}\right) \tag{5.33}$$

$$f_{\mathrm{C}}=\gamma H_{\mathrm{Z}}-\frac{F_{\mathrm{C}}}{L}=\sigma_{\mathrm{m}}\frac{ti}{b+1}\left(1.5i-\sin\alpha-\frac{3ti}{b+2}\right) \tag{5.34}$$

$$f_{\mathrm{M}}=\gamma H_{\mathrm{Z}}-\frac{F_{\mathrm{M}}}{L}=\sigma_{\mathrm{m}}\frac{ti}{b+1}\left(\frac{8ti}{b+2}+4\sin\alpha_2-4i-2\sin\alpha_1\right) \tag{5.35}$$

where $i=H/L$ was the ratio of the block thickness H to the block length L, and $t=h/H$ was the ratio of the distributed height of the boundary horizontal stress to the block thickness H, which represented the arch thickness.

The results of Eqs. (5.33)–(5.35) were the products of σ_{m} and dimensionless variables. It concluded that the efficient bearing load f was the multiple of the maximum boundary horizontal stress σ_{m}, and the variable f could be simplified as:

$$f=\gamma H_{\mathrm{Z}}-\frac{F}{L}=K_{\mathrm{S}}\sigma_{\mathrm{m}} \tag{5.36}$$

The parameter K_{S} could be calculated by the function after the variable σ_{m} in Eqs. (5.33)–(5.35), and K_{S} was defined as the load-bearing coefficient to represent the load-bearing capacity of the structure.

The symmetrical stress arch and the squeezed arch were the load-bearing structure formed by arching distributed stress. The rotating angle α was taken as 5 degrees, and the distribution of the boundary horizontal stress was set as: $b=1$ linearly distributed, $b=2$ nonlinearly distributed. According to Eqs. (5.33), (5.34), the coefficient K_S could be calculated under different block size i and arch thickness t. As shown in Fig. 5.19, the coefficient K_S varied as the following regular: (1) The load-bearing coefficient of the symmetrical stress arch was higher than that of the squeezed arch. (2) The load-bearing coefficient increased with increasing block size i. (3) With the arch thickness t increasing, the load-bearing coefficient increased before reaching the maximum value and

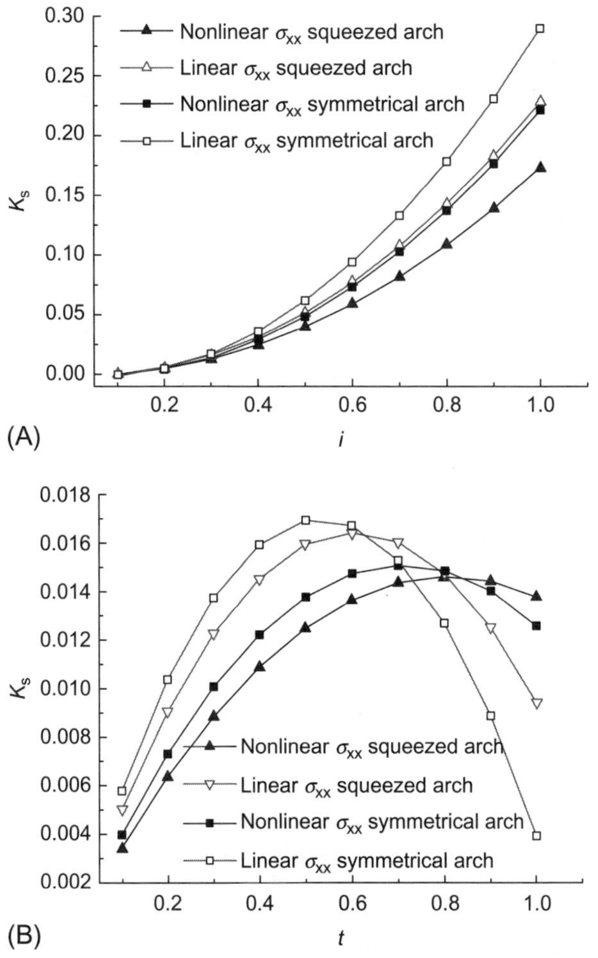

FIG. 5.19 Variation law of the load-bearing coefficient K_S of the stress arch structure. (A) Different block size i ($t=0.5$) and (B) different thickness t of the stress arch ($i=0.3$).

then decreased. (4) When the boundary horizontal stress was linearly distributed, the load-bearing coefficient of the structure was higher.

When the fractured interval was fixed, the symmetrical stress arch and the squeezed arch structure of the thick strata kept a higher load-bearing capacity and played the main carrying function in multiple layers. The distribution of the stress affected the load-bearing capacity of the fractured strata structure. When the boundary horizontal stress was linearly distributed, the structure could reach the maximum load-bearing capacity with a smaller thickness of the stress arch.

For the hinged structure, the blocks kept surface contacting in small range or point contacting, in general condition, the thickness of the stress arch $t < 0.5$. According to Eq. (5.35), when the parameters satisfied $0.5t + \frac{\sin\alpha_2 - 0.5\sin\alpha_1}{i} > 1$, the load-bearing coefficient $K_S > 0$. Considering the example that $t = 0.4$, $\alpha_1 = 5$ degrees, $\alpha_2 = 15$ degrees, the hinged structure could carry load efficiently when the block size i was less than 0.3. Hence, the load-carrying capacity was determined by the block contacting state, rotating angle, and block size.

According to the mining footage of the No. 1 panel of the Shanwan Mine, the working face was advanced 10 m each time in the simulation. As shown in Fig. 5.20, the immediate roof I_1 initially caved with the working face advancing 30 m, the global pressure arch was formed in the B_1–B_5 strata, and the principle stress was transferred to the arch foots.

The basic roof B_1 incurred sliding instability with the working face advancing 50 m, two types of stress filed were formed in the overlying strata, the outer was the global pressure arch of nonfractured zone in the surrounding rock, the arch top was mainly formed by the horizontal principle stress, the load of the upper strata was acted on the arch top and delivered through the concentrated horizontal stress to the arch waist. The inner was the symmetrical stress arch in each strata under the inner boundary of the global pressure arch. The horizontal principle was distributed in the arch top in the mid-span of each strata, and enough horizontal stress was the essential condition to form the pressure arch.

The rear hanging blocks of the basic roof slid with unequal displacement by the friction and the step structure was formed. The principle stress in the main key block appeared arching characteristic by delivering from the coal face to the mined out area. The monitoring data of the vertical stress on each strata was shown in Fig. 5.21. After initial caving of the immediate roof, the lower stress area was formed in the B_1 and B_2 strata under the inner boundary of the global pressure arch; the B_6 and B_7 strata beyond the pressure arch kept an in situ stress state. The global pressure arch carried part of the load of the upper strata and self-weight, and the arch foot was located in the increased stress area of the surrounding rock.

The global pressure arch developed upward and reached the B_7 strata after basic roof caving. The increased load applied on the pressure arch led to the higher vertical stress in the arch foot. Under the inner boundary of the global pressure arch, the lower stress area was extended and the resistance of the support was increased for carrying the weight of this area. The roof weighting was determined by the basic roof near the mining area and performed as a normal weighting stage.

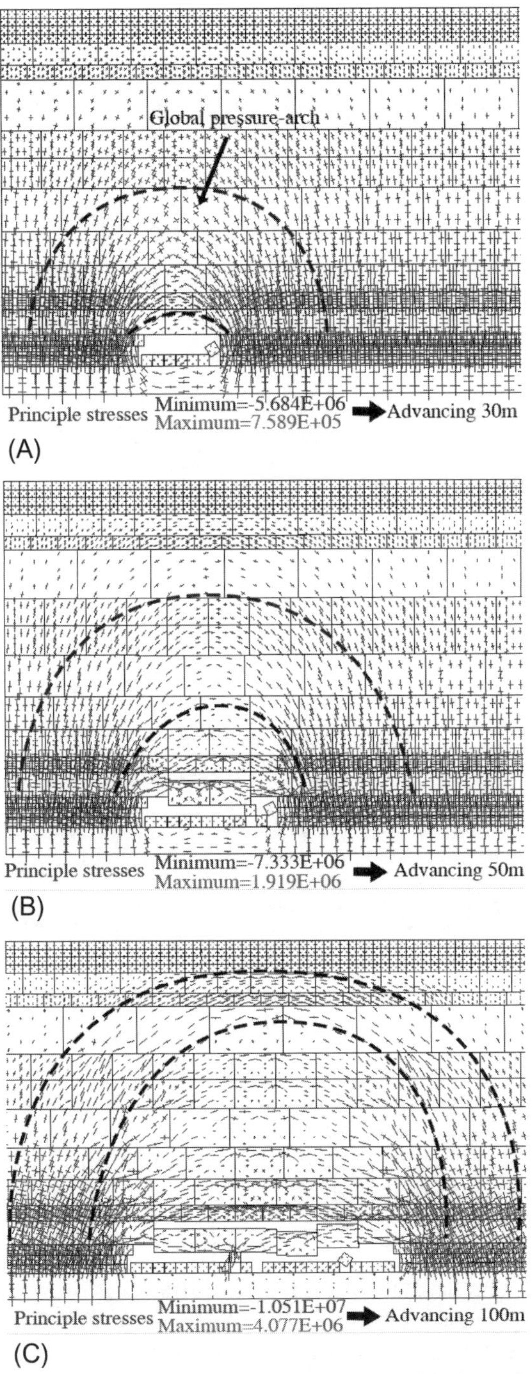

FIG. 5.20 Evolution process of the pressure arch and instability of the lower roof. (A) The immediate roof caving, (B) the basic roof caving, and (C) the key blocks of the basic roof and global pressure arch.

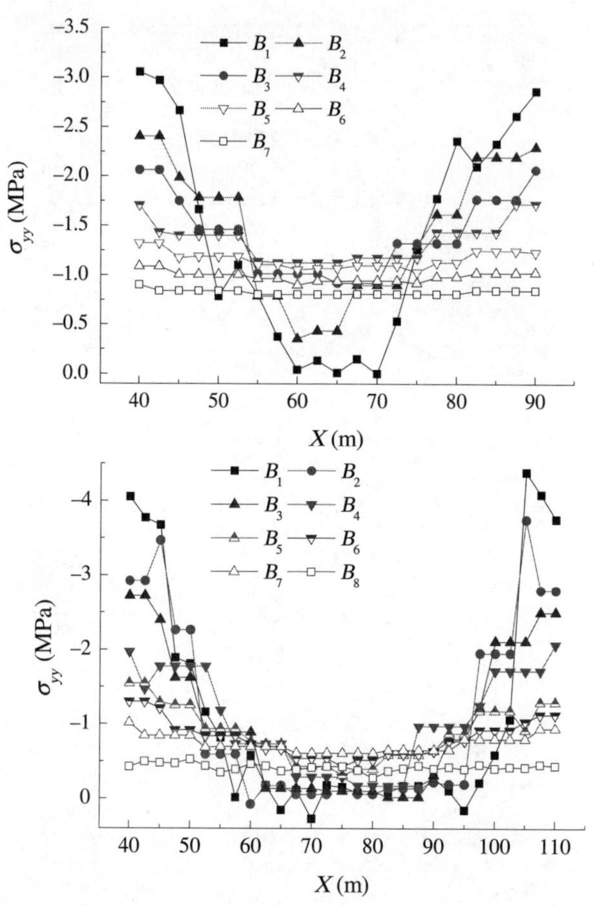

FIG. 5.21 Variation law of the vertical stress σ_{yy} on each strata. (A) The immediate roof caving and (B) the basic roof caving.

As shown in Fig. 5.22A, the roof weighting was caused by the breaking of the B_2 strata, with the working face advancing, the symmetrical stress arch in each strata under the global pressure arch occurred instability in turn, the weighting interval varied from 10 to 30 m, and the long/short and strong/weak weighting was induced periodically. The arching distribution of the principle stress caused the difference of the fractured position between each strata; the span of the strata increased gradually. The instability of the upper strata with a smaller span affected the larger range, and the strong weighting with a short interval was induced.

The equilibrium structure of the fractured strata was formed by the squeezing and gripping of the rotated blocks, as shown in Fig. 5.22C–E. A significant arching effect of the principle stress appeared in the B_4 strata. The squeezed arch structure was formed with the front arch foot located in the bedrock at the

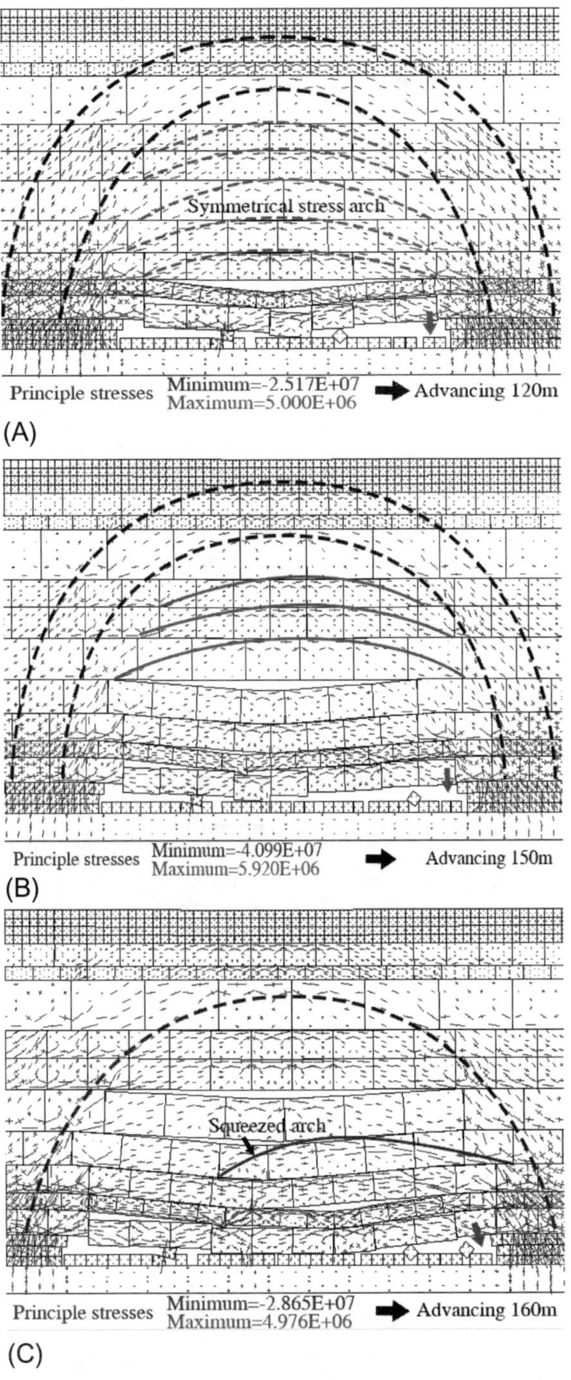

FIG. 5.22 Evolution law of the structure of overlying strata.(A) The strata B_2 breaking, (B) the strata B_3 and B_4 combined breaking, (C) the strata B_5 breaking,

(continued)

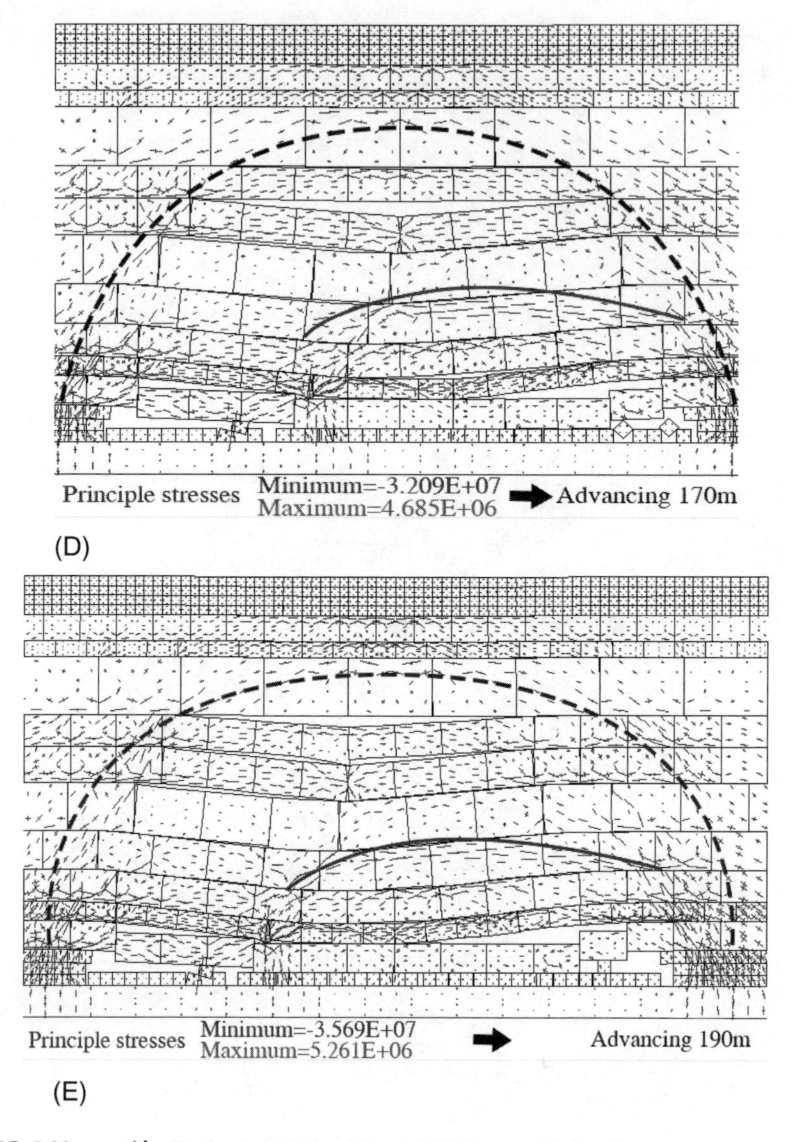

FIG. 5.22—cont'd (D) the strata B_6 breaking, and (E) the strata B_7 breaking.

mining face side and the latter arch foot located in the compacted rock waste. The B_2 and B_3 strata were broken periodically and the hinged structure was formed. The principle stress was delivered as the half arch in the front block and kept horizontal delivering in the rear block. The resistance of the support was increased by the rotation of the middle and lower strata, and the weak weighting with a long interval was induced.

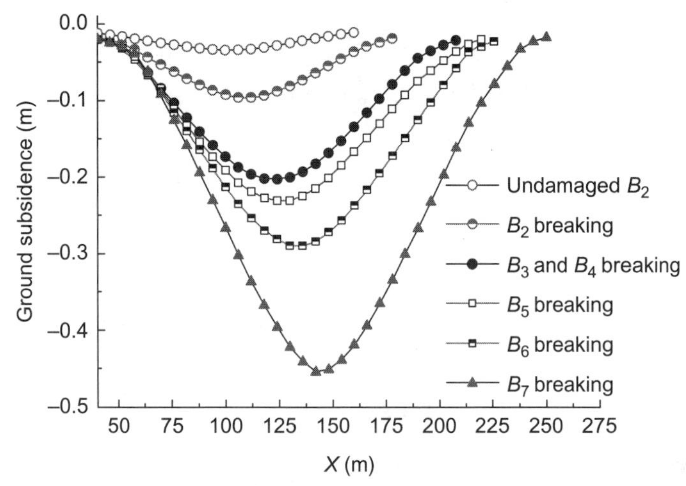

FIG. 5.23 Subsidence characteristic curves of the overlying strata breaking.

As shown in Fig. 5.23, with the working face advancing 150 m, the B_3 and B_4 strata were broken together. The outer boundary of the global pressure arch developed into the loose sand layer, which resulted in the subsidence of 0.20 m on the ground. With successive instability of the symmetrical stress arch in each strata under the inner boundary of the global pressure arch, the subsidence caused by the single strata breaking increased gradually. The subsidence for the B_7 strata breaking was greater than the accumulated value of the B_1–B_4 strata breaking. The instability of the symmetrical stress arch in the upper strata was the main cause of ground subsidence.

As shown in Fig. 5.24A, the global pressure arch has covered the loose sand layer and the subsidence basin was formed. As shown in Fig. 5.24B, a new global pressure arch was reformed between the working face side and the caving compacted rock. The squeezed arch structure was formed in the B_3–B_5 strata under the inner boundary of the global pressure arch. In the hinged structure of the B_2 strata, the rotation of the block led to the sliding instability of the basic roof. The alternated instability of the squeezed arch structure and the hinged structure induced the long/short and strong/weak weighting again.

The arching effect of the principle stress ensured the formation of the load-bearing structure of the overlying strata. For underground excavation in the laminated rock mass, a global pressure arch with elevated stress was observed sustaining the weight of the overburden and providing the natural shield over the opening. Below the pressure arch was a group of local voussoir beam arches formed to transfer the weight of each layer to the abutments (He and Zhang, 2015). The macrostress shell also exited in the rock surrounding a fully mechanized top-coal caving face in deep mining, within the low-stress zone inside the stress shell, the voussoir beam structure only bear parts of the load from the strata (Xie et al., 2009; Wang et al., 2015c).

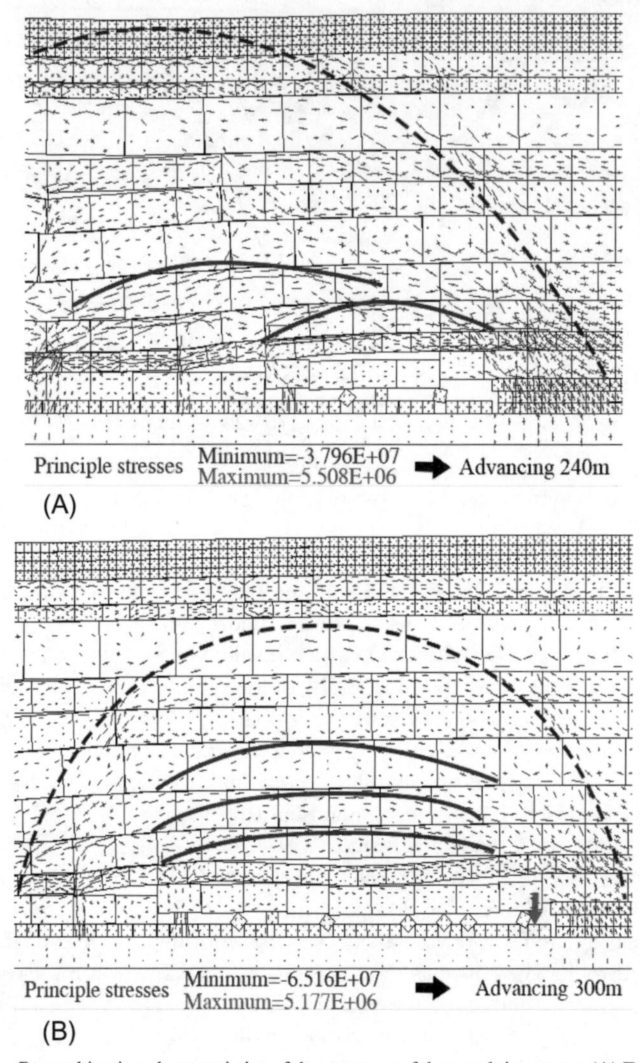

FIG. 5.24 Recombination characteristics of the structure of the overlying strata. (A) Formation of the subsidence basin and (B) the pressure-arch advancing.

Similarly, in the initial stage of shallow coal seam mining, the global pressure arch was formed in the surrounding rock, under the inner boundary of the pressure arch, the load of each strata was carried by the symmetrical stress arch. But the difference was that the pressure arch would develop to the ground, the successive instability of the symmetrical stress arch with unequal span made different weighting interval. After the total instability of the symmetrical stress arch, the latter arch foot of the pressure arch moved into the caving compacted rock, and the alternated instability of the squeezed arch and the hinged structure induced a cyclic occurrence of the long/short and strong/weak weighting.

The equilibrium structure was formed by gripping of the rotated strata block, and the revolution and recombination of the block structure caused the strong and weak changing of the support resistance. Considering the influencing factors of mining height, block size, and rotating angle, a series of mechanical models was established in analyzing the stability of the roof structure (Yang, 2010; Ju and Xu, 2013; Huang et al., 2016), but these models ignored the difference of the load-bearing capacity of the structure. Affected by the contacting state of the blocks and the arching principle stress, only satisfying the limited condition of contacting range, rotating angle and boundary horizontal stress distribution, the hinged structure of the fractured strata could carry load efficiently. The load-bearing capacity of the symmetrical stress arch and the squeezed arch structure depended on the maximum horizontal stress, and the load-bearing coefficient was affected by the block size and boundary stress distribution.

The global pressure arch in the nonfractured zone functioned as a main load-bearing structure in the multiple layer surrounding rock, the single pressure arch in the fractured blocks was located under the global pressure arch. The global pressure arch could be regarded as a protected shell to the single pressure arch. The fractured structure of the roof in the single pressure arch also carried part of the load of the upper strata in the global pressure arch. The instability of the global pressure arch led to caving of the roof in the single pressure arch, and the compound instability of the two types of pressure arch would cause the strong roof weighting.

With shallow coal mining, it was easier to cause the total breaking of the overlying strata and failure of the loose sand layer. The inner stress arches between the coal pillars merging and gradually combined with the external stress arch. This was considered as the main cause of strata instability in shallow partial mining (Cui et al., 2014). For shallow longwall mining, the initial ground subsidence was induced when the pressure arch developed into the loose sand layer, the successive instability of the symmetrical stress arch was the main cause of large scale ground subsidence, the failure of stress arch structure in the upper strata aggravated the formation of subsidence basin.

5.2.5 Conclusions

The revolution of the pressure arch affected the stability of the surrounding rock, and each strata appeared different structure characteristic under the pressure arch. To reveal the evolution characteristics of the overlying strata structure, taking the Shangwan Mine as the engineering background and considering the arching property of the stress, the mechanical models of different mining stages were proposed and the difference of the load-bearing capacity between the structures and its influencing factors were discussed. The conclusions are the following:

(1) The symmetrical stress arch, the squeezed arch, or the hinged structure of the strata blocks can be formed by gripping the horizontal stress. The load-bearing capacity of the structure is affected by the rotating angle, the

contacting range, the boundary stress distribution, and the block size. Only satisfying the limited condition, the hinged structure could carry load efficiently. The symmetrical stress arch and the squeezed arch structure play major roles in load bearing.

(2) The action of the pressure arch in the surrounding rock makes a different fractured position and span of the strata. Under the inner boundary of the pressure arch, the successive instability of the symmetrical stress arch induces alternated strong and weak roof weighting. A new pressure arch is reformed when the caving rock is compacted in the mined out area. The recombining evolution of the squeezed arch and the hinged structure make cyclic strong and weak roof weighting.

(3) During shallow coal mining, the pressure arch can develop into the loose sand layer and induce initial subsidence. Further subsidence is caused by the successive instability of the symmetrical stress arch. The instability of the load-bearing structure of the arching stress in the upper strata is the main cause of strong roof weighting and forming the subsidence basin.

The fractured strata structure along the striking direction of the working face is representative, which reflects the macroregulation of strata movement and has been concerned in the strata control of mining. The research in this study is mainly about this direction. However, under the action of mining stress and joint, there are differences between the mechanical characteristics and engineering response of the rock in each direction, so it is not enough to analyze the striking direction of the working face alone. The fracture state of the strata and the stress evolution rules in different directions need to be further studied. With the extended working face being widely applied in shallow coal mining, the space effect of the strata structure is more significant. The actual mining situation could not be fully reflected by the plane model analysis. In the following work, the evolution rule of the three-dimension pressure arch in the surrounding rock and the mechanical models of the block structure will be studied. The horizontal stress and the contacting state are important factors affecting the arching mechanism of the stress, and the movement of the block is also changed with these factors. The stability of the strata structure under different horizontal stress and contacting condition is also the future research direction.

5.3 Pressure arching characteristics in roof blocks

5.3.1 Introduction

The shallow coal resource is huge in western China, but the fragile ecological environment and mining-induced disasters also seriously restrict the development of the mining area. The Shendong mining area in western China was characterized by large thickness as well as a shallow buried and flat dipping coal seam. It was a typical high efficiency mining district of shallow thick coal mining and the largest underground coal mining area in the world. The fully mechanized longwall mining with large mining height was widely used in this district, and the width of the panel can be reached at 450 m.

The roof weighting and ground subsidence induced by shallow coal mining also affected the development of similar mining engineering around the world (Ma et al., 2009; Zhang et al., 2011; Zhao et al., 2018a). So it is an important problem to reveal the performance of the stepped subsidence and the strong roof weighting during shallow coal mining. The large-scale shallow mining for the full thickness of coal increased the failure range of the thin bedrock roof. The load of the fractured strata was not carried effectively by the roof block structure, resulting in strong mine pressure. So the mechanical model was widely used in analyzing the roof structure and weighting forecast. The applying of the voussoir beam model, the cantilever beam model, the Euler beam model, and the hinged arch bar model also proved practical in many engineering cases (Zhao et al., 2018b). The fully mechanized mining for thick coal with a large mining height would aggravate the shallow strata movement. The moving law and structure types of the roof blocks were special in these cases, and the weighting step and load intensity was characterized by alternating with long and short periods. So the slip instability of the roof blocks would cause the movable column to be retracted sharply and then the supports were crushed.

Aiming at these problems, a great deal of research has been conducted. For example, Ju and Xu used the cantilever and voussoir beam structural models to reveal the periodic alternating law of the weighting step and strata behavior through field observation on the longwall face with a 7.0 m mining height. They pointed out that the key strata broke in advance and step subsidence would lead to the periodical changes of gentle and strong roof weighting (Ju and Xu, 2013). Liu et al. (2015a, b) carried out a simulation experiment on the large height mining of the thick coal in a mine by using $FLAC^{3D}$. The results showed that the height of the fractured zones increased with the mining thickness of the shallow coal seam. Through physical similarity tests, Ren et al. (2013) pointed out that the bedrock fissure zone conducting surface was the main cause of the long and short periodical roof weighting during shallow mining. Zhang et al. (2013) conducted the numerical simulation on a working face in the Shendong mining area by using UDEC, and proposed that if the basic roof could form a stable bearing structure, the overlying strata would present a continuous bending down without slip instability and whole step subsidence. Huang et al. (2016) found that the inclined step rock beam of the key strata was a common structure type in the mining field with a superlarge mining height in the Shendong mining area.

The cantilever beam (Yu et al., 2015; Zhao et al., 2018b) and voussoir beam (Yang, 2010; Li et al., 2016b; Wang et al., 2016a, b, c, d, e, f, g) models were widely applied in analyzing the roof weighting, but these models in previous studies mostly were to estimate the horizontal thrust between roof blocks and the action position by empirical formulas. They did not take into account the effects of the stress distribution on the structure stability. However, the slip instability and rotational deformation were usually all affected by the horizontal stress (Zhao and Song, 2016). Bakun-Mazor et al. (2011) indicated that the horizontal stress distribution along the boundary of the key roof blocks was not necessarily linear and the exact geometry distribution of that stress must be determined by the experimental studies. Tsesarsky (2012) studied the arching law of the compressive stress within the layered sedimentary rocks in shallow mining by using FlAC and analyzed the effects of block size and joint space on the horizontal stress distribution. Shabanimashcool et al. (2014) and Helm et al. (2013) found that the maximum span of the voussoir roof beam prior to the first roof caving depended upon the initial horizontal stress and the roof beams formed a large stable span when they were subjected to the high horizontal stress.

Due to current models of the roof structures giving more focus to the gravity effect on the stability of the roof blocks, the arching law of the principle stress and boundary horizontal stress distribution were not being well recognized. They ignored the influence of the pressure arch structure on the roof movement behavior, the fractured roof of the large height working face under shallow mining for thick coal displayed obvious step subsidence, frequently support crushing accident and commonly ground cracks (Ma et al., 2009; Huang et al., 2016), and these abnormal performances induced by shallow mining were still not being explained reasonably. In this paper, the fully mechanized mining face with a large mining height in the Shendong mining area was taken as the engineering background, and theoretical analysis and numerical simulation were used to analyze the pressure arching effect and the distribution of the horizontal stress in the roof block structure. It was of guiding significance for obtaining the instability and roof weighting characteristics of the pressure arch structure in the key roof blocks to ensure shallow mining safety.

Because there are still many problems that need to be solved in the shallow coal mining, this paper would conduct a further study of the roof weighting of typical shallow coal mining. Taking the fully mechanized mining face with large mining height in the Shendong mining area as the engineering background, theoretical analysis and numerical simulation were used to analyze the arching law of the principle stress and the distribution of the horizontal stress in the roof block structure. Considering that the pressure arch effect in the roof blocks can improve the rationality of the mechanical model, three typical pressure arch models of the roof structure were proposed and the related instability criteria were derived. The results are of importance for mining safety in similar engineering.

5.3.2 Pressure arching characteristics

5.3.2.1 Symmetric pressure arch of two key blocks

As shown in Fig. 5.25, with the mining face advancing from the open-off cut, the cracks were generated at both ends and the mid-span of the basic roof when reaching the limit span, forming trapezoid and arc blocks. The basic roof was periodically broken into multiple removable trapezoid blocks in the middle of the working face. The movements of these arc blocks at both sides were controlled by the trapezoid blocks.

The original fractured blocks F_a and F_b and the periodic fractured blocks P_a and P_b continuously sank and contacted the waste rocks. The hanging blocks

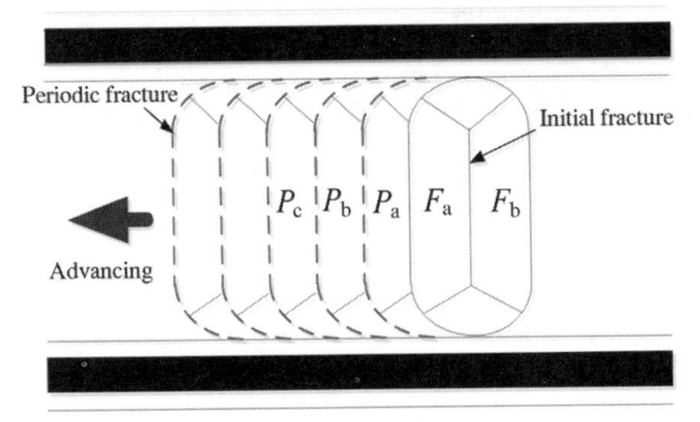

FIG. 5.25 Typical geometry structure of the fractured roof blocks.

were squeezed with the rear blocks to form the fractured structure. The stability of the fractured roof structure depended on the movement of the key blocks while the slip instability of the main block over the support was a direct factor to induce roof cutting and strong strata behavior.

After the full thickness of the coal seam being mined, the caving immediate roof was not enough to fill the goaf, so the large mining height technology provided a movable space for the roof blocks. The hanging blocks showed the step structures after separating from the upper strata. The roof weighting for sliding blocks was stronger than the normal height mining (Huang et al., 2016). The middle of the mining face was the critical control area as roof weighting, so it was feasible to analyze the movement of key roof blocks by using the plane model.

The symmetrical pressure arch model of two key blocks was established, as shown in Fig. 5.26. After the original breaking of the basic roof, the separated roof blocks rotated under their weights and were compressed reciprocally at the abutments in the surface contact state. The friction and horizontal thrust at the abutments directly affected the instability types of the roof blocks, and the internal and boundary stress distribution of the blocks was essential to the stability analysis of the roof structures.

The horizontal stress σ_{ij} at the left abutment of the block (Fig. 5.26C) was distributed as Eq. (5.37) (Wang et al., 2012):

$$\sigma_{xx} = f(x) = a(x-h)^{b} \tag{5.37}$$

where h was the distributed height of the horizontal stress at the left abutment, $f(x)$ was the distribution function of the horizontal stress, and x was the coordinate position.

The horizontal stress σ_{xx} at the coordinate origin point O was the largest value at $x = 0$. The maximum value of the stress $\sigma_m = f(0) = a(-h)^{b}$ according to the stress distribution law in Eq. (5.37), and the horizontal thrust T and its position X_T were respectively given by

$$T = \int_0^h a(x-h)^{b} dx = \frac{ah(-h)^{b}}{b+1} = \frac{\sigma_m h}{b+1} \tag{5.38}$$

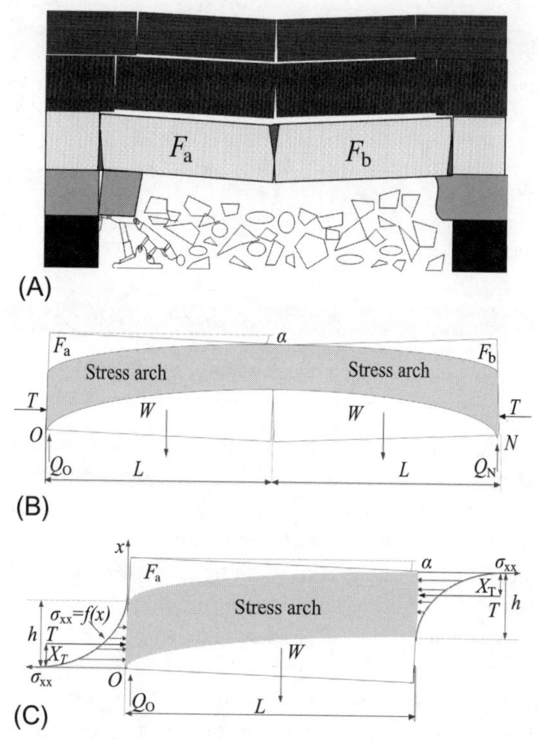

FIG. 5.26 Pressure arch structure of two key blocks of the initial fractured roof. (A) Sketch map of the structure of key blocks, (B) mechanical model, and (C) boundary stress distribution at the abutment.

$$X_T = \frac{1}{T}\int_0^h f(x)xdx = \frac{h}{b+2} \tag{5.39}$$

Assuming that the boundary stress distribution of the contact surface was the same and the horizontal thrust T was equal. If the pressure arch structure of key blocks was in the equilibrium state, taking the moment at the left point O, then $\sum M_O = M_{Q_N} - M_w = 0$, $\sum Y = 2W - Q_O - Q_N = 0$.

So the equilibrium equation was given by

$$Q_N(2L\cos\alpha + 2H\sin\alpha) - W(0.5L\cos\alpha + H\sin\alpha) - W(1.5L\cos\alpha + H\sin\alpha) = 0 \tag{5.40}$$

$$Q_O + Q_N = 2W \tag{5.41}$$

where Q_O and Q_N were the shear force, respectively. H was the block height, α was the rotating angle, W was the bearing load of the roof block, namely $\gamma L(H + H_1)$,γ was the unit weight of the roof rock, and H_1 was the thickness of the upper loading layer.

Arranging Eqs. (5.40), (5.41), then

$$Q_0 = Q_N = \gamma L(H + H_1) \tag{5.42}$$

The roof block F_a was the main key block over the mining working face. The slip instability of the block F_a would lead to the roof step subsidence and strong pressure on the support. However, the instability could be avoided when the contact friction, $T\tan\varphi$, was larger than the shear force Q_O, namely $T\tan\varphi > Q_O$.

Substituting Eqs. (5.38), (5.42) into the above-mentioned inequality, then

$$\sigma_m > \frac{\gamma L(b+1)(H+H_1)}{h\tan\varphi} \tag{5.43}$$

5.3.2.2 *Step pressure arch structure of key blocks*

As shown in Fig. 5.27, with the mining face advancing, the basic roof was fractured periodically. Field observations and simulation experiments showed that the quantity of removable blocks affecting the stability of the mining face was limited, so only the migration of key blocks hanging above the support space was worth considering.

The typical step pressure arch structure of multiple key blocks is shown in Fig. 5.27. The rotating angle α_1 of the main block P_a was larger than the angle α_2 of the rear block. The block P_a contacted with adjacent blocks at the hinged ends, then a semiarch of stress formed in the main block. The rear blocks P_b and P_c with little rotating angle compressed mutually and slid to form a step structure under the action of the horizontal thrust and shear force.

To analyze the mechanics of the key blocks P_a and P_b, taking the moment at the left hinged point A, $\sum M_A = 0$ because the vertical resultant force was zero,

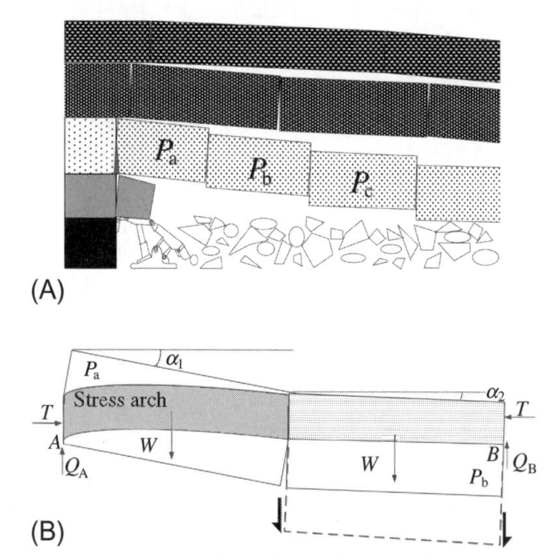

FIG. 5.27 Step pressure arch structure of key blocks of the periodic fractured roof. (A) Sketch map of the structure of key blocks and (B) mechanical model.

let $\sin\alpha_1 + \sin\alpha_2 = A_1$,$\cos\alpha_1 + \cos\alpha_2 = A_2$, then the following equilibrium was given:

$$T(H-LA_1-2X_T)-W(1.5L\cos\alpha_1+0.5L\cos\alpha_2+2H\sin\alpha_1)+Q_B(LA_2+H\sin\alpha_1)=0 \tag{5.44}$$

$$Q_A+Q_B=2W \tag{5.45}$$

Combining Eqs. (5.44), (5.45), then

$$Q_B=\frac{W[0.5L(2\cos\alpha_1+A_2)+2H\sin\alpha_1]}{LA_2+H\sin\alpha_1}-\frac{T[H-LA_1-2X_T]}{LA_2+H\sin\alpha_1} \tag{5.46}$$

$$Q_A=\frac{0.5WL(A_2+2\cos\alpha_2)+T[H-LA_1-2X_T]}{LA_2+H\sin\alpha_1} \tag{5.47}$$

For the main block P_a, $\tan\varphi$ was the friction coefficient of the contacted surface, and if $T\tan\varphi < Q_A$, the block would slide along the working face. According to Eqs. (5.38), (5.39), (5.28), let $\sin\alpha_1\tan\varphi = A_3$, $\frac{2h}{b+2}=A_4$, then the following formula was given:

$$\sigma_m < \frac{0.5\gamma L^2(b+1)(A_2+2\cos\alpha_2)(H+H_1)}{h[L\tan\varphi A_2+H(A_3-1)+LA_1+A_4]} \tag{5.48}$$

Eq. (5.48) can be used as the slip instability criterion for the main block in this structure.

5.3.2.3 *Rotative pressure arch structure of key blocks*

The rotative pressure arch structure of multiple key blocks was shown in Fig. 5.28. The rotating angle of each block was approximately equal, and the horizontal thrusts among these blocks were sufficient to mobilize a frictional resistance to balance the shear force. The roof blocks rotated gradually without slip instability, and the rotative pressure arch with bearing capacity produced a protective effect on the mining space.

The moment at the left hinged point A,$\sum M_A = 0$, and the vertical resultant force was zero, then the equations were given as follows:

$$T(H-3L\sin\alpha-2X_T)+Q_B(3L\cos\alpha+H\sin\alpha)-W(4.5L\cos\alpha+3H\sin\alpha)=0 \tag{5.49}$$

$$Q_A+Q_B=3W \tag{5.50}$$

Combining Eqs. (5.49), (5.50), then

$$Q_B=\frac{W(4.5L\cos\alpha+3H\sin\alpha)}{3L\cos\alpha+H\sin\alpha}-\frac{T(H-3L\sin\alpha-2X_T)}{3L\cos\alpha+H\sin\alpha} \tag{5.51}$$

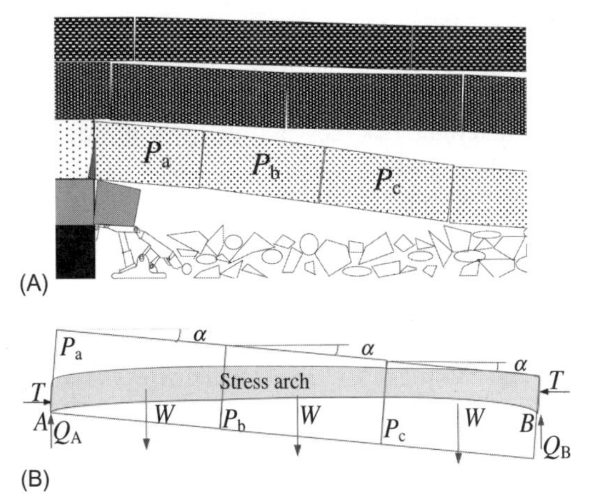

FIG. 5.28 Rotative pressure arch structure of key blocks of the periodic fractured roof. (A) Sketch map of the structure of key blocks and (B) mechanical model.

$$Q_A = \frac{4.5WL\cos\alpha + T(H - 3L\sin\alpha - 2X_T)}{3L\cos\alpha + H\sin\alpha} \tag{5.52}$$

The roof blocks sliding should be avoided to ensure the formation of the structure, namely $T\tan\varphi > Q_A$.

Substituting Eqs. (5.38), (5.39), (5.52) into this inequality, and let $\sin\alpha_1 \tan\varphi = A_3$, $\frac{2h}{b+2} = A_4$, then

$$\sigma_m > \frac{4.5\cos\alpha\gamma L^2(b+1)(H+H_1)}{h[3L(\tan\varphi\cos\alpha + \sin\alpha) + H(A_3 - 1) + A_4]} \tag{5.53}$$

5.3.2.4 Key block stability of initial fractured roof

Taking the No. 1^{-2} coal with a 7.0 m mining height at the Shangwan Mine as the engineering background, the panel was at an average depth of 115.4 m, which dipped 0–3 degrees. The thickness of the coal was 3.8–9.2 m and 7.0 m thick in average. The average thickness of the ground overburden of alluvial sand was 13.7 m, and the average thickness of the bedrock was 102.1 m.

The length of the fully mechanized mining panel No. 51101 was 3654 m, the width was 300 m, and the designed mining height of the panel was 5.2–5.4 m. The total caving method to manage the roof was adopted, which was matched to the German Eickhoff SL-500 shearer and the double column shield hydraulic supports 2 × 4319 kN.

The immediate roof of the mining field was mostly sandy mudstone or siltstone, with a thickness of 0.5–7.1 m, 3.0 m in average. The basic roof was sandstone, with a thickness of 6–15.1 m, 9 m in average, and the compressive strength of that was 16.6–33.8 MPa, 25 MPa in average. The span of the initial roof weighting step was 53.8 m, and there were six times periodic roof weighting in all during the field observation. The span of the roof weighting in the middle of the mining face was 10.9–32.7 m, 20.39 m in average.

Based on the technical parameters of panel No. 51101, the thickness H of the basic roof was 9 m, and the block length L was 26.9 m, according to the initial roof weighting step. The rotating angle α was taken as 4 degrees, the friction coefficient $\tan\varphi$ was 0.5, the bulk density of the roof was 25 kN/m^3, and the compressive strength $\sigma_c$was 25 MPa. As shown in Fig. 5.29, the curves were calculated from Eq. (5.43), which was used to judge the stability of the main block in a symmetrical stress arch structure. The slip instability was avoided when the maximum horizontal stress reached the curve value σ_m. The variable h in Fig. 5.29 was the abutment thickness of the stress arch.

It was found that: (1) When the boundary horizontal stress displayed nonlinear distribution, the stable stress σ_m of the block was higher, so the block with nonlinear boundary stress was easier to slide under the same crustal stress condition. (2) When the abutment thickness h of the pressure arch was 0.1H for the nonlinear boundary stress, the maximum horizontal stress needed to reach 40 MPa to ensure the blocks without sliding while the stress σ_m had exceeded the compressive strength of the blocks under this condition. So, the plastic failure at the hinged end of the blocks occurred, which led to further rotation of the roof blocks. (3) The critical stress σ_m decreased with the thickness h increasing. The decreasing trend was most obvious in the range of 0.1–0.3H, and this trend indicated that it was no longer requiring a high level of stress to avoid the slip instability when the pressure arch abutment reached a certain thickness.

Therefore, the decreasing stress reduced the plastic failure of the block ends and prevented further rotation to increase roof weighting. So, the increasing abutment thickness of the pressure arch could improve the stability of the key blocks.

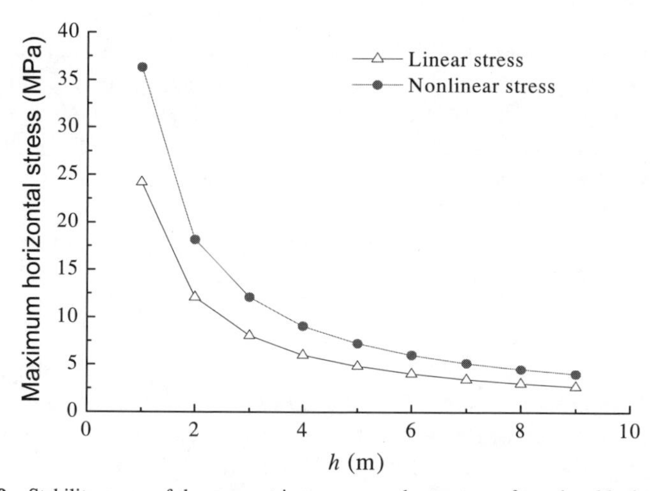

FIG. 5.29 Stability curve of the symmetric pressure arch structure of two key blocks.

5.3.2.5 Key block stability of periodic fractured roof

According to Eqs. (5.48), (5.53), the rotating angle α_1 and α_2 of the key blocks in the step pressure arch structure was adopted respectively as 4 degrees and 1 degree, and the rotating angle α of the key blocks in the rotative pressure arch structure was adopted as 4 degrees. The length L of the block was 20 m according to the periodic weighting step, and the other parameters remain unchanged. The stability judging curves of the step pressure arch and the rotative pressure arch structure were obtained as shown in Fig. 5.30.

It can be found that: (1) The horizontal stress needed to maintain the stability of the multiple key blocks structure was higher than that of the symmetrical pressure arch structure of two key blocks. With the abutment thickness of the pressure arch being less than $0.2H$, the maximum horizontal stress σ_m would exceed the compressive strength σ_c of the blocks, and the plastic failure occurred at the hinged end of the block. (2) The stable horizontal stress of the rotative pressure arch was the largest under the same ground stress condition, and the rotative pressure arch was more difficult to form. (3) The pressure arch with a smaller abutment thickness required higher stress to maintain the stability of the key blocks, and the greater plastic failure would occur at the block ends.

5.3.3 Evolution characteristics of pressure arch

5.3.3.1 Building computational model

Based on the geological condition and the mining field in the Shanwan Mine, the discrete element software UDEC was used to analyze the instability characteristics of the pressure arch of key blocks under the shallow coal mining condition. The numerical test was to simulate the mechanical behavior of the fractured strata blocks, considering the vertical and horizontal virtual joints

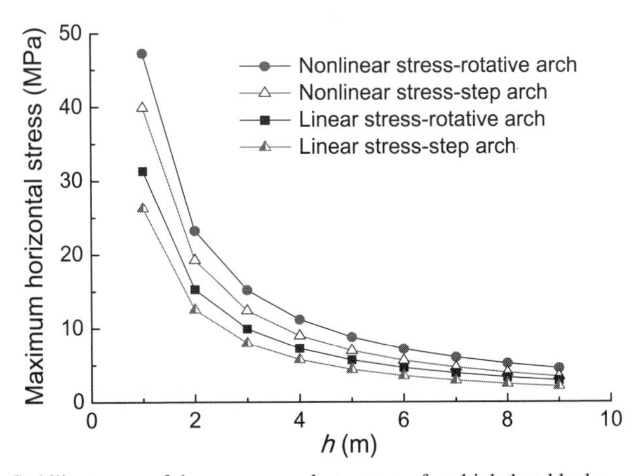

FIG. 5.30 Stability curves of the pressure arch structure of multiple key blocks.

condition, the normal stiffness and tangential stiffness of each joint can be approximately calculated through the joint structure and the deformation of intact rock mass, so the normal stiffness and tangential stiffness was set as 4.0 GPa/m.

As shown in Fig. 5.31, the full mining thickness of the coal seam was 7 m, and the thickness of the immediate roof was 3 m. Considering the falling size of the immediate roof, the virtual joint space was set as 3 m. The thickness of the basic roof was 9 m. Because the periodic weighting length was 10.9–32.7 m in the practical engineering, this numerical test was focused on the stress distribution in the strata blocks during shallow coal mining. So, the virtual block size was set as 15 m in the basic roof according to the weighting length.

The thickness of the strata over the basic roof was set as 30 m to focus on the movement law of the roof strata. According to the field observations and similar material experiment results, the thickness/span ratio of the broken strata decreased with the layer position rising. The fractured strata block in high position was longer than the basic roof block (Yu et al., 2015; Zhao et al., 2018), so the length of the single virtual block was set as 30 m, 25 m, and 20 m in strata ①–③ in Fig. 5.31, respectively.

The model size was 450 m long and 54 m high, and the upper boundary loading was calculated as the bedrock strata of 70 m thickness. The bottom boundary of the model was fixed and the lateral boundaries of the model were fixed in the horizontal direction.

The position to start mining was 50 m away from the left boundary, and the advancing distance was 350 m. The monitoring line was arranged along the *Y* direction at the interval of 1 m in the range of 15–24 m to obtain the evolution law of the stress field and displacement field in the roof structure during mining. The physical and mechanical parameters of the model are listed in Table 5.3, and the Mohr-Coulomb criteria were used in the numerical calculation.

5.3.3.2 Evolution process of key blocks pressure arch

For mining face No. 51101, advancing 10 m each time was adopted in the simulation. As shown in Fig. 5.32A, the compressive stress was positive while the tensile stress was negative. The principle stress showed an obvious arching effect in the initial stage of coal mining. As the advancing distance was 30 m, the immediate roof displayed the initial caving. The symmetrical pressure

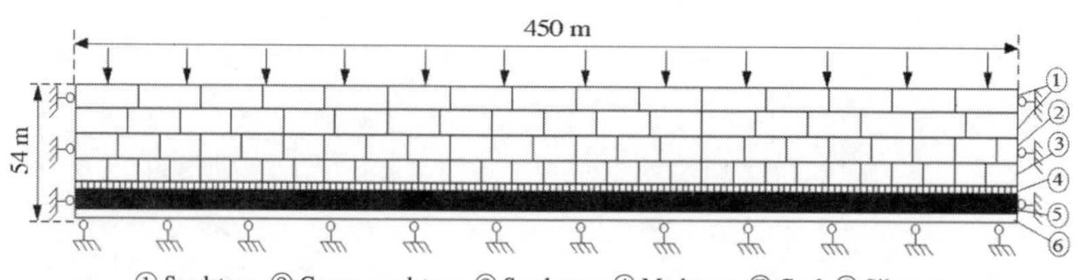

FIG. 5.31 The computational model.

TABLE 5.3 Physical and mechanical parameters of the materials.

Name	Unit weight (kN/m^3)	Elastic modulus (GPa)	Poisson's ratio	Cohesive strength (MPa)	Tensile strength (MPa)	Friction angle (degree)
Coarse sand-stone	24.3	35	0.23	5.5	4	33
Sand-stone	25.0	32	0.24	7.3	4.9	35
Mud-stone	22.4	23	0.15	2.5	1.7	30
Coal	13.1	15	0.29	0.79	0.57	27
Silt-stone	24.6	26	0.22	3.9	2.5	38

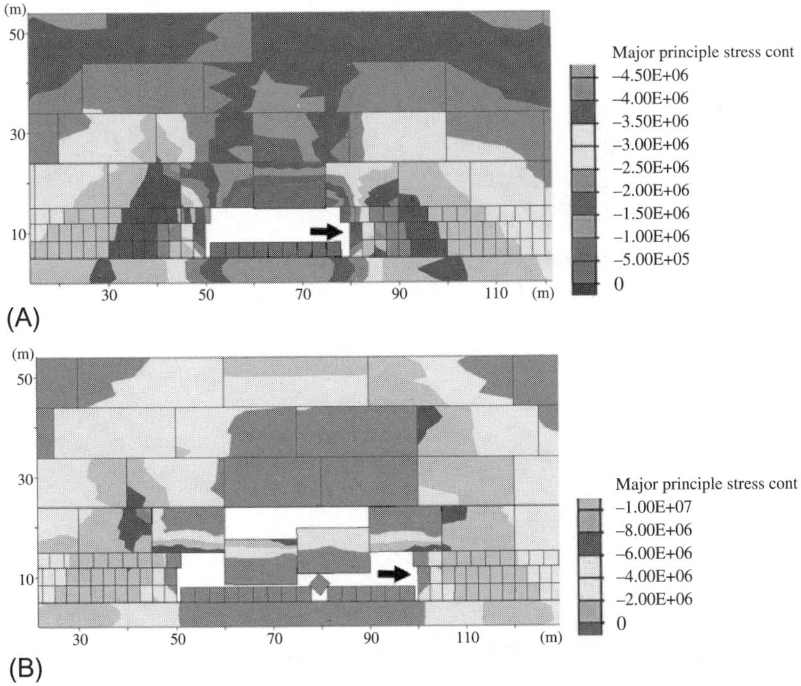

FIG. 5.32 Evolution process of the pressure arch in the initial mining stage. (A) Initial caving of the immediate roof and (B) initial caving of the basic roof.

arch formed in the basic roof and the global pressure arch formed in the overlying strata. Overall, the roof remained stable. The monitoring data showed that the peak value of the major principle stress occurred at the abutment of the pressure arch, and the maximum horizontal stress at both abutments was 1.72 MPa and 1.81 MPa, respectively.

As shown in Fig. 5.32B, the initial caving of the basic roof occurred when the advancing distance was 50 m. The load of the fallen roof blocks was transferred to the surrounding rock by the squeezing and clamping between the blocks. The principle stress was redistributed and delivered through the contacting face. The step pressure arch structure formed in the main key blocks and the fallen roof blocks moved along the stress transferring trace. The peak value of the major principle stress occurred at the surrounding rock of the mined-out area. The maximum horizontal stress at the right arch abutment was 2.41 MPa, but the maximum horizontal stress on the surface of the sliding block was only 0.41 MPa.

As shown in Fig. 5.33A, the overlying bedrock strata was moved totally with the mining face advancing 210 m. The coal mining reached the complete mining stage, and the global pressure arch structure of the bedrock strata was transformed into the single pressure arch structure in each layer. The rotating angle of the basic roof block increased significantly compared with the previous state. The front half pressure arch formed in the hanging basic roof at the mining face side, and the pressure arch top was located in the step sliding basic roof block at

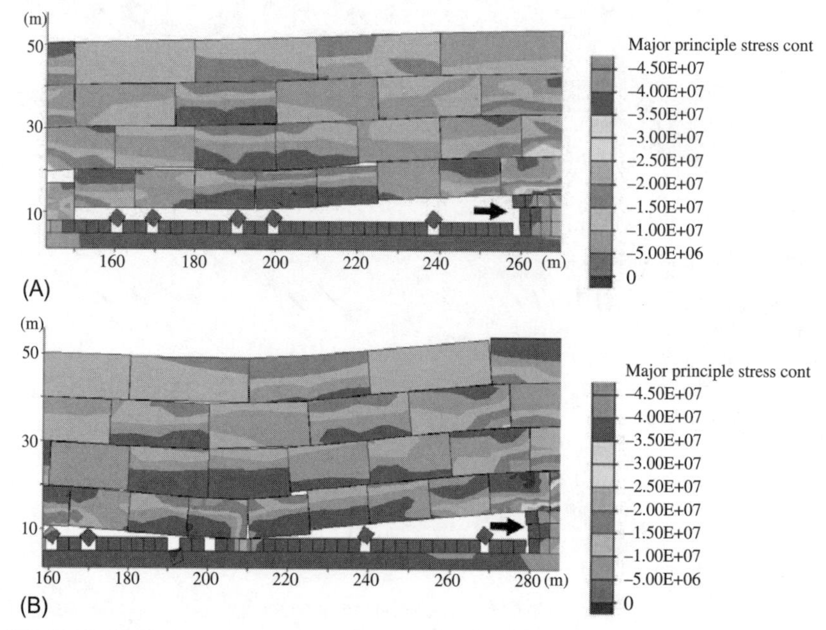

FIG. 5.33 Evolution process of the step pressure arch structure of the basic roof. (A) Step pressure-arch and (B) instability of the pressure-arch of basic roof.

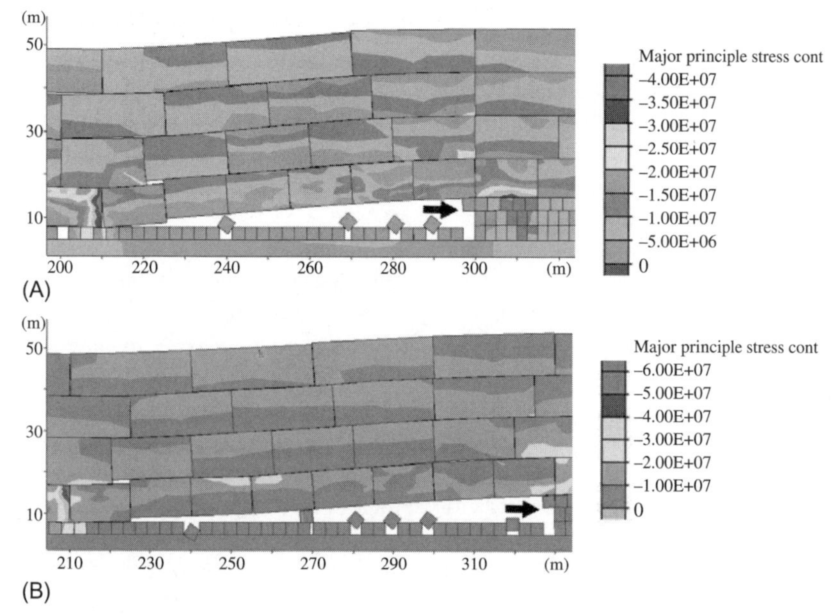

FIG. 5.34 Evolution process of the rotative pressure arch structure of multiple key blocks. (A) Rotative pressure-arch structure and (B) sinking of the pressure-arch of basic roof.

the mid-span of the mined-out area. The hanging basic roof carried the upper strata load and was kept stable by the structure effect of the pressure arch; the roof weight was relatively weak in this period.

It can be seen from Fig. 5.33B, with the mining face advancing 230 m, multiple layers of the bedrock strata rotated and sank toward the mined-out area, the sank roof strata was compacted in the range of abscissa 190–220 m. The pressure arch structure of the basic roof occurred instability under the dynamic action load of the upper strata during the rotation process. The step sliding basic roof blocks in the pressure arch top reached maximum subsidence. The rotated instability of the basic roof in the front pressure arch would increase the resistance of the support at the working face side.

Compared with the stable state of the pressure arch in the basic roof, the instability of the pressure arch in the hanging basic roof led to the load of the upper bedrock strata lose carrier, the additional load acted on the roof support caused the strong periodic roof weighting.

As shown in Fig. 5.34A, with the mining face advancing 250 m, the rotative pressure arch structure was rebuilt in the basic roof, and the peak value of the major principle stress at the left arch abutment was 50.23 MPa. As shown in Fig. 5.34B, with advancing 280 m, the basic roof sank toward the mined-out area and compacted at the mid-span of the pressure arch. The peak value of the major principle stress was 65.40 MPa, and the bearing load of the pressure arch structure of the basic roof was transferred to the caved zone behind the mined-out area of the mining face side.

5.3.3.3 Structure characteristics of symmetrical pressure arch

To explore the distribution characteristics of the symmetrical pressure arch and the block size effect, different models were built considering the block length ranging from 10 m to 20 m, and the length L of the single block was increased 1 m each time. As shown in Fig. 5.35, the major principle stress was distributed as the obvious symmetrical pressure arch in the basic roof when the roof span reached 2 L gradually.

As shown in Fig. 5.36, it was found that the horizontal stress at the pressure arch abutments in the basic roof displayed nonlinear distribution characteristics and with the quadratic function law. The stress σ_{xx} decreased from the bottom to the top along direction of the distributed height h_x, and the maximum horizontal stress was located at the lowest position of the arch abutments at $h_x = 0$.

The horizontal stress at the mid-span of the pressure arch increased linearly from the bottom to the top along the direction of the height h_x, and the maximum horizontal stress σ_m at the mid-span was greater than that at the abutments. The horizontal compressive stress was distributed in the range of $h_x = 4$–8 m, and the mid-span thickness of the arch was less than that of the abutments.

The peak value of the boundary horizontal stress increased with the block length increasing. The maximum horizontal stress was 1.12 MPa, 1.81 Mpa, and 2.38 MPa when the block length was 10 m, 14 m, and 20 m, respectively, while the abutment thickness of the pressure arch reduced gradually with increasing span.

5.3.4 Results and discussion

Shallow coal mining with a large height would induce roof collapse as well as support crushing and ground subsidence. The structure instability of the basic roof and overlying bedrock strata were the main cause of this mining damage, so it was very important to reveal the structure instability characteristics of the hanging roof blocks. Before the roof caving, the arching law of the principle stress made the roof block structure able to bear loading. The instability of the arch structure induced by mining often lead to the roof weighting. The load-bearing structure of the roof blocks could be formed after fully mechanized mining with a large mining height, and the structure characteristics of the hanging blocks were more remarkable and the roof weighting would be unusually strong in the large mining space.

The symmetrical pressure arch structure of the basic roof formed after the initial mining, and the horizontal stress at both sides of the arch displayed nonlinear distribution, which was consistent with the findings in the literature (Wang et al., 2012). But the horizontal stress was linearly distributed at the mid-span of the arch, and the peak value of the horizontal stress was greater than that at both abutments while the mid-span thickness of the arch decreased.

Different views on the thickness of the arch were proposed (Tsesarsky, 2012): (1) When the ratio of the arch thickness h to the roof height H was

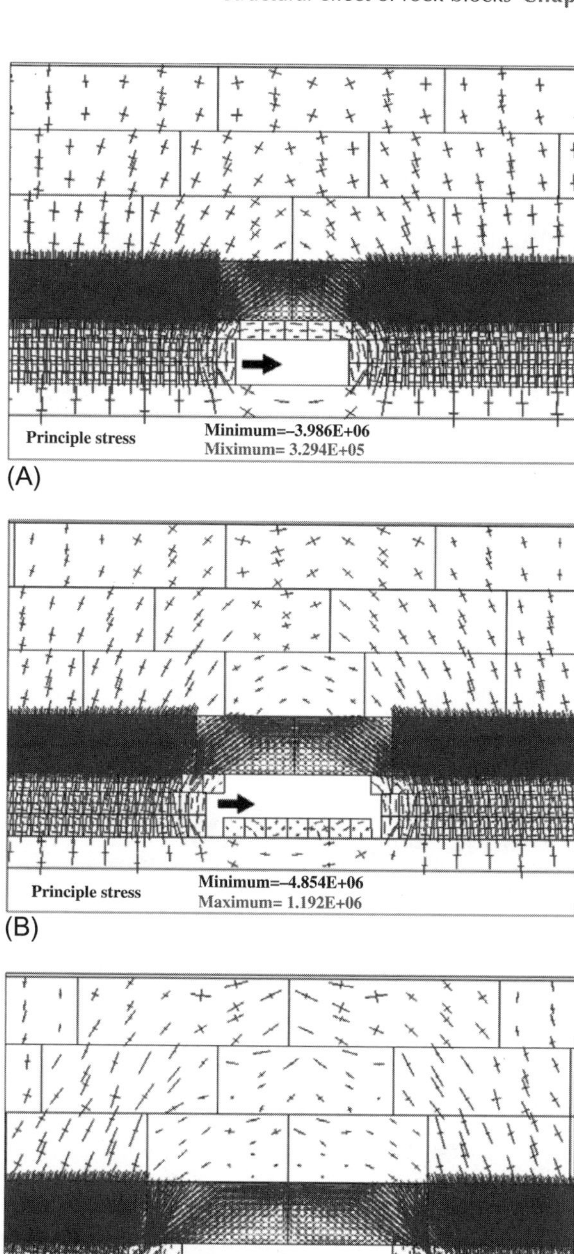

FIG. 5.35 Symmetric pressure arch structure of two key blocks with different block sizes. (A) $L = 10$ m, (B) $L = 14$ m, and (C) $L = 20$ m.

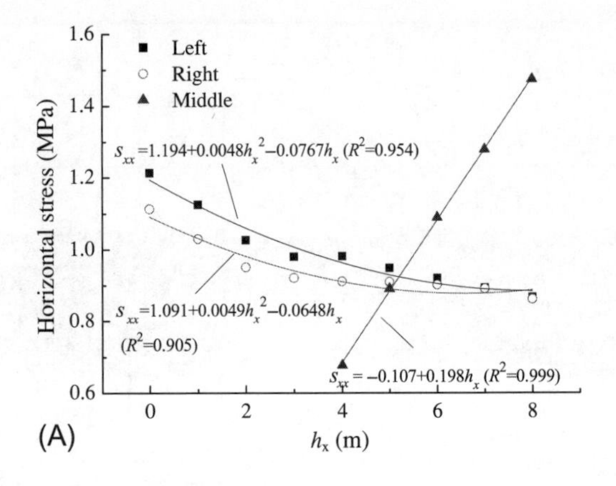

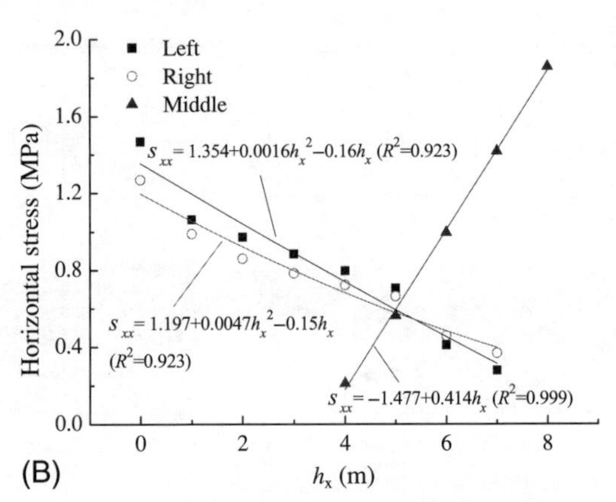

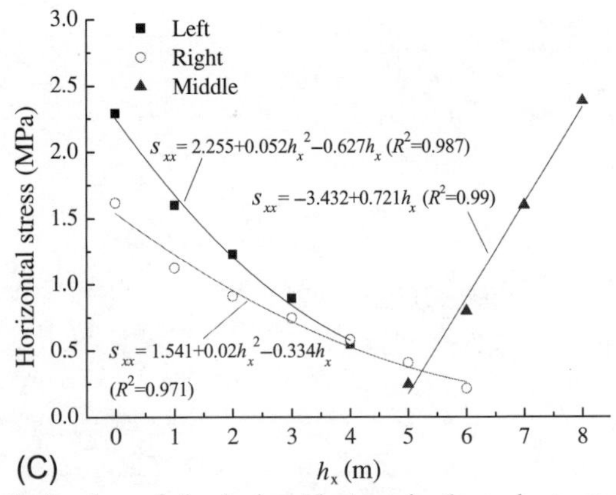

FIG. 5.36 Distribution law of the horizontal stress in the arch structure. (A) $L = 10$ m, (B) $L = 14$ m, and (C) $L = 20$ m.

0.75, the arch structure was at equilibrium, and the ratio $h/H = 0.3$–0.4, which was near failure. (2) When the ratio was $h/H = 0.3$, the arch structure was in the equilibrium state while the ratio $h/H = 0.1$, which was near failure. The results indicated that as the pressure arch reached a certain thickness at the abutments, the required horizontal stress avoiding slip instability obviously reduced, and the increasing thickness of the pressure arch could improve the stability of the key blocks. In addition, the distribution of the principle stress in blocks showed a size effect. With the block length increasing, the abutment thickness of the pressure arch decreased and the mid-span thickness as kept unchanged; however, the peak value of the horizontal stress increased in all.

During the periodic fracture phrase of the roof, the pressure arch structure still existed in the key blocks of the large height mining face, mainly behaving as the step pressure arch structure of multiple key blocks and the rotative pressure arch structure of multiple key blocks. Through mechanical analysis and numerical simulation, it was found that sufficient horizontal stress was necessary to form the rotative pressure arch structure, and the high horizontal thrust at the arch abutments made the load transfer along the arch trace in the blocks. This structure would not be maintained without a strong horizontal constraint, and the horizontal stress on the surface decreased dramatically after the structure instability.

The horizontal stress was relatively lower at the arch abutment of the main key block in the step pressure arch structure. The main key block was easier to slide under the nonlinear distributed boundary stress, and a step structure was formed by the sliding block and the rear blocks. When the semiarch formed again in the main key block over the mining face, the basic roof evolved into a new step pressure arch structure. Field observation showed that step subsidence and strong roof weighting were common, and were induced by shallow mining for the large height working face (Huang et al., 2016; Zhao et al., 2018). The cyclic evolution of the step pressure arch structure was the main reason behind these mining performances. The continuous mining in the large space would weaken the arching effect of the compressive stress in the overlying bedrock strata, and the pressure arch of the bedrock developed upward with the arch abutments moving backward. Moreover, the global pressure arch structure of the bedrock strata was transformed into the multiple single-layer hinged arch structure and the long periodic roof weighting was produced.

5.3.5 Conclusions

Taking the fully mechanized mining face with a large mining height in the Shendong mining area as the engineering background, theoretical analysis and numerical simulations were used to reveal the pressure arching effect and the mechanical evolution characteristics of the key blocks under shallow coal mining. The conclusions are listed as the following:

(1) There are three typical structures: the symmetrical pressure arch of two key blocks, the step pressure arch of multiple key blocks, and the rotative

pressure arch of multiple key blocks during shallow coal mining. The stability of the pressure arch is affected by the horizontal stress distribution, the roof block is easier to slide as the boundary horizontal stresses display nonlinear distribution, and the increasing abutment thickness of the pressure arch can improve the stability of the structure.

(2) In the symmetrical pressure arch structure, the horizontal stresses show nonlinear distribution at both abutments and the linear distribution with a higher peak value at the mid-span of the arch. The boundary horizontal stresses produce an obvious size effect. With the block length increasing, the peak stress increases and the abutment thickness of the pressure arch decreases gradually.

(3) In the period of periodic fracture of the basic roof, sufficient horizontal stress is necessary to form the rotative pressure arch of multiple key blocks. The instability and evolution of this structure are the main causes to induce the common step subsidence and strong roof weighting. The long periodic roof weighting is the outcome of the instability of the global pressure arch structure of the multiple layered bedrock.

Due to the complexity of the engineering geological conditions, the diversity of the roof under shallow coal mining, and the bedrock thickness being a key influencing factor to roof weighing, the characteristics of the pressure arching effect in the roof blocks under different conditions need to be further studied in the future.

5.4 Composite pressure arch in thin bedrock

5.4.1 Introduction

In the western coalfields of China, coal seams are nearly horizontal bedding and shallow buried by the thin bedrock of the overlying strata of the Quaternary loose layer (Zhang et al., 2010). The shallow horizontal coal mining often causes serious strata deformation, such as roof cutting off and large surface subsidence, leading to serious threats to mining safety and the environment in the mining area (Wang et al., 2016c; Qiu et al., 2017). Therefore, it is an urgent issue to be solved to control the stability of the overburdened strata during shallow coal mining in western China.

After the excavation of the underground engineering, the self-supporting pressure arch structure can be formed in the surrounding rock. In the past years, many research efforts have been made to reveal the stress distribution characteristics and evolution process of the pressure arch caused after a small-scale excavation, such as pillar safety and a tunnel or double-arch tunnel construction. To date, some achievements have been obtained, such as the fact that the morphological feature of the pressure arch was affected by the roof thickness, lateral pressure, and construction process (Poulsen, 2010; Chen et al., 2011; Wang et al., 2015c; Li et al., 2016b; Wang et al., 2016d). However, the evolution characteristics of the composite pressure arch due to a large-scale excavation have been rarely reported.

Recently, it has been recognized that a pressure arch structure would be formed in the broken blocks of the roof in the near field in the mining field after the coal seam was mined out (Ju and Xu, 2013; Ju et al., 2015; Yong et al., 2015; Li and Liu, 2017; Jia et al., 2017; Liang et al., 2017; Kong et al., 2018). However, with the increase of bedrock thickness, the stepped subsidence of the roof can be reduced or disappeared during shallow coal mining. In other words, the roof stability in the mining field is closely related to the structural characteristics of the overlying strata. Therefore, the movement characteristics of the broken rocks should be further studied. Xie et al. (2009) studied the stress distribution in fully mechanized top coal caving using physical and numerical tests. The results showed that a far-field pressure arch could be formed in the mining field, and the pressure arch could protect the panel under the periodic weighting of the roof. However, the main pressure arch types and instability criterion of each pressure arch during coal mining were absent.

From the above-mentioned literature, it can be found that some studies displayed the morphological structure and stress distribution of the pressure arch. However, the main pressure arch types and the evolution characteristics of the pressure arches in the broken blocks in the near field as well as the instability criterion of each pressure arch and the interactions of the composite pressure arches in the near field and far field in different mining stages have not been reported. Here, the arching conditions and stability criterion of the composite pressure arch in the overlying strata during shallow coal mining are further studied. This paper focuses on typical pressure arch structures and analyzes the interactions of the near- and far-field pressure arches in different mining stages under thin bedrock shallow coal mining using numerical simulations and physical experiments.

Based on the weighting laws of a mining panel in the Daliuta Coal Mine in the Shendong mining area, the typical mechanical model of the pressure arch in the mining field was established. The principal stress distribution in each mining stage was obtained. The stress development in thin bedrock of the overlying strata of the shallow horizontal coal mining was analyzed and verified by numerical simulations and experiments. The evolution characteristics of the composite pressure arches in the near and far field are of great meaning to the overlying strata stability and mining safety.

5.4.2 Engineering background and pressure arch structure

5.4.2.1 Engineering background

To analyze the roof weighting of thin bedrock under a thick loose layer during shallow coal mining, the No. 22614 panel of the 2^{-2} coal seam in the Daliuta Mine was taken as the engineering background. The inclined longwall retreating mining method was adopted and the full thickness of the coal seam was mined out at one time.

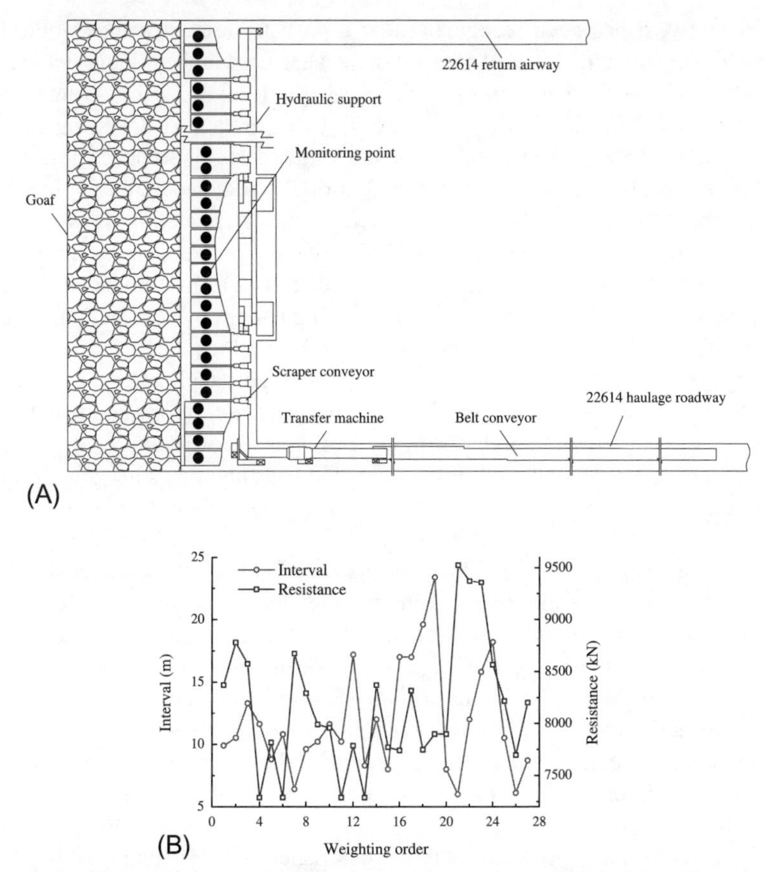

FIG. 5.37 Working face arrangement and roof weighting rule. (A) 22614 working face arrangement and (B) the strata behaviors of No. 22614 working face.

For the No. 22614 panel shown in Fig. 5.37A, the width was 201.7 m and the advancing length was 2436.5 m. The designed mining height was 5.4 m in average, the dip angle was 0–5 degrees, and the average bury depth was 115.4 m. The thickness of the rock strata above the coal seam was 14–47 m, 28.3 m in average. The immediate roof was mostly sandy mudstone or siltstone, with thickness of 1.7–9.6 m. The basic roof was mainly siltstone with a thickness of 13.1–40.4 m. The thickness of the loose sand layer was 83–104.5 m, 96.7 m in average.

As seen from Fig. 5.37B, the variable interval was the advancing distance between twice the roof weighting, the variable resistance was the monitoring resistance of the hydraulic support, there were 27 times roof weighting occurred during the observation period and the advancing distance for each roof weighting varied from 6.1 to 23.4 m, 11.9 m in average. The support was 7290–9526 kN, 8152 kN in average. The strata behaviors of the thin bedrock under a thick loose layer over the 2^{-2} coal seam had the characteristics of a short weighting interval and a strong weighting load, and the longer weighting interval corresponded to stronger strata behaviors. The 21st, 22nd, and 23rd weighting stresses exceeded the support's resistance. The fractured strata kept itself at bearing capacity in the stable structure. The roof weighting was the result of instability and recombination, and the arching effect of the concentrated stress in the rock strata was the symbol of the itself bearing capacity. So it is essential to reveal the pressure arch characteristics in the fractured strata.

5.4.2.2 *Macroscopic pressure arch in far field*

After excavation, the stress in the surrounding rock is transferred to the far-field stable rock. The maximum principal stress σ_1 close to the mining field deflected to form an arch-shaped stress concentrated zone in the surrounding rock. The principal stress σ_1 in the arch is greater than the in situ stress σ_0. Define a judgment index k for the pressure arch in the surrounding rock, which can be calculated by

$$k = \frac{\sigma_1 - \sigma_0}{\sigma_0} \tag{5.54}$$

where σ_1 was the maximum principal stress after coal mining and σ_0 was the in situ maximum principal stress before coal mining.

According to the judgment index k, the stress distribution and pressure arch morphology can be determined, as shown in Fig. 5.38. For $k > 0$, it was in the range of the pressure arch, the low compressive stress zone was in $-1 < k < 0$, and the tensile stress zone was $k < -1$. The outer boundary of the pressure arch was the in situ stress zone. Under its inner boundary, it was a low compressive zone and a tensile zone in which a caving arch was formed in the plastic deformation zone of the roof. The caving arch was the key area to control the surrounding rock. Its weight determined the strata behavior and roof stability, and its range and instability characteristics changed with the evolution of the macroscopic pressure arch in the far field.

5.4.2.3 *Fractured pressure arch in near field*

To investigate the typical pressure arches in the caving roof blocks, the symmetrical pressure arch, the stepped pressure arch, and the rotating-squeezed pressure arch were generalized and the simplified mechanical models were given in Figs. 5.39–5.41, according to their engineering models.

With the working face advancing, the roof was suspended to the ultimate span, and cracks developed at both the coal wall ends and the mid-span. The broken blocks of the separated layer rotated, sank, squeezed, and bit together. The principal stress deviated within the caving zone and the maximum principle stress concentrated and distributed like a symmetrical arch, based on the voussior beam model (Sofianos, 1996; Diederichs and Kaiser, 1999; Zhao et al., 2018). The friction and horizontal thrust between the contact surfaces

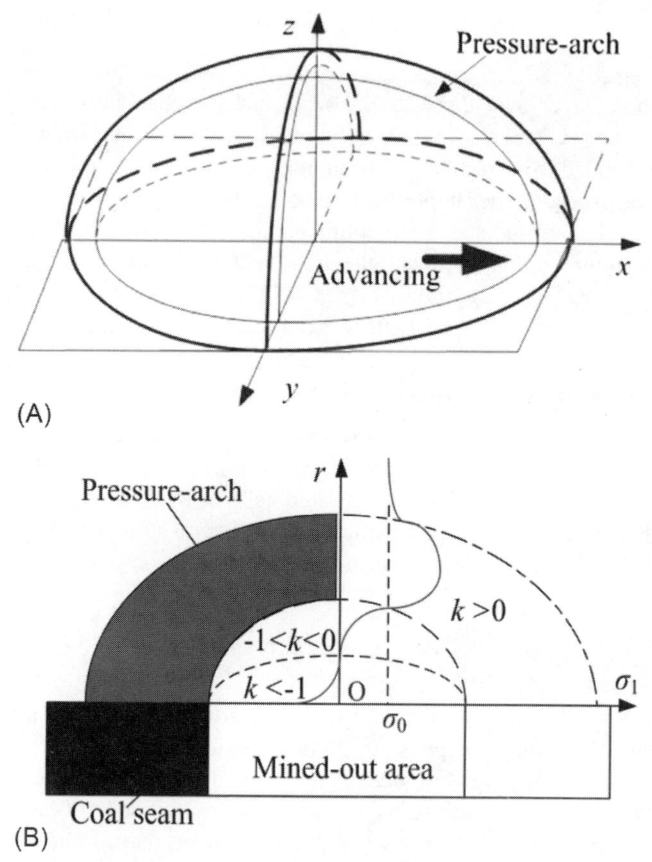

FIG. 5.38 The pressure arch distributed in the surrounding rock. (A) Engineering model and (B) mechanical model.

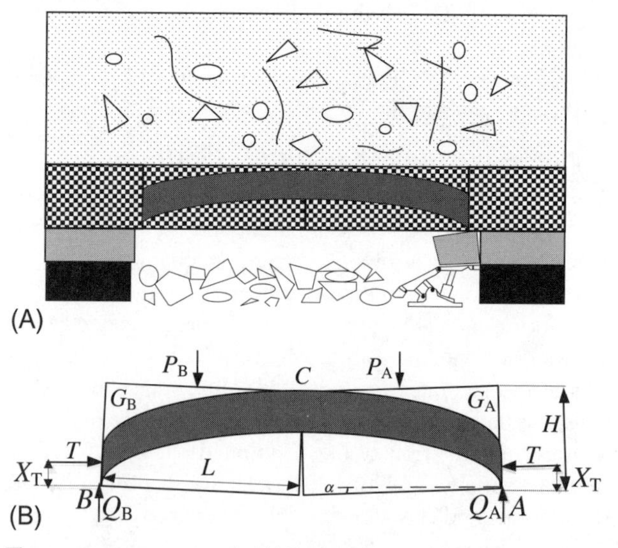

FIG. 5.39 The symmetry pressure arch. (A) Engineering model and (B) mechanical model.

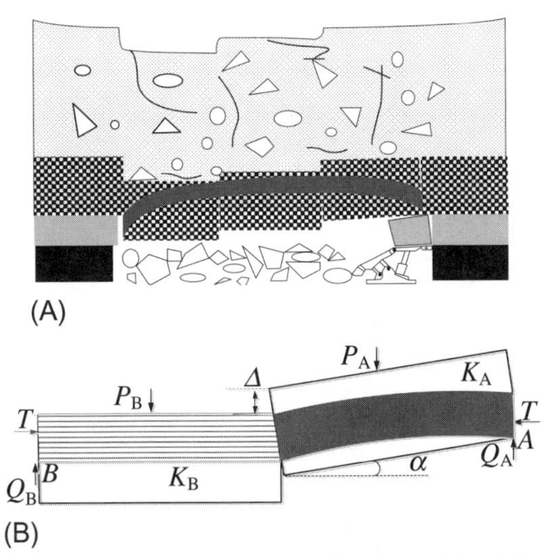

FIG. 5.40 The stepped pressure arch structure. (A) Engineering model and (B) mechanical model.

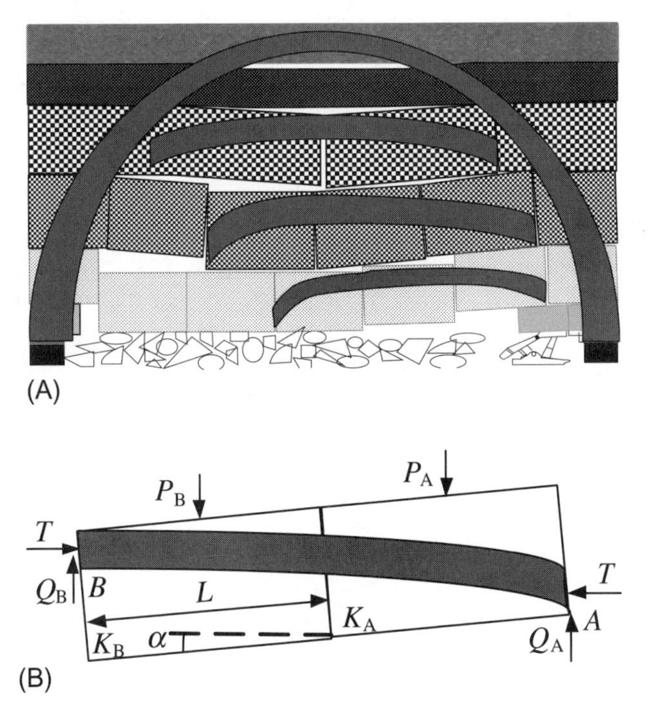

FIG. 5.41 The composite pressure arch. (A) Engineering model and (B) mechanical model.

determined the stability of the broken blocks. The mechanical model of the symmetry pressure arch was built as shown in Fig. 5.39.

Assuming that the roof block was rigid and the contact surfaces were non-cohesive, the stress of the contact surface was the same while the horizontal thrust force was T, $P_A = P_B = P$ and the symmetrical pressure arch was in the equilibrium state. Taking a moment on the left hinge point B, then $\sum M_B = M_{Q_A} - M_P = 0$, $\sum F_Y = 2P - Q_A - Q_B = 0$, the following equation can be obtained.

$$Q_A(2L\cos\alpha + 2H\sin\alpha) - P(0.5L\cos\alpha + H\sin\alpha) - P(1.5L\cos\alpha + H\sin\alpha) = 0 \tag{5.55}$$

$$Q_A + Q_B = 2P \tag{5.56}$$

where Q_A and Q_B were the shear force of the contact surface, respectively. L was the length, and H was the height of the broken block. α was the rotation angle; P was the bearing load of the broken block, which was $\gamma L(H + H_1)$ and contains the blocks' own weight; H_1 was the thickness of the load layer on the block; and γ was the average unit weight of the overburden layer. Combining Eqs. (5.55), (5.56), then

$$Q_A = Q_B = \gamma L(H + H_1) \tag{5.57}$$

Sliding of block K_A would cause a roof stepped subsidence and strong weighting on the working face. To avoid the broken blocks sliding, the frictional force $T\tan\varphi$ (φ was the friction angle) between the contact surface of the broken blocks must be greater than the shear force Q_A, namely $T\tan\varphi > Q_A$. Substituting Eq. (5.57) into an inequality, then

$$T > \frac{\gamma L(H + H_1)}{\tan\varphi} \tag{5.58}$$

As shown in Fig. 5.40, the stepped structure was formed by the friction of the broken blocks in the mining near field, and the first section of the principal stress arch was formed in block K_A near the coal wall side. Taking the block K_A as the object, assuming that the overburden load and the supporting force on the two blocks were equal, $P_A = P_B = P$. Δ was the settlement of the block and X_T was the action height of the horizontal thrust force T. Setting block thickness H, length L, and taking the moment of B point $\sum M_B = 0$, then it can be obtained

$$Q_A = 0.5P + T\frac{H - L\sin\alpha - \Delta - 2X_T}{L} \tag{5.59}$$

Due to the fact that the broken blocks would slide down when the shear force Q_A exceeded the friction $T\tan\varphi$ between the contact surface, it should meet Eq. (5.60) to keep the stability of the broken blocks.

$$T > \frac{0.5PL}{L(\sin\alpha + \tan\varphi) + \Delta + 2X_T - H} \tag{5.60}$$

As the bedrock thickness increased, the macroscopic pressure arch in the far field and the multilayer fracture pressure arch in the near field could be formed in the overlying strata of the mining field. While the sliding blocks kept partial contact by the friction, the load was delivered between the contacting face and the principle stress was concentrated in the load transferring path; the stepped pressure arch was then formed in the basic roof. The symmetrical pressure arch was formed in the upper overlying strata, the blocks rotated and compressed reciprocally in the relatively small space, and the rotating-squeezed pressure arch was formed in the meso-position, which was clamped by the upper bending rock and the inferior supporting rock.

Fig. 5.41 shows the mechanical model of two broken blocks forming the rotating-squeezed pressure arch. Taking a moment on point B $\sum M_B = 0$, assuming that the overburden load and the supporting force on the two blocks were equal, $P_A = P_B = P$. Then

$$Q_A = \frac{2PL\cos\alpha + T(H - 2L\sin\alpha - 2X_T)}{2L\cos\alpha + H\sin\alpha} \tag{5.61}$$

Assuming the formation of the above structure had no sliding between two blocks, namely $T\tan\phi > Q_A$, substituting Eq. (5.61) into the inequality, then

$$T > \frac{2PL\cos\alpha}{2L\cos\alpha(1+\tan\varphi) + H(\sin\alpha\tan\varphi - 1) + 2X_T} \tag{5.62}$$

5.4.3 Computational model and similar experiment

5.4.3.1 Building a computational model

Based on the geological conditions of the Daliuta Mine, the discrete element software UDEC was used to analyze the movement of the thin bedrock under the shallow horizontal coal mining.

As shown in Fig. 5.42A, the single basic roof Model 1 was established with a size 600 m long and 68 m high. The thickness of the coal seam was 5 m and full thickness mining technology was used. The thicknesses of the floor, the immediate roof, the basic roof, and the loose layer above the bedrock were 10 m, 5 m, 20 m, and 90 m, respectively. For this model, a 0.93 MPa load was applied at the top of the model, substituting the weight of the rock and soil layer of 62 m. Considering the effect of the horizontal stress, the initial lateral pressure coefficient was set to be 1.5. The mining starting point was 50 m away from the left boundary, and the advancing distance was 10 m for each mining stage. The bottom boundary of the model was fixed and the lateral boundaries of the model were fixed at horizontal directions.

As shown in Fig. 5.42B,Model 2 was established to analyze the influence of the overburden thickness on the pressure arch, which was 450 m long and 70 m high. The thickness of the immediate roof was 5 m, and the overburden above the immediate roof was 50 m. The load 1.35 MPa was applied at the top of the model, substituting the loose layer of 90 m. Other conditions were the same as Model 1.

The numerical test was to simulate the mechanical behaviors of the fractured blocks, considering the vertical and horizontal virtual joint conditions. The

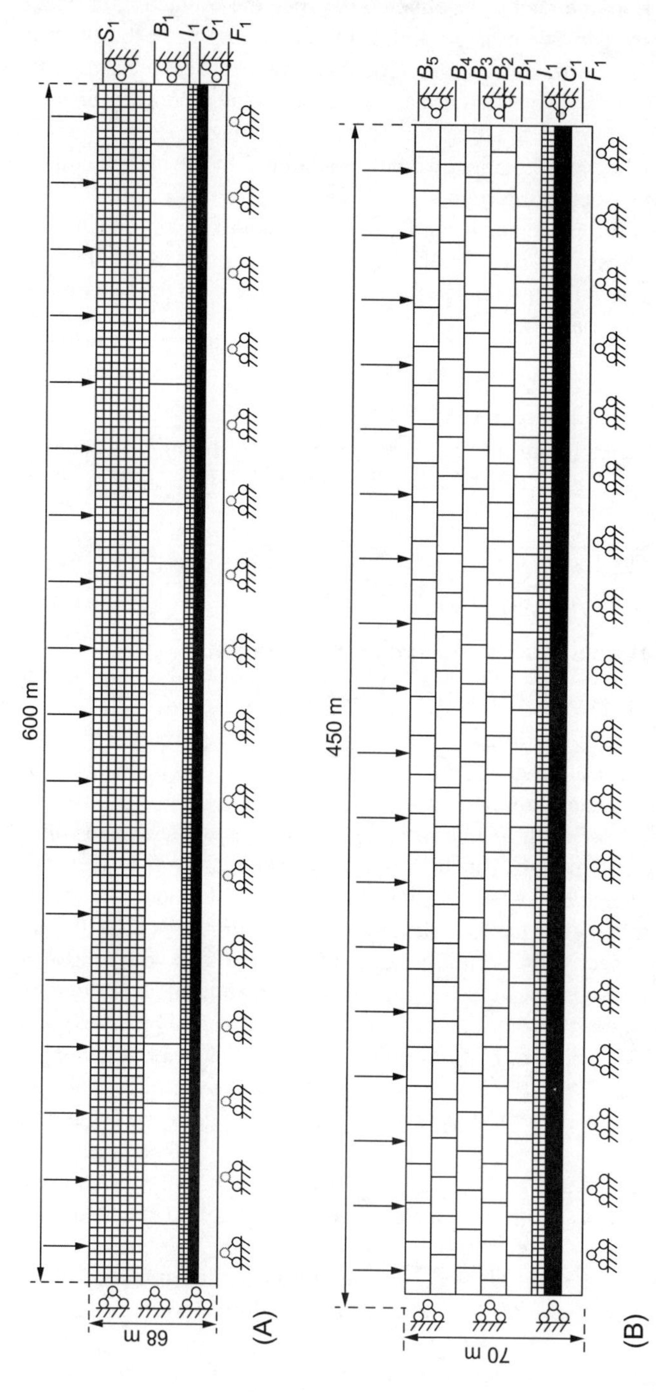

FIG. 5.42 The computational model. (A) The single basic roof model (*S1*, soil, *B1*, sandstone, *I1*, mudstone, *C1*, coal, *F1*, siltstone) and (B) the multilayer overburden model (*B2*, *B4*, *B5*, sandy mudstone; *B3*, coarse sandstone; *B1*, sandstone; *I1*, mudstone; *C1*, coal; *F1*, siltstone).

TABLE 5.4 Material properties of coal and strata.

Name	Unit weight (kN/m^3)	Elastic modulus (GPa)	Poisson's ratio	Cohesive strength (MPa)	Tensile strength (MPa)	Friction angle (degree)
Soil	19.20	0.08	0.32	0.10	0.04	26
Sandy mudstone	24.00	29.00	0.23	3.70	2.10	32
Coarse sandstone	24.30	35.00	0.21	5.50	4.00	33
Sandstone	25.00	32.00	0.24	7.30	4.90	35
Mudstone	22.40	23.00	0.15	2.50	1.70	30
Coal	13.10	15.00	0.29	0.79	0.57	27
Siltstone	24.60	26.00	0.22	3.90	2.50	38

physical and mechanical parameters of the models were listed in Table 5.4. The Mohr-Coulomb constitutive model was used for the soil/rock and continuous yield constitutive for the discontinuities (Lal et al., 2017; Li et al., 2017a; Meier et al., 2017). The normal stiffness and shear stiffness of joints can be approximately calculated through the joint structures and the deformation of the intact rock mass. The normal stiffness and shear stiffness were set as 4 GPa/m, and the friction angles of the joints were set as 30 degrees.

5.4.3.2 Similar materials experiment

The self-developed similar materials platform combined with a multifunction servo control was used in the test; it was developed to simulate different stress field conditions. The size was 3.565 m long, 0.640 m wide, and 2.915 m high. Nonlinear loading could be exerted independently by lateral loading racks and a top hydraulic system, and the range of the load was 0.1–100 kN.

Taking the same conditions in the numerical model, the thickness of the coal seam was 5 m, the thickness of the floor was 16 m, and the thickness of the overlying strata was 54 m. The external load was applied to the top of the model substitute the loose layer 90 m. The lateral pressure coefficient was set to be 1.5, which was applied by the lateral loading racks.

According to the similarity criteria and test requirements, the geometric similarity ratio was $C_L = 1:100$, the time similarity ratio was $C_t = 1:10$, the density similarity ratio was $C_\gamma = 1:1.56$, and the stress similarity ratio was $C_\sigma = 1:156$. The model size was 275 cm long, 30 cm wide, and 76 cm high.

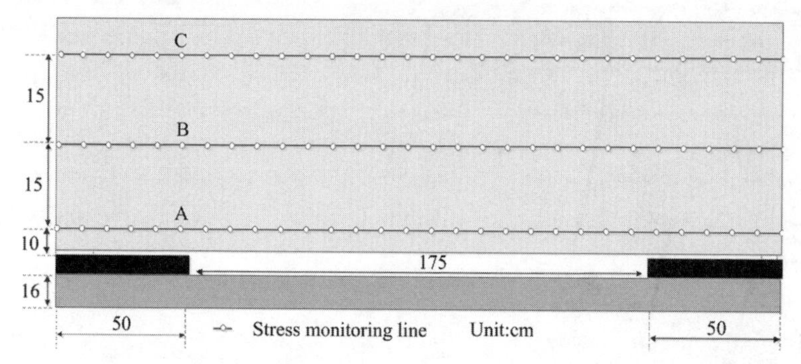

FIG. 5.43 Schematic diagram of monitoring lines in the experiment.

The mining starting point was 50 cm away from the left boundary, and the advancing distance was 6 cm at each mining stage, 175 cm in total.

According to the composition of the similar materials, sand was selected as the aggregate, gypsum as the main cementitious material, lime as an auxiliary cementitious material, borax as a retarder, and mica powder as the weak plane among the different strata. The fine-grained river sand with a particle size less than 0.05 mm was chosen as the aggregate in the model experiment, gypsum was used as the main cementing material, and calcium carbonate was used as the auxiliary cementing material (Zhang et al., 2017). To ensure the reliability of the material ratio in a similar simulation test, according to the proportion numbers 337, 455, 573, and 673, a small amount of river sand, calcium carbonate, and gypsum was made to make cylindrical specimens of the size ϕ50 mm $\times$ 100 mm. Taking the specimen of proportion number 337 as an example, the compression specimen was damaged after 83 s of loading, and the peak strength of the specimen was 0.48 MPa.

From the bottom to the top, three stress monitoring lines were set in the model, at 10 cm, 25 cm, and 40 cm away from the coal seam, as shown in Fig. 5.43. To obtain the evolution laws of the principal stress during different mining stages, the high sensitivity BX120-3CA-type strain gauges were buried at an interval of 10 cm along the monitoring lines, and the TST-3822 strain gauge analysis system was used to monitor the data. The CTS-622-type total station was used to measure the coordinates of the measured points to obtain the displacement field during mining. The similar material ratio of the model was shown in Table 5.5.

5.4.4 Results and discussion

5.4.4.1 Structures of symmetrical and stepped pressure arches

As shown in Fig. 5.44, with the working face advancing 30 m, the principal stress of the hanging rock was transferred from the mid-span to both undisturbed sides, and the increased principal stress formed the symmetrical pressure

TABLE 5.5 Similar material proportion of the model.

Lithology	Thickness (cm)	Compressive strength (MPa)	Actual density (kg/m^3)	Proportion number
Coarse sandstone	10	31	2430	355
Mudstone	8	23	2240	437
Sandy mudstone	3	21	2300	537
Fine sandstone	9	36	2550	337
Siltstone	20	39	2600	328
Mudstone	4	23	2240	437
Coal	5	20	1500	573
Siltstone	16	39	2600	328

arch. The mid-span vault and both arch feet were the horizontal principal stress concentration zone.

As shown in Fig. 5.45, relying on the effect of the horizontal thrust and friction, the sliding blocks of the roof formed a hanging stepped structure (Fraldi and Guarracino, 2010). The principal stress among the blocks was transferred from the edge of the working face to the caving zone in the goaf. Considering that the increased stress in the pressure arch was larger than that in the primary rock, the principal stress increased to form a pressure arch in the blocks, of which the front arch foot was located on the top of the coal wall and the rear arch foot was located in the crushed blocks.

The principal stress distribution curves of the vault and two arch feet of the pressure arch in the broken zones were shown in Fig. 5.46; the horizontal axis of the diagram named H_x was the distance of the point away from the basic roof bottom. In the two pressure arches, the principal stress showed similar distribution, it increased gradually at the mid-span from bottom to up, and reached the maximum value at the vault of the mid-span, the reason of maximum principal stress at mid-span more obvious was the large deformation and compressing effect of roof at midspan. It also decreased gradually at two arch feet from the bottom to the top, and reached the maximum value at the lowest point of the arch foot. Overall, the stress change and the maximum value of the symmetrical pressure arch were smaller than that of the stepped pressure arch.

5.4.4.2 Stress distribution of rotating-squeezed pressure arch

With the working face advancing in Model 2, the distribution of the principal stress in the overlying strata was shown in Fig. 5.47. With increasing the

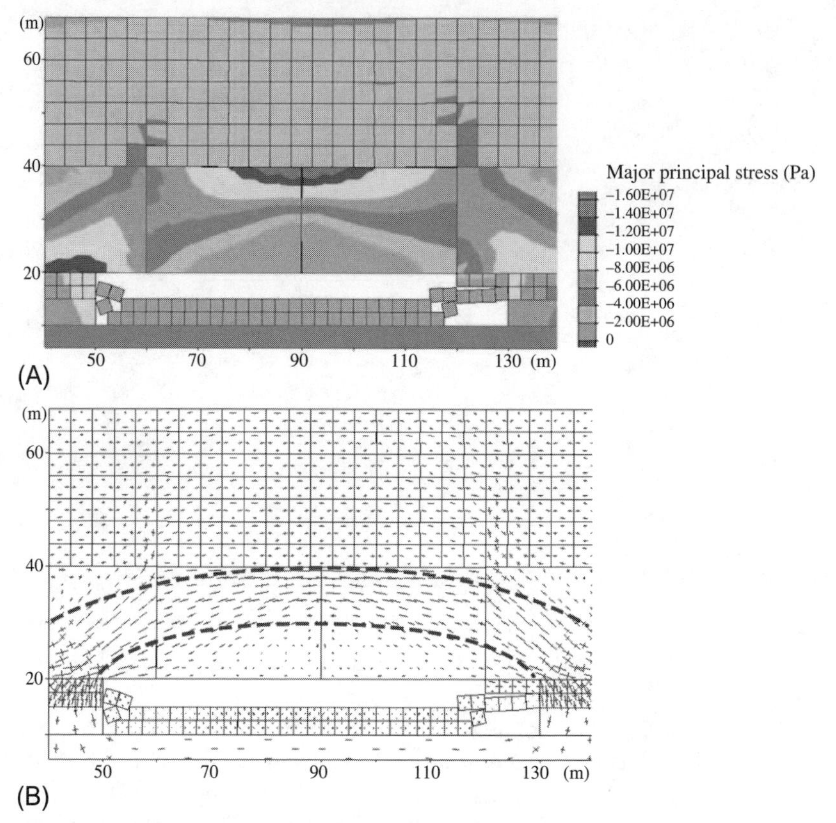

FIG. 5.44 The symmetrical pressure arch structure. (A) The maximum principal stress distribution and (B) traces of the principal stress.

overburden thickness, the composite pressure arches of the near and far field were formed in the surrounding rock. The distribution of the maximum principal stress was shown in Fig. 5.47A at the working face advancing 100 m. After slipping down and contacting with the waste rock, the basic roof formed the stepped pressure arch. Each stratum above the basic roof formed independent symmetrical pressure arches, which formed a superimposed symmetrical pressure arch in the far field. There were three principal stress concentrated zones, that is, the vaults and both arch feet of each symmetrical pressure arch.

As mining continued, the multilayer symmetrical pressure arch of the overlying strata lost stability, as shown in Fig. 5.47B. The broken blocks rotated and contacted with the waste rock, and the principal stress in the surrounding rock redistributed and formed a rotating-squeezed pressure arch. The latter arch foot was transferred to the compacted waste rock in the caving zone. The hanging overburden in the goaf formed the vault zone of the pressure arch, and the front

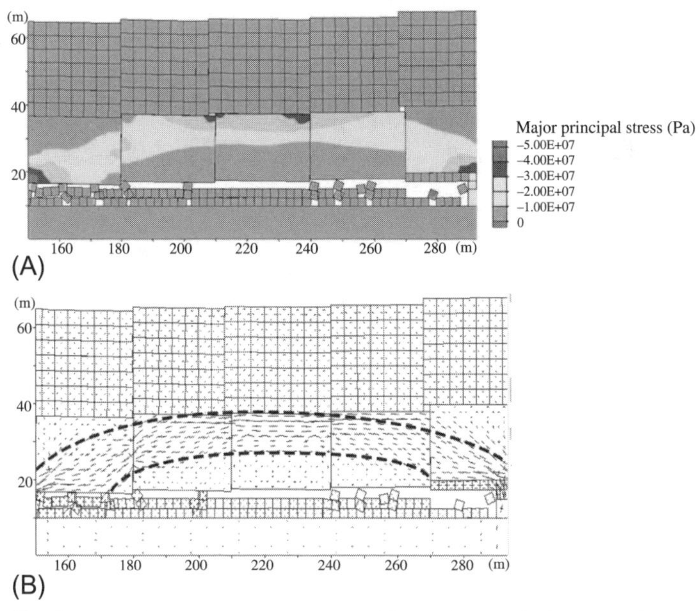

FIG. 5.45 The stepped pressure arch structure. (A) The maximum principal stress distribution and (B) traces of the principal stress.

arch foot was located in the rock around the coal wall. This pressure arch in the far field had self-bearing capacity and protected the stepped pressure arch of the basic roof.

5.4.4.3 Experimental verification on strata fracture structure

As seen from Fig. 5.48A, the immediate roof caved and after the working face advanced 32 m and the broken blocks formed a hinged structure, this phenomenon was similar to the result of immediate roof caving in Fig. 5.44. Along with the mining process, the overlying strata became unstable gradually, forming a symmetrical pressure arch over the caving arch in the mining field (Fig. 5.48B). The upper strata showed a large fracture interval, and the lower roof formed a short block and emerged sliding. After the working face advanced 113 m, the upper fractured overburden rock formed a rotating-squeezed pressure arch structure, and the lower basic roof formed a stepped pressure arch (Fig. 5.48C). Compared with the multiple strata composite pressure arch in the numerical simulation, the fractured roof structure was more obvious near the mined-out area for the fragmentation of similar material. The strata structure at the working face side coincided with the rotating-squeezed pressure arch structure in the numerical simulation.

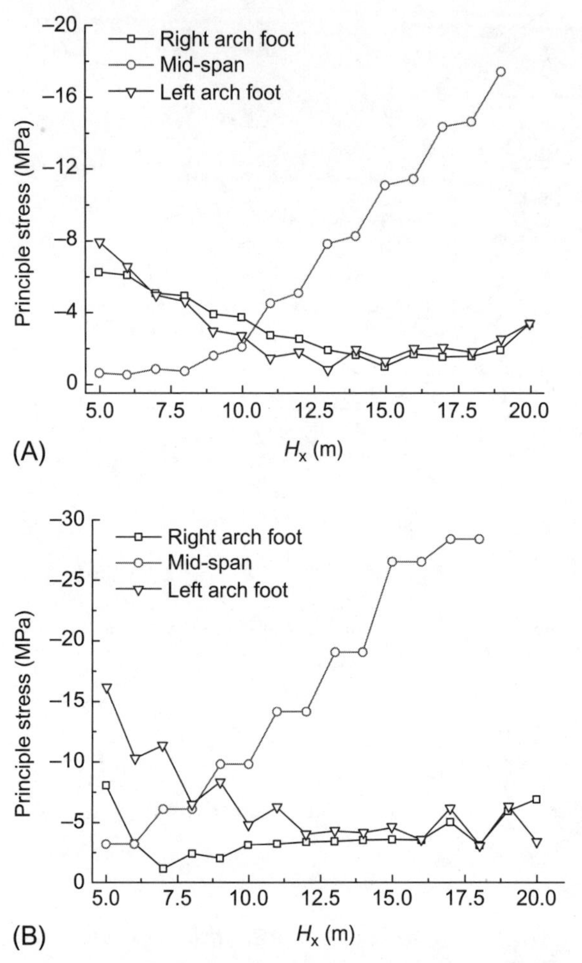

FIG. 5.46 Distribution of the maximum principal stress of the pressure arch. (A) The symmetrical pressure arch and (B) the stepped pressure arch.

With the working face advancing 70 m, the maximum principal stress of the strata in each monitoring line was shown in Fig. 5.49A. Data from monitoring lines 1 and 2 showed that the principal stress in the near field decreased while line 3 showed that the principal stress increased at the upper hanging overburden in the far field, forming a vault zone of the macroscopic pressure arch.

As shown in Fig. 5.49B, by selecting monitoring points that were 40 cm away from the left boundary of the model in the three monitoring lines, the principal

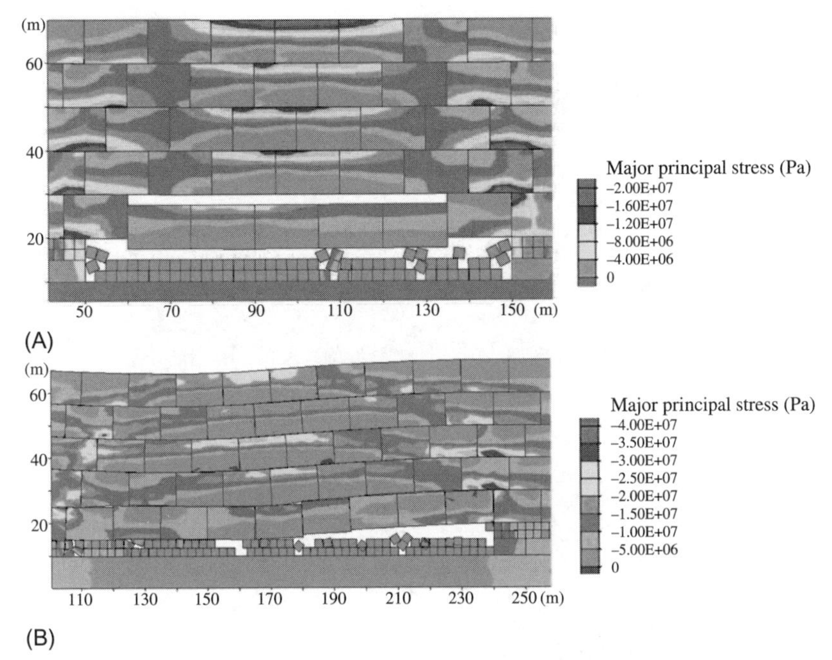

FIG. 5.47 The composite pressure arch in the near and far field. (A) The symmetrical pressure arch and stepped pressure arch and (B) the rotating-squeezed pressure arch and stepped pressure arch.

stress at each monitoring line changed with the advance of the working face. The principal stress of the overlying strata in each layer increased gradually in the initial mining stage, and reached its maximum value when the working face advanced 90 m. The macroscopic pressure arch in the far field lost stability after all the overburden breaking. As a result, the bearing capacity and the principal stress of the undisturbed surrounding rock in the far field reduced.

After excavation, the principal stress of the overlying strata deflected and the load of the hanging rock was transferred to stable surrounding rock in the mined-out area. The pressure arch was formed as a self-bearing structure, of which the structure stability greatly affected the safety of the underground engineering (Chen et al., 2011). The structural characteristics of the strata were different along with the mining process. Relying on the rotating and squeezing among the broken blocks, a bearing pressure arch structure was formed by the broken overburden strata in the near field. And a macroscopic pressure arch could be formed by the stress concentration zone in the surrounding rock in the far field, which was able to bear the load of its own and the loose layers.

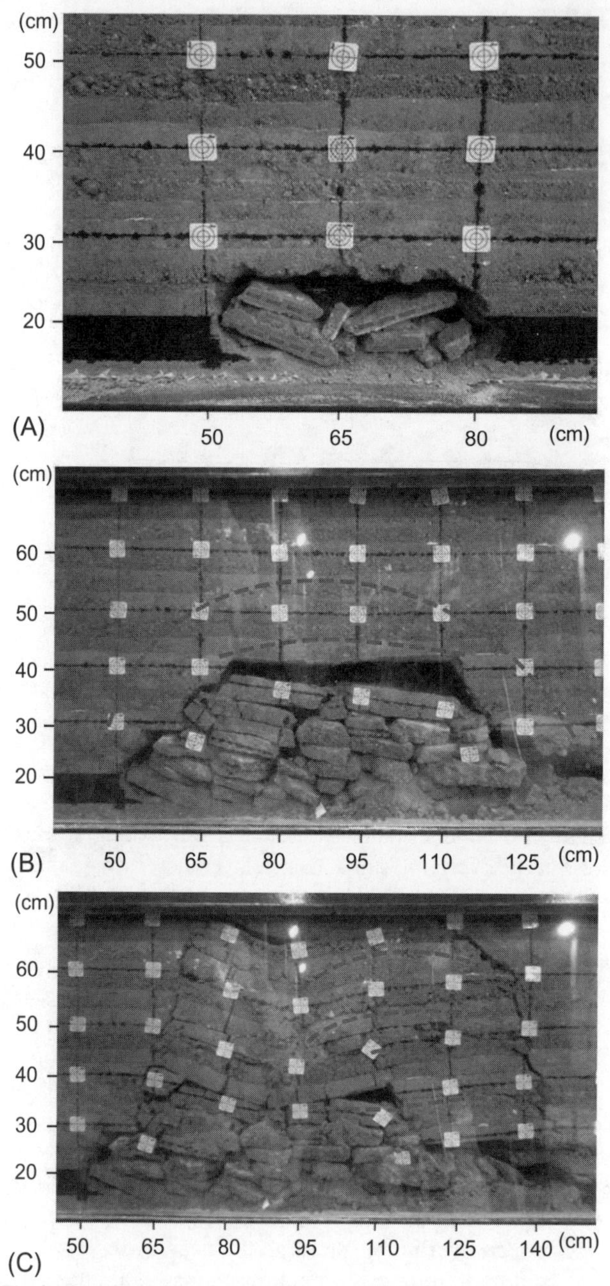

FIG. 5.48 The fracture structure evolution in the overburden rocks. (A) The slipping and rotating of the broken blocks, (B) the symmetrical pressure arch, and (C) the rotating-squeezed pressure arch.

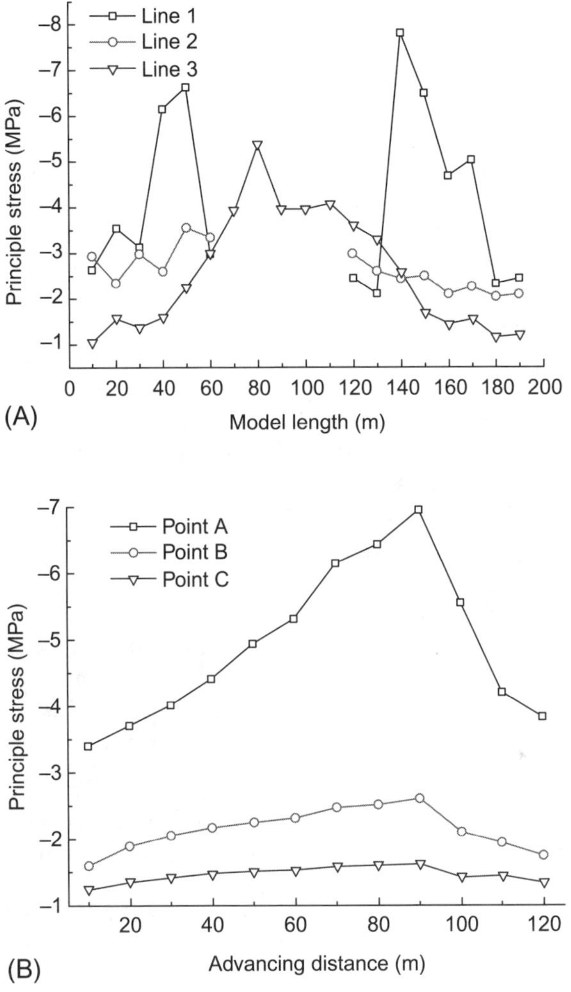

FIG. 5.49 The maximum principal stress distribution in the surrounding rock. (A) Advancing 70 m and (B) different advancing distances.

For the thick bedrock under deep mining conditions, the stable macroscopic pressure arch in the far field was formed in the overburden, of which the arch body moved forward with the working face advancing (Xie et al., 2009; Wang et al., 2018). For overburden totally broken during the thin bedrock mining, it was difficult to form the macroscopic pressure arch in the far-field surrounding rock. The single basic roof slide could cause the stepped subsidence. The principal stress of the stepped braking blocks was transferred to both goaf sides and formed a stress concentration zone as an arch shape. So, the instability of the stepped pressure arch could cause the large area roof cutting off, strong weighting in the working face, and serious surface subsidence.

With an increase of the bedrock thickness, the structural characteristics of the overburden rock changed gradually. A pressure arch in the caving zone was formed below the macroscopic pressure arch in the far-field surrounding rock. The upper hanging strata formed a symmetrical pressure arch, of which the arch feet were located in the undisturbed surrounding rock. The macroscopic pressure arch could be formed along with the overburden load being transferred to the far-field surrounding rock.

The self-bearing capacity of the overburden in the mining field depended on the composite pressure arch structure. As a load-bearing structure for thin bedrock beneath a thick loose layer under shallow coal mining, the stepped pressure arch had lower stability as it could slide easily, resulting in a short weighting interval and strong weighting load on the working face. With the increase of bedrock thickness, the bearing capacity of the composite pressure arch in the overburden rock increased, and the rotating-squeezed pressure arch in the fractured rock could bear the load of the upper loose layer and protect the safety of the working face.

5.4.5 Conclusions

To investigate the characteristics of the composite pressure arch, Daliuta Coal Mine was taken as the engineering case, and the mechanics models of different pressure arches were established under different mining conditions. The evolution characteristics of the pressure arches in the near and far field were analyzed and verified by the similarity material test and numerical simulation. It was found that:

(1) For thin bedrock with a single base roof, the stepped pressure arch can be formed during coal mining. The principal stress of two arch feet decreases gradually with arch height increasing. Instability of the broken blocks induces a short weighting interval and a strong weighting load on the working face.
(2) With the bedrock thickness increasing, the structural characteristics of the pressure arch can be changed. A composite pressure arch can be formed in the far field in the multilayer overlying strata. Its arch feet were located in the undisturbed surrounding rock, and it can bear the loose layer load.
(3) The periodical fractured strata can form a rotating-squeezed pressure arch structure, which can bear the loose layer load, protect the stepped pressure arch, and reduce the roof weighting load. The latter arch foot of the macroscopic pressure arch can be transferred to the compacted waste rock.

The stability of the pressure arch in the broken blocks is crucial for mining safety and environmental protection. With the increase of bedrock thickness, the overburden rock deformation becomes complicated. The evolution rules of the composite pressure arch can be determined by the size of the overburden caving zone and the spatial structure characteristics in the mining field.

5.5 Pressure arch performances in thick bedrock

5.5.1 Introduction

During shallow coal mining with thick bedrock in the western area of China, the overlying strata shows the features of alternate instability, and the strong (weak) roof weighting on the working face often induces support crushing accidents. The stability of the surrounding rock is the key problem that restricts mining safety (Zhang et al., 2011; Xu et al., 2014). With the working face advancing, the pressure arch in the surrounding rock can be formed and the caving zone below the pressure arch can expand gradually (Wang et al., 2015c). Determination of the forming conditions and arching characteristics of the pressure arch in the mining field and analyzing the nonuniform weighting mechanism on the working face during shallow coal mining with thick bedrock are the urgent problems to be solved.

As we all know, the pressure arch widely exists in the surrounding rock after coal mining, and the pressure arch helps to keep the roof stable during coal mining. Ren and Qi (2011) found that the load of the hanging strata in the mining field can be transmitted to the surrounding rock in the far field by the pressure arch, and the size as well as the inner and outer boundaries of the pressure arch affect the load transferring laws of the roof strata. Due to the fact that the pressure arch is a macrobearing structure in the surrounding rock and different morphologies of the pressure arch have different bearing capacities, some mechanics index can be used to determine the bearing capacity of the pressure arch. The caving zone is formed by the instability of the pressure arch, and the weight of the caving zone is the load source of the supporting equipment on the working face. Therefore, the interaction between the pressure arch and the caving zone is an important basis for the design of the support resistance.

The macroscopic pressure arch functions as a self-supporting structure in the bedrock of the mining field, and the size affects the bearing capacity. The evolution characteristics of the pressure arch have an important influence on the scope of the caving zone and the weighting laws on the working face. Based on the strata behaviors of a typical working face in the Shangwan Mine in the Shendong mining area, the mechanical model of the pressure arch in the surrounding rock was constructed and the identifying indicators of the arching parameters were proposed in this paper. The evolution characteristics of the pressure arch and the caving zone under the thick bedrock condition were analyzed and verified by the physical test of similar materials and numerical simulations during shallow coal mining. The size of the pressure arch and the distribution characteristics of the load-bearing zone were obtained. The nonuniform weighting mechanism of the working face was revealed. The results are of great significance to mining safety in a similar engineering practice.

The principal stress in the surrounding rock after coal mining has the characteristics of self-adjusting and forming a pressure arch. As a bearing structure, the pressure arch protects the stability of the mining field. The pressure arch can provide a reliable basis for determining the support resistance of the working face. Xie et al. (2009) studied the distribution characteristics of the pressure arch in deep mining by a similar material test and numerical simulation. They found that there was a macroscopic pressure arch around the mining field that bore the hanging strata load, and the instability of the strata below the pressure arch caused the periodic weighting on the working face. Cui et al. (2014) analyzed the instability mechanism of the overburden strata being induced by the shallow coal mining. They thought that the pressure arch in the near and far field converged under the mining disturbance, and the failure of the thick and hard rock strata led to the disappearing of the pressure arch and the instability of the overburden rock. Wang et al. (2015c) established the mechanical model of the pressure arch and analyzed the instability mechanism of the pressure arch in the mining field with different dip angles. They found that with the increase of the coal seam dig angle, the failure mode of the pressure arch was gradually from compression, shear to tensile failure, and the local damage could cause the overall structural instability finally. The above studies show that the instability of the pressure arch can lead to the formation of the caving zone. However, during shallow coal mining, the forming conditions and evolution characteristics of the pressure arch in the overburden rock with the thick bedrock are still scarcely reported.

Under shallow coal mining, the instability of the overburden caving zone often brings about a sharp increase in support resistance, which is an important threat to mining safety. Based on the laws of mining pressure observation in the fully mechanized working face, Ju and Xu (2013) and Huang et al. (2016) revealed the alternated regularity of the interval and strength of weighting by using the combined mechanics models of the voussoir beam and the cantilever beam. They pointed out that the premature rupture of the key strata and the step subsidence would result in the weighting of the alternated regularity of strong (weak) periodically. According to the strength and thickness of the overburden rock in the Shendong Mining Area, Liu et al. (2015a, b) classified the characteristics of the overburden strata in shallow coal mining, and analyzed the bearing characteristics and load transfer laws of the overburden strata in the caving zones. Soni et al. (2007) summarized the instability characteristics of thin bedrock mining by monitoring the instability of the overlying strata during shallow coal mining in the Kampudi coal mine in India. Helm et al. (2013) and Salmi et al. (2017) studied the instability mechanism of the overburden strata under shallow coal mining in Edinburgh, UK. The instability of the overlying strata in shallow coal mining has been extensively studied. However, the influences of the pressure arch bearing capacity on the range of the overlying strata are scarcely reported, and the interaction between the pressure arch and the caving zone still needs further study.

There are different viewpoints on the pressure arch boundary in the surrounding rock and the loosening area under the pressure arch. Through numerical simulation of the roof load distribution in the goaf by DDA software, He and Zhang (2015) found that the loosening area under the pressure arch could not bear the overburden pressure. They also found that the principal stress in this region was mainly horizontal stress and the vertical stress was close to zero, and the separated rocks tended to slip off. Wang et al. (2016a, b, c, d, e, f, g) applied the pressure arch theory to predict the caving range of the deep tunnel roof, and took the zero tensile stress curves as the inner boundary of the pressure arch. Kong et al. (2018) analyzed the formation and evolution of the pressure arch in the surrounding rock by numerical simulation. Based on lateral and vertical stress distribution, they obtained the morphological characteristic curves and established a method of identifying the pressure arch range. Overall, the loosening rocks under the pressure arch threatens mining safety, and the support resistance is determined by the caving zone movement. So it is an urgent issue to determine the loosening area and reveal the instability mechanism of the pressure arch under shallow coal mining.

There are many results of the morphology and stress distribution of the pressure arch. However, there is a lack of effective indicators to identify the size and bearing capacity of the pressure arch, and the interaction between the pressure arch and the caving zone during coal mining needs to be further studied. Therefore, this study focus on analyzing the evolution characteristics of the pressure arch in thick bedrock during shallow horizontal coal mining, tries to propose mechanical criteria to evaluate the pressure arch size and bearing capacity, and researches the interaction between the pressure arch and the caving zone during shallow coal mining by physical tests and numerical simulations.

5.5.2 Engineering background

The Shangwan coal mine is located in the Inner Mongolia province in northern China. Taking the No. 51104 working face of the 1^{-2} coal seam in the Shangwan Mine as the engineering background, the longwall and full thickness mining technology were adopted in the working face, and the caving method was used to manage the roof. The working face width was 301 m, the dip angle was 0–5 degrees, and the average depth was 115.4 m.

The rock formations in the Shendong mining area were mainly formed during the Cretaceous, Jurassic, and Quaternary periods, among the Quaternary period was sandy soil layer, there were eleven kinds of lithology in the rock strata in this area, the Cretaceous strata was mainly sandstone. The average thickness of the coal seam was 6.5 m (Bai et al., 2016).

As seen from Figs. 5.50, there were 20 times the weighting occurred during the observation period, and the weighting interval varied at 9.4–32.3 m, 16.7 m in average. The support resistance was 5571–8975 kN, 7107 kN in average, and the rated working resistance was 8638 kN. During coal mining with the thick bedrock and thin loose layer, the strata behaviors had the periodically strong (weak) features, and the weighting appeared to be the alternated regularity of

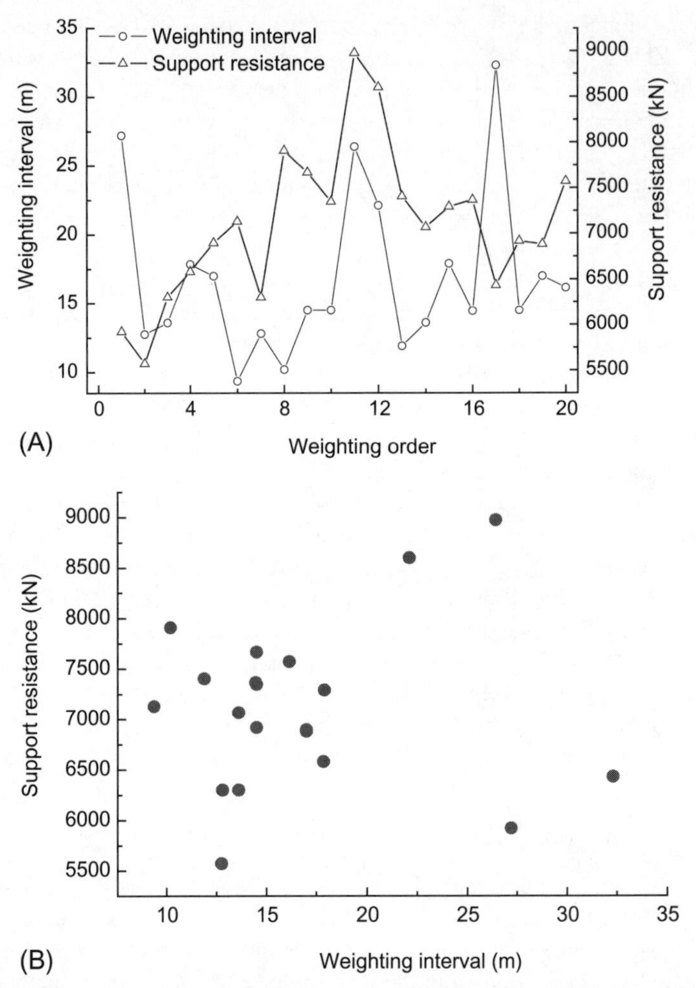

FIG. 5.50 Strata behaviors of the working face. (A) Weighting order and (B) weighting interval support resistance.

long (short) periodically. For part of the periodic weighting, the short interval corresponded to the strong weighting, and the normal weighting alternated with the strong (weak) weighting.

5.5.3 Pressure-arch analysis and experimental methods

5.5.3.1 Theoretical analysis

After shallow coal mining, the load of the overburden rock was transferred to the stable surrounding rock in the mining field. As shown in Fig. 5.51, the maximum principal stress σ_1 was deflected and the concentrated zone around the

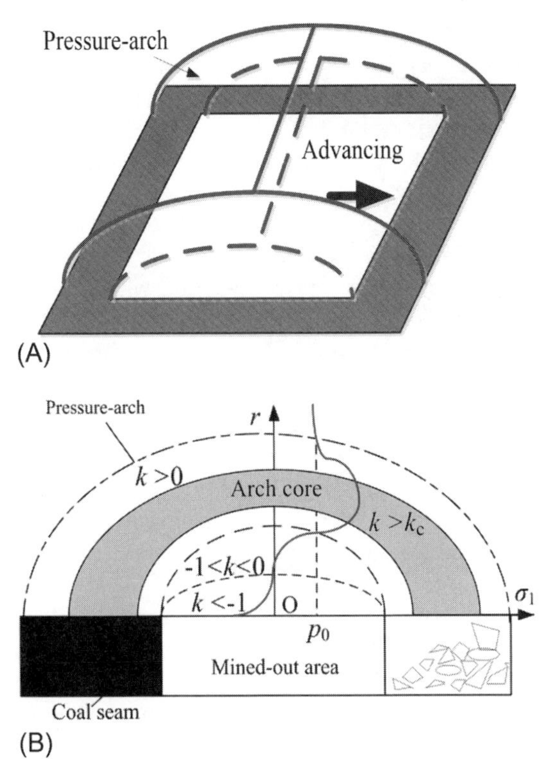

FIG. 5.51 The pressure arch in the surrounding rock. (A) Engineering model and (B) mechanical model.

mining field. The pressure arch was formed in the concentration zone of the compressive stress, of which the inner principal stress σ_1 surpassed the in situ stress p_0, which was

$$\sigma_1 > p_0 = \gamma h \tag{5.63}$$

where γ was the average volume-weight of the rock mass and h was the burial depth.

Outside the outer boundary of the pressure arch was the in situ stress zone, and inside the inner boundary of the pressure arch was the lower pressure zone and tension stress zone. The caving arch in the unloading area of the overburden rock was the key zone for rock control, of which the weight determined the strata behavior and the roof stability in the mining field. The spatial morphology and movement rules changed with the evolution of the three-dimensional pressure arch. To define k as the arching index of the pressure arch in the surrounding rock, which could be calculated by

$$k = \frac{\sigma_1 - \sigma_0}{\sigma_0} \tag{5.64}$$

where σ_1 was the maximum principal stress after coal mining and σ_0 was the in situ maximum principal stress before coal mining. For $k > 0$, it was the range of the pressure arch, and the lower compressive stress zone when $-1 < k < 0$ and the tensile stress zone when $k < -1$.

The index k reflected the stress concentration degree of the surrounding rock, and the average value of index k of each position was another index to evaluate the performance of the stress concentration in the rock. Taking identifying indicator k_c as the average value of k, the indicator k_c was the ratio of index k in each rock unit to the whole rock volume, which could be used to analyze the bearing capacity of the pressure arch. When $k > k_c$, it was located in the nucleus of the pressure arch, of which the thickness of the pressure arch determined the self-bearing capacity in the surrounding rock; when $0 < k < k_c$, it was located in the outside zone of the nucleus of the pressure arch.

5.5.3.2 Building computational model

Taking mining of the 1^{-2} coal seam in the Shangwan Mine as the engineering background, the computational model was established by FLAC^{3D}. The thickness of the coal seam was 7 m, the thickness of the siltstone floor was 10 m, and the thickness of the sandstone roof was 102 m. As shown in Fig. 5.52, the model size was 400 m long, 352 m wide, and 137 m high. The load of 0.45 MPa was applied to the top of the model, substituting for the 30 m loose layer weight. The bottom boundary of the model was fixed and the lateral boundaries of the model were fixed in the horizontal direction.

The position of start mining was 150 m away from the left boundary, the advancing distance was 300 m for each mining stage, and the inclined length was 240 m. Two roadways were excavated along the x-axis direction 150 m away from both sides of the border; their width was 6 m, the height was 5 m, and the excavation of the coal seam was simulated by using the null model. The physical and mechanical parameters of the coal and rock mass are listed in Table 5.6, and the Mohr-Coulomb criteria was used in the numerical calculation.

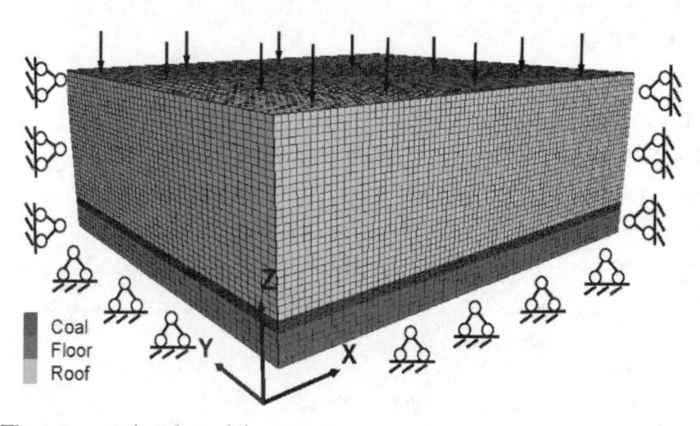

FIG. 5.52 The computational model.

TABLE 5.6 Material properties of coal and roof strata.

Name	Unit weight (kN/m^3)	Elastic modulus (GPa)	Poisson's ratio	Cohesive strength (MPa)	Tensile strength (MPa)	Friction angle (degree)
Sandstone	25.00	36.50	0.22	2.60	1.50	30
Coal	13.10	12.70	0.29	1.20	0.60	27
Siltstone	24.60	37.90	0.20	4.50	3	40

5.5.3.3 Similar materials experiment

Taking mining of the 1^{-2} coal seam in the Shangwan Mine as the engineering background, the thickness of the coal seam was 7 m, the thickness of the floor was 16 m, and the thickness of the overlying strata was 97 m. The self-developed similar simulation platform with combined multifunction servo control was used in the test (Wang et al., 2016a, b, c, d, e, f, g); the size was 3.565 m long, 0.640 m wide, and 2.915 m high. The nonlinear load was applied independently by the lateral loading rack and the top hydraulic system to simulate different stress conditions, and the range of the load was 0.1–100 kN. The external load was applied on the model top, substituting the 40 m loose layer above the overlying strata, and the lateral pressure coefficient was set as 1.5, which was simulated by the lateral loading rack.

According to the similar criteria and test requirements, the geometric similarity ratio was $C_L = 1 : 100$, the time similarity ratio was $C_t = 1 : 10$, the density similarity ratio was $C_\gamma = 1 : 1.56$, and the intensity and stress similarity ratios were $C_\sigma = 1 : 156$. The model size was 275 cm long, 30 cm wide, and 120 cm high. The position of the start mining was 50 cm away from the left boundary, and the advancing distance was 6 cm for each mining stage, 180 cm in total. According to the composition of similar materials, sand was selected as the aggregate, gypsum as the main cementitious material, lime as an auxiliary cementitious material, borax as a retarder, and mica powder as the weak plane among the strata.

The fine-grained river sand with a particle size less than 0.05 mm was chosen as the aggregate in the model experiment, gypsum was used as the main cementing material, and calcium carbonate was used as the auxiliary cementing material. To ensure the reliability of the material ratio in a similar simulation test, according to the proportion number 337, 455, 573, and 673, a small amount of river sand, calcium carbonate, and gypsum was made to make cylindrical specimens of the size 50 mm × 100 mm. The uniaxial compressive test of the specimen was carried out on the servo pressure tester, as shown in Fig. 5.53. Taking the specimen of proportion number 337 as an example, the

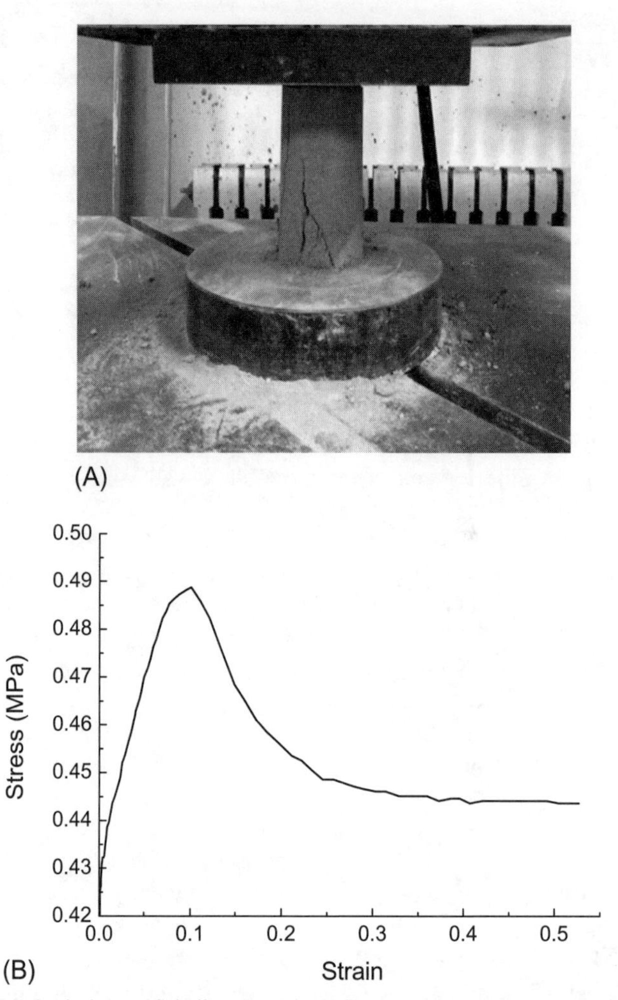

FIG. 5.53 Mechanical tests of similar material. (A) Test specimen and (B) stress-strain curve.

compression specimen was damaged after 83 s of loading and the peak strength of the specimen was 0.48 MPa.

As shown in Fig. 5.54, there were three stress monitoring lines being set in the model at 10, 45, and 80 cm away above the top of the coal seam; the monitoring and acquisition system of the stress data was mainly completed by the computer. The horizontal distance among the sensors was 10 cm. The unit strain gauge was made of a 3 cm × 3 cm × 3 cm polyurethane cube. On the three sides of the cube diagonally adjacent, three 45-degree strain flowers were pasted, and the vertical strain gauges were pasted along the horizontal and vertical directions on other surfaces. The specification of the strain flower is BX120-3CA (sensitive gate size is 3 mm × 2 mm). The monitoring of the principal stress in the model was achieved through the main direction strain sensor,

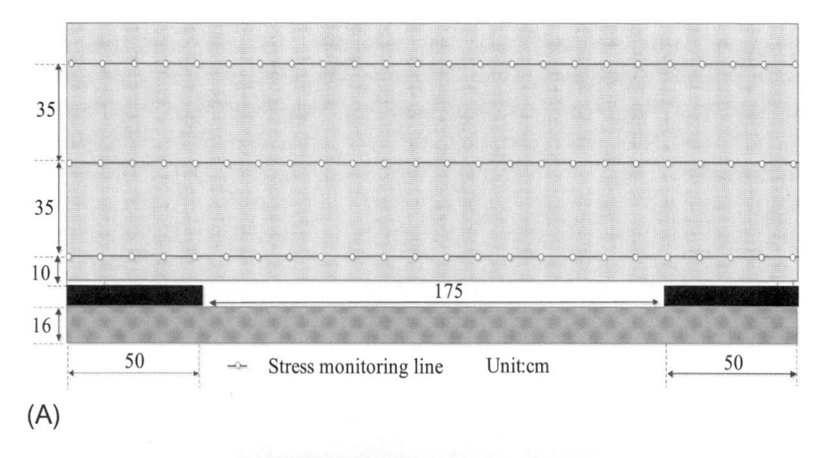

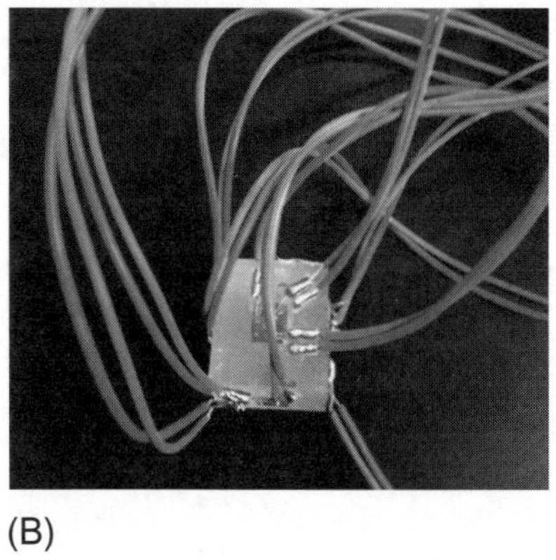

FIG. 5.54 Schematic diagram of monitoring lines and strain sensor. (A) Stress monitoring lines and (B) strain sensor.

and the TST-3822 strain gauge analysis system was used to monitor the principal stress of the similar materials in the model. The CTS-622-type total station was used to measure the space coordinates of the measured points to obtain the displacements of the overlying strata. The similar material ratio of the model is shown in Table 5.7, in which the first digital of the proportion number represented the ratio of sand and cementitious material while the second and third digital represented the ratio of two cementitious materials.

5.5.4 Results and discussion

5.5.4.1 Arching characteristics of principal stress

Choosing the middle position of the working face as the research object, selecting the cross-section at $y = 176$ m, and the vector field of the principal stress distribution was shown in Fig. 5.55. The maximum principal stress in the

TABLE 5.7 Similar material proportion of the model.

Lithology	Thickness (cm)	Compressive strength (MPa)	Actual density (kg/m^3)	Proportion number
Sandy mudstone	7	21	2350	537
Coarse sandstone	8	37	2430	355
Fine sandstone	7	36	2550	337
Siltstone	10	39	2600	328
Medium sandstone	14	27	2300	455
Coarse sandstone	7	37	2430	355
Sandy mudstone	4	21	2350	537
Fine sandstone	15	36	2550	337
Siltstone	8	39	2600	328
Mudstone	5	23	2240	437
Fine sandstone	8	36	2550	337
Sandy mudstone	4	21	2350	537
Coal	7	20	1500	573
Siltstone	16	39	2600	328

mining field was the vertical direction, and the principal stress in the mining disturbed zone was obviously deviated after the initial coal mining. As the shadow zone displayed the principal stress deviation, with the load of the hanging rock transferred to the surrounding rock at both sides of the goaf, the principal stress showed as increasing.

As shown in Fig. 5.55, with the working face advancing, the stress deviation zone expanded gradually. The direction of the principal stress in the lower part of the principal stress deflection zone displayed horizontal, the compressive stress was deviated toward the surrounding rock of the mined out area, assuming

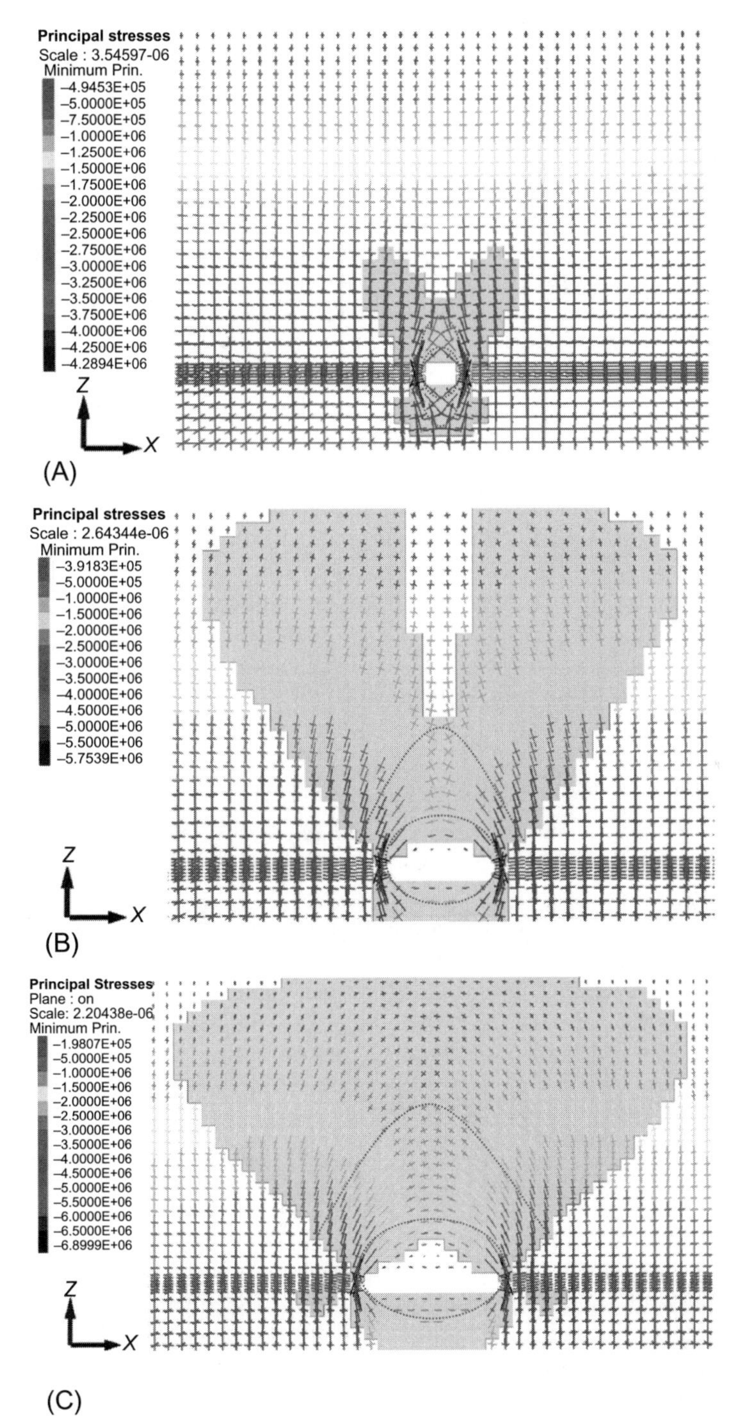

FIG. 5.55 Vector field of the principal stress in the surrounding rock. (A) Initial mining stage, (B) the pressure arch in the near-field, and (C) the pressure arch in the far-field.

that the deflection zone of compressive stress above the mined-out area was the pressure arch area, and the pressure arch in the near-field formed. Above the stress deflection zone was the in situ stress zone, and the pressure arch in the near field bore the upper rock load and transferred it to the stable surrounding rock. As the span of the goaf area increased, the whole direction of the principal stress of the upper overlying strata in the mining field deviated and the pressure arch in the far field formed.

As shown in Fig. 5.56A, selecting the top of the coal seam as the origin point, the vertical stress in the mid-span was analyzed in the mining field. The principal stress of the overlying strata within 0–15 m above the coal seam deviated from the vertical direction after the initial mining stage, and the vertical stress of the deviation zone decreased. After the working face advanced 50 m, the range of the principal stress and the horizontal stress in the deflection zone increased, and the vault of the pressure arch in the near field formed (Fig. 5.56B).

After the working face advanced 100 m, the unloading zone existed within 0–15 m in the mid-span of the overlying strata below the pressure arch, where all the stress indicators of the overburden rock reduced. In the range of 10–55 m, the maximum principal stress of the overburden rock was the horizontal stress, and the pressure arch situated in the horizontal stress increased zone. In addition, the pressure arch of the disturbed surrounding rock in the far field was in the range of 90–104 m (Fig. 5.56C).

5.5.4.2 Characteristic parameters of pressure arch

After the horizontal principal stress of the mid-span overburden rock exceeded the in situ stress, the pressure arch of the disturbed surrounding rock in the far field formed. As shown in Fig. 5.57A, taking the working face advancing 100 m as an example, the distribution of the arching index k in the mining field was obtained by using the built-in FISH language programming in FLAC^{3D} software and the average k_c was 0.21. The arching index of the roof strata in the near field was $k < 0$, and the arching index of the upper hanging overburden rock and the undisturbed surrounding rock in both sides of the goaf was $k > 0$. There was a symmetrical arching core in the pressure arch, where the arching index was greater than 0.21. The arching index gradually decreased from the core area to the inner and outer boundaries of the pressure arch, and the maximum value was located in the vault.

As shown in Fig. 5.57B, according to the arching index distribution characteristics of the pressure arch, the surrounding rock zone could be partitioned. There was the pressure arch in the disturbed surrounding rock, of which the arching core was near the goaf and which could bear the high stress. The mining unloading zone was below the inner boundary of the pressure arch, and the caving zone was located at the bottom of the unloading zone.

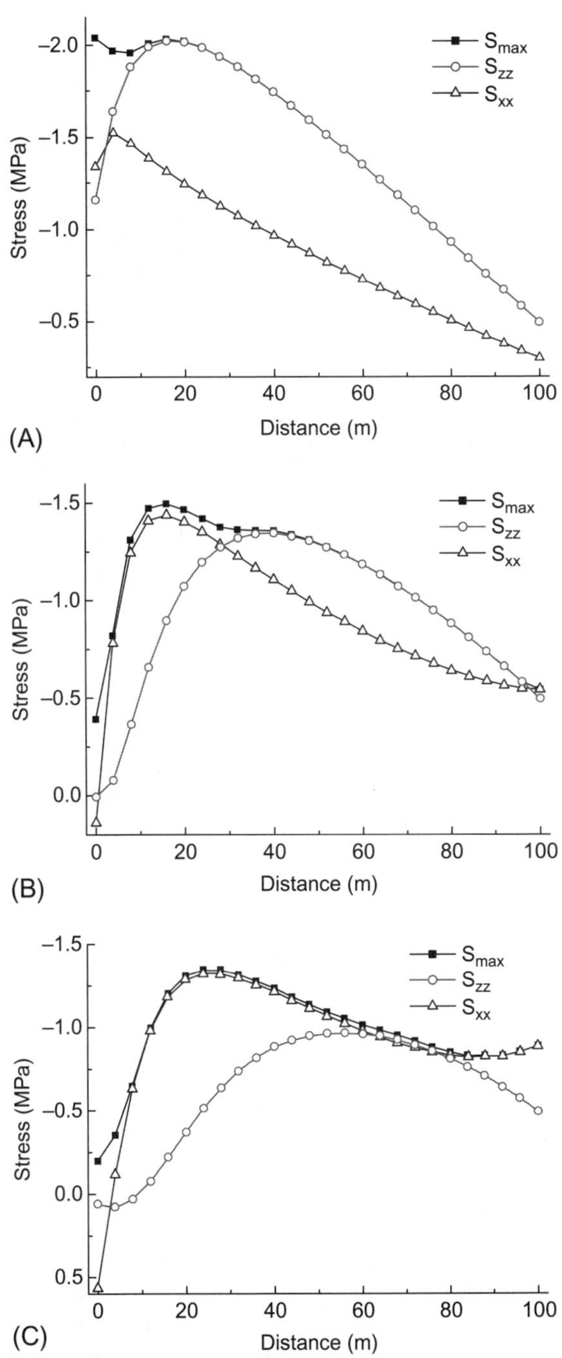

FIG. 5.56 Stress distribution characteristics in the overlying strata. (A) Initial mining stage, (B) the pressure arch in the near-field, and (C) the pressure arch in the far-field.

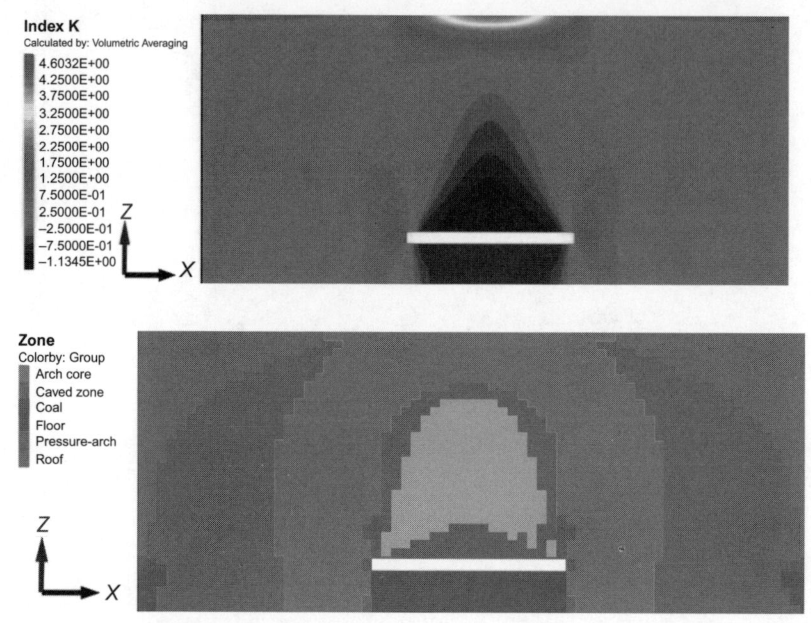

FIG. 5.57 The arching index and pressure arch in the surrounding rock. (A) The arching index distribution and (B) the pressure arch in the surrounding rock.

To analyze the relationship between the bearing capacity of the pressure arch and the development of the caving zone, taking the middle position of the inclined direction of the working face as the research object, the width of the pressure arch core and the failure volume of the strata caused by coal mining could be monitored. As shown in Fig. 5.58, in the beginning of the pressure arch formation in the surrounding rock, the width of the pressure arch core continued to increase, which was kept constant after the pressure arch formation. When the width of the pressure arch core decreased, the failure volume of the strata caused by coal mining increased. The destructive volume of the rock strata displayed the nonuniform periodic variation as the working face advanced.

5.5.4.3 Relationships between pressure arch and caving arch

With the working face advancing 95 m, the caving arch of the overburden broken zone in the near field is shown in Fig. 5.59A. The immediate roof and the basic roof showed the stepped structure. The marked red line was the squeezed-arch structure by the rotation of the central broken blocks, and the upper abscission layer of the overlying strata was the symmetrical hinged structure.

As shown in Fig. 5.59B, the upper marked range was the macroscopic caving arch of the overburden broken zone when the working face advanced 140 m, an arch trace failure line developed along the inner boundary of the pressure arch for the concentrated stress, the caving arch was formed for the function

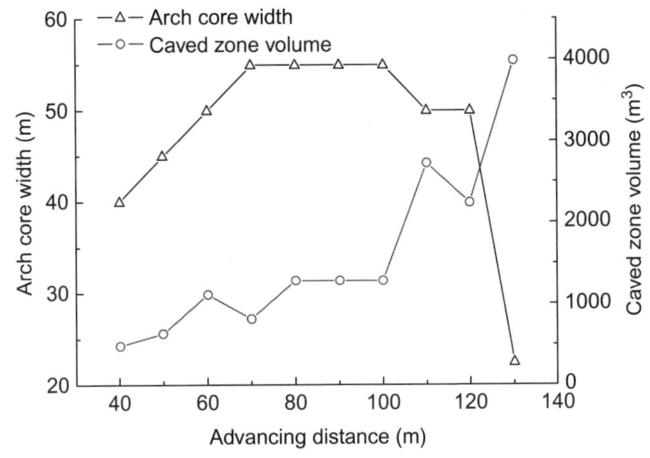

FIG. 5.58 The width of the pressure arch core and failure volume of the strata caused by coal mining.

of pressure arch in the bedrock. The lower marked line represented the squeezed-arch structure by the rotation of the broken blocks in the near field, which bore the load of the upper broken rocks. With the working face advancing 180 m, as shown in Fig. 5.59C, the covering zone and the bending settlement zone of the overburden rock formed below the marked red line, the latter arch foot of the pressure arch of the broken rocks was situated in the compacted zone, and the front arch foot was located in the undisturbed overburden rocks above the coal wall.

The distribution characteristics of the maximum principal stress of the broken strata in each monitoring line are shown in Fig. 5.60A after the working face advanced 90 m. The monitoring data showed that the zone of the increased principal stress formed the pressure arch in the surrounding rock. The arch foot was located in the lower rock strata in the near field, and the maximum principal stress of the arch foot was 13.67 MPa. The vault of the pressure arch was located in the upper hanging overburden rock in the far field, and the maximum principal stress of the vault was 4.59 MPa.

As shown in Fig. 5.60B, with the working face advancing 140 m, the front arch foot of the pressure arch in the surrounding rock transferred to the undisturbed surrounding rock. Owing to the range of the hanging strata expanding, the load bored by the pressure arch increased, and the maximum principal stress of the arch foot arrived at 14.13 MPa. The vault of the pressure arch expanded simultaneously, and the maximum principal stress of the vault became 6.04 MPa.

The macroscopic pressure arch was the macrobearing structure of the overburden rock in the mining field, and the bearing capacity of different forms was diverse. The pressure arch size of the surrounding rock in the far field affected

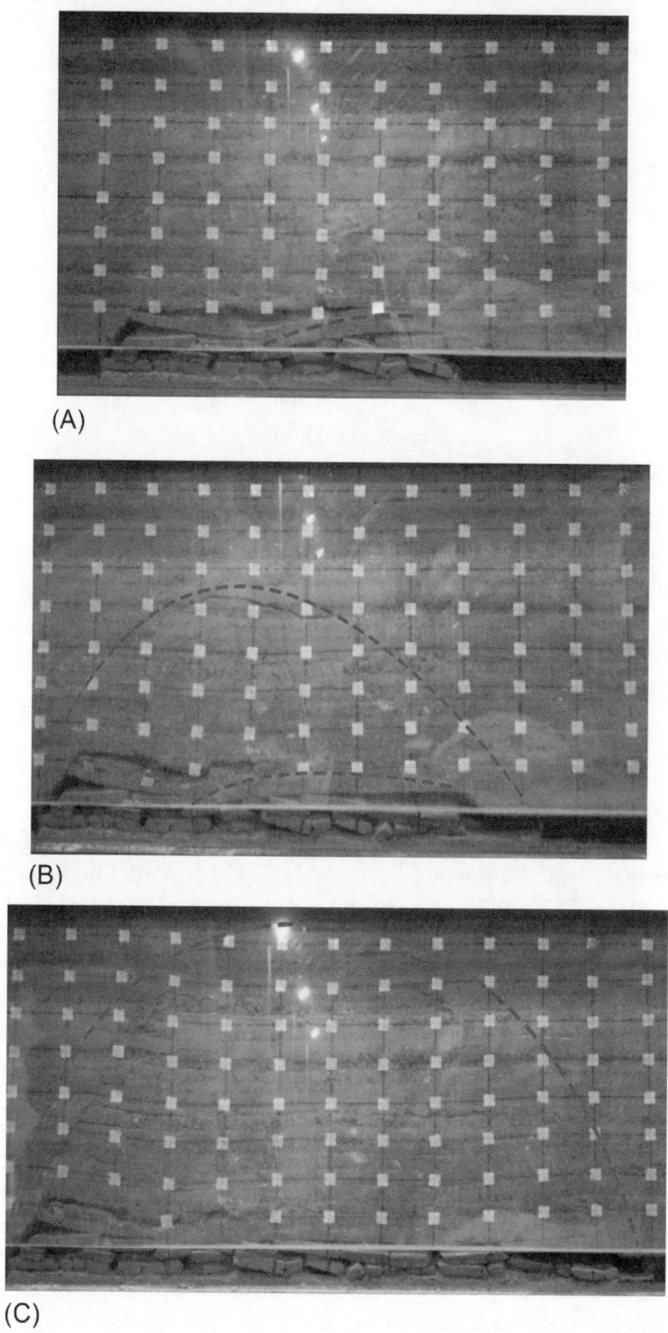

FIG. 5.59 Evolution characteristics of the pressure arch and the caving zone. (A) The pressure arch in the near-field, (B) the caving arch zone, and (C) the inner boundary of the caving arch.

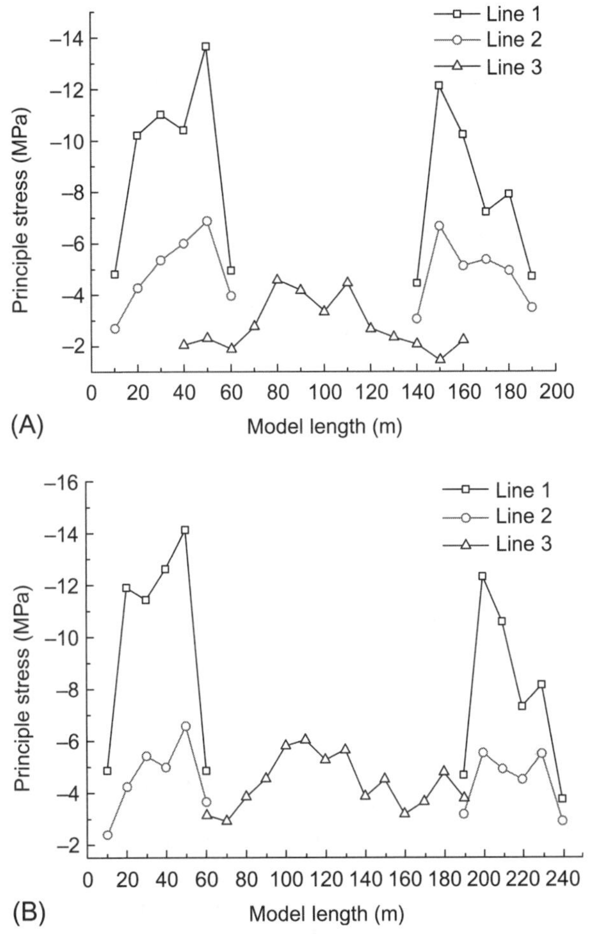

FIG. 5.60 The distribution of the principle stress with the working face advancing in the surrounding rock. (A) Advancing 90 m and (B) advancing 140 m.

the range of the overburden broken zone in the near field (Xie et al., 2009; Wang et al., 2015a, b). The alternating instability of the caving zone led to the occurrence of the strong (weak) weighting periodically during shallow coal mining (Ren and Qi, 2011). However, there was a lack of mechanical criteria for the pressure arch size and bearing capacity, and the evolution relationship between the pressure arch and the caving zone failed to be established.

Because the coal mining disturbance made the principal stress in the upper hanging overburden rock deviate to the stable surrounding rock, the vault of the pressure arch formed in the zone of the increased horizontal principal stress in the mid-span hanging overburden rock, and the arch shoulder and arch feet of the pressure arch were located in the undisturbed surrounding rock. There existed a core bearing zone in the pressure arch that could be accurately delineated by the identifying indicators of the pressure arch in the surrounding rock. The greater the width of the pressure arch core area, the stronger the pressure arch bearing capacity. The destructive volume of the broken strata during coal mining displayed the nonuniform periodic variation with the working face advancing.

At the initial mining stage, the pressure arch in the near field formed in the overburden broken zone, and the weighting on the working face was relatively weak as a result of the small zone instability of the caving arch. After the macroscopic pressure arch forming, the bearing load of the support became larger, and the sliding instability of the caving arch would cause strong weighting. The bearing capacity of the pressure arch in the surrounding rock affected the range of the caving zone. According to the width of the core area of the pressure arch, the bearing capacity of the overlying strata and the weight of the rock strata in the caving zone could be quantitatively analyzed.

5.5.5 Conclusions

To analyze the evolution characteristics of the pressure arch during shallow coal mining, taking the Shangwan Mine as the engineering background, the mechanics model of the pressure arch in the mining field was built, and the mechanical indicators of the pressure arch and its load-bearing zone were put forward. Then, the mechanical model was verified and analyzed, and the nonuniform weighting mechanism of the shallow coal seam with thick bedrock was revealed by analyzing the evolution characteristics of the pressure arch and the caving zone. The conclusions are the following:

The increased zone of the horizontal principal stress in the mid-span hanging overburden rock formed the vault of the pressure arch, and the arch feet and waist of the pressure arch were located in the surrounding rock. The formation of the pressure arch indicated the load of overlying strata above the mined-out area was effectively transferred to the surrounding rock; it is useful to evaluate the load-bearing area of the overlying strata. There existed a core bearing zone in the pressure arch, and the arching indicators of the pressure arch were greater than the average value.

The greater the width of the pressure arch core, the stronger the pressure arch bearing capacity. When the width of the pressure arch core decreased, the failure volume of the strata caused by coal mining increased. The indexes of the pressure arch shaped could be applied in the area division of bedrock strata. The load carried by the working face support also could be calculated based on the pressure arch area and the failure area volume. The destructive volume of the rock strata displayed the nonuniform periodic variation as the working face advanced, so the roof weighting behaved as a long/short weighting interval and a strong/weak weighting strength, identifying the changing law of the pressure arch provided a way to forecast roof weighting.

At the initial mining stage, the broken blocks were the main bearing structure in the near field, the pressure arch effect occurred in the fractured roof strata when the broken blocks carried the load effectively, and the weighting on the working face was weak. The strong weighting of the working face was induced for the sliding instability of the caving arch, and the caving arch of the fractured bedrock was located under the inner boundary of the pressure arch. The latter arch foot of the pressure arch transferred to the compacted zone when the fractured bedrock was compacted in the mined-out area. The front arch foot transferred to the stable surrounding rock in front of the coal wall, and the fractured bedrock under the pressure arch was the load source of roof weighting.

Because the mining height, the length of the working face, and other factors impact the pressure arch, the evolution characteristics of the pressure arch in the surrounding rock under different mining conditions need to be further studied. The formation conditions of the pressure arch under different lateral pressure coefficients are also a future research direction.

5.6 Elastic energy of pressure arch evolution

5.6.1 Introduction

The stability of the overlying strata during shallow coal mining, such as the large-scale roof falling and step-like ground subsidence, is the key problem that can restrict mining safety (Ju and Xu, 2015). The self-bearing structure of the pressure arch can form in the overlying strata after coal mining, and this structure can support the load of the upper strata and soil layer. So, the weighting intensity of the working face is determined by the caved rock in the unloading zone under the inner boundary of the pressure arch. The elastic energy is accumulated in the pressure arch under the concentrated stress, and the released energy for mining is the internal cause of rock failure (Wang et al., 2017). So it is necessary to reveal the distribution characteristics of the stress field and energy field in the mining field, and to analyze the stability of the overlying strata during shallow coal mining based on the evolution characteristics of the pressure arch.

Currently, most scholars focused on the pressure arch of the surrounding rock was mainly the small excavated space such as underground cavern and roadway, and the support was used to stabilize the roof by forming the self-bearing pressure arch structure (Huang et al., 2002; Li, 2006). The pressure arch in the overlying strata during coal mining usually affected the roof stability, and the factors such as buried depth, mining conditions, overburden property, and the horizontal stress also influenced the distribution of the pressure arch (Wang et al., 2015c). With the working face advancing, the inner and outer boundaries of the pressure arch in the mining field were constantly changed, and the pressure arch was often located in the area of the concentrated stress. The stress field and energy field in the surrounding rock were redistributed and affected each other, so the stability analysis of the overlying strata was inadequate if neglecting the correlation of these two indexes. The large mining height and long working face were typical characteristics of shallow coal mining in China. There were differences of the stress distribution and failure characteristics along the strike and dip of the mining field, and the abnormal roof weighting control was also a technical problem during shallow coal mining.

The arching effect of the principal stress exists in the underground engineering due to excavation. The pressure arch theory was widely used in the underground tunnel and mining engineering. For example, Poulsen (2010) proposed a method for calculating the coal pillar load by the pressure arch theory, and the method was verified by the numerical calculation based on shallow coal mining. Kim et al. (2013) investigated the arching effect on a vertical circular shaft by using experimental tests and theoretical analysis, and they quantified the distribution of the lateral earth pressure by a three-dimensional arching effect. Xie et al. (2009) conducted numerical and physical tests to investigate the stress distribution in the rock surrounding during the fully mechanized top-coal caving. The results showed that a macrostress shell composed of high stress existed in the surrounding rock, the working face was protected by the stress shell, and the strata of the low-stress zone inside the stress shell produced periodic roof weighting. However, the definition of the stress shell was the lack of a mechanical criterion. Ren and Qi (2011) studied the stress field during shallow coal mining through field observations and numerical simulations. They pointed out that the bearing structure of the pressure arch could be formed in the overlying strata while the pressure arch was not always stable in the whole mining process, and the mining height and the length of the working face had influences on the pressure arch. But the instability mechanism and the specific form of the pressure arch have not been concerned. So the above studies showed that there was a global pressure arch in the overlying strata in the mining field, but there were differences in the evolution characteristics of the pressure arch under different mining conditions. For some aspects, such as the forming condition, instability mechanism, and spatial distribution characteristics of the pressure arch during shallow coal mining are still lack of specific research.

He and Zhang (2015) found that a global pressure arch was formed in the surrounding rock and a loosened zone beneath the pressure arch was formed in the roof after underground engineering excavation. The strata within the loosen zone did not sustain the overburden load and the strata separated from each other. Wang et al. (2016a, b, c, d, e, f, g) applied pressure arch theory to prediction the collapse scope of a deep-buried tunnel. A curve with a tensile stress of zero was chosen as the inner boundary of the pressure arch. Below the inner boundary of the pressure arch was the cracked rock in a tension state, and the unstable surrounding rock in this scope could cause roof collapse. The above mentioned studies indicated that the instability of the loose strata below the pressure arch should be paid attention to the failure mechanism of this zone. The elastic energy stored in the primary rock was suddenly released after coal mining. The released energy was transformed into dissipated energy for the plastic deformation and rock failure process, and the stress in the surrounding rock was partly concentrated after redistribution. So rock failure was the result of the energy release during mining loading and unloading. Huang and Li (2014) studied the conversion of strain energy in triaxial unloading tests on marble. They found that the initial confining pressure and unloading rate affected the failure mode and strain energy conversion of rock, and the prepeak conversion rate of the strain energy was increased with an increase of the unloading rate. The rock specimen incurred failure with flying fragments after much strain energy being released after the postpeak stage. Zhang and Gao (2015) conducted axial loading-unloading experiments to investigate the energy evolution of water-saturated sandstone samples. The results indicated that the saturation process would decrease the prepeak energy accumulation while it could increase the dissipated energy but reduce the magnitude and rate of the postpeak energy release. Shabanimashcool and Li (2015) calculated the potential energy stored in the pressure arch of the roof beam and the released energy owing to deflection of the beam based on the theory of minimum potential energy. However, this research was limited in analyzing the stability of the pressure arch in a single roof strata. Rezaei et al. (2015) established a mechanical model based on the strain energy balance to determine the mining-induced stress over gates and pillars. They also analyzed the height of the distressed zone and the coefficient of stress concentration by using the proposed model, but the relation of the mining-induced stress distribution and strain energy revolution was not established.

Because it was not perfect enough about the spatial pressure arch and elastic energy distribution under shallow coal mining, this study would conduct further work based on the monitoring data of the roof weighting of a typical shallow coal mining. So, typical shallow coal mining in the Shendong mining area was taken as the engineering background, and the load distribution characteristics of the overlying strata were analyzed based on the monitoring data of the support resistance. Through theoretical analysis and numerical simulations, the formation and changing processes of the pressure arch in the overlying strata were studied, and the instability mechanism of the overlying strata during shallow coal mining was revealed by focusing on the accumulation and release law of the elastic energy in the pressure arch.

5.6.2 Engineering background

The data of the roof weighting in the Shangwan Mine were observed from the No. 51101 and No. 51104 panels in the No. 1^{-2} coal seam. The average thicknesses of the coal seam of No. 51101 and No. 51104 panels were 6.70 m at an average depth of 115.40 m, which dipped 0–3 degrees. The longwall retreating mining method was adopted in each panel, and the full thickness of the coal was mined at one time. The full seam top-coal-caving method was used to manage the roof, and the panel widths were 240 m and 300 m, respectively. The immediate roof was mostly sandy mudstone or siltstone and the main roof was sandstone.

Based on the peak value P_m of the support resistance during the initial weighting and 18 times the periodic weighting, the panel roof weighting characteristics along the strike were obtained as shown in Fig. 5.61.

It can be seen from Fig. 5.61A that the periodic weighting strength increased gradually with the working face advancing, and the strong and weak periodicity occurred in the roof weighting. The support resistance wholly increased with increasing the panel width. The strong weighting of the working face with the length of 240 m and 300 m occurred after the fifth time of the periodic roof weighting while the impacted load of the roof existed in the large mining space.

The distribution characteristics of the average value P_a and peak value P_m of the final support resistance along the seam dip direction are shown in

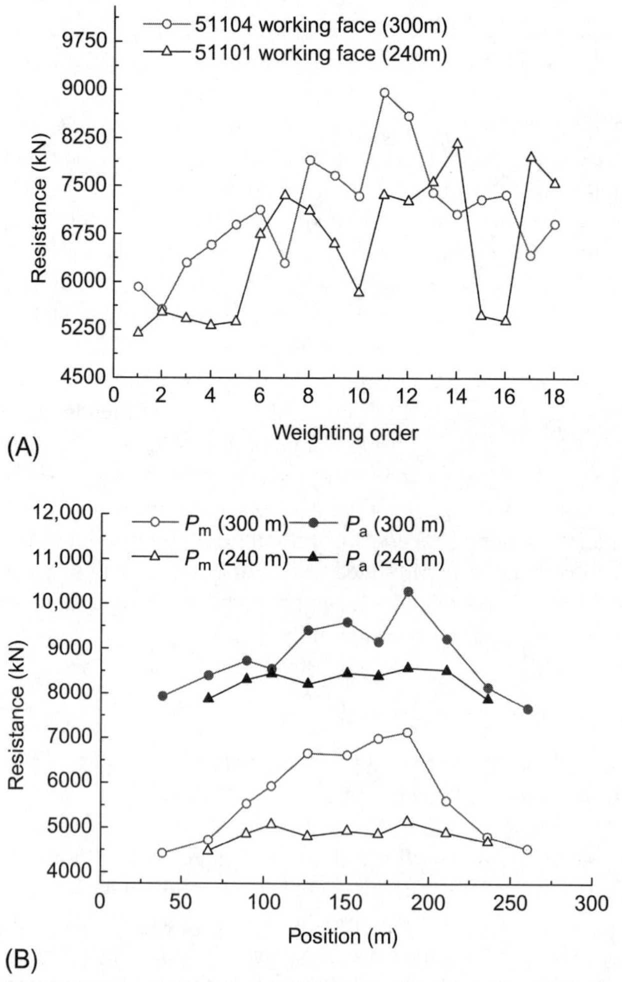

FIG. 5.61 The panel roof weighting characteristics along the seam strike and dip directions. (A) Along strike of the mining field and (B) along dip of the mining field.

Fig. 5.61B. The support load and the position of the working face were parabolic, and the support resistance in the middle of the working face was greater than that on both sides. However, the load at the head- and tailgate sides was the smallest and was distributed as a symmetrical arch as a whole. The roof weighting appeared most violently in the middle of the working face and became gentle toward both sides.

The support resistances along the strike and dip of the working face indicated that the size effect affected the load distribution of the overlying strata and the roof weighting, namely the action load of the roof was stronger with the larger mined-out space. The distribution characteristics of the support load in different directions depended on the spatial failure of the overlying strata.

5.6.3 Pressure-arch analysis and computational model

5.6.3.1 *Pressure-arch analysis*

As shown in Fig. 5.62, the load of the overlying strata was transferred to the stable surrounding rock in the mining field after coal mining. The load of the hanging strata above the mined-out area was transferred to the surrounding rock. This transferring action formed the arch top, and the increasing stress

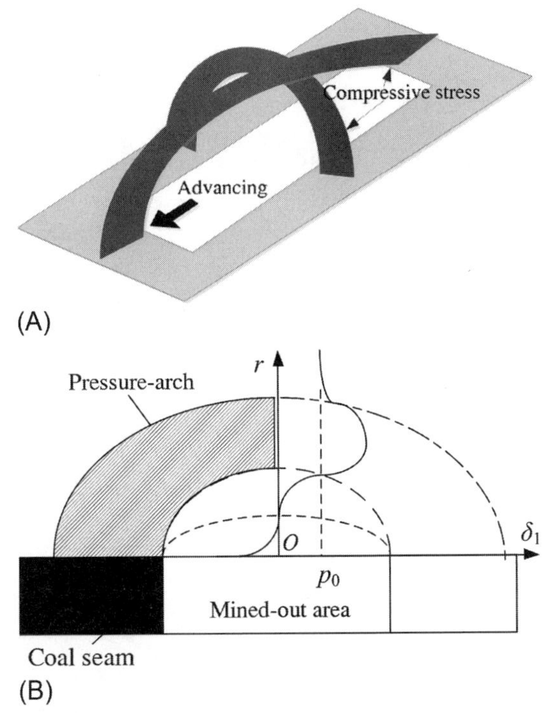

FIG. 5.62 The pressure arch in the overlying strata of the mining field. (A) Sketch of spatial distribution of pressure-arch and (B) sketch of major principle stress distribution.

contributed to the formation of the arch waist in the surrounding rock. The load of the upper strata and the loosened soil layer was carried by the rock mass of the pressure arch. The external range of the outer boundary of the pressure arch was the zone of the primary rock stress while the zone of the reduced compressive stress and the zone of the tensile stress were located under the inner boundary of the pressure arch.

The rock mass in the deep stratum could be regarded as the elastic medium under the compacted conditions, and a large amount of elastic strain energy was accumulated in the rock under long-term geological action. The energy release during coal mining caused the deformation and failure of the surrounding rock, and the stress was redistributed after the surrounding rock lost its balance state. The elastic energy was accumulated again in the zone of the concentrated stress.

The sudden energy release was the internal cause of rock failure (Solecki and Conant, 2003), and the strain energy of the rock element in the principal stress space can be expressed as:

$$U^{\mathrm{e}}=\frac{1}{2E}\left[\sigma_1^2+\sigma_2^2+\sigma_3^2-2v(\sigma_1\sigma_2+\sigma_2\sigma_3+\sigma_1\sigma_3)\right] \tag{5.65}$$

where σ_1, σ_2, and σ_3 was the principle stress of each direction, E was the elastic modulus, and v was the Poisson's ratio.

If the stress of the primary rock was in the triaxial anisobaric state, p_0 was the distant vertical stress, and σ_1, $\sigma_2=\sigma_3=\lambda p_0$ (see Fig. 5.63). The elastic strain energy U_1^{e} of the surrounding rock before excavation was given by

$$U_1^{\mathrm{e}}=\left[(1+2\lambda^2)p_0^2-2v(2\lambda p_0^2+\lambda^2 p_0^2)\right]/2E \tag{5.66}$$

If the strike length of the working face was L, the westergrad stress function being applied (Du et al., 2015), considering the plane strain state and the principal stress distribution of the surrounding rock in the mined-out area, was given by

$$\sigma_1=\frac{p_0}{2}\sqrt{\frac{L}{r}}\cos\frac{\theta}{2}\left(1+\sin\frac{\theta}{2}\right) \tag{5.67}$$

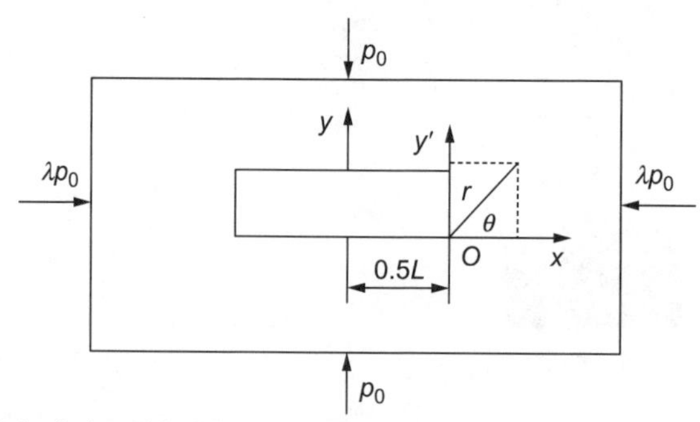

FIG. 5.63 Mechanical model of the surrounding rock.

$$\sigma_2 = \frac{p_0}{2}\sqrt{\frac{L}{r}}\cos\frac{\theta}{2}\left(1 - \sin\frac{\theta}{2}\right) \tag{5.68}$$

$$\sigma_3 = \upsilon p_0\sqrt{\frac{L}{r}}\cos\frac{\theta}{2} \tag{5.69}$$

where L was the span along the strike direction of the panel and r was the vertical distance from the coal seam top.

Taking the boundary of the mined-out area in the mining field along the horizontal direction as an example, $\theta = 0$ degree, the elastic strain energy of the surrounding rock after coal mining was given by

$$U_2^e = \left[(0.5 - 0.5\nu - \nu^2)\right]p_0^2 L/2Er \tag{5.70}$$

Assuming that the lateral stress coefficient $\lambda = 0.5$, the variation of elastic energy ΔU^e of the surrounding rock was given by

$$\Delta U^e = U_2^e - U_1^e = p_0^2\left[(0.5 - 0.5\nu - \nu^2)\frac{L}{r} + 2.5\nu - 1.5\right]/2E \tag{5.71}$$

The pressure arch of the surrounding rock in the mining field was the increasing zone of the principle stress, and the deformation of the surrounding rock was mainly elastic before this principle stress exceeded its strength; the elastic energy was accumulated in the pressure arch. The pressure arch of the surrounding rock was in the dynamic evolution process during coal mining and the energy in the strata was also cyclically accumulated and released, so the unloading zone of the strata was the action result of the pressure arch and the elastic energy. The caved arch of the overlying strata was the cause of roof weighting and disaster, so it was crucial to reveal the instability mechanism of the overlying strata to investigate the evolution characteristics of the pressure arch and elastic energy.

5.6.3.2 Computational model

As shown in Fig. 5.64, based on the engineering background of the typical shallow coal mining of the Shendong mining area, the numerical model was established by using the FLAC^{3D} software. The thickness of the coal seam was 7.0 m, the mining height was 6.0 m, the thickness of the siltstone floor was 20.0 m, and that of the roof was 100.0 m. Focusing on the evolution characteristics of the stress field and the energy field of the overlying strata, the difference of the rock property between the strata was ignored and the roof strata was unified as the sandstone.

The model size was 400 m long, 352 m wide, and 137 m high. The load 1.8 MPa instead of the weight of the 12 m soil layer was applied to the top of the model while the bottom boundary of the model was fixed and the lateral boundaries of the model were fixed in the horizontal direction.

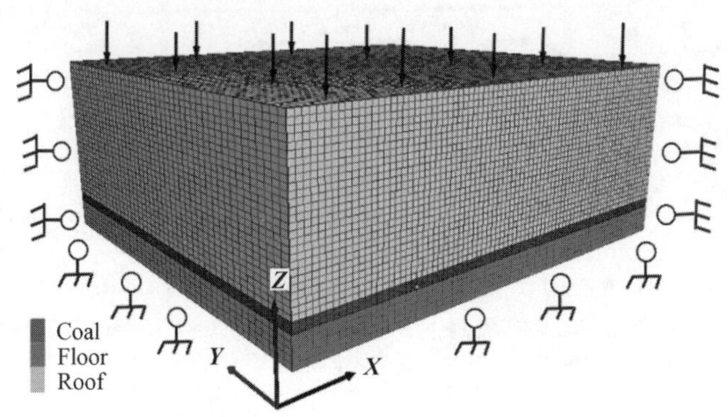

FIG. 5.64 The computational model.

The position to start mining was 50 m away from the left boundary while the advancing distance along the strike was 300 m and the length of the working face along the trend was 240 m. The roadway was excavated 50 m away from both side boundaries along the direction of the *X*-axis while the width of the roadway was 6 m and the height was 5 m. The physical and mechanical parameters of the model are listed in Table 5.8, and the Mohr-Coulomb criteria was used in the numerical calculation.

5.6.4 Simulation results and discussion

As shown in Fig. 5.65, the major principal stress direction of the overlying strata was deviated after the coal mining, and the vertical stress in the roof decreased while the vertical stress in the lateral surrounding rock increased. The stress in the unloading zone below the deviated principal stress zone decreased, and the major horizontal principal stress was tensile stress in this zone. The principal stress of the overlying strata near the mining field was deviated into an arch, and the zone of increasing principle stress was only formed at two sides of the mined-out area. The continuous circle of the pressure was not formed

TABLE 5.8 Physical and mechanical parameters of the materials.

Name	Unit weight (kN/m^3)	Elastic modulus (GPa)	Poisson's ratio	Cohesive strength (MPa)	Tensile strength (MPa)	Friction angle (degree)
Sandstone	25.0	36.5	0.22	2.6	1.5	30
Coal	13.1	12.7	0.29	1.2	0.6	27
Siltstone	24.6	37.9	0.20	4.5	3.0	40

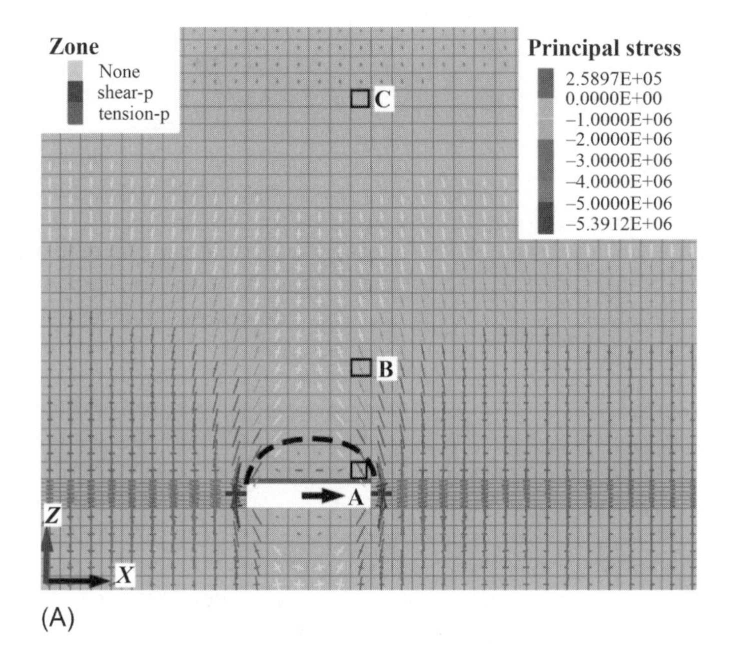

(A)

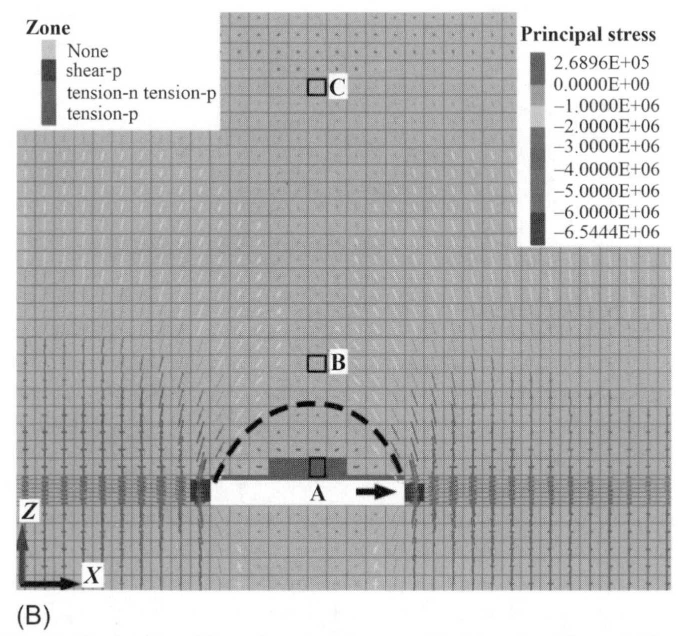

(B)

FIG. 5.65 Distribution law of the major principle stress. (A) Advancing 30 m and (B) advancing 50 m.

because of the smaller horizontal stress in the upper strata. The inner boundary of the deviated stress zone moved upward with the working face advancing 50 m. The unloading area expanded and some of the units incurred tensile failure.

The elastic energy of the unit was monitored using the built-in calculation function and the FISH language in FLAC^{3D} while the lower roof unit A, the middle roof unit B, and the top roof unit C were regarded as the monitoring objects. The changing law of elastic energy of the monitoring unit with the working face advancing 50 m is shown in Fig. 5.66. The roof unit A was in the concentrated stress zone before the coal below it was mined. The elastic energy of this unit before the working face advanced 20 m was in the rising stage, and a large amount of elastic energy was released sharply after the working face advanced 20 m. The unit B was located in the middle position of the roof and the energy in unit B increased gradually after the working face advanced 10 m. However, the energy in unit B increased a little after the working face advanced 20 m and was totally released after the working face advanced 30 m. The unit C was far from the mined area and the energy was obviously kept rising after the working face advanced 10 m.

Overall, coal mining would lead to an energy release in the roof. Some differences existed in the energy evolution characteristics of the strata in different positions under the influence of stress redistribution. The elastic energy of the unit was in the rising stage under the action of the concentrated stress, and the unloading zone of the roof strata near the mining area was formed by the sustained elastic energy release.

As shown in Fig. 5.67, the three-dimensional distribution characteristics of the pressure arch foot were obtained by extracting the monitoring data of the major principal stress of the roof 2.5 m away from the coal seam. The stress reduced in the roof of the mined-out area, and the zone of the concentrated stress was formed in the undisturbed surrounding rock, which was the pressure arch foot of the overlying strata. The exposed area of the overlying strata increased with the working face advancing, and the load carried by the pressure arch and the peak principle stress of the arch foot increased gradually.

The pressure arch was the zone of the concentrated stress in the overburden strata, and the elastic energy was accumulated in the pressure arch. If applying the criterion that the principle stress in the pressure arch was higher than the stress of the primary rock to judge the arch geometry, the plane shape and spatial distribution characteristics of the pressure arch could be obtained by programming with the FISH language in FLAC^{3D}.

As the central section view of the working face shown in Fig. 5.68, the continuous arching circle of the concentrated principle stress was formed in the surrounding rock when the working face advanced 100 m. The arch foot moved forward with the working face advancing 150 m while the inner boundary range of the pressure arch expanded.

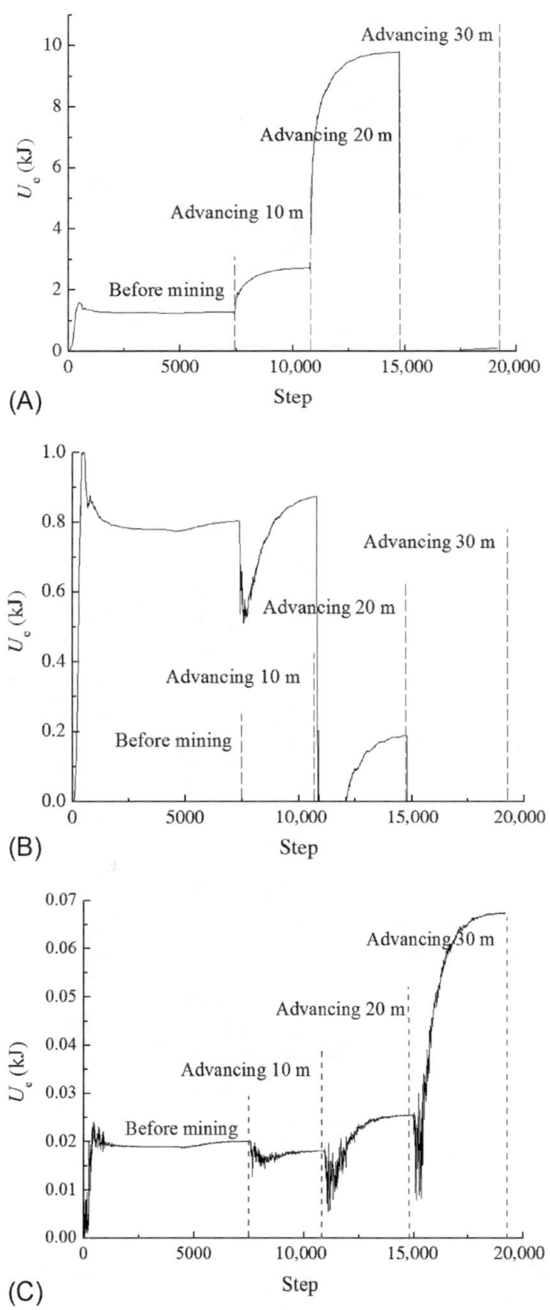

FIG. 5.66 Changing law of elastic energy of the monitoring unit. (A) Unit A, (B) unit B, and (C) unit C.

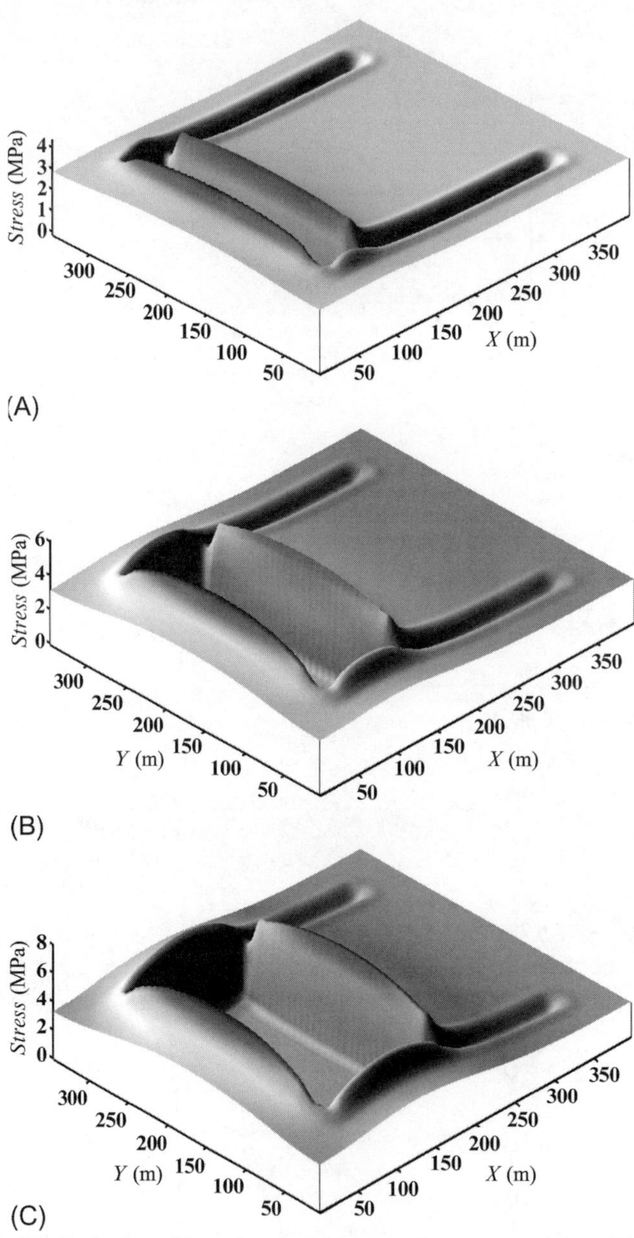

FIG. 5.67 Distribution law of the major principle stress at the pressure arch foot. (A) Advancing 50 m, (B) advancing 100 m, and (C) advancing 150 m.

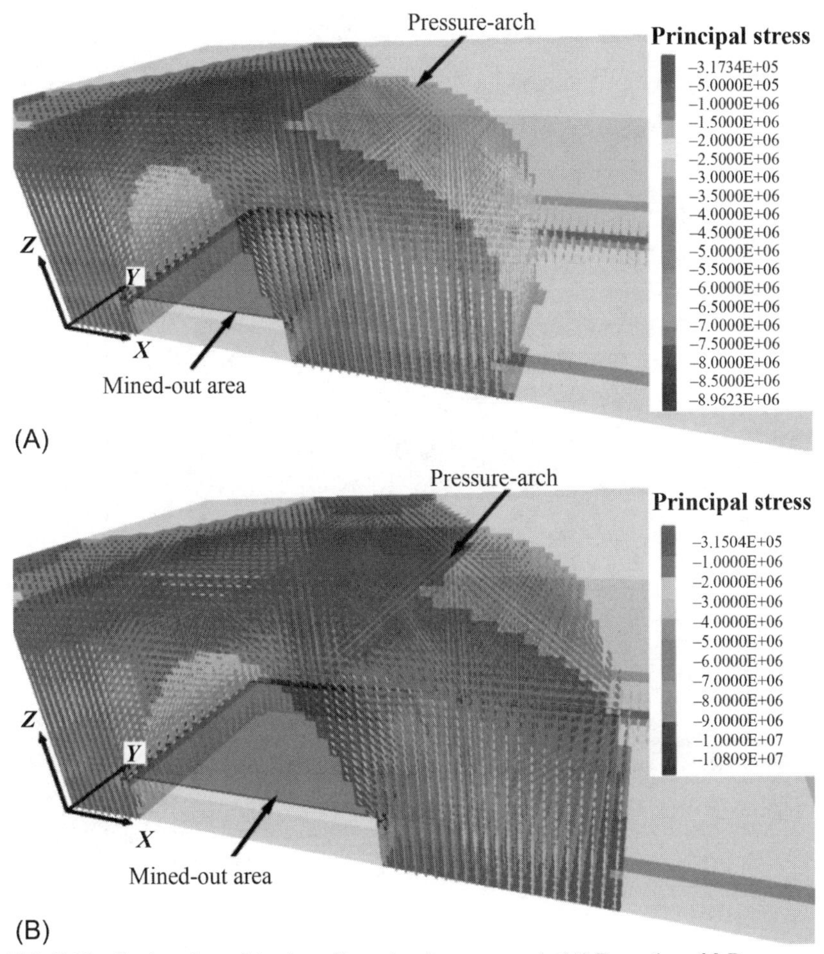

FIG. 5.68 Cutting view of the three-dimensional pressure arch. (A) Formation of 3-D pressure-arch and (B) expansion of 3-D pressure-arch.

The cutting map of elastic energy in the middle of the working face is shown in Fig. 5.69. The inner boundary of the pressure arch was formed after the energy was released when the working face advanced 90 m, and the zone of energy release under the inner arch expanded with continued mining. The roof strata in the zone of the released energy and the reduced stress were destroyed when the horizontal principle stress reached the tensile strength of the surrounding rock.

As shown in Fig. 5.70, the caved arch of the destroyed strata in the unloading zone was formed under the inner boundary of the pressure arch. The continued coal mining caused the pressure arch foot to move forward, and the inner

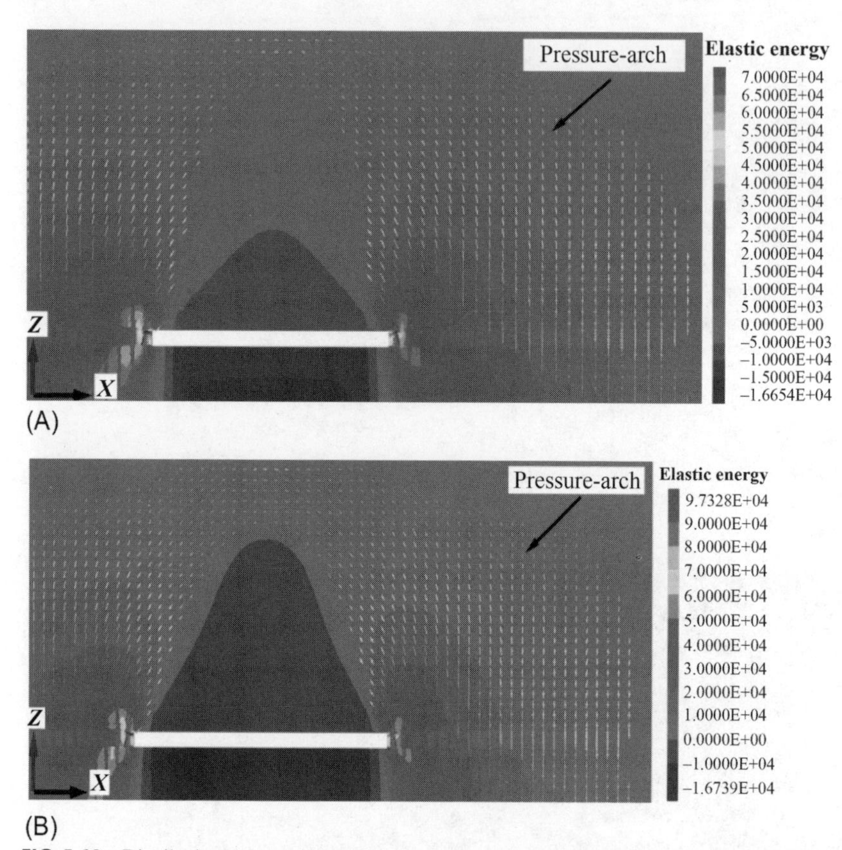

FIG. 5.69 Distribution characteristics of elastic energy in the middle of the working face. (A) Formation of the energy release zone and (B) expansion of the energy release zone.

boundary of the pressure arch was extended while the caved arch range expanded. The rock mass at two arch feet was under the action of the concentrated stress. The obvious vertical shear zone was formed in the roof strata at the side of the working face, and the caved arch would slide along the fractured zone.

Taking the roof units D and E that were 5 m and 25 m away from the coal seam as the monitoring objects, with the working face advancing 110 m, the variation law of the elastic energy of the roof under the action of the pressure arch could be analyzed.

As shown in Fig. 5.71, when the working face advanced 110 m, the elastic energy of 29.27 kJ in the unit D was released after coal mining, which resulted in the unit being destroyed. The energy accumulated in the unit again under the concentrated stress in the arch foot, and the elastic energy of unit D was fully released for the arch foot moving forward with each mining stage. The peak

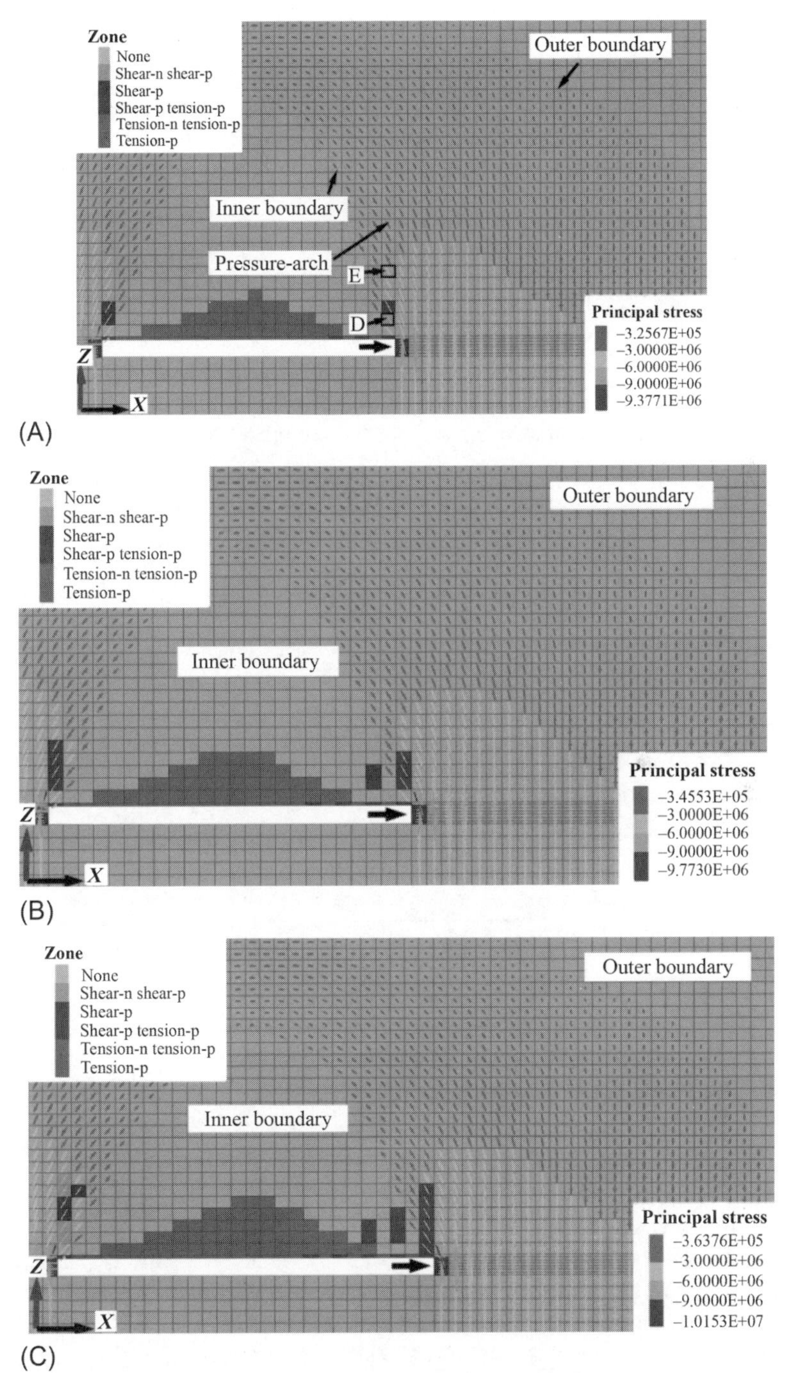

FIG. 5.70 Evolution characteristics of the pressure arch and the caved arch. (A) Advancing 110 m, (B) advancing 120 m, and (C) advancing 130 m.

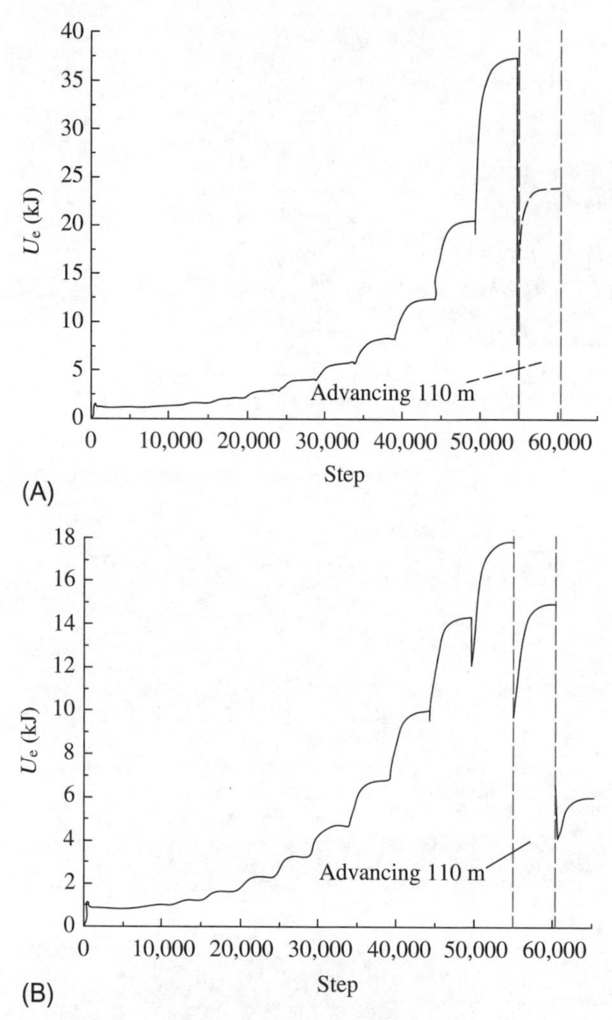

FIG. 5.71 Changing rule of the elastic energy of the monitoring units. (A) Unit D and (B) unit E.

elastic energy of unit E was relatively small. The released energy was 8.70 kJ after coal mining, and the energy was released again for the pressure arch transfer, but the residual energy still remained in the unit.

Compared with the energy monitoring data during the working face advancing 30 m, the energy value was increased totally in the later mining stage, which resulted from the increased load being carried by the pressure arch and the larger strain of the strata. Under the concentrated stress, the energy in the units in the deep position of the surrounding rock was stored with the working face advancing. The concentrated stress was relatively lower near the mining area of the surrounding rock, so the stored energy was lower in this position. Due to the limit of the lower strata, the energy release was inhibited in the upper strata, the units in the middle position of the surrounding rock was not released completely, part of the elastic energy was remained in these units.

The distribution characteristics of the principal stress and elastic energy along the trend of the working face after advancing 150 m are shown in Fig. 5.72. The global pressure arch also existed in the strata along the trend of the working face, and the zone of released elastic energy was below the inner boundary of the pressure arch. The caved arch for shear failure was formed in the unloading zone, which caused the distribution of the support resistance in the middle of the working face to be greater than that on both sides.

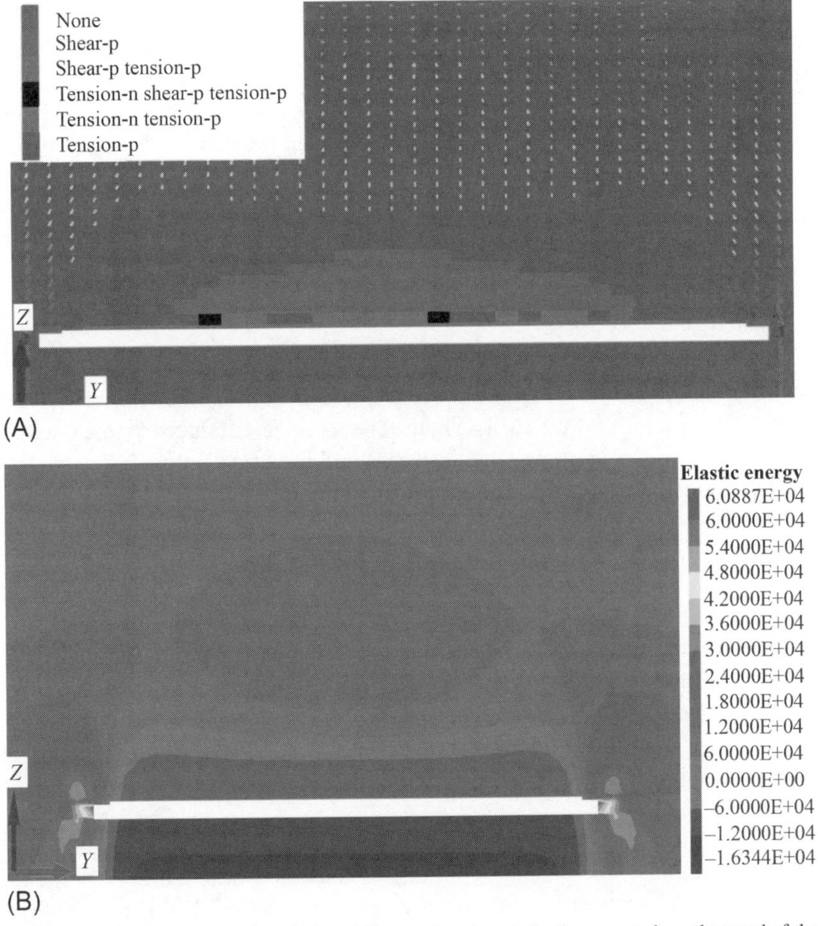

FIG. 5.72 Distribution characteristics of the caved arch and elastic energy along the trend of the working face. (A) Caved arch and pressure-arch and (B) elastic energy distribution.

According to the pressure arch distribution and energy variation, it was known that the elastic energy was mainly stored in the pressure arch. The energy value in the arch waist and arch feet was relatively higher. The mining action could induce the energy to be released suddenly at the arch foot.

The instability of the loosened roof below the pressure arch could threaten the safety of the working face (Li, 2006) while the global pressure arch could provide natural protection in the mining field during coal mining. Such factors as the in situ stress ratio, buried depth, and dip angle could affect the formation and stability of the pressure arch, and the initial stress condition was a key factor to form the pressure arch (He and Zhang, 2015; Abdollahipour and Rahmannejad, 2013). The stress shell of the surrounding rock under deep fully mechanized top-coal caving conditions could carry and transmit the overburden load, and its internal and external shape changed with the working face advancing. So the loading state and the structural characteristics of the strata in the loosened zone under the stress shell determined the roof weighting characteristics of the working face (Xie et al., 2009). The pressure arch could also be formed in the roof under shallow coal mining conditions while the mining height and the length of the working face had influence on the stability of the pressure arch.

Because the initial horizontal stress under shallow coal mining was relatively small, the continuous pressure arch of the overlying strata could not be formed in the initial mining stage. However, the global pressure arch could be formed after the horizontal principal stress exceeded the stress of the primary rock.

The compressive tests and the rating loading and unloading tests showed that the sudden release of elastic energy would induce rock failure (Wang et al., 2012; Huang and Li, 2014). So, the energy release induced by the unloading action of coal mining was the main reason for the roof failure in the mining field. Furthermore, the sliding instability of the caved arch of the overlying strata was the main supporting task of the working face during shallow coal mining.

The feet of the pressure arch were usually the zones of the concentrated stress, which determined the failure characteristics of the roof along the trend and the height of the sliding zone along the strike in the mining field. So, the stress concentration and the elastic energy value of the pressure arch foot were important indices for predicting the sliding instability of the roof.

5.6.5 Conclusions

To ensure mining safety during shallow horizontal coal mining, taking the typical coal mining in the Shendong mining area as the engineering background, the evolution characteristics of the pressure arch and the elastic energy of the overlying strata were studied, and the distribution characteristics of the pressure arch in different mining stages were obtained. Furthermore, the failure mechanism of the overlying strata was revealed based on the accumulation and release of the elastic energy. The conclusions are listed as follows:

(1) The elastic energy of the roof is released during coal mining, and the continuous pressure arch forms when the horizontal principal stress in the midspan of the roof exceeds the stress of the primary rock. The inner boundary of the pressure arch may be determined based on the zone of the fully released elastic energy.
(2) The elastic energy is mostly accumulated in the pressure arch, and the stored energy at the arch foot is the highest. The energy being suddenly released after coal mining can result in rock failure. The concentrated stress and the released energy in the arch foot all increase with the working face advancing, and the height of the shear failure zone of the roof also increases.
(3) The unloading zone is located under the inner boundary of the pressure arch, and the load of the caved arch in this zone is the root of the roof weighting. The pressure arch feet can determine the failure types of the roof along the trend and the height of the sliding roof zone along the strike of the mining field. The sliding instability of the caved arch in the shear failure zone at the arch feet can cause strong roof weighting.

The strata stability depends on the evolution law of the pressure arch, and enough horizontal stress is the necessary condition for the formation of the continuous pressure arch during shallow coal mining. So the characteristics of the pressure arch under different conditions need to be further studied in the future.

5.7 Height predicting of water-conducting zone

5.7.1 Introduction

Driven by the increasing world population and industrialization, the demand for energy has developed rapidly (Booth, 2007) and has caused a series of serious environmental problems (Li and Zhou, 2006; Howladar and Hasan, 2014; Wang and Meng, 2018). Underground coal mining causes surface subsidence, water loss, damage to infrastructure, etc., especially in some eco-environmentally fragile areas. Ground control theory (the study of the behavior and control of the rock roof due to mining) is essentially used to solve these problems (Peng, 2008). "Ground control" includes roof control, rib control, floor control, pillar design, longwall-face shield design, overburden failure, and surface subsidence (Trueman et al., 2009; Peng, 2015; Li et al., 2015; Xu et al., 2016; Wang et al., 2016a, b, c, d, e, f, g; Zhao et al., 2018). Further, the height of the fractured water-conducting zone (FWCZ) is a key parameter in ground control for coal mines operating under a body of water (Guo et al., 2012a). Accordingly, the height of the FWCZ must be predicted prior to mining.

Methods of predicting the height of the FWCZ caused by longwall mining in general require only geological mining conditions, using numerical simulation, physical testing, in situ investigations, and/or microseismic monitoring techniques (Adhikary and Guo, 2015; Ghabraie et al., 2015; Yu et al., 2015; Cheng et al., 2017b). Peng (2006), an American academic, believes that the caved zone ranges in thickness from 2 to 8 times the mining height, and that the fracture zone ranges in thickness from 30 to 50 times the mining height and is related to the lithology of the overlying rock. Majdi et al. (2012) presented five mathematical approaches for estimating the height of the destressed zone (that is, the height of the caving zone plus fracture zone), and he argued that while the short-term height of the destressed zone ranges from 6.5–24 times the mining height, it is 11.5–46.5 times the mining height in the long term. Wu et al. (2017) proposed a method of determining the height of the FWCZ using the radial basis function neural network (RBFNN) model in MATLAB. Xu et al. (2017) developed a trapezoidal fractured model of overburden strata movement induced by coal mining to explain and prevent water leakage and water inrush during mining processes. These earlier works have offered good insight into the methods of predicting the height of the FWCZ under general geological and traditional mining conditions.

However, as coal mining technology and equipment are improved in China to meet the growing demand for coal, high-intensity longwall mining is becoming increasingly common. The available theoretical methods of predicting the height of the FWCZ from high-intensity mining in thick coal seams are poor, and few studies have been carried out on the internal mechanism of overburden failure in such mines. In the present study, the processes and mechanisms of overburden failure transfer (OFT) are proposed, based on the ground control ideas of Peng (2006). The mechanism by which the FWCZ develops based on OFT was analyzed by establishing suitable models; a new theoretical method is proposed for predicting the height of the FWCZ in high-intensity mining conditions. The proposed theoretical method, numerical simulation method, and engineering analogy method were used to predict the height of the FWCZ, and the results were compared with in situ measurements reported in the literature.

5.7.2 High-intensity mining in China

Coal is the main source of electrical energy in China. Coal production from individual longwall panels has risen from millions of tons to more than 10 million tons per year in recent years. For example, the Daliuta Coal Mine was the first large modern underground coal mine in China with a longwall panel producing an annual average of 10 million tons. This has since been matched in many mines or mine groups in Shendong, Wanli, Yushen, Huating, Datong, Shanxi, Yili, and other mining areas (Guo and Wang, 2017). The increase in the number of high-intensity longwalls is due to their high production rate, and thus research on the theory and technology of high-intensity mining has increased in recent years.

"High-intensity mining" refers to a coal mining method used in China. Fan (2014) defined high-intensity mining as a mining method characterized by a large mining area, a fully mechanized longwall with high production, and large panel sizes with rapid face advance and large mining height in thick coal seams. Guo and Wang (2017), on the basis of previous studies, defined high-intensity mining as a high-yield and high-efficiency coal mining method of thick coal seam (more than 3.5 m) mining (top-coal caving mining or large mining height, large panel width (more than 200 m), fast face advance rate (more than 5 m a day), high output per single unit (usually 5–10 Mt./a, minimum 3 Mt./a), small coal seam depth/thickness ratio (less than 100), and severe failure of overburden and surface.

High-intensity mining in thick coal seams as defined above is characterized by the use of advanced equipment and mining technology, simple geological and mining conditions, shallow mining depth, large coal seam thickness, large panel sizes, fast face advance, and high panel production and efficiency (Peng, 2017). The main mining methods are fully mechanized mining to full seam height or fully mechanized top-coal caving. In addition, data shows that the Tashan Coal Mine underwent a successful industrial trial in 2010 to assess the mining of a thick coal seam (average thickness 18.44 m) using fully mechanized top-coal caving. The project identified the key technology and equipment required for fully mechanized caving mining in a very thick coal seam, and the problem of mining a 14–20 m coal seam was successfully solved in 2014. Panel width was 200–302 m (average 262 m); length 1379–4966 m (average 2803 m); maximum daily face advance 15.75 m (average generally above 8 m); and annual output of a working face about 13.43 million tons. These figures indicate that the technology and equipment for high-intensity mining in China are mature.

However, high-intensity mining usually causes unique ground pressure behavior, such as large-scale support failure and extensive rib spalling. The large mined-out space left after high-intensity mining leads to more violent roof movement and strong ground pressure behavior, which eventually causes more than usually severe damage to the overlying strata and surface. For example, the fracture zone caused by high-intensity mining sometimes extends as far as the surface or into the fracture zone of overlying strata, and is thus directly connected to surface cracks. Therefore, there are only two zones above the high-intensity mined-out area: a caved zone and a fractured zone. This is unlike the three zones (i.e., caved, fractured, and continuous bending zones) following traditional underground coal mining. The issues of surface subsidence, step cracks, desertification, soil erosion and degradation, environmental pollution, and the bearing capacity of land affected by high-intensity mining are most severe in eco-environmentally fragile areas (Guo et al., 2012b; Cheng et al., 2017a). Therefore, it is vital that the development mechanism of overburden failure and the height of the FWCZ caused by high-intensity mining be studied.

5.7.3 OFT influence on FWCZ development

As the coal in a high-intensity longwall panel is being extracted slice by slice, the surrounding strata are forced to move toward and fill the goaf left by the extracted coal. This process induces a series of actions: rock strata movement, surface subsidence, abutment pressure concentration at both ends of the longwall face, etc. (Peng, 2008). Put simply, when a site has been undermined, overburden is in a state of failure, and the space-time progression of the failed overlying strata and the crack evolution characteristics of both the coal and rock are extremely complicated (Wang and Tian, 2018). The essence of this failure process is that each stratum above the goaf fails because it is no longer supported as the working face advances, and the failure state of the strata is transferred to the overlying stratum. Eventually the bulking properties of the collapsed rock limit the upward propagation of the overlying strata failure.

5.7.3.1 Processes of overburden failure transfer

As a longwall panel of sufficient width and length is excavated, the overlying strata are disturbed. The immediate roof is affected most severely, then the effect lessens toward the surface. Fig. 5.73 shows the process of overburden failure transfer above the goaf in response to high-intensity longwall mining below it. Before extracting a working face, the overburden is in the state of primary rock stress (Fig. 5.73A); after set-up entry, stratum 1 is in a state of suspended integrity (unsupported but intact) (Fig. 5.73B). Afterward, as the face advances, stratum 1 is in a state of suspended rupture (unsupported and fractured) (Fig. 5.73C). Finally, this stratum fails (Fig. 5.73D). The overhanging rock following the failure and collapse of stratum 1 is again stable (overhanging stability) (Fig. 5.73E). As the face advances further, the overhanging section of stratum 1 is in a state of overhanging rupture (Fig. 5.73F). Finally, the overhanging rock fails and collapses (Fig. 5.73G).

Therefore, when overburden failure is transferred to stratum n–1, in the following analysis the overburden has been divided into three zones in the direction of the advancing face: (i) unsupported failure zone, (ii) overhanging-broken failure zone, and (iii) primary rock stress zone. Thus there are two zone boundaries, one between each of these three zones (Fig. 5.73H).

In other words, the three zones of disturbance in the overlying strata processing upward due to high-intensity longwall mining are shown in Fig. 5.73I. In the caved zone, the strata fall to the mine floor and break into irregular, laminar shapes of various sizes in the process. The fractured zone above the caved zone contains strata that have broken into blocks whose sizes are governed by vertical and subvertical fractures and horizontal cracks due to bed separation. Above the fractured zone is the sagging zone, in which the strata and alluvium deform gently without causing any major fractures that cut through the thickness of the strata; each stratum behaves essentially as a continuous, intact medium. The height of the FWCZ in this study was taken to be equivalent to the combined height of the caved and fractured zones above the goaf. Therefore, when

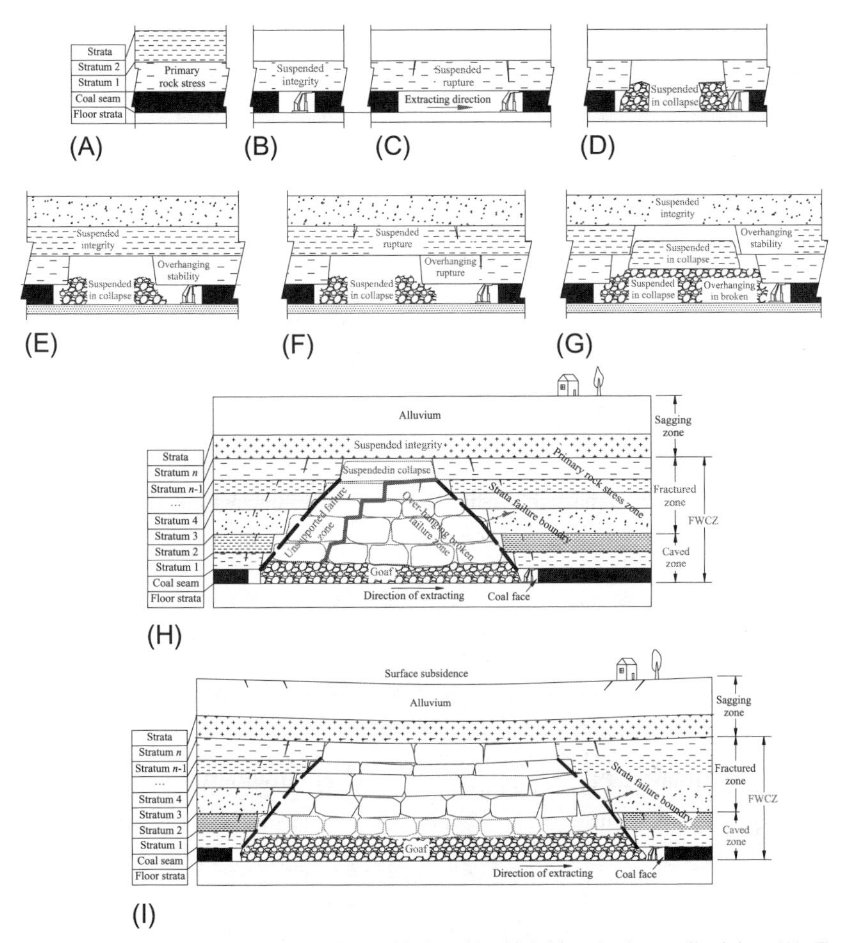

FIG. 5.73 Progressive OFT above the goaf induced by high-intensity longwall mining. (A)–(I) show the states of the overburden strata at different mining stages.

the failure of overburden transfers to stratum n–1, each of strata 1 to n–1 has failed. The FWCZ height (H_f) is given by

$$H_f = m + \sum_{i=1}^{n} h_i \tag{5.72}$$

where m is the mining thickness and h_i is the thickness of stratum i.

5.7.3.2 *Division of OFT into stages*

From the above, failed strata 1 to n–1 fill the goaf as the coal face advances. The separation distance between strata n–1 and n is given by

$$\Delta_{n-1,n} = m - \sum_{i=1}^{n-1} h_i(K_i - 1) \tag{5.73}$$

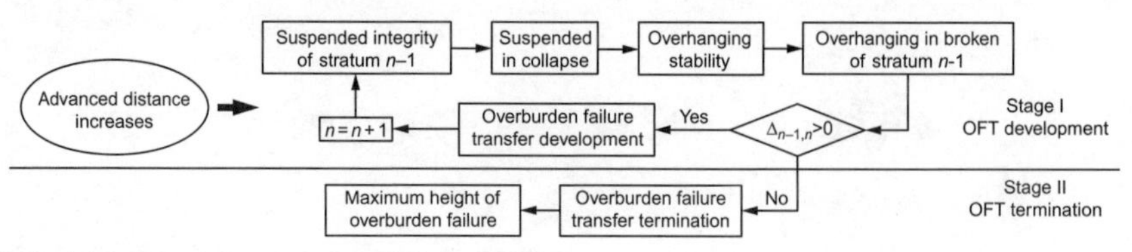

FIG. 5.74 Flow chart and stage division of OFT.

where $\sum_{i=1}^{n-1} h_i$ is the total thickness of strata 1 to n–1, and K_i is the bulking factor of the rock.

When $\Delta_{n-1,\ n} > 0$, OFT is in stage I, the OFT development stage. At this point, the failure of overburden transfers to stratum n–1. Overburden failure then continues to transfer to higher strata and $\Delta_{n-1,\ n}$ decreases as the face advance distance increases.

When $\Delta_{n-1,\ n} \leq 0$ (negative values of $\Delta_{n-1,\ n}$ are taken to be 0), failed stratum n–1 no longer transfers to stratum n, and thus the transfer process ceases at stratum n–1. Then OFT moves into stage II, the termination stage, wherein the overburden failure development reaches its maximum height. Therefore, once the mined-out range is large enough (reaching supercritical mining), OFT is in the termination stage. The two OFT stages are illustrated in Fig. 5.74.

Caving begins in the lowest stratum in the immediate roof and propagates upward into the fractured zone. The fallen strata bulk out and their overall volume increases. The gap between the top of the caved rock piles and the uncaved rock continues to decrease with further face advance (OFT development stage). When the gap vanishes, the overburden failure transfer stops (OFT termination stage).

5.7.3.3 Development characteristics of FWCZ

The stratum bends due to self-weight and overburden pressure. When the pressure exceeds the tensile strength of the rock, cracks and horizontal separations are generated. If major fractures extend through the entire thickness of the stratum, a water-conducting fracture is formed. As the face advances, the failure of stratum 1 is transferred to stratum 2 (i.e., OFT is in the development stage) and stratum 2 is then classified as the FWCZ. When $\Delta_{i,\ i+1} \leq 0$, OFT is in the termination stage, and upward development of water-conducting fractures ceases. The height of the FWCZ is then the sum of the heights of failed strata 1 to i above the goaf. Rock above stratum i is in the sagging zone.

Further advance of the working face transfers the stresses in a failed stratum to the stratum above it. The maximum unsupported (suspension) length and overhang length of each layer of rock depend on whether the stratum belongs to the FWCZ, taking the lithology, thickness, and strength of each stratum into account.

5.7.4 Development mechanism of FWCZ based on OFT

5.7.4.1 Maximum unsupported and overhang lengths

Fig. 5.73H–I shows the n strata in total overlying the coal seam. Initially, stratum 1 is unsupported but not fractured. As the face advances further, the suspended rock fails if stratum 1 reaches its maximum unsupported length ($D_{s1\ max}$), leaving stratum 2 unsupported until it also fails at $D_{s2\ max}$, and so on until the process reaches stratum n.

When a stratum fails, one end is fixed inside the rock mass, and the opposite (free) end overhangs like a cantilever. As the face continues to advance, the cantilevered rock fails at $D_{o2\ max}$, as illustrated in Fig. 5.75.

5.7.4.2 Failure criteria of stratum

The height at which unsupported or cantilevered rock fails is obtained by judging whether each stratum above the goaf has reached its maximum unsupported length ($D_{s\ max}$) or overhang length ($D_{o\ max}$). The failure criterion for unsupported strata is given by

$$D_s \geq D_s \max \tag{5.74}$$

and the overhang failure criterion is given by

$$D_o \geq D_o \max \tag{5.75}$$

5.7.4.3 Mechanical models of OFT

5.7.4.3.1 Model of unsupported strata

When stratum i ($i = 1, 2, 3, 4, \ldots, n-1$) is unsupported, the two opposite sections of stratum i are fixed inside the rock mass, which is the basis for the mechanical model. That assumes that the unsupported stratum is adequately represented by a uniformly loaded, elastic, isotropic, homogeneous beam of unit width fixed at both ends, as shown in Fig. 5.76.

For stratum i in Fig. 5.76, G_i is the weight of the unsupported beam representing stratum i; $D_{si\ max}$ is its maximum unsupported length; q_i is the

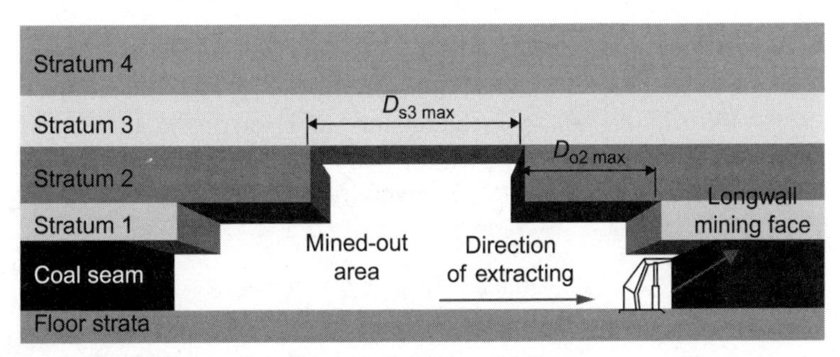

FIG. 5.75 The conditions of partial strata above the mined-out area.

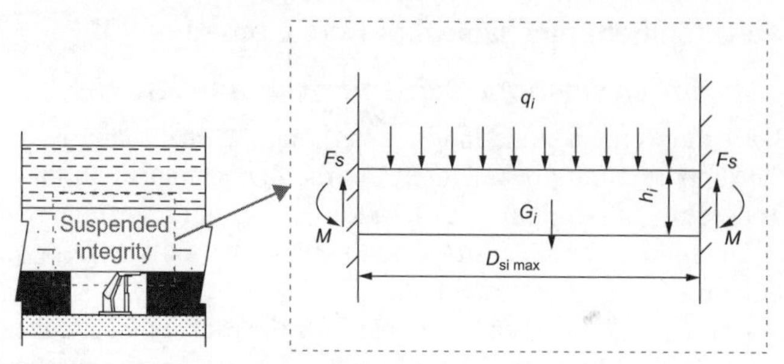

FIG. 5.76 Mechanical model of unsupported strata.

accumulated load on the unsupported part of stratum i; M is the bending moment; and F_s is the shear force. A beam fixed at both ends fails if $\sigma_{\max} \geq R_T$, where R_T is its tensile strength and $\sigma_{\max}$ is the maximum normal stress acting on it. Therefore, $D_{si\ \max}$ is given by

$$D_{si}\max = h_i\sqrt{\frac{2R_T}{q_i + k_i}} \tag{5.76}$$

in which

$$q_i = \frac{E_i h_i^3(\gamma_i h_i + \gamma_{i+1}h_{i+1} + \cdots + \gamma_n h_{n+1})}{E_i h_i^3 + E_{i+1}h_{i+1}^3 + \cdots + E_n h_n^3} \tag{5.77}$$

where E_i is the elastic modulus of stratum i; and k_i is the gravity load on stratum i, given by $k_i = \frac{G_i}{L_{ki}\max} = \gamma_i h_i$, in which γ_i is the bulk density of stratum i.

5.7.4.3.2 Model of overhanging strata

After the immediate roof caves and falls into the goaf, the broken rock no longer transmits horizontal force in the direction of mining. Therefore, in the caved zone, after stratum i has collapsed, one end is fixed in the rock mass and the other end overhangs the fallen rock. The model is simplified by assuming that the beam is a cantilever of unit width (Fig. 5.77), which has been widely used in the analytic study of overlying rock strata movement of underground coal mining.

In Fig. 5.77, G_i' is the weight of the overhanging stratum i; x is the unsupported length of stratum i + 1 after stratum i fails; $D_{oi\ \max}$ is the maximum overhang length for stratum i; and q_i' is the accumulated load on the stable overhang part of stratum i. Stratum i fails when the maximum normal stress exceeds the tensile strength of the rock. $D_{oi\ \max}$ is given by

$$D_{oi}\max = h_i\sqrt{\frac{R_T}{3(q_i' + k_i)}} \tag{5.78}$$

In the fractured zone, adjacent blocks from each broken stratum are still in full or partial contact, across either the vertical or subvertical fractures. This

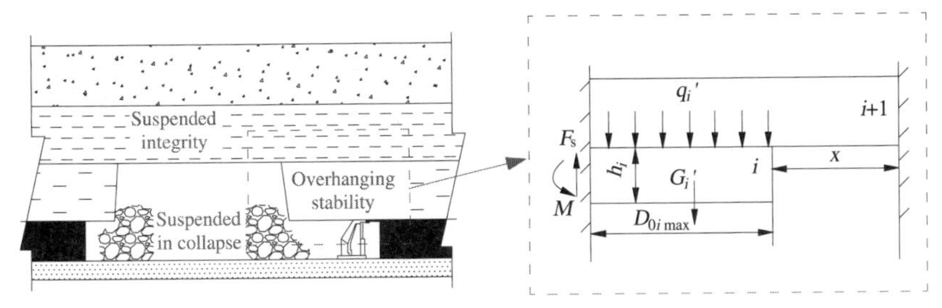

FIG. 5.77 Mechanical model of overhanging strata.

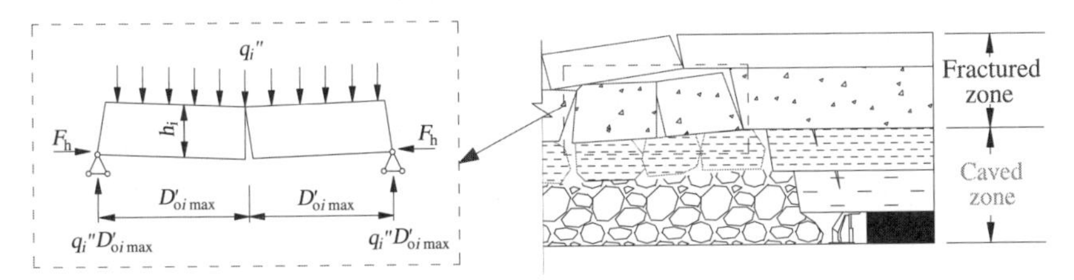

FIG. 5.78 Three-hinged arch model of adjacent blocks.

produces a horizontal force (F_h) transmitted through the blocks of broken rock, preventing their free movement. These were called force-transmitting beams by Peng (2006, 2008). A three-hinged arch model of adjacent blocks in the fractured zone is shown in Fig. 5.78.

In Fig. 5.78, q_i'' is the accumulated load on adjacent blocks; F_h is the horizontal force; and $D'_{oi\,max}$ is the length of a fallen block of rock. For equilibrium, $D'_{oi\,max}$ is given by

$$D'_{oi\mathrm{max}} = \sqrt{\frac{2F_h h_i}{q_i'' + k_i}} \tag{5.79}$$

from which it is clear that $D'_{oi\,max}$ is proportional to F_h.

The criteria for collapsed unsupported and overhanging strata are obtained from Eqs. (3)–(7).

From Fig. 5.73, and combining Eqs. (5.74) to (798), the height of the FWCZ above the goaf is affected by many factors: face advance distance, thickness of individual strata and coal seam, stability of unsupported rock, amount of loading imposed on overhanging strata, tensile strengths of individual strata, elastic modulus, bulking factor, etc.

5.7.5 Example analysis and numerical simulation

5.7.5.1 Example analysis

5.7.5.1.1 General situation

The No. 8100 longwall panel at the Tongxin Coal Mine in the Datong Coal Mine Group was taken as an example. The average mined thickness is 15.3 m, the dip of the coal seam is 2–3 degrees, and the length of the working face is 1406 m, the dip length is 193 m, and the vertical depth of the coal seam is 403–492 m below ground. The extraction speed of the panel is 5 m/d, and annual output reaches 10 million tons. The No. 8100 longwall panel has the characteristics of high-intensity mining, such as large mining thickness, large panel size, and high production and efficiency, and thus it belongs to the high-intensity mining panel. Overlying strata are mainly hard sandstone, with an immediate roof of mudstone 0.8–6.5 m thick; the main roof is medium-hard gritstone 2.2–8.3 m thick. Fig. 5.79 shows the stratigraphic columns for 200 m of overlying strata.

The bulking factor of the immediate roof is 1.18, which has been measured in the field. A study by Guo et al. (2002) proposed that the residual bulking factor (K) of the overlying strata decreases logarithmically as follows:

$$K = K_d - 0.017 \ln h \ (h < 100 \text{ m}) \tag{5.80}$$

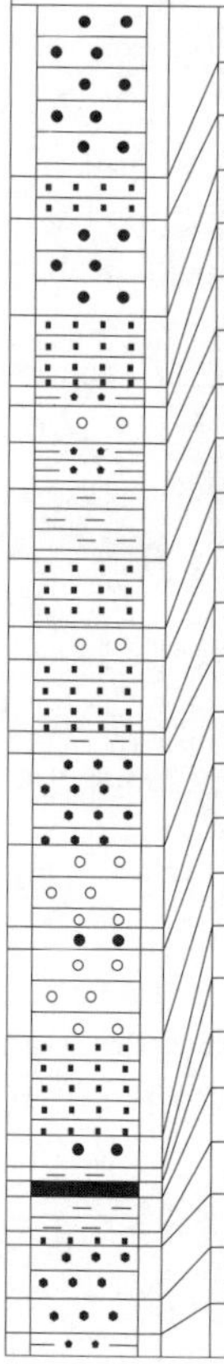

Column		Lithology	Thickness (m)	Distance from coal seam roof (m)
	25	Gritstone	25.4	200.1
	24	Fine sandstone	6.2	174.7
	23	Gritstone	14.3	168.5
	22	Fine sandstone	10.7	154.2
	21	Silty sandstone	2.9	143.5
	20	Conglornerate	5.1	140.6
	19	Silty sandstone	6.9	135.5
	18	Siltstone	10.5	128.6
	17	Fine sandstone	10.3	118.1
	16	Conglornerate	4.6	107.8
	15	Fine sandstone	10.7	103.2
	14	Siltstone	3.2	92.5
	13	Medium sandstone	13.7	89.3
	12	Conglornerate	12.0	75.6
	11	Gritstone	3.5	63.6
	10	Conglornerate	12.9	60.1
	9	Fine sandstone	14.8	47.2
	8	Gritstone	4.3	32.4
	7	Siltstone	2.4	28.1
	6	Shan 4 coal	2.1	25.7
	5	Siltstone	5.3	23.6
	4	Fine sandstone	2.1	18.3
	3	Medium sandstone	7.7	16.2
	2	K3 sandstone	5.3	8.5
	1	Silty sandstone	3.2	3.2

FIG. 5.79 Stratigraphic columns.

where K_d is the bulking factor of the immediate roof and h is the vertical distance of an overlying stratum to the coal seam roof.

When $h > 100$ m, the overlying strata are categorized as hard rock and thus K is small, generally between 1.10 and 1.20 (Peng, 2006). In Fig. 5.79, the 14th layer (siltstone) is 92.5 m above the coal seam roof. From Eq. (5.80), for this stratum $K = 1.103$. Therefore, the value of K for the 15th layer and higher should not exceed 1.103. They were all assigned a value of 1.100.

5.7.5.1.2 Calculation of maximum unsupported and overhang lengths

The maximum unsupported length and overhang length of each stratum were calculated by combining Eqs. (5.76)–(5.78) and Fig. 5.7. $D_{s\ max}$ and $D_{o\ max}$ are shown in Table 5.9.

The overburden failure processes following the 8100 longwall extraction were as follows:

- The unsupported length of stratum 1 (silty sandstone) collapsed when the face had advanced about 33 m. The separation between strata 1 and 2 was 14.72 m ($\Delta_{1,\ 2} = 14.72$ m).
- Unsupported stratum 2 then collapsed when the face had advanced about 65 m; $\Delta_{2,\ 3} = 14.08$ m
- By analogy, by the time the face had advanced 262 m, overburden failure would be transferred to stratum 10. That is, strata 1–9 would all have failed, and $\Delta_{9,\ 10} = 9.9$ m. In fact, the FWCZ lies above the mined-out area, as shown in Fig. 5.80.

5.7.5.1.3 Height calculation of FWCZ

To clearly illustrate the relationship between the separation distance and the face advance distance, a plot of separation distance vs longwall face advance distance based on Eq. (5.73) and the above calculations is shown in Fig. 5.81. Fig. 5.81 shows that during development of the longwall face, the separation between the roof and floor in the development drifts was at its maximum value of 15.3 m. As the face advanced, the separation between the strata decreased linearly. The above calculations indicate that when the face had advanced 294 m, $\Delta_{10,\ 11} = 8.67$ m. When the face had advanced 638 m, the separation distance between stratum 22 and stratum 23 was reduced to $\Delta_{22,\ 23} = -0.12$ m (revised to 0 to reflect the actual situation). That is, the failure of stratum 21 was not transferred to stratum 22.

To visualize the process of overburden failure as the face advanced, the overburden failure height versus the longwall face advance distance is plotted in Fig. 5.82.

Fig. 5.82 indicates that when the advance distance was 294 m, OFT was in stage I, the development stage, and the overlying strata failure height was 75.4 m. When the face has advanced 638 m (and revised $\Delta_{22,\ 23} = 0$ as

TABLE 5.9 Physicomechanical parameters of rocks.

	Lithology	Thickness (m)	Bulk density (kNm^{-3})	Volume expansion coefficient	Tensile strength (Mpa)	Elastic modulus (GPa)	$D_{s\ max}$ (m)	D_o max (m)
25	Gritstone	25.4	25.37	1.100	5.42	20.12	104.18	42.53
24	Fine sandstone	6.2	27.54	1.100	8.64	35.87	58.92	24.05
23	Gritstone	14.3	25.24	1.100	5.34	21.31	51.34	20.96
22	Fine sandstone	10.7	26.82	1.100	8.11	36.12	55.56	22.68
21	Silty sandstone	2.9	26.51	1.100	4.14	18.56	29.40	12.00
20	Conglomerate	5.1	27.15	1.100	3.92	28.42	24.83	10.14
19	Silty sandstone	6.9	25.98	1.100	5.81	18.46	36.65	14.96
18	Siltstone	10.5	25.20	1.100	4.52	23.17	36.67	14.97
17	Fine sandstone	10.3	26.51	1.100	7.87	36.01	47.10	19.23
16	Conglomerate	4.6	26.95	1.100	4.23	28.64	33.33	13.61
15	Fine sandstone	10.7	27.17	1.100	7.93	35.21	51.67	21.09
14	Siltstone	3.2	24.58	1.103	4.45	23.48	32.41	13.23

13	Medium sandstone	13.7	25.52	1.104	7.01	29.62	50.59	20.65
12	Conglomerate	12.0	27.10	1.106	4.34	28.74	40.98	16.73
11	Gritstone	3.5	23.89	1.109	5.24	19.98	37.65	15.37
10	Conglomerate	12.9	27.35	1.110	4.34	28.43	42.94	17.53
9	Fine sandstone	14.8	25.62	1.114	8.20	35.62	61.19	24.98
8	Gritstone	4.3	24.21	1.121	4.82	20.32	39.73	16.22
7	Siltstone	2.4	25.78	1.123	4.25	23.35	23.39	9.55
6	Shan 4 coal	2.1	10.36	1.125	1.27	4.20	19.09	7.79
5	Siltstone	5.3	26.45	1.126	4.97	23.64	28.24	11.53
4	Fine sandstone	2.1	27.12	1.131	7.81	35.54	29.55	12.06
3	Medium sandstone	7.7	26.73	1.133	6.14	29.57	34.31	14.01
2	K3 sandstone	5.3	25.44	1.144	7.68	36.21	38.37	15.66
1	Silty sandstone	3.2	26.31	1.180	5.47	18.35	32.48	13.26

Notes: The maximum unsupported length and overhang length are expressed by $D_{s\ max}$ and $D_{o\ max}$, respectively.

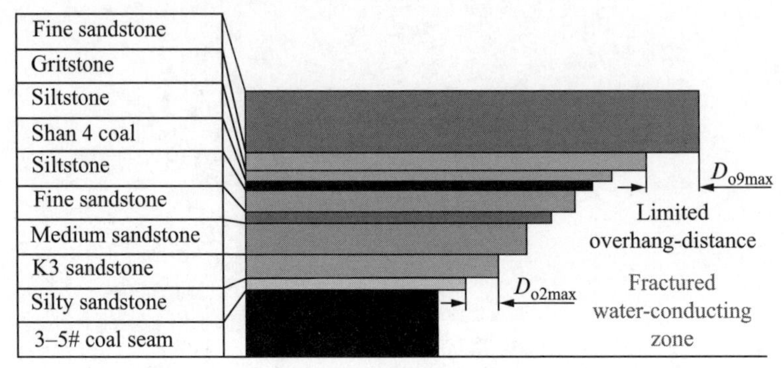

FIG. 5.80 Maximum overhang length of overlying strata above the 8100 working face.

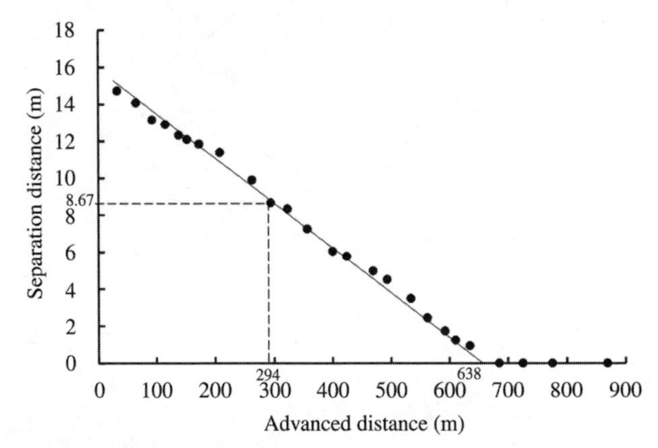

FIG. 5.81 Separation distance versus longwall face advance.

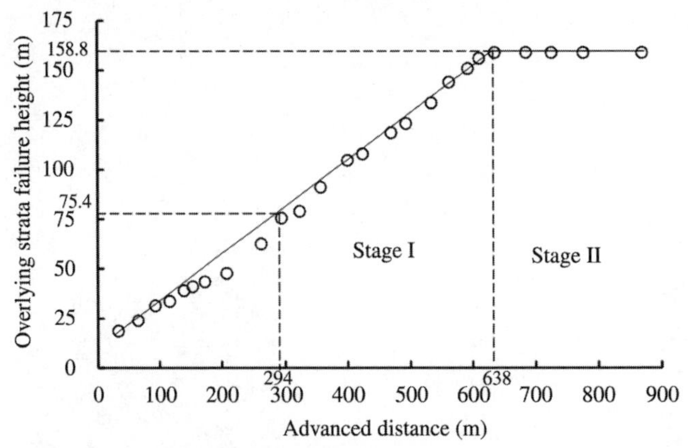

FIG. 5.82 Overburden failure height vs longwall face advance.

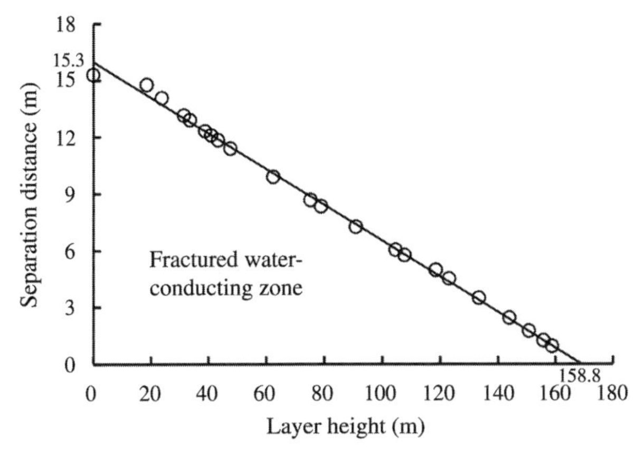

FIG. 5.83 Relationship between separation distance and layer height.

discussed above), the OFT moved into stage II, the termination stage, and the longwall face reached supercritical mining. Therefore, the maximum overlying strata failure height is about 158.8 m.

In addition, to explain the relationship between the separation distance and layer height more intuitively, separation vs layer height is plotted in Fig. 5.83.

Therefore, water-carrying fractures do not propagate above stratum 22. In Fig. 5.83, the lower left triangular area is the FWCZ. The intersection of the curve and the horizontal axis in Fig. 5.83 marks the maximum height of the FWCZ (158.8 m).

5.7.5.2 Numerical simulation of FWCZ height

According to the above analysis, the height of FWCZ at the No.8100 HIM panel has been predicted by elastic bending theory. In order to verify the rationality of the analysis result (158.8 m), the numerical simulation is used to predict the height of the FWCZ at the same HIM panel and compare with the theoretical analysis result mutually.

5.7.5.2.1 Numerical simulation model

FLAC3D™ software (Itasca, Minneapolis, MN, USA) was used to generate a 3D numerical model simulating the stress field and rock failure conditions at the working face from actual mining data for the 8100 longwall, Tongxin Coal Mine (Fig. 5.84). Model dimensions were 1200 m × 593 m × 280 m high. The x- and y-axes represented the strike and dip directions, respectively; the z-axis was in the vertical direction. Cover depth was 460 m, and extraction thickness was 15 m. The thickness of the floor was 65 m.

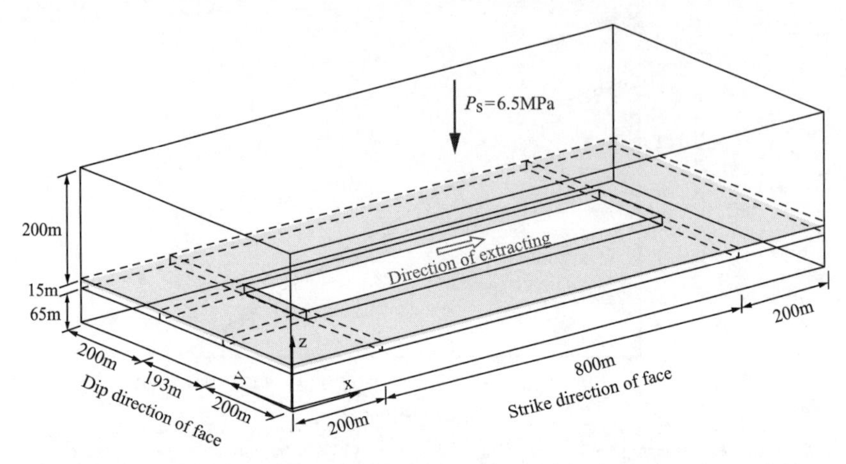

FIG. 5.84 Numerical simulation model for high-intensity mining of superlarge longwall face.

The mining scheme modeled was as follows: Longwall face 193 m in the *y*-direction; total face advance 800 m in the *x*-direction. The mining excavation space was located at the center of the model. Coal pillars 200 m wide were intentionally left to minimize boundary effects in both the *x*- and *y*-directions. Excavation was in 50 m steps for a total of 16 steps.

Because this numerical simulation model was only 280 m high while the depth of the cover was 460 m, an initial vertical principal stress (P_s) was applied to the top of model to simulate the weight of 260 m of overlying rock. P_s was calculated from

$$P_s = \gamma H \tag{5.81}$$

where γ is the bulk density of rock (average 25 kN/m^3 in China), and H is overburden thickness (m).

Eq. (5.81) gives $P_s \approx 6.5$ MPa for these values. Because an elastic, isotropic model was assumed, the horizontal stress components were calculated by introducing Poisson's ratio. The base and lateral boundaries of the model were on rollers to simulate the initial stress field. Fixed base and lateral boundary conditions were used in subsequent excavation simulations. The Mohr-Coulomb yielding criterion was adopted in the excavation simulation (Meng et al., 2016). The rocks were assumed to be elastoplastic, as indicated by laboratory tests. The mechanical parameters used in the simulation are shown in Table 5.10. Please note that the numerical simulation was conducted in the static mode in FLAC3D.

5.7.5.2.2 Simulation results analysis

Here, the numerical modeling focused on the height of the fractured water-conducting zone. Therefore, the permeability distribution by conducting the

TABLE 5.10 Physicomechanical parameters of coal and rocks in numerical simulation.

Lithology	Bulk modulus (GPa)	Shear modulus (GPa)	Cohesion (MPa)	Internal friction angle (degree)
Conglomerate	5.16	3.42	2.1	36
Gritstone	6.28	5.46	5.50	36
Shan 4 coal	3.71	1.91	2.20	21
Siltstone	4.33	3.20	1.77	35
Medium sandstone	4.15	2.72	2.10	35
K3 sandstone	5.28	4.46	2.50	36
3–5# coal seam	3.71	1.91	1.20	21
Silty sandstone	5.33	3.20	1.60	32
Fine sandstone	3.28	2.46	2.50	36

seepage simulation through the surrounding rock mass was not considered. FLAC3D indicated that a plastic zone was induced by the high-intensity mining. The rock failed when in a state of shear or tension. The height of the FWCZ, in the present study, was taken as equivalent to the height of the destressed or plastic zone. Fig. 5.85 is a 3D perspective view and transverse sectional view of the plastic zones and stress distributions when the longwall face had advanced 50, 150, and 350 m.

Longwall mining would generate a goaf and then result in a plastic zone (shear or tension) and a destressed zone inside the surrounding rocks (Fig. 5.85). The results of the numerical simulation showed that when the working face advanced 50 m, the plastic zone and destressed zone distribution in the surrounding rock (Fig. 5.85A) was $H_f = 43$ m. The size of the plastic zone and destressed zone inside the rock mass increased upward with the advance of the working face (Fig. 5.85A–C). For example, $H_f = 135$ m when the working face advanced 350 m. With continued extraction of panel 8100, a vertical cross-section at $Y = 296.5$ m was chosen for the analysis of results (Fig. 5.86).

Fig. 5.86 shows that the plastic zone no longer increased after the face had advanced 650 m ($H_f = 165$ m, Fig. 5.86A) and 800 m ($H_f = 165$ m, Fig. 5.86B). That is, the panel reached the supercritical mining stage at the

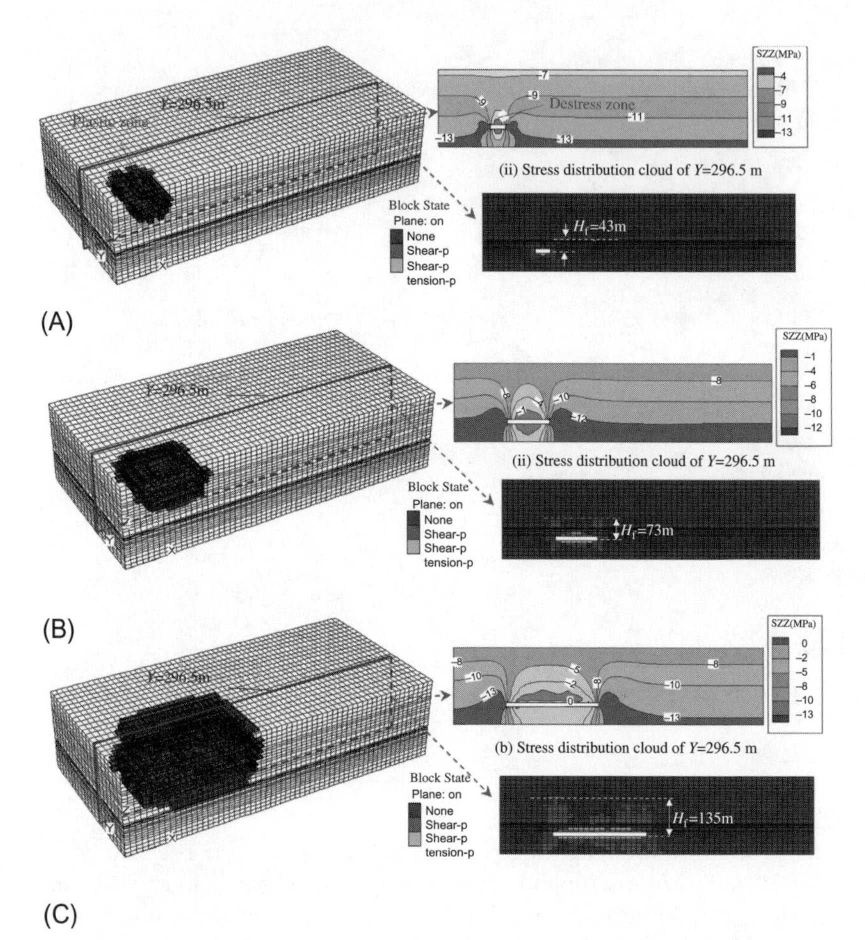

FIG. 5.85 Simulated 3D perspective view and transverse sectional view of plastic zone and stress distribution of panel 8100 after face has advanced (A) Extracted 50 m, (B) extracted 150 m, and (C) extracted 350 m.

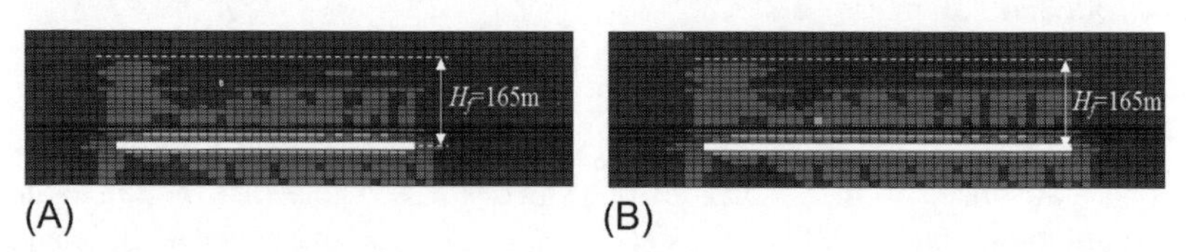

FIG. 5.86 Traverse section view of plastic zone when the panel #8100 had advanced 650 and 800 m. (A) Extracted 650 m and (B) extracted 800 m.

650 m face advance. In summary, mining was predicted to cause stress redistribution inside the rock mass, and the roof strata above the mined-out zone is destressed. The stress distributions at the 650 m and 800 m face advance are shown as Fig. 5.87.

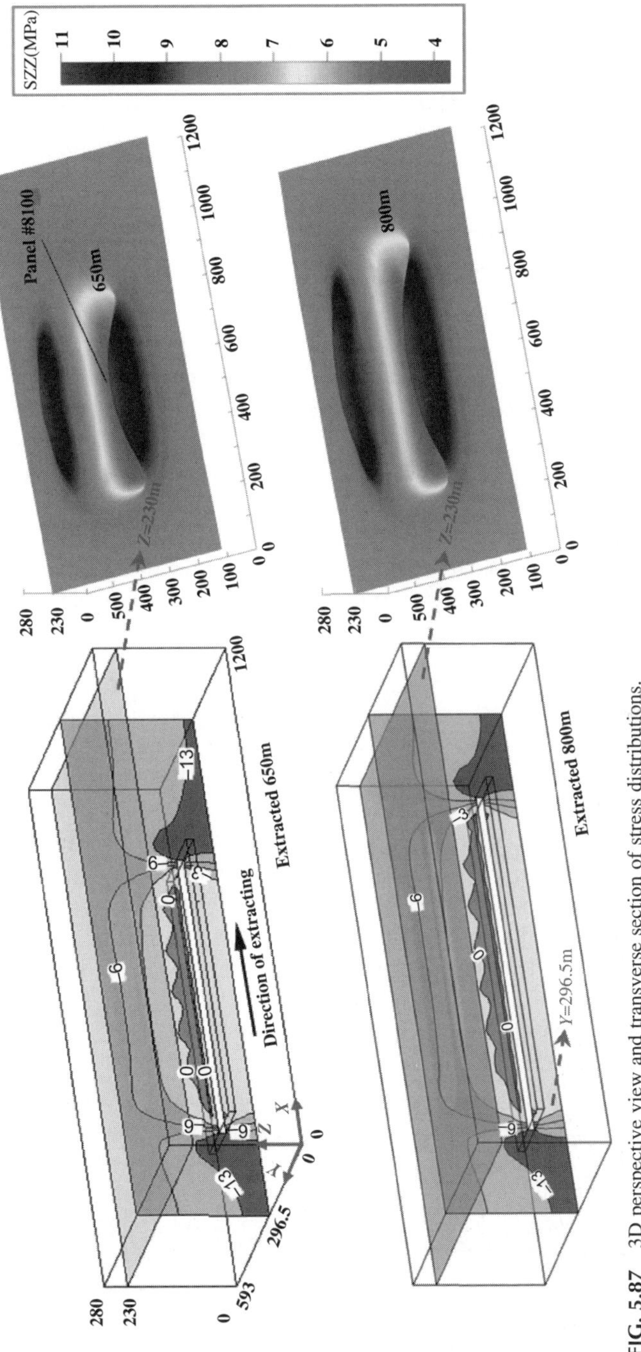

FIG. 5.87 3D perspective view and transverse section of stress distributions.

Fig. 5.87 shows the simulated horizontal cross-sections at $z = 230$ m as mining took place. The stress distributions were almost the same in the vertical direction, which indicated that the destressed zone was completely developed, the panel had reached the supercritical mining stage, and the FWCZ was fully developed and stabilized. The height of the destressed zone was about 165 m.

5.7.6 Engineering analogy

5.7.6.1 Predicted FWCZ height

A total of 138 observations of the height of the FWCZ in different mining areas with different geological conditions and mining methods are reported in the literature (Hu et al., 2017). In situ measurements carried out in geological and mining conditions similar to those in the present study (20 in all) are set out in Table 5.11.

Table 5.11 lists the height of the FWCZ and mining thickness, lithology, mining methods, dip angle of the coal seam, and mining depth. The main differences between the panel 8100 in this study and those in Table 5.11 are mining thickness, seam dip, and mining depth. The general range of FWCZ height is from 63.6 to 161 m. The predicted value of 158.8 m lies within this range.

5.7.6.2 Overall comparison and analysis

The proposed theoretical method and the numerical simulation were compared with measured data to arrive at a prediction of the height of the FWCZ. Comparisons with in situ measurements reported in the literature (Zhang et al., 2014) are listed in Table 5.12.

Table 5.12 shows that the proposed theoretical prediction was within the range of measured values, with a relative error of 5.8%–6.5%. The relative error of the numerical simulation result was 1.1%–12%; the relative error obtained by comparison with measured data was 5.2%–29.4%. Thus, the predicted values obtained by the theoretical method and by numerical simulation coincided with measured values within reasonable error ranges. The large error range obtained when comparing the measured values with those in Table 5.3 is unlikely to satisfy the needs of onsite mining engineering.

In these analyses, comparisons with measured data from other mines were not found to be strongly applicable, although they may be useful as a reference when predicting the height of the FWCZ. However, the result of numerical simulation (165 m) was in close agreement with the theoretical calculation (158.8 m), and both were consistent with the measured values (150–170 m). This verifies the rationality of the all-inclusive OFT method of predicting the height of the FWCZ (i.e., considering the longwall face advance distance, the maximum length of unsupported rock over the goaf, the maximum length of overhanging rock, and the amount of separation between strata).

TABLE 5.11 Measured height of FWCZ: sample data.

Coal mine	Panel	Mining thickness (m)	Lithological character	Mining method[a]	Dip angle of coal seam (degree)	Mining depth (m)	Height of FWCZ (m)
Baodian	1303	8.7	Mid–hard	FMC	4–15	352–517	71
Baodian	5306 (2)	6.9	Mid–hard	FMC	5	335–398	69.7
Dongtan	1305	8.7	Mid–hard	FMC	6	560–640	78.8
Nantun	9310	5.3	Mid–hard	FMC	12–19	476–607	67.5
Jianzhuang	$4^{-2}101$	6.7	Mid–hard	FMC	2–3	370–608	81.6
Jianzhuang	$4^{-2}102$	8.2	Mid–hard	FMC	2–3	466–590	81.6
Jining3#	1031	6.8	Mid–hard	FMC	/	/	80.2
Tangyang	3109	4.4	Mid–hard	FMC	13	460	65
Xinglong	5306	7.1	Mid–hard	FMC	6–13	/	74.4
Xinglong	1301	6.4	Mid–hard	FMC	6–13	/	72.9
Xinglong	4320	8	Mid–hard	FMC	8	/	86.8
Xieqiao	1121 (3)	6	Mid–hard	FMC	13	534.8	67.9
Xieqiao	7507	4.9	Mid–hard	FMC	5	355–390	63.6
Xieqiao	1211 (3)	5	Mid–hard	FMC	13	490.1	73.3

Continued

TABLE 5.11 Measured height of FWCZ: sample data.—cont'd

Coal mine	Panel	Mining thickness (m)	Lithological character	Mining method	Dip angle of coal seam (degree)	Mining depth (m)	Height of FWCZ (m)
Xinjulong	1302 N	8.6	Mid–hard	FMC	7	850	94.7
Xingfu	3417	7.4	Mid–hard	FMC	15	690–840	70
Xiagou	2801	9.9	Mid–hard	FMC	2	316–347	125.8
Tangshang	T2291	10	Mid–hard	FMC	12	636	161
Tongxin	8100	15.3	Mid–hard	FMC	2–3	403–492	160
Zhuxianzhuang	865	13.4	Mid–hard	FMC	15	480	130.8

[a]FMC, *fully mechanized top caving.*

TABLE 5.12 Height of FWCZ determined by different methods.

	Theoretical method proposed		Numerical simulation		Engineering analogy	
Measured value (m)	*Value (m)*	*Relative error (%)*	*Value (m)*	*Relative error (%)*	*Value (m)*	*Relative error (%)*
150–170	158.8	5.8–6.5	165	1.1–12	63.6–161	5.2–29.4

5.7.7 Conclusions

A new method is proposed for predicting the height of the FWCZ due to high-intensity coal mining, following a study of the No. 8100 high-intensity longwall panel at the Tongxin Coal Mine in the Datong Coal Mine Group, China. The following conclusions are drawn:

The processes of OFT were divided into two stages: the transmission development stage and the transmission termination stage. The state of destruction is transferred from a given stratum to the stratum immediately above it as the working face advances.

The proposed theory determines the maximum length of unsupported rock above the goaf and the maximum overhang length following its failure, for each stratum, to judge the extent of damage experienced by each. These provide the OFT criteria for FWCZ development. On this basis, mechanical models of the unsupported strata and the overhanging strata were established. A new theoretical method of predicting the height of the FWCZ in this form of coal mining is put forward, based on OFT processes.

A numerical simulation model and comparison with published measured FWCZ data were used to predict the height of the FWCZ at the mine. The numerical simulation results and the theoretical calculations both agreed closely with in situ measurements of the height of the FWCZ at the 8100 longwall panel at the Tongxin Coal Mine, Datong Coal Mining Group, verifying the rationality of the proposed methods.

References

Abdollahipour, A., Rahmannejad, R., 2013. Investigating the effects of lateral stress to vertical stress ratios and caverns shape on the cavern stability and sidewall displacements. Arab. J. Geosci. 6 (12), 4811–4819.

Adhikary, D.P., Guo, H., 2015. Modelling of longwall mining-induced strata permeability change. Rock Mech. Rock Eng. 48, 345–359.

Alejano, L.R., Taboada, J., García-Bastante, F., Rodriguez, P., 2008. Multi-approach back-analysis of a roof bed collapse in a mining room excavated in stratified rock. Int. J. Rock Mech. Mining Sci. 45 (6), 899–913.

Bai, Q.S., Tu, S.H., Zhang, C., Zhu, D., 2016. Discrete element modeling of progressive failure in a wide coal roadway from water-rich roofs. Int. J. Coal Geol. 167, 215–229.

Bakun-Mazor, D., Hatzor, Y.H., Glaser, S.D., 2011. Dynamic sliding of tetrahedral wedge: the role of interface friction. Int. J. Numer. Anal. Methods Geomech. 36 (3), 327–343.

Booth, C.J., 2007. Confined–unconfined changes above longwall coal mining due to increases in fracture porosity. Environ. Eng. Geosci. 13, 355–367.

Chen, C.N., Huang, W.Y., Tseng, C.T., 2011. Stress redistribution and ground arch development during tunnelling. Tunnel. Undergr. Space Technol. 26, 228–235.

Cheng, G.W., Chen, C.X., Ma, T.H., Liu, H.Y., Tang, C.A., 2017a. A case study on the strata movement mechanism and surface deformation regulation in chengchao underground iron mine. Rock Mech. Rock Eng. 50, 1011–1032.

Cheng, G.W., Ma, T.H., Tang, C.N., 2017b. A zoning model for coal mining-induced strata movement based on microseismic monitoring. Int. J. Rock Mech. Mining Sci. 94, 123–138.

Cui, X.M., Gao, Y.G., Yuan, D.B., 2014. Sudden surface collapse disasters caused by shallow partial mining in Datong coalfield, China. Nat. Hazards 74 (2), 911–929.

Diederichs, M.S., Kaiser, P.K., 1999. Stability of large excavations in laminated hard rock masses: the hinged analogue revisited. Int. J. Rock Mech. Mining Sci. 36 (1), 97–117.

Du, X.L., Wei, J.S., Qian, G., Song, H.W., 2015. Influence of geometric features of coal mining area on the rock pressure arch. Chin. J. Undergr. Space Eng. 11 (S2), 726–731.

Fan, L.M., 2014. On coal mining intensity and geo-hazard in Yulin-Shenmu-Fugu mine area. China Coal 40, 52–55.

Fraldi, M., Guarracino, F., 2010. Analytical solutions for collapse mechanisms in tunnels with arbitrary cross sections. Int. J. Solids Struct. 47 (2), 216–223.

Gao, M.Z., 2004. Study on bending catastrophe phenomena of anchor roof rock beam. Rock Soil Mech. 25 (8), 1267–1270.

Ghabraie, B., Ren, G., Smith, J., 2015. Application of 3D laser scanner, optical transducers and digital image processing techniques in physical modelling of mining-related strata movement. Int. J. Rock Mech. Min. Sci. 80, 219–230.

Guo, W.B., Wang, Y.G., 2017. The definition of high–intensity mining based on green coal mining and its index system. J. Mining Saf. Eng. 34 (4), 616–623.

Guo, G.L., Miao, X.X., Zhang, Z.N., 2002. Research on rupture rock mass deformation characteristics of longwall goals. Sci. Tech. Eng. 2, 44–47.

Guo, H., Yuan, L., Shen, B., 2012a. Mining-induced strata stress changes, fractures and methane flow dynamics in multi-seam LW mining. Int. J. Min. Sci. Technol. 54, 129–139.

Guo, W.B., Zou, Y.F., Hou, Q.L., 2012b. Fractured zone height of longwall mining and its effects on the overburden aquifers. Int. J. Min. Sci. Technol. 22, 603–606.

He, L., Zhang, Q.B., 2015. Numerical investigation of arching mechanism to underground excavation in jointed rock mass. Tunn. Undergr. Space Technol. 50, 54–67.

Helm, P.R., Davie, C.T., Glendinning, S., 2013. Numerical modelling of shallow abandoned mine working subsidence affecting transport infrastructure. Eng. Geol. 154, 6–19.

Hou, G.Y., Tao, L.G., Li, X.W., Zhou, Y.Q., Han, Y.H., 2000. Study on the bifurcation behavior of truss support system under stratified and fractured roof. Chin. J. Rock Mech. Eng. 19 (1), 77–81.

Howladar, M.F., Hasan, K., 2014. A study on the development of subsidence due to the extraction of 1203 slice with its associated factors around Barapukuria underground coal mining industrial area, Dinajpur, Bangladesh. Environ. Earth Sci. 72, 3699–3713.

Hu, H.F., Lian, X.G., Li, Y., 2016. Physical experiments on the deformation of strata with different properties induced by underground mining. J. Eng. Sci. Technol. Rev. 9 (1), 74–80.

Hu, B.N., Zhang, H.X., Shen, B.H., 2017. Guidelines for the Retention and Coal Mining of Coal Pillars in Buildings, Water Bodies, Railways and Main Wells and Alley. Coal Industry Press, Beijing, pp. 172–180.

Huang, Z.P., Broch, E., Lu, M., 2002. Cavern roof stability-mechanism of arching and stabilization by rockbolting. Tunn. Undergr. Space Technol. 17 (3), 249–261.

Huang, Q.X., Dong, B., Chen, S.S., 2016. Determination of roof pressure law and support resistance in the mining face with super-large mining height in approximate shallow coal seam. J. Mining Safety Eng. 33 (5), 840–844.

Jia, J.L., Cao, L.W., Zhang, D.J., 2017. Study on the fracture characteristics of thick-hard limestone roof and its controlling technique. Environ. Earth Sci. 76 (17), 1–10.

Jiang, H.J., Cao, S.G., Zhang, Y., Wang, C., 2016a. Analytical solutions of hard roof's bending moment, deflection and energy under the front abutment pressure before periodic weighting. Int. J. Rock Mech. Mining Sci. 26 (1), 175–181.

Jiang, B.Y., Wang, L.G., Lu, Y.L., Sun, X.K., Jin, G., 2016b. Ground pressure and overlying strata structure for a repeated mining face of residual coal after room and pillar mining. Int. J. Min. Sci. Technol. 26 (4), 645–652.

Ju, J.F., Xu, J.L., 2013. Structural characteristics of key strata and strata behavior of a fully mechanized longwall face with 7.0 m height chocks. Int. J. Rock Mech. Mining Sci. 58 (1), 46–54.

Ju, J.F., Xu, J.L., 2015. Surface stepped subsidence related to top-coal caving longwall mining of extremely thick coal seam under shallow cover. Int. J. Rock Mech. Min. Sci. 78, 27–35.

Ju, J.F., Xu, J.L., Zhu, W.B., 2015. Longwall chock sudden closure incident below coal pillar of adjacent upper mined coal seam under shallow cover in the Shendong coal field. Int. J. Rock Mech. Mining Sci. 77, 192–201.

Kim, K.Y., Lee, D.S., Cho, J., Jeong, S.S., Li, S., 2013. The effect of arching pressure on a vertical circular shaft. Tunnel. Undergr. Space Technol. 37 (6), 10–21.

Kong, X.X., Liu, Q.S., Zhang, Q.B., Wu, Y.X., Zhao, J., 2018. A method to estimate the pressure arch formation above underground excavation in rock mass. Tunnel. Undergr. Space Technol. 71, 382–390.

Lal, A., Mulani, S.B., Kapania, R.K., 2017. Stochastic fracture response and crack growth analysis of laminated composite edge crack beams using extended finite element method. Int. J. Appl. Mech 9 (4), Article ID 1750061: 1–16.

Li, C.C., 2006. Rock support design based on the concept of pressure arch. Int. J. Rock Mech. Min. Sci. 43 (7), 1083–1090.

Li, J.W., Liu, C.Y., 2017. Formation mechanism and reduction technology of mining-induced fissures in shallow thick coal seam mining. Shock Vibr. 6, 1–14.

Li, G.Y., Zhou, W.F., 2006. Impact of karst water on coal mining in North China. Environ. Geol. 49, 449–457.

Li, S.L., Feng, T., Zhu, Z.H., 2014. Mechanical model of rock beam with both ends fixed and force analysis of roof fracture in cemented filling mining. J. Cent. S. Univ. Technol. 45 (8), 2793–2798.

Li, N., Wang, E.Y., Ge, M.C., Liu, J., 2015. The fracture mechanism and acoustic emission analysis of hard roof: a physical modeling study. Arab. J. Geosci. 8, 1895–1902.

Li, C.L., Wang, S.R., Zou, Z.S., Liu, X.L., Li, D.Q., 2016b. Evolution characteristic analysis of pressure arch of a double-arch tunnel inwater-rich strata. J. Eng. Sci. Technol. Rev. 9 (1), 44–51.

Li, C.B., Caner, F.C., Chau, V.T., Bažant, Z.P., 2017a. Spherocylindricalmicroplane constitutive model for shale and other anisotropic rocks. J. Mech. Phys. Solids 103, 155–178.

Liang, S., Elsworth, D., Li, X.H., 2017. Key strata characteristics controlling the integrity of deep wells in longwall mining areas. Int. J. Coal Geol. 172, 31–42.

Liu, X., Song, G., Li, X., 2015a. Classification of roof strata and calculation of powered support loads in shallow coal seams of China. J. Southern African Inst. Mining Metall. 115 (11), 1113–1119.

Liu, X.S., Tan, Y.L., Ning, J.G., 2015b. The height of water-conducting fractured zones in longwall mining of shallow coal seams. Geotech. Geol. Eng. 33 (3), 693–700.

Ma, L.Q., Zhang, D.S., Li, X., Fan, G.W., Zhao, Y.F., 2009. Technology of groundwater reservoir construction in goafs of shallow coalfields. Int. J. Mining Sci. Technol. 19 (6), 730–735.

Majdi, A., Hassani, F.P., Nasiri, M.Y., 2012. Prediction of the height of destressed zone above the mined panel roof in longwall coal mining. Int. J. Coal Geol. 98, 62–72.

Marcak, H., 2012. Seismicity in mines due to roof layer bending. Arch. Min. Sci. 57 (1), 229–250.

Mazor, D.B., Hatzor, Y.H., Dershowitz, W.S., 2009. Modeling mechanical layering effects on stability of underground openings in jointed sedimentary rocks. Int. J. Rock Mech. Min. Sci. 46 (2), 262–271.

Meier, C., Wall, W.A., Popp, A., 2017. A unified approach for beam-to-beam contact. Comput. Methods Appl. Mech. Eng. 315 (1), 972–1010.

Meng, Z.P., Shi, X.C., Li, G.Q., 2016. Deformation, failure and permeability of coal–bearing strata during longwall mining. Eng. Geol. 208, 69–80.

Nomikos, P.P., Sofianos, A.I., Tsoutrelis, C.E., 2002. Structural response of vertically multi-jointed roof rock beams. Int. J. Rock Mech. Mining Sci. 39 (1), 79–94.

Pan, Y., Wang, Z.Q., Li, A.W., 2012. Analytic solutions of deflection, bending moment and energy change of tight roof of advanced working surface during initial fracturing. Chin. J. Rock Mech. Eng. 31 (1), 32–41.

Peng, S.S., 2006. Longwall Mining, second ed., pp. 46–60. Morgantown.

Peng, S.S., 2008. Coal Mine Ground Control, third ed., pp. 319–328. Morgantown.

Peng, S.S., 2015. Topical areas of research needs in ground control: a state of the art review on coal mine ground control. Int. J. Min. Sci. Technol. 25, 1–6.

Peng, S.S., 2017. Advances in coal mine ground control. Woodhead Publishing, Cambridge, pp. 68–95.

Please, C.P., Mason, D.P., Khalique, C.M., Ngnotchouye, J.M.T., Hutchinson, A.J., van der Merwe, J.N., Yilmaz, H., 2013. Fracturing of an Euler-Bernoulli beam in coal mine pillar extraction. Int. J. Rock Mech. Mining Sci. 64 (12), 132–138.

Poulsen, B.A., 2010. Coal pillar load calculation by pressure arch theory and near field extraction ratio. Int. J. Rock Mech. Mining Sci. 47 (7), 1158–1165.

Qiu, J.W., Liu, Z.G., Zhou, L., Qin, R.X., 2017. Prediction model of gas quantity emitted from coal face based on PCA-GA-BP neural network and its application. J. Power Technol. 97 (3), 169–178.

Ren, Y.F., Qi, Q.X., 2011. Study on characteristic of stress field in surrounding rocks of shallow coalface under long wall mining. J. China Coal Soc. 36 (10), 1612–1618.

Ren, Y.F., Ning, Y., Qi, Q.X., 2013. Physical analogous simulation on the characteristcs of overburden breakage at shallow longwall coalface. J. China Coal Soc. 38 (1), 61–66.

Rezaei, M., Hossaini, M.F., Majdi, A., 2015. Determination of longwall mining-induced stress using the strain energy method. Rock Mech. Rock Eng. 48 (6), 2421–2433.

Salmi, E.F., Nazem, M., Karakus, M., 2017. The effect of rock mass gradual deterioration on the mechanism of post-mining subsidence over shallow abandoned coal mines. Int. J. Rock Mech. Min. Sci. 91, 59–71.

Shabanimashcool, M., Li, C.C., 2015. Analytical approaches for studying the stability of laminated roof strata. Int. J. Rock Mech. Mining Sci. 79, 99–108.

Shabanimashcool, M., Jing, R.L., Li, C.C., 2014. Discontinuous modelling of stratum cave-in in a longwall coal mine in the Arctic area. Geotech. Geol. Eng. 32 (5), 1239–1252.

Sofianos, A.I., 1996. Analysis and design of underground hard rock Voussoir beam roof. Int. J. Rock Mech. Min. Sci. 33 (2), 153–166.

Solecki, R., Conant, R.J., 2003. Advanced Mechanics of Materials. Oxford University Press, London, pp. 147–160.

Soni, A.K., Singh, K.K.K., Prakash, A., Singh, K.B., Chakraboraty, K.A., 2007. Shallow cover over coal mining: a case study of subsidence at Kamptee Colliery, Nagpur, India. Bull. Eng. Geol. Environ. 66, 311–318.

Trueman, R., Lyman, G., Cocker, A., 2009. Longwall roof control through a fundamental understanding of shield-strata interaction. Int. J. Rock Mech. Min. Sci. 46, 371–380.

Tsesarsky, M., 2012. Deformation mechanisms and stability analysis of undermined sedimentary rocks in the shallow subsurface. Eng. Geol. 133 (4), 16–29.

Wang, X., Meng, F.B., 2018. Statistical analysis of large accidents in China's coal mines in 2016. Nat. Hazards 92, 311–325.

Wang, X., Tian, L.G., 2018. Mechanical and crack evolution characteristics of coal-rock under different fracture-hole conditions: a numerical study based on particle flow code. Environ. Earth Sci. 77, 297.

Wang, S.R., Hagan, P., Cheng, Y., Wang, H., 2012. Experimental research on fracture hinged arching process and instability characteristics for rock plates. Chin. J. Rock Mech. Eng. 31 (8), 1674–1679.

Wang, S.R., Li, N., Li, C.L., Cao, C., 2015a. Distribution characteristics analysis of pressure-arch in horizontal stratified rocks under coal mining conditions. TehnickiVjesnik Tech. Gazette 22 (4), 997–1004.

Wang, F.T., Zhang, C., Zhang, X.G., Song, Q., 2015b. Overlying strata movement rules and safety mining technology for the shallow depth seam proximity beneath a room mining goaf. Int. J. Min. Sci. Technol. 25 (1), 139–143.

Wang, S.R., Li, N., Li, C.L., Zou, Z.S., Chang, X., 2015c. Instability mechanism analysis of pressure-arch in coal mining field under different seam dip angles. DYNA Ingeniería e Indus. 90 (3), 279–284.

Wang, S.R., Liu, X.L., Zou, Z.S., Luo, J., Wang, E.J., Xue, X.M., Xiao,H.G. 2016a. A combined multi-functional experimental platform under servo control loading conditions, China patent 201620230001.X, 30 Nov.

Wang, Y.C., Jing, H.W., Zhang, Q., Luo, N., Yin, X., 2016b. Prediction of collapse scope of deep-buried tunnels using pressure arch theory. Math. Prob. Eng., (3-4), 1–10.

Wang, S.F., Li, X.B., Wang, D.M., 2016c. Void fraction distribution in overburden disturbed by longwall mining of coal. Environ. Earth Sci. 75 (2), 1–17.

Wang, G., Luo, H.J., Wang, J.R., Tian, F.C., Wang, S., 2016d. Broken laws of key strata on strata behaviors in large height fully-mechanized face of nearly shallow coal seam. J. China Univ. Min. Technol. 45 (3), 469–474.

Wang, J.C., Zhang, J.W., Li, Z.L., 2016e. A new research system for caving mechanism analysis and its application to sublevel top–coal caving mining. Int. J. Rock Mech. Min. Sci. 88, 273–285.

Wang, F.T., Tu, S.H., Zhang, C., Zhang, Y.W., Bai, Q.S., 2016f. Evolution mechanism of water-flowing zones and control technology for longwall mining in shallow coal seams beneath gully topography. Environ. Earth Sci. 75 (19), 1–16.

Wang, S.R., Li, C.L., Wang, Y.G., Zou, Z.S., 2016g. Evolution characteristics analysis of pressure-arch in a double-arch tunnel. Tehničkivjesnik Technical Gazette 23 (1), 181–189.

Wang, S.R., Zhao, Y.H., Zou, Z.S., Jia, H.H., 2017. Experimental research on energy release characteristics of water-bearing sandstone alongshore wharf. Polish Maritime Res. 24 (S2), 147–153. https://doi.org/10.1515/pomr-2017-0077.

Wang, S.R., Wu, X.G., Zhao, Y.H., Hagan, P., 2018. Mechanical performances of pressure-arch in thick bedrock during shallow coal mining. Geofluids, 2018, Article ID 2419659: 1–13.

Wu, Q., Shen, J.J., Liu, W.T., 2017. A RBFNN-based method for the prediction of the developed height of a water-conductive fractured zone for fully mechanized mining with sublevel caving. Arab. J. Geosci. 10, 172.

Xie, G.X., Chang, J.C., Yang, K., 2009. Investigations into stress shell characteristics of surrounding rock in fully mechanized top-coal caving face. Int. J. Rock Mech. Min. Sci. 46 (1), 172–181.

Xu, J.L., Ju, J.F., Zhu, W.B., 2014. Supports crushing types in the longwall mining of shallow seams. J. China Coal Soc. 39 (8), 1625–1634.

Xu, C., Yuan, L., Cheng, Y.P., Wang, K., 2016. Square-form structure failure model of mining–affected hard rock strata: theoretical derivation, application and verification. Environ. Earth Sci. 75, 1180.

Xu, D.J., Peng, S.P., Xiang, S.Y., 2017. A novel caving model of overburden strata movement induced by coal mining. Energies 10, 476–489.

Yang, Z.L., 2010. Stability of nearly horizontal roof strata in shallow seam longwall mining. Int. J. Rock Mech. Min. Sci. 47 (4), 672–677.

Yin, Y.Q., Li, P.N., Di, Y., 2015. Instability and jumping phenomenon of rock structure. Chin. J. Rock Mech. Eng. 34 (5), 945–952.

Yong, Y., Tu, S.H., Zhang, X.G., Li, B., 2015. Dynamic effect and control of key strata break of immediate roof in fully mechanized mining with large mining height. Shock Vibr., 2015, Article ID 657818: 1–11.

Yu, B., Zhao, J., Kuang, T.J., Meng, X.B., 2015. In situ investigations into overburden failures of a super-thick coal seam for longwall top coal caving. Int. J. Rock Mech. Min. Sci. 78 (29), 155–162.

Zhang, Z.Z., Gao, F., 2015. Experimental investigation on the energy evolution of dry and water-saturated red sandstones. Int. J. Min. Sci. Technol. 25 (3), 383–388.

Zhang, J.C., Peng, S.P., 2005. Water in rush and environmental impact of shallow seam mining. Environ. Geol. 48 (8), 1068–1076.

Zhang, D.S., Fan, G.W., Liu, Y.D., Ma, L.Q., 2010. Field trials of aquifer protection in longwall mining of shallow coal seams in China. Int. J. Rock Mech. Mining Sci. 47 (6), 908–914.

Zhang, D.S., Fan, G.W., Ma, L.Q., Wang, X.F., 2011. Aquifer protection during longwall mining of shallow coal seams: a case study in the Shendong Coalfield of China. Int. J. Coal Geol. 86 (2–3), 190–196.

Zhang, W., Zhang, D.S., Wu, L.X., Wang, X.F., 2013. Numerical simulation on dynamic development features of mininginduced fractures in overlying strata during shallow coal seam mining. Electron. J. Geotech. Eng. 18, 5531–5543.

Zhang, H.W., Zhu, Z.J., Huo, L.J., 2014. Overburden failure height of superhigh seam by fully mechanized caving method. J. China Coal Soc. 39, 816–821.

Zhang, X.G., Kuang, X.M., Yang, J.H., Wang, S.R., 2017. Experimental study on mechanical properties of lightweight concrete with shale aggregate replaced partially by nature sand. Electron. J. Struct. Eng. 17 (1), 85–94.

Zhao, Y.H., Song, X.M., 2016. Stability analysis and numerical simulation of hinged arch structure for fractured beam in super-long mining workface under shallow seam. Rock Soil Mech. 37 (1), 203–209.

Zhao, T.B., Guo, W.Y., Tan, Y.L., Yin, Y.C., Cai, L.S., Pan, J.F., 2018. Case studies of rock bursts under complicated geological conditions during multi-seam mining at a depth of 800 m. Rock Mech. Rock. Eng. 51, 1539–1564.

Zhao, Y.H., Wang, S.R., Hagan, P., Guo, W.B., 2018a. Evolution characteristics of pressure-arch and elastic energy during shallow horizontal coal mining. TehnickiVjesnik Tech. Gazette 25 (3), 867–875.

Zhao, Y.H., Wang, S.R., Zou, Z.S., Ge, L.L., Cui, F., 2018b. Instability characteristics of the cracked roof rock beam under shallow mining conditions. Int. J. Min. Sci. Technol. 28 (3), 437–444.

Zhou, H., Xu, R.C., Lu, J.J., Shen, Z., 2015. Study on mechanisms and physical simulation experiment of slab buckling rockburst in deep tunnel. Chin. J. Rock Mech. Eng. 34 (S2), 3658–3666.

Further reading

Huang, D., Li, Y.R., 2014. Conversion of strain energy in triaxial unloading tests on marble. Int. J. Rock Mech. Mining Sci. 66 (1), 160–168.

Li, Z.L., Dou, L.M., Cai, W., Wang, G.F., Ding, Y.L., Kong, Y., 2016a. Mechanical analysis of static stress within fault-pillars based on a voussoir beam structure. Rock Mech. Rock Eng. 49 (3), 1097–1105.

Li, X.M., Wang, Z.H., Zhang, J.W., 2017b. Stability of roof structure and its control in steeply inclined coal seams. Int. J. Min. Sci. Technol. 27 (2), 359–364.

Wang, J.C., Wang, Z.H., 2015. Stability of main roof structure during the first weighting in shallow high-intensity mining face with thin bedrock. J. Mining Saf. Eng. 32 (2), 175–181.

Index

Note: Page numbers followed by *f* indicate figures and *t* indicate tables.

A

D

E

F

G

H

L

M

N

O

P

R

S

T

U

V

W